Springer Textbooks in Earth Sciences, Geography and Environment

The Springer Textbooks series publishes a broad portfolio of textbooks on Earth Sciences, Geography and Environmental Science. Springer textbooks provide comprehensive introductions as well as in-depth knowledge for advanced studies. A clear, reader-friendly layout and features such as end-of-chapter summaries, work examples, exercises, and glossaries help the reader to access the subject. Springer textbooks are essential for students, researchers and applied scientists.

A. R. Bhattacharya

Structural Geology

Springer

A. R. Bhattacharya
Department of Geology (Retired)
University of Lucknow
Lucknow, India

ISSN 2510-1307 ISSN 2510-1315 (electronic)
Springer Textbooks in Earth Sciences, Geography and Environment
ISBN 978-3-030-80794-8 ISBN 978-3-030-80795-5 (eBook)
https://doi.org/10.1007/978-3-030-80795-5

This Springer imprint is published by the registered company Springer Nature Switzerland AG
The registered company address is: Gewerbestrasse 11, 6330 Cham, Switzerland

Dedicated to my parents, in heaven,
and to my wife, Krishna.

Preface

In the course of a long career in teaching and research in structural geology, I felt the need of writing a textbook on this subject mainly to incorporate many things that are as yet not seen in many textbooks. Also, the need to cover the entire updated spectrum of the subject in a balanced way and in simple language was in my mind.

The book is primarily written for undergraduate students who want to take up structural geology at higher levels. It emphasizes basics and fundamental concepts with special reference to field and practical aspects such that there is a balance between theory and practice. This would help students grow their own thinking faculty and to learn the art and science of developing models from their own observations and interpretations. At relevant junctures, the relation of structural geology with allied branches has been discussed, and as such several topics have implications for engineering projects and exploration of minerals, petroleum and groundwater. The book should thus develop excitement and interest in students and readers to explore more on the subject. With all these features, plus many more, the book should be of use to students, academicians, practitioners and professionals.

To read the book, two different lettering styles have been adopted. Words in italic are those that are primarily meant for students for understanding and learning. Those in bold are mainly for headings and subheadings and for highlighting or emphasizing a particular aspect for all categories of readers. Most of the words in all these three styles also appear in Glossary as a refresher to the readers.

The real trigger to write the book came from my students who made me competent enough to write a book of this level. I feel lucky that my teachers at the Department of Geology, University of Lucknow, had enhanced my interest in geology with their excellent teachings. I express my deep sense of gratitude to my former teacher and mentor, the late Professor R.C. Misra, for his erudite guidance and benevolent attitude. I gratefully thank my former teacher, the late Professor S.N. Singh, for laying down a strong foundation of basics in structural geology. I emotionally remember, and owe much to, my former colleague, the late Professor A.K. Jauhri, as it was he who provided me the inner courage and energy to take up this work.

The manuscript of the book has been reviewed (wholly or partly in the form of one or more chapters) by several experts. I am especially grateful to Professor J.P. Burg for reviewing all the chapters of the manuscript and to Professor R.H. Groshong, Jr. for reviewing chapters covering half of the manuscript. Their reviews greatly enhanced the quality of the manuscript and brought it up to this level. At the same time, I am highly grateful to Professors R.D. Hatcher, Jr., Terry Engelder, Peter Hudleston, Reinhard Greiling, Chris Hilgers, R.S. Sharma and D.C. Srivastava for reviewing one or more selected chapters. Their erudite reviews have made high-quality improvements in the manuscript. On behalf of Springer Nature, the manuscript has been reviewed by Professor Soumyajit Mukherjee and two unknown reviewers. Their critical comments and excellent suggestions have significantly enhanced the quality of the manuscript and brought it up to this stage. I express my sincere gratitude to all of them. Professor Dhruba Mukhopadhyay helped me during the initial laying out of the structure of the book, for which I am highly grateful to him. I am especially grateful to Professor

R.D. Hatcher, Jr. for his constant encouragement and help in several ways during the preparation of the manuscript. The revised version of the manuscript has been meticulously read by Mr. D.D. Bhattacharya and Dr. Amit K. Verma; they pointed out several overlooks and inconsistencies. I thank both of them for this work.

Professors Terry Engelder, Cees Passchier, Reinhard Greiling, Stefan Schmid and T.K. Biswal have been kind enough to provide me a few technically high-quality photographs of structures, some of which are used in the book. I express my sincere thanks to all of them for this help. Several friends helped me procure relevant study material and reprints, for which special mention may be made of Professors D.C. Srivastava, Soumyajit Mukherjee, T.K. Biswal and Dr. Amit K. Verma. I sincerely thank all of them.

I have benefited much through discussions with several structural geologists during field visits to a number of geological terrains. Special mention may be made of Professors R.D. Hatcher, Jr. (Appalachians of the USA), Klaus Weber (south Germany and the Himalaya), J.L. Urai (Belgium), Chris Hilgers (NW Germany, Belgium and Holland), S.P. Singh (central India), the late Dr. Hans Ahrendt (Alps of Italy, Jura of Switzerland and the Himalaya) and Dr. Axel Vollbrecht (Alps of Italy and Jura of Switzerland). Some relevant field photographs from most of these terrains are also included in this book at suitable places. I express my sincere thanks to all of them.

I am thankful to the Humboldt Foundation of Germany for awarding me an Alexander von Humboldt Fellowship, under which I visited Germany a few times for research work. Some of the results of my research work carried out under Humboldt Fellowship are reflected in the book. A few fabric diagrams appearing in this book were prepared at Göttingen University, while some of the photomicrographs, also appearing in this book, were prepared at RWTH (Aachen University of Technology), for which I express my grateful thanks to Professors Klaus Weber and J.L. Urai, respectively.

I owe much to the Department of Geology, University of Lucknow, that provided me a platform for teaching and research in structural geology. And this book is a reflection of what I gained from this department. Although the book has been written after my retirement, the department continued to extend me all facilities as and when I needed. The faculty members, research workers and members of the non-teaching staff always cooperated with me in this respect. I sincerely thank all of them. I am especially thankful to Professors Vibhuti Rai and Rameshwar Bali for their help and cooperation in several ways.

For 1 year (2014–2015), I got a book writing grant from the Department of Science and Technology (DST), Government of India, New Delhi, under its USERS programme. I express my grateful thanks to the DST for this grant. This programme was carried out at Sri Jai Narain Post Graduate College, Lucknow, for which I express my cordial thanks to the college authorities.

Dr. Amit K. Verma has very sincerely helped me at all stages of preparation of the manuscript. He especially helped me in procuring reprints and preparing most of the line diagrams and coordinated with me at several critical junctures. I sincerely thank him very much for all this. Mr. A.K. Gupta prepared a number of line diagrams for which I thank him very much. Drs. Narendra K. Verma and Gautam K. Dinkar helped me in several ways; they, as well as Dr. Prabhas Pande, provided me a few relevant photographs, some of which are used in the book. I sincerely thank all of them. Profs./Drs./Messrs. Talat Ahmad, Rajesh Kumar Srivastava, S.C. Tripathi, J.S. Mehta, Anshumali Sharma and N.K. Tewari are especially thanked for their help in several ways. Omission of any name in this list is merely by mistake, not by intention. Lastly, but not the least, my family, especially my lovely grand-daughter Noyonika, persistently provided me the motive force and created the congenial environment for writing the book.

With great pleasure, I convey my grateful thanks to Springer Nature, USA, for publishing my book. At the same time, I am happy to place on record my very nice experience with the Springer Nature team, mainly headed by Drs./Messrs. Zachary Romano, Aaron Schiller,

Herbert Moses, Zoe Kennedy and Nirmal Selvaraj. They meticulously guided me through all the stages of manuscript preparation, and it is because of them and their team members that the manuscript could be brought to the final stage so easily. I understand that occasionally I had put them on vexing points, but they were so patient and helpful that I always found the ball in my court. I express my very sincere thanks to all of them.

Lucknow, Uttar Pradesh, India A. R. Bhattacharya

About the Book

This textbook is a complete, up-to-date and highly illustrated account of structural geology for students and professionals and includes fundamentals of the subject with field and practical aspects. The book aims to be highly reader friendly, containing simple language and brief introductions and summaries for each topic presented, and can be used both to refresh overall knowledge of the subject and to develop models for engineering projects in any area or region. The book is presented in 20 chapters and divided into 3 parts: (A) Fundamental Concepts; (B) Structures: Geometry and Genesis; and (C) Wider Perspectives. For the first time as full chapters in a textbook, the book discusses several modern field-related applications in structural geology, including shear-sense indicators, and deformation and metamorphism. Also uniquely included are coloured photographs, side by side with line diagrams, of key deformation structures not seen in other books before now. Boxes in each chapter expand the horizons of the reader on the subject matter of the chapter. Detailed significance of the key structures, not to be seen in any textbook so far, provides a better grasp to students. Questions at the end of each chapter provide a self-test for students. Glossary at the end of the book is a refreshing aspect for readers. Though written primarily for undergraduate and graduate students, the text will also be of use to specialists and practitioners in engineering geology, petrology (igneous, sedimentary and metamorphic), economic geology, groundwater geology, petroleum geology and geophysics and will appeal to beginners with no preliminary knowledge of the subject.

Contents

About the Author

A. R. Bhattacharya is a Retired Professor of Geology from the University of Lucknow, India. He had obtained his B.Sc., M.Sc. and Ph.D. degrees from this university, and later he joined the faculty of the Department of Geology where he taught, and carried out research work in, structural geology for about four decades. His research interests include geometry and genesis of structures, shear zones, fabric development and evolution of structural complexes. Some of his notable contributions include a new classification of folds, two new parameters of fold geometry, an empirical equation for natural folds, a mathematical model of fold development, quantitative modelling of shear zones and a graphical method for distinguishing different fold generations. His work on the Himalaya has thrown newer lights on the structure and deformational processes in the Kumaun-Garhwal sectors that have far-reaching implications for understanding the evolution of mountain belts. His researches on the Bundelkhand craton of central India, especially on the identification of crustal scale shear zones, structural elucidation of quartz reefs, mesoscopic fold geometry and deformational processes in the internal domains of rocks, are of pioneering nature for this craton. One of his research papers (published in *Tectonophysics*, vol. 387, 2004) was selected as 'Top 25 Hottest Papers' by Science Digest, and it was ranked 15th. He has co-edited two research-level books. He has been an Alexander von Humboldt Fellow of Germany under which he visited Germany a few times, having worked at Göttingen University and at Aachen University of Technology. In addition, he had also visited the USA, France, England, Italy, Switzerland, Belgium and Holland for academic/research purposes. He has a vast geological field experience of the Himalaya and the peninsular shield of central India, as well as of several structurally important regions of the world, mainly the Appalachians of the USA, Alps of Italy, Jura of Switzerland, Variscan Belt, Rhine Graben and NW Schist Belt of Germany and Palaeozoic-Mesozoic Belts of Belgium and Holland.

Part I

Fundamental Concepts

1 Introduction to Structural Geology

A. R. Bhattacharya, *Structural Geology*, Springer Textbooks in Earth Sciences, Geography and Environment,
https://doi.org/10.1007/978-3-030-80795-5_1

Abstract

Standing on a rocky terrain, one may occasionally notice beautiful structures in rocks. The variety of geometrical shapes of the structures poses great inquisitiveness to an onlooker. What are these structures, how are these formed and what does their occurrence indicate are some of the questions that haunt his/her mind. Answers to these questions are hidden in a discipline of geology called *structural geology*, which encompasses the study of all aspects of the deformation structures such as their geometry, field relations, geographic distribution, genesis and related aspects. Since the structures in the context of structural geology are formed by deformation, these are also called *deformation structures* or *tectonic structures*. They occur on scales ranging from tens of kilometres to those that can be seen only under a microscope. Faults, folds and joints are some common examples of structures. Structural geology is closely related to another discipline of geology called *tectonics*, which includes deformation on a much larger scale, i.e. from regional to lithospheric. Structural geology and tectonics commonly go together and even overlap when studied on the same scale. Structures are sometimes associated with economic minerals and hydrocarbons. The type and orientation of structures are important in civil engineering work such as dams, tunnels and bridges. Structural geology has implications for academic, economic, societal and environmental aspects.

This chapter presents a panoramic view of the wonderful world of structural geology so that the reader systematically starts gearing up for an in-depth study of this discipline.

Keywords

Structural geology · Tectonics · Deformation structures · Scale of structures · Methodology of structural geology · Remote sensing · Significance of structural geology

1.1 What Is Structural Geology?

Structural geology encompasses the study of all aspects of the deformation structures such as their geometry, field relations, geographic distribution, genesis and related aspects. Rocks after their formation are subjected to stresses inside the earth. When the amount of applied stresses exceeds the strength of a rock, the latter yields to or accommodates the stresses by undergoing deformation that is manifested in some typical structures called *deformation structures*. In structural geology, we commonly mean structures that are formed due to deformation by externally applied stresses. These structures are preserved for millions of years. Common examples of deformation structures are folds, faults, joints and fractures.

Structural geology is closely related to another branch of geology called *tectonics* that includes the study of lithosphere and deeper mantle rocks also. Considering the earth as a dynamic planet, tectonics includes the study of deformation, bending, rupturing, movements and related processes of the lithosphere including the formation of mountains and ocean basins and evolution of the continents and of the crust.

Structural geology is an important discipline of geology that acts as a tool to specialists and practitioners in engineering geology, petrology (igneous, sedimentary and metamorphic), economic geology, groundwater geology, petroleum geology and geophysics.

1.2 What Are Structures?

1.2.1 Tectonic Structure

In this book, the word **structure** is a password. Structures, in the context of structural geology, are formed by deformation of rocks. As such, these are also called **tectonic structures**. To a layman, or even to a beginner, it may sound unrealistic to note that structures in rocks are formed under hard rock, not soft rock, deformation conditions.

Rocks start their journey as sediments eroded from the continents. These sediments are transported by agencies like rivers, wind and glaciers and are ultimately deposited in a basin in the form of layers (Fig. 1.1). The basin may be a sea, river, desert or glacier. The layers are thereafter buried to deeper levels where they are subjected to pressure and, to some extent, temperature that lead to lithification due to which the layers become hard and compact and are thus converted to layers of rocks. This process is called induration. At depth, the rocks are subjected to stresses mainly due to overburden as well as tectonic stresses that are oriented in some definite directions. Thereafter starts another chapter in the journey of rocks caused mainly by deformation. The rock layers may initially be tilted (Fig. 1.2) and then, depending upon the nature and type of stresses, are followed by formation of different types of structures. The type of structures, mainly their geometrical shapes, depends upon the type and orientation of the externally applied stress on initially flat layers. Some common structures formed in rocks, such as folds, faults, thrusts, joints, lineation and foliation, are briefly described below.

A *fold* is a bend in rock formed due to compressive stresses acting parallel to, or across, the originally flat surfaces of rock. The fold can be an anticline (Fig. 1.3) in which the rock layers are convex upward towards the centre.

Fig. 1.1 Sedimentary layering in loose, unconsolidated sands on the banks of River Ganga, near Rishikesh, India. The sands are deposited layer by layer by the river during flood times. The layering pattern is visible due to variations in grain size and texture of sands in each layer. (Photograph by the author)

Fig. 1.2 After induration of rocks, they are subjected to stresses inside the earth, thus forming structures in rocks. In this photograph, the rocks are exposed as inclined beds. This inclination, or tilting, is an effect of deformation

There can be a syncline also in which case the rock layers are concave towards the centre. Both anticline and syncline may occur together, i.e. adjoining one another (Fig. 1.4).

A *fault* is a fracture along which two blocks of rock have moved past each other (Fig. 1.5). A *thrust* is also a fault, called reverse fault, in which the hanging wall has moved up relative to the footwall (Fig. 1.6). A *joint* is a fracture along which there has been little or no transverse displacement of rock (Fig. 1.7). *Lineation* is any fabric or orientation in a rock developed in a linear fashion (Fig. 1.8). *Foliation* is a planar fabric given by the preferred orientation of minerals showing platy or tabular habit. Foliation can be seen with unaided eye in hand specimen and in outcrops in the form of alternating compositional layers (Fig. 1.9). Sometimes, the minerals that constitute compositional layers are too small to be seen only under a microscope (Fig. 1.10).

Structures in rocks occur on varied sizes. Some are visible with naked eyes, while others are either too small or too large to deter proper study with unaided eyes. For structural studies, therefore, a system of scales of geological structures is in practice. *Megascopic structures* are developed on the scale of several kilometres to tens of kilometres. *Mesoscopic structures* occur on a small outcrop to hand specimen scale. *Microscopic structures* can be studied only under a microscope.

Fig. 1.3 Field photograph of a fold in limestone. This fold is an anticline in which the rock layers are convex upward towards the centre. Loc. near Hampteau Hotton, Belgium. The author happily stands for scale

During fieldwork, a structural geologist commonly studies structures on outcrop to hand specimen scales, i.e. on mesoscopic and macroscopic scales. He/she may use a hand lens for more clarity of a structure, but still the structure falls in the category of mesoscopic scale; the hand lens in this case has been used only to have a better or clearer look of the structure or parts(s) of it.

1.2.2 Nontectonic Structure

Earth materials can deform without being involved in tectonic processes. Stresses can be imposed upon them by, for example, overburden loading, sedimentary processes, shear from overriding sediment-fluid flows and glaciers. Structures thus formed are called *nontectonic structures*.

Sedimentary processes commonly produce a variety of structures called *sedimentary structures*. A common example is the layering pattern (Fig. 1.1) shown by loose, unconsolidated sediments. This layering pattern is developed after the sediments are deposited in a basin by any agency such as wind or water. By their general look, this layering pattern resembles a layered rock. However, the term **bed** is commonly used for lithified, hard rocks. A bed is a unit of a sedimentary rock, just like the unit of a building is a brick, or the unit of an organism is a cell.

Deep on the ocean floor, occasionally huge mass of unconsolidated sediments flow down the deep interiors of oceans due to gravity. These are called *slump structures*. Of the variety of structures formed by slumping, some look like deformation structures such as folds, faults and thrusts, but these are all nontectonic structures and are sometimes called *soft-sediment deformation*.

Fig. 1.4 Photograph of a syncline (left) in which case the rock layers are concave upward at the centre, together with an anticline (right). The rocks constitute a sedimentary sequence of the fold and thrust belt of the Andes Mountains exposed near Lake Argentino, Patagonia region, Argentina. (Photograph courtesy Narendra K. Verma)

Fig. 1.5 Field photograph of faults developed in pink granite of Bundelkhand craton, central India. One can see that a marker layer (light coloured) is successively displaced along a plane, the fault plane. (Photograph by the author)

Fig. 1.6 Photograph of a thrust fault in the Canadian Rockies near Alberta. In the photograph, one can see that a stack of folded Palaeozoic carbonates and sandstones is being transported by a low angle thrust. (Photograph courtesy Narendra K. Verma)

Fig. 1.7 Field photograph of joints developed in quartzite of Bundelkhand craton, central India. (Photograph by the author)

Fig. 1.8 Field photograph of lineations in the form of mullion structure developed in sandstone alternating with shale. Loc. North Eifel, Germany. (Photograph by the author)

Fig. 1.9 Field photograph showing foliation in crystalline-metamorphic rocks of Southern Appalachians, USA. This is gneissic foliation defined by alternating dark and light-coloured bands (running left-right). (Photograph by the author)

Fig. 1.10 Foliation as seen in thin section under a microscope. The foliation is given by a preferred orientation of micaceous minerals that are seen as different colour bands in the photomicrograph

Glaciers are huge bodies of ice that slowly move down the slope due to gravity. Glaciers can produce features that resemble tectonic structures. As such, the toe of a glacier is an area of enormous stress. The accumulated stress is accommodated by the formation of several types of structures that resemble faults (Fig. 1.11), folds, thrusts, etc. Formation of structures by glaciers is called *glaciotectonics*. But these are all nontectonic structures.

1.3 The Ultimate Cause of Deformation

Deformation structures, as we have discussed above, are formed by tectonic stresses. Now the question arises as to wherefrom these stresses are generated. It is only in the late 1960s and early 1970s that we got a satisfactory answer to this nagging question in the form of *plate tectonics* that has virtually transformed the concepts of tectonics. Plate tectonics is based on the concept that the earth's outer rigid shell, called *lithosphere*, is broken into several blocks, called *plates*. The plates are of different sizes—seven or eight very large, six or seven medium-sized and a few smaller ones—that move over the molten upper part of the mantle, called *asthenosphere*. The lithosphere is about 100 km thick and includes the crust and the outer mantle of the earth. The plates are constituted of oceanic lithosphere as well as of continental lithosphere; the latter is thicker than the former. The plates

Fig. 1.11 Structures formed in ice sheet at the toe region of Perito Moreno Glacier in the Andes Mountains, Argentina. The photograph showing a vertical section of the ice sheet displays features analogous to tectonic structures such as faults, joints and folds that are developed due to stresses created by glacial creep and differential loading. Seepage of oxygenated water and soil filling in these structures accentuate their visibility. (Photograph courtesy Narendra K. Verma)

Fig. 1.12 Earthquake at Kalyanpuram, South Andaman, India. This Sumatra-Andaman earthquake—the Great Tsunami—of December 26, 2004, was of magnitude 9.1 causing Grade 5 damage to a newly constructed RCC framed structure (photograph taken on January 09, 2005). The earthquake rendered total collapse of the ground floor under the influence of strong ground motions causing the first floor coming down almost to the ground level, underneath which a car is seen crushed. Geologically, the area consists of Andaman Flysch, which in the present site is under a veneer of soil. (Photograph courtesy Dr. Prabhas Pande)

move in constant motion, at rates varying from 5 to 10 cm per year, and interact with each other such that their boundaries can be convergent, divergent or transform. At convergent plate boundaries, one plate goes down, i.e. subducted, below another, and therefore the crust is destroyed. At divergent plate boundaries, due to release of pressure, molten material, called *magma*, rises up from below to form new crust, and therefore the crust is created. This is accompanied by volcanic activity and shallow earthquakes. In the third type of plate motion, two plates slide past each other along deep fractures called *transform faults*, and as such the crust is neither created nor destroyed. These types of margins are therefore called *conservative*. However, due to rubbing of the crust, transform boundaries create earthquakes.

The theory of plate tectonics postulates that the interiors of plates are undeformed. The plate boundaries, on the other hand, are regions of major deformational processes that result in volcanism, earthquakes and formation of mountain ranges. An important example is the mighty Himalayan Mountain that was formed as a result of collision of the north-moving Indian plate against the Asian plate.

1.4 What if We Fail to Do Structural Work!

Structural work provides specific insight, and sometimes plays a decisive role, in several development projects. Failing to do structural work may sometimes lead to negative, may be disastrous, outcome! Let us take the example of *joints* that are the most common structures in rocks. In size, joints may range from microscopic to tens of metres, or more. Joints never occur alone; they occur in one or more sets, each set containing numerous joints. Presence of joints makes a rock mechanically weak, i.e. the rock is not cohesive, and therefore it cannot bear load for long. Also, joints are avenues for channelization of water and fluids. Wet rocks are weaker than dry rocks. If a civil engineer ignores the presence, and detailed study, of joints in his/her chosen site, the structure or superstructure whatever he/she builds thereupon will be under risk and may collapse sooner or later.

Let us take another example, i.e. *faults*. Faults also range in size from microscopic to tens or hundreds of metres. Presence of faults develops anisotropy to a rock, which thus becomes weak. Faults, like joints, also constitute avenues for channelization of water and fluids. Wet rocks are weak to overlying load. Therefore, if the presence of faults is ignored, or not studied carefully, the civic structures raised in a site may encounter danger over time.

The above and several other examples establish a strong position for structural geology in civil engineering, thereby making it an integral constituent of *engineering geology*. Structural work is a routine practice in engineering constructions like dams, tunnels, bridges and buildings.

The above analogy applies to the exploration of economic minerals too. Occurrence of several economic minerals such as gold and copper and several other metallic minerals, petroleum and gas, as well as groundwater is structurally controlled. For example, mineralizing fluids find faults as easy channels and are ultimately localized/precipitated at lower pressure zones. Faults are also hunting sites for petroleum and gas. Folds too are suitable sites for petroleum and gas, which remain physically stable at the limbs and hinge zones of folds.

Structures like faults play an important, rather crucial, role in the study of *earthquakes*. The latter release untold energy locked up in rocks inside the earth. Faults constitute the most important avenues along which this energy is quickly transferred to the ground surface that causes damage in the form of earthquakes (Fig. 1.12). The degree of damage at any particular spot or area depends on the rock types and their structure. Structural geologists can help in a detailed study of faults, especially in the identification of active faults and measurement and monitoring of stresses in rocks. In collaboration with engineering geologists, structural geologists can prove to be of great help in designing civic structures.

1.5 Methodology of Structural Geology

Because deformation structures occur on various scales and modes, we have to adopt different methodologies that enable us to properly study the structures from as many angles as possible, of which the following are important: field studies, laboratory studies, experimental modelling/simulation, numerical modelling, geophysical studies and remote sensing.

1.5.1 Field Studies

Structural geology is a field-based discipline. Structures first need to be studied in an outcrop in field with the following common practices: (i) identify the rock type or lithology of the structure; (ii) identify the structure: if needed, hand lens may be used; (iii) establish the field relations with the host rock; and (iv) work out the shape, size and attitude of the structure(s). By attitude, we mean the orientation of the structure with respect to horizontal and vertical planes and geographical directions (see Chap. 2). One of the primary aims of field studies is to prepare a geological map of an area or region. A geological map shows geographical distribution of the various rock types of an area. Preparation of a geological map is called **geological mapping**, or simply **mapping**.

1.5.2 Laboratory Studies

The data and samples collected from field need to be carefully studied in laboratory. The data mostly related to shape and orientations of a structure constitute *geometric analysis*. The shape of the structure throws light on how it was formed, i.e. by ductile processes (e.g. folding) or by brittle processes (e.g. faulting and fracturing). The attitudes of the various structures as recorded in field are marked on a lithological map of the area and are called a *structural map* that shows the distribution of various structures in different parts of the area together with their orientations. The data may be represented by suitable plots of which the *stereographic plots* (see Appendix A) are most widely practised.

Structures of rocks of an area or of a region help in *kinematic analysis* that gives an idea of quantitative deformation of a rock involving *strain analysis*. In other words, the kinematic analysis enables us to understand how the rock behaved to the deforming environment by changing its shape, size and orientation. The field data also help in *dynamic analysis* that throws light on the forces that have caused *stress* in the rock. Every rock has strength as its intrinsic property. Because of its strength, a rock resists to the external stress. Results of dynamic analysis may be extended to interpret the movement pattern of crustal blocks or plates.

1.5.3 Experimental Modelling/Simulation

Experimental studies are primarily aimed at understanding the processes of formation of natural structures with analogue models under controlled laboratory conditions. Although the results do not show the actual process of formation of structures, they simulate to a close approximation of the stages of formation of the structures that we can see in the laboratory.

Model experiments have advanced our knowledge on folding, faulting and thrusting. The common practice is to use silicon putty, modelling clay, honey and sand to simulate lithospheric layers. The material is placed in the form of layers in a *deformation box* designed specifically for the experiment. While using a particular material, say clay or silicon putty, one can change the viscosity of the layers that simulate rheological contrast in rock layers. The layers are then mechanically deformed by contraction, extension or strike-slip mode or by any other desired mode. As deformation continues, stages of formation of the structure can be monitored. The structure thus formed is now subjected to interpretation. With sand as experimental material, the experiment is called *sandbox experiment*. Another type of laboratory study pertains to *centrifuge experiments* in which the material is subjected to centrifugal force under which it deforms the way a body deforms under gravity. Such experiments help us to speed up the deformation in order to reach the desired structures.

Deformation of natural rocks and soils under the influence of applied stress and strain rates has been in practice since long by engineering geologists and civil engineers for testing their construction material, generally natural rock, for stability purposes. The experiment is carried out in a specially designed rig in which the rock sample is subjected to compression or tension. Stress can be applied in one direction in which case it is called *uniaxial compression* or *uniaxial tension*, or in two or three directions in which case the deformation is said to be under *biaxial compression/tension* or *triaxial compression/tension*, respectively.

1.5.4 Numerical Modelling

The structural database acquired from field, laboratory and experimental studies may be subjected to *numerical modelling* that requires mathematical simplifications and/or application of computers. Numerical modelling is a sophisticated approach to structural geology due to rather easy availability of computers and has limitations in its application. The main reason is that when the data are subjected to mathematical jargon, we have to make many simplifications to suit the mathematical equations. Also, we have to feed only those data and information that suit the available program or code of the computer.

1.5.5 Geophysical Studies

Geophysical studies reveal a wealth of information on the interior of the earth and especially the structure of the crust where rock deformation occurs. Seismic, gravity and magnetic properties are most important in the context of rocks. In structural geology, however, *seismic methods* have more relevance than others. The common method is to acquire *seismic reflection data* by an artificial sound source placed on the ground surface. Common sound source is an explosive device due to which the sound waves travel across the various rock layers. The return time of the sound waves for different layers is correlated with lithology and depth. Interpretation of all these data helps in unravelling the structure of subsurface rocks (Fig. 1.13).

In onshore exploration, the sound source is put on the ground surface, while in offshore exploration, it is put on the sea floor or can be treated with air/water guns or towed on a cable behind a ship. In onshore areas, the reflected sound signals are detected by *geophones* that are put on the ground surface, while in offshore areas, the signals are received by *hydrophones* towed behind a ship. The signals are digitized so that the data can be processed by computers. Thus, the structure of the rocks such as the layering pattern of beds (inclined or otherwise), folds, faults, thrusts and unconformities can be established.

Structural geologists also sometimes make use of *gravity* and *magnetic data* for some specific purposes such as in unravelling the nature of the continental and oceanic crust. *Magnetic anisotropy* of rocks may quantify the finite strain of rocks. Lithostratigraphy of subsurface rocks as established by magnetic studies leads to establishing a *magnetostratigraphy* that throws significant light on the geological history of an area. The results, in combination with other data, help structural geologists to establish deformation history of an area.

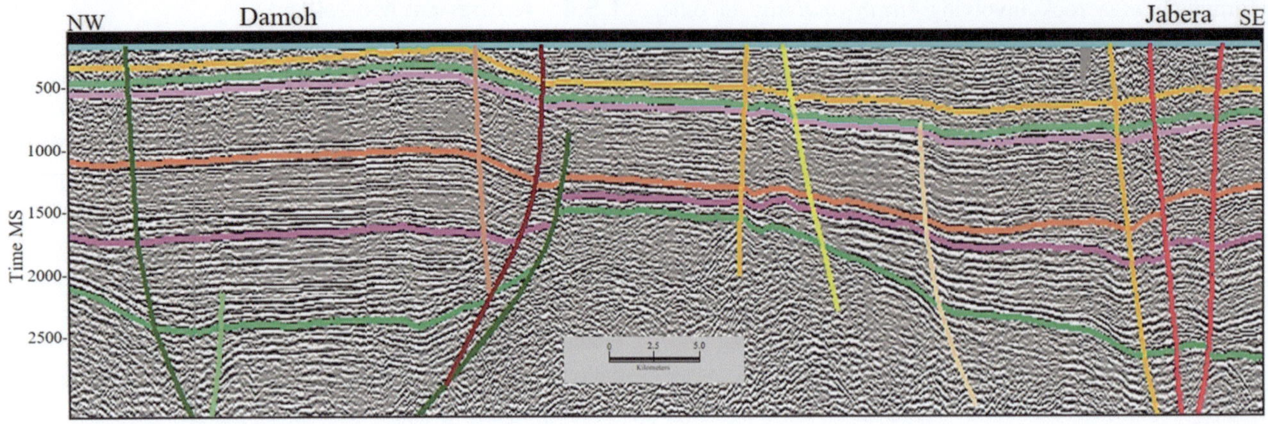

Fig. 1.13 Seismic section unravels the structure of subsurface rocks. The section shows interpreted seismic line through Damoh and Jabera structures in the Vindhyan basin close to the Narmada-Son Lineament, central India. Green horizon is Vindhyan base depicting the sedimentary basin floor. Several steep faults constituting the deformation framework can be seen. Yellow horizon represents an angular unconformity. 'Time MS' is time in milliseconds. (Seismic section with kind permission from Oil and Natural Gas Corporation, India)

Fig. 1.14 NASA image showing the arcuate trend of the Himalayan mountain chain at its southern front

1.5.6 Remote Sensing

Use of remote sensing methods in structural studies is increasing these days, especially with the availability of high-resolution satellite images, called *imageries*, and *aerial photos* that are made by aeroplanes. Nowadays, due to availability of high-resolution imageries, not only large-scale but even outcrop-scale structures of the rocks of an area can be known. If we make a sketch of the structure from the imagery, field studies can follow successfully without wasting time and energy. If the same is available in aerial photos, we are then able to get even a clearer picture of the structure. The follow-up fieldwork, known as ground check, may thus bring more refined and accurate data than provided by imageries. Imageries also help in precisely measuring the rate and direction of displacement of continents or of any point on the earth's surface as well as motion of plates. All this greatly helps in understanding the pattern of evolution of the continents and oceans in general and of some selected regions in particular.

And, on top of the above developments is **Google Earth** that virtually makes you 'see' rocks and their structures exposed on *any* part of the earth, and that too sitting at home. With the aid of satellites, Google Earth provides you the photographs of practically every point of the earth. With computer-controlled resolution, one can scale up and down the pictures. On scaling up, one can see the pictures on the size of continents, mountains and rivers, while on scaling down, one can see small places and even outcrops or rocks and the structures therein.

In addition to Google Earth, there are a few other agencies also, e.g. **NASA** (Figs. 1.14 and 1.15) and **Natural Earth**, that provide earth images on a public domain. We have presented here two examples of satellite images: one showing the arcuate trend of the Himalayan mountain chain at its southern front (Fig. 1.14), and the other showing structures of rocks north of the mountain chain (Fig. 1.15).

Fig. 1.15 NASA image showing a system of en échelon strike-slip faults (yellow rectangle) north of the Himalayan mountain chain. (Since this portion of the image stands uncoded, the location/area is unknown)

1.6 Where Is Structural Geology Today?

The major objectives of structural geology are to understand the geometry of different structures of rocks, the processes by which they were formed and how and why these structures are localized in some specific areas or zones. All these ultimately lead to explaining how rocks behave to the forces imposed by the earth.

In field, we can see a structure but cannot decide how it was formed. In this respect, things have changed with time. Now, we have several indirect methods that singly or in combination can throw light on the genesis of the structure. Remote sensing serves as an important tool with the use of images provided by satellites. Laboratory experiments through analogue materials can show the stages of formation of structures. But do they tell the real story? Here comes the question of time, i.e. geological time. In other words, can the stages or the processes of deformation lasting for a few hours or weeks in the laboratory really mimic the long interval of formation of a particular structure in geological time, ranging from several thousands to millions of years? Further, can the size of experimental models explain the real story of natural structures with sizes of kilometres or tens of kilometres? Nowadays, computers are of help especially in numerical modelling with the given parameters of material and some other physical properties. But since the computer will work according to its code, the results would be a matter of interpretation or simplification. Ultimately, fieldwork comes out to be the most important tool in structural geology as it provides the basic data about a structure and its associations. The above-mentioned methods may, however, add precision and additional information on the structure studied.

Box 1.1 Advent of Modern Structural Geology

In the last about 100 years, structural geology has advanced in several phases. Till the 1950s, the major emphasis remained focused possibly on the understanding of the structure of an area or region and outlining the events that gave rise to the structures or the structural architecture. The 1960s appear to be a turning point when Professor John G. Ramsay (1931–2021) published his book *Folding and Fracturing of Rocks* in 1967. This book can be considered as a landmark in structural geology. The book opened up several new avenues that gave rise to new thinking to structural geologists especially on stress, strain, fracturing and folding. An offshoot of the book is an understanding of rock deformation. Lisle et al. (2019) highlighted some major concepts of the book that have since been developed as major research aspects in structural geology: progressive deformation, finite strain determination, shear zones and folding analysis. Hobbs (2019) highlighted that the book has great impact on the development of structural geology during the last five decades.

In general, Ramsay's book had a mushrooming effect with the involvement of several disciplines such as mathematics, physics, mechanics and engineering together with experimental studies of structures with analogue substances. During the 1980s, all these aspects came in the fold of computers for handling large number of data and their simulation and solution. Undoubtedly, Ramsay's book can be described as a harbinger of modern structural geology.

So, all that we have discussed above can be described as modern structural geology. I am personally amazed to see how structural geology changed in my lifetime from a simple geological compass to the exciting world of computers, satellites and much more . . . .

1.7 Significance of Structural Geology

1.7.1 Academic Significance

- Study of structures throws light on the nature of the forces that are the primary causes for formation of structures. Also, analysis of the orientation of structures indicates the directions along which the forces have acted.
- Interpretation of the nature of deformation of rocks as exposed on the earth's surface throws light on the depth where the rocks were deformed. Structural geology can therefore unravel the deformation history of rocks.
- Structural studies, if carried out in conjunction with tectonics over a large region or a sector of the crust, can ultimately help in throwing light on the evolution of crust and distribution of land and sea.

1.7.2 Economic Significance

- Structural geology is important in the study of economic minerals of commercial value. Structures of rocks control localization of economic minerals of an area. Structural geology can therefore be successfully applied to mineral exploration.
- For the reason mentioned above, some structures such as faults and folds constitute hunting grounds for economic geologists.
- Faults can be pathways for hydrocarbons as well as for groundwater.
- Folds also constitute suitable structures for localization of economic minerals under certain situations.
- In petroleum exploration, the age-old *anticlinal theory* for localization of petroleum has yielded petroleum and gas successfully in several parts of the world. Anticlines are therefore believed to be potential sites for petroleum localization.

1.7.3 Societal Significance

- Structural geology is of great help to our society also. It works closely with civil engineers in several ways.
- Rocks selected for civic construction work should be strong enough to bear the imposed stresses. Presence of joints, foliation, cracks, micro-faults, etc. has adverse effects on the strength of a rock. The construction materials therefore need to be studied from structural angles.
- Structural geology plays a great role in selecting suitable site for dams, tunnels, bridges, pavement and various types of civic construction work. Therefore, in civic projects, structural study of the selected site is a common practice.
- In hilly terrains, construction of roads should be done carefully by taking into consideration the structural features of rocks.

1.7.4 Environmental Significance

- In recent years, environmental consideration especially for towns and human habitation has become an important agenda. The major objective is to check an area/region for environmental stability. In practically all environmental projects where development of an area or township is mooted, structural geologist generally finds his/her place as an active member. In such projects, structural geologist carefully studies the area from structural viewpoints. He/she reports the presence of some structures, if any, that may be detrimental for the project. In such cases, the area may be described as environmentally unstable. Structural geologists thus play an important role in the decision-making of several environmental projects.
- Structural geologist plays a great role in identifying *active faults* or *potentially active faults* of an area. With such faults, an area can be described as environmentally unstable, and, as such, this calls for a rethinking on development work in the area. Structural geologist can suggest remediation processes in such areas.
- Besides active faults, environmental stability of an area also implies that the area should be free from *neotectonic activities* for which the structural geologists are urgently needed for a detailed study.
- Structural geology thus plays a great role in environmental studies.

1.8 Summary

- *Structural geology* encompasses the study of all aspects of deformation structures such as their geometry, field relations, geographic distribution, genesis and related aspects. 'Structures' in structural geology generally connote *deformation structures* or *tectonic structures*.
- Structural geology is closely related to *tectonics*, which also deals with deformation structures but on a much larger scale, say a region or a segment of the crust.

- Structures in rocks are developed on various sizes. *Megascopic structures* are developed on several kilometres to tens of kilometres. *Mesoscopic structures* occur on a small outcrop to hand specimen scale. *Microscopic structures* can be identified only under a microscope.
- Common practice of studying deformation structures includes field studies, laboratory studies, experimental/ analogue modelling, numerical modelling, geophysical studies and remote sensing.
- Structural geology has academic, economic, societal and environmental significance.
- From the academic point of view, structural geology is concerned with deformational processes of rocks that explain how the lithosphere responds to stresses. From the economic point of view, structural geology helps in locating minerals of economic value including petroleum, natural gas and groundwater. From societal viewpoints, structural geology helps engineers to select suitable sites and construction materials for dams, tunnels, bridges, roads and other civic construction works. Structural geology has environmental commitments too as it helps in selecting environmentally congenial areas for human habitation, industries, agriculture and other uses.

Questions

1. What is structural geology? How does it differ from tectonics?
2. Mention where structural geology overlaps tectonics.
3. How do tectonic structures differ from nontectonic structures?
4. Why is it necessary to use the concept of scales in the study of deformation structures?
5. Describe, with examples, what could happen if we do not use structural work in development projects.
6. What are the important aspects of deformation structures for study in laboratory?
7. Give the benefits of experimental studies in structural geology.
8. How is remote sensing helpful in structural geology?
9. Explain why deformation structures of an area need to be studied by more than one methodology.
10. Briefly explain how structural geology is of help in environmental studies.

2 Attitudes of Structures

A. R. Bhattacharya, *Structural Geology*, Springer Textbooks in Earth Sciences, Geography and Environment,
https://doi.org/10.1007/978-3-030-80795-5_2

Abstract

While in field, the foremost thing a geologist should do is to note down the *attitude* of the rock or of any structure he/she intends to study. Attitude is a fundamental geometrical attribute that describes how a plane or a line is oriented on the surface of the earth. Attitude of any geological structure undoubtedly constitutes the starting point of any geological work of whatsoever nature. Measurement of the attitude of structures especially involves the concepts of *direction*, *planar structures*, *linear structures*, *dip of beds*, *strike line*, *bearing* and *back-bearing*. The *direction system* used by a structural geologist is either the *conventional system* or the *azimuth system*. The attitude of planar structures is defined by *dip of beds* and *strike line*, while those of linear structures by *plunge* and *pitch*. *Bearing* and *back-bearing* are used to locate any object or yourself in a topo-sheet. The most fundamental geological field instrument, *clinometer compass*, is described including its use in field. In general, the chapter provides the basic concepts of attitudes of structures, thus enabling a beginner to carry out independent field work in structural geology.

Keywords

Attitude of structures · Direction · True dip · Apparent dip · Strike · Plunge · Rake · Pitch · Clinometer compass

2.1 Introduction

Structural geology is basically a field-based discipline, and its study is accomplished in a few stages. The first is the collection of data in field. This is followed by processing of field data in the laboratory by means of suitable methods and techniques. All these lead to preparation of several types of diagrams, projections and structural maps. The next stage generally includes interpretation of the information or results that are made available from field and laboratory investigations. Understanding of structural details of any area or region, e.g. deformation pattern, deformation history and modes/mechanisms of rock deformation, requires basic data that are collected from field only.

Attitude is a fundamental geometrical attribute that describes how a plane or a line is oriented on the surface of the earth. Planes and lines may be oriented as inclined, vertical or horizontal with respect to the ground surface and may also show some geometric relations with the surface. In this chapter, we shall discuss some elementary concepts and methods pertaining to the attitudes of structures.

2.2 What Is 'Direction'?

In structural geology, *direction* is one of the most commonly used words. Direction can be defined as the *orientation of an object with respect to north*. On the surface of the earth, north has been taken as a key reference for all direction systems.

The most scientific and practical method to locate true north is the use of a magnetic needle. If freely suspended, a magnetic needle always points towards magnetic north, and therefore this method can be used to locate north from any part of the earth's surface. Since the magnet always remains horizontal when stationary (excepting near any magnetic body or close to the magnetic pole), the direction given by the magnet is always shown or indicated in the horizontal direction. Thus, if a plane is inclined towards say NE, the direction of its inclination is indicated or shown by a horizontal line (Fig. 2.1).

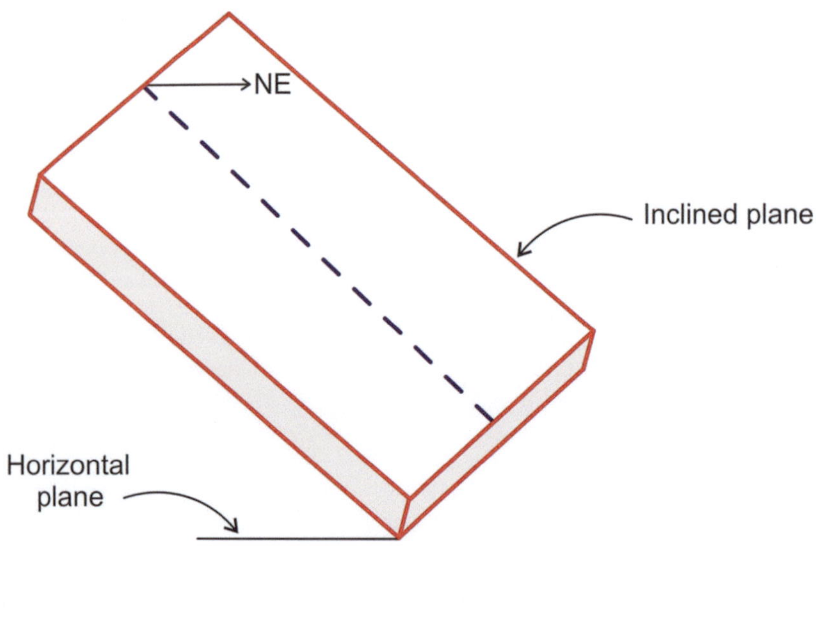

Fig. 2.1 The direction of any object, say a plane, is always indicated or shown by a horizontal line. For the inclined plane shown in the figure, the direction of inclination is NE and it is always shown by a horizontal line

2.3 The Direction System

Direction can be represented by two systems: conventional and azimuth.

2.3.1 The Conventional System

In the *conventional system*, the direction can be read with reference to a circle of 360° angle in which N (0°, 360°), E (90°), S (180°) and W (270°) are called *primary directions* or *cardinal directions*. The other directions can be expressed as, say, N30°E, E42°S, S28°W, W38°N, etc. These are read as 30° east of north, and so on. The bisector directions N45°E, E45°S, S45°W and W45°N are simply expressed as NE, SE, SW and NW, respectively. Note that N45°E should not be written as E45°N though geometrically both appear to be the same. The reason is that of the two directions N and E, N is the original or primary direction from which E is derived; that is why N is written first. Likewise, the direction along the bisector of S and E is written as SE and not ES because of the two directions S is the primary direction from which E is derived. These four directions, called *secondary directions*, are thus written as NE, SE, SW and NW. Another set of directions as dividers of these directions can thus be written as NNE, ENE, ESE, SSE, SSW, WSW, WNW and NNW. These are called *tertiary directions*.

2.3.2 The Azimuth System

In the *azimuth system*, north is taken as the starting point, i.e. 000° or 360°. For measuring any direction, one has to move clockwise, i.e. eastwards, and read the angle by which one has moved from 000°, say 067°, 142°, 214°, 307°, etc. The azimuth system is thus very simple because a direction can be expressed in short. As such, this system is highly practical and can be used easily for recording data during structural fieldwork. It is also much better for computer use too. I suggest geologists to use the azimuth system in their routine fieldwork.

2.4 Attitude of Planar Structures

In structural analysis, orientation of rocks as they occur in field is very important. Rocks can be considered to constitute planes, and these planes contain structural features within them as well as on their surfaces. Planes in rocks are penetrative features and thus can be represented in three dimensions. These are called *planar structures* such as bedding, fault planes, foliations, joints, dikes and sills.

Attitude of a planar structure is mainly given by its dip and strike. *Dip is the amount of inclination of a bed with the horizontal plane as measured in a vertical plane*. Orientation of a bed ABCD has been shown in Fig. 2.2. If a bed is horizontal, its amount of dip will be 00°, and if it is vertical, the amount of dip will be 90°. The bed will be described as inclined when it takes any amount between 0° and 90°. The *direction of dip* is always shown/indicated in the horizontal plane, as mentioned above, because we always read the direction by magnetic needle, which moves freely only when it is horizontal. *A strike line or strike of a bed is the line formed by the intersection of the bed with the horizontal plane*. A strike line always bears two important properties: (i) it always exists on the ground surface, and (ii) it is a straight line.

2.4.1 True Dip and Apparent Dip

We have earlier defined dip as the inclination of a bed with horizontal. Let us again consider a bedding plane ABCD (Fig. 2.3) in which AEB is the unit, right-angle triangle. Angle ABE is the dip of the bed, and BC is the strike line.

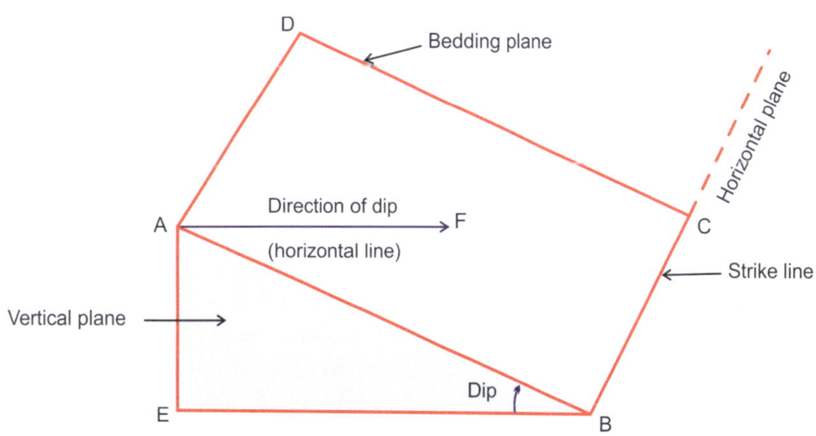

Fig. 2.2 Attitude of an inclined bedding plane ABCD. AEB is a vertical plane. The angle ABE is a dip. The bed dips in the right-hand direction. The direction of dip is shown by AF, which is a horizontal line. The bed intersects the horizontal plane along BC; this intersection is called strike line or strike

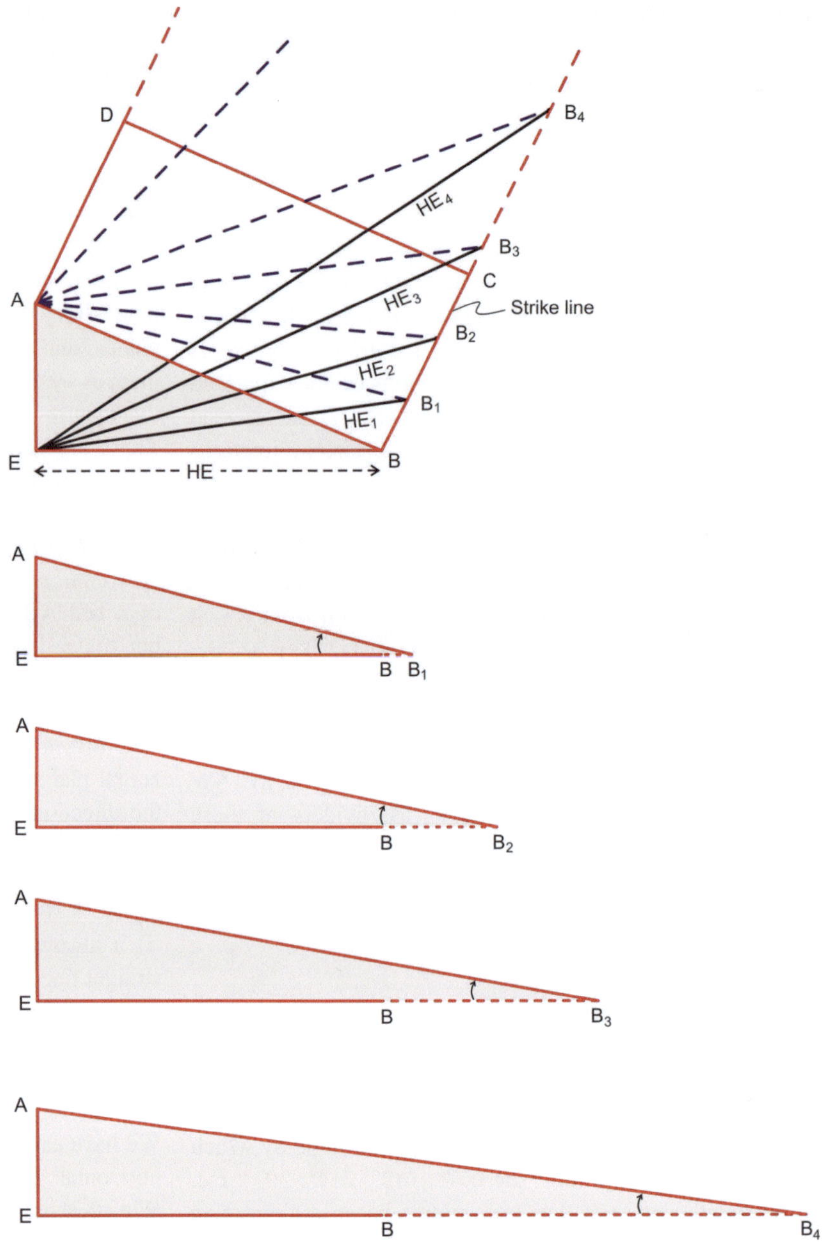

Fig. 2.3 True dip and apparent dip. ABCD is a dipping bed in which AEB is a right-angle triangle, which we consider as the unit triangle. The bed intersects the horizontal plane along BC, which is thus the strike of the bed. If we measure dip in some other directions, say towards B_1, B_2 . . ., the amount of dip progressively becomes smaller than measured along ABE and ultimately becomes 0° along AD, which is a direction parallel to the strike of the bed. The amount of dip measured along the vertical unit triangle is the highest and is called true dip of the bed. The amounts of dips measured along all other directions are lower than the unit triangle and are called apparent dips of the bed. The amount of dip measured along AD, i.e. parallel to the strike line, is zero

Now, if we measure the dip of the bed along a line, say AB_1, we get a different triangle AB_1E. Since the line EB_1 is longer than EB, the angle AB_1E would also be less than ABE. Likewise, we can measure the dip of the bed along some other line, say AB_2. In this case, the line EB_2 would still be longer than EB_1, and therefore the angle AB_2E would be less than angle AB_1E. In this way, we can measure the dip of the bed along any desired lines, say AB_3, AB_4, etc. In each case, as we move away from B, the length of the lines EB_1, EB_2, EB_3, etc. progressively becomes longer than EB, and so the angles AB_1E, AB_2E, AB_3E, etc. would progressively become smaller than ABE (Fig. 2.3, see successive lower figures from the unit triangle). Obviously, the angle ABE of the unit triangle would show the highest amount of inclination of the bed, and this is known as *true dip. True dip can thus be defined as the amount of inclination of a bed with the horizontal plane as measured in a vertical plane at right angle to the strike, and it is the highest amount of inclination of a bed.* From this, it follows that the inclination of a bed measured in any other direction than the true dip direction will always be smaller in amount than the true dip. This is known as *apparent dip. An apparent dip can therefore be defined as the amount of inclination of a bed with the horizontal plane as measured in any direction other than its true dip, and it is always smaller in amount than the true dip.*

The distance between E and B is called *horizontal equivalent* (HE) that can be considered as the equivalent length if the (inclined) bed (AB) is projected on the horizontal surface.

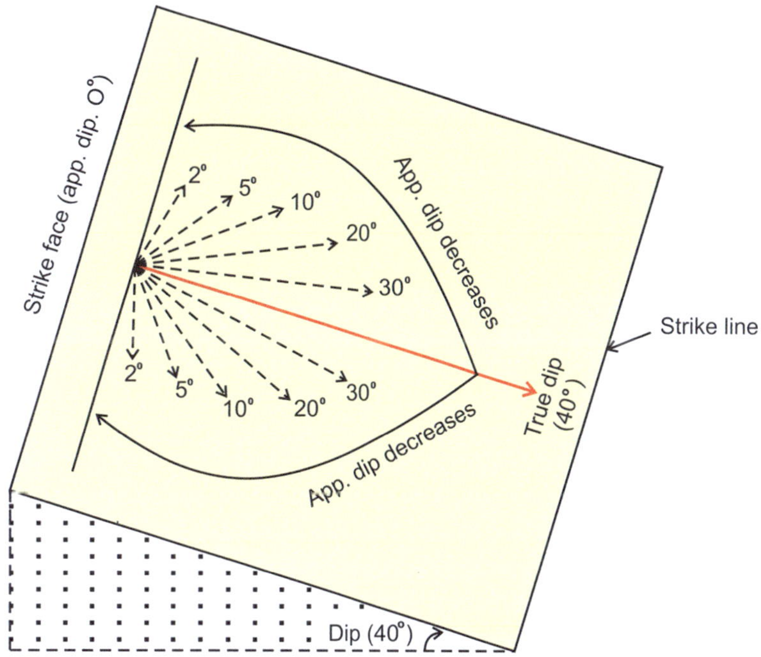

Fig. 2.4 The diagram shows an inclined plane (in yellow colour) with true dip of 40° in the direction shown by red line on the plane. Note how apparent dip progressively decreases on either side of the true dip direction. This is another way of expressing how as the horizontal equivalent increases in different directions the amount of dip decreases (see also Fig. 2.3)

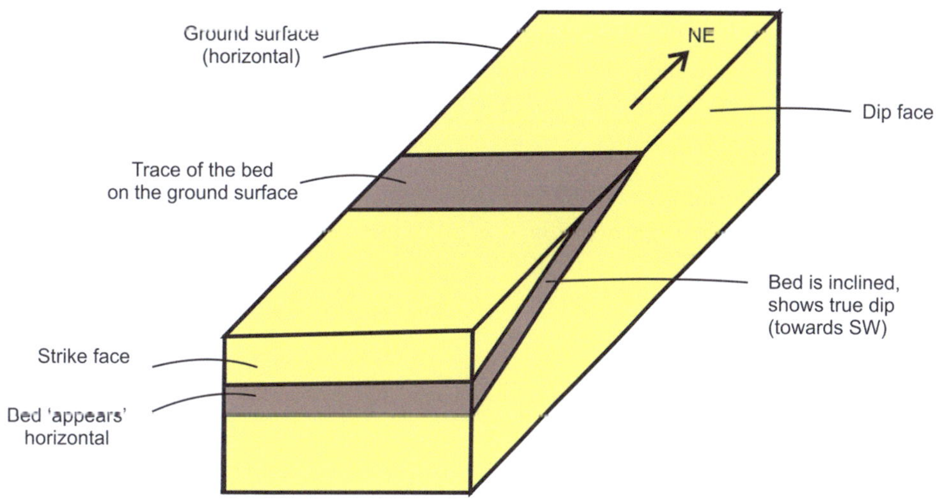

Fig. 2.5 A dipping bed (in brown colour) 'appears' horizontal along the strike face of an exposed rock. See also Figs. 2.3 and 2.4. The face on the right-hand side shows the bed in inclined position. This face is therefore called the dip face. If we gradually move towards the strike face, the bed 'appears' progressively less inclined as is apparent from Figs. 2.3 and 2.4. Ultimately, along the strike face, the same inclined bed 'appears' horizontal

From above, it follows that the horizontal equivalent EB in the unit triangle becomes progressively longer as we take measurements from B onwards along the strike line. It is therefore clear that the vertical plane AEB is perpendicular to the strike. The other vertical planes AEB_1, AEB_2, AEB_3, etc. are not perpendicular to the strike, and the dip measured in each of these vertical planes is called *apparent dip*. At the same time, the amount of inclination of the bed in all such vertical planes (AB_1E, AB_2E, AB_3E, ...) progressively reduces as we move from B onwards along the strike. In other words, the amount of dip shows a progressive reduction as we move away from the true dip.

The above scenario has further been explained diagrammatically in Fig. 2.4. On a bedding plane, all the above situations have been represented in which the amount of dip can be seen to progressively reduce from the true dip direction (highest amount) on both sides on the bedding surface till it reduces, on both sides of the true dip line, to 0° along the strike line. It implies that along the strike line, the amount of dip of a bed is 0°, i.e. the bed when viewed on the strike face *appears* horizontal, although it is a dipping bed. This situation has further been made clear in three dimensions in Fig. 2.5 in which one can notice that a dipping bed (in brown colour) appears horizontal at the strike face, as represented by the front surface in the diagram. This gives a simple rule: *all dipping beds when viewed along the strike face 'appear' horizontal*. Enigmatic situation like this is sometimes faced by geologists during fieldwork. In all such cases, the geologist should immediately check whether it is the strike face or the dip face. Not only this, if it is dip face, then he/she should go for the true dip face where the amount of dip will show the highest value.

2.5 Attitude of Linear Structures

Unlike planar structures that are penetrative, there are some structures in rocks that are commonly non-penetrative and as such are noticed only on the rock surfaces. However, some of them may penetrate up to a few millimetres in the rock. These features show linear attitudes and are therefore called *linear structures* (Chap. 15). As such, they can be represented in two dimensions. Common examples of linear structures include lineation, which is the linear alignment of mineral constituents of a rock. Fold axis, grooves and slickensides also constitute linear structures. Intersection of two planar structures of a rock constitutes a linear structure called intersection lineation. The attitude of a line or linear structure can be described in terms of plunge and pitch, as described below.

2.5.1 Plunge

Most rocks contain linear structures or lineations as commonly noticed on their surfaces. The attitude of a lineation is given by its *trend* and *plunge*. Consider an inclined plane ABCD of a rock (Fig. 2.6a). PQ is a linear structure contained in this plane. The projection of PQ on a horizontal plane AEFD is the line PR, called the *trend line*. Since the latter is a line, it can have two orientations on its either side: (i) its orientation with respect to the geographic north, called the *trend* of the linear structure, and (ii) one in which the linear structure is inclined along the plane ABCD; this gives the *plunge* of the linear structure. In Fig. 2.6a, the trend of the line is given by N38°E (in conventional system) or 038° (in azimuth system). The angle between PQ and PR is the plunge of the linear structure (Fig. 2.6b). *The plunge is defined as the angle made by a linear structure with its projection on the horizontal plane.* The attitude of any linear structure is thus given by two quantities, the direction of its trend and the amount of its plunge. In Fig. 2.6a, the attitude of the plunge is 40°N38E (in conventional system) or 40°/038 (in azimuth system). The inclination of plunge of any linear structure may vary from 0° to 90°.

In Fig. 2.6a, the linear structure shown by PQ may have any orientation in the plane ABCD. It may be parallel to the direction of true dip of the plane or parallel to any other direction falling in the wide range of apparent dips. If the orientation of the linear structure is parallel to the true dip of the plane, the amount and direction of plunge of the linear structure coincide with those of the true dip of the plane. For any direction of apparent dip, the amount of plunge varies between 0° and 90°. If the linear structure is parallel to the strike (i.e. AD) of the plane ABCD, the amount of plunge

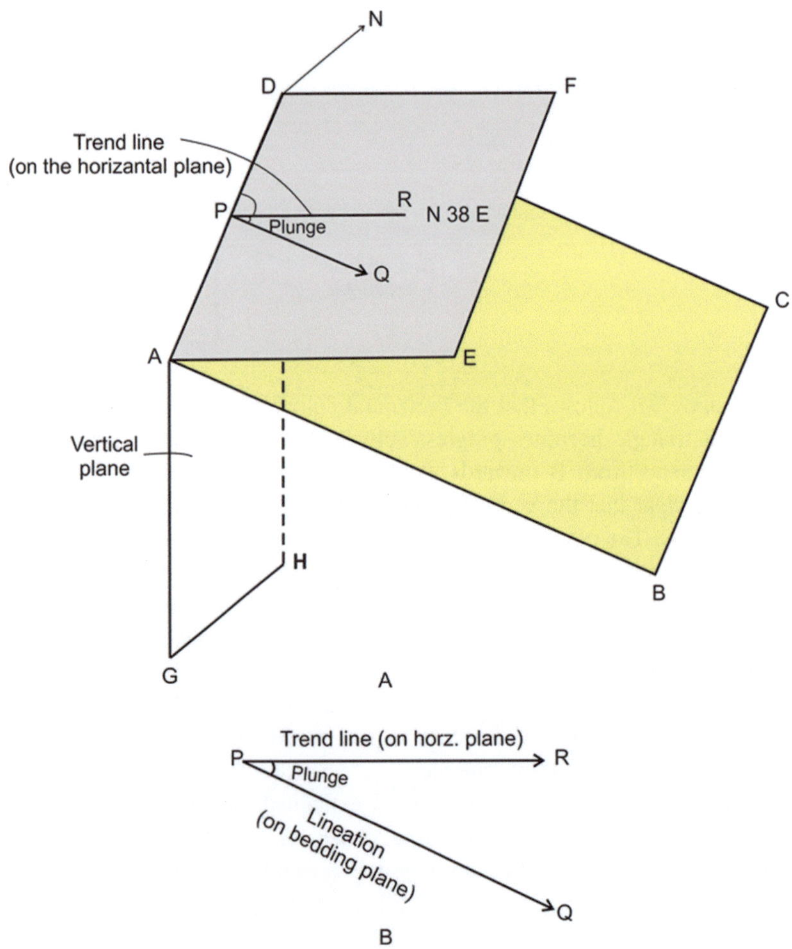

Fig. 2.6 Plunge of a line PQ on a bed. (**a**) ABCD is a dipping bed. PQ is a direction in which plunge is to be measured while PR is the projection of this line on to the horizontal plane AEFD. ADHG is the vertical plane along AD. (**b**) Plunge is given by the angle subtended by PQ with its horizontal projection PR

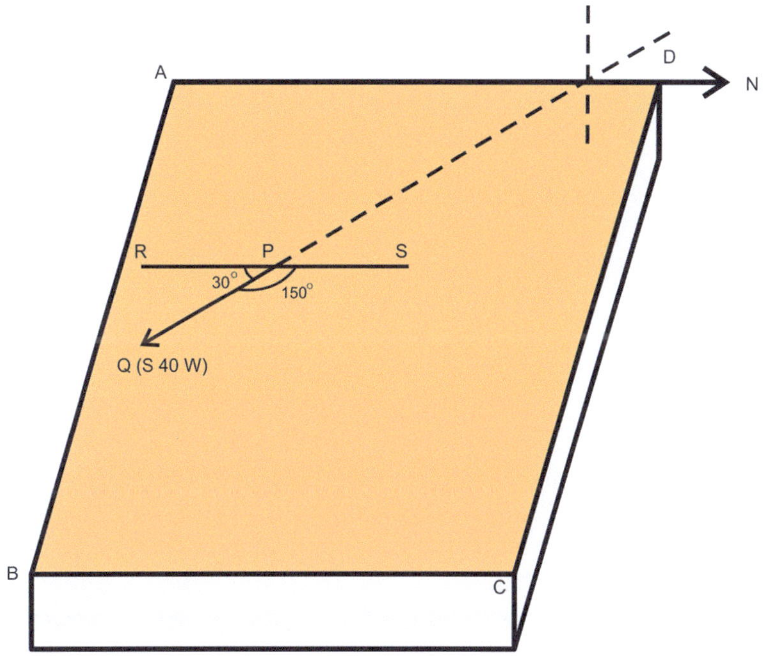

Fig. 2.7 Pitch of a linear structure PQ contained in a dipping plane ABCD. RS is the strike of the linear structure on the inclined plane. The acute angle RPQ (30°) gives the pitch of the linear structure

of the linear structure is 0° and its direction will be the same as the strike of the plane. Further, if a linear structure is horizontal, its plunge is 0°.

2.5.2 Pitch

For measuring the attitude of a linear structure, instead of measuring the plunge, we may measure its attitude with reference to the strike of the plane. In Fig. 2.7, ABCD is an inclined plane containing a linear structure PQ. The strike of the inclined plane is given by RS. The *acute angle* made by the linear structure with the strike of the plane is called the *pitch* (also called the *rake*) of the linear structure. PQ makes two angles with the strike RS, i.e. 30° and 150°. The pitch is always recorded by the acute angle (30°) only. In Fig. 2.7, angle RPQ is the rake of the line PQ.

2.6 Common Field Instrument

The starting point of most geological fieldworks is the measurement of attitude of the beds. For structural studies, however, a variety of structures and features of beds are required to be noted and measured in field for which a *clinometer compass* is an essential field instrument. The clinometer compass is a fundamental field instrument that mainly measures directions and dips of beds and is therefore meant for a beginner.

It may be mentioned here that following the basic principles of clinometer compass, a variety of compasses have been made by various manufacturers. Since they bear brand names, their description is beyond the scope of this book.

The clinometer compass (Fig. 2.8) is a simple instrument that contains a circular dial with a glass casing. The circular dial contains two circular scales. The outer scale is divided into 360 divisions that represent geographic directions. Each division is equal to one degree. The major geographic directions N, S, E and W; their subdivisions NE, SE, SW and NW; and their further subdivisions, i.e. NNE, ENE, ESE, SSE, SSW, WSW, WNW and NNW, are marked on this outer circle. The inner circle is divided into 90° on either side of the zero line. There is a magnetic needle and a pendulum both pivoted at the centre. The magnetic needle moves freely when the instrument is kept horizontal, while the pendulum moves freely when the instrument is kept vertical. When the instrument is in vertical position, the pendulum coincides the zero line of the inner circle. There is a handle or arm perpendicular to the plane of the dial. In the reading position of the compass when the north is towards the front of the observer, this handle vertically rotates up to 90° to the left-hand side of the observer only and not on the right-hand side. The instrument is made of non-magnetic material excepting the magnet.

2.6.1 Why Are E and W Interchanged in a Geological Compass?

In a geological compass, something that strikes a beginner is that the directions E and W are written in an interchanged or reverse manner. In the clinometer, this convention has been

Fig. 2.8 Photograph of a clinometer compass. (Courtesy Department of Geology, University of Lucknow)

made so that the direction of any object can be read directly from the compass. Hold the compass in your hand by keeping it horizontal (because the direction is always read in a horizontal plane). If you rotate the instrument to your left, the directions written on the dial *appear* to rotate to your right, and so also if you rotate it to your right, the directions *appear* to rotate to your left. This automatically creates some confusion; that is, if you align the instrument to read the direction of an object which is, say, actually towards the eastern direction, the instrument will give you a western direction. Therefore, in order to remove this confusion, the directions E and W are written in a reverse manner in the clinometer dial. This convention thus ensures that whatever the direction you read on the dial on rotation of the instrument, you always read the *actual direction* and you need not adjust this direction because of the so-called reversal phenomenon as described above.

Further, to remove the confusion of reversal of direction in a clinometer compass, it is also possible that we can reverse the N and S directions instead of E and W. This would also ensure direct reading of any direction from the clinometer. But since N and S are cardinal directions from which all other directions are derived, the position of these two directions is not disturbed, and only E and W directions are reversed in the clinometer compass. Also, for knowing the position of an object, we read N. Therefore, the direction N is not disturbed (interchanged) for the purpose of directly reading the position of an object.

Box 2.1 Measurement of Attitude by Clinometer Compass

One of the most common uses of clinometer compass is the measurement of the amount and direction of dip of an inclined plane, say a bed. For measuring the amount of dip, place the arm of the compass over the bed (Fig. 2.9). The pendulum moves freely in the vertical plane to rest at a point. Note that the magnetic needle does not work in this position as it works only when the instrument is horizontal. Now read the angle from 0°; this gives the amount of dip. The reader is here reminded of the fact that the dip of an inclined plane, as we have defined earlier, is measured in the vertical plane. That is why we measured here the angle of inclination of the bedding with the pendulum, which is kept in a vertical position during measurement; otherwise, on a horizontal plane, the pendulum will be at 0° position.

For measuring the strike of the bed, place the arm over the bed and rotate it till the pendulum comes at 0° position. Mark a line on the bed parallel to the arm. This line is parallel to the strike of the bed. Note down the directions of this line. Since this is a line, there will be two opposite directions differing in amount by 180°.

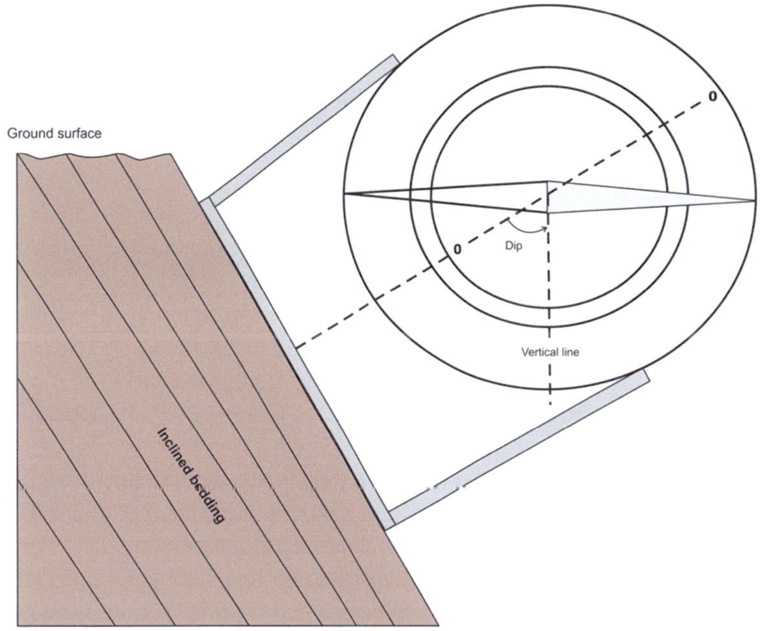

Fig. 2.9 Measurement of dip of inclined strata by clinometer compass (see text)

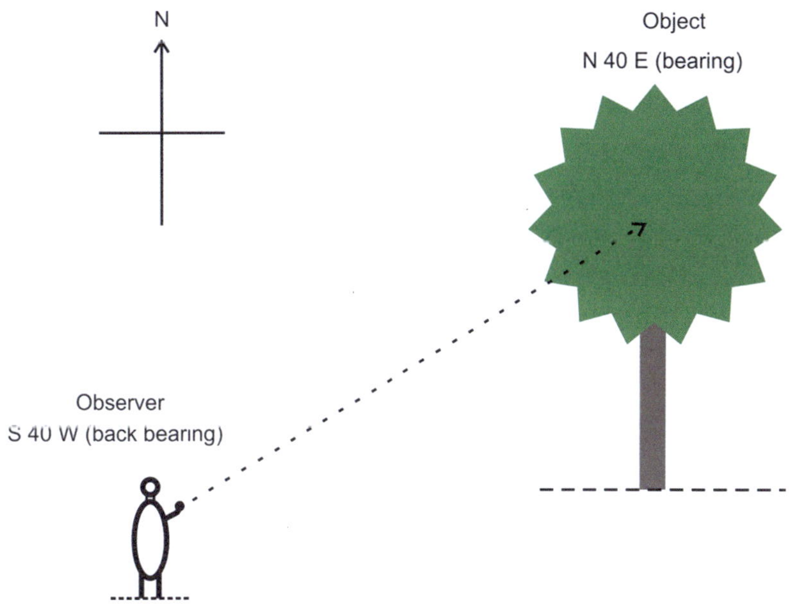

Fig. 2.10 Bearing and back-bearing. The observer reads the position of the object as N40°E; this is called bearing. Conversely, the position of the observer with respect to the object is S40°W; this is called back-bearing

2.7 Bearing and Back-Bearing

During geological fieldwork, one commonly needs two special types of readings from the compass: bearing and back-bearing. While in use, the handle of the compass should always be kept towards the left-hand side of the observer. In the frontal part of this handle, there is a slit to view the object for determining its direction. At the centre of the slit, there is a thin wire to precisely orient the object. If the object is oriented lengthwise, this wire can be aligned parallel to its length. If the object has no definite shape, the wire can be aligned at the centre of the object. In all such operations, ensure that the compass remains horizontal throughout. Once the object has been oriented, read the position of north in the outer dial. This will give the direction of the object *with respect to the observer* and is known as *bearing* (Fig. 2.10). If you read south direction in this position, you get the direction of the observer *with respect to the object*, and this is known as *back-bearing*. The bearing of an object can also be read directly by aligning the length of the handle towards the object and then reading the position of north. If you read the position of south, you get the back-bearing. This method gives the bearing or back-bearing quickly.

2.8 Summary

- Structural geology being basically a field science, collection of data on the attitude of beds and the structures therein constitutes the starting point of any structural work. A structure in rocks can be described in terms of some planes or lines whose orientation with respect to the ground surface (inclined, vertical or horizontal) constitutes what is called the *attitude*.
- *Direction is the orientation of an object with respect to north*. North is taken as a reference direction from which all directions have been derived. North is read by the needle of a compass because a horizontally suspended magnetic needle always points to north on one side and south on the other.
- The *direction system* involves (i) *conventional system* in which all directions are first subdivided into primary or cardinal directions (N, S, E, W), then into secondary directions (NE, SE, SW, NW) and then further into tertiary directions (NNE, ENE, ESE, SSE, SSW, WSW, WNW and NNW) and (ii) *azimuth system* in which north is defined by 000° as well as 360°. The other major directions, i.e. east, south and west, take up the azimuths of 090°, 180° and 270°, respectively. All other directions take up values between 000° and 360°.
- A structure can be planar (i.e. penetrative) or linear (i.e. non-penetrative). Attitude of planar structures is represented by the *amount of dip* and *direction of dip*, while that of linear structures contained in a plane is represented by *plunge* and *pitch*.
- Attitude of structures is measured with a geological compass called *clinometer compass*. In a geological compass, the directions E and W are interchanged or reversed so that the direction of any object can be read directly from the compass. The geological compass also helps to read *bearing* and *back-bearing* of an object. The former is the direction of an object with respect to the observer, while the latter is the direction of the observer with respect to the object. Bearing and back-bearing are commonly needed during fieldwork.

Questions

1. What do you mean by attitude of a structure? Explain with an example.
2. Define 'direction'. Describe how the direction system is derived.
3. Define true dip and apparent dip. Explain the difference between the two with the help of suitable diagrams.
4. What is strike line? Give its important properties.
5. Why are the directions E and W interchanged in a clinometer compass? Why are not N and S considered for this interchange?
6. Why the direction NE is not written as EN?
7. Is the direction N47°E correctly written? Give reasons.
8. Are the directions ENN, WSS, SEE and NWN correctly written? Give reasons in each case.
9. What do you mean by azimuth system of directions? Why is this system preferred during fieldwork?
10. Give the corresponding azimuth values of ENE, NNW, SW and W.

3 Stress

A. R. Bhattacharya, *Structural Geology*, Springer Textbooks in Earth Sciences, Geography and Environment,
https://doi.org/10.1007/978-3-030-80795-5_3

Abstract

A layman may laugh at you when you say that a rock feels the pinch of a force imposed on it! Yes, it does feel irrespective of the amount of force. In fact, the rock gets disturbed when a force is imposed upon it. The disturbance thus developed in a rock is called *stress* which is expressed as the force acting per unit area of a rock. If the applied stress persists uniformly till its amount exceeds the strength of the rock, the latter undergoes deformation, thus developing *strain*. Depending upon the specific conditions in the earth's crust, stress can be of various types such as hydrostatic stress, differential stress, deviatoric stress and lithostatic stress. The stress that is locked in the rocks when they were formed in the geological past is called *palaeostress* that has implications for the amount of deformation (strain) in rocks as well as for the directions in which the stresses had acted upon the rocks. Precise knowledge of the *present-day state of stress* inside the earth's surface is important in geology and in engineering geology, especially for various engineering and mining projects as well as for our preparedness for earthquakes. This chapter highlights some aspects of stress as relevant to structural geology.

Keywords

Stress · Principal stress axes · Uniaxial stress · Biaxial stress · Stress ellipse · Stress ellipsoid · Mohr stress diagram · Palaeostress · Stress tensor · Stress inside the earth

3.1 Introduction

Rocks in the earth's crust are under the burden of the overlying rocks as well as of their own weight. A rock mass is thus always being affected by externally applied forces. These forces develop *stress* within the rock. If the amount of stress developed in a rock exceeds its strength, the rock then yields and undergoes deformation; that is, it changes its shape and/or size to accommodate the induced stress. Although the rock deforms due to stress, the primary cause of stress is the forces that have acted upon the body. So, any study on stress should also take into account the nature of forces that act upon a body. Estimation of stress in rocks is a complicated subject. Although estimation of stress in rocks located below the ground surface up to depths of a few hundreds of metres is possible, it would give the present-day or contemporary stress only and not the actual stress that was operative when the rock was formed or deformed. Study of stress is important especially for getting an idea of direction of application of forces during deformation as well as in the estimation of strain in rocks. In this chapter, we shall discuss some basic concepts of stress in rock that are necessary for structural studies.

3.2 Force

A force is that which changes or tends to change the state of rest or of uniform motion of a body. The force that acts upon a body is applied from some external agency that either makes or tends to make the body move from its original position or stops or tends to stop a moving body. In the first case, the body is said to have been affected by an acceleration, which is measured as the force acting per unit time and tends to increase the motion of the body. A force is best expressed by *Newton's second law of motion*, i.e.

$$F = ma \tag{3.1}$$

where F is a force, m is the mass of a body and a is the acceleration produced in the body due to the force F. In the second case, the body is said to have been affected by a retardation, which is the force acting per unit time and tends to decrease, and finally stops, the motion of the body. A force is a *vector quantity*, while the mass is a *scalar quantity*. A vector quantity has a magnitude and a direction such as weight, velocity and gravity, while a scalar quantity has a magnitude only such as mass, speed and heat.

Considering a rock mass as a deformable solid or fluid, the forces acting upon it can be of two types, *body forces* and *surface forces*.

1. **Body forces** act upon a body and are therefore given by the volume or size of the body. Such forces act on every point of the body and are therefore measured in three dimensions. Thus, the bigger a body is, the larger its body force is such as the planetary bodies. Common examples of body forces include gravity and magnetic force. Effect of gravity produces weight to a body.
2. **Surface forces** act upon the boundaries of a body, i.e. along the area where the forces are in contact with the body. If we push a small wooden block by hand, the block moves for a distance. The distance moved is dependent upon the forces that the block has received along the contact area only. Further, instead of being displaced, a body, say a rock mass, may get deformed. In this case, the external forces have developed stress in the rock mass, and in order to accommodate the internally developed stress, the rock mass gets deformed. The deformation is thus an expression of the stress that the rock mass has developed. Surface forces are therefore relevant to structural geology

because rocks deform by the application of some external forces along a contact plane.

3.3 Types of Force

In structural geology, contact forces are more common, while the direct role of force at a distance has not yet been significantly highlighted and can thus be considered negligible. Balanced forces cause stress, while unbalanced forces produce stress plus displacement. The common types of forces (Fig. 3.1) are briefly described below.

Confining pressure (Fig. 3.1a) at a point is one that is exerted equally from all sides. In the earth's crust, the confining pressure is the lithostatic pressure developed due to the load of the overlying rocks. Since this pressure is equal on all sides of a body, it can be compared with the hydrostatic pressure produced by fluids. In the context of the rocks in the earth's crust, use of confining pressure is more common. *Shear* (Fig. 3.1b) can be compared to the effect produced when the cards in a pack are displaced past each other. Development of shear in a rock takes place by the formation of a couple, also called shear couple, caused by relative movement of rock material past one another. Shear can be produced by a shear stress. *Bending* (Fig. 3.1c) involves deformation of a beam when a load is applied across it. The beam deforms plastically. Bending occurs when the bending stress exceeds the yield strength of the material but below the ultimate strength. Because of the bending load, both compressive and tensile stresses develop in the beam. These stresses are of opposing nature (depending on, say, whether a fold is an anticline or a syncline), and as such there is a neutral surface in the beam where there is no bending stress. Depending upon the direction and location of the transverse stress, the beam assumes a variety of shapes after deformation. In structural geology, bending mainly has implications for fold mechanics. Some of the folds in crustal rocks have been interpreted to have formed by bending (see Chap. 8). *Torsion* (Fig. 3.1d) involves deformation of a body when it is twisted by applying a stress in one direction at one end while the other end remains either motionless or twisted in opposite direction. In other words, torsion involves application of couples that act in parallel planes but in opposite directions and about the same longitudinal axis of rotation. In crustal rocks, situations favouring the development of torsion practically do not exist. As such, torsion as a stress in rocks is an unlikely process. A torque (Fig. 3.1e) is a force that tends to rotate a body. A *torque* may increase or decrease the speed of rotation of the body. Since rotational forces seldom develop in rock masses of the earth, application of torque in structural geology is therefore negligible to absent.

3.4 Stress

Force acting upon a body or a rock develops stress within it. The body is then said to be under stress due to the applied force. If F is the amount of force acting upon a unit area A of a body, then the stress (σ) developed in the body due to this force is given by

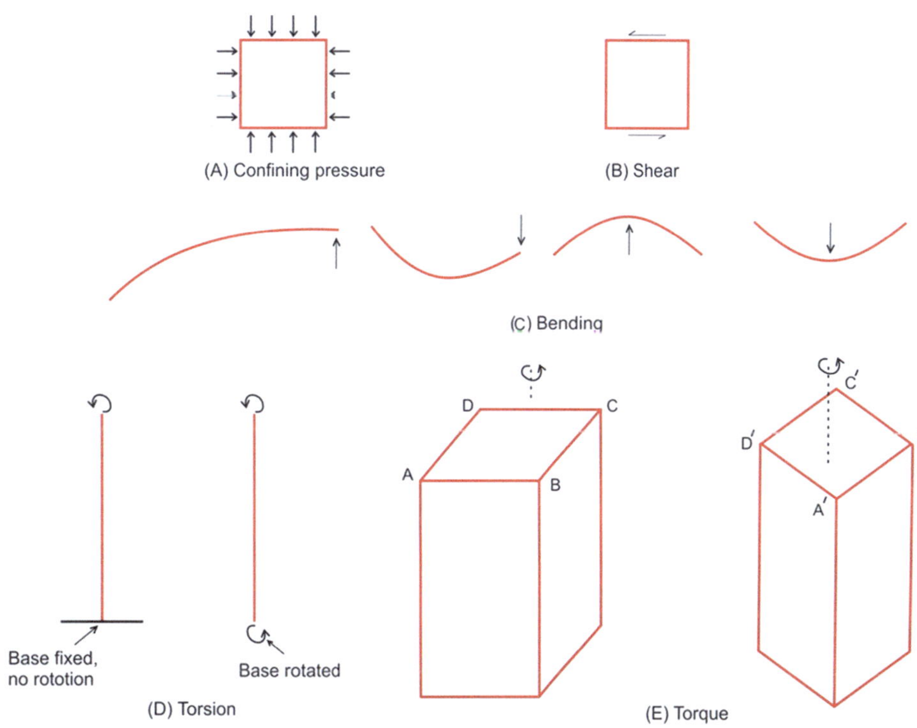

Fig. 3.1 Common types of forces. (**a**) Confining pressure. (**b**) Shear. (**c**) Bending. (**d**) Torsion. (**e**) Torque. (See text for details)

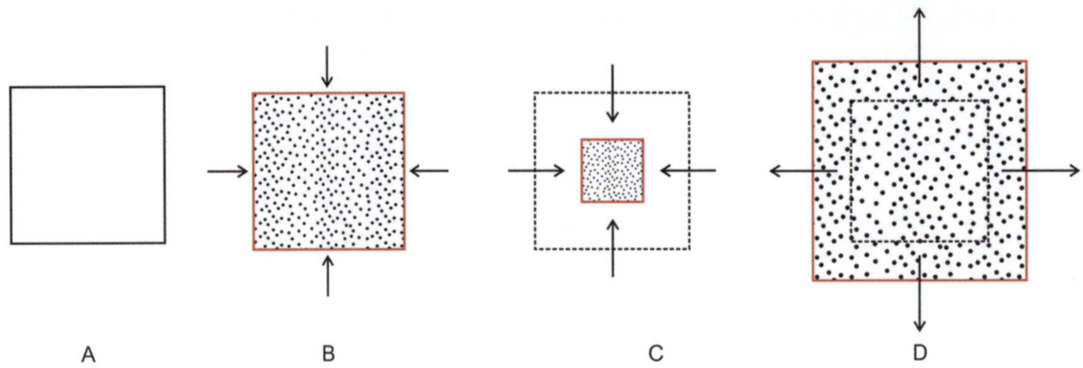

Fig. 3.2 Stress acting upon a body as explained diagrammatically. (**a**) A body without any externally applied stress continues to remain in its original shape. (**b**) The body under the influence of equal amount of stress acting from all sides develops stress within it. (**c**) If the applied stress persists uniformly and its amount exceeds the strength of the body, the latter undergoes dilatation. Since the stress is compressive, the body shows reduction of volume, which in two dimensions causes reduction in area as shown in **c**. (**d**) If the applied stress is tensile, the body would show increase of volume, which in two dimensions causes increase in area as shown in **d**

$$\sigma = F/A \tag{3.2}$$

Unlike force, a stress has no directional significance but has a point of application. The stress developed in a rock interacts with the constituents of rock and tends to develop physical change or deformation (Fig. 3.2) to the rock, which we call *strain*. Thus, stress causes strain in a rock. If stress is the cause, then strain is the effect. Obviously, if a rock is not subjected to any stress, it continues to remain in its original shape and volume (Fig. 3.2a). If the body is under the influence of equal amount of stress acting from all sides, it is said to be under stress (Fig. 3.2b). If on the other hand the stress overcomes the strength of the rock, the latter succumbs to the stress and shows changes in its shape or volume, and we then say that the rock is strained or deformed. A stress will cause a change in shape as shown in Fig. 3.2c (compressive stress) and in Fig. 3.2d (tensile stress).

In the above definition of stress, we have assumed that the distribution of stress is uniform within the body. In practical sense, it is generally not so, and the stress developed in a body due to a force F may vary from cross section to cross section. In such case, the stress at a point can be given as

$$\sigma = \lim_{\Delta A \to 0} \Delta F/\Delta A \tag{3.3}$$

It may be noted here that although the above two equations, i.e. Eqs. (3.2) and (3.3), provide values of stress, there are subtle differences. The former equation provides the value of stress developed in a body due to a force F assuming that the distribution of stress is *uniform* within the body. The latter, on the other hand, gives value of stress at a point assuming that the distribution of stress, in practical sense, is *not uniform*.

The stress that causes deformation to a rock can be estimated for many minerals via experimental mechanics. The study and analysis of stress help in throwing light, among others, on the directions of the applied forces and explaining the geometry, stages and pattern of deformation of a rock mass. *Stress analysis* helps in the designing of structures such as dams and tunnels. It is also used in several geological phenomena such as plate tectonics, earthquakes, volcanism and landslides.

3.5 Units of Stress

The unit of stress can be derived from the unit of force. In the International System of Units (SI), the unit of force is newton (N). One newton is defined as the force required to produce an acceleration of 1 m/s to a mass of 1 kg. In the CGS system, the unit of force is dyne; 1 dyne = 10^{-5} N. The most commonly used unit of pressure is *bar* and that of stress is *megapascal* (MPa). Atmospheric pressure is expressed in *pascal*. On the surface of the earth, 1 atmospheric pressure is equal to 100,000 Pa, which for the sake of simplicity is taken to be equal to 1 bar. 1 MPa is equal to 10 bars, and so 100 MPa is equal to 1000 kilobars or kba. Thus, any further conversions can be easily made.

3.6 Tensile Stress and Compressive Stress

In stress analysis, it is important to know the orientation of stress with respect to the body. We can imagine a plane upon which a stress is being applied at a point. This enables us a two- or three-dimensional study of stress, which can be of three important types (Fig. 3.3): (a) *Tensile stress* (Fig. 3.3a) that develops when the stress acts along a plane in opposite directions when the body is in equilibrium: as a result, the

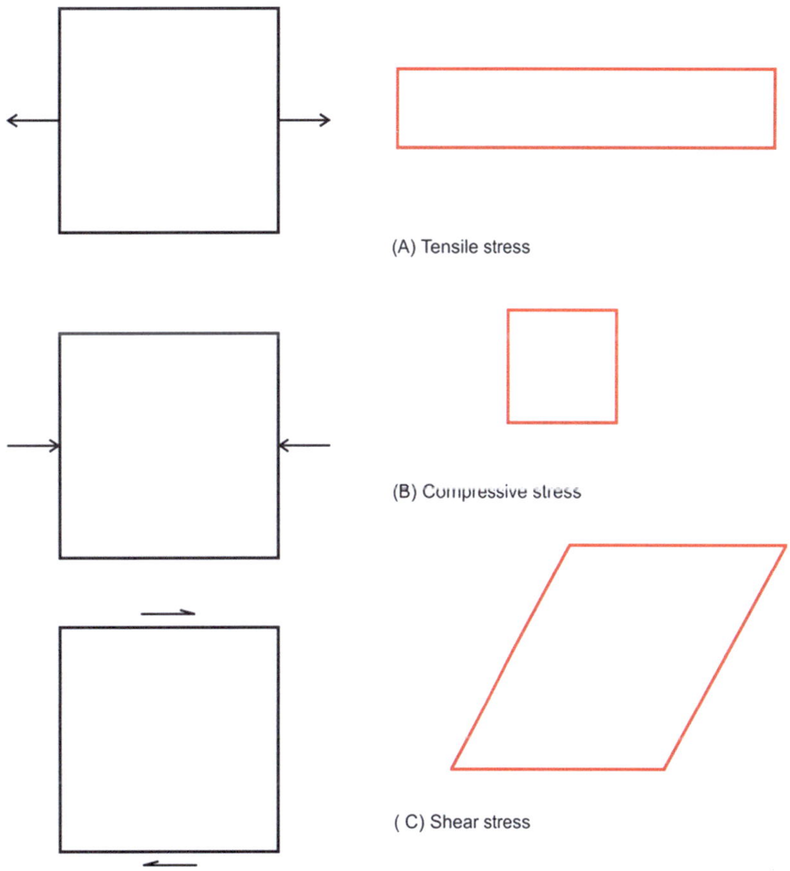

Fig. 3.3 Diagrammatic sketches to show the three common types of stresses. (**a**) Tensile stress acting in opposite direction; the body gets stretched in the direction of stress. (**b**) Compressive stress acting towards each other; the body gets shortened. (**c**) Shear stress. A square changes to a rhomb

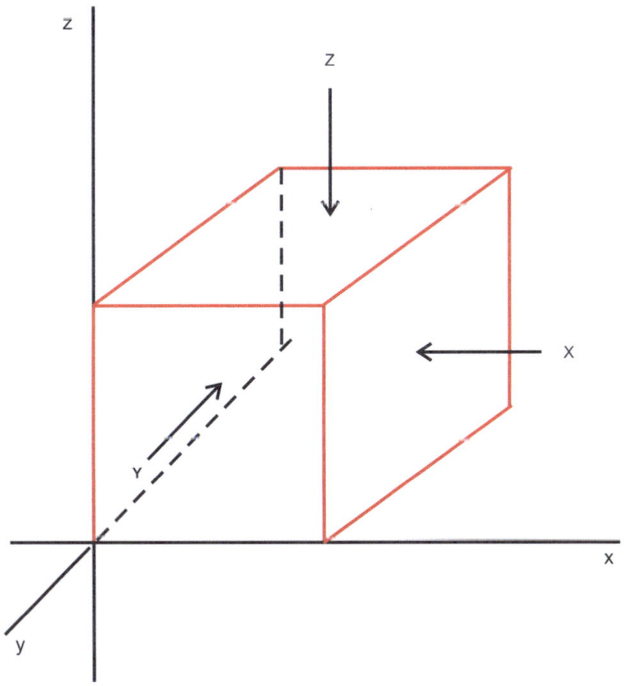

Fig. 3.4 Principal stress axes. Assuming that the cube is under stress from all directions, we can imagine that all the stresses are represented by only three stresses acting parallel to each of the coordinate axes *x*, *y* and *z*. Each of these stresses is called a principal stress. Thus, the stress acting parallel to *x*-axis is designated as X, while the others are Y and Z

body gets stretched. Tensile stress is assigned a negative (−) sign in structural geology. (b) *Compressive stress* (Fig. 3.3b) that develops when the stress acts perpendicular to the plane when the body is in equilibrium: it means that an equal and opposite stress has been active. As a result, the body gets shortened. Compressive stress is assigned a positive (+) sign (because in geology tensile stress is least likely to exist in nature). Since each of the above two types of stresses, i.e. compressive and tensile, acts perpendicular to the plane of the body, each is called direct stress or normal stress. (c) Shear stress (Fig. 3.3c) that develops when the stress acts in opposite directions but not along the same plane or line: as a result, the body shows angular changes of its original planes and thus changes its shape. A square for example is changed to a rhomb, while a rectangle is changed to a parallelogram.

3.7 Principal Stress Axes

Analysis of stress becomes easy if we imagine that all the stresses acting upon a body are represented by only three stress components acting parallel to each of the coordinate axes *x*, *y* and *z* (Fig. 3.4). The normal stress acting parallel to each axis is called a principal stress. The principal stress is normal stress acting on a plane that has zero shear stress. Of these three stresses, one is considered to take a maximum value, while others take a minimum and an intermediate value. The coordinate axes along which the stresses act are called *principal axes of stress*. The stress acting parallel to *x*-axis is designated as X, while the other two stresses are designated as Y and Z. This convention enables to know the orientation of any principal stress in the coordinate system. It is also assumed that shear stresses acting along each of these principal axes of stress are zero. As such, each principal stress represents the normal stress.

3.8 Uniaxial Stress

If a stress acts along a linear body (Fig. 3.5a), say a rod or a bar, we say that the body is subjected to forces that act along one direction only, i.e. along its length. In this case, two of the three principal stresses are not acting and can therefore be considered to be of zero value. This type of stress is called a *uniaxial stress*, and the bar is said to be under uniaxial stress. If σ_1 is the principal stress acting along the length of the body, then $\sigma_1 \neq 0$, $\sigma_2 = 0$ and $\sigma_3 = 0$. If the amount of force is F and the cross-sectional area of the rod is A, the uniaxial stress thus developed in the rod is given by $\sigma = F/A$. If the forces act in opposite directions along the length of the rod, the latter is said to be under uniaxial tensile stress (Fig. 3.5b) that causes increase of length of the rod. With forces acting towards each other, the rod is said to be under uniaxial compressive stress (Fig. 3.5c) that causes shortening of length of the rod. In structural geology, according to the *convention of sign* as mentioned above, the tensile stress is assigned a negative (−) sign while the compressive stress a positive (+) sign.

3.9 Stress Ellipse

It is possible to know the two-dimensional stress at a point if the normal stress (σ_n) and shear stress (τ) are known. In this case, these two stress components act on a plane of any orientation passing through the point. An infinite number of planes are thus possible. We can however consider a simpler case in which all the normal stresses acting on the planes arc represented by either a compressive stress or a tensile stress. In such case, if we represent all the stresses acting on the body by lines, the longest and shortest lines trace an ellipse (Fig. 3.6) called the *stress ellipse*. This ellipse can be traced if two of the stresses acting perpendicular to each other are

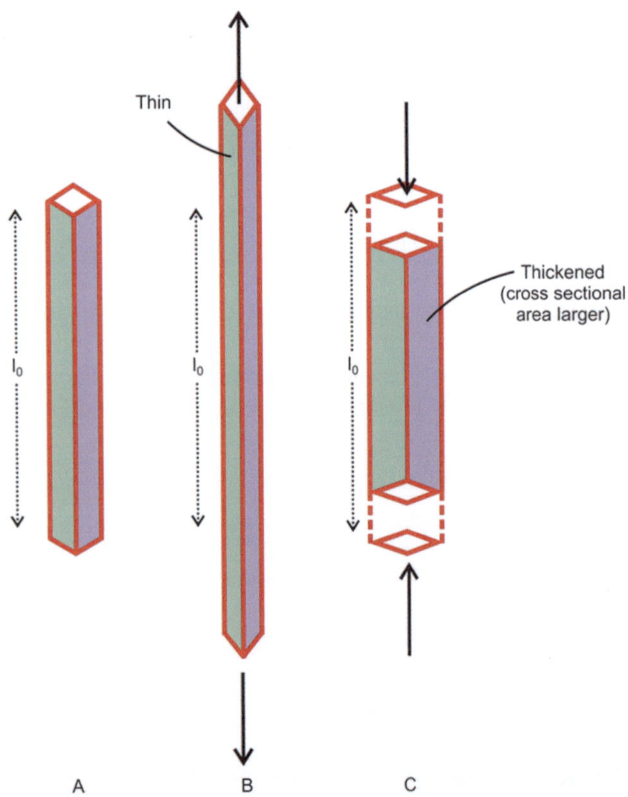

Fig. 3.5 Uniaxial stress showing stress acting along the length of a bar. (**a**) Original length (l_o). (**b**) The bar is under uniaxial tensile stress causing increase of its length. (**c**) The bar is under uniaxial compressive stress causing shortening of its length

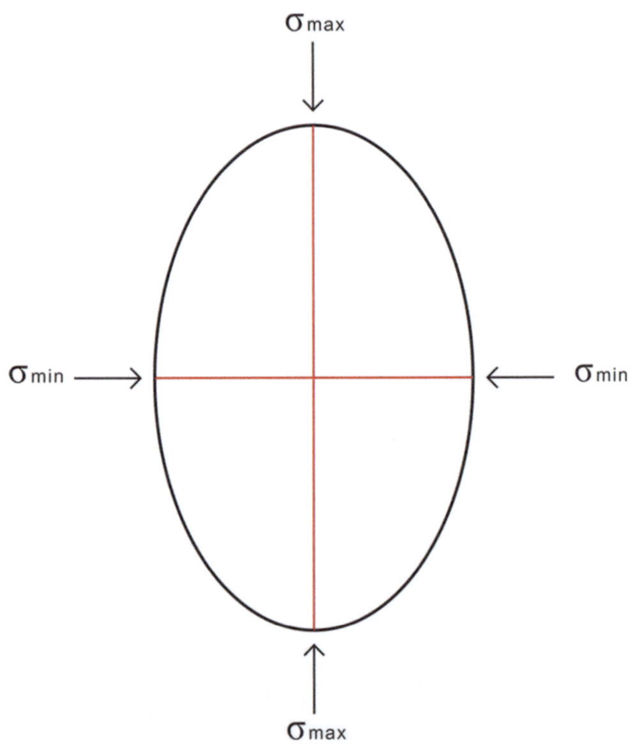

Fig. 3.6 Stress ellipse. The long and short axes represent the maximum stress and the minimum stress, respectively. The stresses are compressive in this case but can be tensile also

known. *A stress ellipse is thus a graphic method of showing relationships between the maximum and the minimum stresses.*

3.10 Biaxial Stress

When all the stresses act upon a body, the state of stress is known as *biaxial stress* or *two-dimensional stress*. A biaxial stress system is represented in two directions: a normal stress and a shear stress. Therefore, of the three principal stresses, one is equal to zero, and as such the biaxial stress is represented by

$$\sigma_1 \neq 0, \sigma_2 \neq 0, \sigma_3 = 0$$

A biaxial stress state in a body is thus developed when it is subjected to only two stresses σ_1 and σ_2 acting on its plane while the normal stress σ_3 is zero. Since all the stresses act in one plane, the biaxial stress is also called *plane stress*.

Application of biaxial stress in crustal rocks is as yet a matter of debate. As such, the concept of biaxial stress in structural geology is limited to geomechanics only, i.e. laboratory testing of materials mainly for engineering purposes.

3.11 Biaxial Stress on a Plane

We now consider a case when a force (F) acts at an angle to the plane (Fig. 3.7a). In this case, the force can be resolved into two components: (a) *direct* or *normal stress* that acts perpendicular to the plane and is denoted by σ (sigma) and (b) *shear stress* that acts parallel to the surface and is denoted by τ (tau) (Fig. 3.7b). Since the body is in equilibrium, we can imagine that σ is being affected by an equal and opposite component N. The shear stress τ can further be resolved into two components τ_1 and τ_2 at right angles to each other but in the same plane (Fig. 3.7c). Thus, we have resolved a force F acting on a plane at a point P into three stresses σ, τ_1 and τ_2.

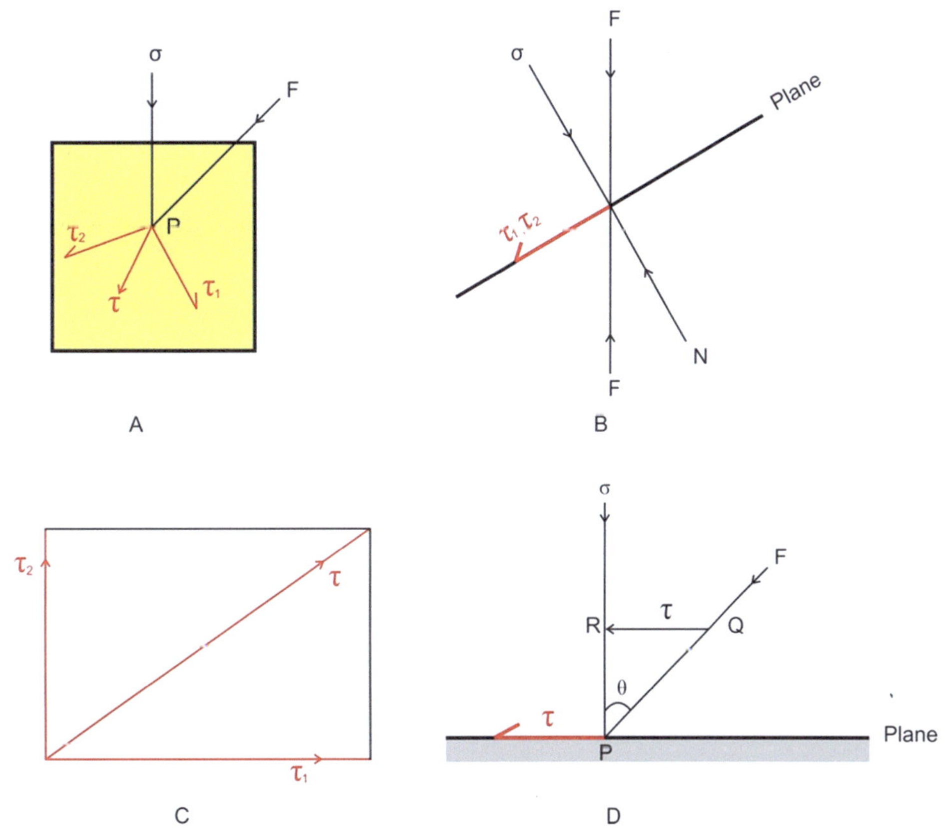

Fig. 3.7 Biaxial stress acting on a plane. (**a**) A force F acting at some angle to the plane is resolved into two stresses: one acting perpendicular to the plane, called normal or direct stress (σ), and the other acting parallel to the plane, called shear stress (τ). On the plane, τ can further be resolved into two mutually perpendicular stresses τ_1 and τ_2. (**b**) Another way to show the perpendicular position of the normal stress to the plane while the two shear stresses τ_1 and τ_2 remain along the plane. If the body is in equilibrium, the normal stress σ is subjected to an equal and opposite stress N acting at the same point. (**c**) The shear stress τ can be resolved into two mutually perpendicular stresses τ_1 and τ_2 by the law of parallelogram. Thus, the force F has been resolved into three mutually perpendicular stresses σ, τ_1 and τ_2. (**d**) The stress system shown in B is reproduced in a simple way on a horizontal plane. The force F acting on the plane is resolved in two mutually perpendicular stresses: normal stress (σ) and shear stress (τ)

Let us consider the above stress system in a rather simple way. The normal stress (σ), shear stress (τ) and applied force (F) are shown in Fig. 3.7d. From any point Q along the line of applied force, a perpendicular can be drawn on the line of stress to meet at R. In the triangle QRP, length QR then represents the shear stress (τ) while the lengths QP and RP represent the applied force (F) and the normal stress (σ), respectively. From the triangle QRP, we thus get the following:

$$\begin{aligned}\sin\theta &= \frac{\tau}{F}\\ \cos\theta &= \frac{\sigma}{F}\\ \tan\theta &= \frac{\tau}{\sigma}\end{aligned} \tag{3.4}$$

Geometrical representation of 2D stress on a plane thus helps in getting values of some unknown parameter if other two parameters are known.

In the coordinate system, the 2D stress can be represented by abscissa (x) and ordinate (y) as shown in Fig. 3.8. In the case of a normal stress (σ), each face of the plane can be considered to be represented by a set of two equal and opposite stresses, σ_{xx} and σ_{yy} (Fig. 3.8a), while in the case of a shear stress, the plane can be considered to be under the influence of two equal stresses, σ_{xy} and σ_{yx} (Fig. 3.8b), acting on the sides of the plane. The 2D stress σ_{ij} acting on the plane can then be represented by four components as shown by the following matrix:

$$\sigma_{ij} = \begin{bmatrix}\sigma_{xx} & \sigma_{xy}\\ \sigma_{yx} & \sigma_{yy}\end{bmatrix} \tag{3.5}$$

The state of two-dimensional stress acting on a plane such that there are only four components, as expressed above, and of which there are only three independent components, is called *plane stress.*

If, instead of designating the two orthogonal axes as x and y, we represent them by x_1 and x_2, the state of 2D stress at a point can be expressed by the following simple matrix:

$$\sigma_{ij} = \begin{bmatrix}\sigma_{11} & \sigma_{12}\\ \sigma_{21} & \sigma_{22}\end{bmatrix} \tag{3.6}$$

This matrix has only four components. Since $\sigma_{12} = \sigma_{21}$, the matrix has only three independent components.

3.12 Mohr Two-Dimensional Stress Diagram

Christian Otto Mohr, a German engineer, developed a graphical method in 1882 that represents in two dimensions the relationship between shear stress and normal stress. From the graph, it is possible to compute the values of shear stress, normal stress and angle of failure at the point when a crack is developed, i.e. when failure occurs. Mohr's theory can thus be applied to predict the failure of brittle materials.

In a *Mohr diagram*, the normal stress σ_n and shear stress τ_s are represented by the abscissa and ordinate of a graph. The known values of the greatest principal stress σ_1 and the least principal stress σ_3 are plotted on the axis of normal stress. According to Mohr's theory, it is possible to represent the state of failure by paired values of normal stress and shear stress on any plane with any orientation within the body by constructing a circle passing through σ_1 and σ_3. This is called *Mohr circle* (Fig. 3.9). It is also known as Mohr stress diagram. A circle is drawn passing through the values of the greatest principal stress (σ_1) and the least principal stress (σ_3) along the x-axis. The difference between the greatest and the least principal stresses, i.e. ($\sigma_1 - \sigma_3$), is called *differential*

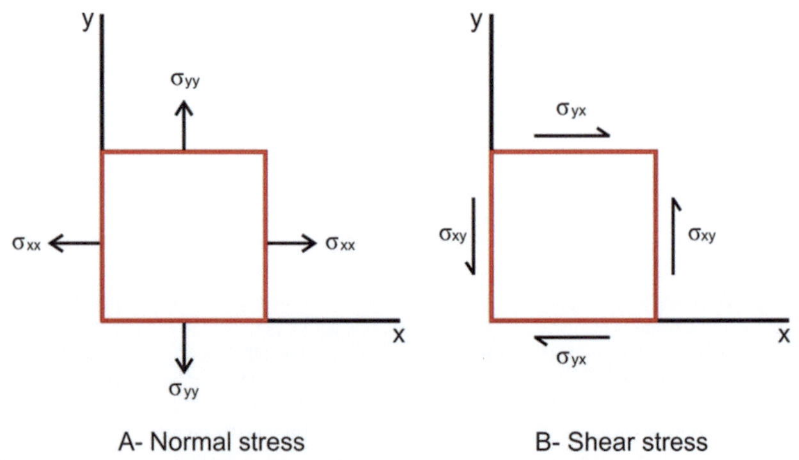

Fig. 3.8 Representation of two-dimensional stress on a plane in the coordinate system. (**a**) Normal stress, (**b**) shear stress (see text)

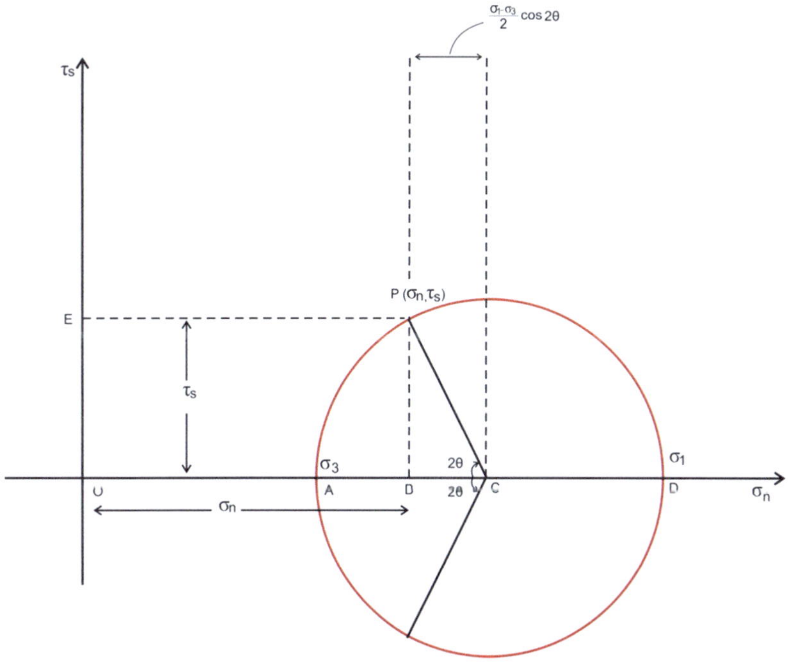

Fig. 3.9 The Mohr circle or the Mohr stress diagram. See text for details

stress and is represented by the diameter of the circle. If the angle between the shear fracture and the axis of normal stress is plotted on the circle, a point P where a fracture occurs is obtained on the circle. This is also known as the *fracture criterion*.

According to Mohr's theory, the coordinates of this point (σ_n, τ_s) should give the values of σ_3, normal stress and shear stress at the time of failure. 2θ is double the angle between σ_3 and the plane. From P, perpendiculars are drawn on the axes of τ_s and σ_n. Graphically, the values of normal stress and shear stress can be represented by lengths of PE and PB, which can be computed as below.

The diameter of the circle is ($\sigma_1 - \sigma_3$) so that its radius PC is 1/2 ($\sigma_1 - \sigma_3$). The centre C is {1/2 ($\sigma_1 - \sigma_3$), 0}. In the triangle PBC:

$$\mathrm{PB} = \tau_s = \mathrm{PC}\ \ \sin 2\theta = 1/2(\sigma_1 - \sigma_3)\sin 2\theta$$

Further, the x-coordinate of P is

$$\begin{aligned}\mathrm{OB} = \mathrm{OC} - \mathrm{BC} &= 1/2(\sigma_1 - \sigma_3) - \mathrm{PC}\ \ \cos 2\theta \\ &= 1/2(\sigma_1 + \sigma_3) - 1/2(\sigma_1 - \sigma_3)\cos 2\theta\end{aligned}$$

The x-coordinate of C should give the value of confining pressure at failure, i.e.

$$C = 1/2(\sigma_1 + \sigma_3) \tag{3.7}$$

Thus, at the point of failure, the values of shear stress, normal stress and confining pressure are given by

$$\tau_s = 1/2(\sigma_1 - \sigma_3)\sin 2\theta \tag{3.8}$$

$$\begin{gathered}\sigma_n = 1/2(\sigma_1 + \sigma_3) - 1/2(\sigma_1 - \sigma_3)\cos 2\theta \\ C = 1/2(\sigma_1 + \sigma_3)\end{gathered} \tag{3.9}$$

The Mohr circle, thus, gives a relationship between shear stress, normal stress and orientation of any plane. Accordingly, if the values of any two variables are known, the value of the third variable can easily be known.

From above, the radius r and the centre (C) of the circle σ_c are

$$r = 1/2(\sigma_1 - \sigma_3) \tag{3.10}$$

$$\sigma_c = 1/2(\sigma_1 + \sigma_3) \tag{3.11}$$

With these values, the equation for the Mohr circle is

$$\begin{gathered}(\sigma_n - \sigma_c)^2 + \tau_s^2 = r^2 \\ \{\sigma_n - 1/2(\sigma_1 + \sigma_3)\}^2 + \tau_s^2 = \{1/2(\sigma_1 - \sigma_3)\}^2\end{gathered} \tag{3.12}$$

On solving Eq. (3.12) for τ_s, we get

$$\tau_s = \pm\sqrt{\sigma_n(\sigma_1 + \sigma_3) - \sigma_n^2 - \sigma_1\sigma_3} \tag{3.13}$$

The positive and negative signs of Eq. (3.13) represent the top and bottom half of the Mohr circle, respectively; the top will represent right-lateral and the bottom left-lateral shear stress.

3.13 Three-Dimensional Stress

Stress that can be geometrically expressed and analysed in three dimensions is called *three-dimensional stress*. Since rocks exposed on the surface of the earth extend at depth, use of three-dimensional stress helps us interpret the forces acting at a particular point of the earth.

3.13.1 Stress Ellipsoid

A stress ellipsoid is a convenient way of graphically representing the state of three-dimensional stress acting at a point. If we assume that the normal stress acting upon the three planes of a body is represented by the lengths of the axes of a triaxial ellipsoid such that the shear stresses acting upon all the planes are zero, the ellipsoid thus constructed is called a *stress ellipsoid* (Fig. 3.10). The three mutually perpendicular planes in the ellipsoid are called the *principal greatest normal stress axis* (σ_1), *principal intermediate normal stress axis* (σ_2) and *principal least normal stress axis* (σ_3) such that $\sigma_1 \geq \sigma_2 \geq \sigma_3$. The state of stress acting at a point is then completely given by the directions and the sizes of the principal stresses.

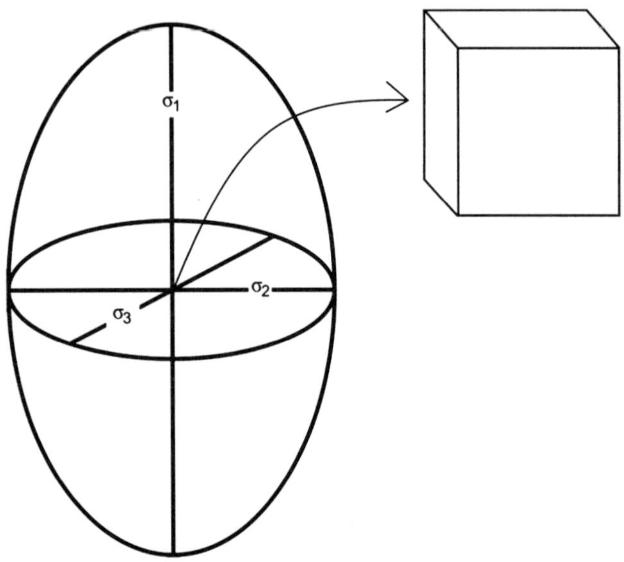

Fig. 3.10 Stress ellipsoid that represents the three-dimensional state of stress at a point that is represented by the infinitesimal cube located at the centre of the ellipsoid. σ_1, σ_2 and σ_3 are three mutually perpendicular surface stresses acting on the three principal planes. These are called the principal greatest normal stress axis, principal intermediate normal stress axis and principal least normal stress axis, respectively. Along these planes, normal stress has the greatest value while shear stress is zero. The stresses σ_1, σ_2 and σ_3 can be compressive (+ve sign) or tensile (−ve sign)

The plane that contains each of these axes is called a *principal plane*. Along the principal planes, normal stresses have greatest values while shear stress is zero. Each plane consists of two principal normal stresses. The difference between any two stresses gives the *differential stress* along that plane. Since σ_1 and σ_3 are the greatest and least stresses, ($\sigma_1 - \sigma_3$) gives the maximum differential stress. All planes other than the principal planes in the stress ellipsoid are *shear planes* because along such planes there is always a component of shear. Thus, we have only three planes along which the value of shear stress is zero.

3.13.2 Three-Dimensional Stress at a Point

We now consider three-dimensional stress acting at a point. For this, we imagine an infinitesimally small cube, i.e. one whose further smaller size is not visible. Let the cube be under stress by the action of a system of forces that can be represented by a single force F acting at the centre of the body. Let us now consider this cube on a three-dimensional coordinate system (Fig. 3.11) in which the three faces are parallel to each of the orthogonal axes x, y and z. Each of the six faces of the cube can thus be considered to be under the influence of the force. If we assume that the body is homogeneous and the effect of stress is also homogeneous, and also that the body is in equilibrium, the effect of each stress on a particular face is counterbalanced by an equal and opposite stress acting on the opposite face.

As we have explained earlier, a force acting at an angle to a plane would develop stresses that can be resolved into three components, one normal stress acting perpendicular to the plane and two shear stresses that are perpendicular to each other but act in the same plane. Thus, for the face, say ABCD, the force F will develop a normal stress σ_y and two shear stresses τ_{yx} and τ_{yz}. For another face CDEG, the stresses will be a normal stress σ_z and two shear stresses τ_{zy} and τ_{zx}. Likewise for the face ADEH, there will be a normal stress σ_x and shear stresses τ_{xz} and τ_{xy}. Since the body is in equilibrium, the stresses acting on each of these three faces will be counterbalanced by equal and opposite stresses acting on the opposite three faces of the cube. Thus, the state of stress σ_{ij} acting on the cube has been resolved into nine components and can be given by the following matrix or tensor form:

$$\sigma_{ij} = \begin{bmatrix} \sigma_x & \tau_{xy} & \tau_{xz} \\ \tau_{yx} & \sigma_y & \tau_{yz} \\ \tau_{zx} & \tau_{zy} & \sigma_z \end{bmatrix} \tag{3.14}$$

Each of the above stresses is called *stress components*. These nine components completely define the state of stress

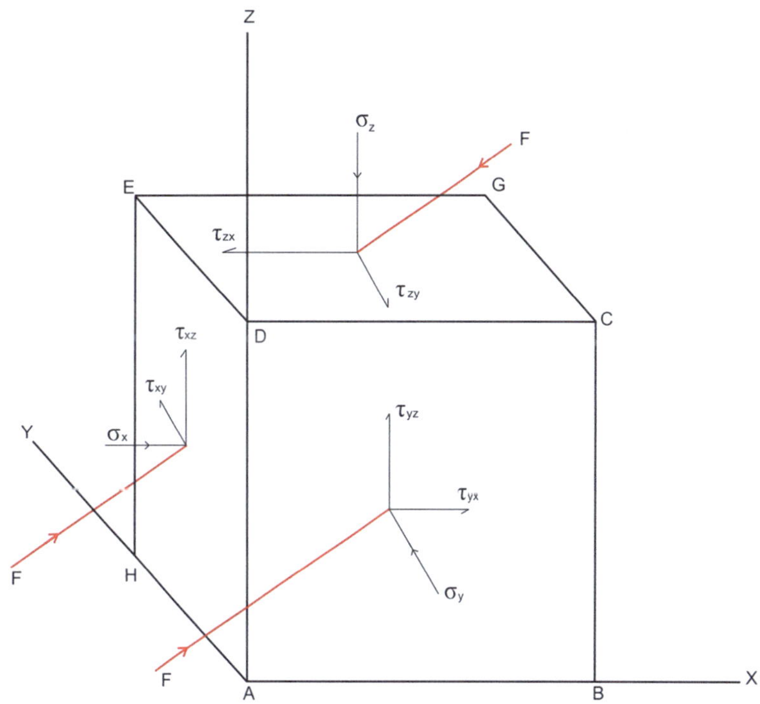

Fig. 3.11 Three-dimensional stress acting on an infinitesimally small cube. Note that on each face, a force *F* is resolved into three mutually perpendicular stresses of which one is a normal stress acting perpendicular to the plane while the other two are shear stresses acting parallel to the plane. Thus, with three stresses acting along each face, the three faces of the cube constitute a system of nine stresses (see text for further details)

in the cube in three dimensions. Since the cube is in equilibrium, each of the shear stresses acting on either side of an edge must balance each other and therefore there is no rotational movement, and therefore

$$\begin{aligned}\tau_{xy} &= \tau_{yx}\\ \tau_{yz} &= \tau_{zy}\\ \tau_{zx} &= \tau_{xz}\end{aligned} \tag{3.15}$$

Thus, there are only six independent components—three normal stresses σ_x, σ_y and σ_z and three shear stresses τ_{xy}, τ_{yz} and τ_{zx}—that completely define the three-dimensional stress acting at a point.

If we rotate the cube so that the coordinate system (*x*, *y* and *z* axes) gets aligned parallel, respectively, to the greatest, intermediate and least principal stresses, there will be no shear stress on the faces of the cube ($\tau = 0$ for all faces), and the stress tensor (i.e. Eq. 3.14) takes the form

$$\begin{bmatrix}\sigma_x & 0 & 0\\ 0 & \sigma_y & 0\\ 0 & 0 & \sigma_z\end{bmatrix} \tag{3.16}$$

Equation (3.16) represents a stress tensor with zero shear stresses and contains only three normal stresses, called *principal stresses*, acting perpendicular to the faces along which shear stresses are zero.

3.14 States of Stress

Depending upon the specific conditions, stress can be of various types; some common types are briefly described below.

3.14.1 Hydrostatic Stress

A state of *hydrostatic stress* develops when all the principal stresses are equal, i.e. $\sigma_1 = \sigma_2 = \sigma_3$. In this case, all the principal stresses are compressive (Fig. 3.12), and there is no shear stress as is the case with fluids. Since all the stresses are compressive, hydrostatic stress tends to reduce the volume.

3.14.2 Differential Stress

In a system with three unequal stresses, σ_1, σ_2 and σ_3, the difference between the greatest and the least stresses is called *differential stress*, σ_d, given by

$$\sigma_d = \sigma_1 - \sigma_3 \tag{3.17}$$

Crustal rocks are under the effect of differential stress, which plays a great role in rock deformation. A rock gets deformed when the amount of differential stress exceeds its strength.

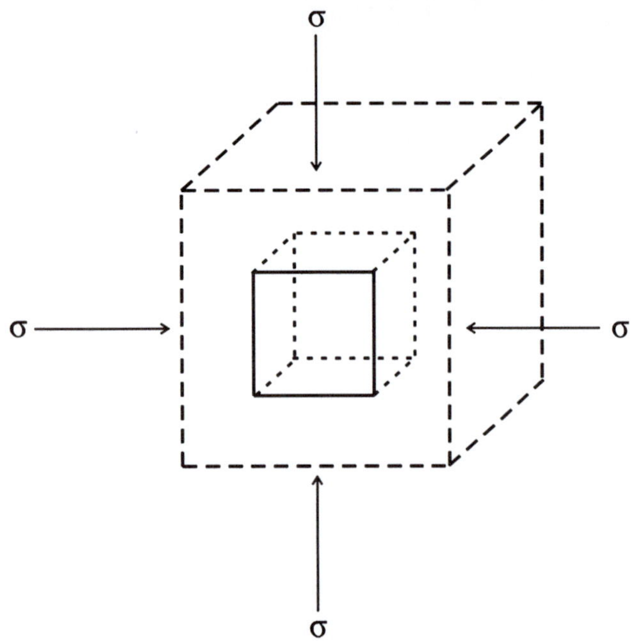

Fig. 3.12 Hydrostatic stress. All the principal stresses are equal and compressive that tend to reduce the volume of the body

3.14.3 Deviatoric Stress

Consider a system where all the principal stresses are unequal, i.e. $\sigma_1 \neq \sigma_2 \neq \sigma_3$, and all are compressive. In such a case, we can imagine a *mean stress*, σ_m, which is given by

$$\sigma_m = (\sigma_1 + \sigma_2 + \sigma_3)/3 \quad (3.18)$$

Now, the body can be considered to be under the influence of three unequal principal stresses (Fig. 3.13)—($\sigma_1 - \sigma_m$), ($\sigma_2 - \sigma_m$) and ($\sigma_3 - \sigma_m$)—the average of which constitutes a mean stress (σ_m). The actual principal stresses thus mark departure under the influence of a system of these three stresses. Each of these is known as *deviatoric stress*. *A deviatoric stress is thus a component of a stress that expresses the difference between a normal stress and the mean stress*. It can be both a two-dimensional stress and a three-dimensional stress. Deviatoric stresses tend to change the shape of a body, which is then said to have undergone distortion.

Fig. 3.13 Hydrostatic stress and deviatoric stress in two dimensions. (**a**) A hydrostatic stress (σ) causes a change (reduction) in volume but no change in shape of a body. (**b**) Deviatoric stresses $\sigma_1 - \sigma_m$ and $\sigma_3 - \sigma_m$ cause a change in shape of a body

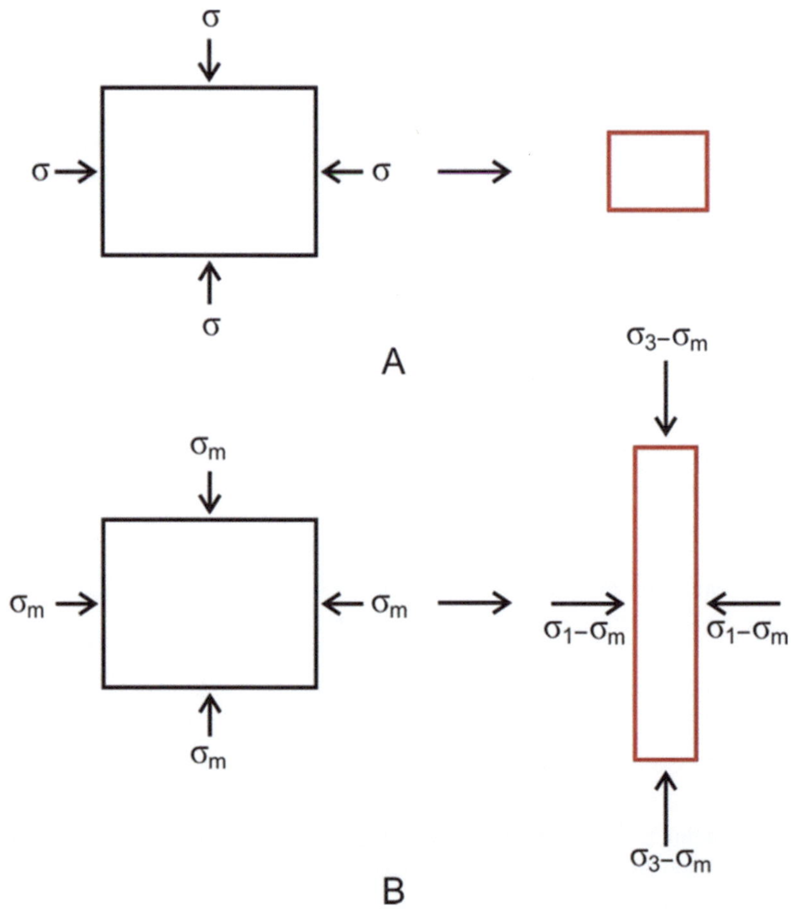

Box 3.1 What Is a Tensor?

A *tensor* in general represents a set of numerical quantities that can describe the physical state of a material. A tensor relates the various geometrical vectors that, when related to a coordinate system, constitutes an organized frame for a set of physical properties of the body. Although tensors are individually independent of a coordinate system, they can be expressed in a coordinate system during a computational work. Tensors thus constitute geometric objects that relate the vector and scalar quantities of an object and can be represented as a multidimensional array of numbers. It may be noted that vectors and scalars are also tensors. Study of tensor is based on the use of a fixed reference frame, which is the coordinate system that uses indices for its notation. In the context of a tensor, a *dimension* represents the range of the indices.

In structural geology, tensors are important in providing a mathematical framework in relation to certain physical properties of rocks such as stress and strain in particular and rock mechanics in general.

Ranks of Tensors

Since a tensor represents a multidimensional array of numbers, it is often required to specify the number(s) by which this multidimensionality is represented. A *rank* (or *order* or *degree*) *of a tensor* represents the number of dimensions or indices required to define the multidimensionality of an array. A rank describes the tensor's dimensions. These numbers are called *components of the tensor*. The rank of a tensor is therefore equal to the total number of dimensions or indices that are needed to define each component. A *tensor of rank zero* is one that has only a single number of dimensions, e.g. a scalar quantity that is unrelated to any axes of reference. A scalar has no indices at all. A *tensor of rank one* is one that has only one dimension, e.g. a vector. A vector requires one axis of reference, and with each axis three numbers or components are associated. *A tensor of rank two* is one that needs two components to define the array of its multidimensionality, e.g. stress and strain. A stress requires two axes of reference, and with each axis nine numbers or components are associated. If the two components σ_{ij} of a stress are related to two vectors l_i

Box 3.1 (continued)

and k_j in a linear manner, then the components σ_{ij} form a matrix as given below:

$$\sigma_{ij} = \begin{bmatrix} \sigma_{11} & \sigma_{12} & \sigma_{13} \\ \sigma_{21} & \sigma_{22} & \sigma_{23} \\ \sigma_{31} & \sigma_{32} & \sigma_{33} \end{bmatrix} \tag{3.19}$$

In general, a rank of a tensor represents the number of indices required to determine the components. As we have indicated above, a rank of zero is represented by a scalar, a rank of one by a vector, a rank of two by matrix (stress), a rank of three by hyper-matrix and so on.

Box 3.2 Concept of Traction

Traction implies the distribution of forces acting on any specific surface of a solid substance, say a rock mass. The concept of traction appears to have not yet been widely used in structural geology, though there are several situations where the concept finds its utility. Pollard and Fletcher (2005) have used this concept to elucidate deformation in rocks. They highlighted the role of a *traction vector* that gives the distribution of forces, or lack of it, acting on any arbitrary surface within a rock mass. Traction vector can explain several geological situations efficiently such as what were the forces distributed on the surfaces of a fault that would cause it to slip or what were the forces distributed on the surfaces of a dike that would cause it to open. In the concept of traction, the earth's surface has been recognized as a traction free surface.

Let us come back to a rock which represents a material continuum as its space is filled up with some material. A rock can be considered heterogeneous as well as homogeneous depending upon the scale of our observation. On a smaller scale, it may show several constituents, say rock fragments, large grains or grain aggregates, so that it looks heterogeneous. On a larger, say outcrop, scale, the same rock may look homogeneous apparently with no discontinuities, thus reflecting a material continuity. The concept of traction vector, which also presupposes a material continuum, can thus be extended to rocks also.

3.14.4 Lithostatic Stress

Rocks in the earth's crust are always under stress caused by the weight of the overlying burden that gives rise to *lithostatic stress*. Since these rocks are in equilibrium, i.e. physically stable, we can assume that at each point the stress is acting uniformly from all sides. In other words, the principal stresses are equal, all the principal stresses are compressive and there is no shear stress.

3.15 Palaeostress

3.15.1 Nature of Palaeostress

Palaeostress refers to the stress locked in the rocks in the geologic past. It is a branch of structural geology whose target is characterizing stress systems acting in the past from their record in deformation structures, singularly from fault-slip data (Simón 2019, p. 124). Palaeostress analysis is concerned with the directions along which stresses were active during rock deformation and, in certain cases, gives the magnitude of stress of the geologic past. With a beginning in France in the 1970s, this branch is nowadays practised all over the world. Excerpts from a review by Simón (2019) on 40 years of palaeostress analysis are highlighted below.

Palaeostress analysis is based on four basic models of relationships between brittle structures and stress/strain axes (Fig. 3.14): (a) conjugate faults, (b) overall discontinuous deformation, (c) Wallace-Bott's principle and (d) orthorhombic or 'biconjugate' fault pattern.

The *conjugate fault model* (Fig. 3.14a) is based on the conceptual relationship between conjugate fault systems and stress axes, as was established by Anderson (1951). The *overall discontinuous deformation model* (Fig. 3.14b) is based on inferring the strain axes from an assembly of faults, stylolites and tension gashes that jointly represent the bulk of deformation of a rock body (Arthaud 1969). *Wallace-Bott's principle* assumes that fault slip parallels the maximum resolved shear stress, which depends on the attitude of the fault plane and the stress ratio $R = (\sigma_z - \sigma_x)/(\sigma_y - \sigma_x)$, where σ_z is the vertical principal stress and $\sigma_y > \sigma_x$ (Fig. 3.14c). From slip data on a minimum of four independent faults, the four parameters that define the reduced stress tensor (three Euler angles for the orientation of stress axes, and R) can be obtained. *Orthorhombic or 'biconjugate' fault pattern* (Fig. 3.14d) is based on the analysis of faulting in three-dimensional strain fields (Reches 1978). Considering all the above models, palaeostress analysis is based on a number of assumptions (Simón 2019, p. 126):

(a) The stress state is homogeneous within the studied rock body.
(b) All fault slips are related to a single stress tensor.
(c) All deformation is accommodated by discrete slip on faults, i.e. blocks bounded by the fault planes are rigid and show no significant rotation.
(d) Slip accumulates on randomly oriented pre-existing fractures.
(e) The studied rock volume is large compared to the scale of each individual fault, and fault displacements are small with respect to fault dimensions.
(f) Movement on each fault is independent of the other ones, i.e. either they are not coeval or they do not interact to each other.

Simón's review emphasizes that palaeostress is a powerful tool for understanding ancient stress fields recorded in rocks, and as such it has implications for lithospheric dynamics and present-day plate kinematics.

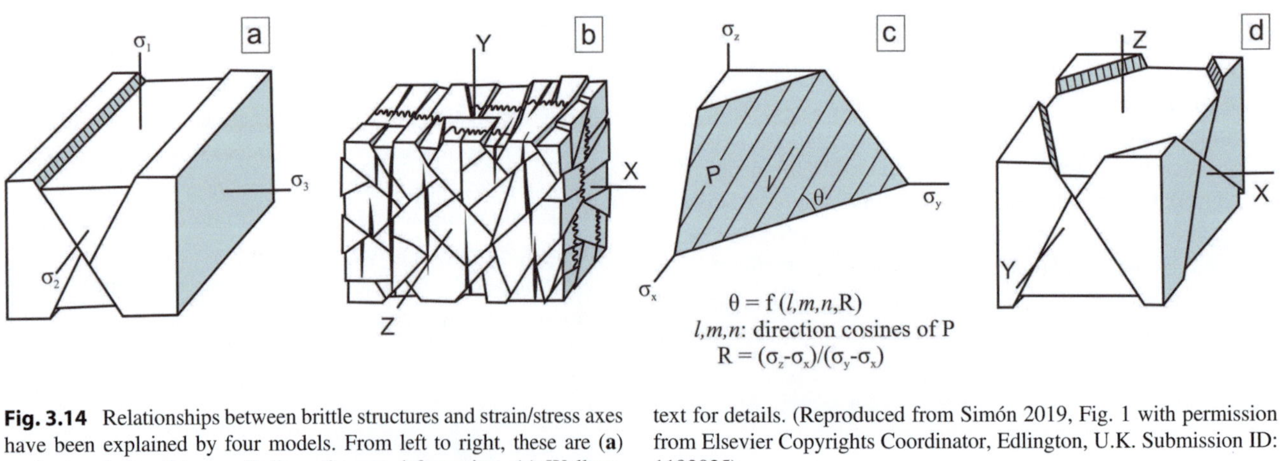

Fig. 3.14 Relationships between brittle structures and strain/stress axes have been explained by four models. From left to right, these are (**a**) conjugate faults, (**b**) overall discontinuous deformation, (**c**) Wallace-Bott's principle and (**d**) orthorhombic or 'biconjugate' fault pattern. See text for details. (Reproduced from Simón 2019, Fig. 1 with permission from Elsevier Copyrights Coordinator, Edlington, U.K. Submission ID: 1193025)

3.15.2 Estimation of Palaeostress

Estimation of palaeostress is sometimes also called *palaeopiezometry*. The features/signatures in rocks that help in estimating the palaeostress are called *palaeopiezometers*. Palaeostress has been estimated by various workers by devising several methods. The various methods use both macroscopic (folds, faults, fractures, joints, veins) and microscopic (grain size in dynamically recrystallized rocks, calcite twins, deformation lamellae in quartz, pressure solution, microfractures, etc.) features of rocks. Estimation of palaeostress has been done by several workers (see Srivastava et al. 1998; Simón 2019, and the references therein) using different methods. Although description of all the methods is not possible here, some common tools or features used in palaeostress estimation are highlighted.

3.15.2.1 Faults

Faults are common structures used in palaeostress analysis. Movement on fault planes occasionally gives rise to slickenside lineations that help in reconstructing the orientation of the fault slip. The stress-slip relationship for the fault could be established by applying the *Wallace-Bott hypothesis* that states that slip along a fault plane should be parallel to the direction of the greatest shear stress. It is thus possible to establish the orientation of all the principal stress axes. It is assumed that most large faults have signatures on small scales or outcrop scales. The above hypothesis can be applied for a population of smaller faults, and the results thus obtained can be generalized for the larger faults. Although the Wallace-Bott hypothesis provides an important tool for palaeostress analysis, it has some limitations. For example, the hypothesis requires that the fault has undergone only one phase of deformation and that it has undergone no rotation. Field evidences, however, reveal that many fault planes show more than one set of slickensides and evidence of rotation. Despite this, the hypothesis stands as a reliable tool for investigating the stress-slip relation of faults.

3.15.2.2 Grain Size

Deformation of rocks under certain conditions promotes recrystallization. When recrystallization takes place during concurrent deformation, it is called *dynamic recrystallization*. Recrystallization in the absence of deformation is called *static recrystallization*. During dynamic recrystallization, a strained grain releases its internal strain energy by the formation of smaller, strain-free grains. As deformation continues, smaller, strain-free grains form more in number and the rock ultimately becomes finer grained. The process of recrystallization is accentuated by increase of temperature.

Dynamic recrystallization involves appearance of new dislocation densities, and the degree of recrystallization depends much on the differences in dislocation density across the grain boundaries. Thus, dynamic recrystallization is a stress-sensitive process. In fact, it is the differential stress that dominates during concurrent deformation. This in turn suggests that the size of the subgrains formed due to dynamic recrystallization is a function of the differential stress. The size of the dynamically recrystallized subgrains is related to the flow stress by the following equation:

$$\sigma = Ad^{-m} \tag{3.20}$$

where σ is the differential stress in megapascals, d is the grain size in μm and A and m are constants for the material (Mercier et al. 1977; Twiss 1977; Ord and Christie 1984).

The above relation implies that during dynamic recrystallization, the average grain size undergoes reduction with increase in differential stress. Thus, the grain size of a dynamically recrystallized rock can serve as a means to estimate palaeostress in such rocks. The method as such demands further refinement. For example, the values of the flow stress for a given grain size have been found to be different by different workers. Also, we are not sure whether the grain size reduction is due to differential stress only; several other factors may also play their part. Despite all these shortcomings, the method is significant especially in the absence of any other technique available at hand.

3.15.2.3 Calcite Twins

Some minerals such as calcite develop twins due to deformation. If simple shearing takes place parallel to twin planes in a calcite crystal, the mechanism is called *twin gliding*. The calcite twins thus developed are the result of stress that produces bends in the crystal lattice structure. Twin gliding can take place at very low shear stress of 10 MPa (Twiss and Moores 2007, p. 500). By geometric construction of the twin plane and the shear direction, it is possible to infer the orientation of the principal stresses that gave rise to twin gliding.

Yamaji (2015) pointed out that although mechanical twinning along calcite *e*-planes is used for palaeostress analyses, the orientations of twinned and untwinned *e*-planes are known to constrain not only stress axes but also differential stress, *D*. The orientations lose the resolution of *D* if the twin lamellae were formed at *D* greater than 50–100 MPa. This can have distortive effects on palaeostress analysis.

3.15.2.4 Veins

Veins are commonly used for palaeostress estimation. Veins represent hydrothermal fluids, which under high fluid pressure result in dilation of pre-existing anisotropy/flaws in a rock, thus forming hydrothermal veins (Yardley 1986; GDH Simpson 1998). Veins are often associated with mineral deposits, some of which are of high commercial value such as gold.

Recently, Lahiri et al. (2020) used quartz veins for estimating palaeostress in mineralized and non-mineralized zones of Gadag, falling in a part of Dharwar craton of South India. The mineralized zones show occurrences of gold. The authors collected data on orientation and spacing of veins. By plotting the orientation data for every outcrop, the authors obtained two types of distributions: cluster distribution (when $P_f < \sigma_2$) and girdle distribution (when $P_f > \sigma_2$), where P_f is fluid pressure. Their palaeostress analysis reveals that the veins were emplaced when fluid pressure (P_f) varied from $P_f > \sigma_2$ (intermediate principal stress) to $P_f < \sigma_2$.

3.16 Stress Tensor

Use of tensor (see also Box 3.1) can be extended to our understanding of stress (see Means 1976; Oertel 1996, for details). Stress is a tensor of rank 2. Since stress is a force per unit area, we can assume that a unit cube of material is affected by forces acting on it in three dimensions (Fig. 3.15). We know the surface area of each face. If we know the value of the force acting on the body, we can calculate the amount of stress acting on the cube in three dimensions. The state of stress acting on the cube can be resolved into nine components that form a tensor, i.e. stress tensor.

If we use eigenvalues σ_{ij} for stress notation, then σ_1, σ_2 and σ_3 represent the principal axes of stress representing the maximum, intermediate and minimum principal stress axes. The 3D stress tensor (in the coordinate system x, y, z) can then be given by the following expression:

$$\sigma_{ij} = \begin{bmatrix} \sigma_{xx} & \sigma_{xy} & \sigma_{xz} \\ \sigma_{yx} & \sigma_{yy} & \sigma_{yz} \\ \sigma_{zx} & \sigma_{zy} & \sigma_{zz} \end{bmatrix} \tag{3.21}$$

If we choose coordinates x_1, x_2 and x_3, the state of stress acting on the cube can be resolved into nine components as shown in Fig. 3.16. The 3D stress matrix in this case takes the following form:

$$\sigma_{ij} = \begin{bmatrix} \sigma_{11} & \sigma_{12} & \sigma_{13} \\ \sigma_{21} & \sigma_{22} & \sigma_{23} \\ \sigma_{31} & \sigma_{32} & \sigma_{33} \end{bmatrix} \tag{3.22}$$

Since the body is in equilibrium and there is no shear stress on principal planes,

$$\sigma_{12} = \sigma_{21}, \sigma_{13} = \sigma_{31} \text{ and } \sigma_{23} = \sigma_{32},$$

and therefore the number of independent stress components is six.

If $\sigma_{13} = \sigma_{31} = \sigma_{23} = \sigma_{32} = \sigma_{33} = 0$, the stress tensor can then be expressed in 2D in the following form:

$$\sigma_{ij} = \begin{bmatrix} \sigma_{11} & \sigma_{12} \\ \sigma_{21} & \sigma_{22} \end{bmatrix} \tag{3.23}$$

A stress tensor is taken to be symmetric, i.e.

$$\sigma_{ij} = \sigma_{ji}.$$

Further, a symmetric matrix can be diagonalized, i.e.

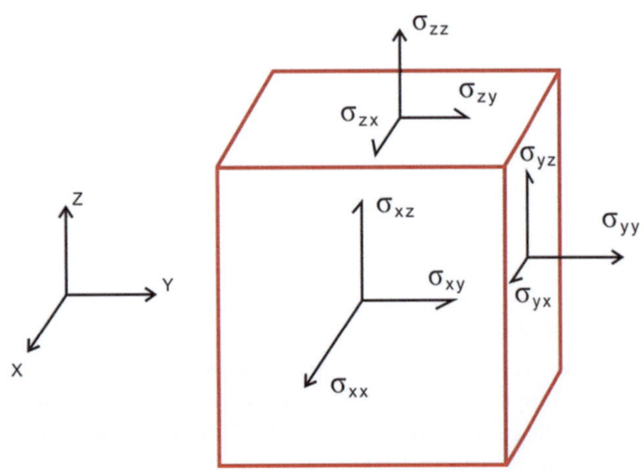

Fig. 3.15 Representation of stress tensor acting on a cube with eigenvectors on Cartesian coordinates X, Y, Z

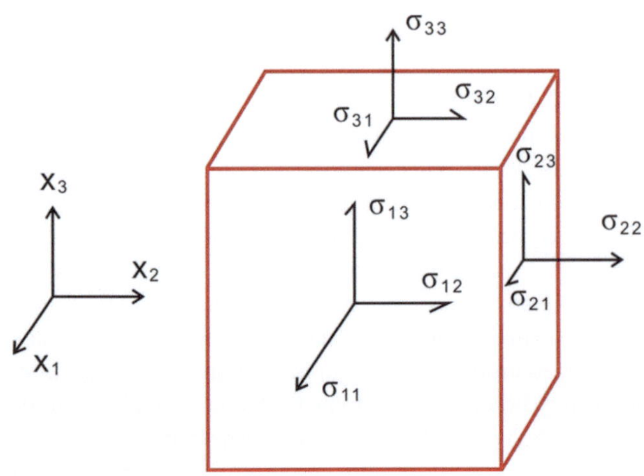

Fig. 3.16 The state of stress acting on a cube as represented in coordinate system x_1, x_2, x_3

$$\sigma_{ij} = \begin{bmatrix} \sigma_1 & 0 & 0 \\ 0 & \sigma_2 & 0 \\ 0 & 0 & \sigma_3 \end{bmatrix} \quad (3.24)$$

In this matrix, the off-diagonal terms of stress tensor are zero and the diagonal elements of the tensor, i.e. σ_1, σ_2 and σ_3, represent the three principal stresses (note that shear stresses are equal to zero).

3.17 Stress Field

As highlighted earlier, the state of stress in a rock mass inside the earth's surface is compressive and by nature it is hydrostatic, i.e. equal in all directions. This seems to be an oversimplification as the stresses are not always of equal amount in all directions. A generalized state of stress in rocks is called *stress field*. In fact, it is the stress field that determines what types of deformation the rock mass should show. Knowledge on stress field thus throws light on the deformation behaviour of the crust at any particular time.

Several workers have attempted estimation of stress field in some selected parts of the crust. Here, we present an example of the regional stress field of the Indo-Australian plate highlighting the results of the work of Cloetingh and Wortel (1986). The authors considered the Indian plate as elastic with a nominal thickness of 100 km and divided the spherical surface of the Indian plate in grids with the maximum grid size of 5 degrees. The results suggest a high level of regional stress field in the Indo-Australian plate. There is a concentration of compressive stresses of the order of 3–5 kbr in the Ninetyeast Ridge area. The maximum compression acts horizontally in NW-SE-oriented direction. The areas west of the Indian peninsula have an approximately N-S-directed tensional and E-W-oriented compressional stress field. There is more symmetric distribution of stress and deformation in the oceanic lithosphere with respect to the Indian peninsula. The Indian plate is undergoing a considerable net resistance at the Himalayan collision zone, and the exceptionally high level of the present-day regional stress field of the Indian plate is a transient feature that results from the unique dynamic situation in which the Indian plate now finds itself.

3.18 Stress History

In any sedimentary terrain, rocks have a long history from the time of sedimentation and burial to the time when they are seen exposed on the surface of the earth as a lithified, and may be deformed, rock. Accordingly, the stress condition also changes all through the long history of the rock from burial to the present-day form, and all this is called *stress history*. Of all, the joints are possibly the commonest structures that are formed at various stages from burial of sediments till uplift and erosion. However, the joints formed in such basins are formed under different stress conditions as the host rock continued to change its state (from unconsolidated mass to a lithified mass) as well as its location. For example, during sedimentation and diagenesis, the area witnessed increase of stresses, while during uplift and erosion, the stresses were gradually released. It is thus possible to understand stress history of any sedimentary basin or any sedimentary terrain by careful study of joints.

3.19 Stress Inside the Earth

3.19.1 Nature of Stress Inside the Earth

Earth below the surface is a storehouse of stress. Our knowledge of underground stress is important for certain engineering aspects. During excavation, for example, knowledge of pre-existing in situ stresses is necessary, especially for the stability of the excavation (Jaeger et al. 2007, p. 399). However, precise estimation of stress from below the earth's surface is rather difficult. Here, all the stresses are compressive; other types of stresses are considered negligible. Anderson (1951) suggested that below the earth the state of stress is everywhere equal to vertical. Also, estimation of horizontal stress is difficult, and therefore our knowledge on complete information on the state of stress on the earth's surface is as yet far from being satisfactory (Zoback and Zoback 1989).

In general, the in situ stresses below the earth vary as a function of depth, but the nature of the stresses and their variations are difficult to ascertain. The underground stresses are influenced by the topography, tectonic forces, constitutive behaviour of the rocks and local geological history (Jaeger et al. 2007, p. 399).

3.19.2 Basic Stress Types Inside the Earth

Our understanding of the basic stress types in the lithosphere comes from laboratory experiments. Engelder (1993, p. 22) suggested that the governors for earth stress include three general types of mechanisms that lead to failure of intact rock: *crack propagation*, *shear rupture* and *ductile flow* (Fig. 3.17). *Friction* is an additional governor for lithospheric stress if joints or shear fractures are reactivated to slip due to large shear traction. In the middle portion of the brittle intracontinental crust, the least compressive stress, σ_3, is

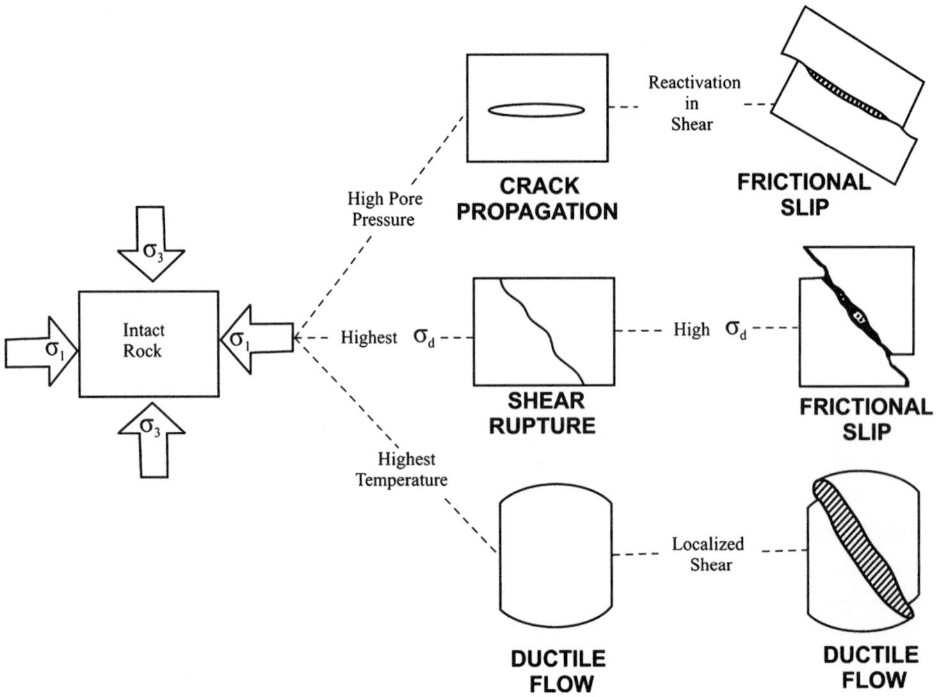

Fig. 3.17 Basic three types of stress inside the earth as revealed from the laboratory test. See text for details. (Slightly simplified from Engelder T, 1993, Stress Regimes in the Lithosphere Fig. 1–6. Reproduced with permission from Princeton University Press. Request ID: 600067225)

commonly horizontal but equally likely vertical in the upper kilometre or two (Nadan and Engelder 2009).

On the basis of rheological properties, Scholz (1990) divides lithosphere into two parts: the *schizosphere* or brittle region and the *plastosphere* or ductile region. Since earthquakes are a manifestation of brittle behaviour, the extent of schizosphere is demarcated by earthquakes. Since schizosphere shows evidences of some ductile deformation mechanisms such as stress solution, the boundary between schizosphere and plastosphere is gradational or blurred.

3.19.3 Causes of Stress Inside the Earth

Stresses inside the earth arise due to several causes of which the following are important.

3.19.3.1 Overburden

The weight of the overlying rocks and soils is broadly called overburden at any point inside the earth. By far, it is the most important factor to develop stress at any point. Obviously, the magnitude of stress inside the earth caused due to overburden increases with depth.

3.19.3.2 Pore Fluid Pressure

Presence of water or any fluid or gas has great effect on the physical-mechanical properties of rocks located inside the earth. Water constitutes the most common fluid inside the earth. Presence of water strongly affects the strength of rocks. Water is believed to be present in the pore space of the rocks. Water reduces the stress that is locked in the fractures, and this in turn proportionately reduces the pore fluid pressure, i.e. the hydrostatic pressure; as such, the normal stress of the rock is reduced (Ranalli 1987, p. 94). In such cases, failure is controlled by the *effective stress* (σ'), which is the principal stresses (considered positive for compressive stresses) minus the pore pressure (P) (Jaeger et al. 2007, p. 98), i.e.

$$\sigma'_1 = \sigma_1 - P, \sigma'_2 = \sigma_2 - P, \sigma'_3 = \sigma_3 - P \tag{3.25}$$

Effective stress significantly affects the deformational processes such as initiation of fractures or initiation of sliding of a pre-existing fault, and all these, on a grand scale, control various geodynamic processes such as faulting, thrust movement and seismic activity.

3.19.3.3 Thermal Stresses

Rocks are susceptible to changes in temperature in several ways. Rocks have variable thermal expansion, and because of this some rocks undergo change in volume to accommodate the change in temperature while others do not. Rocks that undergo change in volume can be considered to have accommodated the imposed thermal stress; such rocks practically do not induce additional stress. Rocks that do not undergo change in volume, on the other hand, can be considered to store thermal stress within them. In the rocks of the upper crust down to a depth of 15–20 km, thermal stresses arise from resistance to the temperature-induced volume change (Ranalli 1987, p. 146). In general, increase of temperature generates a compression, and it has been estimated that a change in temperature by 100 K generates stresses of the order of 100 MPa (Ranalli 1987, p. 148).

The amount of thermal stresses generated in the upper crust is generally very low as compared to stresses generated by some other processes. However, it is believed that thermal stresses can in some cases produce fractures, if not large-scale signatures of deformation.

3.19.3.4 Plate Motion

The state of stress in the lithosphere constitutes a primary cause of plate motion. High degree of stress is generated at plate boundaries where seismic and tectonic processes are active. The state of stress in a plate depends much on the type of relative plate motion, i.e. whether the boundary of two plates is convergent, divergent or transform. A major part of stress in a plate also arises from the plate-mantle interaction, which in turn varies depending upon whether the plate is oceanic or continental. The forces acting on the subducting slab are the largest and control the velocity of oceanic plates (Forsyth and Uyeda 1975). All this also implies that the state of lithospheric stress controls, directly or indirectly, the thermal convection system of the underlying mantle.

3.19.3.5 Burial

Below sea level, burial of sediments is a common process that generates a good deal of stress (burial stress) below the surface. Burial stress acts in vertical direction due to gravity. Maltman (1994) suggested that there is a proportional increase of burial stress with a depth at least up to about 1000 m below seawater. Here, sedimentation and tectonism also commonly go together, and it is the tectonic forces that induce dewatering to generate overpressures. Thus, in regions where burial is active, the crust continues to add stress.

3.20 Significance of Stress

3.20.1 Academic Significance

- Our knowledge of stress in rocks is necessary because it helps in giving an idea of directions in which the stresses had acted upon the rocks and also the amount of deformation, i.e. strain in rocks.
- Knowledge of stress field throws light on the deformation behaviour of the crust at a given time.
- Estimation of stress of the crust has implications for the direction of relative plate motions and seismicity distribution.

3.20.2 Engineering and Economic Significance

- The stresses inside the earth not only are heterogeneously distributed both horizontally and vertically but also keep changing with time. Despite all these, our understanding of the state of stress inside the earth is important, especially for several utility and practical aspects in addition to academic interests.
- In the construction of *dams*, knowledge of the state of stress is helpful in ensuring that the weight of the superstructure should be much below the affordable stress of the earth at that point.
- In *mining*, information on stress is necessary for the construction of tunnels, pits and related engineering work.
- During drilling for *petroleum*, artificial fractures are created at depth to increase the permeability of the petroleum-bearing horizon; this enhances the production of petroleum.
- In *civic construction* work, knowledge of in situ stress of the foundation ensures longer safety of the structures and buildings to be constructed.

3.21 Summary

- A rock mass under the influence of externally applied forces develops stress within it. Stress is thus caused by external forces and is given by the forces acting per unit area of a body.
- Every rock has strength (i.e. the value of stress that causes failure), small or large, due to which it is able to bear some or a large quantity of stress. If the stress developed in a rock exceeds its strength, the rock accommodates the stress by changing its shape or size, which is called *strain*. Thus, if stress is a cause, strain is an effect.
- Considering a rock mass as a deformable solid or fluid, the forces acting upon it can be of two types, *body forces* and *surface forces*. Body forces act upon a unit volume of the body and are therefore given by the volume or size of the body such as gravity or magnetic force. Surface forces act upon the boundaries of a body and thus act along the area where the forces are in contact with the body. Force acting on a unit area of the body is known as *stress*.
- Stress can be tensile, compressive or shear stress. *Tensile stress* develops when the stress acts along the plane in opposite directions. *Compressive stress* develops when the stress acts perpendicular to the plane. *Shear stress* develops when the stress acts in opposite directions but not along the same plane or line; as a result, the body shows angular changes of its original planes and thus to its shape.
- If all the stresses acting upon a body are represented by only three stress components acting parallel to each of the coordinate axes x, y and z, the normal stress acting parallel to each axis is called a *principal stress* and is designated as X, Y and Z, respectively, such that $X > Y > Z$. The coordinate axes along which the stresses act are called *principal axes of stress*.

- If we represent all the stresses acting on a body by lines, the longest and shortest lines trace an ellipse called *stress ellipse* that represents 2D stress. Likewise, the normal stress acting upon three planes can be represented by the lengths of the axes of a triaxial ellipsoid, and this gives a *stress ellipsoid*.
- When all the stresses act in the plane of a body, the state of stress is known as *biaxial stress*. It is a 2D stress that is represented in two directions: a normal stress acting normal to the plane and a shear stress acting along the plane.
- 2D stress can also be determined by *Mohr diagram* that can estimate the values of shear stress, normal stress and angle of failure at the point when a crack is developed, i.e. when failure occurs.
- Stress can be hydrostatic stress, differential stress, deviatoric stress and lithostatic stress.
- Stress locked in rocks in geologic past is called *palaeostress*. Basic principles of estimating palaeostress have been described.
- A generalized state of stress in rocks inside the earth's surface is called *stress field*. Knowledge of stress field throws light on the deformation behaviour of the crust at any particular time.
- The stress condition of a rock mass changes all through the long history of the rock from burial to the present-day form, and all this is called *stress history*.
- The state of stress inside the earth varies as a function of depth and is influenced by the topography, tectonic forces, constitutive behaviour of the rocks and local geological history.

Questions

1. What is stress? Explain why does a body develop stress.
2. What are the differences between body forces and surface forces?
3. What are normal stress and shear stress?
4. What is the difference between tensile stress and shear stress?
5. Outline salient features of uniaxial stress and biaxial stress.
6. What is Mohr diagram? Give its utility in structural studies.
7. With the help of a cube, describe the stress at a point and the state of stress acting on the cube.
8. What is deviatoric stress? What does it signify in the deformation of a rock?
9. Describe how the concept of stress tensor can be extended to structural geology.
10. Describe the causes of stress inside the earth.

4 Strain

A. R. Bhattacharya, *Structural Geology*, Springer Textbooks in Earth Sciences, Geography and Environment,
https://doi.org/10.1007/978-3-030-80795-5_4

Abstract

This chapter takes you to where one can visualize how a stress can produce its effects on an undeformed rock. The amount of deformation that a body has undergone in response to an externally applied stress is called *strain*. The various types of tectonic structures such as folds, faults and boudins noticed in the rocks are manifestations of the strain that the rocks have undergone. Strain may cause change in shape or volume or both. The state of deformation of a rock mass can be referred to three mutually perpendicular lines, and then we can imagine a sphere which after homogeneous deformation is changed to an ellipsoid, called *strain ellipsoid*. The latter in two dimensions can be described a *strain ellipse*. An undeformed body can take up different shapes by homogeneous progressive strain either by (a) *pure shear* or *irrotational shear* when the shape changes in such a way that the material lines do not change their angular relations or by (b) *simple shear* or *rotational shear* when the shape changes in such a way that the material lines have changed their original angular relations. The process of deformation by which a body continues to change its shape and size is called *progressive deformation*. Knowledge of strain is important in geology as, among others, it provides information on the physico-mechanical aspects of rocks. Identification of low-strain and high-strain zones in rocks has bearing in the localization of some mineral deposits.

Keywords

Strain · Homogeneous and heterogeneous strain · Strain ellipse · Strain ellipsoid · Flinn diagram · Pure shear and simple shear · Coaxial and noncoaxial deformation · Strain path · Vorticity · Mohr strain diagram

4.1 Introduction

In previous chapters, we have mentioned that external forces acting on the surface of a body develop stress within it. If the stress thus developed exceeds the strength of the body, the latter undergoes change in its shape and size, which we call deformation. *The amount of deformation that a body has undergone in response to an externally applied stress is called strain.* An elastic body, however, returns back to its original shape and size after the applied stress is removed, and therefore it apparently does not show any signature of deformation.

Crustal rocks are always under stress on their boundaries because of overburden. This stress in the course of time causes strain in rocks and produces tectonic structures such as folds, faults and boudins, as noticed in rocks. However, the type of structure produced as a result of strain depends upon several factors such as type of stress (compressive or tensional), composition of the rock, presence or absence of internal structures (bedding, foliation, etc.), temperature and a few others. Thus, stress causes strain, which in turn produces tectonic structures in rocks. Strain may cause changes in rocks in various geometric forms such as changes in the length of reference lines or angular changes in the original structure and changes in shape, area or volume of the rocks. In this chapter, we shall discuss the concept of strain, while in the next chapter we shall describe some common methods for estimating strain in deformed rocks.

4.2 Components of Strain

Strain in a body can develop in a variety of ways (Fig. 4.1) of which the following four are important: (a) *Translation* that involves movement of a body from one point to another such that there is no change in its shape or volume and therefore it is no strain: In this case any straight line within the body remains parallel to itself throughout the motion. (b) *Rotation* when the body shows a change in its orientation from the original position: Translation and rotation together constitute *rigid body deformation* such that there is no visible strain in the body. (c) *Dilation*, also called *dilatation*, involves change in the volume of a body. As a result of strain, the volume of a body may increase or decrease while the shape remains unchanged. (d) *Distortion* is the change in the shape of a body while the volume remains unchanged. Dilation and distortion together constitute *non-rigid body deformation.* A body is said to show *bulk deformation* when the strain causes changes in both shape and volume.

4.3 Homogeneous and Inhomogeneous Strain

A strain is *homogeneous* when the changes in shape or size of a body have taken place in such a way that the reference lines and planes of the undeformed body do not change their geometric relations after deformation. In this case, the straight lines and the parallel lines of the undeformed body remain straight and parallel after deformation, and the bounding planar surfaces remain planar (Fig. 4.2a). A strain is *inhomogeneous* or *heterogeneous* when deformation of a body has taken place in such a way that the reference lines and planes of the undeformed body have changed their geometric relations after deformation. Thus, the straight lines and the parallel lines of the undeformed body do not remain straight and parallel after deformation, and the bounding planar surfaces become curved (Fig. 4.2b).

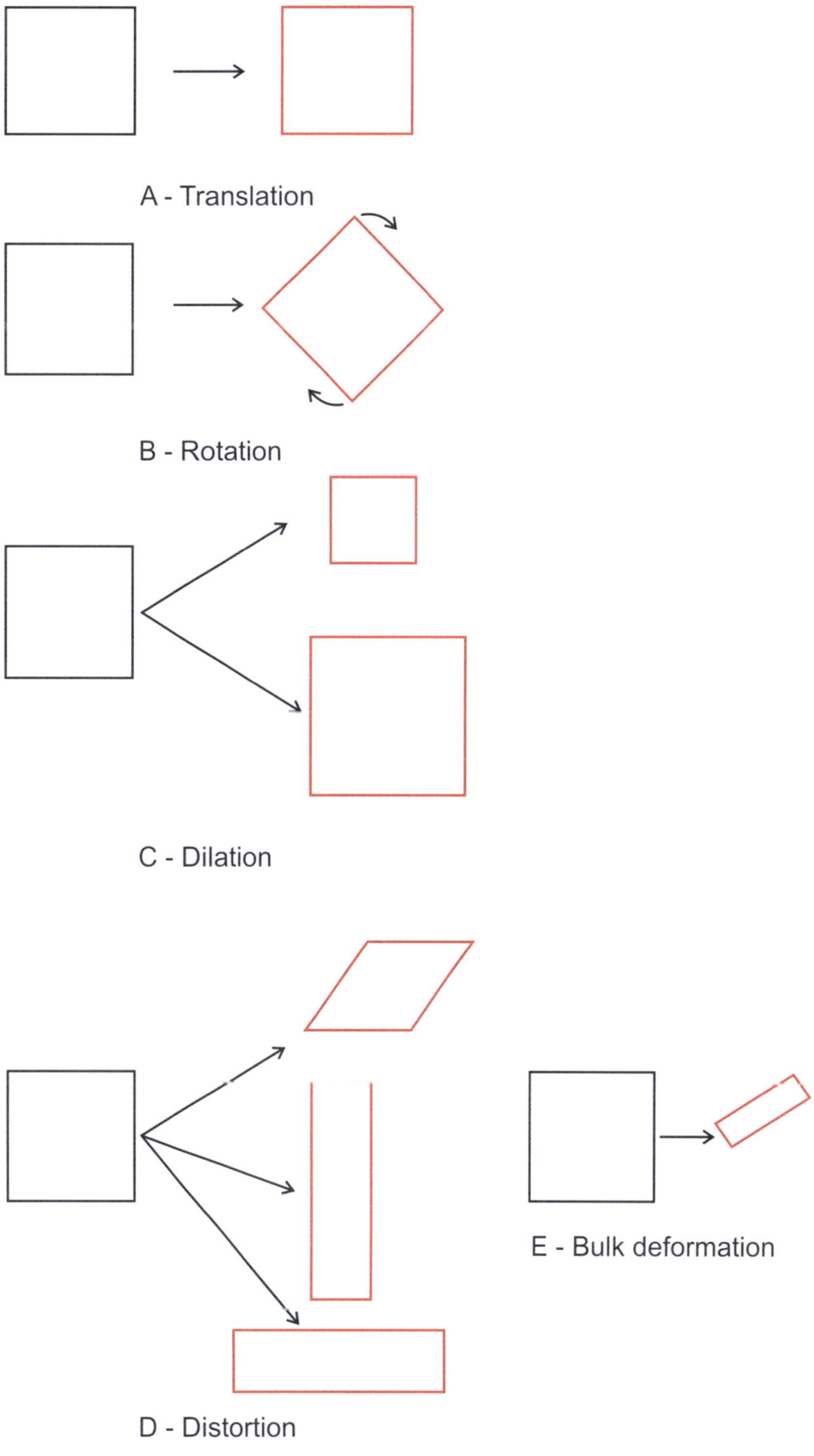

Fig. 4.1 The various ways in which a body can develop strain. (**a**) Translation. (**b**) Rotation. Note that in **a** and **b** there is no visible strain in the body. (**c**) Dilation. (**d**) Distortion. (**e**) Bulk deformation

It may be noted that the concept of homogeneous and inhomogeneous strain is scale dependent. Several large structures that show homogeneous deformation on the scale of kilometres or tens of kilometres commonly show inhomogeneous deformation on the scale of centimetres or metres. The reverse is also true. A large-scale structure extending for several kilometres showing inhomogeneous deformation commonly shows homogeneous deformation on smaller, say centimetre or meter, scales. However, for practical purposes, we try to resolve the larger structures showing inhomogeneous deformation into smaller components showing homogeneous deformation (Fig. 4.3). The smaller

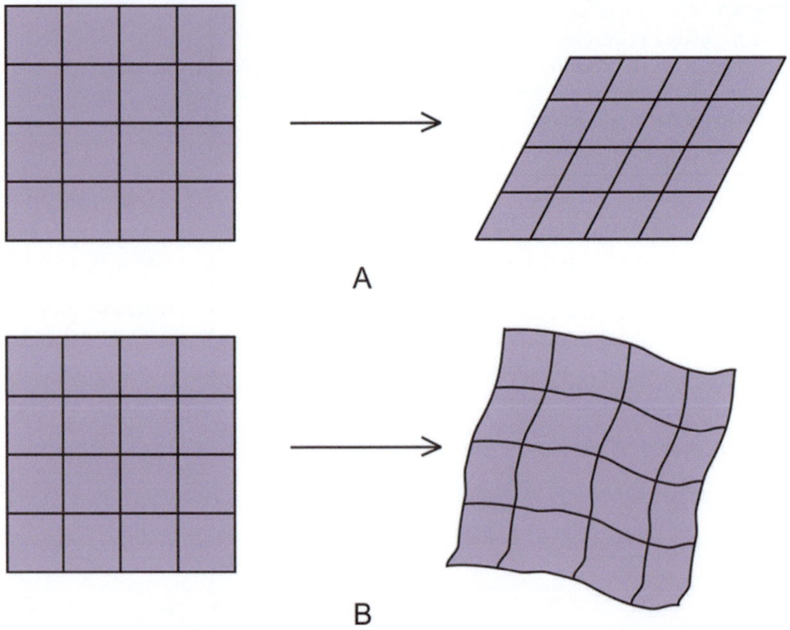

Fig. 4.2 (**a**) Homogeneous strain in which the reference lines and planes of the undeformed body do not change their geometric relations after deformation; that is, the straight lines and the parallel lines of the undeformed body remain straight and parallel after deformation, and the bounding planar surfaces remain planar. (**b**) Inhomogeneous strain in which the reference lines and planes of the undeformed body change their geometric relations after deformation

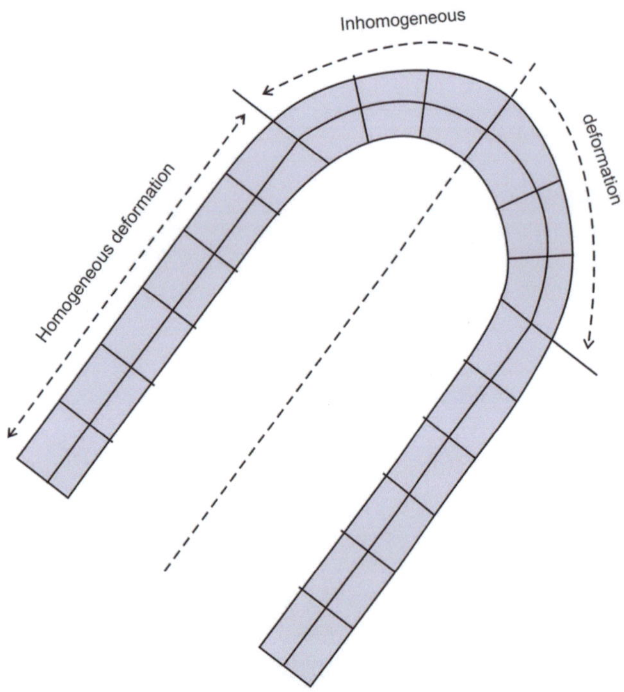

Fig. 4.3 A structure (fold in this case) may not show homogeneous strain throughout. For estimation of strain, the structure can be subdivided into domains of homogeneous and inhomogeneous strain. The inhomogeneous domain can further be conveniently subdivided into homogeneous subdomains to enable strain estimation

components in turn allow estimation of strain easily. The strain thus estimated from the smaller components can be extrapolated for the larger structure, and the wider implications can thus be worked out.

4.4 Measures of Strain

As we have mentioned above, a body can take up strain in a number of ways, mainly as a change in length (linear strain), angle (angular strain) or volume (volumetric strain). Consequently, a number of geometrical measures have been worked out to estimate strain as given below.

4.4.1 Linear Strain

Linear strain is the quantitative change in the reference lines of a body after deformation and can be expressed by the following measures.

4.4.1.1 Extension

Let us consider a reference line of length l_o in the undeformed body. After deformation, let this line assume a length l_f (Fig. 4.4). The linear change due to deformation of the body along this line is thus $(l_f - l_o)$. The *extension* is then defined by the ratio of the linear change to the original length of the body. In other words, extension gives the change in unit length of a line due to deformation of a body and is given by

$$e = \left(l_f - l_o\right)/l_o \qquad (4.1)$$

With multiplication by 100, we get

$$\%\ \text{extension} = \left[\left(l_f - l_o\right)/l_o\right] \times 100 \qquad (4.2)$$

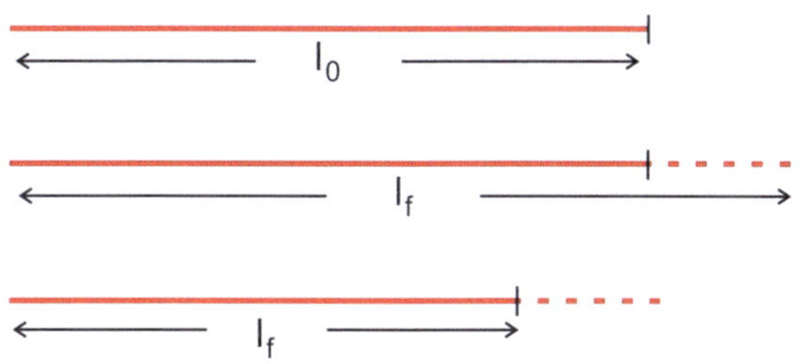

Fig. 4.4 Linear strain as given by deformation of a linear object. Due to deformation, an original length (l_o) can undergo increase (extension) or decrease (shortening) of its original length

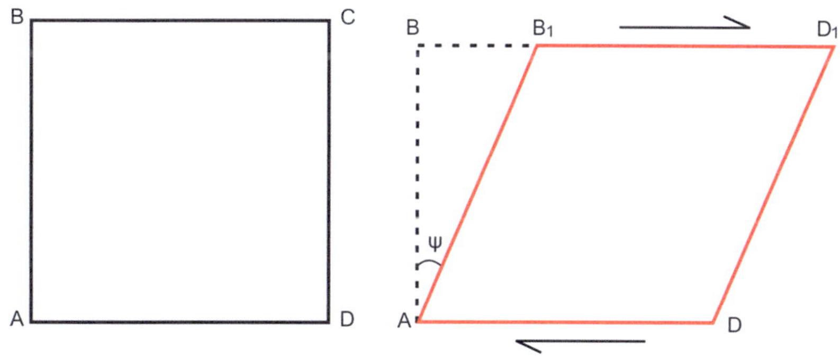

Fig. 4.5 Shear strain. Two originally parallel lines AB and DC are sheared by angle ψ. The shear strain is given by tan $\psi = BB_1/AB$

If the value of e is positive, the strain is described as elongation, while with a negative value the strain is shortening.

4.4.1.2 Stretch

Stretch (S) is a measure to express the final length of a reference line in the deformed state of a body. It is given by the ratio of the final length (l_1) (l_f) to the original length (l_o), i.e.

$$S = \left(l_f/l_o\right) = (1+e) \tag{4.3}$$

From (4.3), $l_f = l_o \times S$. This implies that if we know the value of S, we can compute the final length.

4.4.1.3 Quadratic Elongation

The squared value of stretch gives another measure called quadratic elongation (λ), i.e.

$$\lambda = S^2 = (1+e)^2 = \left(l_f/l_o\right)^2 \tag{4.4}$$

Thus, the value of quadratic elongation can be obtained by getting the ratio of the final length and the original length and then squaring it. If $l_f = 1$, $\lambda = (l_f)^2$. Quadratic elongation can thus also be defined as the squared value of the final length of a deformed body whose original length was unity, i.e. 1.

4.4.1.4 Natural or Logarithmic Strain

Stretch of a line can also be expressed by a parameter called *natural strain* or *logarithmic strain* ε_n. If the original length l_o changes to l_f after deformation, ε_n is given by

$$\varepsilon_n = \ln\left(l_f/l_o\right) = \ln S = \ln(1+e) \tag{4.5}$$

where ln is the natural strain. The natural strain is thus the natural logarithm of stretch. The logarithmic strain is generally used for large strains; in such cases, we can use a log scale. Its use is helpful in comparison of strains because the log scale is proportional to the changes in lengths.

4.4.2 Shear Strain

The measures of linear strain as described above give only the changes in the length of the reference lines of a body after deformation. However, during deformation, the shape of a body may change in such a way that instead of changes in the length of lines, the orientation of the original line may change. In other words, after deformation, the final line may make an angle with the original line. This angle is called *angular shear* (ψ) because it is during shearing of two parallel planes that two mutually perpendicular lines can be deflected and thus make an angle (Fig. 4.5).

Let us consider two planes AD and BC parallel to each other. Let AB be a line perpendicular to the planes. Due to shearing, the point B shifts to B_1. Assuming homogeneous deformation, the shift of the line AB from B to B_1 maintains the angular shift ψ. The displacement BB_1 is compatible with the angular shear ψ. From the triangle ABB_1,

$$\tan\psi = BB_1/AB \tag{4.6}$$

In (4.6), tan ψ is called *shear strain* (γ), and thus $\gamma = \tan\psi$. Thus, if we know the position of the final line after deformation, we can determine the angular shear (ψ) and the shear strain (tan ψ). Shear strain is described as positive (+) when the line deflects in the clockwise direction, and it is negative (−) when the deflection is anticlockwise. The value of shear strain is zero when ψ is zero, i.e. when there is no shearing and the body is in undeformed state; it is infinity when ψ is 90°. As such, the value of shear strain ranges between zero and infinity.

Further, from (4.6),

$$BB_1 = AB \quad \tan\psi \tag{4.7}$$

As mentioned above, BB_1 gives the displacement caused due to shearing. Therefore, if we know the value of shear strain, we can calculate the amount of displacement any original object has undergone as a result of shearing.

4.5 Volumetric Strain

During deformation, as mentioned above, a body not only undergoes changes in line lengths and angular shear, but it may also undergo change, decrease or increase, in volume. The volume strain (ΔV) is given by the ratio of the change in volume to the original volume of a body. If V_O is the original volume of a body and V_1 is the final volume that it acquires after deformation, volumetric strain is then given by

$$\Delta V = (V_1 - V_O)/V_O = V_1/V_O - 1 \tag{4.8}$$

Volume of a sphere of unit radius is $V_O = 4/3\pi\,(1)^3 = 4/3\pi$. Volume of an ellipsoid of semi-axes r_1, r_2 and r_3 is

$$V_1 = \frac{4}{3}\pi r_1 \cdot r_2 \cdot r_3$$

Therefore from (4.8),

$$\Delta V = \frac{\frac{4}{3}\pi r_1 \cdot r_2 \cdot r_3}{\frac{4}{3}\pi} - 1 \tag{4.9}$$

$$\text{or } \Delta V = r_1 \cdot r_2 \cdot r_3 - 1$$

Since the ellipsoid is formed from the unit sphere, the semi-axes of the ellipsoid, i.e. r_1, r_2 and r_3, can be considered as extension along x-, y- and z-axes. Thus, $r_1 = (1 + e_x)$, $r_2 = (1 + e_y)$ and $r_3 = (1 + e_z)$. From (4.9), the value of volume change thus caused by deformation can be given as

$$\Delta V = \left[(1 + e_x)(1 + e_y)(1 + e_z)\right] - 1 \tag{4.10}$$

It may be noted that during deformation, the volume of the original rock may increase or decrease. Accordingly, in (4.10), ΔV may take up a positive or negative value, respectively.

4.6 Strain Ellipse

It is often convenient to refer to the state of deformation of a rock mass by considering two or three mutually perpendicular reference lines. Thus, we can imagine a sphere which is converted to an ellipsoid (Fig. 4.6) if subjected to homogeneous strain. If we cut this sphere along any plane, we get a circle. In this case, the circle is converted to an ellipse on

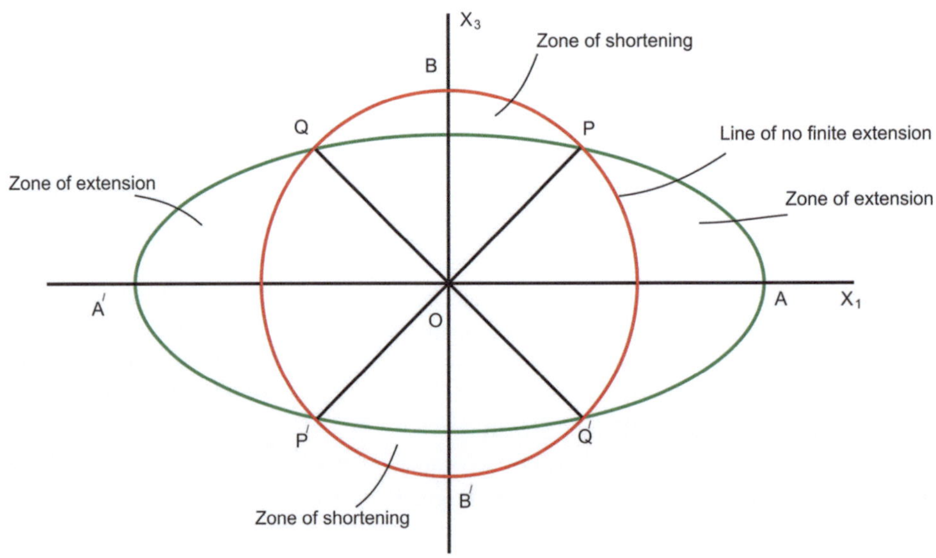

Fig. 4.6 Strain ellipse. A strain ellipse (green colour) superimposed on a circle (red colour) under plane strain condition. Lines PP′ and QQ′ are lines of no finite extension. These lines divide the strain ellipse into two zones: zone of extension and zone of shortening. In the zone of extension (POQ′A and QOP′A′), the radii of the ellipse are extended. In the zone of shortening (POQB and P′OQ′B′), the radii of the ellipse are shortened

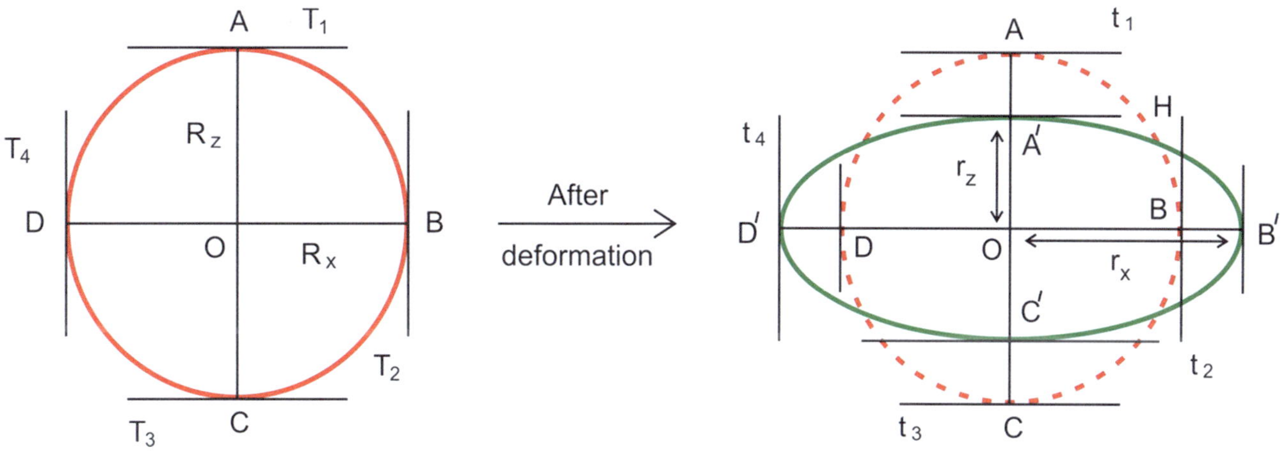

Fig. 4.7 If a circle is deformed to an ellipse, the shear strain parallel to each of the principal axes of strain is zero. The reason is that the tangents drawn perpendicular to the radii of the circle are parallel to those perpendicular to material lines of the ellipse. There is therefore no rotation of the principal strain axes

deformation. We can now imagine that deformation along some particular line is the greatest, and the least along a line that is perpendicular to the former. We can also refer these two or three lines to Cartesian coordinates. The above examples can be considered in two dimensions, which we refer to as strain ellipse, as well as in three dimensions, which we refer to as strain ellipsoid.

A *strain ellipse* describes the state of strain in two dimensions when a material circle changes its shape to an ellipse due to deformation by homogeneous strain. The ellipse can be referred to two mutually perpendicular axes X and Y. It is assumed that strain along one of these axes, X, is the greatest, while along the other the strain is the least. These two axes are then called *greatest principal strain axis* and *least principal strain axis*, respectively.

Because of the typical geometric relations established when a circle is changed to an ellipse, a strain ellipse shows some specific properties as highlighted below.

1. If we orient the circle and the ellipse on the Cartesian coordinate, we can obtain the values of strain measures such as extension and stretch along any selected line. For the circle (Fig. 4.6), since the radius maintains a uniform value, extension and stretch are zero. However, for the ellipse, we can obtain the values of extension and stretch along any selected line.
2. The superimposition of a strain ellipse on the initial circle (Fig. 4.6) clearly separates zones of extension from those of shortening. The line of separation of these two zones is called *line of no finite extension*. In the zone of extension, the radii of the ellipse are extended, while in the zone of shortening the radii of the ellipse are shortened.
3. If a circle is deformed to an ellipse (Fig. 4.7), the shear strain parallel to each of the principal axes of strain is zero. The reason is that the tangents drawn perpendicular to the material lines of the circle and those of the ellipse are parallel. There is therefore no rotation of the principal strain axes. These two principal strain axes are thus also called *axes of zero shear strain*.

4.7 Strain Ellipsoid

4.7.1 Principal Strain Axes

A *strain ellipsoid* describes the state of strain in three dimensions when a material sphere changes its shape to an ellipsoid due to deformation by homogeneous strain. The ellipsoid can be referred to three mutually perpendicular axes X, Y and Z (Fig. 4.8). It is assumed that strain along one axis (X) is the greatest, while that along the other axes is the least (Z) and intermediate (Y), such that $X \geq Y \geq Z$. Accordingly, these three axes are called *greatest principal strain axis*, *intermediate principal strain axis* and *least principal strain axis*, respectively. The lengths of these three principal axes define the *principal longitudinal strains*. Along the X-, Y- and Z-axes, these longitudinal strains, respectively, are $(1 + e_1)$, $(1 + e_2)$ and $(1 + e_3)$ such that $(1 + e_1) \geq (1 + e_2) \geq (1 + e_3)$ (Fig. 4.9). The principal quadratic extensions along these axes are λ_1, λ_2 and λ_3 in which $\lambda_1 \geq \lambda_2 \geq \lambda_3$.

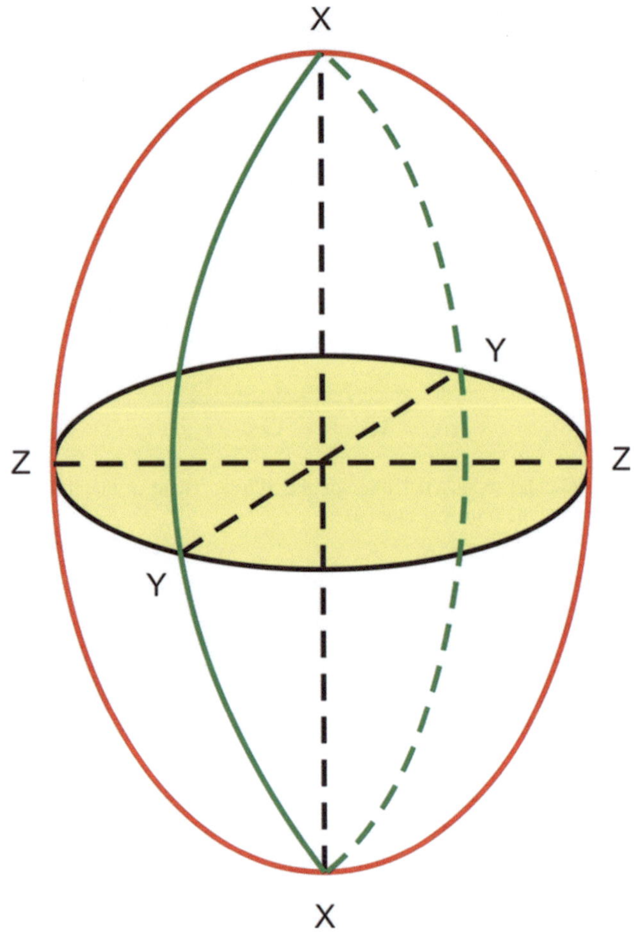

Fig. 4.8 Strain ellipsoid showing the three mutually perpendicular principal strain axes

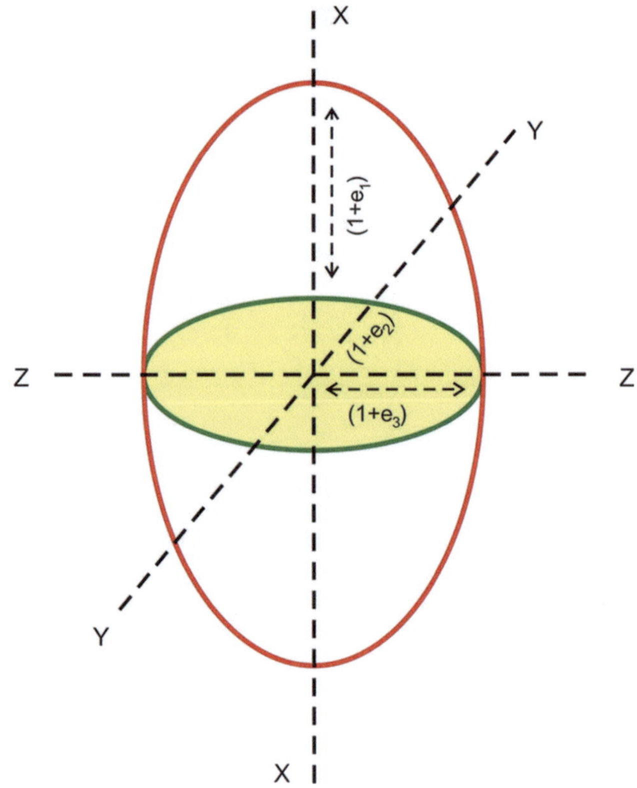

Fig. 4.9 Longitudinal strains along the three principal strain axes X, Y and Z are $(1 + e_1)$, $(1 + e_2)$ and $(1 + e_3)$, respectively

A strain ellipsoid is assumed to have formed when a sphere is subjected to a homogeneous strain. The volume of a unit sphere is $4/3\pi$. The volume of the strain ellipsoid whose radii are λ_1, λ_2 and λ_3 is

$$4(\lambda_1\lambda_2\lambda_3)^{1/2}/3\pi \tag{4.11}$$

We can assume a situation when extension along X is fully counterbalanced by shortening along Z so that there is no extension along Y, and therefore the deformation involves no volume change. This special case of deformation is called *plane strain*.

It may be mentioned here that strain and stress ellipsoids behave in an inverse way if the deformation was homogeneous. Thus, the direction of the greatest principal strain (X) corresponds to the direction of the least principal stress (σ_3) and that of the least (Z) and intermediate (Y) principal strain axes corresponds, respectively, to the greatest (σ_1) and intermediate (σ_2) principal stress axes. This inverse relationship commonly helps in interpreting the directions along which the stresses were applied to the rock mass or to the area.

4.7.2 Equation of Strain Ellipsoid

A sphere of unit radius is represented by the following equation:

$$x^2 + y^2 + z^2 = 1 \tag{4.12}$$

On homogeneous deformation, this unit sphere becomes an ellipsoid. We orient this ellipsoid on three mutually perpendicular axes such that the longest axis of the ellipsoid constitutes the x-axis while the shortest and the intermediate axes lie along the z- and y-axes, respectively. If the radii of this ellipsoid are r_1, r_2 and r_3 in which $r_1 \geq r_2 \geq r_3$, the ellipsoid can then be represented by the following equation:

$$\frac{x^2}{r_1^2} + \frac{y^2}{r_2^2} + \frac{z^2}{r_3^2} = 1 \tag{4.13}$$

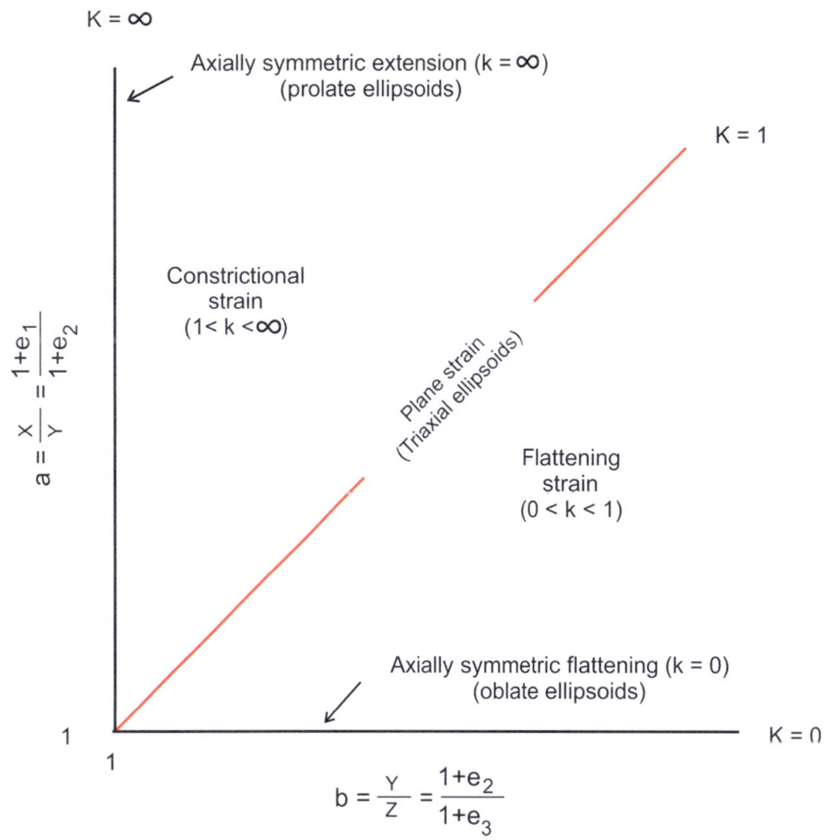

Fig. 4.10 Flinn diagram. It is a plot of axial ratios *X*/*Y* and *Y*/*Z* that shows fields of five different strain states: axially symmetric flattening ($k = 0$), flattening strain ($0 < k < 1$), plane strain ($k = 1$), constrictional strain ($1 < k < \infty$) and axially symmetric extension ($k = \infty$)

If the reference coordinate axes *x*, *y* and *z* are rotated, they coincide with the three mutually perpendicular axes of strain (see Fig. 4.9). This can be thought of as a unit sphere distorted into an ellipsoid, whose equation (Ramsay 1967, p. 124) is given by

$$\frac{x^2}{\lambda_1} + \frac{y^2}{\lambda_2} + \frac{z^2}{\lambda_3} = 1 \tag{4.14}$$

4.7.3 Shape and Ellipticity

The *shape* and *ellipticity* of a strain ellipsoid can be determined if we know the values of stretch along the greatest (s_1) and the least (s_3) principal axes of strain. The ellipticity *R* is given by

$$\begin{aligned} R &= (s_1/s_3) = (1+e_1)/(1+e_3) = (l_1/l_0)/(l_3/l_0) \\ &= l_1/l_3 \end{aligned} \tag{4.15}$$

The ellipticity is a useful parameter as it helps in quantitatively distinguishing two or more strain ellipsoids and thus helps in comparing the state of deformation of rocks of two or more areas. However, complications arise if the rock has not been subjected to homogeneous deformation throughout.

4.8 Representation of Strain States

Derek Flinn devised a method in 1962 that helps in understanding the states of strain for heterogeneously deformed rocks by means of a diagram known as *Flinn diagram*. The diagram shows how an initially spherical object takes up different ellipsoidal shapes during progressive deformation. The diagram thus helps identify different states of strain during progressive deformation. The method assumes constant volume during deformation.

The Flinn diagram (Fig. 4.10) is basically a plot of axial ratios *X*/*Y* and *Y*/*Z* that are, respectively, represented by two parameters *a* and *b*:

$$a = X/Y \tag{4.16}$$

$$b = Y/Z \tag{4.17}$$

Here *X*, *Y* and *Z* ($X > Y > Z$) are, respectively, the value of stretch along *X*, *Y* and *Z* principal axes of strain. From Eq. (4.3)

$$X = (1+e_x), Y = \left(1+e_y\right), Z = (1+e_z) \tag{4.18}$$

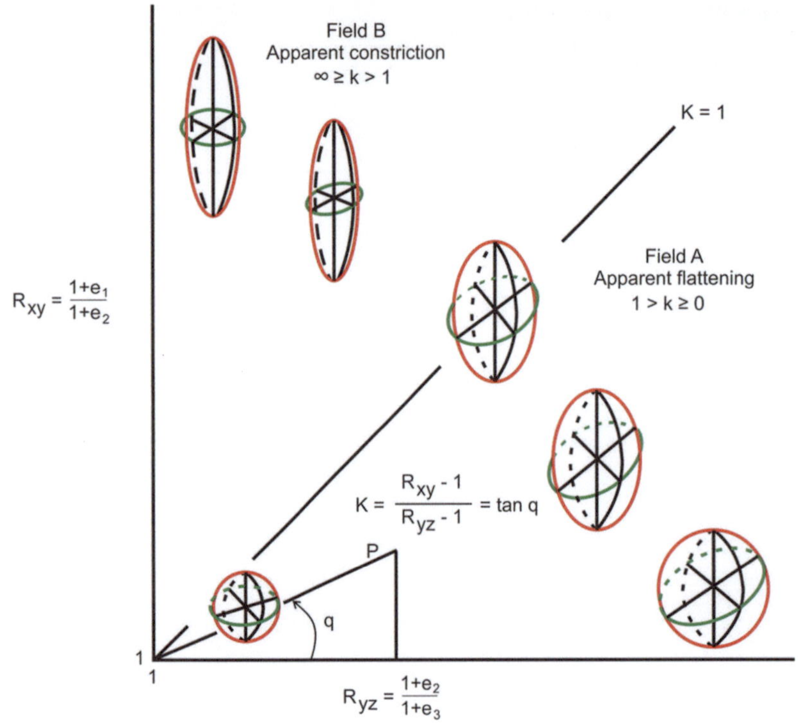

Fig. 4.11 Different types of ellipsoids falling in various strain fields in the Flinn plot. The $k = 1$ line represents plane strain. Colour scheme by the author. (Reproduced from Ramsay and Huber 1983, Fig. 10.8 with permission from Elsevier Senior Copyrights Coordinator, Edlington, U.K. Submission ID: 1198143)

A graph is plotted with b as abscissa and a as ordinate. Since the values of a and b cannot be less than one, the origin of the graph is taken as (1, 1). A parameter k is then identified such that

$$k = \frac{a-1}{b-1} \tag{4.19}$$

The value of k ranges from 0 through 1 and infinity (∞). In the graph, k takes up a value of 0 along abscissa and ∞ along ordinate. By drawing the bisector, $k = 1$, the graph gets divided into two broad fields. The field between $k = 0$ and $k = 1$ is called *flattening strain* and between $k = 1$ and $k = \infty$ is called *constrictional strain*, while the line $k = 1$ is called *plane strain*. From the plot, it is thus possible to identify the five states of strain (Fig. 4.10) and so five types of strain ellipsoids as described below:

1. $k = 0$: uniaxial strain; oblate uniaxial ellipsoid; pancake-shaped ellipsoid
2. $0 < k < 1$: flattening strain; oblate ellipsoid
3. $k = 1$: plane strain; plane strain ellipsoid
4. $1 < k < \infty$: constrictional strain; prolate ellipsoid
5. $k = \infty$: uniaxial strain; prolate uniaxial ellipsoid; cigar-shaped ellipsoid

The ellipsoids representing the different strain fields in the Flinn plot thus show three different shapes (Fig. 4.11). The *oblate ellipsoid* represents a flattening strain in which there is a uniform shortening in the *Z*-direction and equal extension in all directions at right angles to it; this shape corresponds to a *pancake*. The *prolate ellipsoid* represents a constrictional strain in which there is a uniform extension along the *X*-direction and equal shortening in all the directions at right angles to it; this shape corresponds to a *cigar* or a *rugby ball*. The *plane strain* is represented by a *triaxial ellipsoid* in which extension along the *X*-direction is equally counterbalanced by shortening along the *Z*-direction so that the *Y*-direction remains unchanged.

A merit of the Flinn diagram is that we can know the different shapes of the ellipsoids by identifying only one parameter k. The nature of strain, e.g. flattening, constrictional or plane strain, during any stage of deformation can also be known. However, the diagram does not throw light on some important aspects of deformation. For example, since the diagram is based on the ratios of two principal stretches (*X*/*Y* and *Y*/*Z*), it does not give an idea of the size of the ellipsoids. Further, the rotational aspect of material lines is not indicated in the diagram, and therefore it cannot separate pure shear and simple shear or coaxial deformation and noncoaxial deformation.

Although the Flinn method assumes constant volume during deformation, it can be extended to consider volumetric changes also. In the Flinn graph, $k = 1$ represents plane strain, which implies that any change along X is equally compensated by a change in Z such that there is no deformation along Y, i.e. $Y = 1$. It follows that when there is volume change, we need to locate the position of the line that separates the flattening and constrictional fields. This is done in the following way.

For plane strain, since $Y = 1$, Eqs. (4.16) and (4.17) become

$$a = X \quad (4.20)$$

$$b = 1/Z \quad (4.21)$$

The line $k = 1$ in this case separates the flattening field from the constrictional field (Fig. 4.11).

Since there is no stretch along Y-axis, $(1 + e_y) = 1$. This is the case in the plane strain condition when $k = 1$. Eq. (4.10) for volume change thus reduces to

$$1 + \Delta V = (1 + e_x)(1 + e_z) \quad (4.22)$$

4.9 Pure Shear and Simple Shear

An undeformed body can take up different shapes by homogeneous progressive strain in two ways with reference to material lines or reference axes. In one case, the shape can change in such a way that the material lines do not change their angular relations. In other words, the orientations of the principal strain axes X, Y and Z have not changed even though their lengths have changed. This is a case of pure shear, which is a type of irrotational strain (though there are subtle differences between pure shear and irrotational shear as highlighted below as well as in Sects. 4.10 and 4.11). In another case, the shape of the body can change in such a way that the material lines have changed their original angular relations. In other words, the orientations of the principal strain axes X, Y and Z have changed together with change in their lengths. This is a case of *simple shear*, which is a type of *rotational strain* (though there are subtle differences as highlighted below as well as in Sects. 4.10 and 4.11).

The irrotational strain and rotational strain can be better explained in two dimensions. Figure 4.12a geometrically describes an example of irrotational strain by showing the stages of how a circle is deformed to an ellipse. It is assumed that deformation is homogeneous and that each change in shape has taken place in successive deformation events by the addition of smaller uniform units, i.e. by *incremental strain*. It may be noted that the orientation of the principal strain axes X and Z has not changed—they always remain perpendicular to each other—even though their lengths have changed. Since the deformation is progressive, each next ellipse is

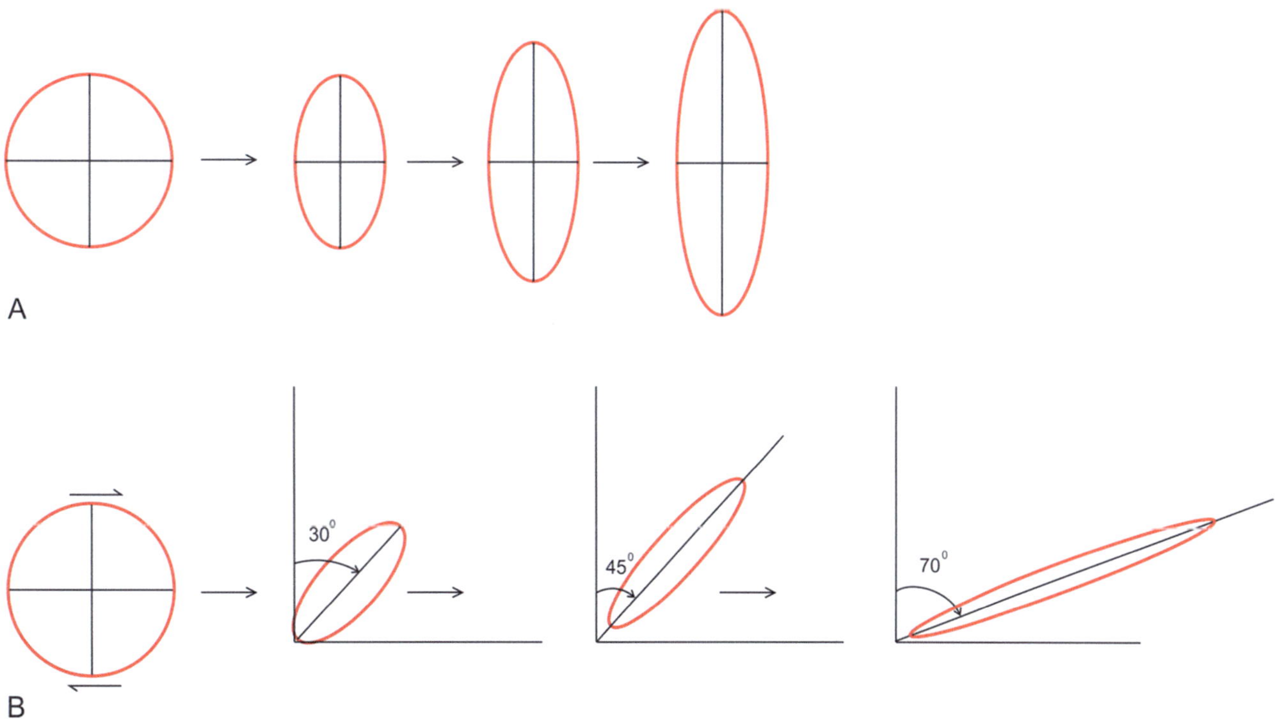

Fig. 4.12 Pure shear and simple shear. (**a**) Pure shear does not involve rotation of the principal strain axes during progressive deformation; this is also sometimes called irrotational strain. (**b**) Simple shear involves rotation of the principal strain axes during progressive deformation; this is also sometimes called rotational strain

more deformed than the previous one. In this case, X refers to the direction of extension, while Z is the direction of shortening.

The rotational strain has been explained in Fig. 4.12b by showing how a circle is deformed to an ellipse by homogeneous strain. Here also, we assume that each change in shape of the original body has taken place by the addition of smaller units, i.e. by incremental strain. In rotational strain, unlike irrotational strain, the orientation of the principal strain axes X and Z continues to change with progressive deformation. Any ellipse thus formed as a result of one incremental strain would give rise to the next ellipse, which is more deformed than the previous one. One can also notice that the lengths of the principal strain axes continuously change with progressive deformation. Here again, X refers to the direction of extension while Z is the direction of shortening.

4.10 Coaxial and Noncoaxial Deformation

The terms *coaxial deformation* and *noncoaxial deformation* are also sometimes used in the context of irrotational strain and rotational strain. Coaxial (vorticity = 0) and noncoaxial (vorticity >0) refer to strain paths (progressive strain), not strains. In the irrotational strain and rotational strain, the rotation of principal strain axes is referred to the arbitrarily chosen coordinate axes. In coaxial deformation and noncoaxial deformation, on the other hand, the rotation of the principal strain axes is referred to the principal axes of incremental strain. Use of the terms coaxial and noncoaxial, thus, has more relevance and significance to the actual field situations of rocks. In fact, any structure may bear imprints of coaxial or noncoaxial deformation.

The nature of orientation and rotation of material lines constitute an important aspect of distinction between pure and simple shear deformations (Hanmer and Passchier 1991). During progressive pure shear, there are two material lines that do not change length, while during progressive simple shear one of the two material lines of zero stretching rate remains attached to the flow plane. This implies that in progressive pure shear the instantaneous stretching axes remain attached to the material line while in progressive simple shear the low plane remains attached to the material line.

In natural rocks, signatures of coaxial and noncoaxial deformations can be observed in structures developed on all scales ranging from field to microscopic. We present here one common example each from field (fold) and microscopic (porphyroclast) showing signatures of coaxial and noncoaxial deformations. The fold in Fig. 4.13 bears signature of coaxial deformation. The black dashes approximately representing the axial trace are vertical suggesting no rotation of the fold, while in the fold in Fig. 4.14 the axial trace is inclined suggesting rotation of the fold. By the same analogy, in Fig. 4.15, the black dashes representing the long axis of the porphyroclast are parallel to the surrounding foliation suggesting no rotation of the grain. In Fig. 4.16, on the other hand, the long axis of the porphyroclast is inclined to the surrounding foliation suggesting rotation of the grain.

Fig. 4.13 Signature of coaxial deformation as shown by a multilayer fold developed in sandstone bed of Southern Appalachians, USA. The axial plane (black dashes) of the fold is vertical suggesting no rotation of the structure. (Photograph by the author)

Fig. 4.14 Signature of noncoaxial deformation as shown by an asymmetric fold in sandstone. Due to noncoaxial deformation, the axial plane of the fold has undergone rotation. The black dashes show the approximate trace of the fold axial plane. Loc. NE of Bhowali, Kumaun Lesser Himalaya. (Photograph by the author)

Fig. 4.15 Signature of coaxial deformation in thin section as shown by a plagioclase porphyroclast. The black dashes show the approximate trace of the long axis of the porphyroclast. This marker line is parallel to the orientation of the surrounding fabric, thus suggesting no rotation of the porphyroclast. (Photomicrograph courtesy Amit K. Verma)

Fig. 4.16 Signature of noncoaxial deformation in thin section as shown by an asymmetrically disposed plagioclase porphyroclast. The black dashes show the approximate trace of the long axis of the porphyroclast. Note that this marker line is inclined and is asymmetrical with respect to the surrounding fabric, suggesting rotation of the porphyroclast during progressive deformation. (Photomicrograph courtesy Amit K. Verma)

4.11 Strain Path

The various states of strain that a body acquires from undeformed to deformed state during progressive deformation are referred to as *strain path*. Strain path is often referred to as *deformation path* because it is sometimes possible to trace back, at least partly, the states of strain (deformation). As mentioned earlier, the body acquires a deformed state by addition of each incremental strain. It must be mentioned here that the strain path does not necessarily provide information on strain states during progressive deformation because it is possible that during deformation the body may have undergone different types of deformation, e.g. extensional or compressional tectonics.

Since a body can acquire a deformed state by two main processes, i.e. coaxial and noncoaxial, strain paths can be of two types, i.e. coaxial strain path and noncoaxial strain path.

4.11.1 Coaxial Strain Path

A body is said to have undergone a *coaxial strain path* if during progressive deformation the principal strain axes of the body coincide with the principal axes of incremental strain. In other words, the strain axes of the body continue to remain parallel as the body takes up different shapes with progressive deformation. The coaxial strain path can be shown by a simple example of deformation of a circular body in two dimensions (Fig. 4.12a). Due to an imposed strain, the body takes up an elliptical shape. If the body is subjected to a coaxial strain path, the material lines (long and short axes) of the body and those of the surrounding area maintain a constant geometrical relation. The lengths of the material lines of the body continue to change under strain, but their relations with those of the matrix do not change. For this type of strain path, we assume that one axis of the strain ellipsoid remains unchanged while the other two undergo changes in their lengths. Further, an increase in one material line is sympathetically associated with a decrease in the other, and as such there is no change in volume during straining. This type of strain path is also commonly called *pure shear*. At this stage, it is important to note that the use of coaxial strain path as a synonym of pure shear should be done with caution. The main reason is that sometimes it is opined that constant volume is also possible even if all the three axes undergo changes in their lengths.

4.11.2 Noncoaxial Strain Path

A body is said to have undergone a *noncoaxial strain path* if during progressive deformation the principal strain axes of the body do not coincide with the principal axes of incremental strain. In this case, the strain axes of the body rotate, i.e. do not remain parallel, as the body takes up different shapes with progressive deformation. The noncoaxial strain path can be shown by a simple example of deformation of a circular body in two dimensions (Fig. 4.12b). Due to an imposed strain, the body takes up an elliptical shape. During progressive deformation, the material lines (long and short axes) of the body remain perpendicular to each other though they continue to change their lengths with each increment of strain but at the same time the body rotates with respect to the material lines of the matrix. We assume that one axis of the strain ellipsoid remains unchanged while the other two undergo changes in their lengths such that an increase in one material line is associated with a sympathetic decrease in the other, and as such there is no change in volume during straining. Thus, during noncoaxial strain path, the material lines of a body remain perpendicular to each other but keep changing their orientation with incremental strain. This type of strain path is also commonly called *simple shear*.

4.12 Progressive Deformation

During deformation, the amount of strain acquired by a body goes on increasing continuously as long as the deformation persists. The body thus continues to take up different shapes (Fig. 4.17) and sizes as deformation proceeds. The strain ellipses also continue to change their shapes accordingly. *The process of deformation by which a body continues to change its shape and size is called progressive deformation.* It is believed that the final strain achieved, and thus the final shape taken by the body, is the cumulative effect of several deformation events, and each deformation event took place by increment of some smaller strain, i.e. incremental strain. In this case, we assume that the incremental strain is within finite limits. In Fig. 4.17, the finite strain at time t_1 may be considered as incremental strain for practical purposes. Finite strain at t_1 was achieved next to the time (t_0) when the body was in an undeformed state. When the deformation is over by the addition of incremental strain uniformly, the body acquires the final shape or volume which is actually seen in field today. *The amount of strain estimated from a deformed rock at the time of observation is called finite strain.* In common practice, use of the word 'strain' generally refers to 'finite strain' unless it is specified otherwise.

4.13 Vorticity

During rotational strain, as we have mentioned above, the principal axes of strain as well as all material lines rotate during deformation. However, all material lines do not rotate

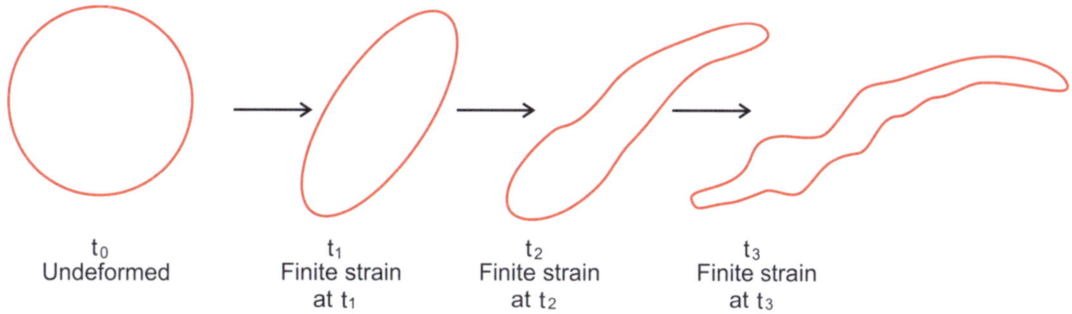

Fig. 4.17 Progressive deformation showing how an originally undeformed body takes up different shapes. Measurement of strain at any instant (say t_2 or t_3) gives the amount of finite strain at that instant. If say t_3 represents the final shape of the body at the time of measurement in field, the strain measured gives the finite strain at that spot

in the same rate. This phenomenon is expressed by a parameter called *vorticity that gives the average rate of rotation of all the material lines of a body under progressive deformation in relation to the coordinate axes*. For a noncoaxially deformed rock, vorticity thus constitutes an important measure of the amount of rotation as compared to distortion. Since the coordinate axes constitute a fixed reference frame, it is therefore possible to know the exact amount of rotation taken up by the material lines during deformation. During progressive simple shear (rotational strain), vorticity maintains an amount of unity. However, when rotation is much higher for any object, vorticity may exceed the amount of unity. During progressive pure shear, on the other hand, vorticity is zero because the deformation is coaxial as the coordinate axes do not rotate with progressive deformation.

Several workers have worked on the quantitative aspects of rotation by identifying a few parameters as well as by formulating equations that describe the various types of rotation (e.g. Ramberg 1975; Ramsay 1967; Ghosh and Ramberg 1976; Ghosh 1993). A simplified version of vorticity as based on these works is presented below.

Vorticity can be expressed in terms of coordinate axes. If we consider the velocities along the x- and y-axes, the deformation may cause a rotation perpendicular to the xy-plane, i.e. along the z-axis (Fig. 4.18a). It may be noted that the deformation in this case has caused rotation only and no internal deformation. Similarly, we can consider deformation along other planes, i.e. yz- and xz-planes (Fig. 4.18b); in such cases, the rotation would be, respectively, along x- and y-axes. If u, v and w are the velocities along the x-, y- and z-axes, respectively, the three-dimensional rate of rotation along the three axes is given by

$$W_x = \frac{1}{2}\left(\frac{\delta w}{\delta y} - \frac{\delta v}{\delta z}\right) \quad (4.23)$$

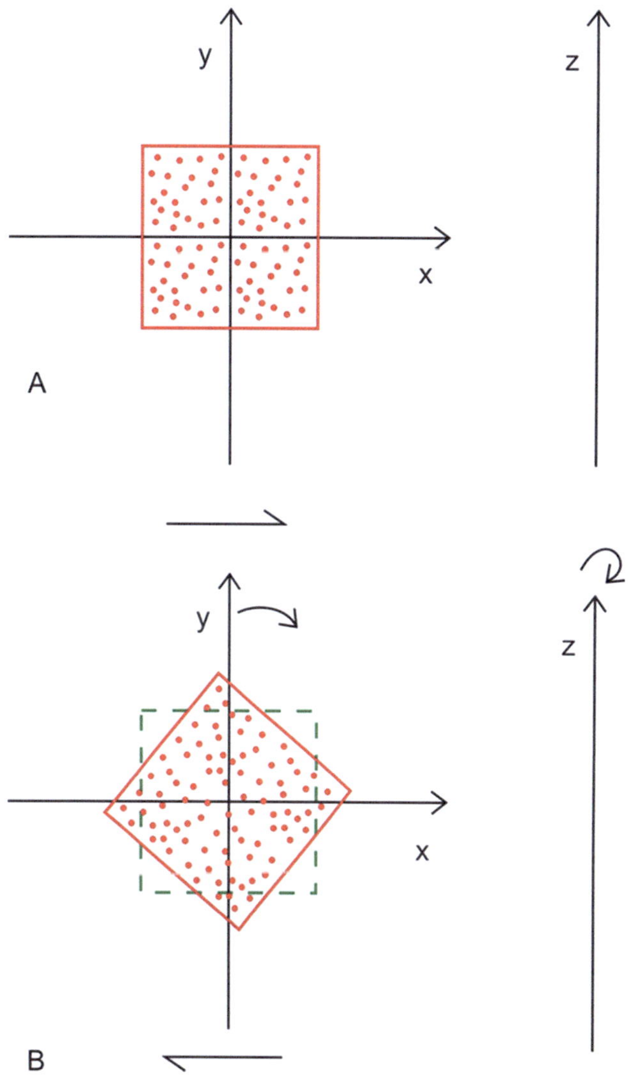

Fig. 4.18 Vorticity as explained in two dimensions. (**a**) A square oriented along x- and y-axes. If the square is rotated clockwise, i.e. towards the x-axis, the axis of rotation (z) lies perpendicular to xy-plane. (**b**) The square on being subjected to an angular shear takes up a new position in the xy-plane. This causes a rotation along an imaginary line z oriented perpendicular to the xy-plane

$$w_y = \frac{1}{2}\left(\frac{\delta u}{\delta z} - \frac{\delta w}{\delta x}\right) \quad (4.24)$$

$$w_z = \frac{1}{2}\left(\frac{\delta v}{\delta x} - \frac{\delta u}{\delta y}\right) \quad (4.25)$$

The above three equations simplify to

$$w = 2w_z = \frac{\delta v}{\delta x} - \frac{\delta u}{\delta y} \quad (4.26)$$

Each of the components $2w_x$, $2w_y$ and $2w_z$ is known as *vorticity* for the plane concerned. Vorticity is represented by W. If the deformation is irrotational or coaxial, i.e. the deformation is accompanied by no rotation, the value of W is zero.

Since rotation is a characteristic feature of noncoaxial deformation, it is therefore possible to quantify the degree of noncoaxiality by the degree or amount of rotation. This concept has been explained by Truesdell (1954) by suggesting that the degree of noncoaxiality can be expressed by a *kinematical vorticity number* as given by

$$W_k = \frac{W}{\left[2\left(\dot{\varepsilon}_1^2 + \dot{\varepsilon}_2^2 + \dot{\varepsilon}_3^2\right)\right]^{1/2}} \quad (4.27)$$

where $\dot{\varepsilon}_1, \dot{\varepsilon}_2$ and $\dot{\varepsilon}_3$ are principal linear strain rates and W is the amount of the vorticity vector. For rotational shear, the absolute value of W_k is 1, while for irrotational shear it is 0. The degree of noncoaxiality, thus, increases with increase in the kinematical vorticity number. We may, thus, have an intermediate stage also, called *general shear* (noncoaxial) for which the value of W_k lies between 0 and 1. To sum up:

$$\text{Pure shear (coaxial)} : W_k = 0 \quad (4.28)$$

$$\text{General shear (noncoaxial)} : 0 < W_k < 1 \quad (4.29)$$

$$\text{Simple shear (noncoaxial)} : W_k = 1 \quad (4.30)$$

Box 4.1 Strain Tensor in the Context of Structural Geology

Some elementary ideas of tensor have been presented in Chap. 3 (Box 3.1). Here, we confine only to strain tensor. We begin with some important types of tensors as summarized by Allmendinger et al. (2012), pp. 81–82:

- *Zero-order tensor*: A *scalar quantity* that can be represented by a single number that is independent of the coordinate system, e.g. temperature, mass and density.
- *First-order tensor*: A *vector quantity* which is a physical entity with a magnitude and direction that remain the same in all coordinate systems, e.g. velocity, displacement and temperature gradient.
- *Second-order tensor*: It is a physical quantity represented by nine numbers and is independent of coordinate system, e.g. stress, strain and thermal conductivity.

Box 4.1 (continued)

Strain is a three-dimensional quantity that can be fully described by either six or nine quantities or components. With six different components, such as stress, the tensor is described as a *symmetric tensor*, and with nine components, such as finite homogeneous strain, it is described as an *asymmetric tensor*. A *strain tensor* is therefore an asymmetric tensor.

In the context of three-dimensional strain, commonly there is a component of rotation also. For the sake of convenience, we may therefore consider strain in two ways. In one way, we can view strain as a symmetric quantity, especially when we consider deformation in the form of a strain ellipse. A strain ellipse is always symmetrical if it bears two reference lines perpendicular to each other. Therefore, any change in a marker line, i.e. distortion or deformation, is symmetrical with the reference lines or the principal strain axes.

In another way, we can view strain as an asymmetric quantity, especially when we consider the component of rotation (rotational strain). In fact, in most cases of deformation, rotation is always there (without rotation the deformation becomes symmetric). When we consider rotation, one problem immediately crops up. What was the original position of the rock in its undeformed state? As a matter of fact, we do not have any precise knowledge of the marker lines or planes in the undeformed state of the rock and any assumption can be considered as guess only. Therefore, while considering deformation as an asymmetric tensor, we need to take into account many simplifying assumptions, and all this makes the exercise highly arduous, yet imperfect. As such, we are not going to discuss asymmetric tensor in further detail.

An asymmetric tensor is completely given by the following nine coefficients by a matrix of the following type (Ramsay 1967, p. 123):

(continued)

Box 4.1 (continued)

$$\begin{vmatrix} a & b & c \\ d & e & f \\ g & h & i \end{vmatrix} \quad (4.31)$$

This tensor includes both symmetric and asymmetric components of deformation. The symmetric part can be expressed in terms of six independent quantities showing symmetrical relationship as given below:

$$\begin{vmatrix} a & \frac{b+d}{2} & \frac{c+g}{2} \\ \frac{b+d}{2} & e & \frac{f+h}{2} \\ \frac{c+g}{2} & \frac{f+h}{2} & i \end{vmatrix} \quad (4.32)$$

If the rotational component of deformation is considered, we need to take into account three independent quantities with the signs of the quantities changed:

$$\begin{vmatrix} 0 & \frac{b-d}{2} & -\frac{g-c}{2} \\ -\frac{b-d}{2} & 0 & \frac{f-h}{2} \\ \frac{g-c}{2} & -\frac{f-h}{2} & 0 \end{vmatrix} \quad (4.33)$$

The tensor as represented in the above form is called *skew-symmetric matrix* (Ramsay 1967, p. 124), and it gives the rotational part of the strain.

Application to Structural Geology

Application of tensor to strain studies, like those of stress (Chap. 3), holds potential to advance our knowledge on strain in rocks. Ramsay (1967) applied the concept of tensors to some structural aspects including strain (also stress, as described in Chap. 3). From structural viewpoints, strain tensors are discussed in the books of Means (1976), Nicolas and Poirier (1976), Ranalli (1987), Nicolas (1987), De Paor (1996) and Allmendinger et al. (2012). Strain tensor can explain some important structural aspects as given below.

Velocity field broadly means the distribution of velocity in a given area, and as such it can be expressed in the spatial and time coordinates, i.e. $V(x,y,z,t)$. The analogy of this concept in structural geology is that rocks are believed to behave as a fluid under high temperature and pressure as existing at depth. Velocity field thus advances our knowledge in *vorticity*, *steady flow* and *irrotational flow* as applied to deformation of rocks.

Box 4.1 (continued)

Flow apophyses (Tikoff and Fossen 1993) are eigenvectors of deformation matrix and can be considered to represent the strain axes. Due to rotation, the changing strain field from, say, incremental shortening to incremental extension during folding can be expressed by eigenvectors of the deformation matrix.

4.14 Mohr Strain Diagram

We have described Mohr diagram for stress (Chap. 3), which is in the form of a circle in which normal stress is plotted as abscissa and shear stress as ordinate. Likewise, it is also possible to construct a Mohr diagram for representing the state of finite strain. This is known as *Mohr strain diagram* in which, in contrast to stress diagram, the reciprocal of quadratic elongation, λ' ($= 1/\lambda$), is plotted as abscissa and modified shear strain, γ' ($= \gamma/\lambda$), as ordinate. Assuming that there is no change in the area, the states of strain can be represented by a circle (Fig. 4.19). The radius of this circle is $(\lambda'_2 - \lambda'_1)/2$, and the centre C is $(\lambda'_1 + \lambda'_2)/2$, (0).

The angle (measured clockwise) for any material line CP with the x-axis is $2\varphi'$. This line intersects the circle at P, whose coordinates are given by (λ', γ'). The point P represents the state of strain in the circle. Ramsay (1967, pp. 69–70) and Ramsay and Huber (1983, pp. 93–95) presented the mathematical solutions for describing the state of strain with reference to angles measured in the strained condition as given below:

$$\lambda' = \frac{\lambda'_1 + \lambda'_3}{2} - \frac{\lambda'_1 - \lambda'_3}{2}\cos 2\phi' \quad (4.34)$$

$$\gamma' = \frac{\lambda'_1 - \lambda'_3}{2}\sin 2\phi' \quad (4.35)$$

The values of λ' and γ' give the states of strain for the body. The angular shear strain ψ is given by the inclination of the line joining P with the origin (0, 0) and is expressed as

$$\psi = \tan^{-1}\gamma - \frac{\gamma'}{\lambda'} \quad (4.36)$$

It is therefore possible to construct a Mohr circle from deformed objects. From this circle, the values of principal longitudinal strains and their orientations can be evaluated in the deformed plane.

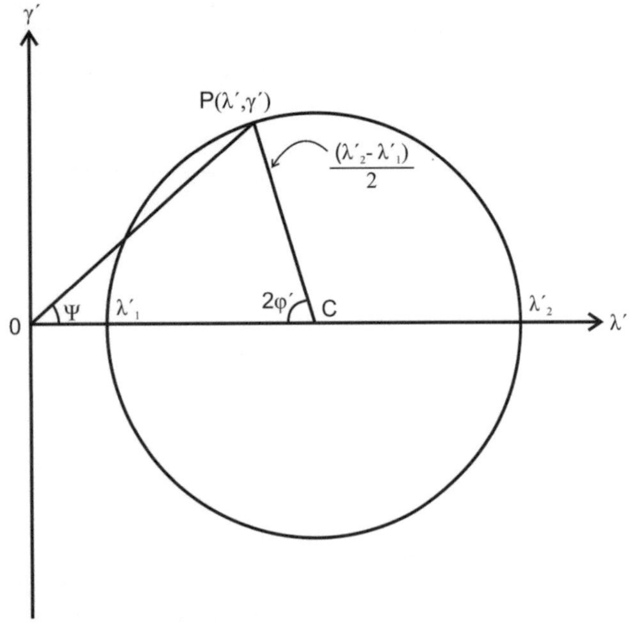

Fig. 4.19 Mohr strain diagram. The circle, called Mohr circle for finite strain, represents the state of homogeneous finite strain. See text for details

4.15 Strain History

In most deformed terrains, the nature of strain does not remain the same all through the deformation history. Systematic analysis of strain carried out in different parts of an area helps in understanding the various states of strain shown by the rocks. With the available data, it is possible to frame a sequence showing the various states of strain undergone by the rocks starting from their undeformed state. All this helps in working out the *strain history*, also called *deformation history*, of rocks.

There is no specific methodology for working out the strain history of deformed rocks. It all depends upon the deformation features and the corresponding strain signatures. For example, careful study and analysis of strain may reveal that the rocks of an area may show a change from initially pure shear to a later simple shear mode of deformation. In another case, the analysis may reveal a change from an initial rotational strain to translation. Likewise, there may be an area that has initially undergone compressional tectonics followed by extensional tectonics. Further, there are cases of introduction of some new material such as veins, dikes and mineral fibres. Development of faults, fractures and tension gashes may also take place at some of the intervening stages of deformation. Formation of fractures and related structures may suggest a brittle regime of deformation, but at some later stages the same features may show bending suggesting appearance of ductile regime. Also, structures formed at some stage may indicate a coaxial strain path, but the same structures may show evidence of rotation suggesting introduction of a noncoaxial path of deformation at some later stages. To summarize, it can be said that understanding of strain history is a complicated process that demands careful study of the strain signatures and deformation features of the rocks.

4.16 Significance of Strain

- Knowledge of strain provides a good deal of information on the physico-mechanical aspects of rocks and a quantitative picture of deformation of rocks. All these have implications for stress field and directions of stress application.
- Strain studies help in comparing the states of deformation of rocks, and thus the behaviour of rocks to the imposed stresses, of two or more localities or regions.
- Strain studies also throw light on whether a rock was brittle, ductile or ductile-brittle at the time of deformation.
- Localization of some mineral deposits or fluid systems can be traced on the basis of occurrence of low-strain and high-strain zones in rocks.

4.17 Summary

- *Strain* is the amount of deformation that a body has undergone in response to an externally applied stress. In crustal rocks, strain is developed as they are always under stress due to overburden. Strain in rocks is manifested in various tectonic structures such as folds, faults and boudins.
- A body can develop strain by *translation, rotation, dilatation, distortion* or *bulk deformation*.
- A body is said to have undergone *homogeneous deformation* if the parallel lines of the undeformed body remain straight and parallel after deformation and the bounding planar surfaces remain planar. In *inhomogeneous* or *heterogeneous deformation*, the reference lines and planes of the undeformed body have changed their geometric relations after deformation.
- The state of deformation of a rock mass can be considered by imagining a sphere converted to an ellipsoid if subjected to homogeneous strain. This gives a *strain ellipsoid* that can be referred to three mutually perpendicular axes: *greatest principal strain axis, intermediate principal strain axis* and *least principal strain axis*. A strain ellipsoid on cutting along some plane gives a *strain ellipse* with only two strain axes: *greatest principal strain axis* and *least principal strain axis*.

- A convenient diagram that helps in understanding the states of strain for inhomogeneously deformed rocks is a *Flinn diagram* that shows how an initially spherical object takes up different ellipsoidal shapes during progressive deformation. The Flinn diagram is basically a plot of axial ratios *X*/*Y* and *Y*/*Z*.
- An undeformed body can take up different shapes by homogeneous progressive strain in two ways with reference to material lines or reference axes: (a) *pure shear* or *irrotational shear* when the shape changes in such a way that the material lines do not change their angular relations and (b) *simple shear* or *rotational shear* when the shape changes in such a way that the material lines have changed their original angular relations. If the rotation of the principal strain axes is referred to the principal axes of incremental strain, the terms pure shear and simple shear are designated as *coaxial deformation* and *noncoaxial deformation*, respectively.
- The various states of strain that a body acquires from an undeformed to the deformed state during progressive deformation are referred to as *strain path* or *deformation path*. The process of deformation by which a body continues to change its shape and size in response to the imposed deformation is called *progressive deformation*. Systematic analysis of strain carried out in different parts of an area helps in understanding the various states of strain shown by the rocks from their undeformed state; this provides the *strain history*, or *deformation history*, of the rocks.
- Although the principal axes of strain as well as all material lines rotate during rotational strain, all material lines do not rotate in the same rate. This phenomenon is expressed by a parameter called *vorticity* that gives the average rate of rotation of all the material lines of a body under progressive deformation in relation to the coordinate axes.
- A convenient way of representing the state of finite strain is *Mohr strain diagram* in which the reciprocal of quadratic elongation, λ' $(= 1/\lambda)$, is plotted as abscissa and modified shear strain, γ' $(= \gamma/\lambda)$, as ordinate. Assuming that there is no change in the area, the states of strain can be represented by a circle.

Questions

1. Explain the concept of strain in rocks and describe the various ways in which rocks can develop strain.
2. What is the difference between finite strain and incremental strain?
3. How would you distinguish linear strain from shear strain?
4. Give the utility of strain ellipse and strain ellipsoid in structural studies.
5. What is Flinn diagram? Describe how is it constructed and what does it signify?
6. Give the concept of pure shear and simple shear. How does this concept differ from coaxial deformation and noncoaxial deformation?
7. How would you identify the signatures of coaxial and noncoaxial deformations in rocks?
8. What is strain path? Describe the common types of strain paths shown by deformed rocks.
9. What do you mean by vorticity? What does it signify?
10. What is Mohr strain diagram? How does it differ from Mohr stress diagram?

5 Estimation of Strain

A. R. Bhattacharya, *Structural Geology*, Springer Textbooks in Earth Sciences, Geography and Environment,
https://doi.org/10.1007/978-3-030-80795-5_5

Abstract

Nowadays, a statement such as weakly, moderately or strongly deformed rock sounds rather vague to a structural geologist as it does not give any idea of by how much the rock is deformed. Over the last about half century, this vagueness has been mitigated to a great extent by estimation of strain, i.e. the amount of deformation, from deformed rocks. This is done by identifying some objects in a deformed rock, called *strain markers*, whose original shape or size is known. Strain estimated at the time of measurement is called *finite strain* that can be considered as a cumulative effect of several deformation events, and the amount of strain added in each deformation event may be described as *incremental strain*. A variety of methods have been developed for estimation of one-, two- and three-dimensional strain in deformed rocks. Some commonly used methods are described in this chapter.

Keywords

Finite strain · Incremental strain · Strain markers · One-dimensional strain · Two-dimensional strain · Three-dimensional strain · Image analysis techniques · Automation of strain

5.1 Introduction

Estimation of strain from deformed rocks is nowadays a common practice in structural geology. Development in this field gained momentum in the 1960s, and thereafter the literature continued to be flooded with material on this aspect. Undoubtedly, the books of Ramsay (1967) and Ramsay and Huber (1983, 1987) greatly advanced our knowledge on strain estimation and provided a strong basis for the development of this aspect. Notable works of Dunnet (1969), Elliott (1970), Talbot (1970), Dunnet and Siddans (1971), Lisle (1977, 1985, 2010), De Paor (1981, 1988), Mulchrone (2013), Mulchrone et al. (2005), Heilbronner and Barrett (2014) and several others significantly added to this advancement. Recently, McCarthy et al. (2019) reviewed various methods devised by various workers for estimation of 2D and 3D strain.

Strain can be estimated as one-, two- and three-dimensional strain. Elementary methods of one-dimensional or linear strain are already described in Chap. 4 (Sect. 4.4). In this chapter, we have described some common methods of estimating 2D and 3D strain. Only a few commonly used methods are described in detail, while others are briefly described in respect of their principle, brief methodology and significance. However, more detailed description, like a laboratory manual, is beyond the scope of this book for which the reader is suggested to consult the respective works. In addition to the methods described in this chapter, there are several others also for which the reader is suggested to consult the two volumes of Ramsay and Huber (1983, 1987).

5.2 Strain Estimation from Deformed Rocks

The strain measured on a deformed rock can be thought of as the cumulative effect of the geometrical changes that the rock has undergone during its deformation history. The amount of strain given by a rock at the time of measurement is called *finite strain*. This finite strain of a rock can be thought of as the cumulative effect of several deformation events, and the amount of strain added in each deformation event may be described as *incremental strain*.

Estimation of strain from rocks is done by *strain markers* that are objects in a deformed rock seen in hand specimens or in thin sections, whose original shape or size is known. During deformation, such objects may show changes in several ways, e.g. changes in line lengths, angles, shapes, sizes or volumes. In all such cases, we know, respectively, the actual ratios of lines, angles, shapes, sizes or volumes of the strain markers we have selected. Some common strain markers include oolites, fossils, pebbles, vesicles and amygdules in volcanic rocks.

5.3 Recent Advances in Strain Analysis

The book *Folding and Fracturing of Rocks* by John Ramsay (1967) laid down several basics of strain analysis that motivated other workers to develop this aspect of structural geology mainly as a result of application of newer techniques. The present status of strain analysis has considerably grown since then (see McCarthy et al. 2019). Some recent developments are highlighted below.

5.3.1 Extension of 2D to 3D Methods

Most of the available work on strain analysis pertains to the estimation of two-dimensional (2D) strain as compared to relatively much less work on 3D. However, in order to bridge the gap, several workers (e.g. Milton 1980) developed methods whereby 2D methods can be extended to 3D too.

5.3.2 Algebraic Method

The algebraic methods (e.g. Shimamoto and Ikeda 1976) mostly use mathematical and statistical analysis. Application of such methods grew in order to minimize the errors and shortcomings in some available methods.

5.3.3 Numerical Algorithm

Numerical algorithm methods are employed to accurately determine the best-fit ellipse or ellipsoid (e.g. Robin 2019) by using several planar surfaces of differing orientations using orthogonal as well as non-orthogonal sections.

5.3.4 Synthetic Strain Markers

Synthetic strain markers are artificially prepared objects that mimic natural strain markers, such as oolites or conglomerates. These objects are subjected to deformation under laboratory conditions, such as homogenous/inhomogeneous strain, pure shear and simple shear. An additional benefit of this method is that the synthetic objects can be packed and oriented in the same way as the natural markers in their outcrop position (see Thissen and Brandon 2015).

5.3.5 Automation of Strain

Recognition of best-fit ellipse or ellipsoid of strain markers requires precise identification of object boundaries and their centres. This is a time-consuming job. To overcome this, efforts are made for accurate automatic segmentation of thin sections as well as automated extraction using image processing or GIS-based techniques.

5.3.6 Anisotropy of Magnetic Susceptibility (AMS)

The *anisotropy of magnetic susceptibility* (AMS) (e.g. Borradaile and Henry 1997; Borradaile and Jackson 2010; McCarthy et al. 2015) has been shown to have great potential in tectonic analysis. In general, the AMS is believed to be related to the layer parallel shortening of a rock. This implies that the magnetic fabrics/subfabrics can be used for orientation studies of the constituent magnetic minerals of a rock. As such, this method finds great value in strain analysis.

5.3.7 SURFOR Method

The SURFOR method was first proposed by Panozzo (1984) and was extended in detail by Heilbronner and Barrett (2014). The method is based on the changing length of the projection of line segments as the outline is rotated (Panozzo 1984) and quantifies the fabric on the basis of shape, size and orientation of 'surfaces' that can be any linear element such as fractures or grain boundaries (Heilbronner and Barrett 2014).

5.3.8 Electron Backscatter Diffraction (EBSD)

Electron backscatter diffraction (EBSD) is a technique for the study of microstructure of polycrystalline materials. Basically, it is a scanning electron microscopy (SEM) using X-ray diffraction (XRD). Polished specimen of a rock is placed in an SEM chamber that faces a diffraction camera that gives an electron backscatter diffraction pattern. Study of the diffraction pattern provides information on the microstructure and grain orientation that in turn helps strain analysis and quantification of fabrics of the rock under investigation.

5.3.9 X-Ray Computed Tomography

X-ray computed tomography is a method wherein one can 'see' each individual object in three dimensions and study many more objects by the use of a photographic camera, an optical scanner, electron backscattering image, X-ray maps, etc. or any combination of these (see Robin and Charles 2015, and the references therein). This method, also called *CT analysis*, has now become a common practice in 3D strain analysis.

5.3.10 The Present Status

From above, it appears that the present status of strain analysis is much different from the Ramsay system mainly due to the use of computers as a handy tool. Nevertheless, for the basic principles, credit should go to the Ramsay type of methods. Currently, use of technology coupled with mathematical treatment of data seems to be a common trend in strain analysis.

5.4 Two-Dimensional Strain

Two-dimensional (2D) strain can be estimated for those geological objects whose initial shape is known such as oolites, reduction spots in slates, conglomerates and some fossils. Estimation of 2D strain is by far more in common use possibly due to the ubiquitous occurrence of suitable strain markers in geological terrains and also a variety of available techniques. Some common methods of 2D estimation of strain are described below.

5.4.1 Axial Plots

Ramsay (1967) suggested a simple method of estimating strain from rocks that show elliptically deformed objects. These objects can be studied directly in the field or from their photographs. The long and short axes of the ellipses are measured. The data are plotted on a graph with long axis as ordinate and short axis as abscissa (Fig. 5.1). If the rock has undergone homogeneous deformation, the data should fall on or about a straight line (Ramsay 1967, p. 193). The inclination of the line with the abscissa (θ) is noted down and its gradient, i.e. tan θ, gives a measure of strain, named strain ratio, whose value ranges from 0 to infinity. The strain ratio thus obtained provides the ratio of the major and minor axes of the strain ellipse. At the inclination of 45°, the strain ratio is equal to 1. It is therefore possible to estimate strain from elliptically deformed objects from a locality or a spot. It may be noted that the strain thus obtained gives the amount of strain in that particular section in which measurements have been taken. If we measure strain from different sections of the same rock, or say in a cut section, the amount of strain would be different. The reason is that in an ellipse, the elongation of a line would be different in different sections.

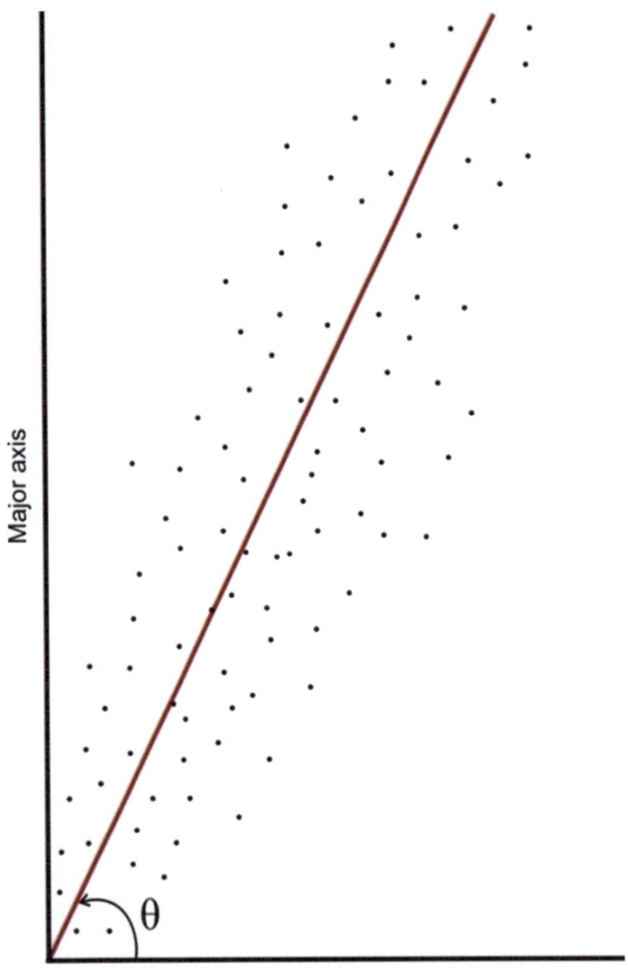

Fig. 5.1 Representation of two-dimensional strain with arbitrarily chosen data with the long and short axes of elliptically deformed objects such as oolites plotted as ordinate and abscissa, respectively. The tangent of the angle (θ) subtended by the best-fit line with x-axis is a measure of the strain expressed as the strain ratio given by tan θ

5.4.2 Centre-to-Centre Method

During deformation, the strain in a rock is manifested in several geometric forms. For example, some large-size components of a rock may show relative changes in their position during deformation. In such cases, the *centre-to-centre method*, also called *nearest neighbour method*, can be applied to estimate the tectonic strain. This method was first proposed by Ramsay (1967), and worked out in detail by Ramsay and Huber (1983), for rocks in which some large-sized components (pebbles, sand grains or oolites) occur in relatively finer grained matrix. For such rocks, there exists great difference in the rheological behaviour, i.e. flow characteristics, between the larger components and the matrix. Estimation of tectonic strain is based on the fact that during deformation, the larger components systematically change their distance (length) and angle with respect to the nearest grains in relation to the strain undergone by the rock.

The centre-to-centre method has been explained here by a simple example in which a few grains of a deformed rock have been selected (Fig. 5.2). Select a grain and its centre

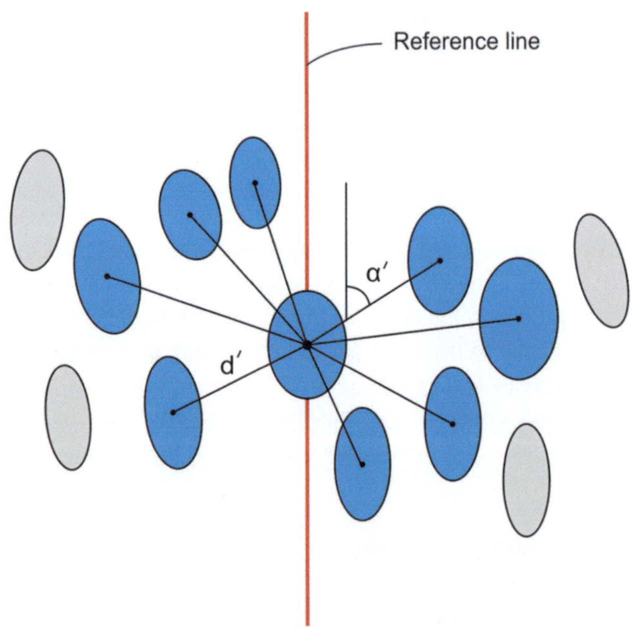

Fig. 5.2 Estimation of two-dimensional strain by centre-to-centre method. Position of deformed grains in a thin section has been shown. With reference to the centre of a selected grain located along the reference line, measure the distance (d') of each centre of the nearest grains (blue colour) and the angle (α') from the reference line. Ignore grains (grey colour) that are not the nearest neighbours of the selected grain. Repeat the process for other grains to obtain more data

with respect to which measure the distance (d'), also called *tie line*, of the centres of the nearest grains (marked in blue colour) and the angle (α') from a reference line. Ignore those grains that are not the nearest neighbour (marked in grey colour), though they may be the nearest neighbour of some other grain for which one has to repeat the same process for obtaining more data. Plot a $d' - \alpha'$ graph (Fig. 5.3) that gives a curve. Measure the maximum (d'_{max}) and minimum (d'_{min}) values. The tectonic strain R_T is then given by the ellipticity of the strain ellipse, which is expressed as the ratio between these two values, i.e.

$$R_T = d'_{max}/d'_{min} \tag{5.1}$$

A merit of the method is that it is applicable for rocks showing large difference in the sizes of larger and smaller components and that the value of d_{max} gives the orientation of the long axis of the strain ellipse. A demerit is that it is not always possible to precisely mark the centre of the larger objects, and the strain thus estimated is not precise.

5.4.3 Elliptical Objects

Ellipsoidal objects (oolites, xenoliths, pebbles, reduction spots, etc.) constitute one of the most important geological strain markers. The populations of these objects can be statistically studied in three mutually perpendicular directions in their cut sections that enable both two-dimensional and three-dimensional strain measurements. Techniques for two-dimensional strain from elliptical objects have been described by several workers of which some important ones are described below.

5.4.3.1 Fry Method

Fry (1979) developed a method, called *Fry method*, for estimating two-dimensional strain that is practical and rapid. It is based on centre-to-centre method and graphically presents a point distribution that gives the finite strain ellipse in the form of a central vacancy. The method can be performed manually as well as using computer programs.

In the Fry method (Ramsay and Huber 1983), the centres of all the particles are obtained on an overlay; the centres must be numbered (Fig. 5.4). Take a second tracing overlay

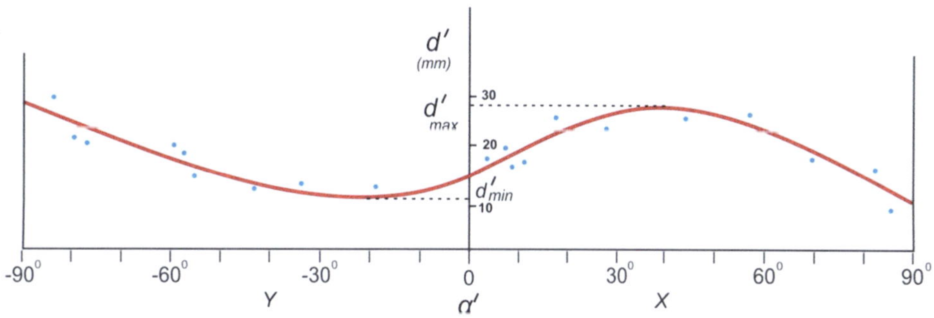

Fig. 5.3 Estimation strain by centre-to-centre method. Plot of d' against α' gives a graph from which maximum (d_{max}) and minimum (d_{min}) are noted down. In this example, the strain ratio $R_T = d'_{max}/d'_{min} = 28/12 = 2.3$

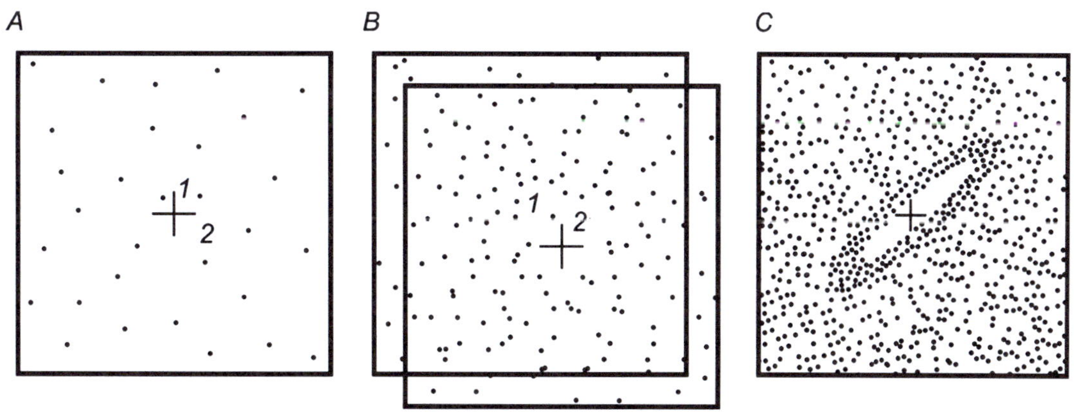

Fig. 5.4 Fry method for estimation of two-dimensional strain. The step-by-step method is represented in (**a**), (**b**) and (**c**) as explained in the text. (Reproduced from Ramsay and Huber 1983, Fig. 7.10 with permission from Elsevier Senior Copyrights Coordinator, Edlington, U.K. Submission ID: 1198140)

and mark a reference point at the centre with two coordinate axes passing through it. Place the second overlay over a central point (1) of the first one. Mark all the other centres on the second overlay (Fig. 5.4a). The reference point is then moved (not rotated) to the second point (2). Mark all the centres in this position (Fig. 5.4b). Continue to move the reference point to all other points (3, 4, 5…) so that all the centres marked on the first overlay are covered. We thus get a dense cluster of unevenly distributed points (Fig. 5.4c). However, one can notice a circular or elliptical region (vacancy) that contains either no point or only a few points. A circular vacancy field indicates that the rock has undergone no strain while an elliptical vacancy indicates strain in the rock. If on the other hand no vacancy field is noticed, no tangible conclusion can be drawn.

The circular or the elliptical vacancy observed in the final overlay (Fig. 5.4c) usually shows heavy concentration of points at the boundaries of the vacancy, and this in fact accentuates the shape of the vacancy. According to Ramsay and Huber (1983), this concentration relates to the commonest closest packing distance between the particles and implies that the initial particle size was fairly constant. A weak concentration on the other hand implies that the initial particles had varied shapes or the packing was not tight.

5.4.3.2 R_f/φ Method

The *R_f/φ method* was first developed by Ramsay (1967) and was then worked out in detail by Ramsay and Huber (1983). This method is applicable to deformed rocks that contain originally elliptical objects such as oolites, pebbles and sand grains. Sedimentary rocks that are formed as channel deposits occasionally contain elliptical pebbles with their longer axes aligned parallel to the direction of stream flow. Such rocks occasionally create problem because after deformation the original elliptical objects are again elliptical in shape (Fig. 5.5). However, some marker planes formed during deformation such as rock cleavage, foliation or schistosity help in relating the new shape and orientation of these objects to the imposed strain. The method thus constitutes an efficient technique of strain estimation for some particular types of deformed rocks.

The R_f/φ method as described below is based on Ramsay and Huber (1983). An originally elliptical object with ellipticity R_i subjected to homogeneous strain will again produce an ellipse but with a changed ellipticity R_f (= long axis/short axis). The shape and orientation of the deformed ellipse will depend upon the original shape and orientation of the ellipse with reference to the strain axes. In Fig. 5.5a, a few elliptical objects are shown in their undeformed state. The ellipticity of each of these objects R_i is 2.0, but their orientation φ with respect to a marker direction x is different. A circle $R_s = 1.0$ has also been shown for reference. A plot (Fig. 5.5a) between R_i (along abscissa) and φ (along ordinate) is made for the entire range of φ. Since all the objects are in undeformed state, each has an initial ellipticity $R_i = 2$, and therefore the points fall along the vertical line $R_i = 2$.

Let, as in Fig. 5.5b, the elliptical objects be subjected to a homogeneous strain with strain ellipse ratio $R_s = 1.5$. The shape (ellipticity, R_f) and orientation (φ') of the ellipses change according to the imposed strain and their initial orientation. The ellipses show full range of orientation, i.e. between −90° and +90°. The R_f/φ' plot shows that the long axes of the ellipses fall around or come closer to the long axis of the strain ellipse. It may be noted that, given the same imposed strain, change in orientation (φ') of the initial ellipse depends on its initial orientation. That is why the ellipses at two positions, at $\varphi = 90°$ and 0°, will not change their orientation; they change their ellipse ratio (shape) only. The ellipse at $\varphi = 90°$ will become less elliptical, while the one at $\varphi = 0°$ will become more elliptical.

With stronger deformation, say at an imposed strain $R_s = 3.0$, the objects become more elliptical (Fig. 5.5c). In this case, the orientations of the long axes of the ellipses are confined to a smaller field (between −15° and +15°) and the R_f/φ' plot becomes a closed curve. Thus, with increase in the intensity of deformation, R_f increases and the range of variation of φ' from the mean value decreases. The range within which the long axes of the ellipses are confined is called *fluctuation F*. When deformation is strong, as in Fig. 5.5c, F becomes short.

According to Ramsay and Huber (1983), although the R_f/φ' plot provides a useful method of estimating strain in elliptically deformed objects with initial elliptical shapes, the method works efficiently only when the undeformed objects are randomly oriented and do not contain any imprint of tectonic disturbance. In such cases, the R_f/φ' plots yield a single curve. However, with different initial shapes (i.e. different values of R_i) and different tectonic strains shown by the elliptical objects, the plot does not give a single curve but spreads to yield more than one curve. As such, the method bears an important significance that it can separate situations when the initially random objects already contained imposed tectonic strain or not.

5.4.4 Panozzo's Projection Method

The *projection method* assumes great significance in strain analysis. The reason is that most of the strain estimation methods, especially those based on *shapes of mineral grains* (e.g. Ramsay 1967; Shimamoto and Ikeda 1976), are applicable for elliptical objects only. However, many rocks contain mineral grains that show departure from elliptical shape, such as polygonal shapes with planar boundaries. For such type of shapes, Panozzo (1984) proposed a method of strain

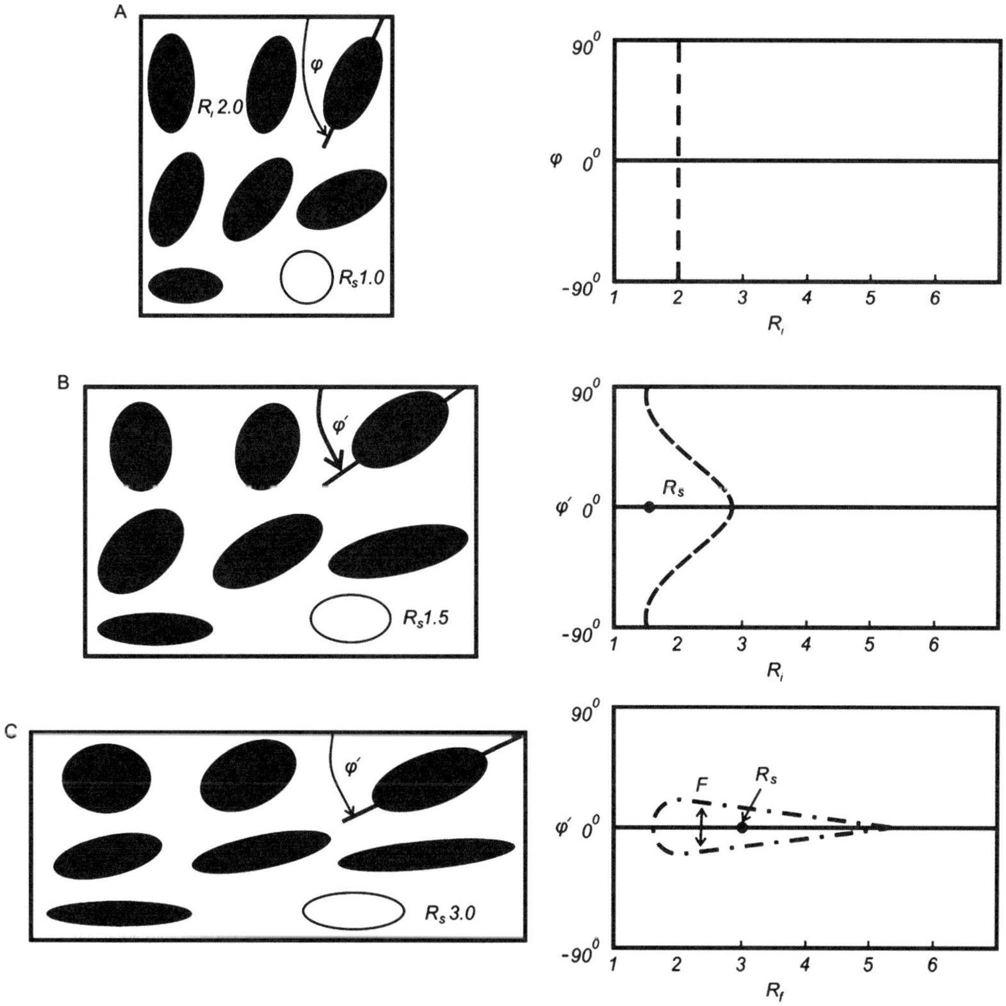

Fig. 5.5 Initially elliptical objects (**a**) after deformation remain elliptical. However, their initial orientation (φ) and ellipticity (R_i) change to say φ' and R_f (**b**, **c**). See text for details. (Reproduced from Ramsay and Huber 1983, Fig. 5.3 with permission from Elsevier Senior Copyrights Coordinator, Edlington, U.K. Submission ID: 1198139)

estimation that is applicable for mineral grains with planar boundaries. The method assumes that the grain boundary surfaces may develop a preferred orientation of the surface elements as a result of homogeneous strain. This change of orientation of surface is a function of strain, and this constitutes the basis of strain analysis of this method.

The starting point is a thin section of a rock that shows polygonal grains with flat, planar boundaries. The grain boundaries are initially drawn as close polygons. The outlines of shapes of the mineral grains are digitized by sets of small straight lines. The lines representing the surface of a grain are digitized by suitable software. With this, the grain boundaries are projected as deformed strain ellipses. In this context, the significance of undeformed and deformed fabrics is important. *The undeformed state is often equated to a state of randomness or isotropy, to an absence of preferred orientation or even to an absence of fabric* (Panozzo 1984, p. 219). The 'undeformed state' is relevant here because it constitutes a reference state for defining the finite strain. The method holds that a set of randomly oriented lines form an isometric polygon that approaches the shape of a circle. The deformation affects the individual line segments according to their orientation. The projection method thus gives an ellipse.

The above-mentioned Panozzo method has recently been successfully applied by us (Bhattacharya and Verma 2020) in the Main Central Thrust (MCT) zone of the Himalaya. Along the MCT, the crystalline rocks of the Higher Himalaya have been thrust over the sedimentary belt of the Lesser Himalaya (Gansser 1964). As such, the MCT is a zone of high concentration of ductile shear strain. Systematic study of quartz-rich rocks of the MCT zone reveals a common occurrence of quartz grains that are elliptically stretched in the direction of mylonitic foliation due to high ductile shear strains. However, at several places of the shear zone, the quartz grains are polygonal showing planar boundaries, thus constituting a locally developed strain-insensitive fabric. For such quartz grains, the available methods of strain estimation are not easily applicable, and as such we have followed the Panozzo method. The amount of strain for each sample is given by the ellipticity of the ellipse generated by 'Fabric 8' software. Our results reveal that rocks that show higher values of strain, as obtained by the Panozzo method, contain more quartz grains with polygonal boundaries, whereas those showing lower values of strain contain relatively less number of polygonal quartz grains.

5.4.5 Hyperbolic Net Method

De Paor (1988) suggested that there is a need for amalgamation of the available methods to develop the capability to handle tectonically imbricated fabrics. He described a hyperbolic net that can be used as a standard stereonet capable to yield valid results from initially uniform and imbricated fabrics. The net can simultaneously handle more than one method of strain analysis. The method takes into account the fact that the final shape and orientation of an ellipse are independent of its initial size and position. They instead depend only on initial shape and orientation together with the shape and orientation of the strain ellipse.

The *hyperbolic net* (De Paor 1981) is a combination of two intersecting families of curves (Fig. 5.6). One family constitutes vertical and horizontal asymptotes and vertices

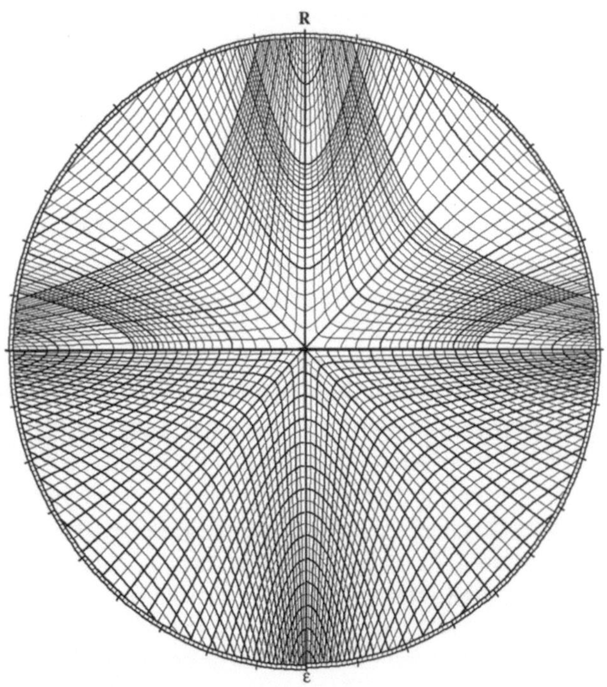

Fig. 5.6 The hyperbolic net. (Reproduced from De Paor 1988, Fig. 1 with permission from Elsevier Copyrights Coordinator, Edlington, U.-K. Submission ID: 1193082)

that lie along the diagonals. This family tracks the progressive changes of shape and orientation of the ellipses during deformation and is therefore called *ellipse trajectories*. The other family has symmetric non-orthogonal asymptotes such that the vertices lie along the horizontal or vertical axes and mark equal increments of strain along the trajectories and is therefore called *strain contours*.

The strain calculation by the hyperbolic net method is based on the assumption that the ellipses deformed homogeneously with each other and with the matrix and that they did not possess a primary sedimentary fabric. A tracing overlay is placed on the net to plot a point that represents an ellipse of axial ratio R_f and orientation φ_f. For step-by-step plotting procedure, the reader is suggested to follow De Paor (1988).

5.4.6 Theta Curves

The R_f/φ method of Ramsay (1967) and that of Ramsay and Huber (1983) assume that the elliptical markers initially had the same eccentricity. After deformation, they assume shapes that yield a characteristic curve on a graph of R_f versus φ. However, in practice, the elliptical markers commonly have variable initial shapes. As such, the R_f/φ data yield a diffused characteristic curve. To overcome this, Lisle (1977, 1994) suggested a method called theta curves. The method yields a set of theta curves (Lisle 1994, Fig. 5.7) for constant intervals of theta for each strain. Assuming that the markers had uniform orientation distribution, a set of curves are formed which divide the data equally.

The basis of the theta curve method assumes that a suite of initial elliptical markers had variable eccentricities (R_i) but identical orientations (Fig. 5.7a). In deformed state, the original orientations change (Fig. 5.7b). Assuming that all orientations (θ) are equally represented in the undeformed condition, the R_f/φ plot will be represented by points that are equally distributed in the cells defined by the theta curves (Fig. 5.7c).

The theta curve method of Lisle (1977) assumes great significance in strain analysis as it is the first computerized technique for determining the strain from R_f/φ data (Yamaji 2008, p. 1463).

5.4.7 Normalized Fry Method

McNaught (1994) developed a method for constructing a normalized Fry plot and determining the orientation and axial ratio of the fabric ellipse. According to the author, grains can be approximated by any shape as long as the centre and area of each grain can be determined, and particularly

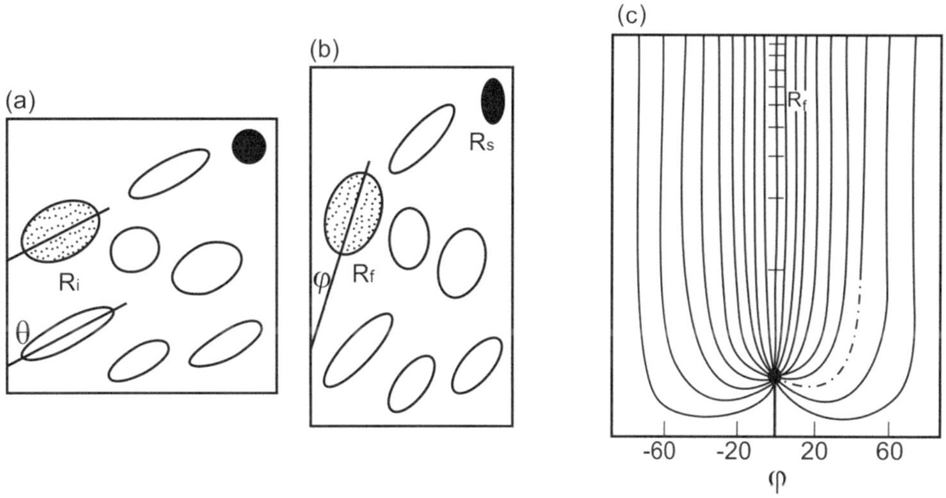

Fig. 5.7 R_f/φ analysis by theta curve method. (**a**) Elliptical markers in their undeformed state with variable R_i but identical orientations. (**b**) The markers in deformed state. (**c**) The R_f/φ plot for the deformed markers shown in (**b**). See text for details. (Reproduced from Lisle 1994, Fig. 2.5 with permission from Elsevier Copyrights Coordinator, Edlington, U.K. Submission ID: 1198370)

irregular polygons are helpful because they can be used to approximate both elliptical and non-elliptical grains.

If the centre and area of each grain in an aggregate are calculated, a normalized Fry plot can be constructed that helps in calculating the least-square best-fit ellipse (see McNaught 1994 for details). In order to avoid the problems associated with the elliptical approximation of grains and the problems associated with the variation in two-dimensional grain size (e.g. Erslev and Ge 1990), McNaught made an attempt to approximate individual grains by polygons rather than ellipses. This helps in constructing a normalized Fry plot.

From a photomicrograph of a thin section, the grain boundaries are traced onto an overlay from which the points along the grain boundaries are entered into a computer file. If these points in series are connected by straight-line segments, they form irregular polygons that approximate each grain in which the points are the vertices of the polygons. Having each grain approximated by a polygon, the centre and area of the polygon are calculated from which a normalized Fry plot is constructed. The average radius of each polygon is defined the way the average radius of an ellipse, i.e. $\sqrt{\text{area} \times \pi}$, is defined.

5.4.8 Intercept Method

Strain estimation often requires quantification of rock fabrics. Since these types of measurements are time consuming, they are not in common use. To mitigate this, Launeau and Robin (1996) developed a method, the *intercept method*, whereby objects such as grains or aggregates of minerals, pores, lineaments, etc. can be identified in an image and thus can be used for quantification of fabric anisotropy on 2D image. The method basically analyses boundaries of objects as population of lines. The various intercept counts are analysed by a Fourier series. The fabric, its symmetry, direction(s) and intensity are quantified by polar plots or rose diagrams. According to the authors, an axial strain ratio can also be calculated from the Fourier components if the objects or lines can be assumed to have deformed passively from an initially isotropic state.

5.4.9 Mean Radial Length Method

Mulchrone et al. (2003) suggested a method for calculating finite sectional strain from distributions of elliptical objects. The method is based on the fact that the mean radial length of a set of uniformly oriented ellipses in the unstrained state is equal to that of a circle. As such, the mean radial length itself evaluates the strain ellipse. In other words, if an ellipse distribution is subjected to homogeneous deformation, the circle of mean radial length is transformed to the finite strain ellipse (Mulchrone et al. 2003, p. 530). The method assumes that the initial distribution of objects is uniformly random with respect to their orientation and that the distribution of the axial ratios is independent of orientations.

5.4.10 SAPE Method

Mulchrone et al. (2005) suggested a SAPE (Semi-Automatic Parameter Extraction) program for extracting information relevant to strain analysis from input digital images that are manually produced by tracing the outlines of selected objects. There exist several manual methods for measurement of strain marker parameters, i.e. aspect ratio/long axis orientation (R_i/φ_i). The authors adopted a manual method first to identify the longest possible line through the target shape and consider this to be the long axis. The short axis is then considered to be the longest possible line perpendicular to the long-axis direction. For the use of SAPE, digital photographs of the samples are imported into the program. Also for each sample, a line trace of marker boundaries is made from the digital sample photographs and then scanned into the bitmap image, and these are used as input image for SAPE.

5.4.11 Method of Point Fabric Patterns

As an improvement to the Fry method, Lisle (2010) developed a technique for finite strain estimation from deformed patterns of points that possess anticlustered properties. This is a computer-based method that provides a number of acceptable solutions for the strain ellipse in a given geological situation. The best-fit strain ellipse is obtained by repeatedly de-straining the point data to identify those strain states that are capable of restoring the data to assumed initial conditions. It has been shown that the data derived from strongly anticlustered point distributions give better results than weakly anticlustered point distributions that give vague, poorly defined vacancy fields on the Fry diagram. Further, in order to detect isotropy of the restored point fabrics by statistical tests, the minimum sample size should be about 75 points.

The above method of strain analysis is performed in three stages (Lisle 2010). (1) The data in the form of coordinates of points in the deformed array are acquired so that the inter-point vector (Fry) diagram is obtained. (2) The data set thus acquired is subjected to a series of geometrical transformations, each of which corresponds to the imposition of a strain ellipse with a particular axial ratio and orientation. (3) The final stage involves comparison of each of the transformed data sets on the Fry plot with those assumed for the pre-deformation configuration of the points. The reader is now suggested to consult Lisle (2010) for obtaining the plots.

According to Lisle, the point patterns can be considered a type of strain marker whose quality can vary greatly. Pronounced anticlustering constitutes a favourable circumstance when the point patterns are geometrically analogous to well-defined circular markers that serve as accurate strain gauges.

5.4.12 Gaussian Blur Technique

The Fry plot sometimes gives a diffused elliptical vacancy that yields inaccurate strain estimation by this method. In order to mitigate this problem, several types of image analysis techniques have been developed (e.g. Erslev 1988; Erslev and Ge 1990; Waldron and Wallace 2007; Lisle 2010; Vinta and Srivastava 2012; Mulchrone 2013). As an example, Vinta and Srivastava (2012) developed a method that is independent of any iteration or analytical solution. They use the Gaussian blur technique that transforms a Fry plot containing discrete points into a continuous area image with a prominent central vacancy. The required Gaussian blur is obtained by successive application of the one-dimensional Gaussian function Eqs. (5.2) and (5.3) in x- and y-directions, respectively. With this, one can get the same Gaussian blur as is produced by a two-dimensional Gaussian function Eq. (5.4).

A Gaussian function in one-dimension is represented by

$$G(x) = \frac{1}{\sqrt{2\pi\sigma^2}} e^{-\frac{x^2}{2a^2}} \tag{5.2}$$

$$G(y) = \frac{1}{\sqrt{2\pi\sigma^2}} e^{-\frac{y^2}{2a^2}} \tag{5.3}$$

In two dimensions, it is the product of two Gaussians:

$$G(x, y) = \frac{1}{\sqrt{2\pi\sigma^2}} e^{-\frac{x^2+y^2}{2a^2}} \tag{5.4}$$

where x and y are the distances from the origin on the horizontal and vertical axes, respectively, and σ is the standard deviation of the Gaussian distribution given by the Gaussian blur radius.

The method involves two additional scripts, one for getting a normalized Fry plot and the other best fitting the ellipse through the central vacancy, and operates through three steps: generation of the normalized Fry plot, application of Gaussian blur and extraction of the central vacancy. Since the method is non-iterative, it is more direct than some other existing techniques.

5.4.13 'Fitting the Void' Method

Mulchrone (2013) suggested an objective method for identifying the shape of the void that develops by analysis of the spatial arrangement of point distributions. The method applies high weighting to points on the inner boundary of the central vacancy developed in a Fry-type plot. For normalization purposes, the average radius of an elliptical object is required which is given by

$$\sqrt{r_{\max} r_{\min}}$$

where $r_{\max}$ and $r_{\min}$ are the long and short axes, respectively. One major problem with the Fry distribution, as pointed out by the author, is that the distribution contains excessive data and

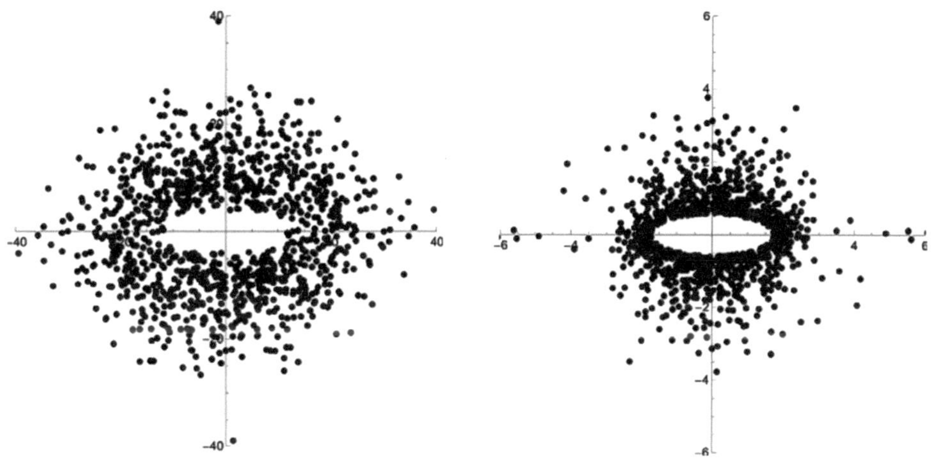

Fig. 5.8 Development of a well-defined central void by analysis of the spatial arrangement of point distributions. (Left) Raw Delaunay triangulation nearest neighbour distribution. (Right) Normalized Delaunay triangulation nearest neighbour distribution. (Reproduced from Mulchrone 2013, Fig. 1 (e and f) with permission granted by Elsevier Senior Copyrights Coordinator, Edlington, U.K. Submission ID: 1193096)

therefore most of the data which relates to far-apart objects is redundant and tends to slow down computational approaches. As such, not only this sparsity accentuates with increasing strain, but the data set also becomes manageable and thus the central void continues to be well defined even after considerably large imposed strains (Fig. 5.8). The method is suggested to work best for closely packed object arrangements.

5.4.14 ACF Method

The term ACF means *autocorrelation function*. This method was first proposed by Panozzo (1992) and later on developed by Heilbronner and Barrett (2014) for studying microstructures and textures of geological materials. *The autocorrelation function describes the correlation of an image with itself as a function of displacement of the image with respect to itself* (Heilbronner and Barrett 2014, p. 389). The bulk ACF as resulted from averaging all ACFs provides the axial ratio $(b/a)_{\mathrm{bulk}}$ of the ellipse.

5.4.15 Automated Image Analysis Technique

Image analysis is the conversion of an image into a number or a set of numbers, i.e. *the extraction of some information* (Heilbronner and Barrett 2014, p. 7). The image of a rock primarily provides the grain size distribution. The grains are separated, and their diameters are measured. The image is converted from colour to greyscale. This greyscale image, after several image processing steps, is then processed by software for obtaining the results. The image analysis techniques are based on automatic segmentation of thin-section images by image processing or GIS-based techniques that produce grain boundary maps. For this, commonly quartz clasts are used as strain markers. Image analysis techniques have been found to be of great importance, among others, in the Fry method in which the accuracy depends much on the sharpness of the central vacancy. The reader is suggested to consult the book by Heilbronner and Barrett (2014) that presents an excellent review of the image analysis techniques for microstructural studies and fabric analysis of rocks.

5.4.16 Strain from Deformed Fossils

5.4.16.1 Wellman Method

Wellman (1962) devised a geometric method to determine strain from *deformed fossils* that initially had a bilateral symmetry before deformation. Bilateral symmetry is shown by brachiopods, trilobites and a few others. It is, however, rare to find such fossils in deformed state and that too in sufficient number. For a sound knowledge of strain, about 10 data are sufficient. Strain estimated from deformed fossils is reliable to a great extent because the fossils in their undeformed state show some fixed angular relations between material lines. These angular relations can be observed and noted from the adjoining undeformed strata for comparison. If in an outcrop showing deformed fossils, at least one fossil is noticed in an undeformed state and then the perpendicular lines represent axes of principal strains, and so also of the strain ellipsoid. Since the deformed fossils selected for strain analysis commonly occur embedded on a bedding plane in the outcrop, the strain thus estimated gives a two-dimensional strain along that particular plane.

A photograph of the deformed fossils is taken and can be transferred on to a tracing that represents the deformed fossils as they occur in the outcrop (Fig. 5.9). It is assumed that all lines that were originally straight remain straight even after deformation, thus suggesting that the deformation is homogeneous. Since original lengths are not known, only the length changes within the bedding plane can be known.

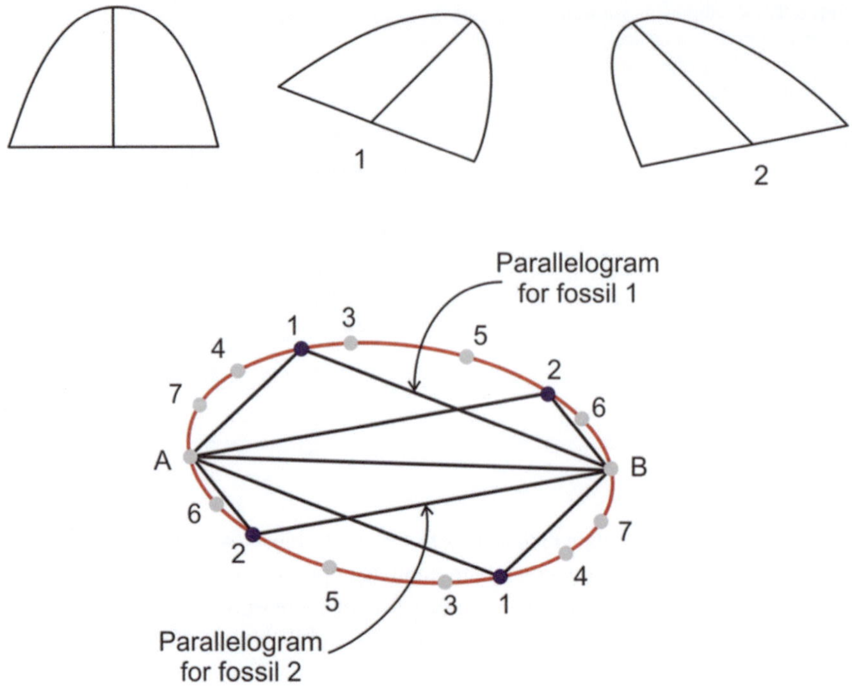

Fig. 5.9 Estimation of strain from deformed fossils by Wellman method. An undeformed bilaterally symmetrical fossil (top left) on deformation takes up a shape shown in 1, while 2 represents the shape of another fossil. The lower diagram is a plot for obtaining an ellipse. See text for details. In the plot, points 1 and 2 (in black colour) represent the selected two fossils in the form of parallelograms, while points 3–7 (in grey colour) are only for demonstration of the method

We have demonstrated here estimation of strain for only two deformed fossils, 1 and 2 (Fig. 5.9). First, draw a reference line (AB) of any orientation. Draw a line parallel to the baseline of the fossil passing from A and B. Draw another line parallel to the medial line of the fossil passing from A and B. Now we get a parallelogram. Obtain another parallelogram for fossil 2. Of the four apices of the parallelogram, two are given by the points A and B while the other two points are located on either side of the line AB. We thus get four points given by the four apices of the parallelogram. Now, repeat this process for other deformed fossils, say 3–7. We thus get 14 points in all. By joining these points, we get an ellipse. The short axis of the ellipse is parallel to the apparent shortening, and the ratio of the two axes gives the amount of strain undergone by the rock.

It may be noted that a perfect ellipse is obtained only if the rock has undergone homogeneous strain during deformation. In most cases, however, the points do not trace a perfect ellipse. In such case, the best-fit graph should be obtained to get the nearest possible ellipse. The ratio of long and short axis of this best-fit ellipse gives the amount of two-dimensional strain in the selected bedding plane.

5.4.16.2 Breddin Graph

In 1956, Hans Breddin developed a method, known as *Breddin graph*, for estimating strain from deformed objects, mainly deformed fossils that show a relationship between angular shear and orientation of the strain ellipse. The method is based on the fact that a pair of perpendicular lines after deformation shows an angular shear (ψ) that varies with the orientation (φ') of the pair of lines with axes of principal strain. The angular shear (ψ) is plotted along the ordinate with both +ve and −ve values, while the orientation (φ') is plotted along the abscissa with both +ve and −ve values. Breddin thus developed a family of curves (Fig. 5.10) from which the value of ellipticity R can be directly read for a particular set of data.

The Breddin graph is a useful and practical method because data for only one or two samples, for example deformed fossils, are sufficient to give an idea of the strain undergone by a rock.

As far as strain from deformed fossils is concerned, a number of workers have proposed methods for different types of fossils for which the reader is suggested to consult the book by Ramsay and Huber (1983).

5.5 Three-Dimensional Strain

5.5.1 Background

Three-dimensional (3D) strain can be estimated when all the three dimensions of a strain marker are seen either in outcrop or in cut sections. This mainly requires identification of a strain ellipsoid. A major problem of 3D strain analysis lies in obtaining necessary data in 3D and that is why several

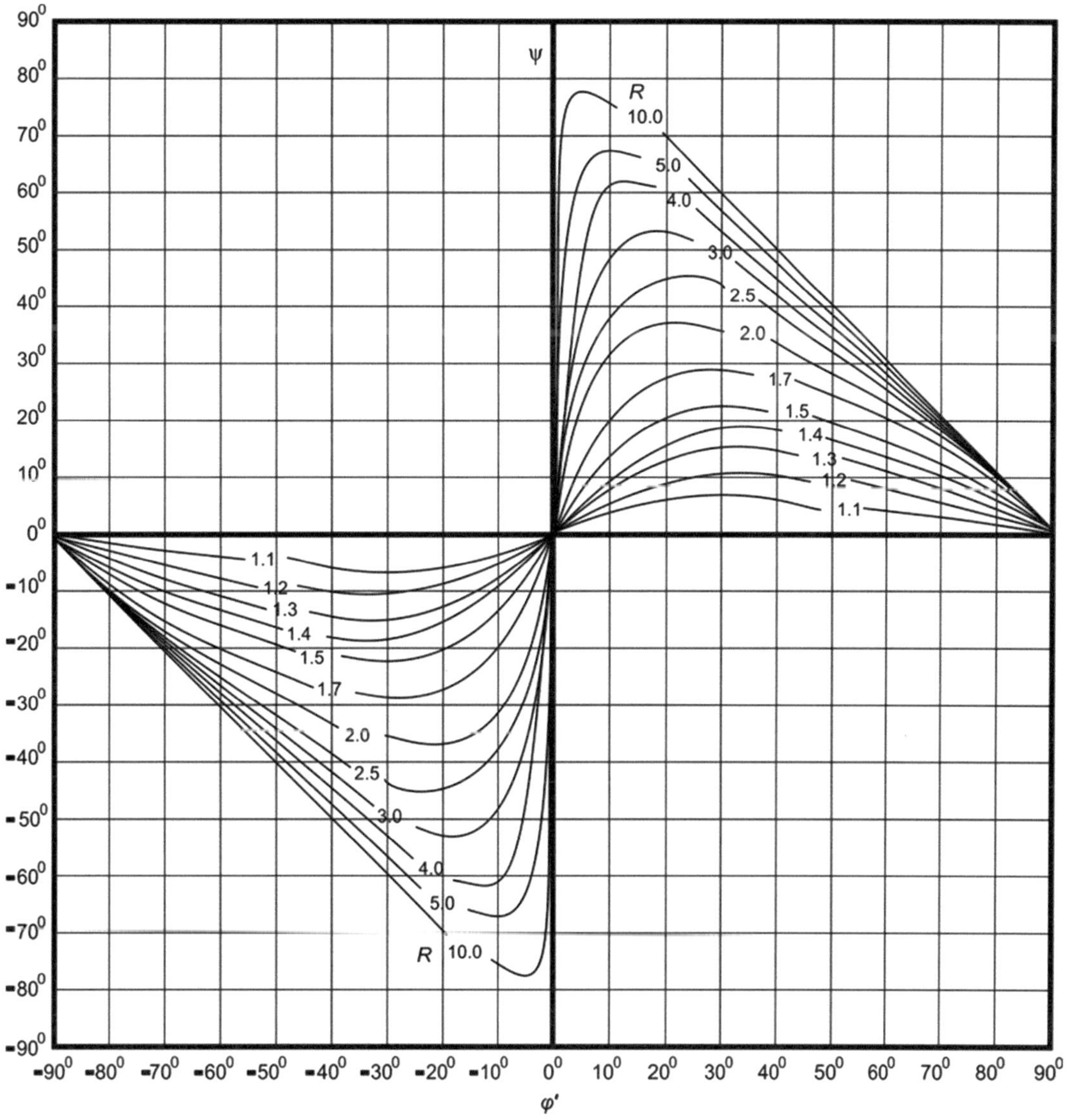

Fig. 5.10 The Breddin graph showing a plot between the values of the angular shear (ψ) and orientation (φ') with respect to the long axis of the ellipse. From the plot, the value of ellipticity R can be directly read. (Reproduced from Ramsay and Huber 1983, Fig. 8.5 with permission from Elsevier Senior Copyrights Coordinator, Edlington, U.-K. Submission ID: 1198142)

available 2D methods cannot be easily extended to 3D strain analysis. However, only a few, e.g. the Flinn method (1962) (described in Chap. 4), can be used to represent strain ellipsoid. The Flinn method was modified by Ramsay (1967) and Ramsay and Huber (1983) by using a logarithmic scale. Hossack (1968) used Nadai-Hsu plot using a polar scale.

Estimation of 3D strain took a new turn in the 1990s when computer-based methods using software became popular. These methods broadly use automated image analysis, quantification of fabric in respect of orientation, and shape and size of grains and grain boundaries. In the following sections, we describe some common methods of estimation of 3D strain especially in respect of their principle, brief methodology and significance.

5.5.2 Ellipsoidal Objects (Cloos Method)

Several types of strain markers are initially spherical in shape such as oolites in carbonate rocks, vesicles in volcanic rocks and reduction spots in slates and concretions. During homogeneous deformation, these objects assume ellipsoidal shapes.

Ernst Cloos (1947) was the first to estimate strain from ellipsoidally deformed objects. He observed in an anticline west of South Mountain, Maryland, USA, that during deformation of limestone, the originally circular oolites are deformed to ellipsoids. He systematically estimated strain from these deformed oolites from different parts of the anticline. In the anticline, the oolites occur as ellipsoidal objects.

He assumed that these ellipsoidal oolites must have an initial spherical shape before deformation. He further assumed that during deformation, there was no volume change. If the initial radius of an oolite is r, its volume is

$$V_0 = \pi r^3 \tag{5.5}$$

Let r_1, r_2 and r_3 be the semi-axes of the ellipsoid; the final volume after deformation is then given by

$$V_f = \pi r_1 \cdot r_2 \cdot r_3 \tag{5.6}$$

Since there is no volume change,

$$V_0 = V_f$$
$$\pi r^3 = \pi r_1 \cdot r_2 \cdot r_3$$
$$r = \sqrt[3]{r_1 \cdot r_2 \cdot r_3} \tag{5.7}$$

Thus, we get the value of the original radius of an oolite. If this radius is considered as the initial length of a sphere, then the other lengths r_1, r_2 and r_3 constitute the final lengths along three axes of the ellipsoid. With these values, it is now possible to get the amount of strain undergone by a deformed oolite by estimating the various strain parameters as given below:

$$\begin{aligned}
\text{Elongation } (e_x) \text{ along } x-\text{axis} &= (r_1 - r)/r \\
\%\ \text{Elongation } (e_x) \text{ along } x-\text{axis} &= [(r_1 - r)/r] \times 100 \\
\text{Elongation } (e_y) \text{ along } y-\text{axis} &= (r_2 - r)/r \\
\%\ \text{Elongation } (e_y) \text{ along } y-\text{axis} &= [(r_2 - r)/r] \times 100 \\
\text{Elongation } (e_z) \text{ along } z-\text{axis} &= (r_3 - r)/r \\
\%\ \text{Elongation } (e_z) \text{ along } z-\text{axis} &= [(r_3 - r)/r] \times 100
\end{aligned}$$

From all these, other parameters of strain such as stretch (S) and quadratic elongation (λ) can be estimated as described earlier (Chap. 4).

5.5.3 Strain by Direct Measurement of Axes

Occasionally, rocks exposed on the surface show ellipsoidally deformed markers such as pebbles in sedimentary rocks or quartzo-feldspathic aggregates/grains in gneissic rocks. Because of erosion, sometimes the rock exposes all the three tectonic sections, viz. XY, XZ and YZ (Fig. 5.11). Here, X represents the direction of maximum elongation, Z is the direction of maximum shortening and Y is the intermediate axis, and $X > Y > Z$. It would be better to select the XY and XZ sections for strain estimation because these sections

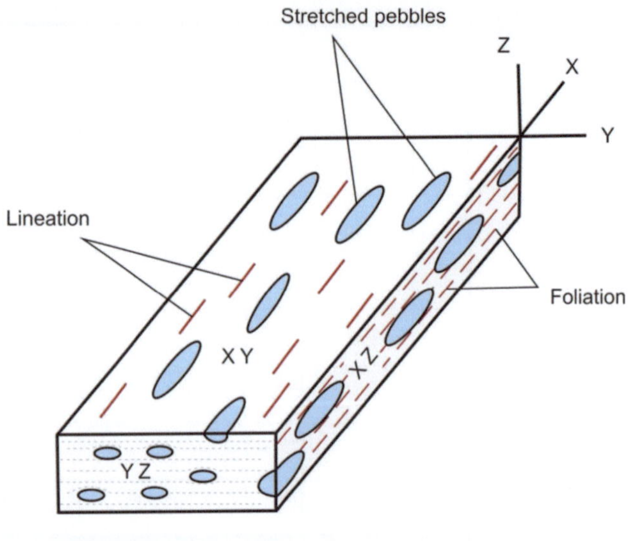

Fig. 5.11 Block diagram showing the three tectonic axes XY, YZ and XZ of a deformed rock. A number of stretched pebbles have been shown. X-axis represents the direction of extension, while Z represents that of shortening. Foliation is contained in the XZ-plane, while XY-plane contains stretched pebbles as well as a few lineations

contain objects showing maximum elongation and maximum shortening, respectively. If sufficient number of ellipsoidal objects are available at a spot, an oriented sample of the rock can be taken from this spot for strain estimation. The sample is then cut in three planes XY, XZ and YZ. From the outcrop or from the photograph of the cut sections, the long axis (X) of the ellipsoidal objects in the XY-plane, the short axis (Z) in the XZ-plane and the intermediate axis (Y) in the YZ-plane are directly measured. Mean value each of X, Y and Z is thus obtained. The three-dimensional strain for the rock is then expressed as a ratio of these three mean values of X, Y and Z. If for example $X = 8$, $Y = 6$ and $Z = 2$ are the mean values, the three-dimensional strain is

$$X : Y : Z = 8 : 6 : 2 \text{ or } 4 : 3 : 1$$

This method would add precision to strain estimation mainly because the planes are precisely cut perpendicular to each other, thus representing the three tectonic planes.

5.5.4 Algebraic Method

Shimamoto and Ikeda (1976) developed a new algebraic method to determine the shape and orientation of the strain ellipsoid by using deformed ellipsoidal objects as strain markers. The method assumes that the originally ellipsoidal objects have been changed to ellipsoidal shapes on being homogeneously deformed with their matrix. They developed basic theories that enable to determine the strain ellipsoid from measurements of lengths and orientations of all principal axes of deformed ellipsoidal objects and to construct the strain ellipsoid from the two-dimensional analysis on three mutually orthogonal planes. Since general finite homogeneous deformation is treated, the method is equally applicable to both rotational and irrotational strains.

According to Shimamoto and Ikeda, it would be more useful and practical to analyse the strains in three dimensions from the two-dimensional strain analysis on plane sections. An averaged strain ellipsoid can be considered as a sphere centred at the origin of a perfectly random orientation of the objects. This implies that the averaged final ellipse on any plane section simulates the shape of strain ellipse on that plane. The authors have mathematically shown that all elliptic sections of an ellipsoid cut by parallel planes are geometrically similar.

Let us assume that an ellipsoid is geometrically similar to the strain ellipsoid. Let the shape matrix be represented by $[C_{ij}]$. If the strain analysis is done on three mutually perpendicular planes, i.e. *xy*-, *yz*- and *zx*-planes, the three ellipse sections of Eq. (5.8) on *xy*-, *yz*- and *zx*-planes, respectively, become

$$[x\ y\ z]\begin{bmatrix} C_{11} & C_{12} & C_{13} \\ C_{21} & C_{22} & C_{23} \\ C_{31} & C_{32} & C_{33} \end{bmatrix}\begin{bmatrix} x \\ y \\ z \end{bmatrix} = 1 \tag{5.8}$$

$$[x\quad y]\begin{bmatrix} C_{11} & C_{12} \\ C_{21} & C_{22} \end{bmatrix}\begin{bmatrix} x \\ y \end{bmatrix} = 1 \tag{5.9}$$

$$[y\quad z]\begin{bmatrix} C_{22} & C_{23} \\ C_{32} & C_{33} \end{bmatrix}\begin{bmatrix} y \\ z \end{bmatrix} = 1 \tag{5.10}$$

$$[x\quad z]\begin{bmatrix} C_{11} & C_{13} \\ C_{31} & C_{33} \end{bmatrix}\begin{bmatrix} x \\ z \end{bmatrix} = 1 \tag{5.11}$$

The ellipses on *xy*-, *yz*- and *zx*-planes, respectively, are

$$[x\quad y]\begin{bmatrix} f_{xy} & h_{xy} \\ h_{xy} & g_{xy} \end{bmatrix}\begin{bmatrix} x \\ y \end{bmatrix} = 1 \tag{5.12}$$

$$[y\quad z]\begin{bmatrix} f_{yz} & h_{yz} \\ h_{yz} & g_{yz} \end{bmatrix}\begin{bmatrix} y \\ z \end{bmatrix} = 1 \tag{5.13}$$

$$[x\quad z]\begin{bmatrix} f_{xz} & h_{xz} \\ h_{xz} & g_{xz} \end{bmatrix}\begin{bmatrix} x \\ z \end{bmatrix} = 1 \tag{5.14}$$

The arithmetic mean of $[C_{ij}]$ over the above-mentioned six possible cases can be used for the construction of the strain ellipsoid.

5.5.5 Adjustment Ellipse Method

Milton (1980) suggested a method by which an adjustment ellipse is determined analytically for each of the three sections. These adjustment ellipses make the three sections compatible that enables determination of the strain ellipsoid. According to the author, an ellipsoid can be expressed by six parameters, i.e. the lengths and orientations of its three principal axes. Likewise, an ellipse in a defined plane can be expressed by three parameters. Three elliptical sections will thus define a nine-parameter figure and not a six-parameter figure. This is because, according to the author, *any* three ellipses may not necessarily come from the same ellipsoid and therefore they may not be compatible. Theoretically, in a homogeneous strain field, the strain ellipses measured on any sections should be compatible, provided that they constitute sections of one strain ellipsoid. However, observational and experimental errors as well as inhomogeneities in strain field may exist. Therefore, the three strain ellipses must be adjusted to make them compatible. By measuring the ellipses on three planes, the author proposed a method based on the full uncertainties in the ellipse matrices by the use of a computer program (see Milton 1980 for details).

As a matter of fact, accuracy of results depends upon the accuracy of the data. One can expect optimum geometry for obtaining the strain ellipsoid from three sections when the sections themselves are the three principal sections of the ellipsoid. In such cases, the principal axes can be measured directly. The author suggests that for a rock exhibiting a fabric with a pronounced cleavage and lineation, the best results can be obtained by cutting three orthogonal sections parallel and perpendicular to the cleavage and lineation.

5.5.6 Autocorrelation Method

Thissen and Brandon (2015) presented a method for 3D estimation of strain using *autocorrelation* properties of deformed rocks. Autocorrelation, according to the authors, is the *correlation of an image with itself as a function of an offset or lag which in two dimensions has both distance and direction*. An initially isotropic geologic material should also have an initially isotropic autocorrelation function (ACF). An isotropic material when subjected to deformation undergoes an anisotropic distortion that can be estimated, apart from the Ramsay and Fry methods, from the ACF. The non-linear, best-fit method can minimize the anisotropy in the ACF.

The authors used both synthetic (by generating a 3D isotropic aggregate of randomly distributed spherical grains) and natural materials (deformed ironstone ooids) for their ACF analysis. The proposed method has been suggested to work best for internally uniform materials that have strong contrast with the matrix and other grains.

Thin-section images and X-ray tomography have been used for studying the grains and the overall fabric. A tomographic image, called a *tomogram*, represents the sample volume within a 3D grid. The 3D tomography data, especially for the grain size, are analysed by suitable software with the ultimate aim of getting an ellipse (2D) or an ellipsoid (3D).

5.5.7 Strain by X-Ray Computed Tomography

Rocks often contain spherical objects (e.g. ooids, reduction spots, conglomerates) that are used to estimate 3D strain if they occur in deformed rocks. The traditional methods of 3D strain analysis (Robin 2002; Launeau and Robin 2005) use different planar sections of varying orientations to obtain an average 'sectional ellipse'. Several such sectional ellipses can then be combined to calculate a fabric ellipsoid. As a matter of fact, this ellipsoid does not reveal the complete three-dimensional shape. To overcome this, Robin and Charles (2015) suggested an alternate method for 3D strain analysis by the use of *X-ray computed tomography*, as briefly described below.

If a solid object is placed between an X-ray source and a detector arrangement, a sequence of 2D X-ray images or 'views' is recorded around the object. A *3D tomogram* is then constructed from the successive views by the use of algorithms. Each point in the tomogram is called a 'voxel' (3D pixel), and each voxel is encoded with its (x, y, z) position inside the object. A greyscale value is assigned corresponding to the X-ray absorption which, in turn, corresponds to the volume in the object. All these give a full three-dimensional image of the object. The image is examined as planar sections through the volume (see Robin and Charles 2015).

By the use of tomography, the authors succeeded to fit ellipsoids to spherical objects as visible in three dimensions and to calculate an average ellipsoid. For this, they used least-square fitting of the algebraic parameters describing an ellipsoid to the data points measured. The authors used this method on meteorite chondrules that are small spherical crystalline objects with a diameter <1–2.5 mm.

5.5.8 Strain from the March Model

Considering that the March model explains rotation of linear or planar strain markers as passive, geometrical elements during progressive 3D homogeneous strains, Mulchrone and Talbot (2016) developed the model for the first time for a best-fit approach to 3D strain estimation. Their method is of use to those who carry out field measurements of planar or linear deformed elements in any rock that has undergone homogeneous strains. It is assumed that the data to be processed has an initially isotropic distribution. The authors have derived the probability density function associated with the March model. This function is used to estimate the best-fit distribution and 3D strain parameters.

The authors have applied their proposed method to the Neoproterozoic rocks of SW Scotland. Assuming constant volume, the best-fit parameters obtained are $X = 1.64$, $Y = 1.50$ and $Z = 0.41$, which yield a Flinn k-value of 0.034. The strain ellipsoid thus obtained gives values $X/Y \approx 1.1$ and $Y/Z \approx 3.4$. Estimation using the March distribution gives the following corresponding values: $X/Y \approx 1.09$ and $Y/Z \approx 3.71$. This remarkable consistency between the two methods based on conceptually different assumptions indicates that the method developed by the authors works reasonably well.

5.5.9 Strain Probe Method

Experimental studies commonly require estimation of finite strain from images or surveys of scattered marker points taken at successive times. Likewise, for estimation of crustal geodetic strain, one requires the measurement of velocities of geodetic stations.

Robin (2019) presented a mathematical treatment that enables the best-fit strain to be known from motions of n marker points distributed over an area or a volume as presented in a Cartesian reference frame. According to the author, the method should be useful when analysing physical deformation models and interpreting repeated geodetic surveys, and although the method focuses on two-dimensional Lagrangian strain, the strain probe can equally be expressed in a Eulerian reference frame.

The experimental part involves imposition of a nominally simple shear strain on a thin layer of a deformable material—such as ice, camphor, norcamphor or octachloropropane—

sandwiched between two glass plates. Since the crystals are birefringent, the deforming grains can be observed with a petrographic microscope. Since grain boundaries migrate through the material, the grains themselves do not provide any record of deformation. But, if specks (5–20 μm) of an opaque material, e.g. fine grinding powder, are embedded in the model material, they serve as markers that were tracked during deformation. The motion of such markers could be related to their strain and orientation and thus helps in getting strain ellipses.

Robin also tracked geodetic velocities of 120 stations in Central Anatolia and calculated the strain. With these two examples, i.e. the motions of scattered markers in experiments and geodetic velocities, he presented a mathematical treatment that led to obtaining a deformation tensor. While the experiment provides an example of a finite strain calculation, the geodetic data are an example of infinitesimal strain.

5.5.10 Visualization Methods

Spitz et al. (2020) presented visualization methods for 3D finite strain that are helpful in characterizing finite strain in 3D numerical solutions in regions of folding and overthrusting. They found that the Hsu diagrams, indicating Nadai strain, ε_S, and Lode's ratio, ν, are particularly useful in visualizing finite strain and its spatial gradients. With the use of Hsu diagram, they provided a graphical representation of finite strain that relates strain magnitude to strain symmetry and thus helps in identifying different deformation regimes. They presented 3D models of power-law viscous flow, suggesting that the lateral variation of deformation style, from folding to overthrusting, is caused by the lateral variation of the thickness and geometry of a detachment horizon and that the strong layers experienced a bulk strike-slip shear deformation that was not imposed by the boundary conditions.

Analysis of finite strain in 3D numerical simulations of power-law viscous folding and overthrusting suggests that (i) the magnitude of ε_S generally increases from folding to overthrusting, (ii) an initial flat ramp geometry of the detachment generates a distinctive pattern of ε_S, (iii) lateral variations of ε_S and ν can be used to identify lateral variations in subsurface structures and (iv) internal strike-slip shearing is generated due to the folding-overthrusting transition. Further, different lateral variations of finite strain are produced due to different lateral variations in geometry. Thus, assuming that the model geometry mimics natural tectonic inheritance, Spitz et al. (2020) suggest that natural observations of extensional and/or strike-slip shear structures do not necessarily indicate a regional scale extension or strike-slip shearing.

Box 5.1 A General Appraisal of the Methods of Strain Analysis

- Most of the available methods of strain analysis pertain to 2D strain. 3D strain estimation methods are relatively much less. A few 2D strain methods can, however, be extended to 3D strain also (e.g. Shimamoto and Ikeda 1976).
- Despite the availability of several different methods of strain estimation, the Fry method appears to hold great potential to calculate the bulk strain of a rock in cases where the particles and the matrix show competence contrast or where the strain is discontinuous, and as such the method can be applied in a wide range of materials such as quartz grains, granite plutons, sand volcanoes, feldspars in gneisses and salt domes (Lisle 1994, p. 35).
- The Fry method has been the choice of several workers till late with the main objective of adding accuracy to the central void (ellipse). Gladly, several recent methods have succeeded in doing so.
- In strain analysis studies, especially 3D, precise detection of grain boundaries is a common problem. Accuracy of the final result depends much on precise identification of the grain boundaries. Grain boundaries are occasionally blurred to varying extents due to cementation, recrystallization and several grain boundary processes including migration of inter-granular fluids. It appears that only a few workers have given serious attention to this problem. Heilbronner (2000) and Heilbronner and Barrett (2014), for example, developed methods for creating grain boundary maps from petrographic thin sections.
- Strain analysis is also beset with one problem on which less attention appears to have been given. Strain estimated by one method may not tally with that estimated by some other method(s) for the same rock mass. The applications or implications of the results may lead to anomalous conclusions. Soares and Dias (2015), for example, made an appraisal of the 2D and 3D strain analysis by the Fry and R_f/ϕ methods as applied to the quartzitic rocks of the NW Iberian Variscides. They observed that the magnitudes of strain obtained from the Fry method are always higher than the ones obtained by R_f/ϕ. Strain quantification in tectonites, according to them, is not a straightforward process because the heterogeneous deformation of the rock fabrics could have been achieved by a combination of intra-, trans- and/or inter-granular mechanisms. Since each method has distinct sensitivities to different mechanisms, the

(continued)

Box 5.1 (continued)

results usually give different strain values to the same rock sample. Soares and Dias are therefore of the opinion that proper interpretation of strain results should be given a priority in structural studies.

- Automation together with application of mathematics seems to be a recent trend of strain analysis. The latter requires a few assumptions. In such cases, the accuracy of the final result will depend much on the type of assumptions made. There are situations when the assumptions are made to suit automation, ignoring the conditions and limitations of the rock. Obviously, the final result may not tell the real story of the rock. This aspect therefore needs to be explored more. Otherwise, while the input of technology/automation will continue to advance, the application of mathematics will remain in the world of 'assumptions'. The net expected result, i.e. accuracy and more accuracy, may still remain a far cry!

5.6 Significance of Strain Estimation

- The strain markers give the amount of deformation undergone by a rock in a quantitative manner.
- Estimation of strain helps in understanding the rheological behaviour of rocks during deformation.
- Comparison of strain in different lithologies of an area or spot helps us to give an idea of the rheological behaviour, i.e. the flow characteristics of the rocks when subjected to an external stress system.
- Strain data can be interpreted to throw light on the depth of deformation and the general physical conditions of the earth's interior.
- In sedimentary terrains, strain analysis helps in restoration of stratigraphic thicknesses (Ramsay 1969).
- Strain data help us compare the amount of deformation undergone by the rocks in different areas or locations.
- Strain analysis helps in understanding the mode of formation of a variety of structures in rocks, e.g. secondary foliations and crystallographic fabrics, because of their supposed relationship to the finite strain in the rock (Lisle 1985, p. 2).
- In general, presence of strain markers suggests that the rock is susceptible to deformation. In other words, their presence indicates that the rock has the ability to absorb the imposed stresses and, in turn, the ability to transfer the stresses to different parts of the rock, thus resulting in deformation.

5.7 Summary

- Estimation of strain in deformed rocks is a quantitative approach to deformation of rocks, and as such it is a common practice in structural geology.
- Strain given by a rock at the time of measurement is called *finite strain* that can be thought of as the cumulative strain developed due to several deformation events, and the strain developed in each deformation event is called *incremental strain*.
- Strain from deformed rocks is estimated by *strain markers* that are objects in a deformed rock seen in hand specimens or in thin sections, whose original shape or size is known. Common strain markers include oolites, fossils, pebbles, vesicles and amygdules in volcanic rocks. During deformation, such objects may show changes in line length, angles, shapes, sizes or volumes.
- Strain can be measured in one, two and three dimensions. Each category includes several methods as described in the chapter.
- Estimation of strain bears great significance in structural geology and has implications for the behaviour of rocks to stresses beneath the earth's crust.

Questions

1. What is a strain marker? Why is this necessary for the estimation of strain in rocks?
2. Can the estimation of strain be carried out on any rock? Give reasons and then add what types of rocks are actually selected for this purpose.
3. What do you mean by one-dimensional, two-dimensional and three-dimensional strains? Why is estimation of such types of strains in practice? Give limitations of each of these types.
4. Give the significance of Fry method.
5. Name a method of strain estimation in which an originally elliptical object remains elliptical even after deformation. Briefly give the principle of such method of strain estimation.
6. Name a method of strain estimation that uses non-elliptical markers. Briefly give the basic principle of this method.
7. What is R_f/φ method? Give merits of this method.
8. Can all types of fossils be used in the estimation of strain? Give reasons and then mention what types of fossils are commonly used for this purpose and why.
9. What are synthetic strain markers? Why are such markers used?
10. Outline the significance of estimation of strain in rocks.

6 Rheology

A. R. Bhattacharya, *Structural Geology*, Springer Textbooks in Earth Sciences, Geography and Environment,
https://doi.org/10.1007/978-3-030-80795-5_6

Abstract

Rocks flow! Sounds unrealistic to a non-geologist. But yes, rocks flow under certain physical conditions though we cannot see it. Study of deformation and flow under different physical conditions is called *rheology*. Rocks change their deformation behaviour under different physical conditions. Rheology thus deals with how the physical factors control the nature of deformation of a material in respect of flow. The rate at which a rock deforms is called *strain rate*. The relationship between strain rate and stress is called *constitutive law* for a substance. The physical parameters of rocks such as rigidity, elasticity and viscosity are called the *intrinsic parameters*. The latter depend upon external factors such as temperature, pressure and time that together are called the *extrinsic parameters*. The mathematical relations existing among all these factors constitute the *rheological equations* or *constitutive equations*. Mathematical expression that relates differential stress, temperature and strain rate is called *flow law* that is expressed by means of specific constitutive equations supported by experimental data. Rheological properties of lithospheric rocks depend much on the temperature gradient. Rocks down to the brittle-ductile transition (15–20 km) are under very high temperature gradient and thus behave as *elastic material*, while those below it behave as *ductile material* due to low temperature gradient. Although rheology is an important aspect of rock mechanics, the discussions contained in this chapter will show that rheology is also an important aspect of structural geology.

Keywords

Rheology · Rock mechanics · Strain rate · Steady-state flow · Constitutive law · Constitutive equations · Continuum mechanics · Isotropic and anisotropic materials · Rheological models · Flow laws

6.1 Introduction

The word *rheology* is derived from the Greek *rheos* meaning 'to flow'. Rheology thus concerns with deformation and flow of rocks. By the term 'flow', we naturally think of something like water, oil, pitch, gas and the like that flow. The flow behaviour of all these substances is extended to rocks also. The main reason is that rocks show different behaviour to deformation under different physical conditions, and under certain conditions they deform by flowage. As such, rheology is concerned with rock deformation and is considered as an important aspect of structural geology.

In previous chapters, we have emphasized that rocks deform when they are subjected to stress. This establishes the fact that stress and strain are related. However, this relationship is dependent upon physical properties of a rock as well as upon some physical factors such as temperature, state of stress and strain rate. In addition, the lithological composition, grain size, temperature, confining pressure as well as time also affect the behaviour of rocks to stress. A rock that deforms, for example, by brittle fracturing at shallow depths of the crust may deform by ductile flow at deeper levels of the crust. Similarly, at lower temperatures, a material that yields to external stress by the formation of fractures at high strain rates may flow at high temperatures and at low strain rates. These and several other examples show that rocks change their deformation behaviour under different physical conditions. Rheology thus deals with how the physical factors control the nature of deformation of a material in respect of flow. Our knowledge on rheology has greatly increased through experiments involving rocks deforming in different media. Study of rheology is therefore important in understanding the deformation of rocks under a wide range of physical conditions.

Rheology constitutes an important aspect of *rock mechanics*. Application of rheology to rock deformation can be seen in only a few works (e.g. Nicolas and Poirier 1976; Kirby 1983; Ranalli 1984, 1987; Jaeger et al. 2007; Allmendinger et al. 2012).

6.2 Strain Rate

Strain rate is the rate at which a rock changes its strain with time. It gives an idea of the rate of deformation of a rock with time. During deformation, the constituents of a rock change their position with time irrespective of whether the rock is undergoing shape change or volumetric change. In the process, the rock may undergo expansion or contraction. In all such cases, the rate at which the changes of constituents or markers of a rock take place with time is an expression of strain rate. If a linear body of original length l_o acquires a length l_t after a time t, the rate of elongation is given by

$$\varepsilon = \frac{l_t - l_o}{l_o} \tag{6.1}$$

The strain rate is then given by

$$\dot{\varepsilon} = \frac{d\varepsilon}{dt} = \frac{d}{dt}\left(\frac{l_t - l_o}{l_o}\right) \tag{6.2}$$

In a similar way, we can measure strain rate with volumetric change.

The strain rate can also be considered when a rock undergoes progressive shear deformation without any volume change. In such case, the rate of progressive shearing is called *shear strain rate*.

Strain is dimensionless because it is the ratio of two lengths. Strain rate is therefore expressed in per second or s^{-1}. Strain rate is measured by deforming the rocks in laboratory. It can also be measured from geological structures formed by natural deformation. Plate tectonics also helps in measuring strain rates. Geological strain rates fall in the order of $10^{-12}\ s^{-1}$ to $10^{-15}\ s^{-1}$, while the average geological strain rate is commonly given as $10^{-14}\ s^{-1}$ (Fagereng and Biggs 2019).

6.3 Steady-State Flow

We have mentioned earlier that an elastic body yields by rupture (yield point) when an external stress reaches a threshold value. Most solids show this property, and therefore materials showing elastic property are believed to be solids. Below the brittle-ductile transition (15–20 km, see Chap. 7), most rocks show a constant rate of flow under constant stress conditions. The material showing this property is said to show steady-state flow, which operates under constant applied stress (σ) conditions ($d\sigma/dt = 0$) in which the strain rate remains constant (*steady-state creep*) (Nicolas and Poirier 1976, p. 135):

$$\dot{\varepsilon} = r/h = \text{constant} \tag{6.3}$$

where r ($= d\sigma/dt$) is the recovery rate and h is strain hardening. Because of flowage, such materials are described as *fluids. Fluids are therefore substances that show constant rate of flow even under small stress.* The term fluid has been used here in the rheological sense because such conditions occur at high temperature and pressure at depth. This is in contrast to the fluids that occur on the surface of the earth (e.g. water) under normal conditions. Therefore, in order to distinguish the meaning of fluid as it occurs on the surface of the earth and at depth, the term *creep* is used for the one that occurs at depth. As such, we can also use the term *steady-state creep* for materials showing a constant rate of flow under constant stress. Viscosity of the concerned fluid is an important controlling factor of steady-state flow. It can now be said that solids show two extreme properties: *elastic* showing instantaneous deformation and *fluids* showing steady-state deformation.

6.4 Transient Flow

Flow is transient when the strain rate changes with *time* under constant stress (Ranalli 1987, p.67). Again, since the flow in this case takes place at high temperature and pressure and at depth, we can also use the word *transient creep* for this type of flow.

6.5 Isotropic and Anisotropic Materials

Rheology, as we have mentioned earlier, basically concerns the flow of a material. Of all the important factors on which rheology, and so also the deformation, depends, e.g. stress, strain and temperature, the mechanical properties of the material are also important. A material is said to be *isotropic* when its mechanical properties are the same in all directions. If the mechanical properties are different in different directions, the material is said to be *anisotropic*, and therefore the strength of such a material will vary in different directions.

Isotropy and anisotropy of rocks control their deformation in a significant way. A mechanically isotropic rock may deform in a way different from an anisotropic rock. Anisotropy of a rock can develop by a variety of ways. Presence of structural discontinuity such as cleavage or schistosity in a rock may render it anisotropic. These discontinuities are formed due to preferred orientation of platy or flaky minerals. The anisotropic planes allow the imposed stresses to pass through the rock. On the other hand, the same rock without internal discontinuities will accommodate the imposed stresses in a different way. All these will be reflected in the type of deformation or the flow characteristics shown by the rock. Likewise, the presence of fractures in a rock also adds to its anisotropy. Thus, a rock showing no fractures at one locality may deform in a way different from another locality where it shows fractures. Factors that commonly add anisotropy to rocks mainly include cleavage, slaty cleavage, bedding, joints, fractures, compositional heterogeneities shown by mineral constituents and presence of platy or flaky minerals.

Isotropy or anisotropy of rocks is also scale dependent. If we consider rocks on a larger scale, say in a part of the crust, then smaller chunks of rocks say on a local scale may behave as an isotropic mass as compared to the entire block of the crust we have selected. On local scale, the rock may show heterogeneities in several ways, but on the larger perspective, these heterogeneities do not significantly add to anisotropy to the overall rheological behaviour of that part of the crust.

6.6 Constitutive Law

Till now, we have emphasized that deformation of rocks depends much upon the state of stress, strain rate and temperature. However, the physical properties of rocks also constitute an important factor that controls deformation. In natural rocks and solids, for example, the same state of stress may show different behaviour of deformation if the materials show different properties. An iron rod will fracture at low temperature if subjected to hammering, while the same rod will flow like a fluid when heated to higher temperatures. Likewise, a rock if hammered yields by fracture at low temperatures, while it will yield by flowage at higher temperatures. The strain rate and stress relations are thus dependent upon the physical properties of a substance. The relationship between strain rate and stress is called the *constitutive law* for a substance. It is so named because the relationships existing between strain rate and stress depend upon the constitution or composition of the substance.

In constitutive law, we generally consider a rock as a continuous medium; that is, it is free from asperities, microfractures, cracks or some other forms of discontinuities. Such a material deforms uniformly in all directions within its boundaries, and as such the concept of *continuum mechanics* can be applied. In doing so, the deformation of the substance can be considered by simple mathematical relations.

6.7 Constitutive Equations

6.7.1 What Is a Constitutive Equation?

A *constitutive equation* in general expresses *a relation between any two physical quantities that are characteristic of a given substance*. In geology, especially when we talk of rock deformation, constitutive equations mathematically represent the relations that stress bears with strain or strain rate. Constitutive equations are grain size insensitive and are therefore formulated by considering a single-valued (mean) grain size of a rock. However, these equations can be combined with some other equations that relate the physical properties of the rock concerned.

The physical properties are important in rheological considerations including constitutive equations because they determine the mechanical state of a material for continuous media in the sense of *continuum mechanics* (= mechanics that consider a rock as a continuous medium). In rheological sense, the physical parameters include properties such as rigidity, elasticity and viscosity. These factors or parameters constitute the *intrinsic parameters* of a material. The intrinsic properties, in turn, depend upon the external factors such as temperature, pressure and time. The latter factors or parameters are called the *extrinsic parameters*. As a matter of fact, the mechanical state of a material is dependent upon the kinematic (e.g. velocity, displacement) and dynamic (e.g. forces) parameters and thus on the intrinsic and extrinsic parameters. The mathematical relations existing among all these factors constitute the *rheological equations* or *constitutive equations*.

6.7.2 Generalized Constitutive Equation

Formulation of constitutive equations requires the basic intrinsic and extrinsic quantities of a material to be known. Within the domain of continuum mechanics, i.e. for a continuous medium, the generalized form of the constitutive equation can be written as

$$R(\varepsilon, \dot{\varepsilon}, \sigma, \dot{\sigma}, u, t, \ldots \{M\}) = 0 \qquad (6.4)$$

where the rheological parameters are strain (ε), strain rate ($\dot{\varepsilon}$), stress (σ), rate of stress ($\dot{\sigma}$), velocity (u), time (t) and intrinsic material parameters$\{M\}$ (Ranalli 1987, p. 16). All these parameters are considered together under R, which is called the *rheological function* for the material. The various rheological parameters of a constitutive equation are determined by observation and experiment; they can also be derived by theory. A variety of constitutive equations can thus be established.

Materials that show the same equation(s) fall in the same rheological class such as elastic, viscous and plastic. The forms of the constitutive equations, however, change depending upon the rheological parameters. A material may show different constitutive equations if the extrinsic parameters change. For example, a rock on the surface of the earth behaves as an elastic material; that is, it shows instantaneous strain that is fully recovered if the stress is released, and that the stress is directly proportional to strain. If the same rock is brought down to great depths of the earth, it flows as a viscous body because of high temperature and pressure. Here, the extrinsic factors (temperature, pressure and fluid pressure) have changed the intrinsic properties of the material; that is, the material has changed from a rigid solid body to a viscous body such that its flow is now governed by the stress that varies directly with the strain rate. The constitutive equation under this condition would be different from what it was for the same material when it was on the surface of the earth. All this clearly suggests that constitutive equations are dependent on the rheological (intrinsic and extrinsic) parameters of a material. In the next section, we consider constitutive equations for elastic, viscous and plastic materials in a simple way.

6.7.3 Constitutive Equations for Elastic Materials

Hooke's law states that strain varies directly with stress. In a stress-strain diagram, this is represented by the straight line in Fig. 6.1. The graph rises till the body remains elastic. Point X represents the position in the graph where the material yields by rupture. Therefore, the point X also represents the *elastic limit* of the material. Since stress is known, elastic strain can be easily determined. Once the rupture point is reached, the graph follows a reversible path along the same straight line such that the strain again comes back to the zero position. (Rocks can, however, show hysteresis loop in which the strain returns to zero on a different path. We are not going to discuss this aspect here, for which the reader is suggested to consult books on rock mechanics.) On release of stress, the body recovers the strain in the same way it got strained, and the recovery of the strain follows the same path. The strain undergone by an elastic body is thus recoverable as well as reversible.

For an elastic material, strain is directly proportional to stress. For linear considerations, if a normal stress σ_n causes an extension e_n, the Hooke's law is represented by

$$\begin{aligned} &\sigma_n \infty e_n \\ \text{or,}\quad &\sigma_n = \mathrm{E}e_n \end{aligned} \tag{6.5}$$

where E is a constant called *Young's modulus*. If on the other hand a shear stress produces a shear strain σ_s, then according to Hooke's law,

$$\sigma_s = 2\mu e_s \tag{6.6}$$

where μ is the *shear modulus* or the *modulus of rigidity*. In Fig. 6.1, the slope of the graph gives the value of Young's modulus or shear modulus as the case may be.

From the above facts, an *ideally elastic substance* is one in which the strain developed due to a load remains constant during loading but the strain developed in the body reduces to zero on removal of the load. Such a substance recovers the total strain once the load is removed.

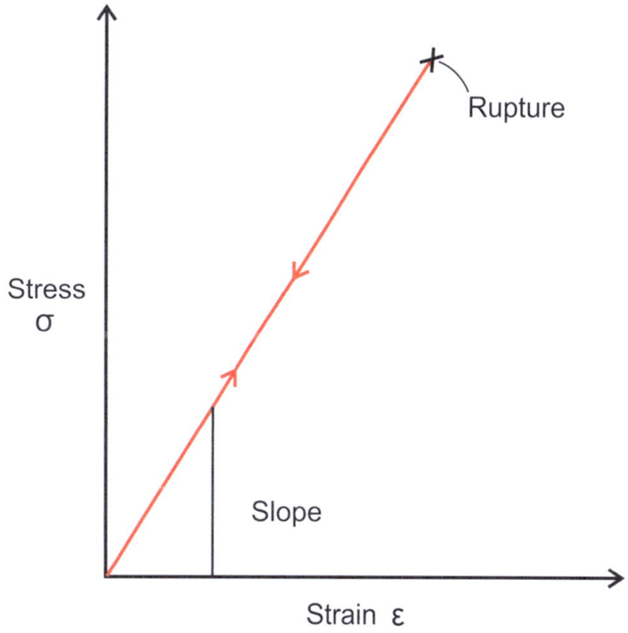

Fig. 6.1 Elastic deformation as represented by the stress (σ)-strain (ε) diagram. At smaller stresses, strain is directly proportional to stress. This is represented by the straight line. Point X represents the position in the graph when the body ruptures after which the graph follows a reversible path along the same straight line till the strain again comes back to the zero position. The slope of the graph gives the value of Young's modulus or shear modulus as the case may be. (See text for details)

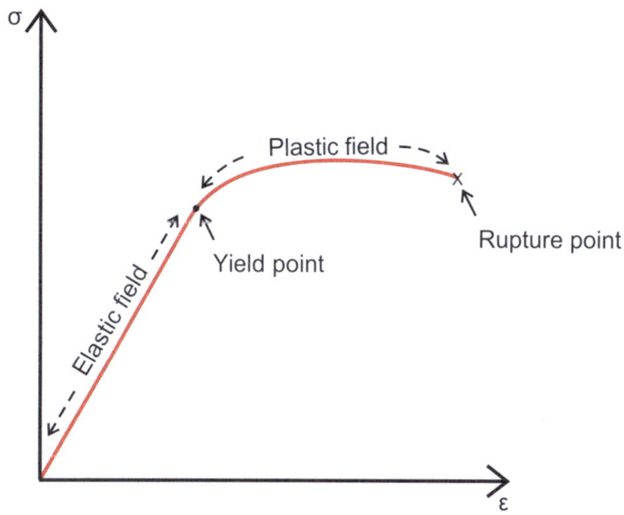

Fig. 6.2 Stress-strain relation for a plastic material

6.7.4 Constitutive Equations for Plastic Materials

The elastic materials as we have seen above do not show any permanent deformation. In such cases, the stress level is commonly low, i.e. well within the yield stress. If the stress level is raised higher, most materials show deformation by flowage. For such materials, the stress-strain graph goes beyond the yield point (Fig. 6.2) till the body yields by rupture. This type of behaviour is characteristic of *plastic materials*. The deformation that occurs in the plastic field is non-recoverable, i.e. permanent, and the deformation occurs without loss of continuity. The plastic materials thus show a graph in which the straight-line portion represents elastic behaviour while the curve above the yield point represents plastic behaviour. Plastic strain is associated with breaking of interatomic bonds and movement of dislocations of the crystal lattices, and therefore plastic strain is non-recoverable.

Most workers believe that plastic flow occurs when the maximum stress difference acquires a critical value, which, in turn, depends upon the rheological parameters of the material. Once the critical value is reached, the material continues to show permanent deformation. The maximum stress difference can be expressed by the Mises-Hencky criterion (Jaeger et al. 2007), which can be stated as

$$(\sigma_1 - \sigma_2)^2 + (\sigma_2 - \sigma_3)^2 + (\sigma_1 - \sigma_3)^2 = 6k^2 \quad (6.7)$$

where σ_1, σ_2 and σ_3 are the greatest, intermediate and least principal stresses, respectively, and k is a parameter of the material. Equation (6.7) lays down the conditions necessary for plastic flow. However, this condition holds well if the material has not undergone strain hardening. In such cases, the material is said to be an *ideal plastic* or *pure plastic*. Pure plastic materials show the following rheological equation:

$$\sigma = \sigma_Y \quad (6.8)$$

where σ_Y is the yield strength.

The above model assumes that yielding by plastic flow is independent of hydrostatic pressure. Hydrostatic or fluid pressure, in fact, promotes discontinuous or elastic deformation that is characteristic of the upper parts of the lithosphere.

6.7.5 Constitutive Equations for Viscous Materials

For viscous materials, steady-state flow occurs under constant stress and there is no shear stress. Once the stress is removed, the material does not return to its original shape because it undergoes *non-recoverable deformation* or *permanent deformation*, and in this case the stress is directly proportional to strain rate. This is true for linear viscosity, as viscosity can be non-linear also. The strain rate is a general term for the derivative of the strain with respect to time.

Figure 6.3 graphically shows the stress-strain rate relation for viscous materials. For idealized viscous or Newtonian substances, the simple constitutive equation showing the linear rheological relation between stress and strain rate is expressed as

$$\sigma = 2\eta\dot{\varepsilon} \quad (6.9)$$

where σ is the stress; η is the Newtonian viscosity, also called *coefficient of viscosity*; and ($\dot{\varepsilon}$) is the strain rate or flow rate. Material showing high viscosity is said to be competent, and it progressively becomes incompetent with progressive decrease of viscosity. The above constitutive equation holds good for a viscous material undergoing constant volume deformation.

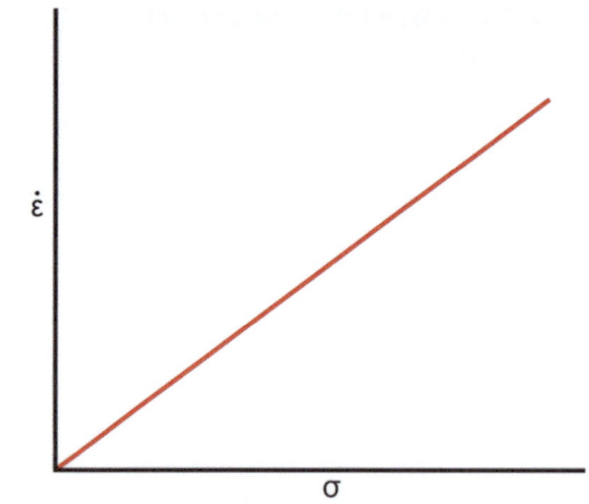

Fig. 6.3 Graph showing rheological relation, i.e. strain rate ($\dot{\varepsilon}$)-stress (σ) relation, for a linearly viscous material. The slope of the graph is related to viscosity (η) in the form $1/2\eta$

6.8 Rheological Models

6.8.1 What Is a Rheological Model?

Most substances behave in their own way with respect to stress and strain rate. These substances range from natural rocks to artificial substances. Most rocks are believed to represent some ideal cases that we consider here under *linear rheological models*. Also, there are substances that represent the combination of one or more types that we consider here under other rheological models. Here, we consider two extreme classes of linear models, i.e. *elastic* or *Hookean* and *viscous*. We also consider here an intermediate model, *plastic*. In addition to these, there may be many more such sophisticated models such as viscoelastic (Maxwell), elastic-plastic (Prandtl), visco-plastic (Bingham) and firmo-viscous (Kelvin or Voigt) models. However, considering the scope of structural geology, we restrict our description to the above-mentioned three ideal models only.

6.8.2 Elastic Model

Elastic substances show instantaneous deformation upon loading, and instantaneous and total recovery upon removal of load. The strain-time graph of an ideally elastic substance has been shown in Fig. 6.4. Elastic deformation is the non-permanent deformation that occurs before it returns to its original form. This non-permanent deformation is large

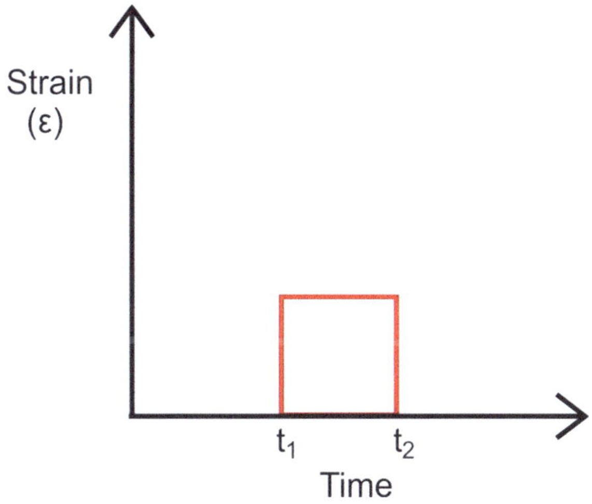

Fig. 6.4 Graph showing strain-time relations of an ideally elastic substance. Load is suddenly applied at time t_1, and the body suddenly shows strain that becomes constant till the load is removed at time t_2 when strain comes down to zero. In other words, the body recovers the total strain when load is removed

and can easily be seen in some substances such as a rubber band. For rocks, on the other hand, this non-permanent deformation is so small that it is invisible and can therefore be considered as infinitesimal.

Most elastic substances obey Hooke's law, i.e. the stress and strain are directly related (Fig. 6.1). The amount of deformation undergone by an elastic substance is thus directly proportional to the imposed stress. The Hooke's law can be stated as

$$\text{stress} \infty \text{strain}$$
$$\text{or stress/strain} = \text{E}$$

The constant E is called Young's modulus.

Hooke's law provides a useful relationship that gives the amount of strain or deformation in any substance if we know the amount of stress. This is possible under laboratory conditions. In the case of rocks, however, the reverse is also true as we can see and measure the strain or deformation undergone by the rock from which we can have an idea of the stress.

Elasticity is a natural property of substances and depends on the intrinsic properties of the substances and the magnitude and duration of application of the applied stress. Crustal rocks have the property of elasticity due to which they bear the loads of long and short durations. Elasticity of rocks has thus great implications for seismology and geodynamics.

Box 6.1 Poroelasticity and Thermoelasticity

Rocks below the earth's surface commonly contain cracks and pores that are filled with fluid phases such as air, water and oil. Since subsurface rocks are commonly under compression, the pore fluids would add *pore fluid pressure* that influences the mechanical behaviour of the rock mass. As a result, the rock may undergo failure by elastic deformation. Pore fluids are dynamic in nature and move according to pressure gradients of the included fluids. This, in turn, brings about local changes in the mechanical behaviour of the rock mass. Thus, the mechanical properties of rocks below the earth's surface are dependent upon the included fluids in the rock masses. The general theory that accounts for this coupled hydromechanical behaviour is *poroelasticity* (Jaeger et al. 2007, p. 168).

Let us consider a porous rock whose macroscopic bulk volume is V_b, the volume occupied by the pore space is V_p and the volume occupied by the solid mineral content is V_m; then (Jaeger et al. 2007, p. 169)

$$V_b = V_m + V_p \tag{6.10}$$

The above equation can be extended (see Jaeger et al. 2007, p. 170) to define the porosity, φ, in terms of the relative amounts of void space and solid components as

$$\varphi = V_p/V_b \tag{6.11}$$

and the void ratio, e, as

$$e = V_p/V_m = \varphi(1-\varphi)$$

A closely related theory is *thermoelasticity* (Jaeger et al. 2007, p. 197) that considers the effect of changes in temperature on the stresses and displacements in a rock body. The theory of thermoelasticity is mathematically and physically analogous to the theory of poroelasticity, with the temperature playing a role similar to that of the pore pressure. An isotropic rock on being subjected to a temperature change will give rise to equal normal strains in three orthogonal directions and no shear strains.

6.8.3 Viscous Model

Under normal conditions, a fluid is in equilibrium because of cohesion of internal molecules. This cohesion is disturbed

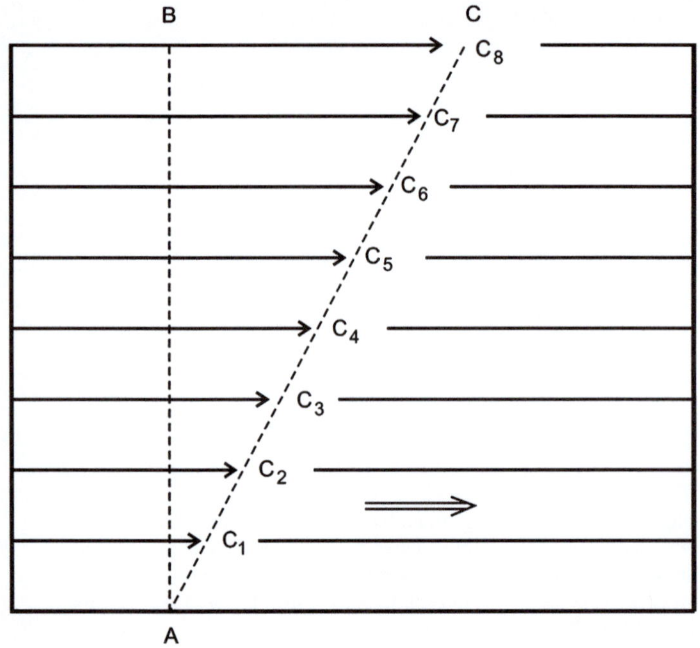

Fig. 6.5 Laminar flow of a fluid can be viewed as a set of parallel layers (C_1, C_2 . . .) moving in the direction of flow. At a particular instant, the displacement is highest along the top layer (from *B* to *C*), and it gradually reduces towards the base where it is theoretically zero

when the fluid flows or tends to flow under the influence of some external agency or gravity. The amount of resistance thus offered by the molecules is a measure of the viscosity of the fluid.

Let us consider laminar flow of a fluid; that is, the fluid can be considered to be layered such that each layer is parallel to each other and also parallel to the surface over which it is flowing. Let the fluid flow, say, due to gravity along a gentle slope. Each layer from base to top is flowing in the direction of flow (Fig. 6.5). During flow, the displacement of a point along the plane from the initial position progressively increases from base to top. Thus, at a particular time, the successive layers from base to top move to points, say, C_1, C_2, . . . C_8 (Fig. 6.5). The displacement is highest along the top layer (from *B* to *C*), and it gradually reduces towards the base where it is theoretically zero. This flow pattern results from the internal friction between the successive layers during fluid motion. *Viscosity is therefore a measure of resistance of a fluid against flow.* For a Newtonian fluid, the rate of shear (D) is directly proportional to shear stress (τ) (Fig. 6.6):

$$\tau = \eta D \tag{6.12}$$

where η is called *viscosity* or *Newtonian viscosity*. Viscosity refers to fluids in the same way as rigidity to solids, and it is dependent upon temperature and pressure.

Application of viscosity to rocks is limited to certain conditions only. As temperature gradually increases below the earth's surface, the viscosity of the material also increases with depth. The viscosity of the mantle is 10^{20} to 10^{22} Pascals, which is extremely large as compared to water (10^{-3} at $T = 20^0$) (Ranalli 1987, p. 71).

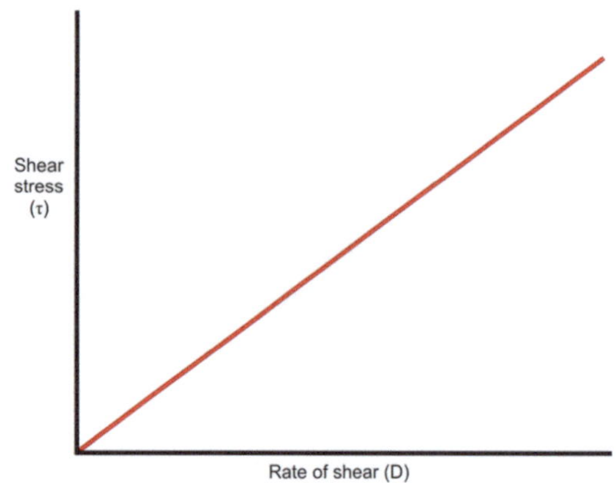

Fig. 6.6 For a Newtonian fluid, the rate of shear (D) is directly proportional to shear stress (τ). The viscosity or Newtonian viscosity (η) is defined by the slope of the graph from *x*-axis

6.8.4 Plastic Model

Plasticity is the property of some rocks to undergo permanent deformation without rupture even after a deforming force is released. The process operates through internal rearrangement of the molecules or constituents of a rock. Since plastic strain develops as a result of breakage or slippage of interatomic bonds, the strain produced in the body is permanent.

Unlike elastic and viscous deformations, the displacement in plastic deformation is not proportional to the applied stress.

As such, plastic models behave as non-linear functions and are sometimes considered to represent an intermediate stage between elastic and viscous substances.

6.9 Flow Laws

6.9.1 What Is a Flow Law?

Deformation of a rock is largely controlled by temperature, differential stress and strain rate. Rocks, therefore, behave differently under brittle and ductile conditions and so also under different range of temperatures. *The general form of mathematical expression that relates differential stress, temperature and strain rate is called flow law*. The law helps us understand the steady-state creep of rocks under different physical conditions by means of specific constitutive equations supported by experimental data. However, since polycrystalline rocks are constituted of distributed grain size that also shows variations in their shapes, it would therefore be more appropriate to formulate flow laws for distributed grain size as well as *composite flow laws* for single-valued and distributed grain size also.

Flow laws have been formulated by several workers for several types of geological materials under varying conditions (e.g. Carter and Avé Lallemant 1970; Heard 1960; Müller et al. 1981; Gottstein and Mecking 1985; Ranalli 1987; Tsenn and Carter 1987; Kohlstedt et al. 1995; Ter Heege et al. 2004). On the basis of these works, some common flow laws are highlighted below.

6.9.2 Flow Laws for Single-Valued Grain Size

Let a homogeneous body be subjected to deformation by simple shear and the deformation is associated with a strain rate $\dot{\varepsilon}$ then the steady-state flow of the body is represented by (Ranalli 1987)

$$\dot{\varepsilon} = A\sigma_s^n \tag{6.13}$$

where A is a function of pressure, temperature and material parameters; n is the stress exponent ($n > 1$); and σ_s is the shear stress acting upon the body. The above equation gives a relation between effective shear stress and effective strain rate and can be considered to represent the *general flow law* for a homogeneous, isotropic material. At higher temperatures and pressures, the flow law is given by a *power law*:

$$\dot{\varepsilon} = A\sigma_s^n \exp(-Q/RT)$$

where Q is an empirical parameter and T is the absolute temperature.

6.9.3 Flow Laws for Distributed Grain Size

Several workers (Ranalli 1984; Michibayashi 1993; Newman 1994; Dijkstra et al. 2002; Ter Heege et al. 2004) have argued that, unlike single-valued grain size, rocks are constituted of distributed grain size that also shows variations in their shapes. Ter Heege et al. (2004) suggested that in a single rock with distributed grain size, different flow mechanisms, unlike that showing single-valued grain size, may operate, say one for coarse-grained rock and another for fine grained. While fine grains may deform by grain size-sensitive (GSS) mechanisms (e.g. diffusion creep), coarse grains may deform by grain size-insensitive (GSI) mechanisms (e.g. dislocation creep). The flow laws thus formulated are called *composite flow laws* that take into consideration the various parameters that characterize the grain size distribution of a rock.

6.9.4 Composite Flow Laws for Single-Valued Grain Size

A polycrystalline rock can be considered as one with a single-valued (mean) grain size as well with a distributed grain size. For such rocks, both the GSI and GSS mechanisms operate as independent and concurrent processes. For such a rock with a single-valued grain size d, the equation for diffusion creep is given by (Ter Heege et al. 2004)

$$\dot{\varepsilon}_{\text{diff}} = \frac{A_{\text{diff}} b D_{\text{diff}} \mu}{KT} \left(\frac{b}{d}\right)^m \left(\frac{\sigma}{\mu}\right) \tag{6.14}$$

For a similar rock, the power law dislocation creep is given by

$$\dot{\varepsilon}_{\text{disl}} = \frac{A_{\text{disl}} b D_{\text{disl}} \mu}{KT} \left(\frac{\sigma}{\mu}\right)^n \tag{6.15}$$

Assuming that both GSI and GSS mechanisms contribute independently to the overall strain rate in a polycrystalline rock, the composite diffusion-dislocation flow law can be obtained by summation of the above two equations (Ter Heege et al. 2004):

$$\dot{\varepsilon} = \left[\frac{A_{\text{diff}} b D_{\text{diff}} \mu}{KT}\left(\frac{b}{d}\right)^m\left(\frac{\sigma}{\mu}\right)\right] + \left[\frac{A_{\text{disl}} b D_{\text{disl}} \mu}{KT}\left(\frac{\sigma}{\mu}\right)^n\right] \tag{6.16}$$

6.9.5 Composite Flow Laws for Distributed Grain Size

The composite flow law is obtained by incorporating a continuous log-normal grain size distribution and is given by Ter Heege et al. (2004) as

$$\dot{\varepsilon}_\sigma = \left[\frac{A_{\text{diff}} b D_{\text{diff}} \mu}{KT}\left(\frac{b}{\exp\left[(3-\frac{1}{2}m)\phi^2\right] d_{\text{med}}}\right)^m\left(\frac{\sigma_\sigma}{\mu}\right)\right] + \left[\frac{A_{\text{disl}} b D_{\text{disl}} \mu}{KT}\left(\frac{\sigma_\sigma}{\mu}\right)^n\right] \tag{6.17}$$

where $\dot{\varepsilon}$ is the strain rate, A is the single-valued grain size, D is the probability density distribution of distributed grain size, b is the magnitude of Burgers vector (m), μ is the shear modulus, K is the shape factor ($K = 4/3\pi$ for spherical grains), T is the absolute temperature, m is the grain size exponent, d is the grain size (diameter) and σ_s is the shear stress. The above composite equation (flow law) is a summation of the diffusion creep (the first term on the right of this equation) and dislocation creep (the second term).

6.10 Rheology of the Lithosphere

6.10.1 Background

Our knowledge on the *lithosphere* vastly grew with the development of plate tectonics. Lithosphere constitutes the mechanically strong outer shell of the earth that responds to stress and undergoes deformation. In general, it extends to a depth of about 100 km and includes the upper and lower crusts and part of the upper mantle. It is thicker below cratons than oceans. We describe below two aspects of rheology of the lithosphere, viz. temperature and rock deformation, both of which seem to have relevance to structural geology.

6.10.2 Lithospheric Rheology in Relation to Temperature

Underneath earth's surface, the temperature increases with depth and as such the earth is characterized by vertical zonation of temperature. As yet, our knowledge on the thermal structure of the lithosphere and the role of temperature in rheology is meagre. The reader is suggested to consult books by Bott (1982), Turcotte and Schubert (1982) and Ranalli (1987).

Temperature is undoubtedly the most important factor of lithospheric rheology as it affects the mechanical properties, and so also the deformation behaviour, of rocks. Although it is an extrinsic property like pressure and time, it strongly affects the intrinsic properties of rocks such as elasticity and rigidity. Temperature in the context of the earth broadly implies heat content of the earth. It is this heat of the earth that strongly controls the geodynamic processes as well as motion of plates. Temperature also depends upon the rate at which heat is produced and the rate at which it is exchanged in the lithosphere. In the lithosphere, heat is transferred by conduction, except in anomalous areas where hydrothermal processes play a significant role (Ranalli 1987, p. 154).

In the case of **continental lithosphere**, a characteristic feature of the earth's interior is the outflow of heat. This outflow of heat is related to, or is a reflection of, the large-scale motions in the mantle. It is believed that convection processes in the mantle are capable of transferring a sizeable amount of heat in the form of outflow from the interior. Temperature causes a change in volume and thus density of a material. Thermal properties of rocks are therefore a function of depth. It is believed that conduction is the main mechanism for transfer of heat in the lithosphere. In this connection, *Fourier's law of heat conduction* as stated below provides a relevant relation (Ranalli 1987) between heat flow and temperature gradient:

$$q_i = -K\,\delta T/\delta x_i \tag{6.18}$$

where q_i is heat flow, K is thermal conductivity and T is temperature. The minus sign indicates that heat flows down the temperature gradient. Considering the fact that in the earth the temperature increases vertically with depth, the surface heat flow is given by (Ranalli 1987)

$$q = -K\,dT/dz \tag{6.19}$$

where z represents depth, which is positive downwards. The minus sign indicates upward flow of heat.

While the continental lithosphere is characterized by the production of heat, the **oceanic lithosphere** shows cooling effect with negligible production of heat because the material moves away from the ridge (Turcotte and Schubert 1982; Bott 1982). Temperature inside the earth is thus dependent upon the heat transfer that affects the rheological behaviour of rocks. Near the surface of the earth, the temperature gradient (*dT/dz*) is very high (about 30° C/km) (Bott 1982; Ranalli 1987). Below the brittle-ductile transition

(15–20 km), the temperature gradient decreases, and this makes the rocks to behave as ductile. There is therefore a direct control of temperature on the rheological properties of the lithospheric rocks.

6.10.3 Lithospheric Rheology in Relation to Rock Deformation

Rheology of the lithosphere basically means the rheology of rocks that depends upon two important factors: intrinsic properties such as rigidity, elasticity and viscosity and extrinsic properties such as temperature, pressure and time. Since both these factors change inside the earth with depth, rocks show variable rheological behaviour with depth. Rocks of the upper crust thus behave differently to stresses from those of the lower crust despite the fact that rocks of both these regions are under loads of long duration. The overall physical conditions of upper crust are dominated by low temperature and low pressure, and as such the rocks show elastic behaviour. The overall physical conditions of lower crust, on the other hand, are dominated by high temperature and high pressure, and as such the rocks here show viscous behaviour. However, between the upper and lower crusts, there exists a transitional zone called *brittle-ductile transition* (BDT) (described in Chap. 7) that appears to be a rheological simplification where all the physico-mechanical properties of rocks are gradational depth-wise. The BDT in general occurs at depths between 10 and 15 km though it varies depending upon the thermal gradient and fluids.

The rheology of the lithosphere that controls its strength is also dependent upon the *differential stress* required to cause failure. The differential stress decreases with increasing temperature, i.e. with depth. Brittle faulting and fracturing are, thus, the dominating modes of deformation in the upper crust where failure occurs by frictional sliding under low-temperature and high-strain-rate conditions. Ductile flow, on the other hand, is the dominating mode of deformation below the BDT where failure occurs by power-law creep under conditions of high temperature and low strain rate.

6.11 Summary

- *Rheology* deals with deformation and flow of rocks. Rocks show different behaviour to deformation under different physico-mechanical conditions, and under certain conditions they deform by flowage.
- If a rock shows a constant rate of flow under constant stress conditions, it is said to show *steady-state flow*.
- The relationship between strain rate and stress is called *constitutive law* for a substance. This law assumes that a rock is a continuous medium that is free from asperities, microfractures, cracks or some other forms of discontinuities. Such materials deform uniformly in all directions within their boundaries, and as such the concept of *continuum mechanics* can be applied to rocks.
- Rheology depends upon the physical parameters of rocks such as rigidity, elasticity and viscosity. These are called *intrinsic parameters* that in turn depend upon the external factors such as temperature, pressure and time that together are called *extrinsic parameters*. The mathematical relations existing among all these factors constitute the *rheological equations* or *constitutive equations*.
- Rheology, and so the deformation behaviour, of a rock also depends upon whether it is *isotropic*, i.e. when its mechanical properties are the same in all directions, or *anisotropic*, if not the same.
- Rocks and artificial substances behave in their own way to stress and strain rate. Rocks that represent ideal cases are considered *linear rheological models* and include two extreme classes, i.e. *elastic* or *Hookean* and *viscous* together with an intermediate model, *plastic*. However, there are substances called *non-linear rheological models* that represent a combination of one or more types.
- Mathematical expression that relates differential stress, temperature and strain rate is called *flow law* that helps us understand the steady-state creep of rocks under different physical conditions by means of specific constitutive equations supported by experimental data.
- Temperature has a direct control on rheological properties of the lithospheric rocks. As such, the most common rheological model of the lithosphere is dependent upon the temperature gradient; that is, rocks down to the brittle-ductile transition (15–20 km) are under very high temperature gradient and thus behave as *elastic material*, while those below it behave as *ductile material* due to low temperature gradient.

Questions

1. What do you mean by rheology in the context of rocks?
2. Enumerate the factors on which rheology depends.
3. Explain strain rate. How does it differ from shear strain rate?
4. What is steady-state flow? How can this concept be applied to rocks?
5. Explain constitutive law and constitutive equations.
6. Discuss the plastic model in the light of rocks.
7. What are Newtonian substances? Give the constitutive equation for such substances.
8. What are isotropic and anisotropic substances?
9. Explain what is meant by flow laws.
10. Describe the rheology of the lithosphere from geological point of view.

7 Concept of Deformation

A. R. Bhattacharya, *Structural Geology*, Springer Textbooks in Earth Sciences, Geography and Environment,
https://doi.org/10.1007/978-3-030-80795-5_7

Abstract

While loading a box of 20 kg weight on your head, have you ever noticed that your face looks distorted? With more weight, your face looks more distorted. A load of, say, 2 kg, on the other hand, will make practically no visible effect on your face. With progressive increase of load, your face progressively starts showing visible effects of distortion. In the language of structural geology with this analogy, the load of 20 kg will produce *strain* or *deformation* on your face because it exceeds your strength or loading capacity. Likewise, rocks too show signatures of deformation when they are under stress. Effect of stress is sometimes visible and sometimes not, depending upon the strength of a rock with respect to the applied stress.

This chapter takes the readers to explore the concept of deformation of rocks with the necessary elementary ideas so that their journey to structural geology may not become an uphill task!

Keywords

Deformation · Homogeneous and heterogeneous deformation · Kinematics of deformation · Dynamics of deformation · Elastic behaviour · Plastic behaviour · Viscous behaviour · Brittle-ductile transition · Creep · Deformation mechanism map

7.1 Introduction

Rocks in the earth's crust are always under the influence of some external forces that tend to change their original shapes. For example, they may bend or develop cracks. The external forces acting upon a rock develop stress (a stress is force per unit area) within the rocks. The stresses are generally of tectonic nature but are also generated by the overburden as well as due to the mass of the rock body itself. As a result, the rocks undergo *deformation*, which is *the change in the original shape, size or volume of a rock mass caused by external stresses*. Deformation results in the formation of structures in rocks such as folds, faults, joints and several other complex structures. Deformation of earth's rocks, if affected over a large region for longer periods, produces large-scale structures including mountains. In this chapter, we will discuss some basic concepts of deformation, while several other aspects of deformation will be discussed in some other chapters of this book.

7.2 Kinematics of Deformation

Deformation primarily involves the processes by which the particle motions are achieved in a rock body. One major objective of structural geology is therefore to understand the motion of rock masses or of their constituent particles. All these are known as *kinematics* that necessarily does not take into consideration the forces acting upon a rock body but mainly deals with the geometry of motion of rock bodies. Depending upon the intrinsic properties of a rock and the nature of the imposed forces, a rock body may show rigid body or non-rigid body deformation.

7.2.1 Rigid Body Deformation

A rock is considered a *rigid body* when during motion the distance between any two points of the body remains unchanged. Motion of a rigid body may be of two types: *translation* (Fig. 7.1a) in which a straight line within the body remains parallel to itself throughout the motion and *rotation* (Fig. 7.1b), in which the body shows a change in its orientation from the original position.

In translation, the motion takes place along a straight line, while in rotation the body rotates around an axis of rotation along which there is no motion. The process involving both translation and rotation is known as *rigid body deformation.*

7.2.2 Non-rigid Body Deformation

A rock is considered a *non-rigid body* when during motion the distance between any two points of the body changes. Motion of a non-rigid body may be of two types: *dilation* (Fig. 7.2a) in which the body shows a change in volume and *distortion* in which the body shows a change in shape (Fig. 7.2b). If during motion both volume and shape change, the body is said to have undergone *bulk deformation* (Fig. 7.2c), and as such the body is then said to be in a deformed state.

7.3 Dynamics of Deformation

The *dynamics of deformation* is the study of stresses acting within a rock body as well as on its boundaries. In fact, it is the stresses that ultimately deform a rock which, in turn, assumes a particular geometry or shape. Dynamic analysis

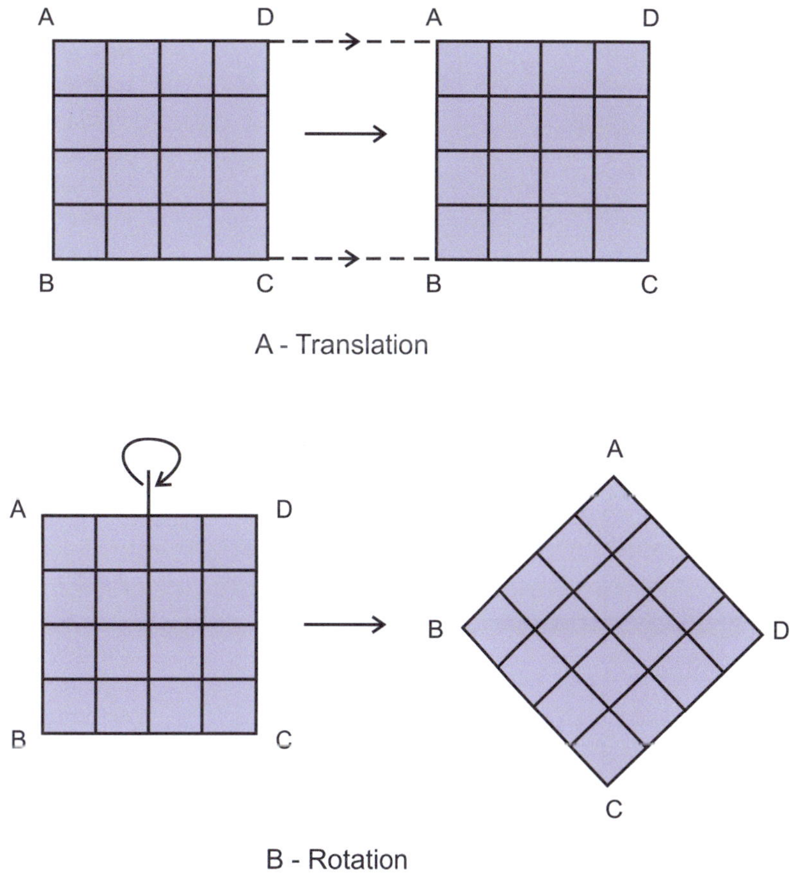

Fig. 7.1 Rigid body deformation. (**a**) Translation. (**b**) Rotation. See text for details

is aimed at understanding the orientations and relative magnitudes of the stresses that have deformed the rock. It involves the study of rock, whether it has behaved as an elastic material deformed by brittle mode with the formation of cracks or fractures or whether the rock has undergone permanent deformation by ductile mode without the formation of cracks. For example, during brittle deformation strain rate is high, while during ductile deformation it is low. Temperature and pressure also play important roles in strain and strain rate.

7.4 Modes of Deformation

Rocks behave (deform) differently to the applied stresses. Crustal rocks, therefore, show different modes of deformation of which elastic, brittle, ductile, plastic and viscous deformations are common, as described below. However, rocks seldom show ideal modes of deformation and tend to follow a deformation path that involves more than one component of some other mode(s). For example, a rock deformed by brittle processes may also involve some or little components of ductile deformation, and vice versa. Inside the crust, the various physico-mechanical conditions of rocks are therefore gradational. This puts a constraint on showing ideal physico-mechanical conditions of deformation at any particular point in the earth's crust.

7.4.1 Elastic Deformation

Within the elastic limit, i.e. within the straight line portion of the stress-strain graph (Figs. 6.1 and 6.2), a rock returns to its original shape and volume after the applied stress is removed. Thus, whatever strain the rock may have undergone during loading is recovered, and therefore this strain is called *recoverable strain*, as is shown by a stretched rubber band that recovers its original shape after it is released. For most rocks, however, the property of showing elastic deformation is very low to negligible as the chances of recovery of stresses are negligible.

7.4.2 Brittle Deformation

Brittle deformation involves failure and change in shape of a rock by the formation of fractures or cracks along which the rock loses its cohesion. It is characteristic of rocks that are elastic. Ideal brittle materials fail well before crossing the elastic limit (Fig. 6.2). Brittle deformation generally occurs at

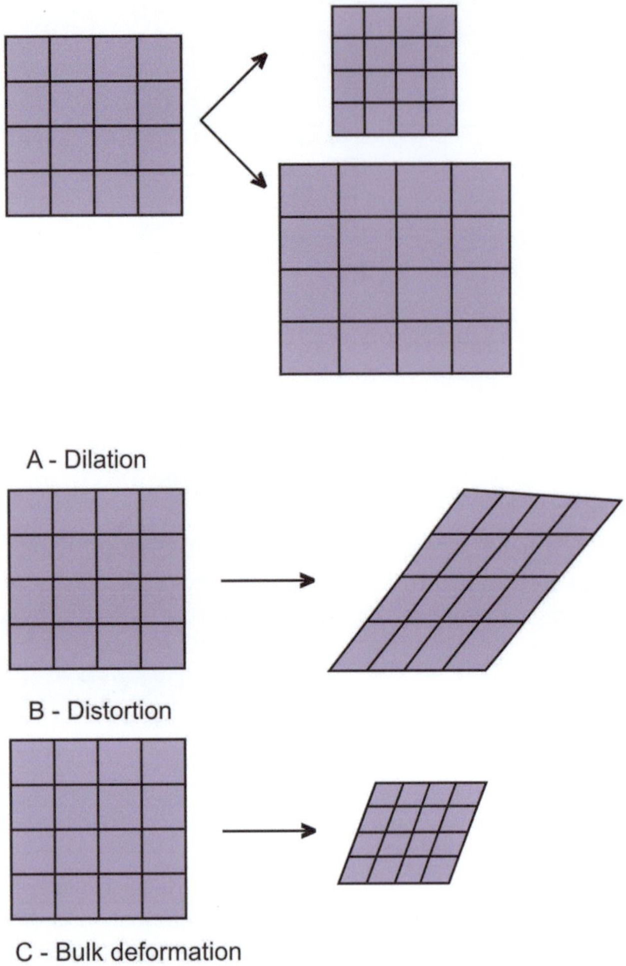

Fig. 7.2 Non-rigid body deformation. (**a**) Dilation in which the body shows change (decrease or increase) in volume. (**b**) Distortion in which the body shows change in shape. (**c**) Bulk deformation in which the body shows change in both volume and shape

conditions of low temperature, low confining pressure and high strain rate under shallow or near-surface conditions.

In crustal rocks, the common structures formed by brittle deformation processes include joints and fractures (Chap. 13) and faults (Chap. 9). A joint is a crack in rock, generally transverse to bedding, along which no appreciable shear displacement has occurred (Engelder and Geiser 1980). Joints (Fig. 7.3) may develop in rock with any orientation. *Fractures* are breaks or discontinuities in a rock due to which the rock has lost its cohesion. The broken parts may show movement across the fracture surface in which case it is called an *extension fracture*. Fractures that occur on microscopic or submicroscopic scales are known as *microfractures* along which a rock is broken down into numerous smaller fragments, a process called *cataclasis* that produces a fine-grained aggregate commonly called *cataclasite*. Another important structure formed due to brittle deformation is *fault* (Fig. 7.4) that involves movement of two parts of a rock parallel to the surface of separation. In the process, the rock loses some cohesion, but still the separated parts remain attached with the main mass. Faulting therefore involves partial loss of cohesion of a rock. Faulting operates on various scales ranging from microscopic to several tens of kilometres.

Rocks occurring at the surface and shallow levels of the crust deform by brittle processes. Rocks and minerals showing brittle behaviour can be broken by the application of stress in a certain manner even if they possess high values of strength. Because of this, diamond which is the strongest of all minerals can be broken into pieces if it is cut in a certain manner. Several strong rocks likewise can be broken into pieces if they possess brittle behaviour.

7.4.3 Ductile Deformation

Ductility is the property of a rock to fail by plastic flow before rupture. Within the ductile limit, a material continues to deform without the development of fractures or cracks till it ruptures. After initial loading, a rock may first undergo elastic deformation represented by the inclined straight part of the curve in Fig. 6.2. For a ductile substance, the graph would go further beyond the elastic limit or the yield point and will follow a curved path (plastic range) till the body ruptures. The highest point of the graph is called *ultimate strength* of the rock. The curved path in the graph, i.e. the curve between the elastic limit and the point of rupture, is an expression of ductile deformation mode of the rock.

Ductile deformation involves accumulation of large permanent strain and failure by homogeneous flow without formation of fractures and cracks. Rocks show several types of structures that are formed due to ductile deformation. A fold (Fig. 7.5) is an excellent example of ductile deformation. To a common man, it is difficult to imagine how a rock flows. In our daily life experiences, we see an ironsmith giving various shapes to a metal. This is possible because the metal possesses ductile properties due to which it is made to flow. The same analogy holds good for rocks also that are able to flow under certain conditions provided that they hold the property of ductility. Rocks possessing ductile behaviour undergo permanent deformation without loss of cohesion. Ductile deformation generally operates at high temperature, high confining pressure, low strain rate and under deeper level conditions.

All rocks do not possess perfect ductility. Rocks possessing ductility may also possess brittleness as reflected in the occurrence of both straight line and curved paths in the stress-strain graphs (Fig. 7.6). A rock possessing high brittleness and low ductility may show a longer straight line followed by a shorter curved path (Fig. 7.6a). The length of

Fig. 7.3 Joints in rocks. Joints are excellent examples of brittle deformation in rocks. B: Bedding. J: Joints. (Photograph by the author)

Fig. 7.4 Brittle deformation in rocks as evidenced by the development of a fault (FF)

the straight path is a reflection of how much brittleness the rock possesses. Material with less brittleness and more ductility may show a shorter straight line followed by a longer curved path (Fig. 7.6b), while a wholly ductile material with no brittleness may show a curved graph starting from the origin with the straight-line portion missing (Fig. 7.6c).

Ductility of a rock depends upon several physical factors of which the temperature, mineral composition of the rock, strain rate and confining pressure are important. Each of these factors has its specific influence on the ductility, and hence deformation behaviour, of a rock. At high temperature, most rocks show ductile behaviour, while at low temperature, they show brittle behaviour. With the presence of minerals like mica and clay minerals, a rock behaves ductile, while the presence of quartz and feldspar makes it behave brittle. Strain rate of a rock influences the mobility of atoms. Thus, at high strain rates, the rocks fail by brittle fracturing, while at low strain rates, they show ductile failure. At low confining pressure as existing on or near the surface of the earth, rocks show brittle behaviour, while at high confining pressure that exists at deeper levels, rocks show ductile behaviour. In addition to these factors, there are several other factors also that influence the ductile or brittle behaviour of rocks such as presence of water, mineralizing fluids, presence of rock cleavage, joints or other planar anisotropies (see also Sect. 7.6).

Some important characteristic features or criteria of ductile deformation include the following: (a) The deformation does not involve loss of cohesion of the body. (b) Deformation takes place in solid state without the appearance of any melt phase. (c) The structures formed show continuity of marker lines without any break. (d) The deformation is more dependent on temperature and strain rate than on stress.

There are certain types of deformation that are sometimes confused with ductile deformation. An example is the formation of *flow bands* in fault zones. Flow bands are formed as a result of cataclasis that involves breaking and crushing of grains along some discrete planes, thus causing loss of cohesion of the original mass. As such, cataclasis is an example of brittle deformation and cannot be included under ductile

Fig. 7.5 Ductile deformation in rocks as exemplified by folds. The scale placed on the structure measures 30 cm. (Photograph courtesy Professor T.K. Biswal)

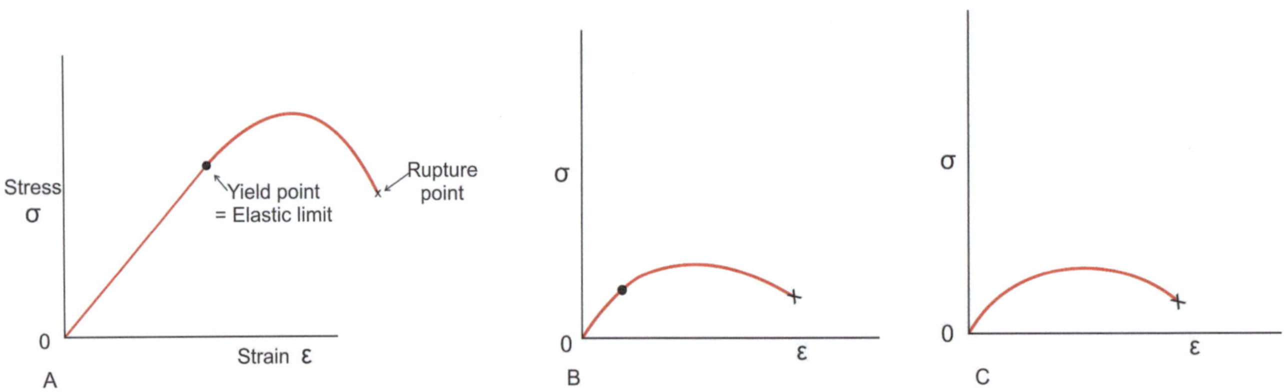

Fig. 7.6 Some common stress-strain graphs shown by rocks containing both elastic and plastic behaviours. (**a**, **b**) If the rock possesses elastic properties, the graph initially starts with a straight line (elastic field) followed by a curved path (plastic field). The length of the straight-line portion is a reflection of how much elastic properties the rock holds: in this case, more in **a** and less in **b**. (**c**) If the rock possesses only plastic, and no elastic, property, the straight-line portion of the graph is missing and the graph rises from the origin with a curved path

deformation even though the flow bands formed by this process look like a product of ductile deformation.

Ductile deformation operates through a variety of mechanisms of which recrystallization, cataclastic flow, diffusional mass transfer and movements on glide planes are important (described in Chap. 16).

7.4.4 Plastic Deformation

This involves deformation of the crystal lattice of the constituent minerals and is therefore called *crystal-plastic deformation* or *intracrystalline deformation* (see Chap. 16).

7.4.5 Viscous Deformation

In this type of deformation, stress is proportional to strain rate implying that viscous deformation is a *time-dependent deformation* and that the rocks accumulate strain with time. During viscous deformation, strain does not bear linear relationship with stress. As such, this type of deformation causes permanent strain.

7.5 Brittle-Ductile Transition

The crust is constituted of several types of rocks. The strength and mechanical properties of these rocks are broadly determined by their dominant mineralogy and the temperature-pressure conditions, which in turn depend upon the depth. As such, rocks behave differently to stresses with increasing depth. At shallow levels of the crust, rocks are at low temperature, low compressive strength and high tensile strength. Under these conditions, rocks behave as elastic materials and release the stresses instantaneously with high strain rate along joints and fractures. These conditions prevail in the upper crust up to a depth of about 15 km below which the rocks deform by ductile processes due to increase of temperature and pressure. The zone where the rocks show transition from brittle to ductile mode of deformation is called *brittle-ductile transition* (BDT), also called *ductile-brittle transition* (DBT) (Fig. 7.7). In general, the BDT represents transition from faulting to flow (Paterson 1978).

The depth of the BDT varies at different parts of the crust. In general, it is believed to lie at depths of 10–15 km. However, the depth of the BDT strongly depends upon the thermal gradient and fluid content that broadly controls mineralogical phase changes. In continental regions of high thermal gradients, the BDT occurs at relatively shallower levels, up to depths of 8–12 km, while in regions of low thermal gradients, it occurs up to depths of 15 km. In regions of continental crust showing recent tectonism, where geothermal gradient is higher, the BDT may move up to shallower levels.

In the BDT zone, the rocks deform by both brittle and ductile processes: those above it by brittle processes and those below it by ductile processes. However, even within the shallower, brittle, zone, the deformation is not always ideally brittle all through its depth. Development of folds is occasionally noticed in the brittle deformation zones of orogenic belts. Such folds are open parallel folds. In some cases, evidences of superposition of homogeneous ductile strain, though mild, are also noticed. For example, in the Himalayan orogenic belt, the rocks of the Siwalik Supergroup (Mio-Pliocene) of the Outer Himalaya have been found to show open parallel folds in sandstones, and rarely some of the folds show flattening strain up to 5% (Bhattacharya 1983). It is possible that such rocks correspond to some deeper levels of the upper crust where pressure-temperature conditions or thermal gradient, or both, may have been locally higher. As such, these rocks may have attained some ductility sufficient to undergo strain superimposition. Likewise, an analogous situation may also exist in the lower crust, i.e. below BDT, where folds in rocks may show progressive increase of strain superimposition with depth, reaching a flowage stage at much deeper levels.

Considering the deformation behaviour of rocks, the crust as well as the lithosphere in general have been modelled (Knipe 1989) on the basis of strength. The strength of the upper crust increases linearly with depth. This is in accordance with **Byerlee's law**, which states that there is a linear correlation between confining pressure (given by depth) and strength of rock. The lower crust, however, shows a reversed pattern and is governed by the flow law of quartz (due to ductile nature of the lower crust). These two types of behaviour are however common at the brittle-ductile transition where both brittle and ductile deformation mechanisms operate.

Below the BDT, joints and fractures are not formed, thus causing slow release of stresses. These conditions impart permanent strain to the rocks, which thus behave as ductile materials and fail by homogeneous flow. Also, the confining pressure is high and therefore the fractures, that are commonly present above the BDT, disappear or get closed. This allows slow release of stresses, and the rocks thus show permanent strain. All this, together with higher temperature, makes the rocks located below the BDT ductile and thus make them deform by flow.

Most of the above characteristics of the BDT zone are believed to be induced by fluid activity (Reynolds and Lister 1987). The main reason is that the magnitude of the difference between pore fluid pressure and mean stress determines the ability of cracks to dilate, and therefore the length of the fractures formed is dependent on the magnitude of the pore fluid reservoir (Lister and Davis 1989, p. 72).

In spite of the above-mentioned features of the BDT, it is still contended that there is a gradual, not abrupt, increase in ductility with temperature and pressure (Paterson 1978; Lister and Davis 1989). Thus, considering the crust as a whole, the existence and the precise depth of the BDT are yet an open field of investigation.

7.6 Factors Controlling Deformation of Rocks

Deformation of rocks primarily depends upon their intrinsic properties that in turn vary with depth from the surface of the earth. Of the various factors that control rock deformation, composition, temperature, pressure and rheology are most significant. All these factors taken together suggest that the location of a rock, i.e. whether on the surface or underneath,

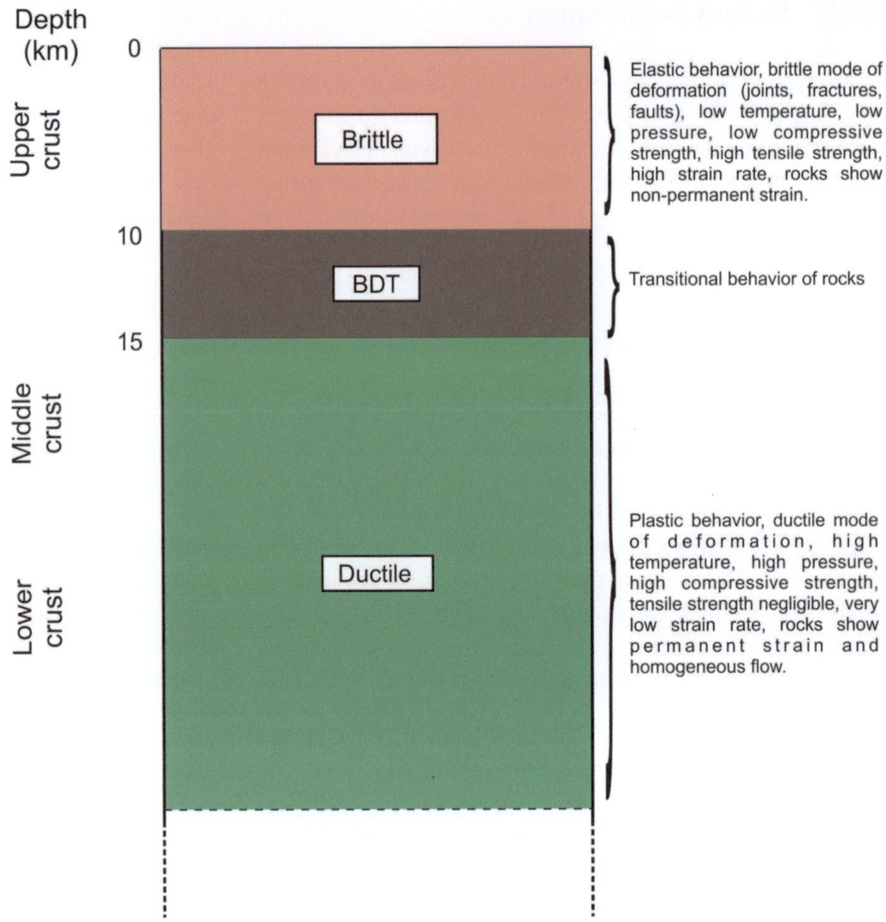

Fig. 7.7 The brittle-ductile transition (BDT) in the crust. Some important properties of upper and lower crusts are highlighted in the light of recent literature

constitutes an important factor of rock deformation. Some common factors that control deformation of rocks are described below.

7.6.1 Composition

Rocks are composed of minerals. The behaviour of the constituent minerals to stress strongly controls the material properties of a rock. Minerals such as feldspars and olivine are brittle, and therefore presence of these minerals imparts brittle character to the rock concerned. Micas and clay minerals are ductile, and their presence imparts ductile character to the rock. Quartz behaves as a brittle mineral in the upper crust, but at higher temperatures, as exist in the lower crust, it is ductile and deforms by crystal-plastic mechanisms and diffusional flow.

7.6.2 Temperature

From the surface of the earth downwards, progressive increase of temperature is a rule, though the rate of increase (= geothermal gradient) varies from place to place. Temperature commonly has a pronounced effect on the deformation behaviour of a rock. At lower temperatures, the rocks release the stresses instantaneously; the strain rate thus becomes high, and the rocks fail by brittle fracturing. Higher temperatures, on the other hand, promote strain to propagate in the rock mass where the strain accumulates; the rocks thus undergo permanent strain and fail by flowage that facilitates movement of ions and their bonds. Low temperature thus promotes brittle deformation, while high temperature promotes ductile deformation.

7.6.3 Pressure

Inside the earth, the confining pressure increases with depth. At low confining pressure as exists at or near the surface, rocks fail by formation of fractures and thus show brittle mode of deformation. At high confining pressure as exists at depth, formation of fractures and their movement are hindered by the surrounding high pressure. As such, the surface of the earth is dominated by brittle deformation processes that progressively give way to ductile deformation processes with depth.

7.6.4 Rheology

Rheology refers to the flow characteristics of rocks, and it dominantly controls strain in rocks. Further, how a rock reacts to stress also depends much on its rheologic behaviour. A rock may instantaneously react to, and release, stress. Such rocks show elastic behaviour and high strain rate and thus fail by brittle deformation, a condition that exists in the upper crust and on the surface of the earth. Rocks, on the other hand, may react slowly to stress due to which strain accumulates in the rocks. This promotes shearing processes and flow mechanisms to operate, and the rocks thus deform by homogeneous flow. Conditions like this exist at lower crust and mantle.

7.6.5 Strain Rate

The rate at which deformation takes place is called strain rate. This rate is dependent upon several factors. Low temperature-pressure conditions promote brittle deformation by the development of fractures that move rapidly to the surroundings. The deformation is thus instantaneous. The strain rate is high in such cases. At high temperature-pressure conditions, on the other hand, fractures do not form, and more time is available during which ions or atoms can move. This makes the deformation to proceed slowly. The strain rate is thus low. Under such conditions, ductile deformation is favoured. High strain rate thus favours brittle deformation, while low strain rate favours ductile deformation.

7.6.6 Planar Features

Presence of layering such as bedding or some other secondary features such as rock cleavage promote deformation (see Box 7.1). These planar features constitute anisotropies that allow the external forces to easily pass through the rock, which in turn deforms to accommodate the imposed stress. Presence of planar features, thus, controls the degree or extent of deformation in a rock.

7.6.7 Orientation of Stress

Whether a rock would undergo deformation, and if so to what extent, also depends on the direction of the applied stress or the orientation of the principal stress. Stress applied in directions intermediate between the two principal stresses would accordingly produce intermediate effects of deformation.

7.6.8 Pore Fluids

Presence of pore fluids is also important in rock deformation. Pore fluids occur in the form of water, chemical fluids, hydrocarbons and gaseous phases. These substances are commonly present in the upper lithosphere, and as such their presence affects brittle fracturing of rocks. Presence of a fluid system causes an additional pressure to rocks called *pore fluid pressure*. The fluids occupy the pore spaces available within the rocks and thus constitute a rock-fluid mass that is relatively at high overpressure than the surrounding rock mass. Once the local fluid pressure exceeds the strength of rock, the latter fails by brittle fracturing and grain boundary sliding.

Box 7.1 Deformation: Massive Versus Strongly Layered Rocks

Deformation of rocks broadly depends upon the amount and orientation of the applied stress. However, the internal structure of the rocks, mainly the layering pattern, also strongly governs the deformation. Because of this, under similar stress conditions, a massive rock and a thinly layered rock will show different behaviour to deformation; the former may not show any visible or weak signature of deformation, while the latter may show visible, occasionally strong, signature.

Let us take the example of a common deformation structure, i.e. folds that are generally formed due to compressive forces acting parallel to layering. Given similar conditions of deformation, a massive rock (Fig. 7.8) would show weak signature of deformation (folds, in this case), while a thinly layered rock (Fig. 7.9) would show profound effects of deformation (folds).

A massive rock, in fact, is constituted of a mess of coarse grains of irregular sizes oriented in a haphazard manner. The externally applied forces, therefore, may not easily or uniformly pass through the rock. The rock, thus, may not develop sufficient stress within it for deformation. As such, the rock either may not show any visible signature of deformation or may show only weak or partial effect of deformation. In a strongly layered rock, on the other hand, the forces may pass through the rock easily, thus bringing the rock under stress that is accommodated by the development of folds.

The above examples can explain why in deformed terrains strongly layered rocks, such as shale, thinly laminated limestone and alternation of shale/sandstone or of shale/limestone, show profuse development of structures (folds) than the associated massive sandstone or massive limestone in the same terrain.

Fig. 7.8 The rock (quartzite) being thickly layered, fold development is weak (it is an open fold)

Fig. 7.9 A rock showing thin layering (of shale and limestone) shows strongly developed folds. (Photograph courtesy Professor T.K. Biswal)

7.7 Time-Dependent Deformation (Creep)

A rock can be deformed under two different experimental set-ups. In one case, the rock can be subjected to varying stress such that the rock is strained at a constant rate. In other words, the stress is changed in such a way that the strain rate remains constant.

In another case, a rock may be loaded with a constant stress such that the strain rate is allowed to vary. Under such conditions, the rock gets strained with time. This *time-dependent deformation* is called *creep* that takes place under constant stress conditions well below the rupture strength of the rock. The rock, thus, undergoes plastic deformation with time without undergoing rupture.

Materials deformed by creep mechanism have been found to show three different types of mechanical responses after initial loading. This is mainly a reflection of "time" in relation to rock strength. Creep experiments show that after initial loading, the strain rate becomes moderately high, which is due to the effect of an elastic deformation. With time, however, the strain rate decreases till it becomes constant. As deformation proceeds with time, the strain rate again increases rapidly. For materials deformed under creep mechanism, the *strain-time relationship* (Fig. 7.10) is represented by a typical curve that reflects three types of mechanical responses shown by rocks: primary, secondary and tertiary creep.

Primary creep is represented by the first non-linear part of the curve. This type of deformation starts immediately after the rock has undergone the elastic response to the applied stresses as represented by the straight-line portion of the curve. During the primary creep, the rock shows a diminishing rate of strain. This type of deformation is also known as *transient creep*.

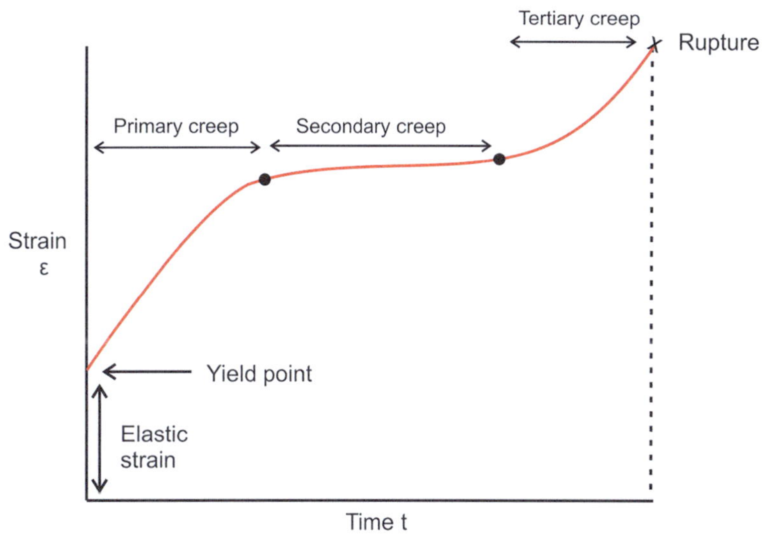

Fig. 7.10 Graph showing strain-time relationship at constant stress (creep experiment). See text for details

Secondary creep is represented by the straight-line part of the curve. During this deformation, the strain rate or creep rate remains constant with time, and as such the strain and time of loading bear a linear relationship. This part of the curve represents plastic deformation and is also known as *steady-state creep*.

Tertiary creep is represented by the rising part of the curve. In this type of deformation, the rock shows rapid fatigue with time and with increasing creep until it fails by rupture. This type of deformation is also known as *accelerating creep*.

The creep or constant stress experiments, thus, show that the deformation pattern shows a definite change or mechanical responses with time: an elastic deformation followed by plastic deformation. This deformation pattern can thus be fitted with an equation of the type

$$\varepsilon = \mathrm{A} + \mathrm{B} \log t + \mathrm{C}t \qquad (7.1)$$

where ε is strain, t is time and A, B and C are material constants. The strain rate is given by

$$\dot{\varepsilon} = \varepsilon / t \qquad (7.2)$$

The creep deformation has great implications for crustal rocks. Below the surface, rocks are constantly under a fixed amount of stress in the form of overburden. Creep experiments show that there is a definite influence of 'time' on rock deformation. A rock can therefore deform plastically even at low levels of stress provided that it is subjected to loading for a long time. It further implies that since crustal rocks mostly remain under the influence of a constant stress (overburden) for a long time, they undergo creep deformation because the strength of the rocks decreases due to loading for a long time. In our daily life also, we see that originally flat horizontal slabs of many ancient monuments show sagging today. This is also an example of creep deformation. Most crustal rocks deform by steady-state creep.

7.8 Deformation Mechanism Maps

Deformation of polycrystalline rocks generally occurs under the influence of more than one mechanism. Depending upon the physical conditions such as temperature, strain rate and differential stress at various stages of deformation, it is possible that different mechanisms may have operated in the deforming system, and at a particular stage some mechanisms may have dominated over others. These physical factors may have controlled the rheological behaviour and macroscopic flow of the rock at various stages of deformation. All these ideas are covered under a *deformation mechanism map* or simply a *deformation map*, which is *a useful diagram that represents the dominant deformation mechanism that has operated at various stages of deformation of a rock under a particular set of conditions* (Ashby 1972; Rutter 1976; Frost and Ashby 1973, 1982). The concept does not take into account brittle fracturing whose effect is superimposed, and thus ignored, by imposing sufficiently large hydrostatic confining pressure during experiment.

The deformation mechanism map (Fig. 7.11) broadly represents a relation between stress and temperature in a graph. The most common type of deformation mechanism map is one in which the various mechanisms of plastic flow, such as dislocation glide, dislocation creep and diffusional flow, are shown on the map by specific fields such that a point on the map refers to the dominant mechanism together with the associated strain rate. In the graph, the ordinate represents

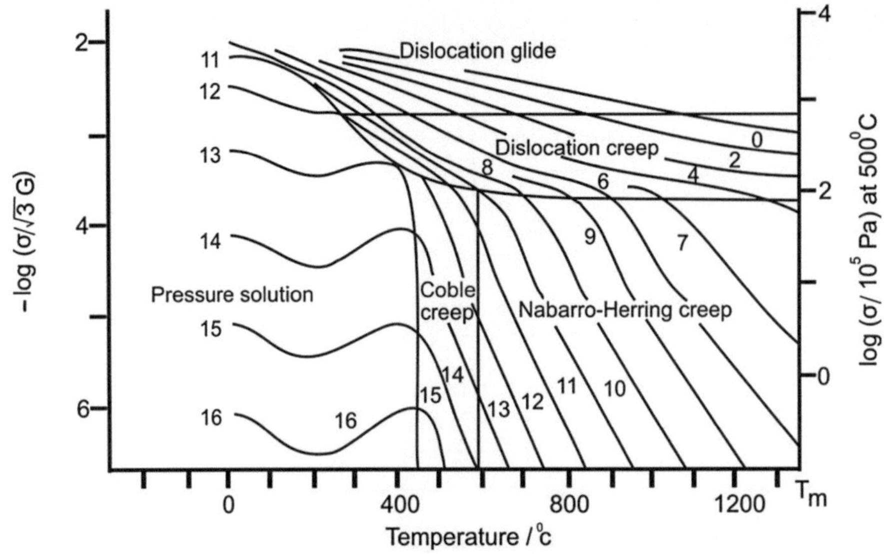

Fig. 7.11 Deformation mechanism map that broadly shows the relation between stress and temperature of a deforming substance. The ordinate represents the stress (σ) as presented in a normalized form by dividing it by the shear modulus (μ). The abscissa represents the homologous temperature, i.e. the ratio of the temperature (T) and the melting point (T_m) of the material. The vertical axis to the left side is in negative while that to the right side is in positive. On the left side, therefore, the numbers decrease upwards with increase of differential stress, while on the right side it is reverse. Since the differential stress increases upward, the contours with higher values show lower strain rates. Dominant deformation mechanisms are shown in the graph in which each field represents only the dominant deformation mechanism under which the material failed. In general, the deformation mechanism map relates three variables: temperature, strain rate and differential stress. (Reproduced from Rutter 1976, Fig. 8, with permission from the Royal Society (U.K.) under Order Licence ID 1172311-1)

the stress (σ), which is presented in a normalized form by dividing it by the shear modulus (μ). The abscissa represents the homologous temperature, which is the ratio of the temperature (T) and the melting point (T_m) of the material. The vertical axis to the left side is in negative while that to the right side is in positive. On the left side, therefore, the numbers decrease upwards with increase of differential stress, while on the right side it is reverse. Further, since the differential stress increases upward, the contours with higher values show lower strain rates.

The deformation mechanism map thus obtained shows the dominant deformation mechanisms. The contours indicate the constant strain rates that are achieved during a particular mechanism. These contours superimpose the deformation fields. It is important to note that each field represents only the dominant deformation mechanism under which the material failed (though however some other mechanisms of plastic deformation may have been active under that stress-temperature condition). From the map, it is thus possible to get the value of stress and temperature for a particular mechanism, and thus the form of the constitutive equation can be derived from the experimental data of the material. The deformation mechanism map, thus, shows the relationship between three variables, i.e. temperature, strain rate and differential stress. As such, if values of two known variables are known, one can get the value of the third variable.

The deformation mechanism map as described above can also be presented in some other forms of which the following two types have also been used with the axes representing (i) shear strain rate and (normalized) shear stress (the contours represent temperature) and (ii) strain rate and temperature (or reciprocal temperature) (the contours represent constant stress).

Deformation mechanism maps are of great significance in metallurgy, where the formation of a particular metal can be observed clearly and therefore the various physical parameters can be measured correctly. However, our understanding of the deformation mechanism maps as applied to rocks is still not clearly known. The maps at best can be considered to be a manifestation of the overall rheologic state and flow regimes of a deformed rock or its mineral constituents. From geological viewpoints, the maps can help in a qualitative way in selecting suitable materials for engineering purposes.

7.9 Deformation and Continuum Mechanics

The concept of *continuum mechanics* is based on the fact that solids and fluids mostly constitute a continuous media containing a finite number of matters. A rock is a solid body, and as such continuum mechanics is applicable for rocks that deform in contrast to the rigid bodies that do not deform. A solid deforms because it possesses shear strength, and therefore it allows shear forces to act on it. Since crustal rocks possess the property of deformation, the concept of continuum mechanics is in use in structural geology (see Means 1976 for details).

While applying the concept of continuum mechanics in structural geology, one problem arises that rocks are seldom homogeneous as they invariably contain cracks, discontinuities, voids, different mineral constituents or mineral aggregates, etc. on all scales ranging from grains to outcrops. It is believed that the application of continuum mechanics provides better predictions than that of the concept of discontinua.

Box 7.2 What Is Continuum Mechanics?

The concept of *continuum mechanics* is based on the fact that solids and fluids mostly constitute a continuous media containing a finite number of matters. A *continuum* may be considered as a unit whose properties are similar to those of the object. The entire space of the object is filled with such units. The continuity of the matter within this object gets disturbed when the matter is subjected to an external load. *Continuum mechanics thus basically deals with the response of solids and fluids to an external load.* It is assumed that the physical properties of the matter that constitute a solid or a fluid are independent of its actual size or the time during which they are measured. In continuum mechanics, the material (solid or fluid) is assumed to be a continuous mass. This branch is thus concerned with the physical behaviour of materials.

Any continuous medium or volume in a rock can be considered as constituted of points or material particles in space and thus can be visualized in the framework of a continuous space. *A continuous medium or a continuum is therefore an idealization which assumes that physical quantities are definable at points in space, considered as lower limits of shrinking elemental volumes* (Ranalli 1987, p. 5).

Continuum mechanics does not involve atomic structure of materials and is therefore independent of the atomic processes. This concept is in contrast to what is advocated in *solid-state physics*, wherein material properties are considered at an atomic level. Rheological behaviour of a rock can thus be considered in two different ways, one taking into consideration the material continua, while the other considering material properties at atomic level. It is believed that while the continuum mechanics is helpful in explaining large-scale or macroscopic behaviour of rocks, the solid-state approach can explain rheology of rocks at the microscopic level. Considering all these aspects together, the continuum mechanics can explain geological processes in a better way.

7.10 Summary

- *Deformation* is the change in the original shape, size or volume of a rock mass caused by external stresses. A rock may undergo *homogeneous deformation* if it has deformed by flowage without the development of any fracture or *inhomogeneous deformation* if it has deformed by the development of fracture without undergoing any flowage.
- Deformation in rocks produces structures such as folds, faults, joints and several other types. Deformation of rocks of a large region, if continued for a longer period, causes large-scale structures including mountains.
- The *kinematics of deformation* involves the processes by which motion of rock masses or of their constituent particles takes place. The *dynamics of deformation* involves the stresses acting within and on the boundaries of a rock mass and is therefore aimed at understanding the behaviour of a rock to an applied stress.
- Since rocks behave differently to an applied stress, a similar stress would produce different effects of deformation in different rocks. Crustal rocks show various modes of deformation of which elastic, brittle, ductile, plastic and viscous modes are common. However, rocks seldom show ideal modes of deformation and tend to show a deformation path involving components of some other mode(s).
- Deformation of crustal rocks is largely dependent upon the strength and mechanical properties of rocks that are broadly determined by their dominant mineralogy and the temperature-pressure conditions, which in turn depend upon the depth. Thus, under conditions prevailing in the upper crust up to a depth of about 15 km, the rocks behave as elastic materials and deform by brittle processes. Below this depth, rocks deform by ductile processes due to increase of temperature and pressure. This zone of transition (15–20 km) from brittle to ductile mode of deformation is called *brittle-ductile transition*.
- Of the various factors that control rock deformation, composition, temperature, pressure and rheology are most significant. Other factors include strain rate, presence of planar features, orientation of stress and pore fluids.
- A rock can be made to deform plastically even at low levels of stress provided that it is subjected to loading for a long time. This is called *time-dependent deformation* or *creep*. Since crustal rocks mostly remain under the influence of a constant stress (overburden) for a long time, they undergo creep deformation because the strength of the rocks decreases due to loading for a long time.

- During deformation of polycrystalline rocks, different mechanisms may have operated in the deforming system such that at a particular stage some mechanism(s) may have dominated over others. This is because of changing physical conditions such as temperature, strain rate and differential stress at various stages of deformation. All these ideas are represented by a *deformation mechanism map* or simply a *deformation map*.
- Since a rock is a solid body that constitutes a continuous media containing a finite number of matters, the concept of *continuum mechanics* is applicable to rocks.

Questions

1. Outline the important criteria of brittle deformation and ductile deformation.
2. Give an example each of brittle deformation and ductile deformation from our daily life.
3. What do you mean by rigid body deformation?
4. What do you mean by creep? Explain why crustal rocks commonly deform by creep mechanism.
5. Folds are believed to be an example of ductile deformation, yet sometimes they show fractures within the structure. Explain why is it so?
6. What do you mean by strain rate? Explain how strain rates vary in crustal rocks.
7. What is brittle-ductile transition (BDT)? Explain deformation of crustal rocks in the light of BDT.
8. What is continuum mechanics? Briefly explain how this concept is of help in structural geology.
9. Briefly explain how temperature affects deformation of rocks.
10. What is a deformation mechanism map?

Part II

Structures: Geometry and Genesis

8 Folds

A. R. Bhattacharya, *Structural Geology*, Springer Textbooks in Earth Sciences, Geography and Environment,
https://doi.org/10.1007/978-3-030-80795-5_8

Abstract

Can rock layers bend? Yes, they can but only when they are ductile, and can thus produce spectacular structures called *folds* that are bends in rocks formed due to compressive stresses acting parallel to, or across, the originally flat surfaces of rock layers. Folds may occur in a single layer or in multilayers. They are developed on all scales ranging from microscopic to kilometric and occur in all types of rocks. They commonly occur in strongly deformed rocks of orogenic provinces. This chapter provides a description of various parts of folds, their classification and theories of their formation. The significance of folds is highlighted.

Keywords

Folds · Geometrical parameters of folds · Fold style · Classification of folds · Fold mechanics · Buckling · Bending · Kinking · Passive folding · Progressive development of fold shapes

8.1 Introduction

Folds are bends formed as a result of permanent deformation of originally planar surfaces of rock layers. They can develop on any planar surfaces in rocks such as bedding or foliation, and their formation reflects ductile behaviour of the layers in which they are developed. They are best developed in sedimentary strata. Folds occur on all scales ranging from microscopic to several kilometres and can develop in practically all types of rock. They can occur as single or isolated structures as well as in groups.

Folds are developed under a wide range of conditions of stress, hydrostatic pressure and temperature. As such, they can occur at various levels of the crust ranging from brittle upper part to ductile lower part. On the surface of the earth, folds commonly occur in regionally deformed terrains of orogenic provinces where they constitute *fold belts.*

8.2 Parts of a Fold

The point on the fold surface where curvature is highest is called *hinge* (Fig. 8.1), and the line on the fold surface that joins the points of the highest curvature is called *hinge line*. Folds are generally studied in their profiles. A *fold profile*, or simply profile, is the form of a fold surface as seen on a plane normal to the hinge line.

Common parts of a fold in profile are shown in Fig. 8.2. In the profile of a fold, the *crest* is the highest point and *trough* is the lowest point; in three dimensions, these correspond to the crest line and the trough line, respectively. The flat or curved surfaces on either side of a fold are called *limbs*. The point on a limb where the curvature reverses is the *inflection point*. The line on the folded surface that separates these curvatures is called the *inflection line*. It may be noted that the crest and hinge point of a fold may or may not coincide. In an upright fold, with horizontal axis and vertical axial plane, both crest and hinge coincide. *Hinge zone* is commonly used for the part of the fold located near the hinge. However, it is necessary to define hinge zone and fold limbs in some definite geometric terms so that their limits are properly known.

The surface formed by joining the hinge lines of the different folded layers is called the *axial plane* or *axial surface* (Fig. 8.2). If the hinge surface is planar, it is called axial plane. Since an axial plane is a planar surface, it can be described by its dip and strike. It must be noted that if the axial plane is not vertical, the crest line and the hinge line may not coincide. The intersection of the axial plane with any surface—fold surface, ground surface, etc.—is known as the *axial trace* (Fig. 8.2) of the fold.

Considering a fold as a wave-like trace (Fig. 8.3), it is possible to represent it in the x–y coordinate system by a simple function of a curve $y = f(x)$. The crest and trough are the points where the curvature is maximum and therefore $dy/dx = 0$, while d^2y/dx^2 reaches its maximum negative value at crest and maximum positive value at trough.

Ramsay (1967, p. 349) suggested a measure in which the *curvature of the fold* surface is compared with that of a (unit) circular arc that passes through both the inflection points (Fig. 8.4a). *Hinge zone is that part of a fold surface whose curvature is greater than that of a (unit) circle whose diameter joins the two inflexion points, and the limb is that part whose curvature is less than that of the circle* (Professor Peter

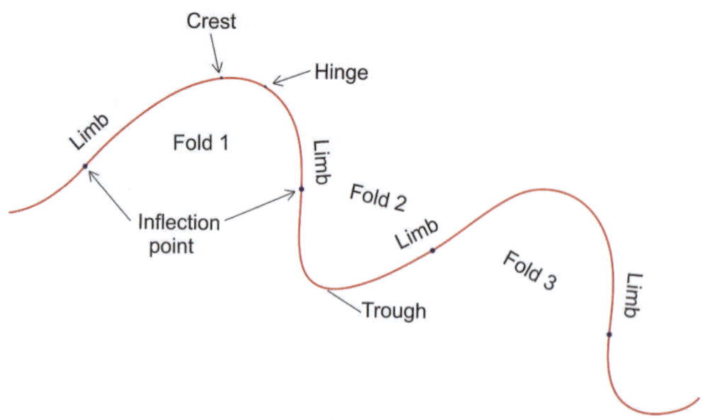

Fig. 8.1 A fold profile showing limbs, crest, hinge, trough and inflection point. Folds 1 and 3 are up-arching showing convexity upwards, while fold 2 is down-arching

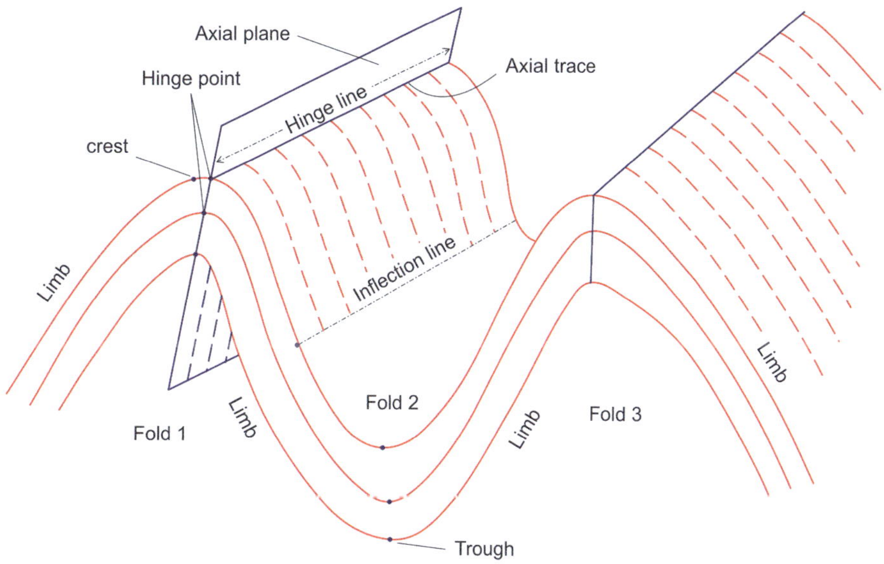

Fig. 8.2 Common parts of a fold showing axial plane, hinge line, inflection line, crest and hinge point

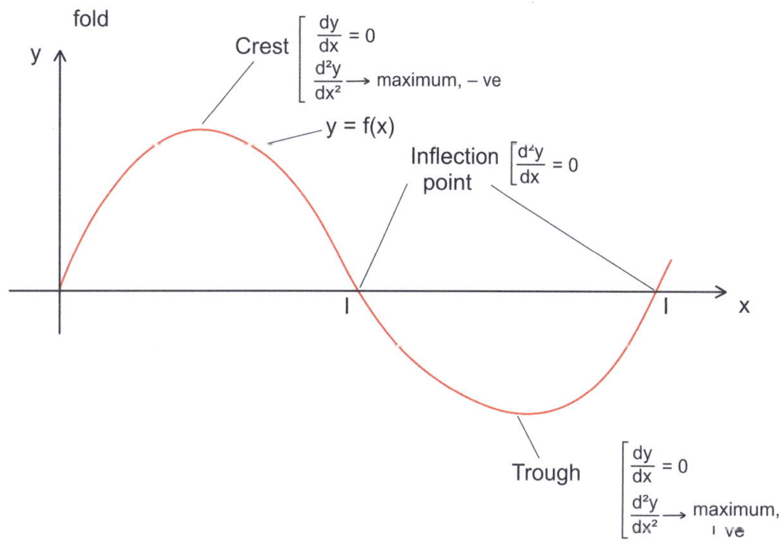

Fig. 8.3 The profile of many folds can be represented as a wavy surface in the form of a simple function $y = f(x)$. The relations existing between the crest, trough and inflection points are shown

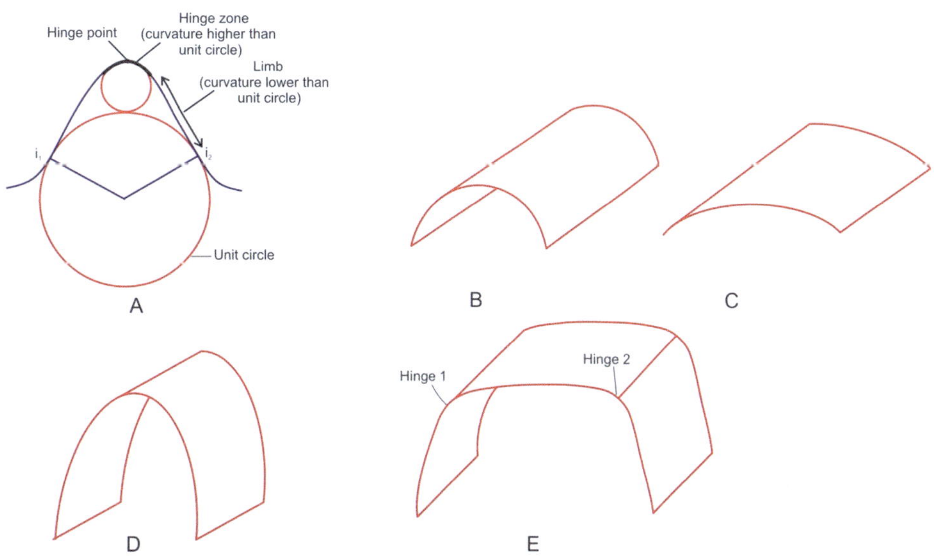

Fig. 8.4 Hinge zone and fold limb. (**a**) Hinge zone is that part of a fold surface whose curvature is greater than that of a (unit) circle whose diameter joins the two inflexion points, and the limb curvature is less than that of the circle. (**b**, **c**) Folds in which the curvature remains unchanged and do not have any hinge. (**d**) A fold in which the point showing the highest curvature constitutes the hinge; such fold is called a single-hinge fold. (**e**) A fold showing two points where the curvature reaches maximum; such fold is called a double-hinge fold

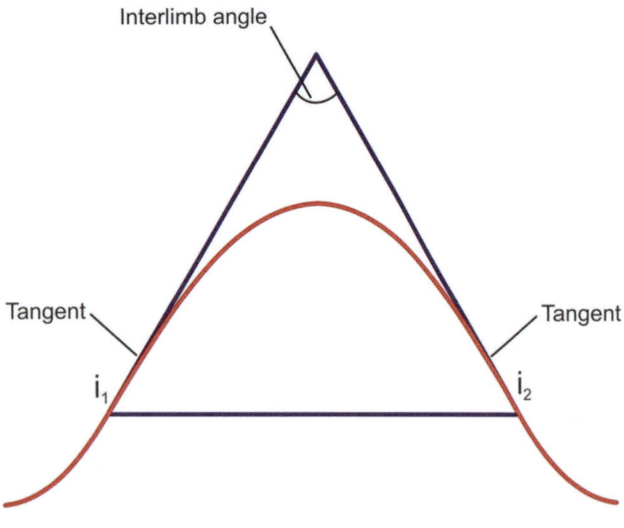

Fig. 8.5 Interlimb angle of a fold is the angle that the tangents to the two limbs subtend at the hinge

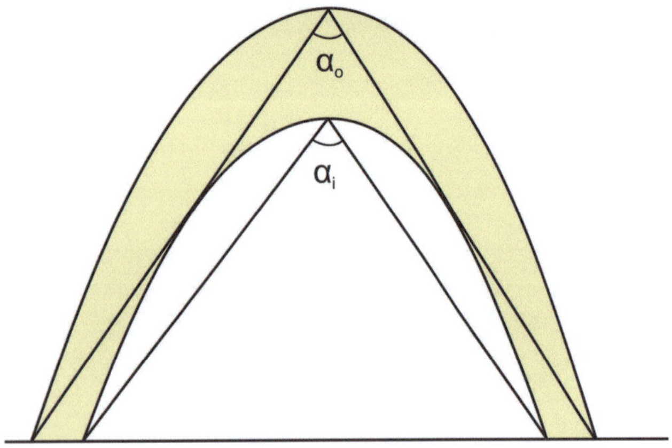

Fig. 8.6 Fold axial angle is the angle subtended by a trace at its hinge. α_i and α_o are axial angles for the inner and outer arcs, respectively. This parameter was proposed by Bhattacharya (1992)

Hudleston, personal communication 2020). Folds in which the curvature remains unchanged do not have any hinge (Fig. 8.4b,c). In nearly all folds, the folded surface shows changes in curvature from point to point; the point showing the highest curvature constitutes the hinge, and such fold is called a *single-hinge fold* (Fig. 8.4d). In yet another type of folds, there are two points where curvature reaches maximum; such fold is called a *double-hinge fold* (Fig. 8.4e).

Interlimb angle (Fig. 8.5) is the angle formed by the two tangents drawn at the inflection point of the opposite limbs. *Axial angle* (Bhattacharya 1992) is the angle that two limbs of a fold subtend at its hinge (Fig. 8.6). The angle is formed by two lines joining hinge with the inflection points. For a folded layer, there is thus an *outer axial angle* (α_o) and *inner axial angle* (α_i), which are the acute angles subtended by the outer and the inner arcs, respectively, at the hinge of the fold. For parallel folds, outer (α_o) and inner (α_i) axial angles are equal.

Depending upon the orientation or attitude of the fold, the hinge line may be horizontal or inclined. If the hinge line is inclined to the horizontal, it is said to *plunge*, and the fold is called a *plunging fold* (Fig. 8.7a). If the hinge line is horizontal, plunge is zero; such folds are called *non-plunging folds* (Fig. 8.7b). Most natural folds are plunging and occur on all sizes ranging from hand specimen to several kilometres (Fig. 8.8). The direction of overturning or leaning of the axial plane of a fold is called the *vergence* (Fig. 8.9) of the fold. Vergence is only for asymmetric folds as it describes the sense of asymmetry. Asymmetry is best defined by relative limb lengths, not by vergence (Professor Peter Hudleston, personal communication 2020).

8.3 Geometrical Parameters of Folds

Description of a fold requires some geometrical aspects to completely specify a fold. Since a fold is a wave-like structure in rock strata, we can apply the characteristics of a wave in section. If we join the inflection points of the

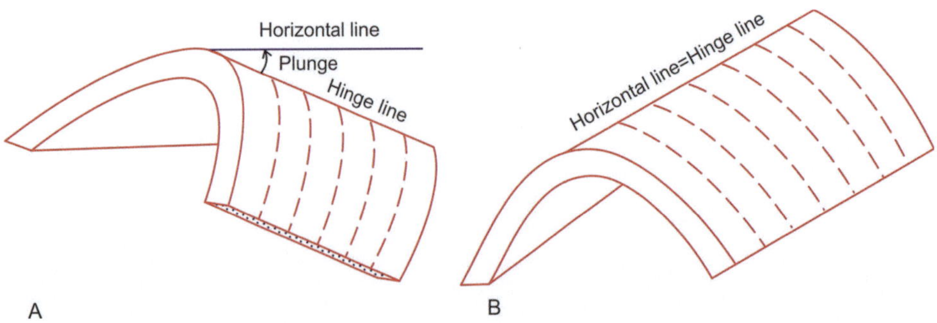

Fig. 8.7 Plunging and non-plunging folds. (**a**) Plunging fold. The plunge is the angle between the hinge line and its horizontal projection line. (**b**) Non-plunging fold in which the hinge line is parallel to a horizontal line

Fig. 8.8 A huge plunging fold in Zagros Mountains, Iran. The arrowhead of the dotted red line shows the approximate direction of plunge of the fold. Beds of the limbs, dipping in opposite directions, are shown by dotted yellow lines. (Aerial photo courtesy Narendra K. Verma)

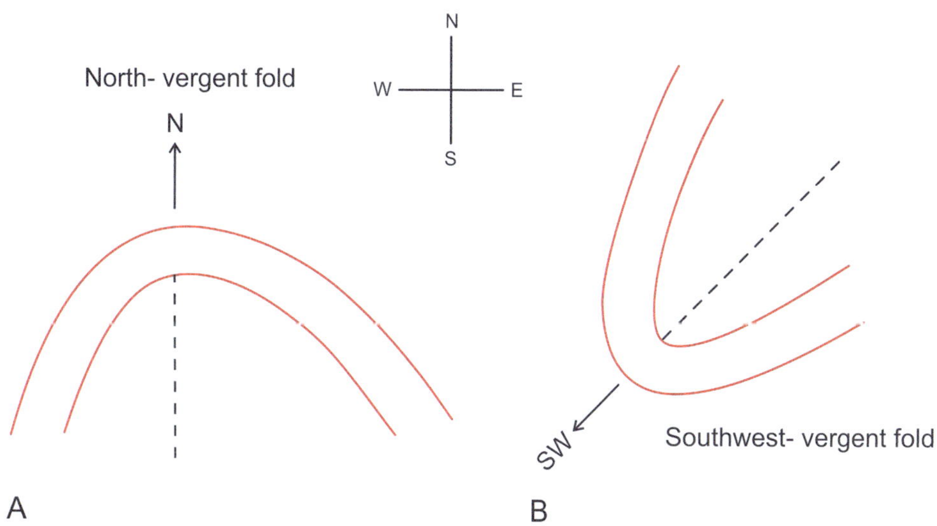

Fig. 8.9 Vergence of a fold. (**a**) North-vergent fold. (**b**) Southwest-vergent fold

adjoining folds by a line, the maximum height of a fold gives the amplitude (Fig. 8.10) of the fold; amplitude is thus half the distance between the crest of an anticline and the trough of an adjacent syncline. Amplitude and wavelength can be better studied for symmetric folds only as they lead to some generalized information on the geometry of folds, which is not so for asymmetric folds. Wavelength and amplitude are relatively more practical parameters of folds (see Box 8.1).

Folds often occur in a set such that they form successive waveforms bounded between two surfaces, each of which is called an *enveloping surface* (Fig. 8.11), which is formed by joining the tangents to the crests and troughs. A *median surface* (Fig. 8.11) is formed by joining the successive inflection lines of each fold. Both enveloping surface and median surface may be planar or curved.

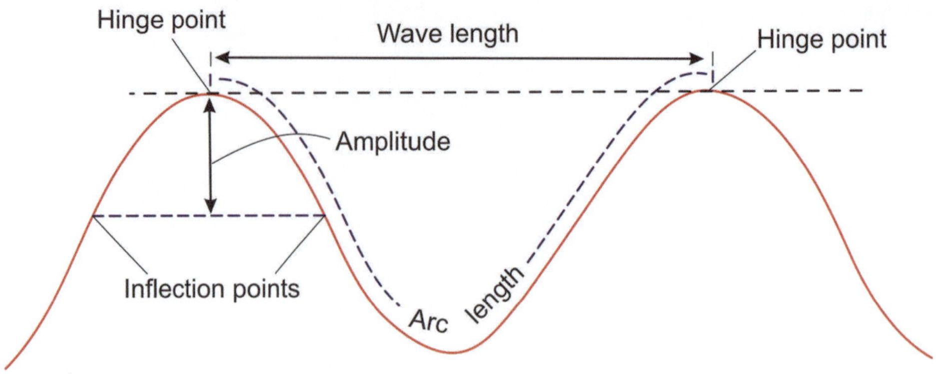

Fig. 8.10 Wavelength, amplitude and arc length of a fold in profile section. The distance between two successive crests or troughs gives the wavelength, the maximum height (in case of anticline) or depth (in case of syncline) of the fold is the amplitude and the length measured along the arc between two hinge points is the arc length

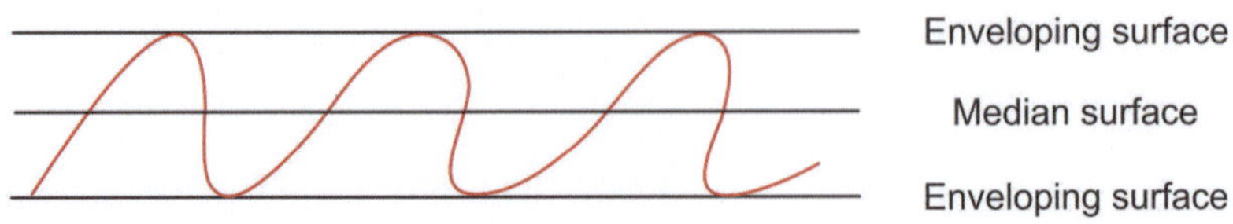

Fig. 8.11 Enveloping surface and median surface of a fold

Box 8.1 Wavelength-Amplitude Ratio: Use in Identifying Fold Generations

Of all parameters of a fold, the wavelength and amplitude can give in a better way the shape of the folded layer on profile section in a plane perpendicular to the fold axis. However, one practical problem is that folds are seldom visible with one anticline and a corresponding syncline. As such, the corresponding hinges and therefore the actual wavelength (*W*) are not always possible to measure. To mitigate this, Hudleston (1973a) suggested a scheme based on a *W/4* unit—"quarter wavelength unit"—with *y*-axis normal to tangent at the hinge point and the *x*-axis normal to *y* through the inflection point.

The above scheme has been found to be of great practical use by me (Bhattacharya 1986) while carrying out wavelength and amplitude measurements on a number of (symmetrical) folds at different localities of the Bundelkhand craton, central India. The Proterozoic craton mainly consists of granites, gneisses, quartzites, amphibolites and various types of schists and has undergone several cycles of magmatism, metamorphism and tectonism. The rocks have undergone several generations of folding as reflected in the rocks of the craton on minor to small outcrop scales.

Data on quarter wavelength of the measured folds were plotted against amplitude (Fig. 8.12). The plots clearly reveal that folds of a particular generation are more or less confined to a definite field or range, suggesting that the folds of a particular generation

Box 8.1 (continued)

show a definite range of wavelength-amplitude ratio that is different from other generations. In the light of field studies and the above plots, it has thus been indicated (Bhattacharya 1986) that the early-formed folds of the craton show relatively lower values of wavelength-amplitude ratio that gradually increases for folds of later generations. The early-formed folds, thus, appear relatively tighter than the later formed folds.

Individual generation of folds can be identified in field by the various relations of fold axial planes and fold axes with other features of folds. However, to arrange the various generations in suitable sequence is not always easy. In such case, the use of the wavelength-amplitude ratio for getting the above plots comes as a handy solution to some extent. The plots also indicate the possible number of generations (e.g. five in the present case study) undergone by the folds. The method can, thus, help in identifying the number of generations of folds, and their possible sequence of formation, in any terrain affected by polyphase deformation.

Since folds are curved surfaces, it is therefore better to consider one more geometric feature, i.e. *curvature* (Fig. 8.13), of a fold. Knowledge of curvature gives better information of fold shape, which in turn may help in understanding the mechanisms of folding. The curvature

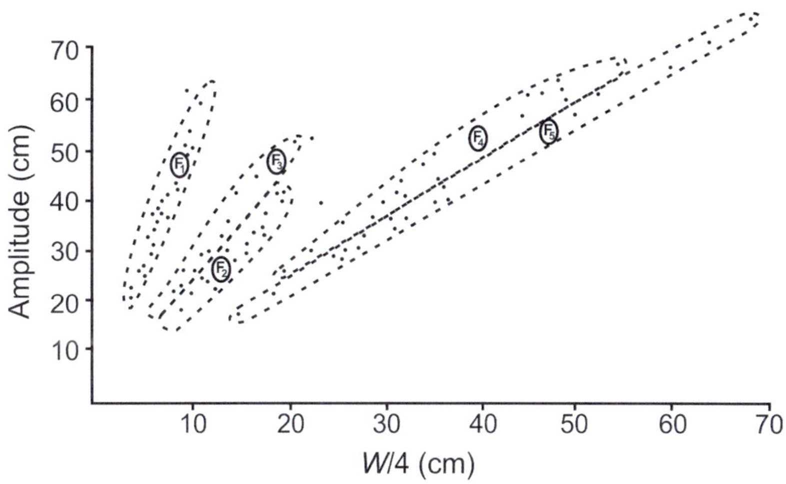

Fig. 8.12 Data on quarter wavelength (*W/4*) and amplitude are plotted for folds of various generations (F_1 to F_5) of the Bundelkhand craton, central India. The plots reveal that folds of a particular generation are more or less confined to a definite field or range. (Reproduced from Bhattacharya 1986, Fig. 3 with permission from Elsevier Senior Copyrights Coordinator, Edlington, U.K. Submission ID: 1193117)

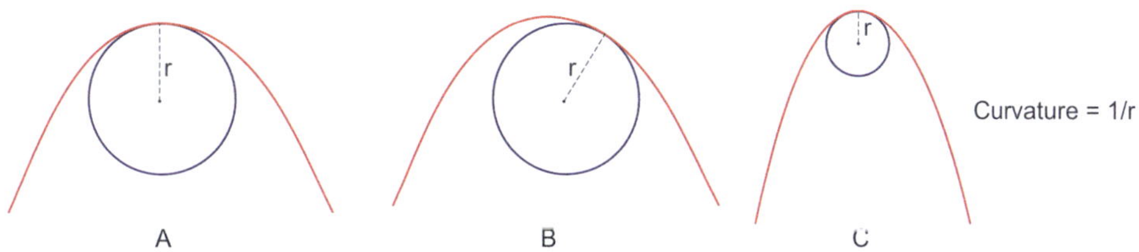

Fig. 8.13 Curvature of a folded surface at a point is the reciprocal of the radius of curvature at that point. (**a**) Curvature at hinge. (**b**) Curvature at a point in the limb of an open fold. (**c**) Curvature at the hinge of a tight fold. Note, in general, that if the radius is large, e.g. an open fold (**b**), the curvature is small, while if the radius is small, e.g. a tight fold (**c**), the curvature is large

(C) of a line is given by the reciprocal of the radius of curvature(r), i.e.

$$C = 1/r \tag{8.1}$$

A straight line has an infinite value of radius of curvature, and therefore its curvature is zero. The maximum curvature of a fold occurs at its crest and trough. The value of curvature is zero at inflection point.

8.4 Fold Style

The style of a fold is a general term to describe the overall geometrical shape or attitude of a fold in terms of a few parameters that help in distinguishing different fold forms. Some common parameters by which the style of folds can be described include cylindricity, symmetry, aspect ratio, tightness and bluntness (following Twiss and Moores 2007) as described below.

8.4.1 Cylindricity

The *cylindricity* means by how much a fold approaches an ideal cylindrical shape. In a cylindrical fold, the shape of a fold in profile remains the same at all points along a hinge line (Professor Peter Hudleston, personal communication 2020). A cylindrical fold can be generated by moving a line parallel to itself in space such that in every part of the folded surface there should be a line having the same orientation as that of the hinge line.

8.4.2 Symmetry

The *symmetry* of a fold is given by how much the shape of the fold on one side of a hinge, in profile section, is identical to that of the other side (Fig. 8.14). A fold is said to be symmetric if the fold shape is a mirror image on either side of the axial plane. In such case, the bisector of the interlimb angle is also the bisector of the median line that joins the two inflex-

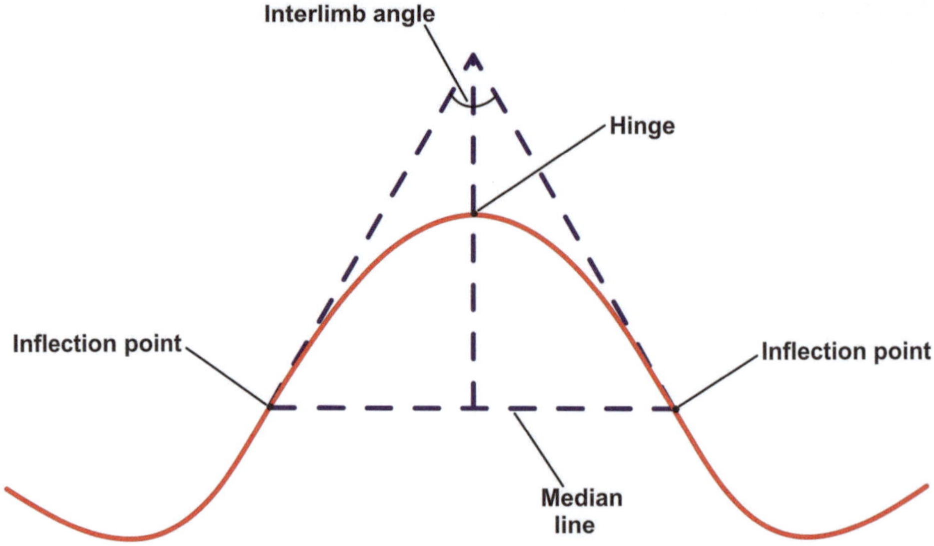

Fig. 8.14 Symmetry as applied to a fold in profile section. The symmetry indicates by how much a fold is identical, or mirror image, on either side of its axial plane

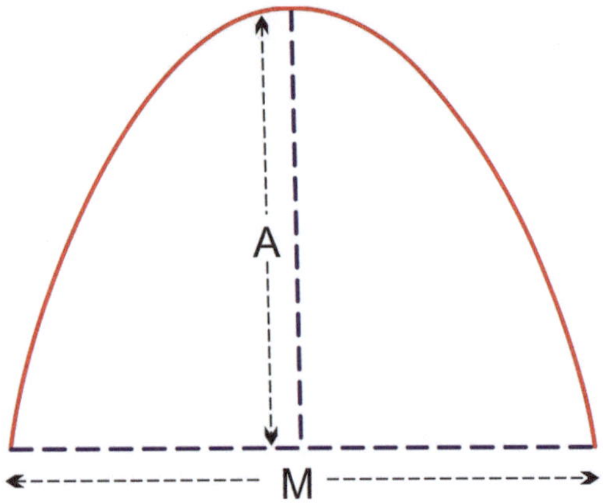

Fig. 8.15 Aspect ratio of a fold is defined as the ratio between the amplitude (A) of a fold along the axial plane and the distance (M) between two adjacent inflection points

ion points, and therefore the limbs of the fold are of equal length. Folds that do not show symmetry are called asymmetric folds. Symmetric and asymmetric folds have been further described below under classification of folds (Sect. 8.5).

8.4.3 Aspect Ratio

The *aspect ratio* (P) is the ratio existing between the amplitude (A) of a fold along the axial plane and the distance (M) between two adjacent inflection points (Fig. 8.15). For a symmetric fold, the latter constitutes the half-wavelength. With increasing values of aspect ratio, the folds are successively named as wide, broad, equant, short and tall. In general, the folds with low values of aspect ratio look wide, while those with high values look tall.

8.4.4 Tightness

The *tightness* of a fold is given by the value of its interlimb angle, i.e. the angle subtended by the limbs at the hinge. With increasing intensity of folding, the interlimb angle decreases and the folds become progressively tighter and are called gentle, open, close and tight. This parameter has been used as a basis for a fold classification (see Fleuty's classification below).

8.4.5 Bluntness

The *bluntness* of a fold refers to the relative curvature of its hinge. It is defined by the ratio (b) between the radius of curvature (R_c) at fold closure and the radius of the circle (R_o) that is tangent to the limbs at the inflection points (Fig. 8.16), i.e. $b = R_c/R_o$. With increasing values of bluntness, the folds are progressively called sharp, angular, subangular, subrounded, rounded and blunt. With the value of bluntness as zero, the fold has perfectly sharp hinges; with unity, the fold is said to be perfectly circular; and with value more than unity, the fold is said to be blunt.

Box 8.2 How Large a Buckle Fold Can Form in the Crust? Any Idea!

Crustal rocks get folded on various scales ranging from several tens of kilometres to microscopic. But how large a fold can develop in crustal rocks? Although this seems to be an interesting aspect, only a little attention has yet been given.

Formation of large-scale anticlines and synclines involves movement of large masses of rocks. Though

(continued)

Box 8.2 (continued)
formation of these structures requires suitable geologic conditions, Ramberg and Stephansson (1964) and Ramberg (1967) highlighted that gravity is also an integral part of their formation. According to them, gravity controls the upward (anticline) and downward (syncline) movement of rock masses. As such, the formation of large folds depends upon the interaction of lateral (layer-parallel) forces and gravity; the former will be opposed by the latter.

Considering the crust composed of granitic rocks with a strength of the order of 5×10^9 dynes/cm^2, crustal thickness of 30 km and substratum density of 3 g/cm^3, Ramberg and Stephansson (1964) and Ramberg (1967) developed the following relation (see Ghosh 1993, p. 265) as the maximum value of wave-length/thickness ratio:

$$\frac{L}{h} = \Pi\sqrt{\left[\frac{5 \times 10^9}{3 \times 981 \times 15 \times 10^5}\right]} \approx 3.5, \qquad (8.2)$$

where L and h are the arc length and thickness of the folds.

The above relation gives a maximum value of 100 km for the waves formed by compression of the crust, suggesting that further large buckle folds cannot form in the crustal rocks (Ghosh 1993). This result should have great implications for developing geotectonic models and for understanding crustal deformation.

The buckling model however does not hold true for folds developed due to lithospheric folding. McAdoo and Sandwell (1985) reported folding of the lithosphere just south of the Bay of Bengal. These are east-west trending folds with wavelength ranging from 130 to 250 km. According to the authors, the stresses required for the formation of these folds exceed the strength of lithospheric rock, and loads approaching the net compressive strength can cause lithosphere to fold up to a wavelength of about 200 km. They assign formation of these folds to the axial compression caused by the collision of India and Asia.

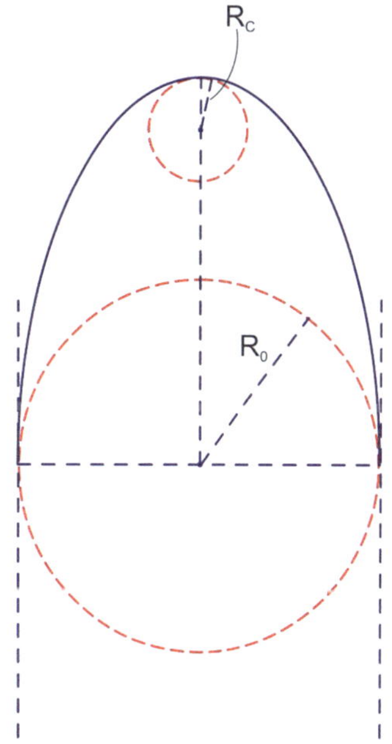

Fig. 8.16 Bluntness of a fold is defined by the ratio between the radius of curvature (R_c) at the fold closure and the radius of the circle (R_o) that is tangent to the limbs at the inflection points

8.5 Classification of Folds

Natural folds occur in a great variety of shapes. They also involve either one layer (single-layer folds) or a number of layers (multilayer folds) in folding. Because of a great variety of features shown by the folds, they have been classified by different workers in several ways. Some folds have been classified on the basis of their geometrical features (geometrical classification) while others on the basis of their formation (genetic classification). We present the geometrical classification only and not genetic ones, as the latter are based on interpretation of the geometrical features of folds, and interpretations may vary from worker to worker.

8.5.1 Classification Based on Interlimb Angle

Interlimb angle (Fig. 8.5) is the angle between the two limbs of a fold and is measured as the angle between two tangents drawn at the inflection points of the two opposite limbs of a fold. Interlimb angle is applicable for folds with rounded crests and curved limbs, and it is a measure of the tightness of folds. Fleuty (1964) proposed a fold classification (Fig. 8.17) based on the amount of interlimb angles. The various folds thus identified are *gentle* with interlimb angle between 120° and 180°, *open* 70° and 120°, *closed* 30° and 70°, *tight* 0° and 30°, *isoclinal* 0°, i.e. with parallel limbs, and *elastica* in which the interlimb angle is negative. Folds showing elastica geometry are also called *fan folds*.

8.5.2 Fleuty's Classification

On the basis of the relationship between the plunge of fold hinge line and the dip of axial surface, Fleuty (1964) devised a graph (Fig. 8.17) showing the fields of the various fold

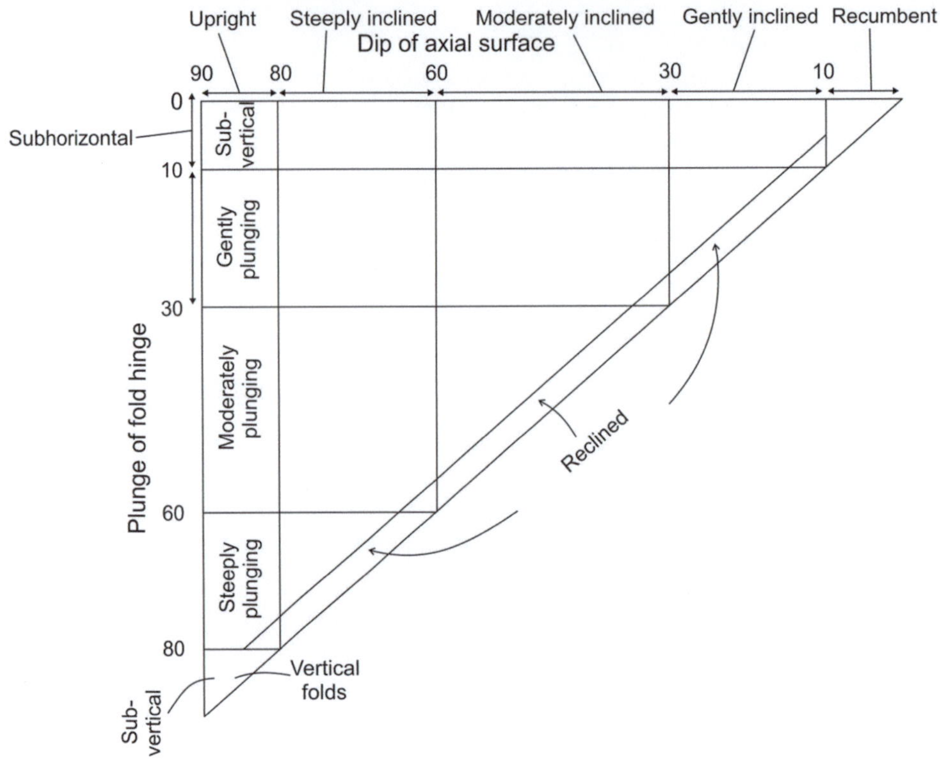

Fig. 8.17 Classification of folds based on the relationship between the plunge of fold hinge and the dip of axial surface (After Fleuty 1964, Fig. 11a. Reproduced with permission from the Geologists' Association, London)

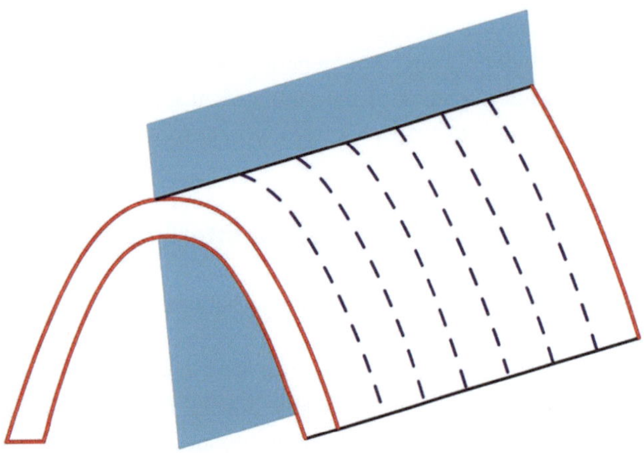

Fig. 8.18 Sketch of an upright fold

types. He thus classified folds into *upright, steeply inclined, moderately inclined, gently inclined* and *recumbent* as well as *sub-horizontal, gently plunging, moderately plunging, steeply plunging*, *sub-vertical* and *reclined.*

An *upright fold* (Figs. 8.18 and 8.19) is one in which the axial surface is vertical.

In an *inclined fold* (Figs. 8.20 and 8.21), the axial surface is inclined; the inclination may be low or high. If the axial surface is horizontal, it is called *recumbent fold* (Figs. 8.22 and 8.23). If the limbs of a fold are parallel, it is an *isoclinal fold* (Fig. 8.24).

In an *overturned fold* (Billings 1972, p. 50; Whitten 1966, p. 617) (Fig. 8.25), the dip of both the limbs is in the same direction and the axial surface is inclined in such a way that the upper limb overrides the lower one; as such, the beds in the lower limb are disposed in an inverted position, while those of the upper limb are in normal position. The overturned limb, called *forelimb*, is steeper than the axial plane, while the normal limb, called *backlimb*, is shallower than the axial plane. This is the geometry for an overturned anticline, while the opposite holds good for an overturned syncline. In a *reclined fold* (Fleuty 1964, Whitten 1966, p. 619) (Fig. 8.17), the axial plane dips between 10° and 80° and the hinge plunges down the dip of the axial plane. A *vertical fold* (Figs. 8.26 and 8.27) is one in which the plunge of the fold hinge and dip of axial surface are nearly vertical.

8.5.3 Classification Based on Plunge of Fold Axis

On the basis of the plunge of fold axes, folds can be classified (Fig. 8.28) into *horizontal fold* (plunge 0°–10°), *plunging fold* (plunge 10°–80°) and *vertical fold* (plunge 80°–90°).

Fig. 8.19 Upright fold in calc-schists of Delhi Supergroup, Rajasthan, India. (Photograph courtesy Professor T.K. Biswal)

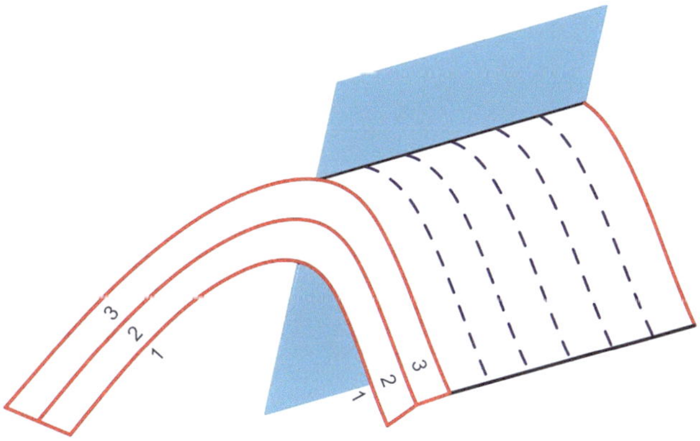

Fig. 8.20 Sketch of an inclined fold. Beds 1, 2 and 3 are from oldest to youngest

Fig. 8.21 Inclined fold in quartzitic rocks of the southern Appalachians, USA. (Photograph by the author)

8.5.4 Classification Based on Limb Curvature

Most folds show curved hinges and limbs. As such, any classification based on curvature (Fig. 8.29) is highly practical and can be easily used in field. An *anticline* (Fig. 8.29a) is a fold that is convex upward towards the centre and the strata dip opposite to each other; as such, the older rocks occur at the centre of curvature. A *syncline* (Fig. 8.29a) is a fold that is concave towards the centre and the strata dip towards the centre of the structure from both sides; as such, the younger rocks occur at the centre of curvature. The area located towards the inner, concave side of a fold is called the *core* of the fold. In most natural rocks, anticlines and synclines occur together, i.e. adjacent to each other (Fig. 8.29a, Fig. 1.4) or in succession showing alternation of a number of anticlines and synclines.

The terms anticline and syncline are used only when the relative age of the rocks at the core is known. However, there are many structures that look like anticlines and synclines, but the age of the rocks at their cores is not known. An *antiform* (Fig. 8.29b) is a fold that looks like an anticline, but the age of the rocks at the core is not necessarily older or is not known. In contrast, a *synform* (Fig. 8.29b) is a fold that looks like a syncline, but the age of the rocks at the core is not necessarily younger or is not known.

If the ages of the rocks of an antiform or of a synform are known, these structures can be named as antiformal syncline or a synformal anticline. In an *antiformal syncline*

(Fig. 8.29c), the strata dip away from the axis of the fold and the rocks at the core are younger; the structure can thus be described as a downward-facing syncline. In a *synformal anticline* (Fig. 8.29c), the strata dip towards the axis of the fold and the rocks at the core are older. A synformal anticline can thus be described as an upward-facing anticline in which it is necessary to know the younging direction, i.e. the direction along the axial plane in which the younger beds occur. It may be noted that anticlines and synclines are relatively simpler structures formed during deformation than antiforms, synforms, antiformal synclines or synformal anticlines. The latter four types are formed when the rocks are affected by several episodes of folding.

A *neutral fold* is one that closes laterally (Fig. 8.30a). A neutral fold can also be a vertical fold (Fig. 8.30b) that closes laterally.

8.5.5 Classification Based on Fold Closure

The continuity of the strike of the rock strata of a fold may close, and the dip of the rocks may be away from (anticline) or towards (syncline) the centre. A *dome* (Fig. 8.31a) is a special type of fold in which the strata dip away from the centre in all directions. In contrast, a *basin* (Fig. 8.31b), or a

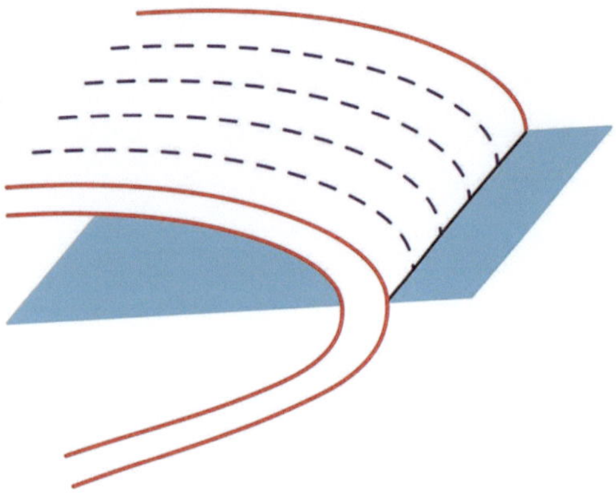

Fig. 8.22 Sketch of a recumbent fold

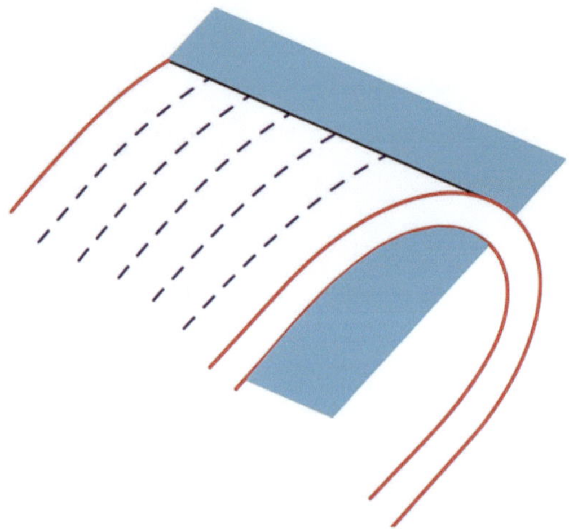

Fig. 8.24 Sketch of an isoclinal fold

Fig. 8.23 Recumbent fold developed in quartzitic rocks of Bundelkhand craton, central India. (Photograph by the author)

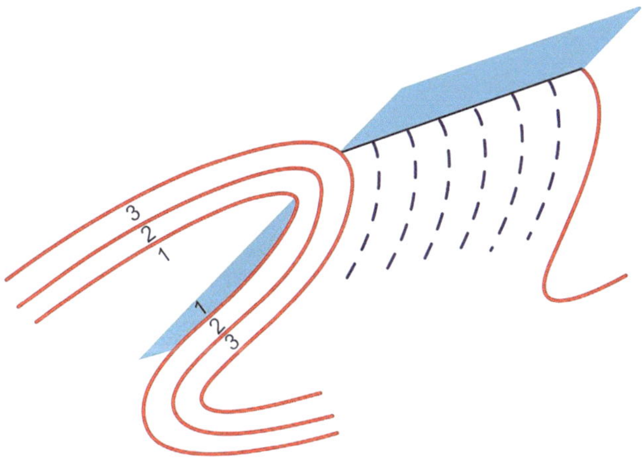

Fig. 8.25 Sketch of an overturned fold

Fig. 8.26 Sketch of a vertical fold

Fig. 8.27 Vertical folds developed in calc-schists of Delhi Supergroup, Rajasthan, India. (Photograph courtesy Prof. T.K. Biswal)

structural basin, is a special type of syncline in which the strata dip towards the centre in all directions. Commonly, the dome-and-basin structures are disposed in elongated forms. In such cases, these are called doubly plunging antiform and doubly plunging synform, respectively.

Sometimes, a dome and a basin close in both directions so that the resultant folds assume elliptical/semi-elliptical shapes in their plan view. Such folds are called a *pericline* (Fig. 8.32). A pericline with anticlinal strata showing outward plunge in two opposite directions is called a *doubly plunging anticline*, while a pericline with synclinal strata showing centrally directed plunge is called a *doubly plunging syncline*. Also, a pericline with anticlinal strata is called an *anticlinal pericline*, while that with synclinal strata is called a *synclinal pericline*.

The crest or trough of the folded surface may pass through different elevations (Fig. 8.33) above a reference horizontal plane. The point where the crest line or the trough line reaches the highest elevation is called *culmination*, and the point where the crest line or the trough line reaches the lowest elevation is called *depression*.

8.5.6 Classification Based on Dip of Folded Strata

The dip of strata of a fold may show a monotonous pattern or local steepening. A *homocline* (Fig. 8.34a) is a structure in which the strata dip persistently in one direction. In field, however, it is not always easy to assign a structure as homocline and the same structure can also be assigned to constitute a simple dipping structure. Detailed field studies sometimes reveal that a homocline is a part of a larger fold of which only one particular limb is exposed or studied, which shows strata dipping in one direction only. As such, most homoclines are sometimes considered to constitute the limb or part of a larger fold, which is not directly or easily visible. Apparently, therefore, a homocline does not show any folding. A *monocline* (Fig. 8.34b) is a local steepening of an otherwise uniformly dipping strata. The local steepening may be caused in horizontal strata or in (lowly to moderately) dipping strata. If a bed includes a horizontal part between two inclined portions, the structure is called a *structural terrace* (Fig. 8.34c).

8.5.7 Classification Based on Angularity of Hinge

Instead of showing curved limbs, some folds show straight limbs that join at hinge, which is angular. *Chevron folds* (Fig. 8.35) have straight limbs and angular hinges and thus look V-shaped. Natural chevron folds show an interlimb angle of about 60° (De Sitter 1958). Chevron folds are also

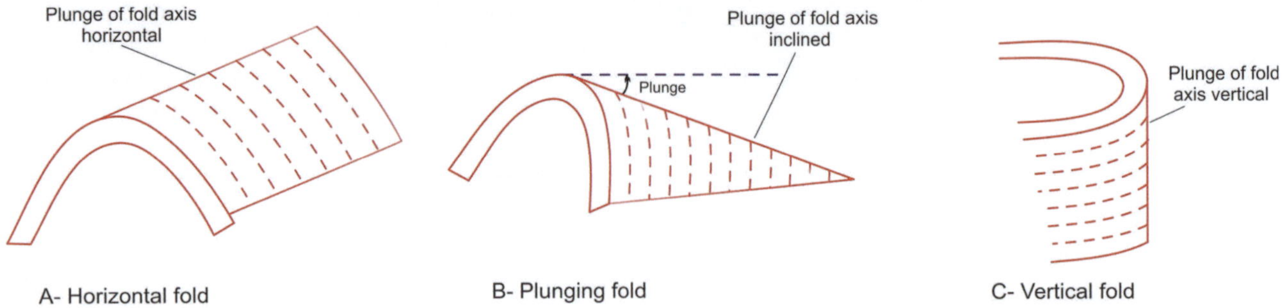

Fig. 8.28 Classification of folds based on the plunge of fold axis. (**a**) Horizontal fold with nearly horizontal plunge (0°–10°) of fold axis. (**b**) Plunging fold with plunge of 10°–80°. (**c**) Vertical fold with nearly vertical plunge (80°–90°)

known as *concertina folds* (Whitten 1966). *Kink folds* (Fig. 8.36) have also straight limbs and angular hinges, but the limbs are of unequal lengths, and therefore these are asymmetrical folds. The steplike smaller limbs of kink folds are called *kink bands*.

Chevron folds and kink folds occur in multilayers and are commonly developed in fine-grained, thin-layered rocks such as phyllites and schists. Chevron folds commonly form in rocks showing alternations of competent and incompetent layers such as sandstone and shale. The terms chevron fold and kink fold have no restriction of size, although most kink bands are indeed small. Chevron folds can be quite large, however, and most have wavelengths much greater than 10 cm (Professor Peter Hudleston, personal communication 2020). The terms chevron folds and kink folds are commonly used for folds in which the distance between the adjacent axial surfaces is up to about 10 cm; for larger folds, the terms *zigzag fold* or *knee fold* are used (Ramsay 1967). An *arrowhead fold* (Fig. 8.37) has an angular hinge and curved limbs.

8.5.8 Classification Based on Symmetry of Folds

A *symmetric fold* can have an axial plane with any dip, while an asymmetric fold is defined by the relative lengths of the fold limbs, without regard to the dip of the limbs (Professor Peter Hudleston, personal communication 2020). A symmetric fold is also called *M-fold* (Figs. 8.38a and 8.39a). Most folds in rocks are *asymmetric folds* that are further grouped as Z-fold and S-fold. A *Z-fold* (Figs. 8.38b and 8.39b) is an asymmetric, Z-shaped fold in which alternate limbs are nearly parallel, whereas an *S-fold* (Fig. 8.38C) is an asymmetric, S-shaped fold in which alternate limbs are parallel and hinge is curved.

8.5.9 Classification Based on Order of Folds

Many large-scale folds leave their imprints on smaller scales at their limbs (Fig. 8.40). The smaller folds may develop as several sets, and folds in each set bear similar geometry as that of the larger fold. The larger folds are called *first-order folds* that may be of regional scale, while those developed on the flanks or limbs are called *second-order folds*. The second-order folds may, in turn, develop another set of smaller folds at their limbs and are called *third-order folds*. In all such cases, *Pumpelly's rule* can be applied, which states that *the small-scale structures generally depict large-scale structures formed at the same time*. The larger fold together with the smaller folds of all sizes is called a *polyharmonic fold*. If the axes of the smaller folds are parallel/subparallel to that of the larger fold, they are called *congruous folds*; if not they are called *noncongruous folds*. In field, the larger folds are not commonly visible, and, instead, smaller folds are rather more easily visible. The Pumpelly's rule can thus help in unravelling the larger structures through the study of the smaller structures.

8.5.10 Cylindrical Folds

Folds are commonly studied in two dimensions in their profile sections because these are more commonly noticed in the field. Their actual form in three dimensions is a matter of interpretation or reconstruction. In three dimensions, folds are commonly grouped into two: cylindrical or cylindroidal folds and non-cylindrical or non-cylindroidal folds. A *cylindrical fold* (Fig. 8.41) shows a geometry of a cylinder that can be assumed to have formed when a line of fixed orientation, which is the fold axis, moves parallel to itself such that the fold surface thus formed contains a set of parallel lines. These

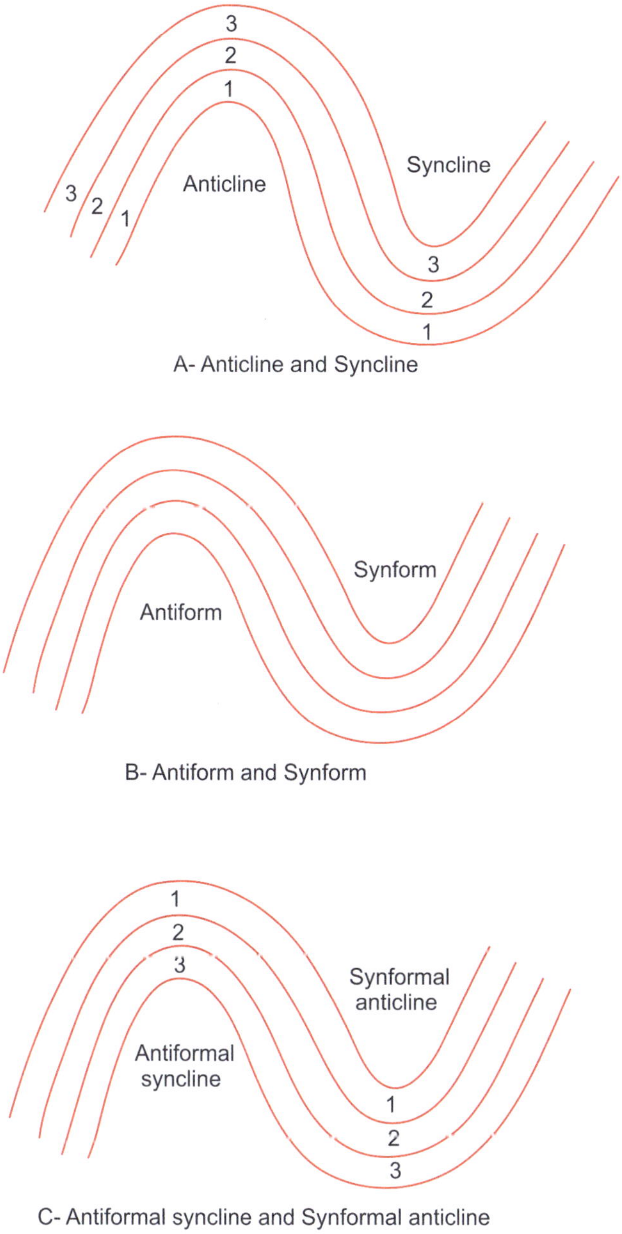

Fig. 8.29 Fold classification based on limb curvature. (**a**) Anticline and syncline formed by folding of beds 1 (oldest), 2 and 3 (youngest). At the centre of curvature, the anticline shows the oldest bed while the syncline shows the youngest bed. (**b**) Antiform and synform look like anticline and syncline, respectively, but the age of rocks at the centre of curvature is not known. (**c**) Antiformal syncline and synformal anticline. In an antiformal syncline, the strata dip away from the axis of the fold and the rocks at the centre are younger. In a synformal anticline, the strata dip towards the axis of the fold and the rocks at the centre are older

lines are actually the hinge lines of the fold at the respective points on the fold profile. Since the curvature of the folded surface thus generated is equal at each point, there can be an infinite number of rectilinear hinge lines that are parallel to each other. *A cylindrical fold is thus characterized by parallel hinge lines that are parallel to the fold axis.* Since cylindrical folds are formed under a set of ideal conditions, such folds are rare in nature. D.M. Ramsay and Sturt (1973) used the term *aberrant folds* for cylindrical folds that slightly deviate from the ideal cylindrical geometry.

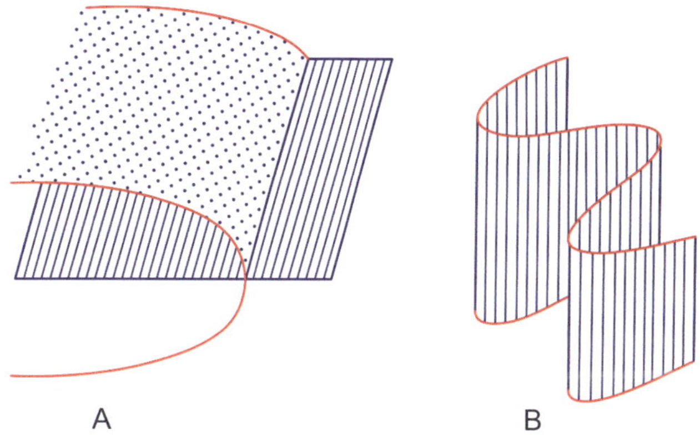

Fig. 8.30 (**a**) A neutral fold that closes laterally (with horizontal axial plane). (**b**) A neutral fold (with vertical axial plane)

Zulauf et al. (2017) suggested that the geometrical parameters of 2D cylindrical folds can be expressed in the same way as for 2D folds and include the following parameters: fold arc length, L; amplitude, A; and wavelength, λ.

Most folds, however, do not show cylindrical geometry and are thus called *non-cylindrical folds*. In rocks, non-cylindrical folds are more common. In a non-cylindrical fold, the hinge lines on a fold surface are curved and are therefore not parallel to each other (Fig. 8.42) as well as to the fold axis. According to Zulauf et al. (2017, p. 128), the dome-and-basin geometry is present in most types of non-cylindrical folds.

A *conical fold* (Fig. 8.43) is a non-cylindrical fold that can be generated by rotating a line in a circular path from one end while keeping the other end fixed. The structure thus generated is cone shaped. The original position of the line forms the fold axis, while the fixed end forms the apex of the conical fold. The fold surface thus formed always maintains a fixed angle, and the fold hinges from the fold surfaces thus formed converge together at a point. A conical fold terminates along its trend (Groshong 2006, p. 111). Conical folds are relatively rare structures. Commonly, they are asymmetrical (Fig. 8.44) as a result of later disturbances.

Recently, Welker et al. (2019) have pointed out that conical folds, if they exist, terminate at a point; the amplitude-to-width ratio and the plunge of the crestal line must remain constant towards the fold terminus. Considering these geometrical attributes, the conical folds, according to the authors,

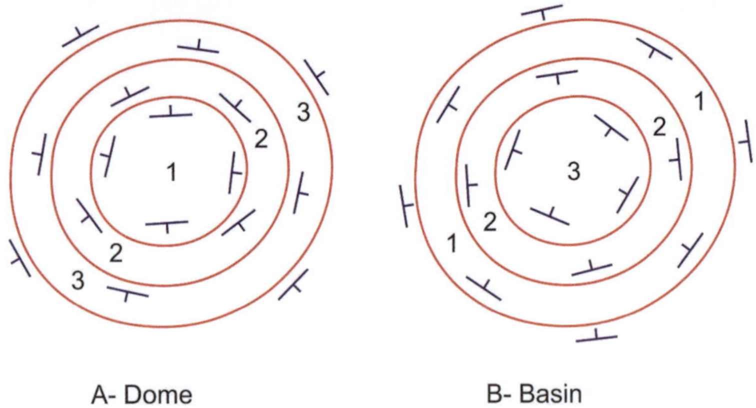

Fig. 8.31 Classification of folds based on fold closure. A dome (**a**) and a basin (**b**) as viewed on a plan view. Bed 1 is the oldest, while 3 is the youngest. Note that in a dome the beds dip away from the centre in all directions so that the older beds occur at the centre. In a basin, the beds dip towards the centre so that the younger beds occur at the centre

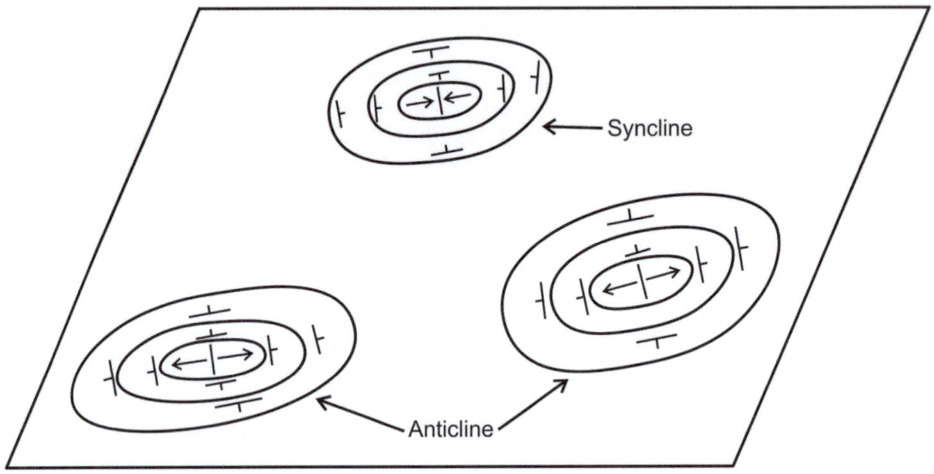

Fig. 8.32 Diagrammatic plan view of pericline. The area exposes two anticlines and one syncline; all are of elliptical shapes. Note the strata in each structure close in all directions. Each of these structures may be considered an elliptical basin or an elliptical dome as the case may be. With anticline, the structure is called an anticlinal pericline, while with syncline, it is synclinal pericline

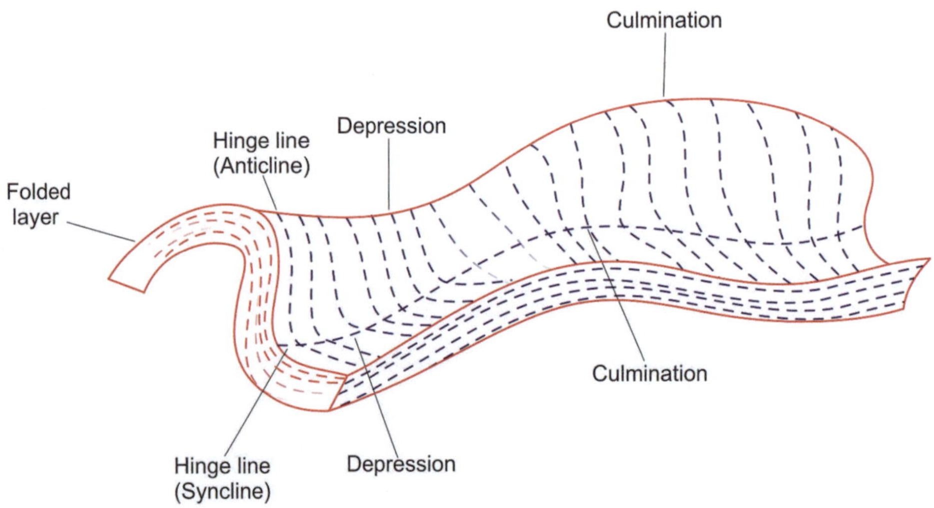

Fig. 8.33 Culmination and depression as related to an anticline and a syncline

represent an unrealistic geometrical form of natural folds, and as such these folds cannot be represented properly by sections of cones. So, there is a word of caution before properly identifying a conical fold.

8.5.11 Fold Systems

Folds occur either as a single structure or in groups. Folds occurring in a group constitute a *fold system.* A *fold train* is formed when a number of folds occur adjacent to one

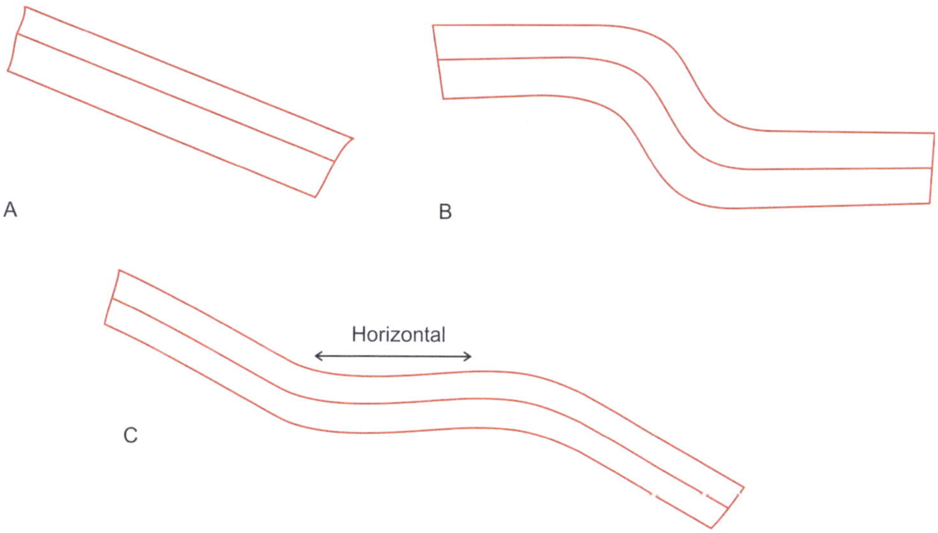

Fig. 8.34 Classification of folds based on the dip of folded strata. (**a**) Homocline. (**b**) Monocline. (**c**) Structural terrace. All folds are viewed on a vertical plane

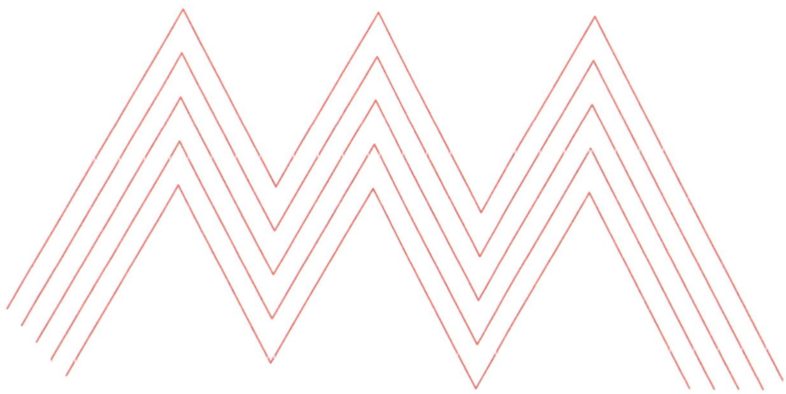

Fig. 8.35 Chevron folds in profile sections. The limbs are straight and equal in length, and hinge is angular

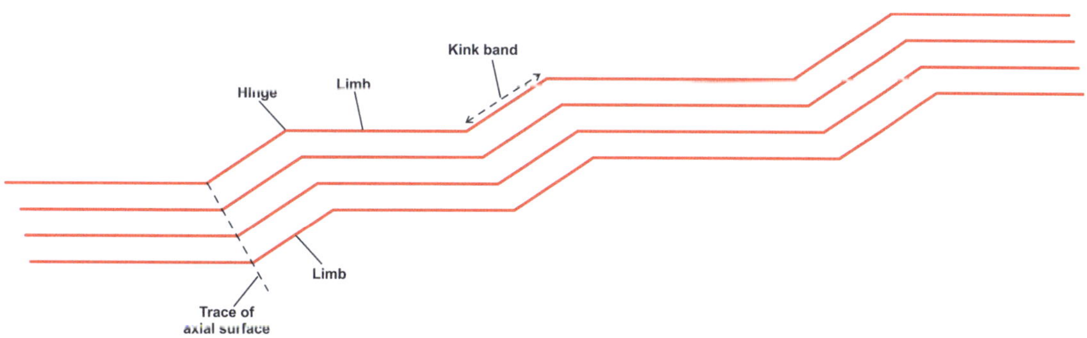

Fig. 8.36 Kink folds in profile sections. The limbs are straight and unequal in length, and hinge is angular. The steplike smaller limbs are called kink bands. The traces of axial surfaces are nearly parallel to each other

another; the individual folds may show the same or different geometry. Some common types of fold systems are described below.

An *anticlinorium* (Fig. 8.45) is a larger anticline that includes a number of smaller folds (anticlines and synclines) on its limbs. Likewise, a *synclinorium* (Fig. 8.45) is a larger syncline containing a number of smaller folds (synclines and anticlines) on its limbs. The larger folds are often called *first-order folds*, while those on the flanks are *second-order folds*. *Harmonic folds* (Figs. 8.46 and 8.47), commonly developed in multilayered rocks, are those in which the fold shape does not change within one lithic unit and is repeated a few times along the axial region, i.e. at every half-wavelength. In cross section, harmonic folds show layers nearly parallel to one

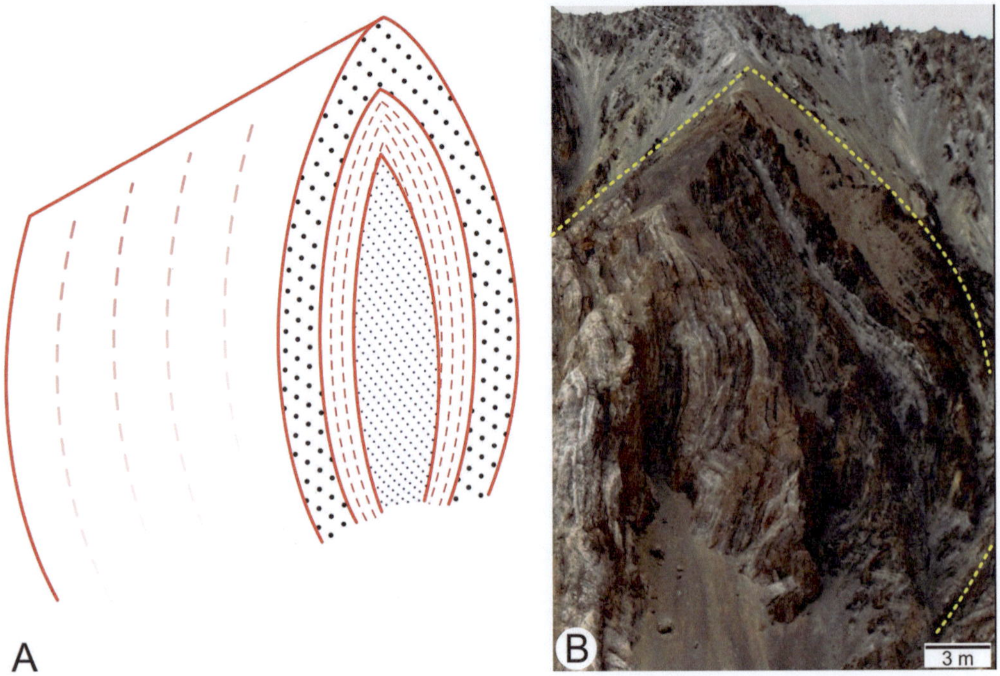

Fig. 8.37 (**a**) Sketch of arrowhead folds. Note that the hinge is angular and the limbs are curved. (**b**) Photograph of arrowhead folds developed in Canadian Rockies. The right-hand limb is curved, but the left-hand limb is affected by later disturbances. (Photograph courtesy Narendra K. Verma)

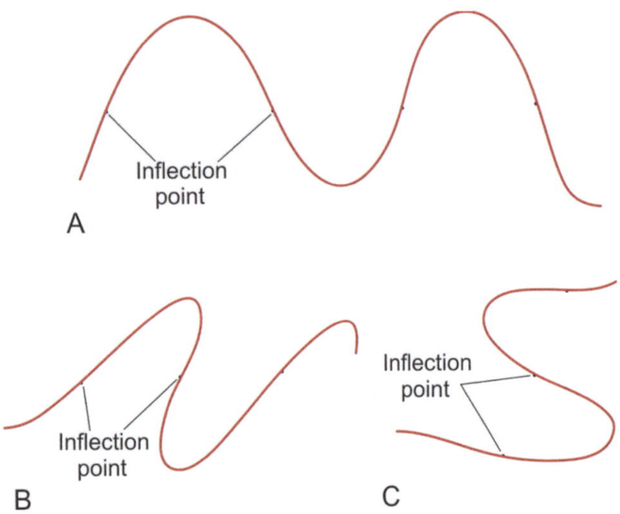

Fig. 8.38 Classification of folds based on symmetry. (**a**) M-fold. This is a symmetrical fold, (**b**) Z-fold. This is an asymmetric fold with clockwise asymmetry. (**c**) S-fold. This is also an asymmetric fold with anticlockwise asymmetry

another (Groshong 2006). *Disharmonic folds* (Figs. 8.48 and 8.49), on the other hand, are those in which the different layers show different fold shapes. Disharmonic folds do not show any systematic geometric relations with the adjoining folds. Individual folds die out shortly, generally within one or two half-wavelengths. Disharmonic folds generally form when the layers behave differently to the compressive stresses due to varying competence and physical/rheological properties of the layers. Disharmonic folds are relatively more common than harmonic folds and can be of varying sizes. *Ptygmatic folds* (Figs. 8.50 and 8.51) occur as veins of quartzo-feldspathic material in high-grade metamorphic rocks and in migmatites. The individual folds show irregular geometry with respect to each other, while the thickness of the folded layer generally remains uniform. The folds are tight with their amplitudes much greater than the wavelength. Ptygmatic folds are generally formed in a competent layer occurring in a more ductile host material. *Parasitic folds* (Figs. 8.52 and 8.53) are smaller folds developed in the limbs or hinge of a fold. These folds are developed during the formation of the main fold. However, these may also form in some later event when folding is coaxial with the earlier developed fold. The geometry of the parasitic folds bears resemblance with that of the host fold. In cases when the larger fold is not seen or exposed, the parasitic folds help in unravelling the geometry of the larger folds, in accordance with Pumpelly's rule, as stated above.

8.5.12 Folds Showing Two or More Axial Planes

The geometry of some folds is such that the fold shows two or more axial planes. A *conjugate fold* is a combination of two asymmetric folds in which the axial planes dip in opposite directions and thus converge. Conjugate folds show two different geometries (Ramsay 1967, p. 358): (a) *conjugate*

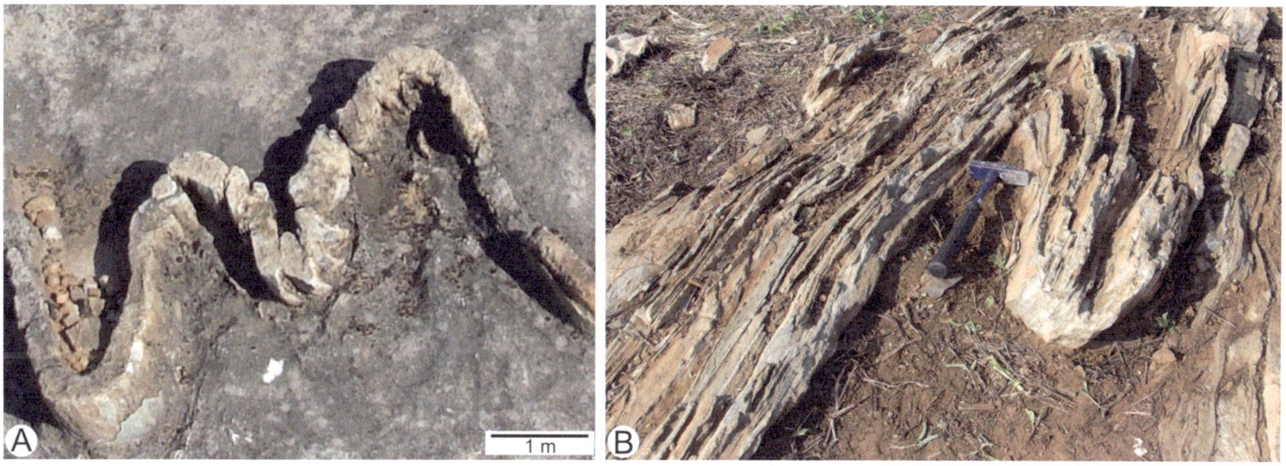

Fig. 8.39 (**a**) M-fold developed in a sandstone layer. (**b**) Z-fold developed in strongly layered calcareous rocks. Both the photos are from Aravalli Supergroup of Rajasthan, India. (Photographs courtesy Prof. T.K. Biswal)

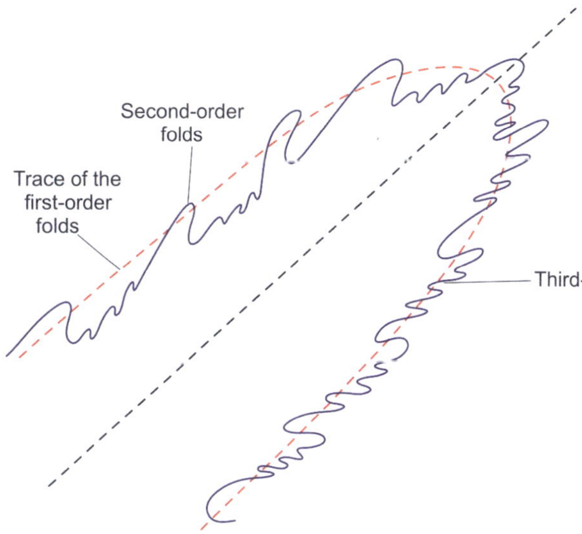

Fig. 8.40 Order of folds. The trace of a large (first order) fold has been shown (by red dashes). On its limbs, two more sets of folds—second order and third order—are developed. The second-order folds here are Z, and third-order are S

kink folds (Fig. 8.54) in which hinge zones are angular and fold limbs are straight formed by the development of kink bands and (b) *box folds* (Fig. 8.55) in which the interlimb angles of the two folds meet at nearly 90° to each other and the hinge zones are rounded. Box fold geometry is commonly shown by moderately thick-layered rocks (Fig. 8.56), while conjugate kink fold geometry is shown by thin-layered, fine-grained rocks.

A *polyclinal fold* (Turner and Weiss 1963, p. 115; Whitten 1966, p. 618) is a group of adjacent folds with subparallel hinge lines, and the fold set consists of more than two axial surfaces with different orientations (Figs. 8.57 and 8.58).

8.5.13 Classification Based on Layer Thickness

The thickness of a folded layer may remain constant or may vary in different ways. On the basis of layer thickness, natural folds commonly show three different geometries: parallel, similar and supratenuous. *Parallel folds* (Figs. 8.59a and 8.60a) are those in which the layer thickness remains constant. In *similar folds* (Figs. 8.59b and 8.60b), each fold surface along an axial plane has the same shape. The curvature of the layers thus remains the same. The thickness of a layer as measured in profile section is maximum in the hinge zone and reduces on the limbs. The geometry of parallel and similar folds is further described below under Ramsay's classification.

In a *supratenuous fold* (Figs. 8.59c and 8.60c), the hinge is thin while the limbs are thicker. Folds of this type are commonly developed when sedimentation and folding take place together, and there is some resistance at the hinge zone during formation of the fold. Supratenuous folds are very rare in natural rocks. Known reports are mostly from marine sediments in seismic sections.

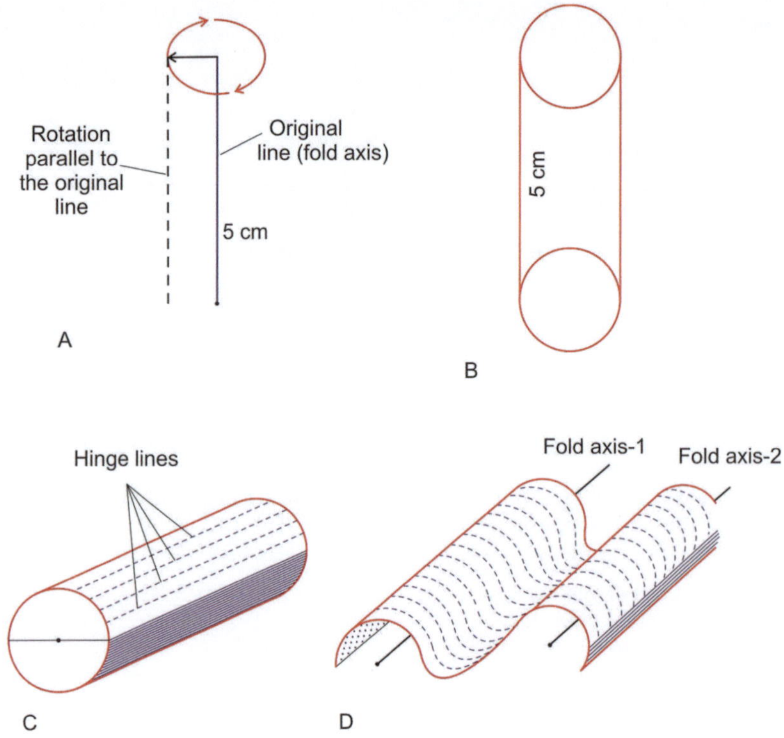

Fig. 8.41 A cylindrical fold shows the geometry of a fold in three dimensions as explained diagrammatically. Rotation of a line around itself (**a**) produces a cylinder-like shape (**b**), which can be considered to constitute a cylindrical fold (**c**). (**d**) Two cylindrical folds occur close to each other. Note that in such folds all hinge lines are parallel to each other as well as to the fold axis

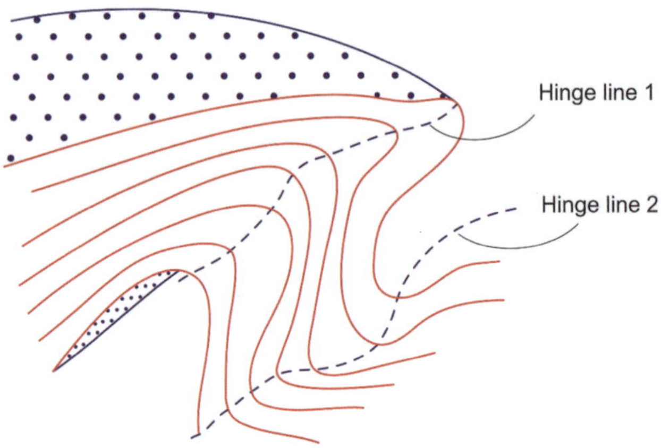

Fig. 8.42 2D diagrammatic representation of a non-cylindrical fold. Note that the hinge lines of the anticlines and the synclines are neither rectilinear nor parallel to each other

8.5.14 Ramsay's Classification

Fold classification based on the behaviour of layering has been found to be highly useful by many workers. Ramsay (1967) proposed a fold classification that relates the layer thickness of a fold with the angle of dip of the folded surface as observed in a profile section. His classification is geometric in nature and gives a complete description of the layering pattern of folds. This classification cannot be directly applied in field. So, either a photograph of the fold is taken or the fold is to be collected from field and then cut along the profile section such that a sketch or photograph is obtained. For the sake of classification, the profile section, i.e. section cut perpendicular to the hinge, is studied while the axial plane remains vertical.

Ramsay's classification takes into account three important parameters of a fold, viz. dip isogons, orthogonal thickness and thickness parallel to fold axial surface.

(a) **Dip isogons** are lines joining the points of equal dip of two successive layers (Fig. 8.61). The dip or slope of a layer is obtained by drawing a tangent at any point on the folded surface. The line drawn parallel to this tangent on the next layer will give a point on the fold surface where the amount of dip is same as that of the other layer. If such tangents are continued to be drawn for other layers of a multilayer fold, one gets a continuous line across the fold. This line will constitute a dip isogon of the value equal to the amount of dip thus obtained.

(b) **Orthogonal thickness** (t_α) of a fold is the perpendicular distance between two tangents on the successive layers where the amount of dip (α) is equal (Fig. 8.62). The parameter t_α can be utilized to give the proportional change in the layer thickness with variation in α by identifying another parameter $t'_\alpha = t_\alpha / t_0$, where t_0 is the layer thickness at hinge (where $t_0 = T_0$).

(c) **Axial surface thickness** (T_α) is the distance between two successive tangents as measured parallel to the axial

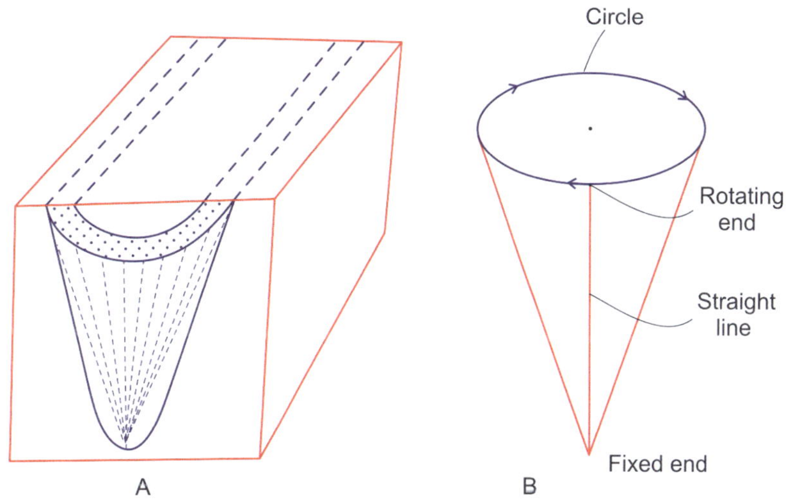

Fig. 8.43 A conical fold can be assumed to form when a line is rotated at one end keeping the other end fixed (**a**). The surface thus generated resembles a cone (**b**)

Fig. 8.44 A conical fold developed in calc-argillaceous rocks of Delhi Supergroup, Rajasthan, India. The structure is slightly asymmetrical due to later disturbances. Diary at the centre of the photograph measures 30 cm. (Photograph courtesy Professor T.K. Biswal)

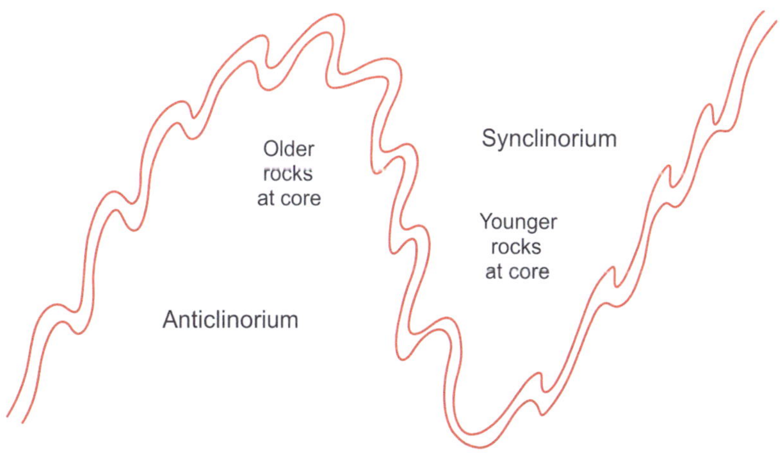

Fig. 8.45 An anticlinorium and a synclinorium, together with the smaller folds on the limbs, constitute a fold system

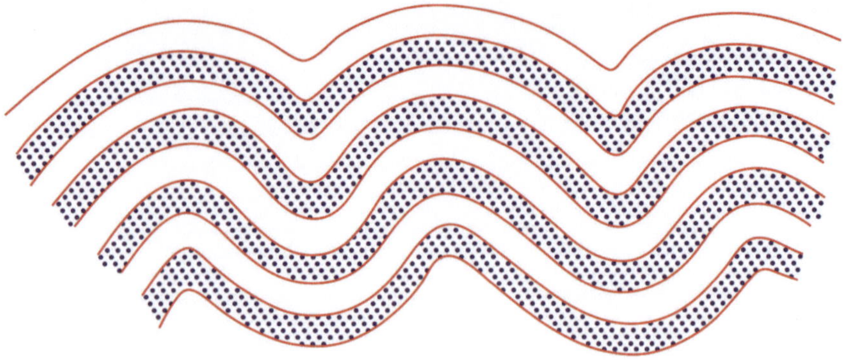

Fig. 8.46 Sketch of harmonic folds

Fig. 8.47 Harmonic folds developed in calc-argillaceous rocks of Delhi Supergroup, Rajasthan, India. (Photograph courtesy Professor T.K. Biswal)

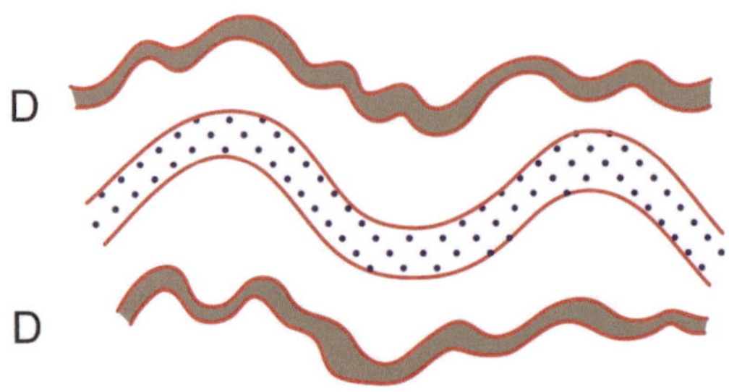

Fig. 8.48 Sketch of disharmonic folds (D) in section

surface of the fold (Fig. 8.62). Further, from Fig. 8.62, the two parameters t_α and T_α are related by

$$t_\alpha = T_\alpha \cos\alpha \tag{8.3}$$

The proportional variation of this distance at different parts of the fold is expressed as $T'_\alpha = T_\alpha / T_0$, where T_0 is the axial surface thickness at hinge. Since at hinge $t_0 = T_0$, Eq. (8.3) changes to

$$t'_\alpha = T'_\alpha \cos\alpha \tag{8.4}$$

Since the above three geometric parameters are dependent, it is therefore possible to clearly explain the nature of variation of the layer inclination α from hinge to limb from these parameters and thus to define the style of the folded layer. With these, Ramsay (1967) proposed three fundamental classes of folds (Figs. 8.63 and 8.64) as described below:

Class 1 folds: Folds in which the dip isogons are convergent towards the inner arc and thus the curvature of the inner

Fig. 8.49 Field photograph of disharmonic folds. (Photograph by the author)

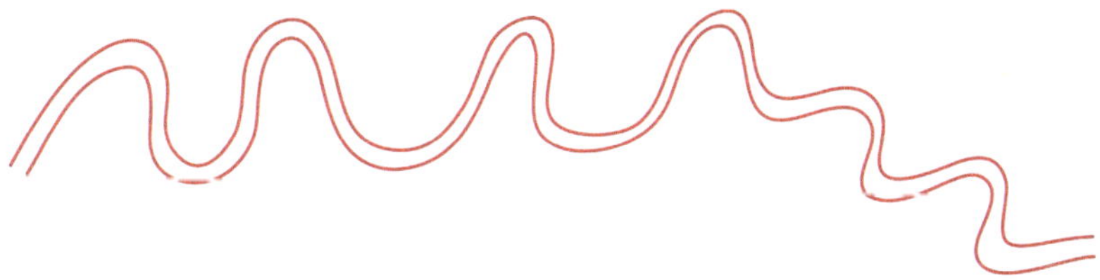

Fig. 8.50 Sketch of ptygmatic folds

Fig. 8.51 Ptygmatic folds developed in calc-argillaceous rocks of Delhi Supergroup, Rajasthan, India. (Photograph courtesy Professor T.K. Biswal)

arc always exceeds that of the outer arc (Figs. 8.63 and 8.64).

Class 2 folds: Folds in which the dip isogons are parallel to the axial surface and thus the curvature of the inner arc is equal to that of the outer arc (Figs. 8.63 and 8.64).

Class 3 folds: Folds in which the dip isogons are divergent towards the inner arc and thus the curvature of the inner arc is always less than that of the outer arc (Figs. 8.63 and 8.64).

Class 1 folds are further subdivided into three subclasses: *Class I A folds*: These folds have strongly convergent dip isogons. The orthogonal thickness of the layers (t_α) is more than the hinge thickness, i.e. $t'_\alpha > 1$. These folds have thinned hinges. *Class I B folds*: These are parallel folds with convergent dip isogons. The orthogonal thickness of the layers always remains constant, i.e. $t'_\alpha = 1$. The layer thickness thus remains constant all through the fold. *Class I C folds*: These folds have weakly convergent dip isogons. The orthogonal thickness of the layers (t_α) in the limbs is always less than the hinge thickness, i.e. $t'_\alpha < 1$. The hinge thickness is more than the limb thickness.

Class 1A folds with thinned hinges are rare in nature and may include *supratenuous folds*. *Class 1B* is a very common type of fold and includes parallel folds. *Class 1C folds* are also common types and could be considered as *modified parallel folds*. *Class 2 folds* have identical shapes of the inner and outer arcs and include the ideal *similar folds*. *Class 3 folds* have much thicker hinges and thinner limbs. These folds are also very common in nature and are considered to have developed by superposition of homogeneous strain on parallel folds during progressive deformation.

Since the above classification is based on geometric features of folds, it is easy to apply. Further, the classification satisfactorily depicts the pattern of variation of the folded layer and thus helps in revealing the geometry of folds in a lucid manner. Although the classification covers a vast range of folds, there are many complex types of folds whose geometry cannot be worked out precisely by the above methods. Such complex folds commonly occur in terrains with highly deformed rocks. However, as pointed out by Ramsay (1967,

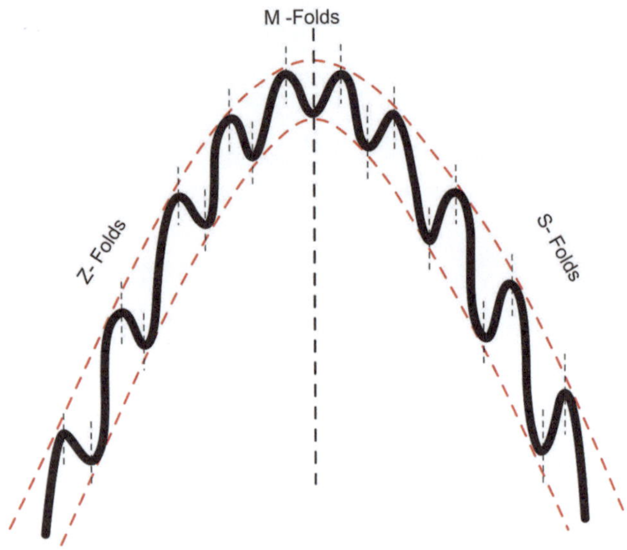

Fig. 8.52 Sketch of parasitic folds

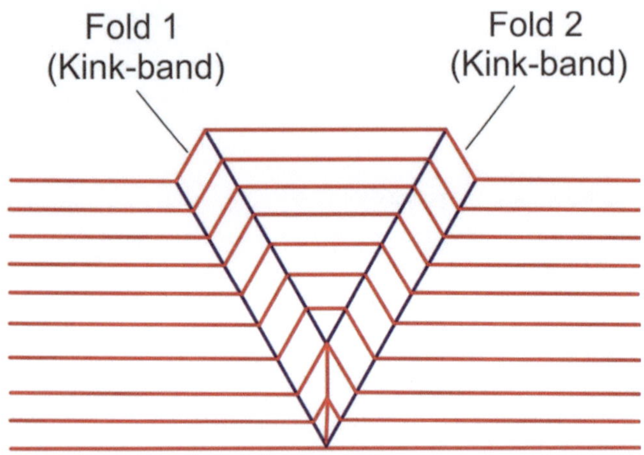

Fig. 8.54 Conjugate kink fold with angular hinge zones and straight fold limbs formed by the development of kink bands

Fig. 8.53 Parasitic folds well developed in one limb of a fold. (Photograph courtesy Professor T.K. Biswal)

p. 369), 'more complex fold shapes may be found that do not fit simply into the broad classification so far established. Although these complex folds are probably very rare, the classification can be expanded to include them'.

8.5.15 Hudleston's Classification

Hudleston (1973a) proposed a classification on the basis of the shape of folds. According to him, since most folds occur in the form of a curve, it is possible to apply simple harmonic analysis or Fourier transformation to study the shape of folds.

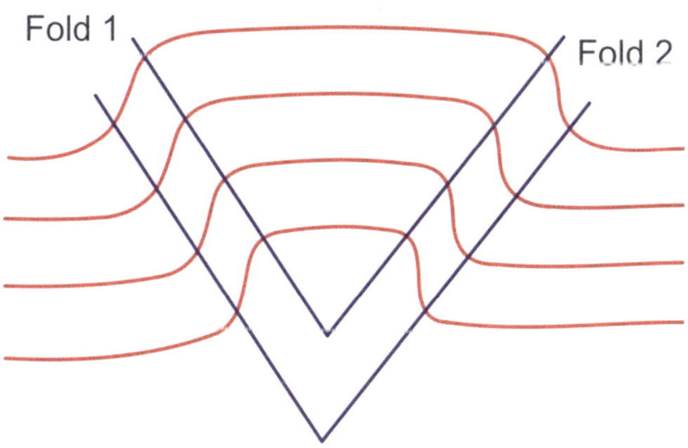

Fig. 8.55 Box folds formed by two folds whose interlimb angles meet at nearly 90° to each other and the hinge zones are rounded

Applying visual harmonic analysis, Hudleston identified 30 ideal fold forms (Fig. 8.65). On the basis of 'shape', all folds fall in six categories (A–F), while on the basis of 'amplitude', the folds fall in five categories (1–5). In the classification, box fold and chevron fold constitute two extremes within which a wide variety of folds are represented.

8.5.16 Classification Based on Axial Angle

Bhattacharya (1992, 2005) proposed a simple classification of folds involving the geometry of the folded layer in profile section. It is mainly based on two new geometric parameters (Bhattacharya 1992): *axial angle* (α) and *thickness ratio* (R)

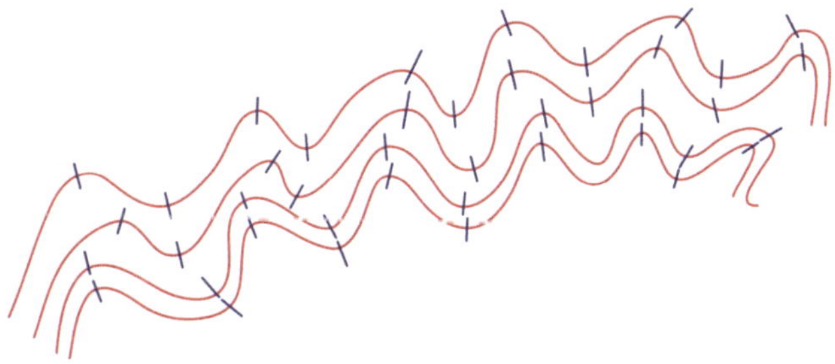

Fig. 8.57 Polyclinal folds in which the fold domain shows more than two axial planes. See text for details

Fig. 8.56 Photograph showing box folds developed in arenaceous rocks in the Tethyan sequence of the Himalayas near Leh, Ladakh, India. The fold is asymmetrical due to later deformations. The fold domain shows two axial planes at nearly perpendicular to each other, and the hinge zone of each fold is rounded. Note that the beds of the right-hand limb have been steepened due to the intervention of a fault. (Photograph courtesy Narendra K. Verma)

Fig. 8.58 Polyclinal folds developed in calcareous beds alternating with sandstone of Delhi Supergroup, Rajasthan, India. Axial surfaces of the folds are shown (yellow lines). (Photograph courtesy Prof. T.K. Biswal)

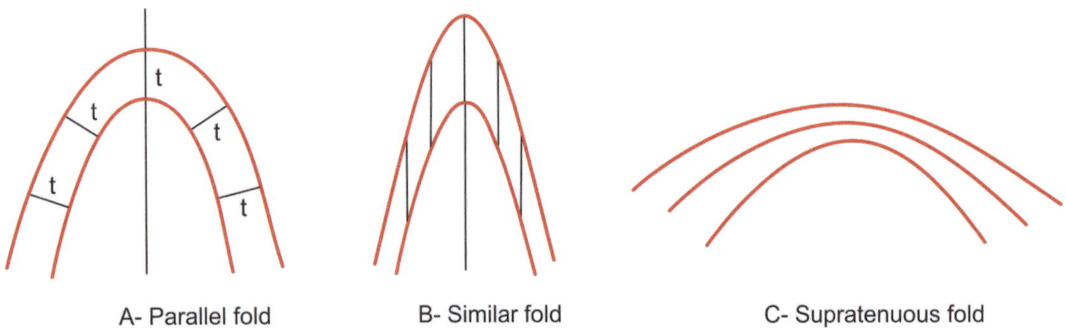

Fig. 8.59 Fold types according to layer thickness: (**a**) parallel fold, (**b**) similar fold, (**c**) supratenuous fold

(Fig. 8.66). Axial angle is the angle for a given (inner or outer) trace or arc of a fold as subtended at its apex (hinge). For a parallel fold, this angle would be of equal value for both the arcs of a fold while for a fold with thickened hinge, on the other hand, this angle would have different values. We can thus identify an inner axial angle (α_i) and an outer axial angle (α_o). In Fig. 8.66, DEF (α_i) and ABC (α_o) are the axial angles for the inner and outer arcs of the fold, respectively, such that a mean value (α) can be considered to represent the axial angle of the fold in question and is given by $\alpha = (\alpha_i + \alpha_o)/2$. The other geometric parameter is thickness ratio (R), i.e. the ratio of hinge thickness (T_h) and limb thickness (T_l). Hinge thickness (T_h) is the orthogonal thickness of the fold along the hinge, i.e. along the axial line (BE in Fig. 8.66), while limb thickness (T_l) is the orthogonal thickness of the limb at their lowest (inflection) points. In Fig. 8.66, A represents the lowest inflection point of the outer curve. A tangent is drawn at A. Another tangent parallel to this one is traced up to the inner curve and meets at D. If this thickness is not equal for both the limbs, a mean thickness may be considered that would represent the limb thickness (T_l) for this fold.

With the above geometric parameters, the single-layer, symmetric folds can be classified into four types (Fig. 8.67): *thickened fold*, *parallel fold*, *supratenuous fold* and *irregular fold* (the term irregular as used here is specific for this classification only, and has no reference to any other fold of this name in the literature). A *thickened* (or *T*) *fold* (Bhattacharya 1992) is one that shows thickening at its hinge. Assuming initiation of folding in a competent layer due to buckling, a thickened fold is commonly believed to form due to superimposition of homogeneous ductile strain (see Ramsay 1962, 1967; Hudleston 1973a; Bhattacharya and

Fig. 8.60 Field photographs. (**a**) Parallel fold. (**b**) Similar-type fold. (**c**) Supratenuous fold. (Photographs: A: Courtesy Professor T.K. Biswal, B: Author, C: Courtesy Gautam K. Dinkar)

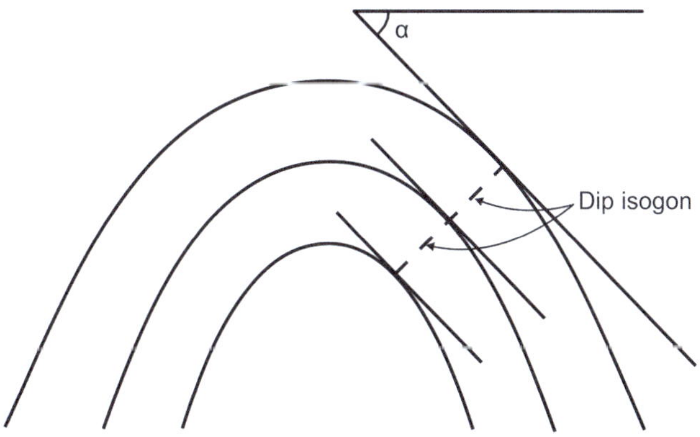

Fig. 8.61 Dip isogons. See text

Siawal 1985; Bhattacharya 1987, 1992; Lan and Hudleston 1996). For such fold, the value of thickness ratio (R) is more than unity and the outer axial angle (α_o) is less than the inner axial angle (α_i), i.e. $\alpha_o < \alpha_i$. A *parallel fold* maintains a uniform layer thickness throughout, and thus the value of R remains unity, i.e. $R = 1$, and both the axial angles (α_o and α_i) are of equal value, i.e. $\alpha_o = \alpha_i$. A *supratenuous fold* is one in which the hinge thickness is less than the limb thickness, and as such the value of R is less than unity, i.e. $R < 1$, and the outer axial angle (α_o) is greater than the inner one (α_i), i.e. $\alpha_o > \alpha_i$. An *irregular fold* does not show any specific values of its parameters and is characterized by (i) irregular shapes, and (ii) zone of thickening and thinning occur irregularly in the fold. As such, the irregular folds do not show any systematic relationship between T_h and T_l as well as between α_o and α_i. Because the first three types of folds (thickened, parallel and supratenuous) show well-defined geometric relations between their parameters, these can be considered here as regular folds in contrast to the irregular folds. For the sake of simplicity, all the above-mentioned folds can be designated as *T*-, *P*-, *S*- and *I*-folds, respectively.

The regular folds also show some definite relationships between α_o and α_i as shown in the graph in Fig. 8.68. All parallel folds fall on the 45° line, all *T*-folds in the field between the 45° line and *x*-axis while all *S*-folds in the field between the 45° line and *y*-axis.

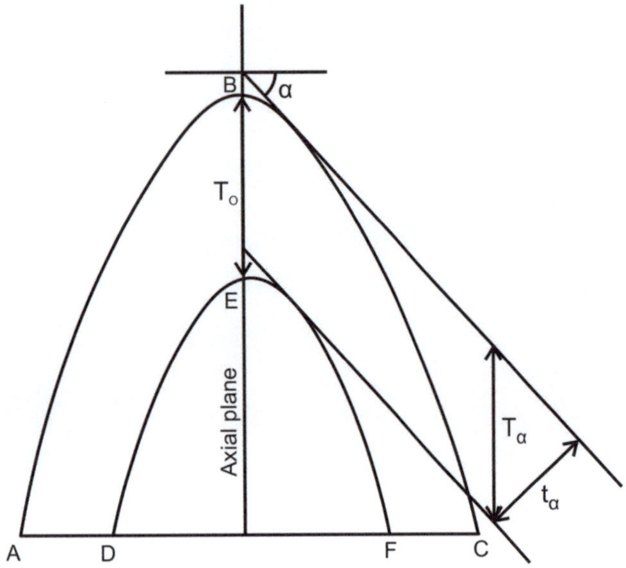

Fig. 8.62 Orthogonal thickness (t_α) and axial surface thickness (T_α)

8.6 Special Types of Folds

In this section, we describe some folds that could not be classified under any of the above schemes.

1. **Sheath fold** (Figs. 8.69 and 8.70a) is commonly developed in ductile shear zones at high strain levels. These are non-cylindrical folds (Fig. 8.69) formed when the early-developed folds rotate during progressive simple shear. Due to rotation, the folds thus developed look elliptical or semi-elliptical in cross section and extend within the rock as tubes (Fig. 8.69) that are parallel to the direction of slip during shear movement. Depending upon ductility, sheath folds may develop in any type of rocks. Many sheath folds have earlier been described as *eyed folds* (Fig. 8.70b).
2. **Intrafolial fold** (Fig. 8.71) is a tight to isoclinal fold with thinned, detached limbs. Such folds occur in ductile shear zones where they form due to intense shear deformation that causes transposition of the limbs.
3. **Rootless fold**: Mostly, the term *rootless fold* is applied to folds in sedimentary strata that pass directly down into unfolded beds, with the transition being abrupt at a zone of detachment, and as such the term detachment fold has become popular for this type of fold (Professor Peter Hudleston, personal communication 2020).

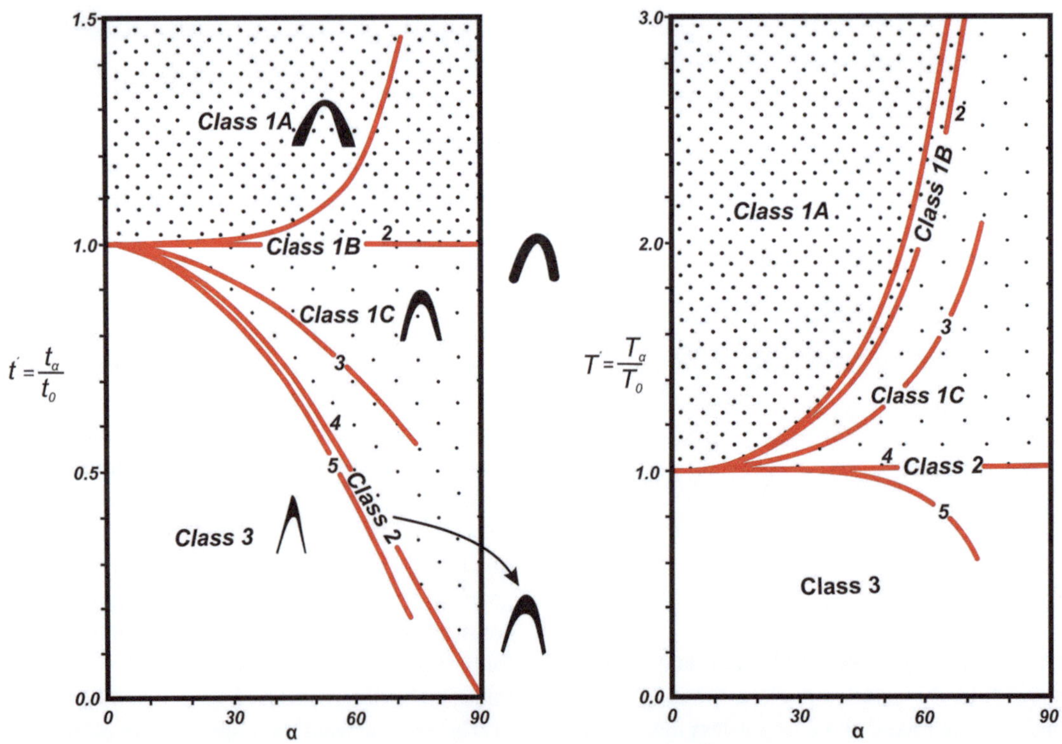

Fig. 8.63 Main types of fold classes in the plot of orthogonal thickness (t'_α) against the angle of dip (α). (Reproduced from Ramsay and Huber 1987, Fig. 17.3 with permission from Elsevier Copyrights Coordinator, Edlington, U.K. Submission ID: 1199999)

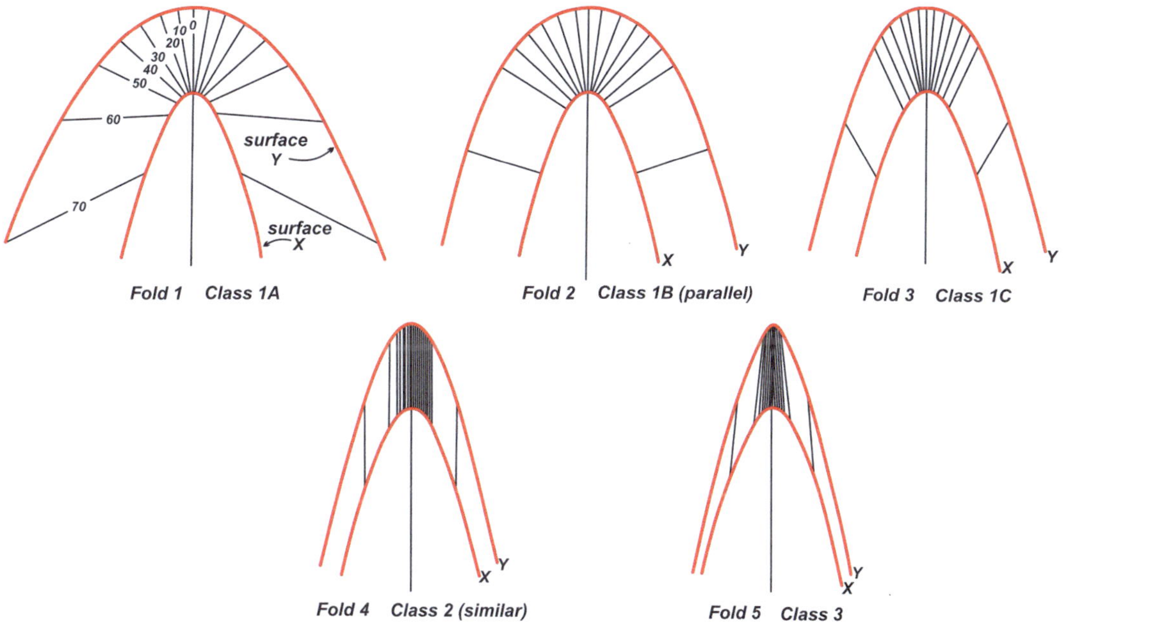

Fig. 8.64 The main fold classes based on dip isogon characteristics. (Reproduced from Ramsay and Huber 1987, Fig. 17.6 with permission from Elsevier Copyrights Coordinator, Edlington, U.K. Submission ID: 1200000)

Fig. 8.65 Hudleston's classification of folds. See text for details. (Reproduced from Hudleston 1973a, Fig. 12 with permission from Elsevier Senior Copyrights Coordinator, Edlington, U.K. Submission ID: 1193118)

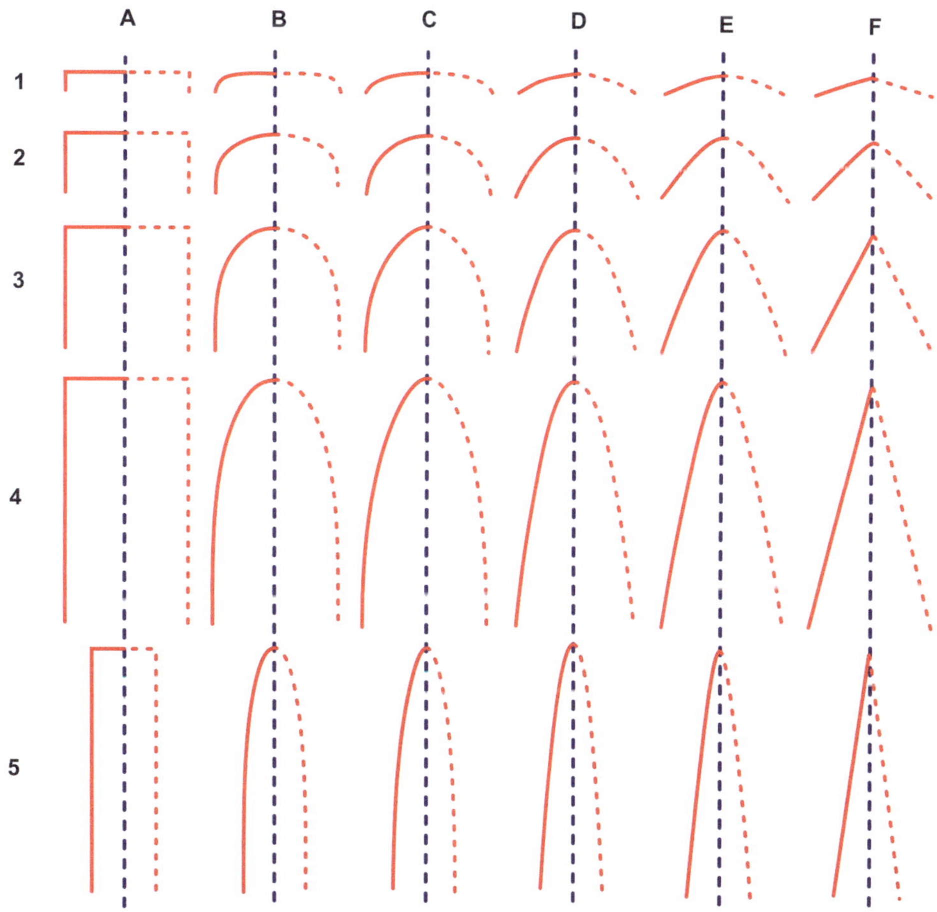

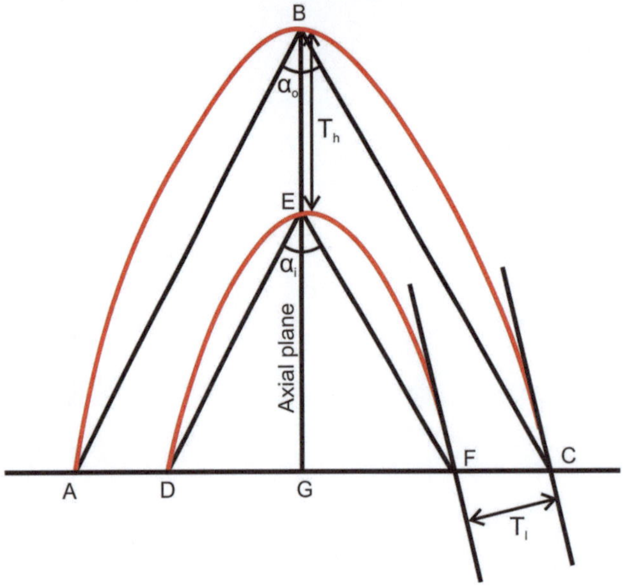

Fig. 8.66 Axial angle and thickness ratio of a fold. See text for details

Fig. 8.67 Classification of folds based on axial angle. See text for details. (Reproduced, with minor modification, from Bhattacharya 2005, Fig. 3 with permission from Japan Society of Geoinformatics)

Fold Type	Thickness Ratio $R = T_h/T_l$	Axial Angles (α_o - outer arc α_i - inner arc)	Fold Profile
T - FOLD (Thickened Fold)	R > 1	$\alpha_o < \alpha_i$	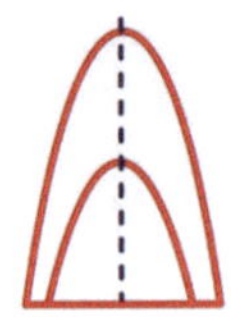
P - FOLD (Parallel Fold)	R = 1	$\alpha_o = \alpha_i$	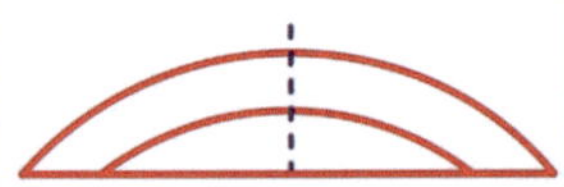
S - FOLD (Supratenuous Fold)	R < 1	$\alpha_o > \alpha_i$	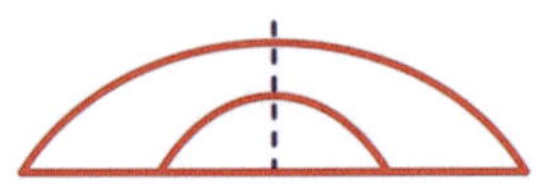
Irregular Fold	Variable Values	Variable Values	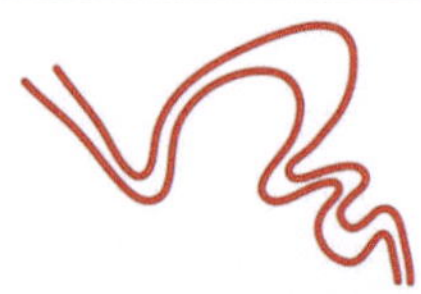

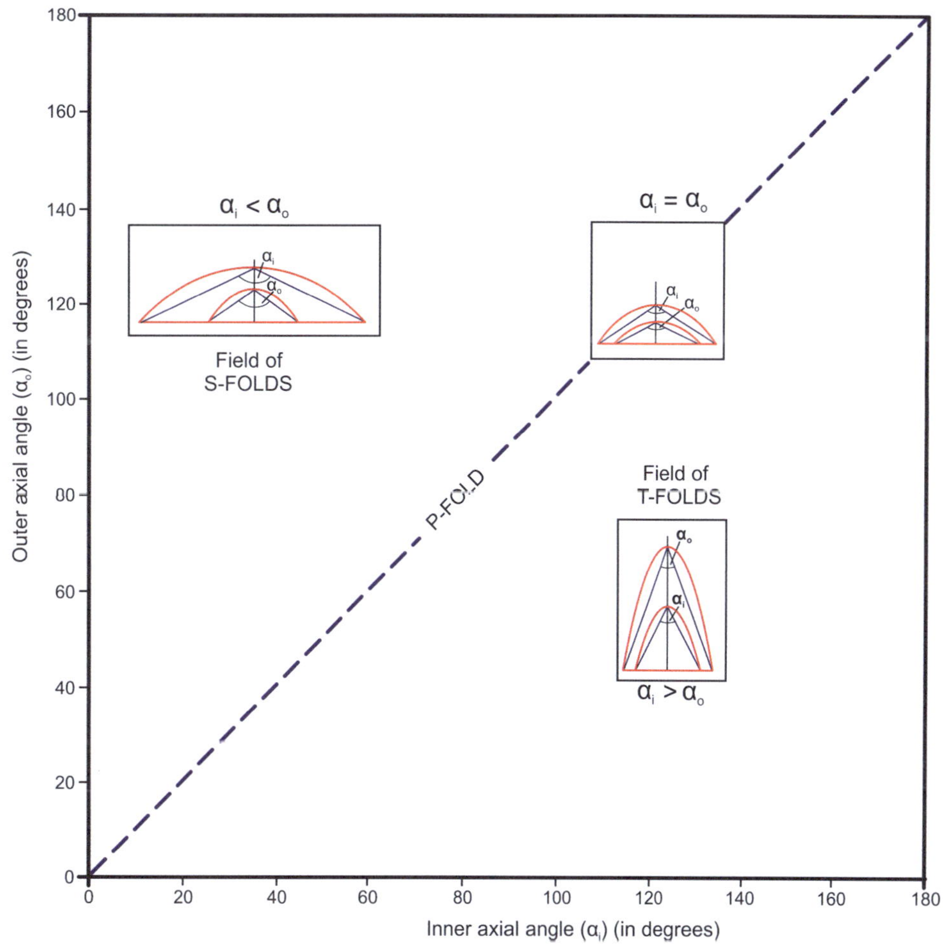

Fig. 8.68 Relationship between α_o and α_i in folds. Parallel folds fall on the 45° line. Fields of *S*- and *T*-folds are shown. (Reproduced from Bhattacharya 2005, Fig. 4 with permission from Japan Society of Geoinformatics)

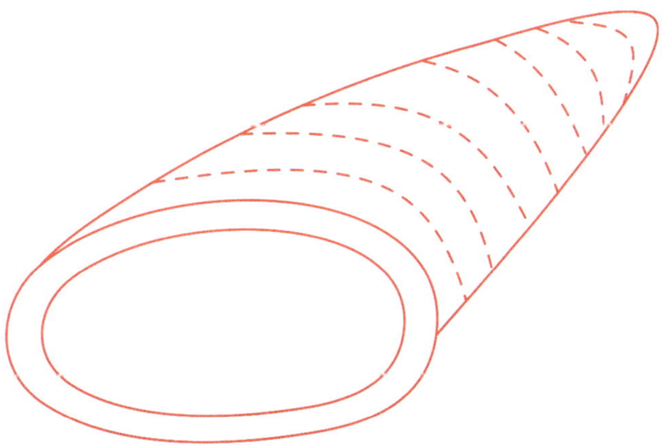

Fig. 8.69 Diagram of a sheath fold (see text)

Box 8.3 Growth Folds

In recent years, rigorous exploration by hydrocarbon geologists led to identifying several types of structures whose formation is related to sediment accumulation.

Box 8.3 (continued)

All such structures together are called *growth structures*. According to Chapman (1983, p. 24), 'growth structures are structures in which the variations in sedimentary thickness show a close relationship to the structure itself and are believed to have formed as a result of deformation that was contemporaneous with sediment accumulation'. Some relevant excerpts from Chapman's work are highlighted below.

Of the various growth structures, the most common structures are anticline and syncline though some other structures such as faults, drape folds, supratenuous folds and monoclines also have common origin. Growth anticlines are those in which the thicknesses of rock units are greater on the flanks than on the crest (Chapman 1983, p. 34). As flanks accumulate more sediment, these are steeper and their steepness increases with depth and conversely the crestal part is thinner than limbs (Fig. 8.72). Because of greater thickness, the flanks undergo greater compaction.

In a basin, the areas of rapid subsidence constitute the flanks of growth anticlines, while those of slow

(continued)

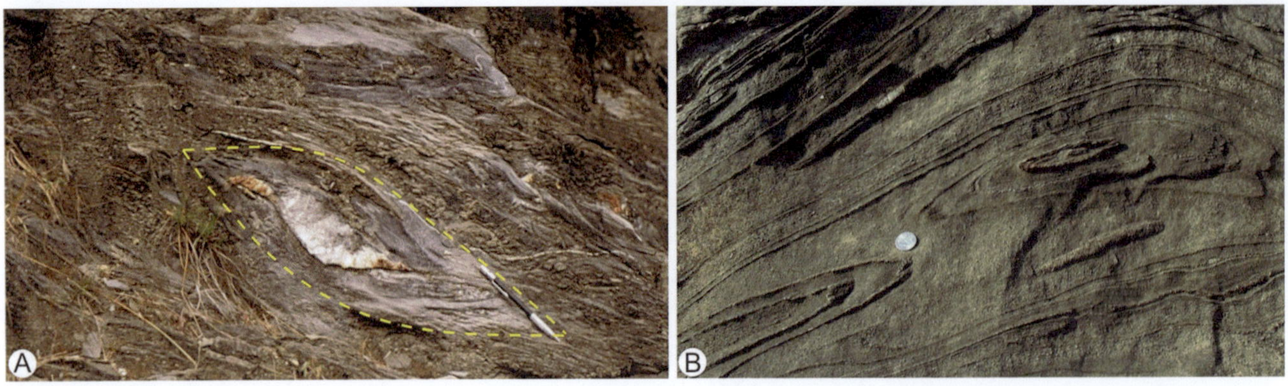

Fig. 8.70 (**a**) Sheath fold developed in mylonitic gneisses of the Main Central Thrust zone, Garhwal Himalayas, India. (**b**) Eyed fold in calcareous rocks of Delhi Supergroup, Rajasthan, India. (Photographs: A: by the author. B: courtesy Professor T. K. Biswal)

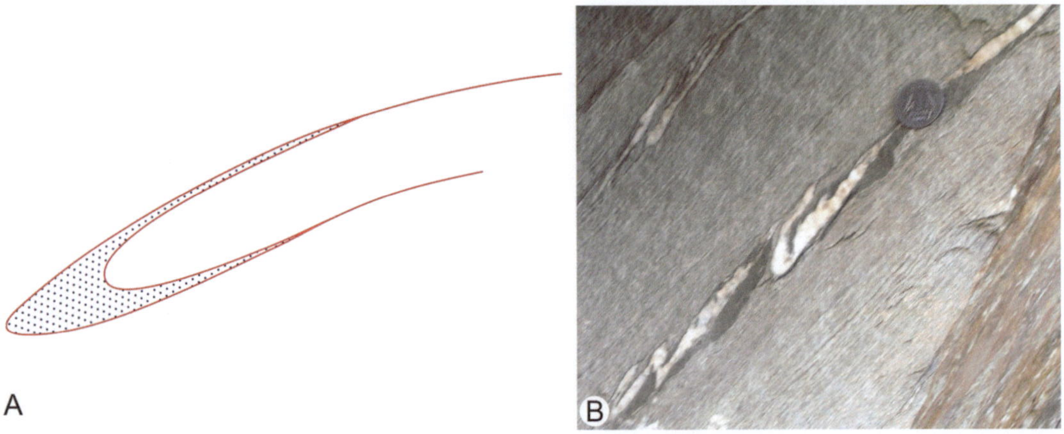

Fig. 8.71 Intrafolial fold. (**a**) Sketch. (**b**) Photograph of intrafolial folds developed in mylonites along the Main Central Thrust, Garhwal Higher Himalayas, near Helong, India. (Photograph by the author)

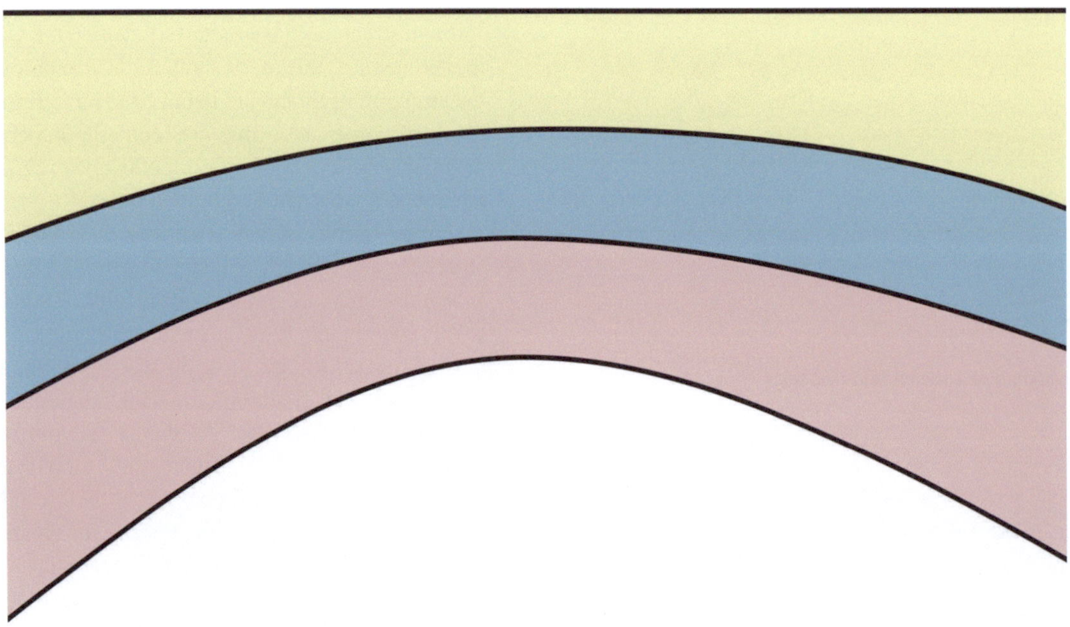

Fig. 8.72 Cross section through a growth anticline

Box 8.3 (continued)
subsidence constitute the crests of anticlines. As such, growth anticlines are associated with active sedimentary basins and have therefore not been reported from surface rocks. Due to close association of growth anticlines with diapirs, many hydrocarbon geologists believe that hydrocarbon-bearing anticlines may have had their history of formation as growth anticlines. In general, growth anticlines are parallel to the depositional strike and in most cases are not associated with any corresponding synclines. These observations suggest that the formation of these anticlines is related to sediment accumulation. The primary cause is differential subsidence relative to base level, i.e. areas of rapid subsidence accumulate greater sediment thickness (thus forming flanks) than those of slow subsidence (thus forming crests). Growth anticlines commonly occur as closed anticlines though they also occur as oval or dome-like.

8.7 Fold Mechanics

Folds show a variety of shapes and sizes and bear a variety of geometric relations with the rocks in which they occur. All these suggest that folds are formed by several mechanisms and that no single mechanism can produce the variety of folds that occur in rocks.

Folds have been studied by several workers from several angles, e.g. geometrical relations existing among the various features of folds (e.g. Ramsay 1962, 1967; Hudleston 1973a; Hudleston and Stephansson 1973; Johnson 1977), mathematical relations existing among the various geometrical parameters of folds (e.g. Ramsay 1962, 1967; Hudleston 1973b; Hudleston and Stephansson 1973; Johnson 1977; Tanner 1989) and experimental deformation of analogue materials whose properties are simulated with natural rocks (e.g. Biot 1961, 1964; Biot et al. 1961; Ramberg 1963, 1964, 1967; Hudleston 1973b; Llorens 2019). The reader is further suggested to consult a review on fold geometry and folding by Nabavi and Fossen (2018) and the references therein.

In the light of these and several other such works, our understanding of fold formation has tremendously advanced by now. In an excellent review, Hudleston (1986) highlighted three important mechanisms of fold formation: buckling, bending and passive folding, as briefly described below.

8.8 Buckling

8.8.1 Basic Ideas

Buckling is a process of formation of folds by the application of stresses parallel to layering of rocks. For buckle folds to form, it is necessary that a competent layer is embedded in a matrix or embedding material with much higher ductility (incompetent material) than the layer. Buckle folds develop in the more competent layer. For buckling, it is necessary that there should be contrast in the competence between the layer and the matrix. Under layer-parallel compressive stresses, the competent layer forms fold(s) due to development of a mechanical instability in the layer. The irregularities existing in the interfaces between the layers are mainly responsible for the formation of folds whose shape and size depend on the viscosities and thicknesses of the layers and the resistance to slip along the interfaces (Hudleston 1973b; Hudleston 1986). If this condition persists, the shape of the bend accentuates to produce a fold with rounded hinge and uniform layer thickness, i.e. a parallel fold. Folds formed by buckling are thus parallel folds.

Buckle folds may be single layered or multilayered. We briefly discuss here mechanics of both these types of folds in the light of the above-mentioned works.

8.8.2 Single-Layer Buckling

8.8.2.1 Stages of Buckling

Hudleston and Treagus (2010) described the major stages of formation of single-layer buckle folds like this. Let us consider that a competent layer with viscosity μ_L and thickness h_O is embedded in a matrix of lower viscosity μ_M (Fig. 8.73a). Initially, under low compressive stress, the layer does not produce folds. As compressive stress is increased, the layer becomes thicker resulting in *layer-parallel shortening*. Experiments show that folds do not form until the layer is shortened by about 20%. Till then, the layer continues to get thicker to accommodate the imposed layer-parallel stress. With further increase of compressive stress, the layer develops mechanical instability due to the presence of irregularities at the interface of the layers that leads to buckling of the layer. The folds thus nucleate (Fig. 8.73b) and grow as they go to select a suitable wavelength. Thereafter, they grow to large amplitude (Fig. 8.73c). At the stage shown in Fig. 8.73b, the layer strain equals the bulk strain. But as the folds grow to large amplitude (Fig. 8.73c), this equality of strain is not maintained. In other words, if $L = \lambda_p$ at stage (b),

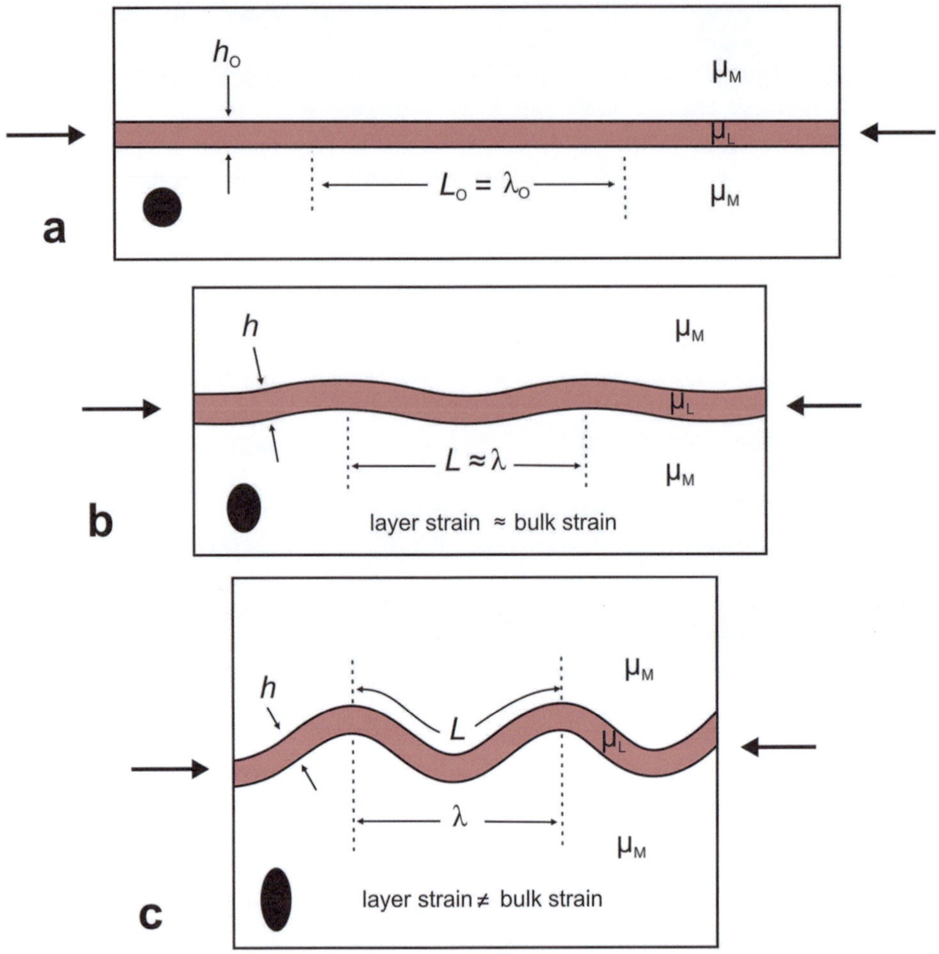

Fig. 8.73 Stages in the development of buckling of a single layer. Stages **a**, **b** and **c** are explained in the text. Note that the changes in the shape of an initial circle into different ellipses reflect the increasing strain involved during successive stages of folding. (Reproduced with mild simplification from Hudleston and Treagus 2010, Fig. 4 with permission from Elsevier Senior Copyrights Coordinator, Edlington, U.K. Submission ID: 1196061)

L in (c) is a rough measure of λ_p at the strain shown in (b) (Hudleston and Treagus 2010, p. 2044). The folds thus formed resemble Class 1B, i.e. parallel folds (Figs. 8.59a and 8.64, Fold 2). In order to accommodate the imposed stress, the matrix material on the other hand fills the space thus produced by folding the competent layer. The incompetent layer, unlike the competent one, produces other classes of folds. The important point is that the layer thickness increases during layer-parallel shortening, and therefore the layer thickness of the buckle fold formed is obviously more than that of the initial thickness of the layer.

Formation of buckle folds involves the following assumptions: (i) Folds form when there is marked contrast in the viscosity of the competent layer and the matrix. (ii) Formation of buckle folds takes place under pure shear and plane strain (no volume change) conditions. (iii) Since the layer thickness remains constant during shortening, there is no component of shear strain within the layer.

8.8.2.2 Dominant Wavelength During Buckling

Let us consider a thin competent layer of viscosity μ_1 and thickness t as embedded in another layer of lower viscosity μ_2. If the layer is subjected to layer-parallel compressive strain, it has been shown that the instability that is developed in the layer produces folds of various wavelengths. These folds grow at different rates. Of these, only one fold initiates buckle folding; it has a wavelength of

$$W_d = 2\pi t \sqrt[3]{\frac{\mu_1}{6\mu_2}} \qquad (8.5)$$

This is called *dominant wavelength* (Biot 1957). Ramberg (1964) called it *characteristic wavelength*. Equation (8.5) has significant implications for buckle folds, some of which are highlighted below:

- The dominant wavelength and the thickness of the folded layer bear a direct relationship. This feature has been

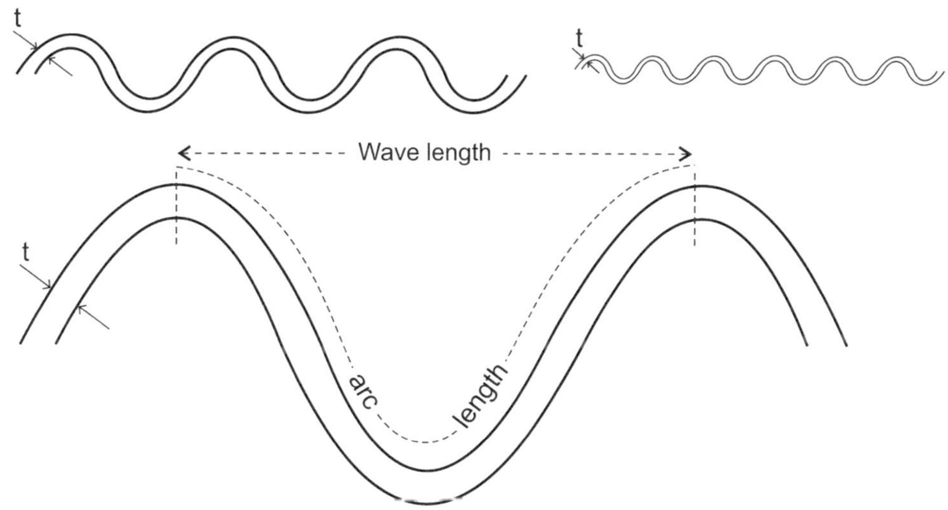

Fig. 8.74 One important conclusion of buckling theory of fold formation is that the thicker layer takes up greater wavelengths and thinner layer takes smaller. For folds of all sizes, the ratio of arc length and thickness remains the same

observed in natural folds also. Currie et al. (1962), for example, studied natural folds developed in strong layers of a multilayer rock sequence and applied prediction for the geometry of common natural folds in respect of the relationship between layer thickness and wavelength.

- The above relation implies that *the thicker layers show larger wavelengths* (Fig. 8.74).
- The fold wavelength is independent of the compressive stress and the strain rate.
- As μ_1 approaches the value of μ_2, the dominant wavelength takes up a value of 3.46 t. This, however, reflects that the theory becomes inapplicable at low viscosity ratios (Professor Peter Hudleston, personal communication 2020).

8.8.3 Multilayer Buckling

8.8.3.1 Nature of Multilayer Buckling

A multilayer may include several competent layers separated by surfaces of anisotropy such as bedding or cleavage or competent layers, e.g. sandstone, alternating with incompetent layers, e.g. shale. In rocks, multilayer folds are more common than single-layer folds. Buckling of multilayer has been extensively studied through experiments, theory and field occurrences (Biot 1961, 1964, 1965; Biot et al. 1961; Ramberg 1961, 1963, 1964, 1967; Currie et al. 1962; Ghosh 1968; Llorens 2019). Some important results of these works are highlighted below:

- Like single layer, the multilayer develops buckle folds due to layer-parallel stress if the layers have marked contrast in their viscosity.
- Experiments on analogue materials and observations on natural folds indicate that buckle folds in multilayer show a variety of shapes ranging from rounded hinge to chevron folds and their shapes mainly depend, among others, upon the competence contrast and the ease to gliding. Ramberg (1964) has shown that folds grow more rapidly with greater difference of competence between the layer and the matrix.

8.8.3.2 Role of Spacing in Multilayer Buckling

Experiments indicate that the spacing between the competent layers of a multilayer controls the type of folds formed during buckling. Ramberg (1964) has shown that buckle folds formed under such conditions are of two types. If the spacing is close and less than the dominant wavelength, the multilayer mechanically behaves as one unit with uniform layer thicknesses. The layers buckle homogeneously, and the folds thus formed are 'harmonic' (Fig. 8.75a). If the spacing is large, i.e. more than the dominant wavelength, each competent layer behaves as an independent unit and thus develops buckle folds whose arc length resembles that of single-layer buckle folds. Folds formed in the thicker layers have a larger wavelength than those formed in the thinner layers. The folds thus formed are 'disharmonic' (Fig. 8.75b). Large spacing of competent layers, in fact, promotes folds to grow at different rates within the multilayer.

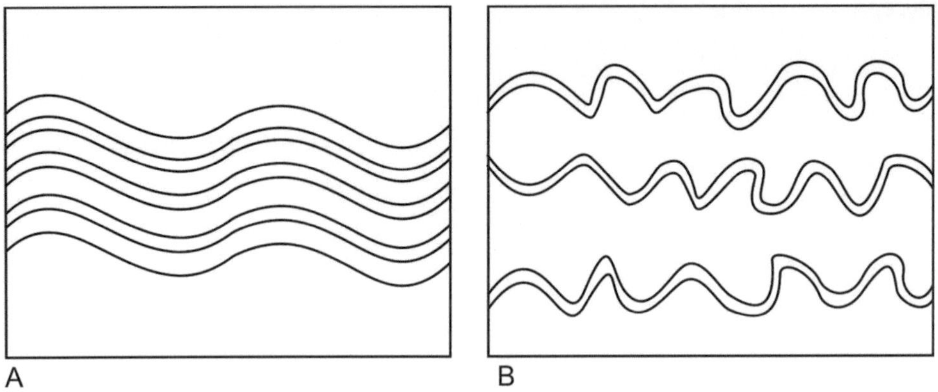

Fig. 8.75 Effect of spacing of competent layers in multilayer buckling. (**a**) Close spacing produces harmonic folds. (**b**) Wide spacing produces disharmonic folds

8.8.3.3 Dominant Wavelength of Multilayer Folds

Work of Biot shows that the dominant wavelength of buckle multilayer folds is not the same as for buckle single-layer folds. Let us consider a multilayer consisting of n competent layers each of equal thickness t and viscosity μ_1, and let the entire slab is embedded in a medium of lower viscosity μ_2. Assuming perfect lubrication between the layers, Biot has shown that the dominant wavelength of multilayered folds as given by

$$W_d = 2\pi t\left(\frac{n\mu_1}{6\mu_2}\right)^{1/3} \tag{8.6}$$

is larger than that of a single-layer fold by a factor $n^{1/3}$.

According to Biot, under above-mentioned conditions, the dominant wavelength W_d (i) increases proportionally with increase in the number of layers in the slab and (ii) remains unaffected by the changes in the viscosity contrast between the layer and the matrix.

8.8.3.4 Characteristics of Buckled Folds

- The arc length/thickness ratio of single-layer buckle folds for one particular lithology remains more or less constant. This implies that a change in layer thickness will change the arc length accordingly. Thicker layers thus show larger folds, while thinner layers show smaller folds. Such folds are also called *periodic folds*.
- Buckle folds show sinusoidal shapes with rounded hinges and are typically parallel folds (Class 1B).
- Laterally, the effect of folding disappears rapidly within a distance equivalent to about one wavelength.
- The buckled competent layer shows two contrasting effects of tangential longitudinal strain (Fig. 8.76). The outer arc of the fold shows layer-parallel stretching (extension), while the inner arc shows layer-parallel shortening (compression). The boundary between these two is a surface called *neutral surface* that shows the effects of neither extension nor compression. The effects of extension on the outer arc include tension cracks and normal faults, while those on the inner arc include thrust faults.

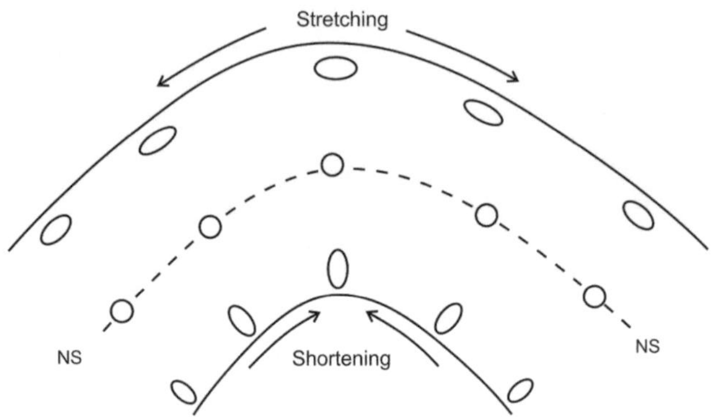

Fig. 8.76 Strain pattern in a buckle fold. While the outer arc undergoes stretching, the inner arc undergoes shortening. Both these strain fields are separated by a neutral surface (NS) along which there is no deformation

8.9 Bending

Folds formed by application of stresses transverse to layering are called *bending folds*. If an iron layer is fixed at one end and is subjected to transverse forces at any point along its length, it bends to form folds. Bending folds may develop in one single layer or in a multilayer. Unlike buckle folds, competence contrast of the layers undergoing bending folds is not necessary. What are instead necessary are the transverse forces that should act at high angles or perpendicular to the layer.

The transverse forces that produce bending folds may act in one direction or in two opposite directions acting at one

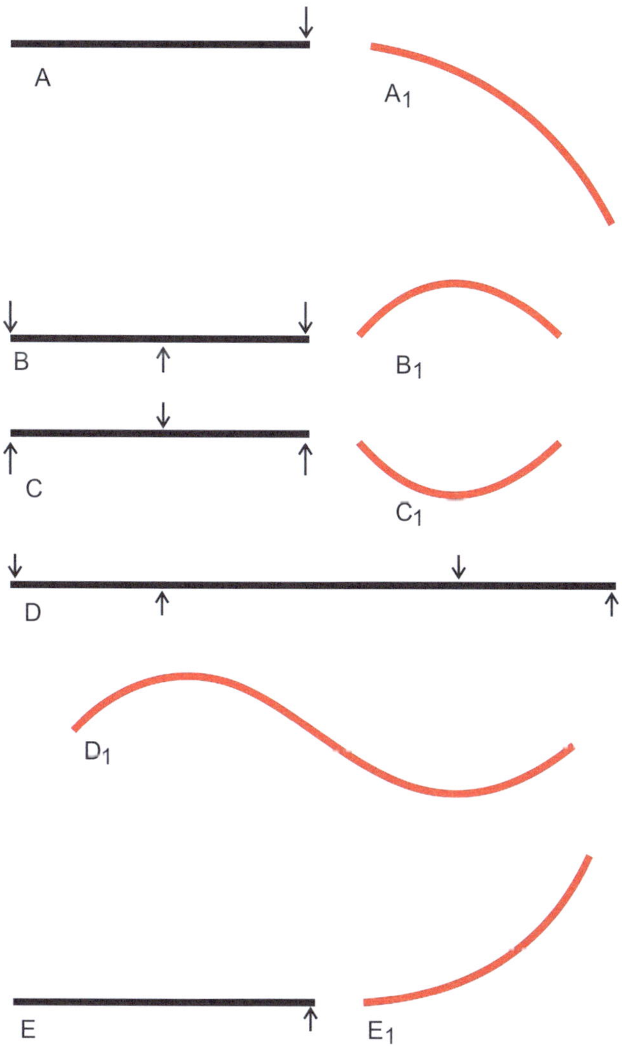

Fig. 8.77 The various ways of application of bending force(s) on an iron bar. The resultant geometry (folds) formed has been shown in each case

point or more along the length of the layer. In each case, a specific type of fold is formed (Fig. 8.77). The overall geometry of folds formed by bending process thus depends largely upon whether the transverse forces act at one point or more and whether in one direction or more. In general, the bending folds do not show tight geometry. With example of an iron bar, some common ways that bending force(s) can be applied have been shown in Fig. 8.77 together with the resultant geometry (folds) that the bar can assume.

The bending folds and buckle folds also differ in their mechanisms of formation. In bending fold, the arching of the layer takes place *parallel* to the direction in which the stresses are applied, while in buckle fold, the arching takes place in the direction *perpendicular* to the direction of application of stresses. Bending folds *do not undergo layer-normal thickening* on application of stresses, while buckle folds *undergo layer-normal thickening* before the layer develops folds.

In rocks, several types of folds have been interpreted to have formed by bending process. These include folds overlying granite intrusive, salt domes, neck folds associated with boudins (Chap. 15), fault-bend folds and fault propagation folds (Chap. 11), forced folds and drape folds (Chap. 9).

8.10 Kinking

Kinking is the formation of kink folds. Such folds are formed by a variety of processes that have been explained by a few models:

Model 1: Paterson and Weiss (1966) produced kink bands in strongly foliated rocks under laboratory conditions by the application of layer-parallel stress. They suggested that formation of kink bands is strongly dependent upon the role of planar anisotropies in relation to the direction of compression.

Model 2: Ramsay (1967, p. 453) suggested that the applied stresses initially develop a buckle that grows into a kink fold by extending the lengths of the kinked portion. The fold thereafter grows sideways by migration of the axial surfaces. He further suggests (p. 453) that the formation of kink bands and conjugate folds involves a maximum shortening of 25%. According to him, kink bands are geometrically related to chevron folds and are probably formed in a similar manner.

Model 3: Ghosh (1968) carried out experiments on analogue models and produced an array of folds by increasing or decreasing the degree of lubrication of layers by greasing. He has shown that when the layers are without lubrication, i.e. when they are in frictional contact, chevron folds with straight limbs and angular hinges are formed. With thorough greasing, on the other hand, sinusoidal folds with curved hinges are produced. He thus demonstrated that lubrication of layers during folding plays an important role in controlling the geometry of folds; kink folds form when the layers are in frictional contact, i.e. without lubrication.

Model 4: Twiss and Moores (2007) suggested that kink bands can form by shearing parallel to the laminations and preservation of the continuity of laminations across the kink band boundaries.

8.11 Passive Folding

Passive folding occurs where the layers simply undergo ductile flow as a passive marker without playing any significant part in deformation. The layers are deformed by shear or flow across the layering, and therefore such folds are also

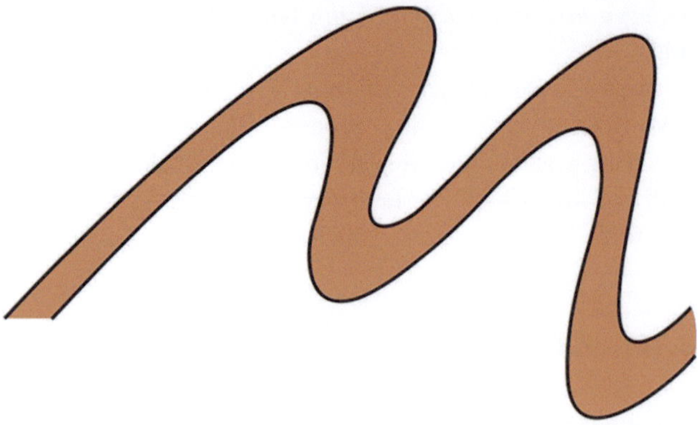

Fig. 8.78 Passive folds. The thickness of the layers changes systematically showing maximum thickness at the hinges

called *passive folds* or *flow folds* (Fig. 8.78). Passive folds commonly occur in shear zones, where the layers do not show any significant competence contrast but are simply subjected to ductile flow. They also commonly occur in high-grade metamorphic rocks, glaciers and salt domes. For passive folds, unlike buckle folds, competence contrasts among the layers and layer-parallel shortening are not necessary and, unlike bending folds, passive folds do not require transmission of forces across the layers.

The dominant *mechanism of passive folds* is ductile flow that produces a systematic curvature in the layers. These folds can form by both homogeneous and heterogeneous ductile strain. Under conditions of heterogeneous shear, the flow is caused along a direction normal or oblique to layering. Both simple shear and pure shear can produce passive folds. Under all these conditions, the layering acts as a passive marker. For passive folds to form, presence of pre-existing curvatures in the layering is a necessity. Objects such as porphyroblasts or pebbles may impart initial curvatures to the layering. The curvatures get accentuated during deformation caused by ductile flow, and the amount of strain in the corresponding positions of the successive layers shows similar values. The thickness of each layer changes systematically, and the fold thus formed approaches Class 2 or similar fold geometry.

8.12 Progressive Development of Fold Shapes

All folds change their shape as they grow excepting when there was either a uniform increase or decrease in area or volume (Professor Peter Hudleston, personal communication 2020). Shape change in a fold is basically a geometrical restructuring of the earlier shape and mostly involves hinge thickening accompanied by a proportionate limb thinning. Progressive development of fold shapes has been explained by a few models as highlighted below.

8.12.1 Model 1

De Sitter (1964) presented a model that quantitatively suggests at what stage a buckle parallel fold becomes a flattened fold. It is known from the buckling theory that under compressive strain a layer initially undergoes layer-normal thickening. With further compression, the layer shows the development of folds that initially show parallel geometry. This assumes that the fold at this stage is represented by concentric arcs. De Sitter suggested that buckling alone can cause shortening of up to 36%. With further shortening, the fold cannot retain its parallel geometry; instead, the imposed compressive strain is accommodated by thickening at the hinge. The fold thus becomes a flattened one. With still further shortening, the limbs tend to become parallel to each other and the fold becomes very tight with highly thickened hinge.

8.12.2 Model 2

Ramsay and Huber (1987) suggested that the originally parallel folds (Class 1B) change their shapes to a flattened geometry (Class 3) by subsequent superimposition of strain in a multilayer. Ramsay and Huber (1987, p. 360) thus posed a pertinent question: 'why we wish to evaluate the flattening components superimposed on fold'. Their answer is: 'It is often necessary to compute the total shape changes in the Earth's crust and folded layers do offer a method for making such calculations. Measuring along the fold arc length will not produce the correct answer to such a computation where we have folds with flattened geometry, but these equations will enable such structures to be restored to a pre-flattened state'.

8.12.3 Model 3

Bhattacharya (1992) presented a model of fold shape development considering the relationship between fold axial angle (α) and thickness ratio (R). The latter is the ratio of hinge thickness (T_h) and limb thickness (T_l) of a (symmetrical) fold. T_h is the orthogonal thickness of the fold (BE in Fig. 8.66) along the hinge, i.e. along the axial line, while limb thickness (T_l) is the orthogonal thickness of the limbs at their lowest (inflection) points. The axial angle (α) is the mean of the inner (α_i) and outer (α_o) curves along the fold axial surface. The parameters R and α are interdependent and control the shape of natural folds. R-α plot (Fig. 8.79) for a number of natural folds studied yields a typical curve (Bhattacharya 1992, Fig. 3) of the type:

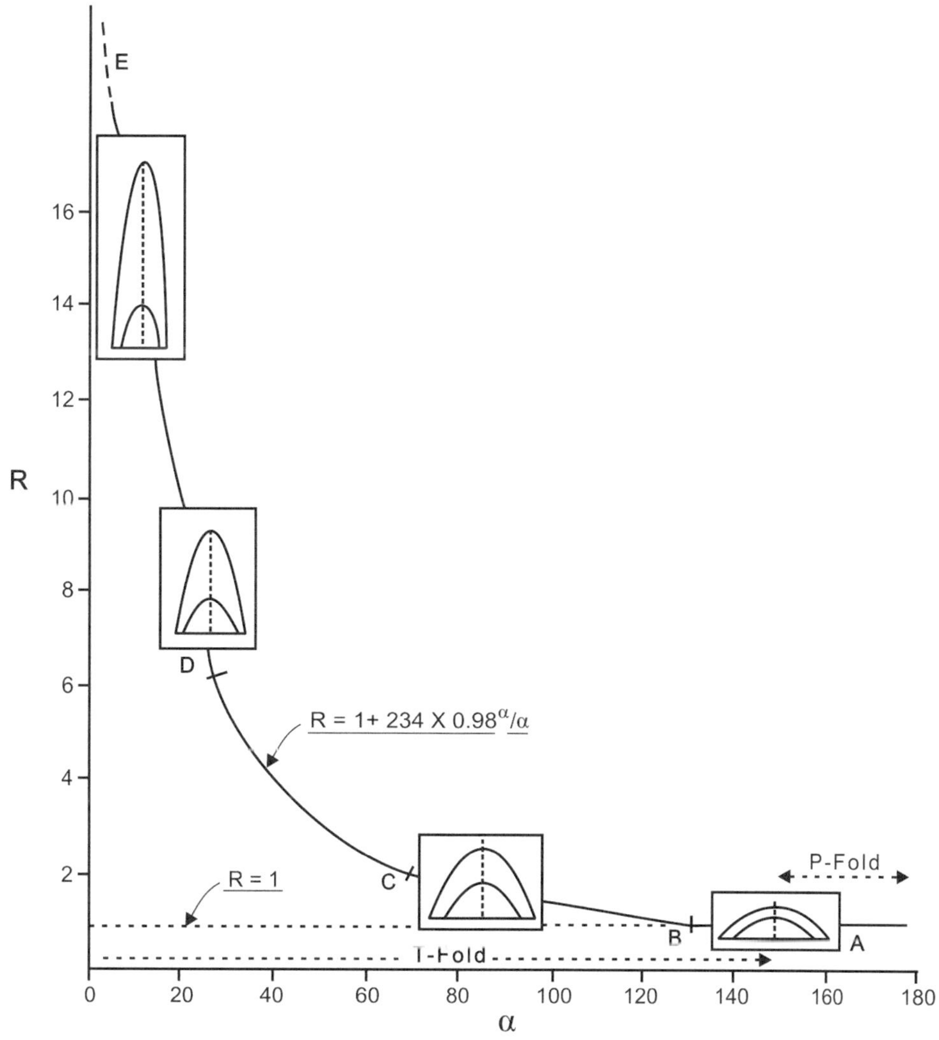

Fig. 8.79 Characteristic curve for natural folds as obtained by a *R*-α plot. *R* is the thickness ratio, and α is the mean axial angle for a fold (see text for explanation). The points A, B, C, D and E are points where the graph changes its shape. Shapes of typical folds developed at various stages of fold formation are shown on the graph to give an idea of sequential changes in fold shapes during progressive deformation. (Reproduced with minor modification from Bhattacharya 1997, Fig. 2 with permission from Japan Society of Geoinformatics)

$$y = 1 + AB^x/x \tag{8.7}$$

where *A* and *B* are parameters; $A = 234.55$, and $B = 0.98419$ (Bhattacharya 1992). Figure 8.79 is the computer-produced best-fit curve with the above-mentioned values of *A* and *B*.

On replacing *y* by *R* and *x* by α, Eq. (8.7) takes the form

$$R = 1 + AB^x/\alpha \tag{8.8}$$

Equation (8.8) is helpful in getting the actual graph for any set of data on *R* and α.

In Fig. 8.79, the various fold shapes developed at different stages of fold formation have been shown. The graph suggests that during progressive ductile deformation, the naturally deformed buckled folds generally retain their parallel geometry until the axial angle is reduced to about 130° or so (the AB portion of the graph). With progressive deformation, increase of hinge thickness is a slow process between axial angles 130° and 70° (BC) but becomes rapid down to angle 30° (CD). Below axial angle 30° (DE), when elevated temperature also comes into play, the folds undergo mass flowage by hinge thickening accompanied by thinning and extension of the limbs.

From a genetic point of view, fold shape development appears to be accomplished at three major stages: an initial *buckling stage* (fold axial angle down to about 130°), followed by a *metastable stage* (between 130° and 70°) and then by a *labile stage* (axial angle <30°). During buckling

Stage	Buckling Stage	Metastable Stage	Labile Stage
Physical Factor(s)	Competency of the rock layer	Influence of pressure, temperature and textural factors of the layer	Elevated pressure and temperature
Fold Mechanism	Buckling	Superimposition of homogeneous ductile strains	Viscous/ductile flowage
Fold Shape		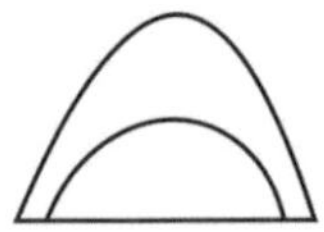	
Mathematical Relation	T = 1	Complicated	Polynomial
Deformation Curve (Diagrammatic only)	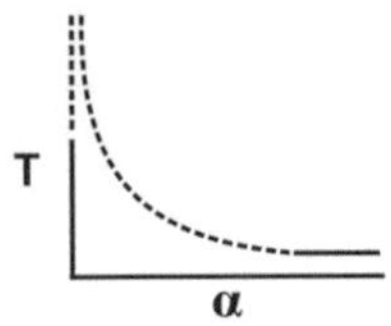	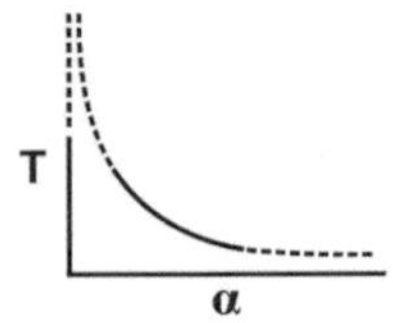	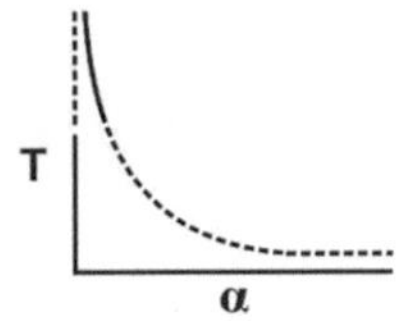

Fig. 8.80 Graphical representation of fold shape changes with progressive deformation. The three major stages of fold shape changes, i.e. *buckling stage*, *metastable stage* and *labile stage*, have been provided with their mechanisms and geometrical and mathematical expressions. (Reproduced with minor modification from Bhattacharya 1997, Fig. 4 with permission from Japan Society of Geoinformatics)

stage, the fold maintains a parallel geometry. With progressive deformation—the metastable stage—the fold accommodates the ductile strain by hinge thickening that progressively increases with decrease of axial angle. The various components of the fold undergo interactions among them to accommodate the imposed compressive stress by systematically changing their shapes with progressive deformation. With further deformation—the labile stage with axial angle $<30^{\circ}$—the fold undergoes flowage by rapid hinge thickening and limb extension as shown by rapid increase of R even with a very small decrease of α, thus making the graph asymptotic along the y-axis. These stages have been graphically shown in Fig. 8.80.

8.12.4 Model 4

Llorens (2019) presented a systematic study of folding of a competent layer deformed in pure and simple shear boundary conditions up to high strains by keeping the following parameters to systematically vary during experimental simulations: viscosity ratio between layer and matrix (m), stress component (n) of the power-law viscous materials and initial layer orientation (β) with respect to the shear plane (in simple shear). The results clearly show that the fold geometry changes with progressive shortening up to 75% in both the cases.

8.13 Significance of Folds

8.13.1 Academic Significance

- Folds are common deformation structures in rocks. Their presence implies that the rocks were subjected to compressional forces at the time of their deformation.
- Folds are formed due to active role of rock layers (excepting passive folds). Occurrence of folds thus implies that the original rocks had a layered pattern.
- Small-scale folds are believed to represent some large-scale structure in an area. As such, even if a large-scale fold is not exposed or is not visible in the field, one may get an idea of the larger fold, e.g. attitude, orientation of the beds and axial plane and nature and type of the fold.
- If folds with different geometries are present in an area, it is possible, though not always, that the area had undergone changing pattern of the directions of compressional forces.
- Folds, if studied along a traverse of an area, give an idea of regional structure of rocks of the area.

- Shape of folds often gives an idea of their occurrence in different parts of an orogenic belt.

8.13.2 Economic Significance

- Folds are considered as repository to a variety of economic mineral deposits.
- Since folds are formed under a specific system of stresses, mineralizing fluids also follow the directions of mass movement. Thus, a systematic analysis of stress pattern through the study of folds can give an idea of the trajectory of mineralizing fluids, if any.
- Petroleum and natural gas tend to be systematically (density-wise) located in different parts of a large anticline. This is also sometimes called as the *anticlinal theory* of localization of petroleum and gas.

8.14 Summary

- *Folds* are bends in rock strata formed due to permanent deformation of originally flat rock layers under the influence of compressive forces acting parallel to, or across, their layering.
- The form of a fold surface as seen on a plane normal to the hinge line is called *fold profile*, while the typical geometrical shape or attitude of a fold is called the *style of a fold*.
- A fold can be specified by some geometrical aspects such as *amplitude*, *wavelength*, *arc length* and *curvature*.
- A fold can be described by some quantitative parameters such as *cylindricity, aspect ratio, tightness* and *bluntness*.
- Natural folds show a variety of shapes. Some folds occur as single layers (single-layer folds), while others involve several layers (multilayer folds). Folds are therefore classified in a variety of ways. Some commonly used schemes of classifying folds have been described.
- Since folds show a variety of geometric relations with the host rocks, genesis of folds occurs by a variety of mechanisms of which buckling, bending, kinking and passive folding are the most common mechanisms.
- *Buckling* is a process of formation of folds by the application of stresses parallel to layering of rocks. Formation of buckle folds requires that a competent layer is embedded in a matrix or embedding material with much higher ductility (incompetent material) than the layer. Buckle folds develop in the more competent layer. *Bending folds* form by application of stresses transverse to layering. *Kinking* is the formation of kink folds. *Passive folding* occurs where the layers simply undergo ductile flow as a passive marker without playing any significant part in deformation. The layers involved in folding do not have significant competence contrast but are simply subjected to ductile flow.
- All folds change their shape as they grow excepting when there was either a uniform increase or decrease in area or volume. Shape change in a fold is basically a geometrical restructuring of the earlier shape. Models of fold shape development are described.
- Folds bear great significance from both academic and economic viewpoints.

Questions

1. What is meant by profile and style of a fold? What is the difference between the two?
2. What do you mean by inflection point and inflection line? Explain your answer with diagrams.
3. What is meant by curvature of a fold? Does the curvature remain the same everywhere in the fold? Where is the highest curvature in a fold?
4. Are the two terms hinge and hinge zone the same for a fold? If not, give the difference between the two.
5. What is the difference between an upright fold and a vertical fold? Explain with the help of diagrams.
6. A fold is SE vergent. What does it mean?
7. What do you mean by cylindricity of folds?
8. What are buckling and bending folds? Explain the difference between the geometry and genesis of these two types of folds.
9. What do you mean by dominant wavelength? Describe the role played by it in fold formation.
10. Why do the folds have economic significance?

9 Faults

A. R. Bhattacharya, *Structural Geology*, Springer Textbooks in Earth Sciences, Geography and Environment,
https://doi.org/10.1007/978-3-030-80795-5_9

Abstract

This chapter takes you to the amazing world where a block of rock mass moves past another such block along a plane of separation called *fault*. A fault is an elongated zone of high shear stress along which two blocks of adjacent rocks have been offset. It is thus a discontinuity or anisotropy due to which the rock loses its cohesion. Being a plane, a fault can be described in geometric terms such as dip and strike. The orientation of stresses acting upon a fault surface controls the geometry of the fault. Rubbing of two blocks of rock masses not only leaves behind a variety of signatures on or in close vicinity of the fault surface but may also genetically modify the host rocks. During fieldwork, a geologist is sometimes surprised to note that in some areas a fault has concealed one or a few beds while in other areas it has got a few other beds repeated. In brief, a fault is able to do or undo many things to the rocks of an area that sometimes leave a field geologist baffled!

In this chapter, the reader may find a detailed description of faults in respect of their geometrical attributes, classification, recognition in field, common rock types found in fault zones and mechanics of faulting. Faults are highly useful structures in the exploration of hydrocarbon, minerals and groundwater. The reader can get all these and many more once he/she takes a plunge into this chapter!

Keywords

Faults · Fault geometry · Classification of faults · Recognition of faults · Fault damage zone · Fault zone rocks · Growth of faults · Fault mechanics

9.1 Introduction

A *fault* is an elongated zone of high shear stress along which the adjoining rocks have been offset. A fault is thus a discontinuity or anisotropy showing visible slip or displacement of rock masses parallel to the fault plane (Fig. 9.1). As such, a fault causes loss of cohesion of the host rock. The slip surface is characterized by more intense deformation than the surrounding rock mass.

The movement or displacement of rocks associated with a fault is an important factor. Some faults show displacements of several kilometres while others a few metres or centimetres only. Likewise, the length of a fault can be several kilometres to a few centimetres. The thickness of a fault plane is also variable ranging from tens of metres to microscopic scales. If the fault is so small that it is visible under a microscope only, it is called a *micro-fault*. However, whatever is the scale of displacement, length or thickness of faults, their presence should be given due attention for study as they control the structure by accommodating large offsets of rocks and are formed under specific mechanisms of deformation.

Faults are excellent examples of brittle deformation of rocks. Rubbing of two blocks of rocks on or near the surface of the earth produces crushed rock material and breccias that occur along the fault surface. Such faults are formed on or very near the surface of the earth. Fault zones are also formed at deep crustal levels where rocks deform by ductile deformation processes. Under such conditions, shear displacement is associated with the formation of *mylonites* and several ductile deformation features along zones called *ductile shear zones*. Faults are thus one of the most common structures developed in all types of rocks occurring from shallow to deeper levels of the crust. In this chapter, we shall discuss faults that are formed by brittle processes only rather than by ductile processes. The main reason is that most geologists commonly link the term fault to include a brittle fault. In general, when we link a fault with brittle deformation mechanisms, we call it a *brittle fault* or simply a *fault*, and with ductile deformation mechanisms, it is a *shear zone*. Faults formed by ductile processes at great depths, i.e. ductile shear zones, are described separately (Chap. 17).

A fault may be represented in rocks by a single or discrete plane (Figs. 9.1 and 9.2) or by a zone, called *fault zone*, where it may include multiple faults occurring together (Figs. 9.3 and 9.4).

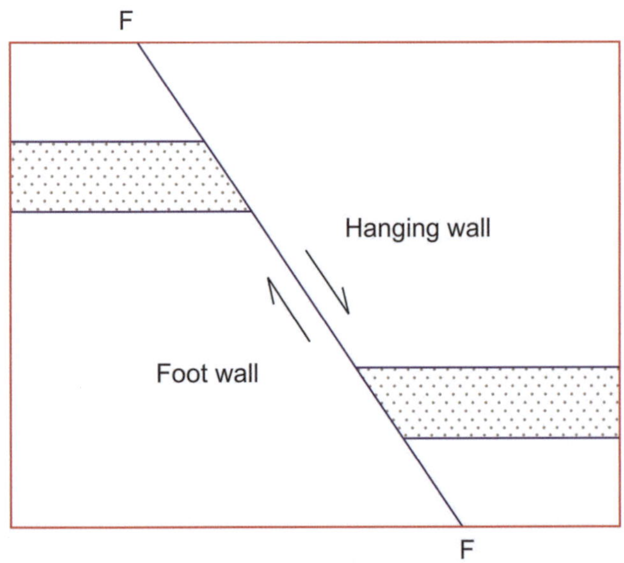

Fig. 9.1 Diagram of a fault (FF) as a discrete plane seen in a section showing the hanging wall, footwall and displacement of a marker bed (dotted)

Fig. 9.2 Fault as a discrete plane developed in a gneissic rock of Bundelkhand craton, central India. Loc. Khilarigaon. (Photograph by the author)

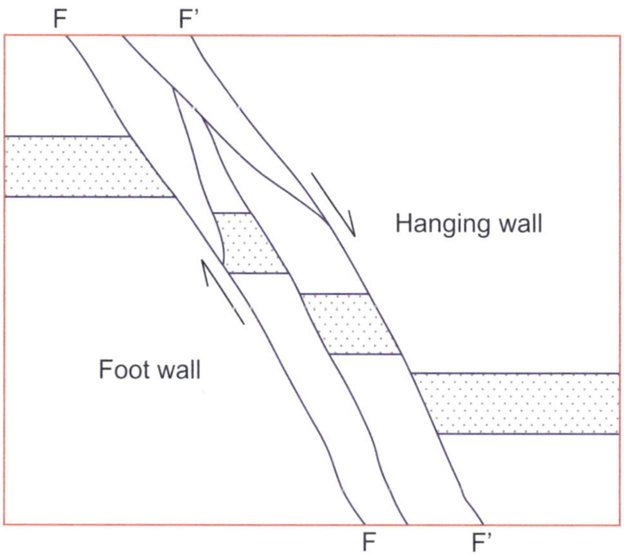

Fig. 9.3 A fault zone formed by several faults. Note that a marker bed not only is displaced across the fault zone but also undergoes sympathetic displacement within the fault zone

9.2 Fault Geometry

9.2.1 Parts of a Fault

A fault is a planar feature, and therefore its parts and geometrical parameters resemble those of a plane. In order to understand the various parts of a fault in three dimensions, it is better to consider a marker plane or bed that has been displaced by a fault (Fig. 9.5). The plane or surface along which the actual movement takes place is called *fault plane* or *fault surface* that divides the rock into two parts called *fault blocks*. If the fault is inclined, the block that rests above the fault plane is called *hanging wall*, while the block below the fault plane is called *footwall*. If the fault plane is vertical, these two terms, i.e. hanging wall and footwall, are irrelevant and, instead, the two blocks are indicated by their geographic positions.

Dip of a fault is the angle that the fault plane makes with the horizontal plane. *Strike of a fault* is the line formed due to intersection of the fault plane with the horizontal plane. The line formed by the intersection of a fault with the ground surface is called *fault line* or *fault trace*.

9.2.2 Geometrical Parameters of a Fault

The angle between a fault plane and the vertical plane is called *hade* (Fig. 9.6), which is therefore the complementary angle of the dip of the fault. For an inclined fault, the displacement component along the horizontal is called *heave*, while the vertical component of displacement is called *throw*. In practice, the throw is more commonly used than the heave. The geometrical attitudes of a line along an inclined fault plane (Fig. 9.7) are identical to those for an inclined plane. Thus, the *plunge* is the angle that a line on the fault plane makes with a horizontal plane, while the *rake* is the angle that a line on the fault plane makes with a horizontal line in the fault plane.

Fig. 9.4 Field photograph of a fault that constitutes a zone, not a discrete plane. The fault has offset a silica vein in the gneissic rocks of the Bundelkhand craton of central India. (Photograph by the author)

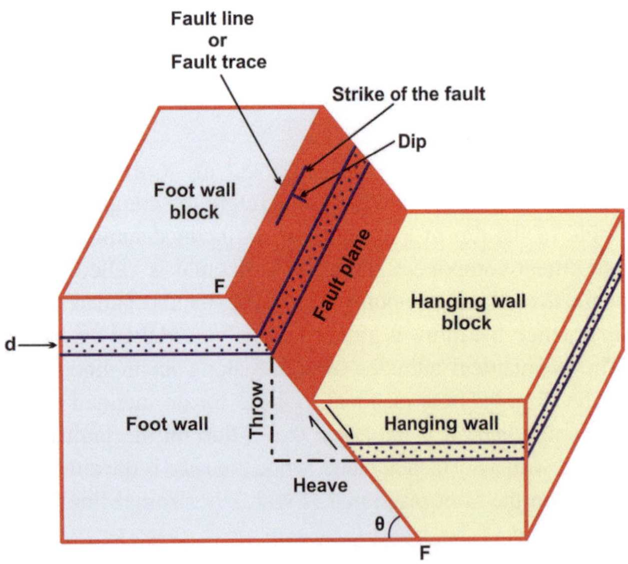

Fig. 9.5 Parts of a fault. FF is the trace of the fault in vertical section. See text for details

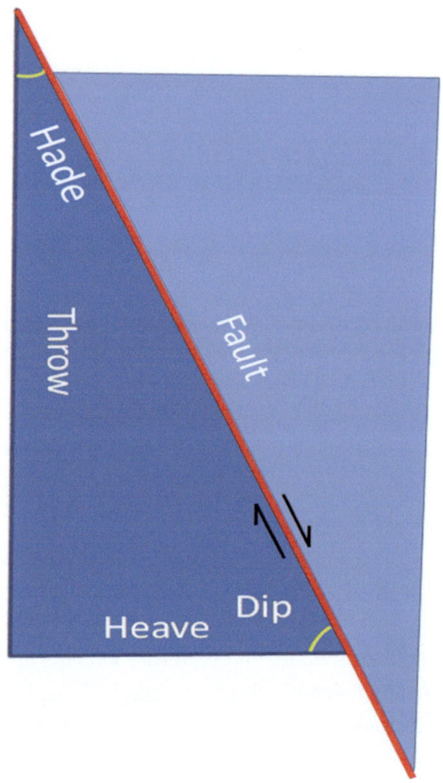

Fig. 9.6 Throw, heave, dip and hade of a fault

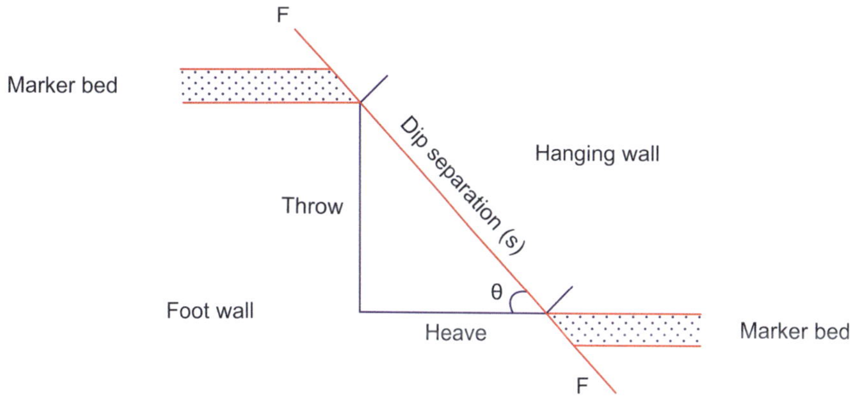

Fig. 9.7 Triangle formed by an inclined fault together with its heave and throw

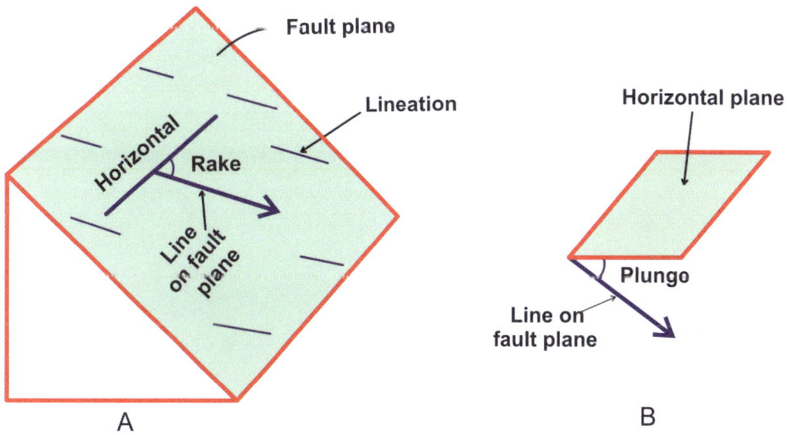

Fig. 9.8 Rake and plunge of a fault. (**A**) Rake is the angle that a line on the fault plane makes with a horizontal line. (**B**) Plunge is the angle that a line on the fault plane makes with the horizontal plane

Box 9.1 Relations Among Some Parameters of a Fault
Geometric representation of an inclined fault plane of a given angle with respect to the horizontal (or vertical) plane enables us to get some measurable quantities of the fault. We consider here three commonly observable parameters: heave, throw and dip separation. These three terms are explained in text. If we construct a fault plane of inclination θ with a horizontal plane (i.e. dip), we get a right-angle triangle with heave (H) as the base, throw (T) as the normal and dip separation (S) as the hypotenuse (Fig. 9.8).

From the triangle:

$$T = S \sin \theta$$
$$H = S \cos \theta$$

Therefore, if we know any two parameters, we can get the third one.

The motion of a fault results in *displacement* or *slip* (Fig. 9.9) of a bed whose attitude is measured in terms of direction and amount. A bed may be displaced in any direction that can be shown by two points that were initially adjacent to each other. The slip between two points along the dip of the fault plane is called *dip-slip*, while the slip along the strike of the fault plane is called *strike-slip*. If the slip has occurred in any direction other than these two, the total displacement of any point is called *net-slip*.

9.2.3 Separation of a Fault

Due to faulting, a bed is separated on either side of the fault plane. The term *separation* (Fig. 9.10) represents the displacement of a bed caused by faulting. The separation can be measured in any desired direction. If the separation has occurred along the dip of the fault plane, it is called *dip separation* (Fig. 9.10a) and is measured in a vertical section perpendicular to the fault plane, and if it has occurred along the strike of the fault plane, it is called *strike separation* (Fig. 9.10b) and is measured on the horizontal plane. The *vertical separation* is the displacement measured in a vertical section perpendicular to the strike of the fault. The term separation can further be used for expressing the displacement of any bed on a horizontal plane in any desired direction; this type of separation is called *geographic separation*. Thus in Fig. 9.10c, the horizontal separation along N-S and along NW-SE directions has been shown. It is therefore clear

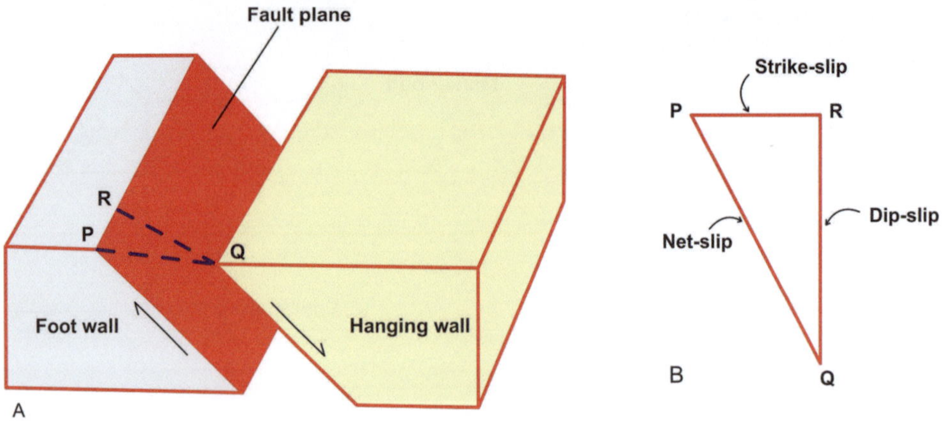

Fig. 9.9 (**a**) Displacement or slip of a bed affected by a fault. *RQ* dip-slip; *PR* strike-slip; *PQ* net-slip. (**b**) Triangular diagram showing the slip components of the fault

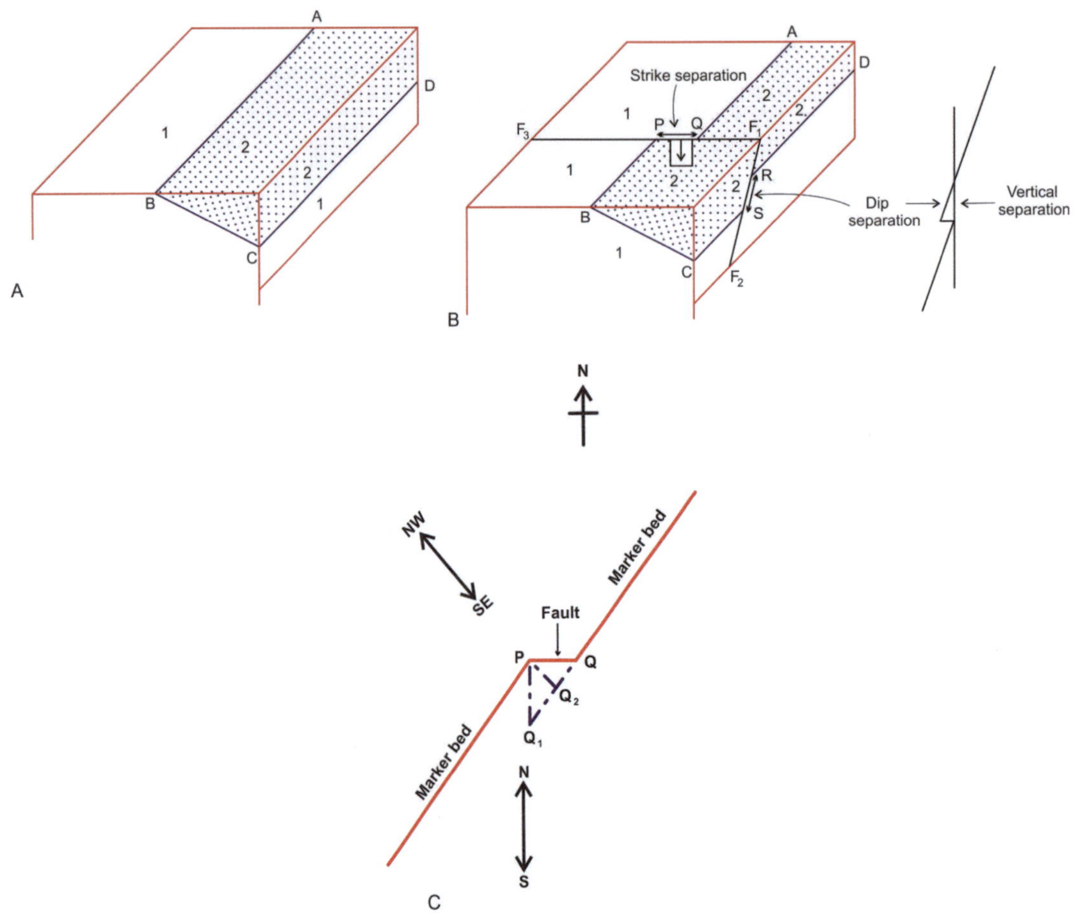

Fig. 9.10 Separation caused by a fault as observed on a geological map. (**a**) Block diagram showing an inclined bed ABCD (dotted) in the horizontal and vertical sections. We have considered here only two beds; ABCD is bed number 2 that overlies bed number 1. (**b**) The bed ABCD is affected by an inclined fault F_1F_2 whose trace in the horizontal plane is represented by F_1F_3. PQ is strike separation, while RS is dip separation. (**c**) Horizontal separation of the above fault can be measured in any desired direction; this is called geographic separation. For example, PQ_1 is the horizontal separation in the N-S direction, while PQ_2 is in the NW-SE direction

that the dip separation, strike separation and vertical separation of a particular fault remain unchanged; only the geographic separation changes with directions.

During drilling operations, some boreholes may encounter a fault; this point is called *fault cut*. In such situations, the normal stratigraphic sequence of the beds is disturbed; some beds are missing or are repeated. That part of the stratigraphic sequence that is missing or repeated is called *stratigraphic separation*. In general, the normal faults cause missing of a part of the stratigraphic sequence, while the reverse faults cause repetition.

9.3 Classification of Faults

Since rocks show a variety of faults, it is necessary to classify them. Faults can be classified on the basis of their geometry and genesis. The former classification is based on the geometry of faults, i.e. their attitude and relation to the adjacent beds, that can be easily seen in field or in hand specimens. The genetic classification takes into consideration the origin and formation of faults, and this is purely based on interpretation. As such, the geometrical classification seems to be more useful and practical than the genetic one. In this chapter, we shall present only the geometric classification of faults. We have considered five different schemes of classification of faults on the basis of translational movement, rotational movement, fault association, orientation of stress axes and slip on fault planes. Figure 9.11 is a flow chart showing classification of faults.

9.3.1 Classification 1: Based on Translational Movement

In translational faults, the marker lines of both the blocks are parallel (Fig. 9.12a), while in rotational faults, the marker lines are not parallel (Fig. 9.12b). We have broadly followed Billings (1972) for describing translational faults that have been further grouped into four categories as given in Fig. 9.11 and as described below.

9.3.1.1 Based on Attitude of Fault Relative to Adjacent Beds

Faults can be geometrically classified on the basis of the attitude of the fault plane with respect to the attitude of the adjacent beds in plan view. In this category, we consider the following types of faults with an example of four beds numbered 1, 2, 3 and 4 in ascending order as shown in the respective diagrams.

In a *dip fault* (Fig. 9.13), the trace of the fault is parallel to the direction of the dip of adjacent beds. In other words, the strike of the fault plane is perpendicular to the strike of the adjacent beds. The *strike fault* (Fig. 9.14) is parallel to the strike of the adjacent beds. For such faults, the net-slip FF_1 is equal to dip-slip. In an *oblique fault*, the strike of the fault is oblique to the strike of the adjacent beds. A *bedding fault* is a fault that is parallel to the adjacent beds, and as such it is a variety of strike fault (see Fig. 9.14).

9.3.1.2 Based on Slip of Fault Plane

Faults are also classified on the basis of net-slip. In a *dip-slip fault* (Fig. 9.15), the net-slip is along the direction of the dip of the fault; the strike-slip component is thus zero. The slip may be in the up (*normal fault*) or down (*reverse fault*) direction of the dip. In a *strike-slip fault* (Fig. 9.16), the net-slip is parallel to the strike of the fault; the dip-slip component is thus zero. Depending upon whether the opposite wall of an observer has moved to the right-hand or left-hand side, a strike-slip fault can be described as *dextral* or *right lateral* if the opposite wall has moved to the right or *sinistral or left lateral* if the opposite wall has moved to the left. An *oblique-slip fault* is one in which the net-slip is parallel neither to the dip or to the strike of the fault plane, i.e. the net-slip is disposed angularly to the dip and strike of the fault, and as such this fault has both a dip-slip component and a strike-slip component.

9.3.1.3 Based on Relative Movement of Beds

Another classification is based on the relative movement of beds along the non-vertical fault plane as observed in a vertical section perpendicular to the fault plane. If the hanging wall has gone down relative to the footwall, the fault is called a *normal fault* (Fig. 9.17), while if the hanging wall has gone up relative to the footwall, the fault is called a *reverse fault* (Fig. 9.18). If the dip of a reverse fault is less than 45°, it is called a *thrust fault* (Billings, 1972, p. 196). If the movement is parallel to the strike of the fault, it is called a *strike-slip fault* (Fig. 9.16) as described above.

9.3.1.4 Based on Amount of Dip of Fault

Since faults can have any amount of inclination, these are also classified according to the amount of dip of the fault plane. A *low-angle fault* (Fig. 9.19) is one in which the dip of fault is less than 45°. With dip around 45°, the fault is called a *moderate-angle fault* (Fig. 9.20). If the dip of fault plane is more than 60°, it is called a *high-angle fault*, and if the dip of fault is 90° it is called a *vertical fault* (Fig. 9.21). In a vertical fault, relative movement by hanging wall or footwall cannot be specified. It can only be specified which block (direction-wise, say northern, western, etc.) has moved up or down relative to the other block along the fault plane. In a *horizontal fault* (Fig. 9.22), the hanging wall and footwall are separated by a horizontal surface (fault), and movement along the fault is specified by describing the direction in which a particular block has moved relative to the other. Most horizontal faults constitute a part of some inclined fault.

9.3.2 Classification 2: Based on Rotational Movement

Instead of movement of a fault along the fault plane in a dip-slip, strike-slip or oblique-slip manner, as described above, a fault may show a rotational movement in which case it is called a *rotational fault* (Fig. 9.23). Due to rotation

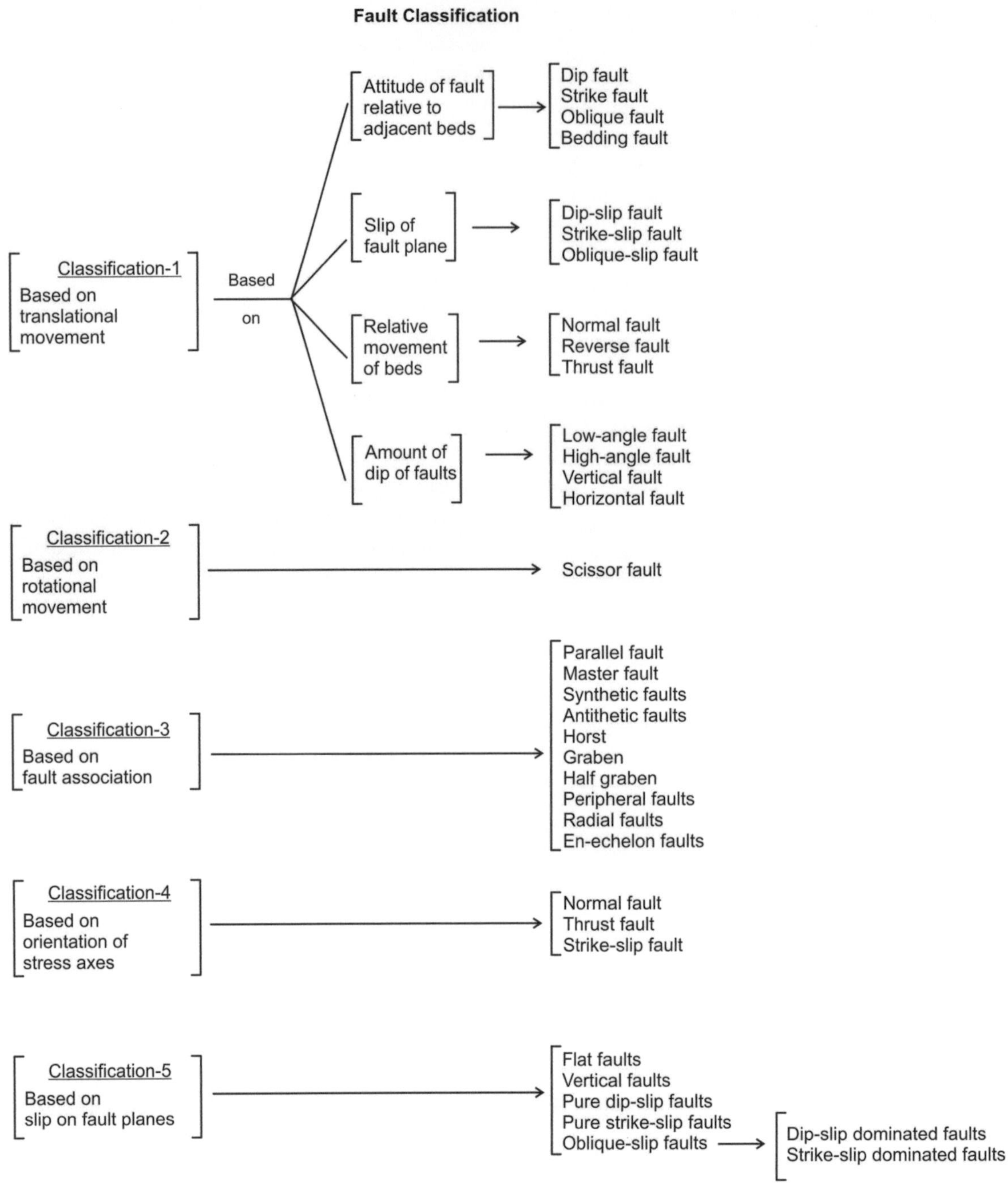

Fig. 9.11 Flow chart showing classification of faults. See text for details

of the fault, the original strike of the beds gets rotated, and therefore the two separated blocks do not show parallelism of strike (Fig. 9.24). In a rotational fault, therefore, the marker lines of both the blocks are not parallel on either block, and the movement is commonly given by the angle by which one block has rotated relative to the other. The axis of movement occurs in a plane perpendicular to the fault plane. The gap thus formed is filled by the material of the host rock.

A *scissor fault* (Fig. 9.25) is one in which the displacement occurs across a point, called *hinge*, such that the displacement on both sides of the hinge is reversed. The amount of displacement at hinge is zero, while it increases from the hinge on either side. In this fault, the displacement increases away from the hinge. Scissor faults are rare in crustal rocks.

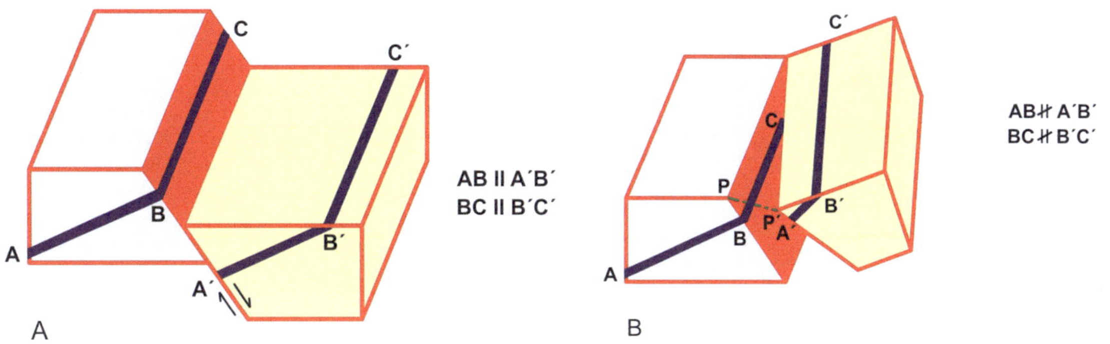

Fig. 9.12 Nature of movement of a marker bed along a fault. (**a**) Translational movement. Note that the original lines AB and BC representing a bed are parallel to the lines A′B′ and B′C′ after faulting. (**b**) Rotational movement. Note that the original lines AB and BC representing a bed are not parallel to the lines A′B′ and B′C′ after faulting. The fault plane is coloured red within which the disposition of a marker bed (black) has been shown

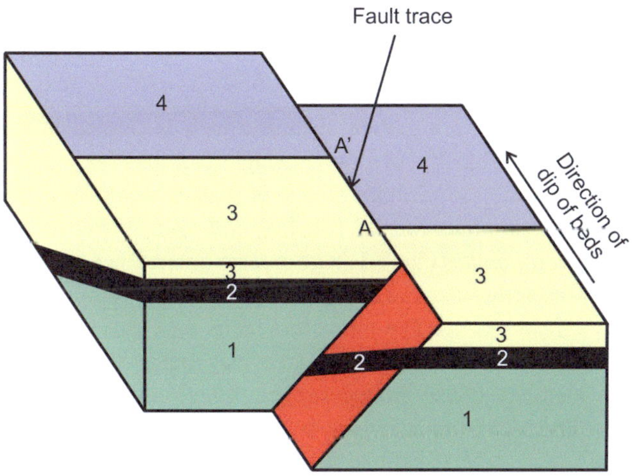

Fig. 9.13 Dip fault. The block diagram shows a succession of four beds, numbered 1, 2, 3 and 4 from base to top, affected by a dip fault. Point A is shifted to A' due to faulting. The trace of the fault is parallel to the direction of dip of beds and is therefore perpendicular to the strike of the beds. Fault plane is coloured red within which the disposition of a marker bed (2) has been shown

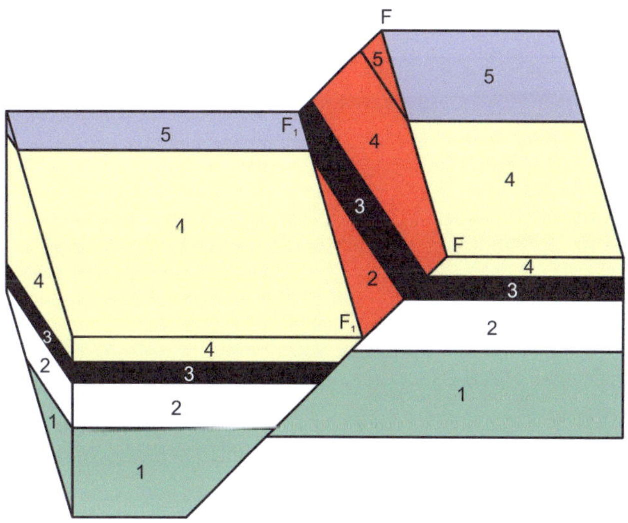

Fig. 9.15 Dip-slip fault. The net-slip FF_1 is along the direction of the dip of the fault and is equal to dip-slip. Strike-slip component is zero. The fault plane is coloured red within which the disposition of a marker bed (2) has been shown

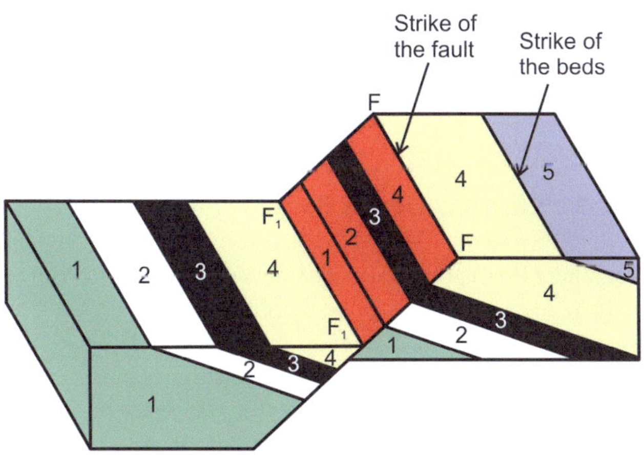

Fig. 9.14 Strike fault. The strike of the fault FF_1 is parallel to the strike of the beds. The net-slip FF_1 is equal to dip-slip. The fault plane is coloured red within which the disposition of a marker bed (2) has been shown

9.3.3 Classification 3: Based on Fault Association

Faults commonly occur in groups. In most cases, the individual faults of the association are of the same type. *Parallel faults* (Figs. 9.26 and 9.27) constitute a set of faults in which the dip and strike of all the faults are the same. The faults thus get arranged in a steplike manner, and as such these are also called *step faults* (Figs. 9.26 and 9.27). A large fault, commonly called a *master fault*, may be associated with a few smaller and parallel faults showing the same sense of displacement as the large fault. Thus, a *synthetic fault* (Fig. 9.28a) is a minor fault that has the same displacement sense and a similar orientation to a related major fault, while an *antithetic fault* (Fig. 9.28b) is a fault that dips in the opposite direction to a related dominant fault or fault set (Peacock et al. 2000).

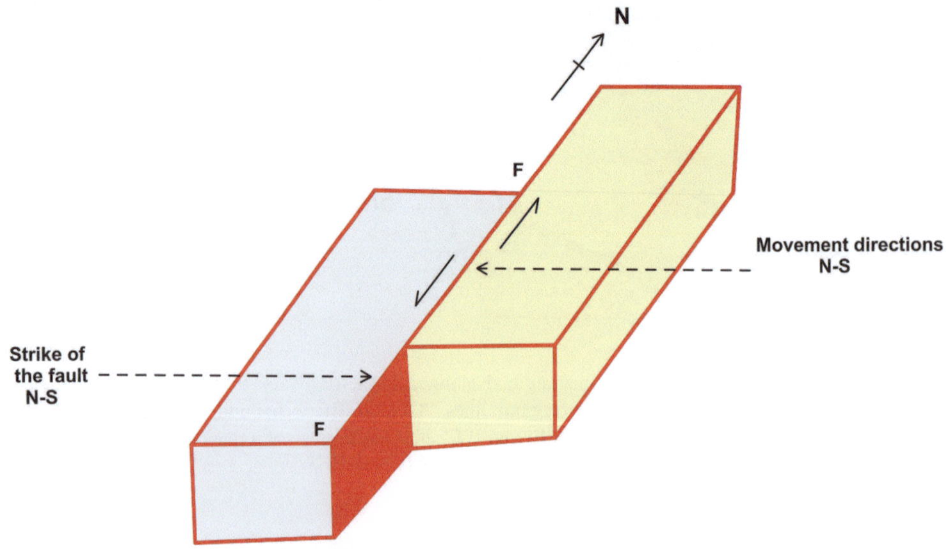

Fig. 9.16 Strike-slip fault. Standing on any block, the opposite block has moved to the left-hand side, and as such it is a sinistral or left lateral fault. The fault plane is coloured red

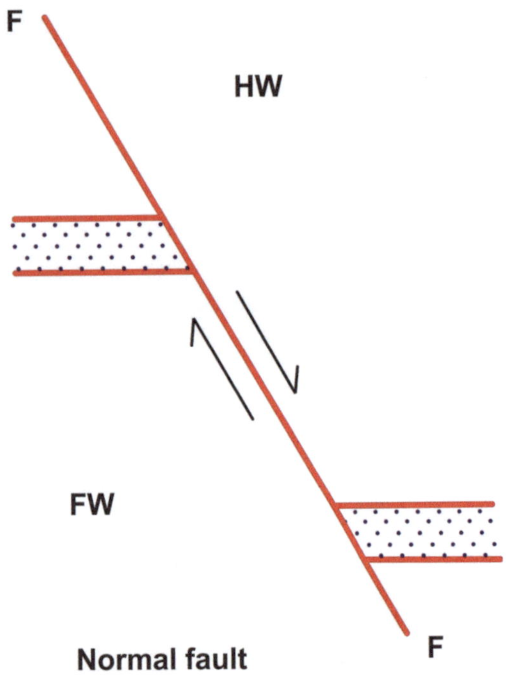

Fig. 9.17 Normal fault. The hanging wall (HW) has gone down relative to the footwall (FW)

Some normal faults occur in a conjugate set in which individual faults on either side dip in opposite direction. Under such a setting, the block that is uplifted between two oppositely dipping faults is called a *horst*, while the block that has been downthrown is called a *graben*. Horsts and grabens (Fig. 9.29) commonly occur adjacent to each other. In a graben, the normal faults dip towards the centre of the structure, while in a horst the normal faults dip away from the centre of the horst.

Sometimes in a graben, a fault trough is formed in the downthrown block when only a single fault has been active such that the sedimentary wedge is thicker along the fault. The structure thus formed is called *half graben* (Fig. 9.30). *Peripheral faults* (Fig. 9.31) are arcuate or circular faults that fully or partly surround a circular area exposing salt diapir or any igneous body. The individual faults look curved on the ground surface. The circular area may be a depressed or elevated part of the ground surface brought to this position by vertical tectonics caused by the circular faults. *Radial faults* (Fig. 9.32) constitute an association in which the faults radiate from some central point. The faults thus converge towards a point. In some cases, a number of smaller faults overlap and are approximately parallel to each other; such faults are called *en echelon faults* (Fig. 9.33).

9.3.4 Classification 4: Based on Orientation of Stress Axes

See Anderson's Theory in Sect. 9.8.

9.3.5 Classification 5: Based on Net-Slip

Angelier (1994) proposed a classification based on the slip on fault planes. The slip can be identified by the presence of slickenside lineations that reveal the sense of relative motion due to the presence of various asymmetric features on the fault surfaces. This gives a variety of *slip fields* on the slip surface. The slip can be in the direction of dip of the fault surface, parallel to the strike or oblique to both the dip and strike directions. Accordingly, the faults can be named as *dip-*

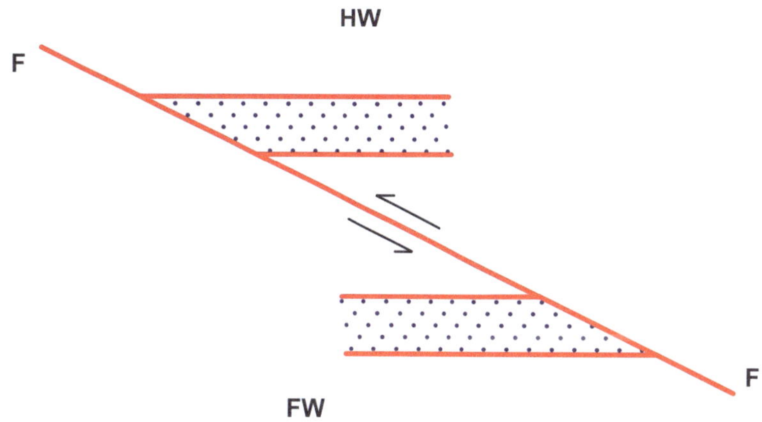

Fig. 9.18 Reverse fault. The hanging wall (HW) has gone up relative to the footwall (FW)

Fig. 9.19 A low-angle fault developed in limestone beds of the Tethys sequence of Kashmir Himalaya, India, south of Sonprayag. (Photograph by the author)

Fig. 9.20 A moderate-angle fault developed in massive sandstone of Krol Group of Lesser Himalaya. Loc.: southeast of Nainital, India. (Photograph by the author)

Fig. 9.21 A vertical fault developed in red sandstone-shale beds of the Rome Formation, Southern Appalachians, at Moores Gap, NW of Knoxville, USA. (Photograph by the author)

Fig. 9.22 A horizontal fault developed in sandstone beds of Krol Group of Kumaun Lesser Himalaya. The fault constitutes a discrete plane and appears to be the continuation of an inclined fault (that can be noticed at the right-hand side of the photograph). (Photograph by the author)

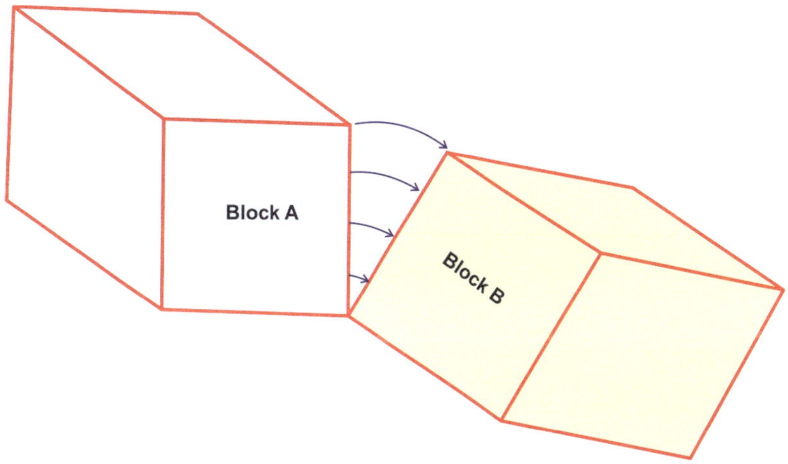

Fig. 9.23 Rotational fault. Block B has undergone rotation in the direction shown by arrows. This rotation is given by the angle made by the two separated edges of the blocks. In the diagram, the gap caused due to rotation is merely a geometric construction

Fig. 9.24 Rotational fault in calcareous rocks of Delhi Supergroup near Ambaji, Rajasthan, India. Note that, due to rotation of the fault, the original strike of the beds gets rotated, and therefore the two separated blocks do not show parallelism of strike (Photograph courtesy Professor T.K. Biswal)

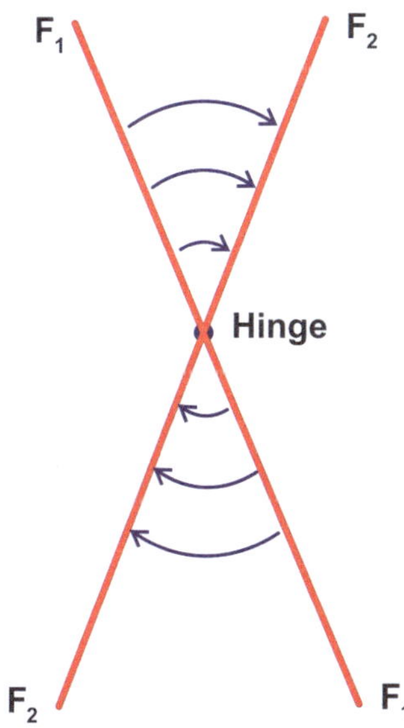

Fig. 9.25 Scissor fault. Two blocks have undergone rotation along two faults F_1F_1 and F_2F_2. The faults rotate along a hinge on either side of which the displacement is reversed and progressively increases with distance from the hinge where displacement is zero

slip fault, *strike-slip fault* or *oblique-slip fault*, respectively. For oblique-slip fault, the degree of obliquity is given by the *pitch*, which is the angle between the slip and the strike of the fault surface. Thus, on the basis of dip-pitch relationship, Angelier (1994) classified faults (Fig. 9.34) by identifying five major domains, as given below. He identified two major components of fault slip: T, which is the transverse horizontal component of displacement, and L, the lateral horizontal component of displacement.

(a) **Flat faults** in which the fault dips less than 10°–15°. Pitch of the slip is insignificant as the strike of the fault is highly variable. These faults lie along the left side of the diagram.
(b) **Vertical faults** in which the slip is predominantly vertical and the fault surface dips more than 75°–80°. Pitch of the slickenside lineation is steeper than 45°. These faults lie along the right side of the diagram.
(c) **Pure dip-slip faults** lie on the uppermost part of the diagram. The ratio of T/L is larger than 4–8.

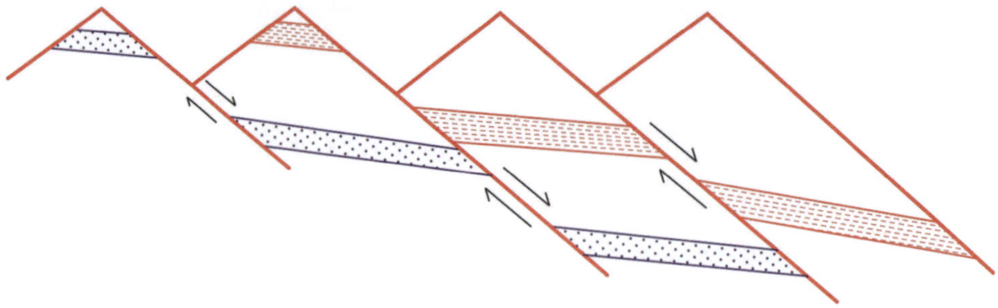

Fig. 9.26 Diagrammatic sketch of parallel faults. A group of faults are arranged in a steplike manner such that the dip and strike of all the faults are the same

Fig. 9.27 Step faults shown by silica bands in a ferruginous quartzite of the Bundelkhand craton, central India. (Photograph by the author)

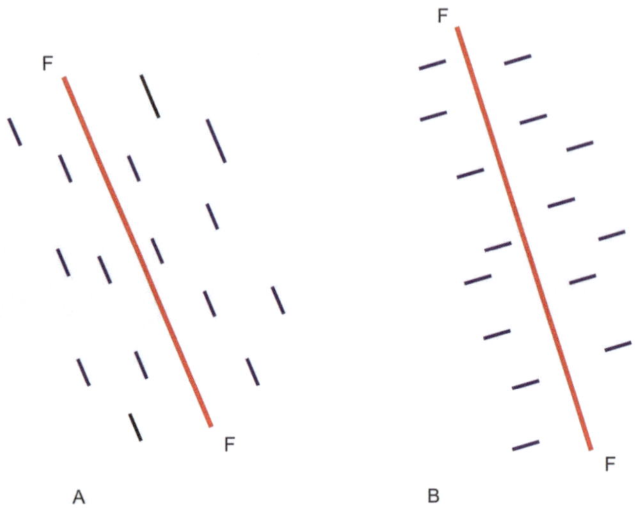

Fig. 9.28 (**a**) Synthetic faults given by a set of smaller faults that are parallel to the main fault or master fault FF. (**b**) Antithetic faults given by a set of smaller faults that are perpendicular to the main fault FF

(d) **Pure strike-slip faults** lie on the lowermost and lower-right parts of the diagram. The ratio of *T*/*L* is larger than 4–8.

(e) **Oblique-slip faults** lie in the central part of the diagram. These faults can further be subdivided into two subareas: a normal-dextral fault in the upper subarea and a dextral-normal fault in the lower subarea.

9.4 Recognition of Faults

Faults are recognized in field by a variety of ways. Although smaller faults can directly be seen in field, the larger faults create some problems. Faults are generally associated with a variety of features that help in locating them. We have grouped here the common methods of recognition of faults mainly on the basis of the features characteristic of fault surface, geological effects of faults on strata, geomorphologic criteria and related features as described below.

9.4.1 Presence of Fault Rocks

Faults are commonly associated with some rocks called *fault rocks*. These rocks are formed during the faulting process and are therefore characterized by their typical textures. Common fault rocks include *breccia* (Fig. 9.35) and *cataclasite* (described below). Invariably, breccias are associated with faults in which case these are called *fault breccias*. *Fault gouge* (Fig. 9.36) is another fault rock which is fine grained and is formed when the original rock is fully pulverized as a result of rubbing during faulting. With progressive movement along the fault surface, the broken pieces undergo pulverization. Thus, a zone of crushed or pulverized rocks or both is developed along the fault surface. The grains are of clay size. Crushing and pulverization are characteristics of brittle faults that are believed to form very near to the surface or at shallow crustal levels. Pseudotachylite (Fig. 9.37) is a very-fine-grained rock resembling glass. It is formed due to

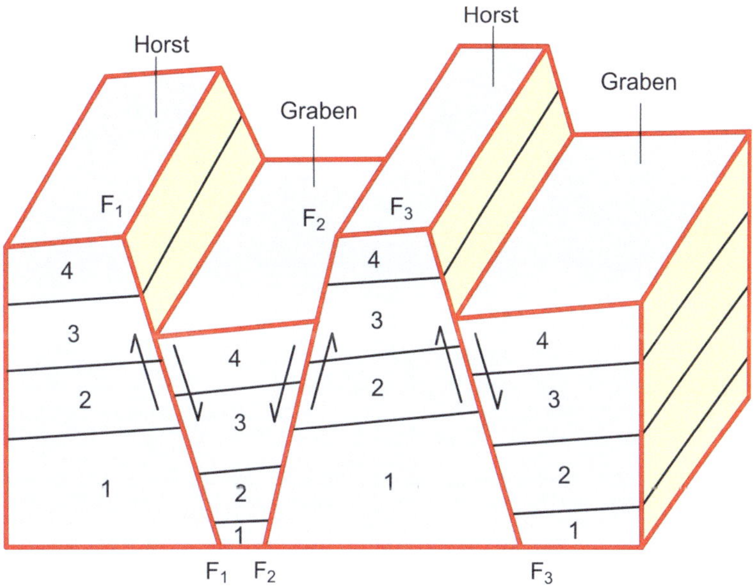

Fig. 9.29 Horst and graben structure. The block that is uplifted between two oppositely dipping faults F_2F_2 and F_3F_3 constitutes a horst, while the block that has been downthrown between two oppositely dipping faults F_1F_1 and F_2F_2 constitutes a graben

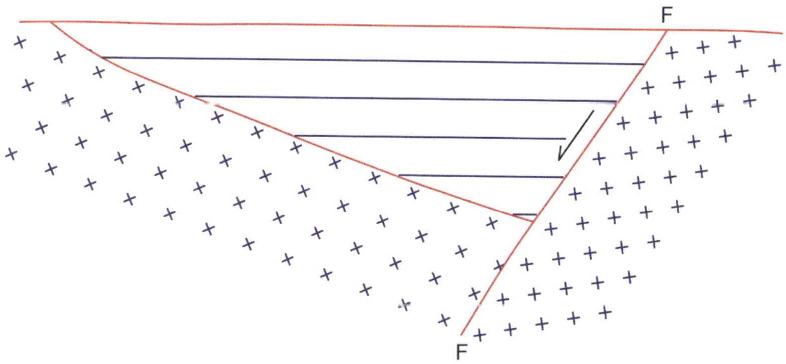

Fig. 9.30 Half graben. The structure is actually a fault trough formed due to downward movement along a single fault FF. As such, the sedimentary wedge is thicker along the fault

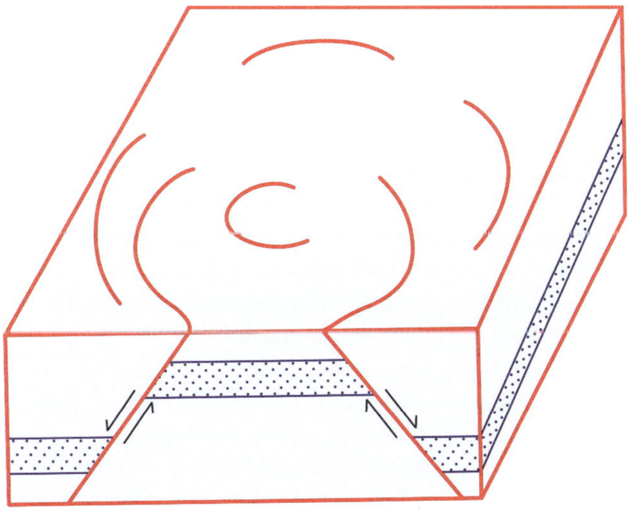

Fig. 9.31 Peripheral faults. The structure includes a number of arcuate or circular faults developed around a salt diapir or igneous body. The faults look circular or arcuate on the ground surface

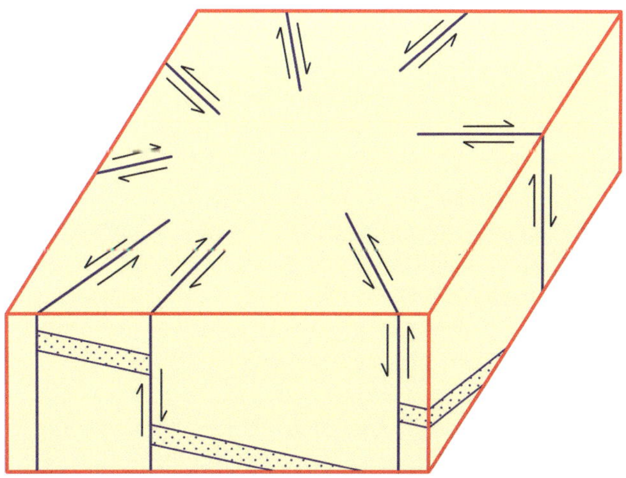

Fig. 9.32 Radial faults. A number of faults radiate from a central point

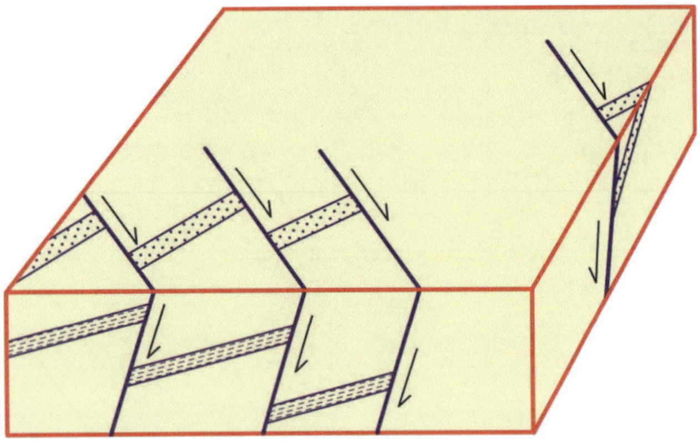

Fig. 9.33 En echelon faults. A few nearly parallel faults overlap each other

high frictional heat sufficient enough to temporarily reach the temperature of melting. As the displacement along the associated fault stops and as the frictional heat disappears, the rock cools down suddenly in the form of glassy material. The molten material thus spreads randomly in the rock mass or in the form of veins; in both the cases, some broken pieces of surrounding rock may be engulfed in the molten mass (Fig. 9.37a,b). The veins may not include pieces of surrounding rocks (Fig. 9.37c) and may show small offsets due to concomitant faulting (Fig. 9.37d). Presence of pseudotachylite is believed to be an evidence of palaeoseismicity (Sibson 1975, 2003).

Box 9.2 Active Faults

When we talk of faults, we generally mean a geological fault, i.e. one that was formed in the geologic past. Thus, a fault may have formed, say, in the Precambrian or Permian or Cretaceous period and so on. Generally, the movement or displacement associated with any fault formed in a certain geologic period may have been locked after its formation; that is, it has not shown any evidence of its movement or activity after its formation. Such faults are simply called faults, and these may be considered as *dormant faults*.

Contrary to the above-mentioned faults, there are some faults that show evidences of movement or activity that are recorded within the historic or prehistoric periods or within the recent times as witnessed by man. Such faults are called *active faults*. Our understanding of active faults is growing these days mainly because of our environmental awareness. Nowadays, study of geoenvironmental stability of an area also includes knowledge of the presence or absence of active faults. For this, evidences from field and historic records are

Box 9.2 (continued)

collected to ascertain the presence or otherwise of active faults in an area. Once the occurrence of an active fault is established, the immediate information one needs is the age or period of activity of the fault. It then becomes imperative to seriously study the geoenvironmental stability of the area and the implications thereof.

In recent years, there has been growing interest in knowing the age or period during which a fault was active. Work on this line has helped in classifying faults according to their period of activity. On the basis of fault activity, faults are nowadays broadly classified under active faults and potentially active faults (see Keller 2001 for details). Active faults are those that have shown activity during the last 10,000 years, i.e. during the Holocene Epoch, while *potentially active faults* have shown their activity between about 1.65 million and 10,000 years, i.e. during the Pleistocene Epoch excluding the Holocene. Faults that have not shown activity during the past 1.65 million years are grouped as *inactive faults*.

Although the criteria for identification of active faults have been worked out, little work is available on what type of faults can be potential for becoming an active fault. Recently, Kaninskaite et al. (2020), while working on the internal architecture and petrophysical properties of some faults of Sicily, Italy, suggested that cemented fault rock is unlikely to be reactivated.

9.4.2 Features or Marks on Fault Surface

Faults are invariably associated with some features or marks on the surface of the rocks. Since these marks are formed due to rubbing of rocks against each other, most of these can be taken as direct proof of faulting. However, this conclusion should be arrived at with caution because during deformation of any kind, beds may undergo some relative movement past each other. Such relative movements are commonly part of some other deformation processes, not necessarily faulting. This confusion can however be resolved by collecting some other evidences of faulting on that spot or locality. Some common features developed on fault planes are described below.

(a) **Slickensides** are polished striated surfaces (Fig. 9.38) formed due to frictional movement during fault motion. The structure occurs as a set of alternating grooves and depressions oriented parallel to the direction of movement along the fault surface. This gives a stepping up and down nature of the slickenside surface (see also Chap. 15).

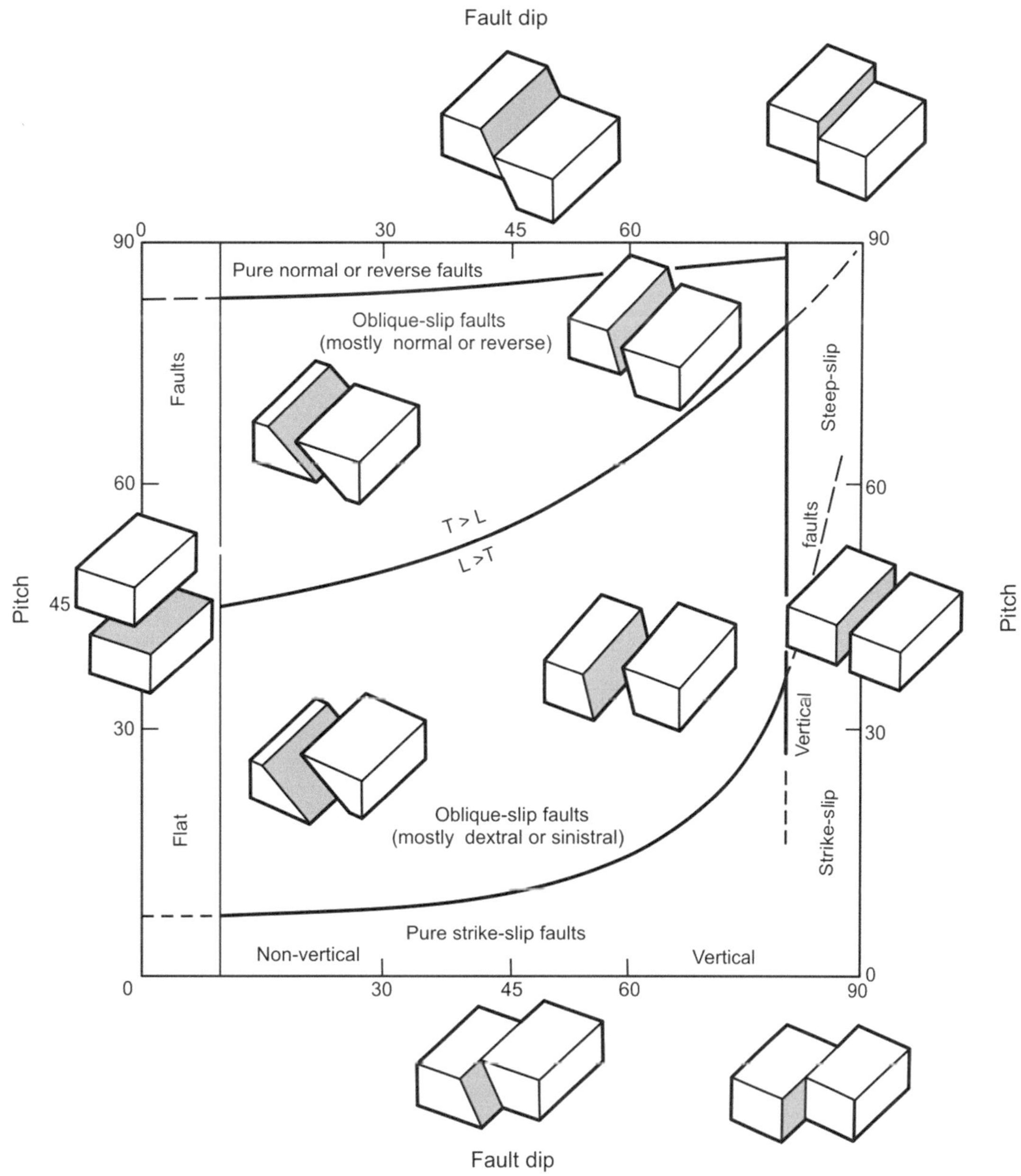

Fig. 9.34 Angelier's classification of faults. The abscissa represents fault dip, and the ordinate represents the pitch of slickenside lineation. See text for details. (Reproduced from Angelier 1994, Fig. 4.15. In: Hancock PL (ed.) 1994. Continental Deformation. Pergamon Press, Oxford, with permission from Elsevier Science & Technology Journals. Request ID: 600072892)

(b) **Slickenlines** are linear scratches developed on a fault surface. Slickenlines actually constitute the grooves of slickensides (Fig. 15.5), which can thus be considered to include slickenlines.

9.4.3 Structures Associated with Fault Surface

Faults are occasionally associated with some structures that are formed during relative movement of fault blocks; some common types of such structures are briefly described below.

(a) **Drag folds** are curved surfaces or flexures in which the strata show upward inclination against the fault in the hanging wall and downward inclination in the footwall (Fig. 9.39). Drag folds are typically asymmetric and non-cylindrical (Fig. 9.40). Since these folds are typically associated with a fault, these are also sometimes called *fault drags*. Presence of drag folds thus suggests relative movement of two blocks along a fault.

(b) **Drape fold** (Fig. 9.41) is a flexure or open fold of the sedimentary layers that conforms to the surface geometry or configuration developed due to the presence of an

Fig. 9.35 Breccia associated with a fault in the Bundelkhand craton, central India. (Photograph by the author)

Fig. 9.36 Fault gouge occurring as a powdery material in a fault zone (demarcated by yellow dashes). The rock is a limestone of the Krol Group of Kumaun Lesser Himalaya, India, exposed northeast of Nainital. (Photograph by the author)

underneath fault. A *drape fold* is thus a fault-related fold. During sedimentation of the cover rocks, the layers under sediment overburden have to take up the shape of the local ground configuration already developed by a fault in the basement rocks. As such, a drape fold is also sometimes called a *forced fold*.

9.4.4 Secondary Crystallization

Since faults are fractures, they constitute channels for fluids below the ground surface. These fluids react with the wall rocks and commonly bring silica (silicification) or calcite into solution, which are later on deposited as veins and lenses near the fault surface (Fig. 9.42). The degree of secondary crystallization however decreases away from the trace of the fault. Silicification is commonly considered as an evidence for the presence of a fault. In addition to silica, some other minerals may also be deposited along faults, thus adding mineralization as an additional criterion. However, some other evidences are also needed to prove the existence of faults. The reason is that fluids may pass through any avenue other than faults.

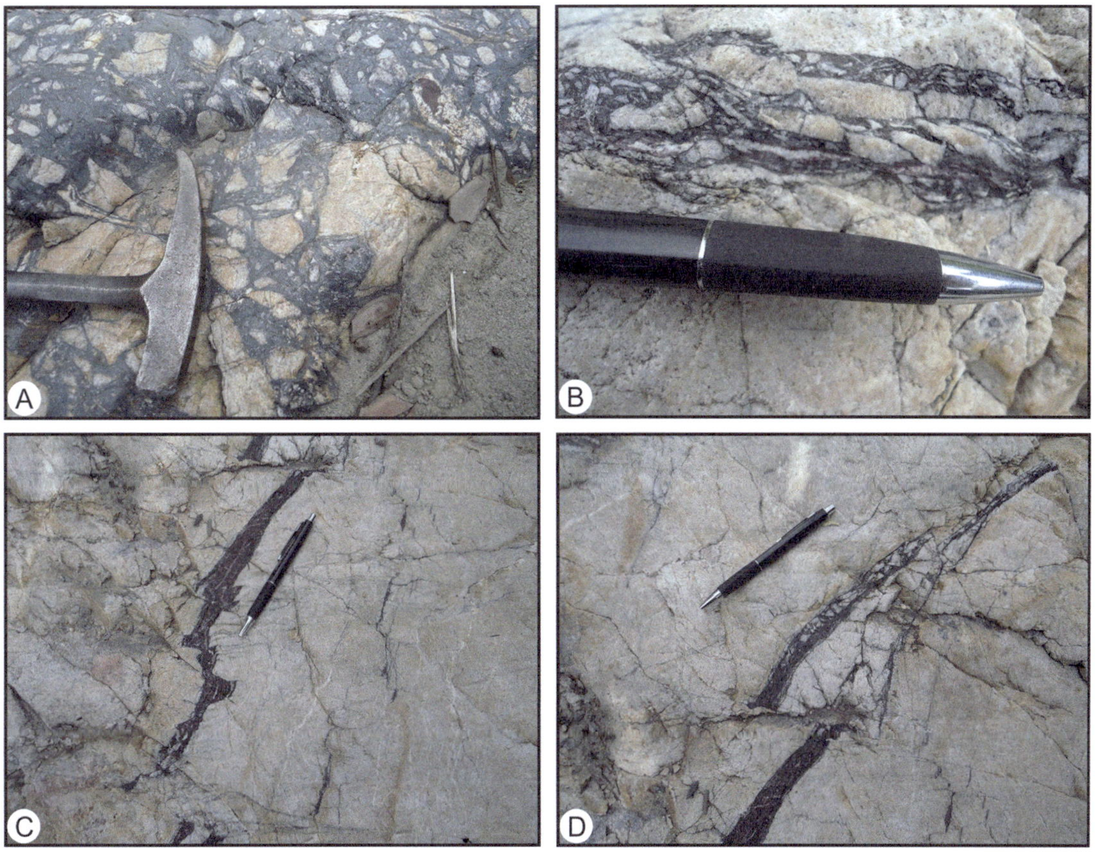

Fig. 9.37 Pseudotachylite (black portions in the photographs). (**a**) The flowing molten mass has spread in a random manner and has engulfed a few broken blocks of the surrounding rock. (**b**) The flowing molten mass occurs in the form of a vein within which a few broken blocks of the surrounding rock can be seen. (**c**) A vein of pseudotachylite. (**d**) The pseudotachylite vein shows offsetting due to concomitant faulting. Loc.: North of Bhimtal on the road to Almora, Kumaun Lesser Himalaya, India. (Photographs by the author)

Fig. 9.38 Slickensides developed in limestone. Note the stepping up and down nature of the slickenside surface. Loc.: Near Hampteau Hotton, Belgium. (Photograph by the author)

Due to brecciation, a part of the associated rock is fragmented. The silica fluids then permeate the zone undergoing brecciation and precipitate as the stress is decreased due to cessation of movement along the fault. Thus, brecciation and silicification can be seen together along a fault (Fig. 9.43).

9.4.5 Behaviour of Strata

9.4.5.1 Abrupt Truncation of Strata

Occasionally, the continuity of a bed may abruptly end along a sharp line (Fig. 9.44) that may represent the presence of a fault. A similar situation may be observed in the geological

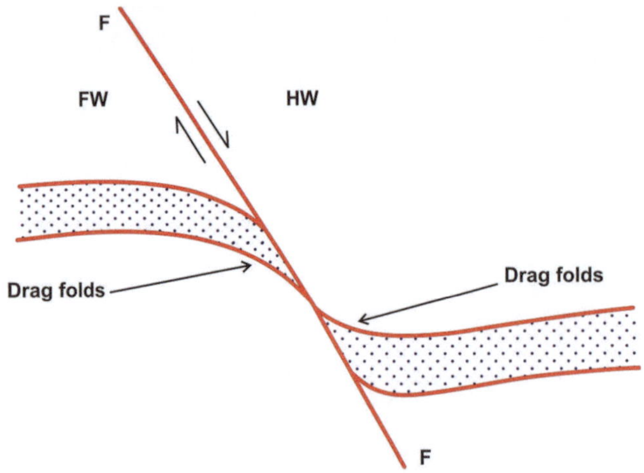

Fig. 9.39 Diagrammatic sketch of drag folds associated with a fault FF. Note that the bed (stippled) shows upward inclination against the fault in the hanging wall (HW) and downward inclination in the footwall (FW)

Fig. 9.40 Drag folds (within the rectangle) associated with a high-angle fault in the Krol Group near Nainital, Kumaun Lesser Himalaya, India

Fig. 9.41 Block diagram to show drape fold developed in the sedimentary layers that overlie an underneath fault

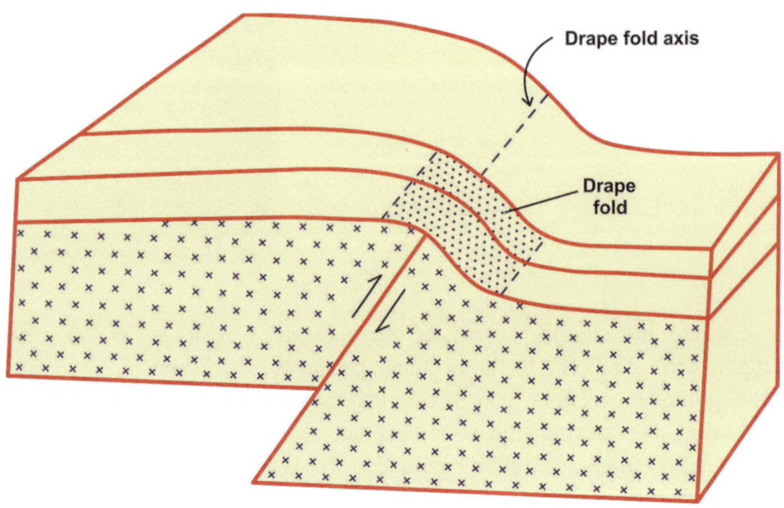

Fig. 9.42 Silicification associated with a fault. A zone of silicification (light-coloured patches) can be seen at the contact of a fault FF (shown by yellow dashes). Along the fault, crushing and pulverization of the footwall rocks can also be seen. The rocks are shales of Lower Krol Formation, northeast of Nainital, Kumaun Lesser Himalaya, India. (Photograph by the author)

Fig. 9.43 Silicification and brecciation occurring together along a fault FF (shown by yellow dashes) developed in carbonate rocks of Lower Krol Formation, near Bhimtal, Kumaun Lesser Himalaya, India. (Photograph by the author)

map of an area in which a particular bed may show truncation along a sharp line. In such cases also, the line of truncation may represent the presence of a fault.

9.4.5.2 Abrupt Truncation of Structure

Due to the presence of a fault, the structure of a rock, say folds, sometimes abruptly truncates along the fault plane (Fig. 9.45). The continuity of the structure may thus be traced on one block only.

9.4.5.3 Juxtaposition of Contrasting Rocks

Sometimes, a fault brings two different rock types disposed on either side of the fault plane (Fig. 9.46). The continuity of each rock type abruptly ends along the fault plane.

9.4.6 Change in Attitude of Strata

9.4.6.1 Sudden Steepening of Strata

A sequence of beds persistently showing low or moderate dips may locally show gradual steepening, or even vertical attitude, of the beds (Fig. 9.47) towards a fault. In such cases, the presence of a fault may be inferred along the zone of highest steepening of beds. Sometimes, beds on both sides of the fault show sympathetic steepening or even curving on opposite sides (Fig. 9.48). In such cases, although a fault is indicated, the fault plane is sometimes difficult to mark precisely.

9.4.6.2 Anomalous Dip and Strike of Strata

In the vicinity of a fault, the strata commonly show anomalous pattern of dip and strike (Fig. 9.49). A line separating the anomalous zones may indicate the trace of a fault. This should however be confirmed by some other observations that indicate the presence of a fault.

9.4.6.3 Abrupt Change in Dip of Strata

In some cases, the dip (both amount and direction) of the strata changes on either side of a fault (Fig. 9.50). The surface that juxtaposes the two blocks with different dips of strata marks the presence of the fault.

Fig. 9.44 Abrupt truncation of strata due to the presence of a fault. The rock is granite gneiss of Bundelkhand craton, central India. (Photograph by the author)

9.4.7 Geomorphologic Criteria

Faults are occasionally associated with some geomorphic features. This is more so with active faults. Some important geomorphic features commonly noticed along faults are described below.

9.4.7.1 Scarps/Fault Scarps

A scarp is a sudden rise of topography as exhibited by a planar slope (Fig. 9.51). The inclination of the slope varies from medium to high.

Scarps are commonly preserved in case of active faults. Presence of a scarp is not always a proof of a fault as several other geologic-physiographic processes can also produce a scarp (for example, when there exists marked difference in the degree of erosion due to the presence of soft rocks in association with hard ones). However, in the case of active faults or rejuvenated faults, there is some degree of certainty for the presence of a fault in which case it may be called a *fault scarp*. In the absence of any proof of a fault, the locally developed steep slope can only be called a *scarp*.

9.4.7.2 Localization of Springs

Occurrence of springs in a row (Fig. 9.52) in a hilly terrain may indicate the presence of a fault. The reason is that a fault is a crack or a fracture plane. The groundwater below the surface is always at high pressure. If a fault is present, the water finds it as an easy channel for movement towards the zones of lower pressure, i.e. ground surface. The water that comes up to the surface forms local springs in a row that follows the trace of the fault. In some areas, *hot springs* are present. Their presence indicates the existence of some deep fractures along which the hot water has moved upwards. Hot springs may thus indicate the presence of a fault.

Fig. 9.45 Abrupt truncation of a folded structure (left-hand side of the photograph) along a fault plane. The rocks are from the Himalayan collisional zone near Leh, Ladakh, Himalaya, India. The simple dipping structure of rocks of the right-hand side block also abruptly ends along the fault plane. (Photograph courtesy Narendra K. Verma)

Fig. 9.46 Two contrasting lithologies are juxtaposed along a fault plane FF in the Bundelkhand craton, central India. Loc.: Talbehat. (Photograph by the author)

Fig. 9.47 Steepening of strata as a fault FF is approached. The strata on the right-hand side progressively become steeper towards the fault surface. Loc.: Bhimtal-Almora motor road, Kumaun Lesser Himalaya, India. (Photograph by the author)

Fig. 9.48 Steepening and curving of strata along a fault surface FF. Loc.: About 20 km NNW of Bhimtal, Kumaun Lesser Himalaya, India. (Photograph by the author)

Fig. 9.49 Anomalous pattern of dip and strike shown by rocks of both the blocks of a fault. Loc.: South of Sonprayag, Kashmir Himalaya, India. (Photograph by the author)

Fig. 9.50 Abrupt change in dip of strata due to the presence of a fault (yellow dashes) in quartzitic rocks of Bundelkhand craton, central India. The dip of the strata, in both amount and direction, is different on either side of the fault. (Photograph by the author)

9.4.7.3 Straightening of Stream Course

On the ground surface, the trace of a strike-slip fault or a vertical fault commonly shows a straight outline. In such areas, the normal course of a river becomes straight as long as it follows the fault line beyond which the river again shows its normal non-straight or serpentine course. Depending upon the length of the fault line, the straightness of the river course may continue for a few hundred metres to several kilometres.

Sometimes, the straight course of a river (Fig. 9.53) in field or in topo-sheet gives indication of a fault. However, in such cases, the presence of the fault should be ascertained in field by other evidences.

Fig. 9.51 Fault scarp. View towards NW onto the SW dipping fault scarp of the Shkoder-Peja Normal Fault (behind a group of six people) that juxtaposes ophiolitic mélange (green meadow) in its hanging wall against the Mesozoic limestones of the High-Karst nappe in the footwall. Albanian Alps near Tropoja. (Photograph courtesy Professor Stefan Schmid)

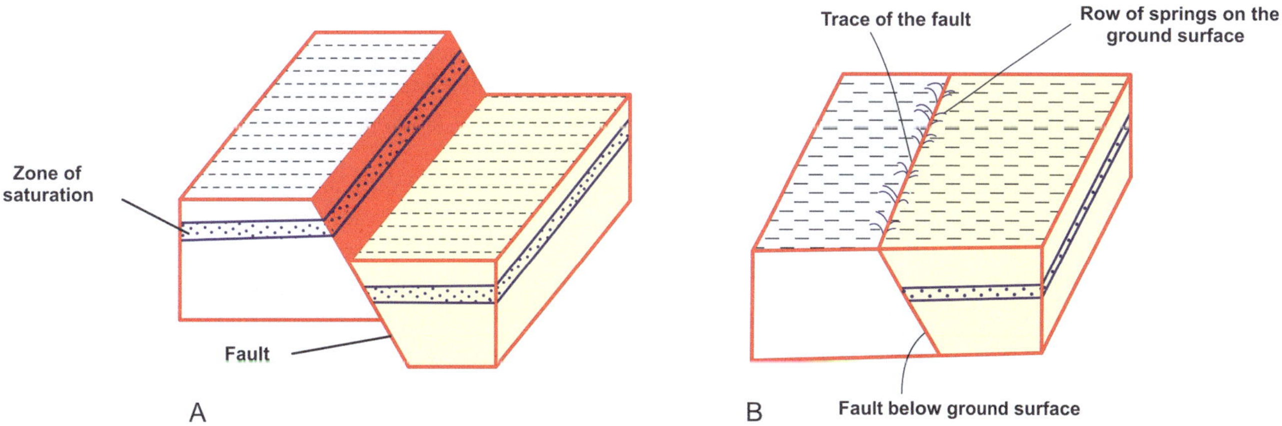

Fig. 9.52 Localization of springs along a linear belt may indicate the presence of a fault. (**a**) A normal fault is developed bringing down a block of rock strata to the right. (**b**) After erosion of the upthrown block (left), a row of springs is occasionally developed along the trace of the fault on the ground surface. The springs originate from where the fault intersects the water table (zone of saturation). The water follows the fault plane and rushes up to the ground surface due to decrease of pressure caused by the development of the fault

9.4.7.4 Sudden Change in Stream Profile

Under normal conditions, a stream follows a smooth profile of its course. If a fault occurs across the stream course, the profile shows a sudden change or break along the fault line. A waterfall is commonly developed (Fig. 9.54) along a normal fault due to the change in stream profile. The reason is that there is a difference in the rate of stream erosion and faulting. If, on the other hand, both these rates are equal, the stream profile would not show any sudden change and consequently the waterfall may not develop.

Fig. 9.53 Straight course of a river sometimes indicates the presence of a fault. Other field evidences such as presence of a steep scarp, juxtaposition of contrasting rock types, sudden changes in dip and strike of strata and slickensides on rock surfaces confirm the presence of the fault. Loc.: Near Raneh Falls on the Ken River, Madhya Pradesh, India. (Photograph courtesy Narendra K. Verma)

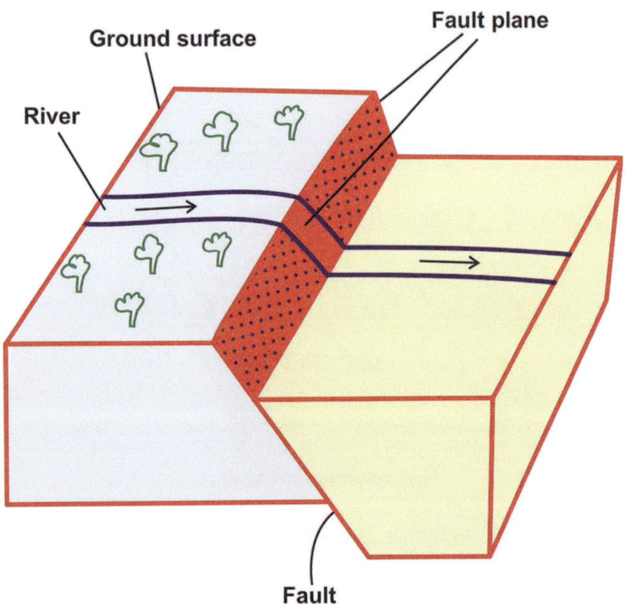

Fig. 9.54 Sudden change in stream profile may occur due to the presence of a normal fault. A waterfall is commonly developed at the point where the stream crosses the fault surface

9.4.7.5 Truncation of a Mountain Front

A mountain chain may suddenly show truncation of continuity at its front that is clearly visible if an alluvial plain is present at the mountain front. It is possible that a fault is present along the line of truncation. In such cases, the rocks of the mountain are seen juxtaposed against the flat plains (Fig. 9.55).

The Himalayan mountain front truncates at high angles against the alluvial plain to its south at several places. Sometimes, a river follows the truncation zone (Fig. 9.56) for some distance. In such cases, that portion of the river course that follows the mountain front can be suspected to be a fault, though some other criteria are also needed for inferring the fault.

9.5 Fault Damage Zone

Although faults are known to be planar surfaces, in reality they do not look ideal in geological or seismic sections. Faults are irregular on all scales and so are their deformation

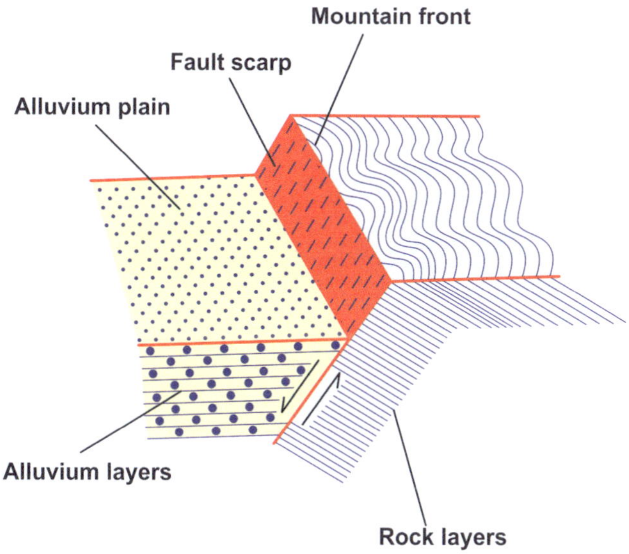

Fig. 9.55 Diagrammatic sketch showing truncation of a mountain chain against an alluvial plain. Note that the rocks of the mountain (showing folded structure in this case) suddenly abut against the fault plane. The upthrown block constitutes the topographic high (mountain) below which the erosionally resistant layers of rock of the mountain abut against the fault surface below the alluvium

products. Close observations suggest that most faults contain a *fault damage zone* (Fig. 9.57) consisting of a material that is the product of brittle deformation caused by the *fault core*. *A damage zone is the volume of deformed wall rocks around a fault surface that results from initiation, propagation, interaction and build-up of slip along faults* (Cowie and Scholz 1992; McGarth and Davison 1995). The nature of damage zones is controlled by various factors including lithology, dip of bedding relative to the slip direction of the fault and stress system (Kim et al. 2004). The damage zone is a heterogeneous mixture of crushed rocks and broken portions of the host rocks. Within the damage zone, some common features associated with faults such as synthetic faults and brecciation are also noticed. Fault damage zone is commonly developed in brittle rocks. The fault core generally constitutes a few millimetre thick surfaces that are responsible for the bulk slip or displacement of the fault.

In field, most faults do not show the development of the fault core ideally. In fact, it may have developed during early stages of faulting, but due to continued displacement along the fault, and the follow-up internal deformation, the deformation products get rearranged within the fault zone. What is

Fig. 9.56 Photograph showing abrupt truncation of the Himalayan mountain front against the alluvial plains of North India. The abrupt truncation produces scarps at several places that can be considered to be due to the presence of a regional scale thrust fault called the Himalayan Frontal Thrust. In the photograph, the Ganga River is seen flowing along the line of this truncation (yellow dashes) at Rishikesh, India. (Photograph by the author)

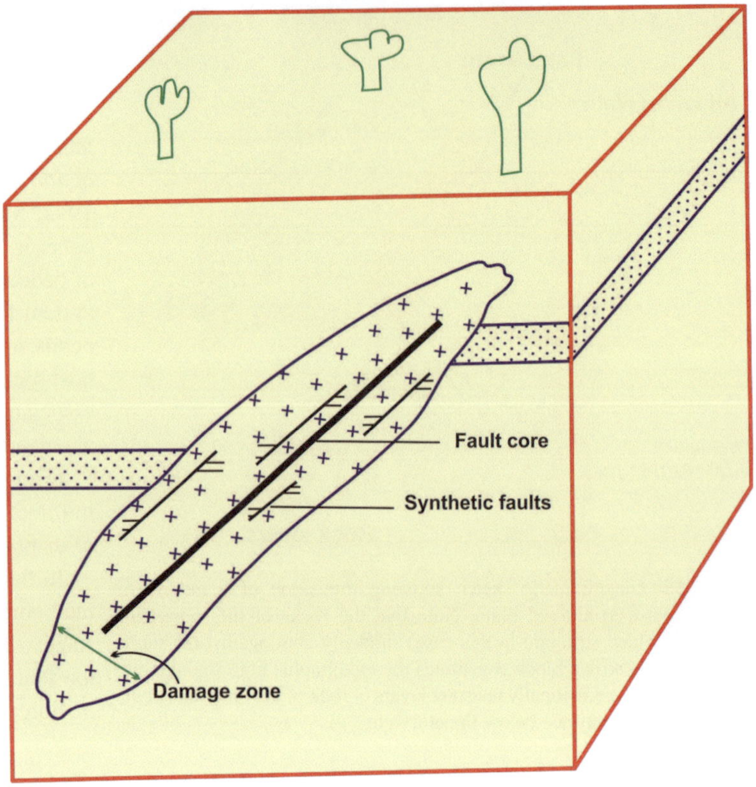

Fig. 9.57 Diagrammatic sketch of a fault damage zone. Note the development of a fault core surrounded by a zone of rocks formed by brittle deformation

Fig. 9.58 Fault damage zone as developed in arenaceous rocks of Kumaun Lesser Himalaya, India. The damage zone is confined between two fault traces (shown by a set of two yellow dashes). Loc.: NW of Bhimtal on the road to Khairna. (Photograph by the author)

thus commonly noticed in field is a heterogeneous mass of deformed rocks in the fault zone with practically no trace of the fault core. In such cases, the entire material contained between the two bounding surfaces of the fault can be considered as damage zone (Fig. 9.58).

9.6 Fault Zone Rocks

Since faults are formed under specific geologic conditions of shearing, they are characterized by the presence of rocks that show specific textures and structures. Sibson (1977) presented a classification of fault rocks that takes into consideration aspects such as cohesiveness, fabric, grain size, matrix and mechanism of formation (brittle or ductile). Sibson's classification has largely been found to be practical as the rock types can be easily identified in field and in thin sections, and also give an idea of their formation at different depths of the crust.

In Sibson's classification, fault rocks are broadly grouped as cohesive and non-cohesive; each group is then subdivided into three series: breccia, cataclasite and mylonite. Further classification is based on fabric (foliated or non-foliated), grain size and matrix. Fault rocks and fault-related rocks have later on been investigated by several workers (e.g. Wise et al. 1984; Peacock et al. 2000), thus suggesting that no unanimous scheme of classification of fault rocks is available as yet. Some common fault rocks from Sibson's classification are briefly described below. (*Mylonite*, a typically shear zone rock, has been described in Chap. 17.)

Breccia is a fault rock dominantly containing more than 30% angular fragments that are set in a fine-grained matrix formed due to pulverization during fault motion. The fragments are of variable sizes, up to several centimetres. *Gouge* is a pulverized rock containing less than 30% fragments and thus appears powdery in outcrops. *Cataclasite* is a fine-grained, non-foliated rock formed by brittle deformation and therefore contains angular fragments. Because of fine-grained fabric, a cataclasite behaves as if formed by ductile processes that make the rock flow, called *cataclastic flow*. Depending upon the proportion of (fine-grained) matrix, a cataclasite is subdivided into three: protocataclasite with 10–50% matrix, cataclasite with 50–90% matrix and ultracataclasite with 90–100% matrix.

9.7 Growth of Faults

A fault after its formation commonly grows in length as well as in width. A common process of growth of a fault is the opening of fractures followed by their joining with the adjacent shear joints with progressive deformation. If deformation does not progress with time, the fault will not grow further and becomes an inactive fault. If deformation continues with time, a fault may grow in width or in length. A fault may become longer if during its growth another fault is available adjacent to it. The end-to-end linkage is a common process of growth of a fault. Further, two faults may join sideways also. With progressive deformation, this process may be repeated with other adjacent faults. All this creates a wide fault zone with time.

Growth of faults is a complicated process. It is difficult to assign any specific process for this phenomenon. Also, it is not clearly known whether all rocks or only some specific rocks show this phenomenon. Pollard and Fletcher (2005, p. 371), for example, indicated that growth of fault is commonly shown by granite and sandstone. In granite, according to them, the process operates by propagation of opening fractures to link adjacent sheared joints that enable the fault to grow in length, whereas in sandstone the process operates by clustering of deformation bands that enable the fault to grow in thickness. Nevertheless, the growth of faults seems to be an interesting aspect for understanding the formation of structures in brittle rocks.

Faults are sometimes subjected to accumulation of stress that enables them to grow after their formation. This growth operates by two distinct mechanisms: stable sliding and stick-slip. In *stable sliding mechanism*, the fault maintains uninterrupted motion, and as such build-up of stress does not take place. This mechanism operates when a segment of a fault comes in contact with groundwater, while other segments of the same fault may show stick-slip behaviour. Water is believed to reduce the accumulated stress in a fault zone. This in turn reduces the hazard of the earthquake. In *stick-slip mechanism*, the fault shows sudden movement in response to accumulation of large stress. Release of stress causes a slip during which there is no motion. Motion of fault is resumed when further build-up of stress exceeds frictional resistance of the fault surface. Stick-slip mechanism explains the occurrence of earthquakes in a simple way. Earthquakes are believed to be the result of release of accumulated stress in rocks. After an earthquake, build-up of stress again continues. According to stick-slip mechanism, an earthquake occurs at slip followed by stress build-up at stick. Repeated earthquakes thus make a fault grow because of repeated breaking of the same segment of the fault. At the same time, some new faults may be created. However, growth of a fault is rather a common phenomenon than creation of new faults in successive seismic events.

Box 9.3 Fault Chronology

Most faults occur as one single surface. However, in some areas, the continuity of a fault is offset by another fault (Fig. 9.59). The cross-cutting relations easily reveal which fault has displaced which one. This gives the *chronology of faults*. However, it is not always clear whether both the faults have formed at the same time (age) or not.

In fact, once a fault has formed, the two associated blocks shift to other places. This disturbs the mass balance on either side of the fault surface. If the rock masses of the two blocks are strong enough to accommodate the newly developed stress field, the two blocks remain in their same position. In the otherwise case, if a new stress field is formed, the shear stress thus developed produces another fault that normally cross-cuts the earlier fault. Thus, the relative age of two or more faults developed at one place can be deduced. Dating of samples collected from each fault surface may give absolute age of the faults.

Fig. 9.59 Faulted faults in shale strata affected by three faults. An earlier fault F_1 abruptly terminates at the NW-SE trending fault F_3, which offsets a NE-SW trending fault F_2. By cross-cutting relations, one can establish the chronology of faults in rocks affected by more than one fault

9.8 Fault Mechanics

Faults are formed under a variety of physico-mechanical conditions in rocks. At shallower levels of the crust, faults are formed by brittle failure of rocks, while at deeper levels ductile conditions dominate. As such, several large brittle faults exposed on the earth's surface are believed to continue at depth in the form of ductile thrusts and vice versa. Also, faults and thrusts are commonly distinguished on the basis of inclination and mode of movement of blocks. With hanging wall moving down with respect to footwall, the structure is a fault, or normal fault, while with the former moving up constitutes a thrust fault, or simply a thrust. This difference in geometry of the structures is a reflection of their genesis. While extensional processes of the crust promote the formation of normal faults, compressional processes constitute congenial environment for the formation of thrusts. This also follows that normal faults are commonly formed by brittle processes, while thrusts require ductile processes. In this chapter, we describe fault mechanics in a generalized way, while thrust mechanism has been described in Chap. 11.

Theoretically, a fault is formed when the shear stress acting along a surface exceeds the cohesive strength and friction along the surface. A fracture of high shear stress, i.e. a *fault*, is thus developed along which a part of the rock mass moves past the other part. A fault assumes different geometrical attitudes that help us to identify different stress patterns responsible for giving rise to the various fault types. Experimental studies coupled with field evidences suggest some specific conditions or criteria for the formation of faults as described below.

9.8.1 Coulomb Criterion of Failure

In 1773, a French physicist, Charles A. Coulomb, suggested that the shear stress τ_s is the sum of the cohesive strength of the material (τ_0) and the coefficient of internal friction (μ) multiplied by the normal stress (σ_n), i.e.

$$\tau_s = \tau_0 + \mu\sigma_n \tag{9.1}$$

The above equation gives the threshold value of shear stress at which a fracture is developed. The coefficient of friction is defined by

$$\mu = \tan\varphi \tag{9.2}$$

where φ is the angle of internal friction given by the angle the line (Eq. 9.3) makes with the axis of σ_n (Fig. 9.60).

Thus, the Coulomb criterion of shear failure can be stated as below:

$$\tau_s = \tau_0 + \sigma_n \tan\varphi \tag{9.3}$$

In (9.3), τ_0 is the cohesive strength of the material when normal stress is zero.

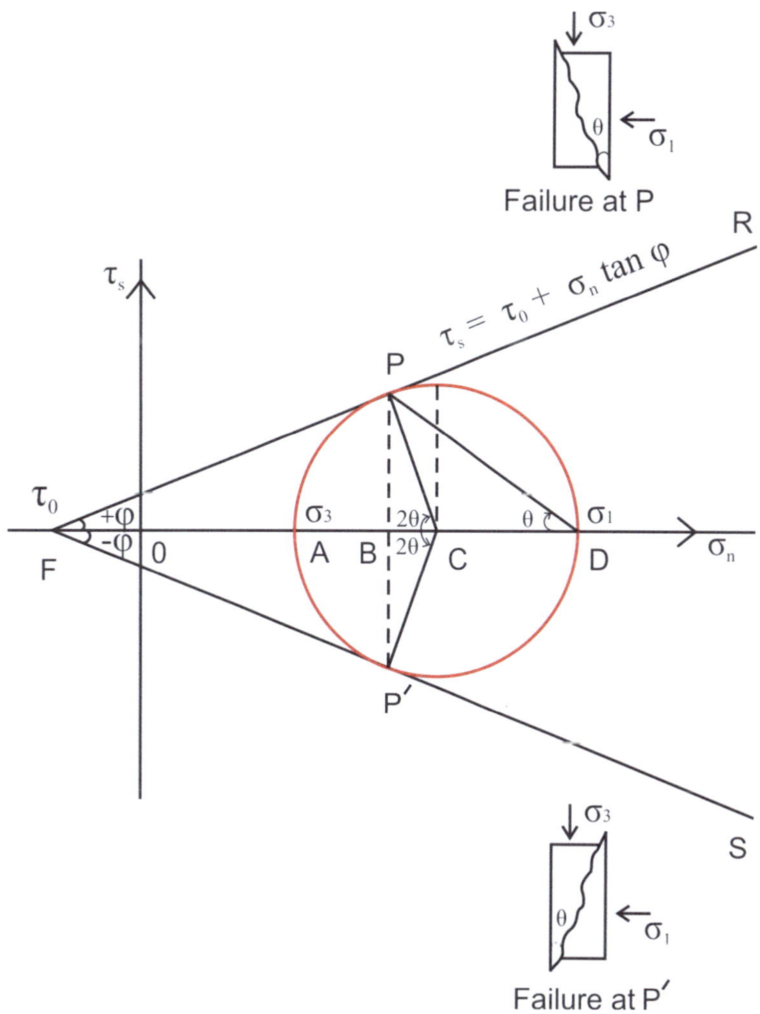

Fig. 9.60 The Mohr-Coulomb criterion for shear failure. The line $\tau_s = \tau_0 + \sigma_n \tan \varphi$ represents the stress condition at shear failure. The two lines meet the axis of σ_n at F where normal stress is zero. The lines FR and FS constitute the envelopes called the Mohr envelope. At failure points P and P′, the relationship between the principal stresses is shown. Note the conjugate position of the two failure planes

The Coulomb criterion suggests that at failure (P and P′) two sets of shear fractures should develop. Therefore, for a given set of values of cohesive strength (τ_0) and coefficient of friction ($\tan \varphi$), Eq. (9.3) gives two straight lines FR and FS on the Mohr stress diagram (Fig. 9.60). These two straight lines constitute the envelopes called the *Mohr envelope*. The diagram thus formed can be considered as the combination of Coulomb criterion and Mohr circle and is known as the *Mohr-Coulomb criterion for shear failure*.

In the Mohr-Coulomb criterion, θ gives the angle between the failure planes and normal stress. At $\mu = 0.0$, $\theta = 45°$; in this condition, the conjugate failure planes are perpendicular. For values of $\mu > 0.0$, $\theta < 45°$.

We have mentioned above that the line FP, as represented by $\tau_s = \tau_0 + \sigma_n \tan \varphi$, makes an angle φ that gives the angle of internal friction. This line, constituting the Mohr envelope, need not be a straight line. If the value of φ remains constant, the Mohr envelope is a straight line (Fig. 9.61a). However, with changing values of φ, the envelope is curved (Fig. 9.61b). In both these types of envelopes, the region within the envelope describes the *state of stable stress*, while that outside the envelope is the *state of unstable stress*.

9.8.2 Anderson's Theory

Anderson (1951) suggested that faults can be related to the principal stress axes and the shear stress. He assumed that there cannot be any shear stress on the surface of the earth. As such, one of the principal stress axes should be vertical while the other two stress axes should be horizontal. He thus considered three possibilities:

1. Maximum principal stress (σ_1) is vertical and the minimum (σ_3) and intermediate (σ_2) principal stresses are horizontal.
2. Minimum principal stress (σ_3) is vertical and the maximum (σ_1) and intermediate (σ_2) principal stresses are horizontal.
3. Maximum (σ_1) and minimum (σ_3) principal stresses are horizontal and the intermediate (σ_2) principal stress is vertical.

In an area, the types of faults formed depend much on the orientation of stress axes. Theory and experimental studies of brittle failure under compressional conditions indicate that at

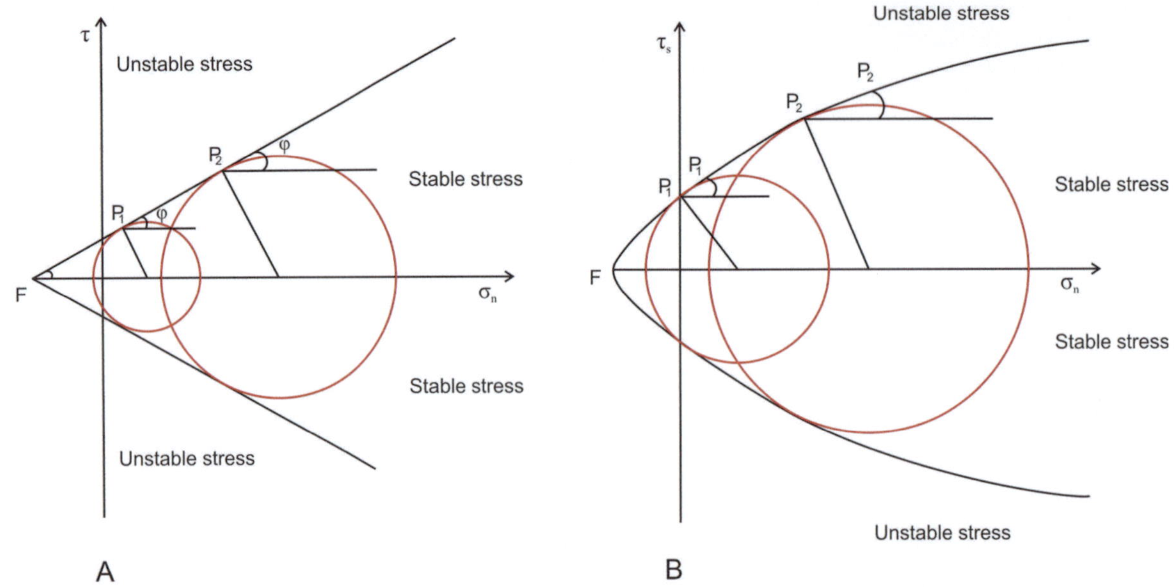

Fig. 9.61 The Mohr envelopes. (**a**) If the angle of friction (φ) remains constant, the Mohr envelope is a straight line. (**b**) If φ is variable, the Mohr envelope is curved. In both the cases, the region within the envelope represents the state of stable stress, while the one outside it represents the state of unstable stress

failure a set of two shear fractures are formed at acute angles (2θ), as indicated in the previous section. The maximum principal stress σ_1 bisects the angle between the conjugate shear fractures (Fig. 9.62); this is known as *Hartman's rule*. The line formed by the intersection of the two shear fracture planes is parallel to the intermediate principal stress axis σ_2. Thus, the maximum principal stress σ_1 makes an angle θ with the fault plane. On the basis of the orientation of stress axes, faults have thus been classified into normal faults, thrust faults and strike-slip faults. Classification of faults on these lines was first proposed by Anderson (1951), as described below:

(a) **Normal faults** are formed when the maximum principal stress (σ_1) is vertical and the minimum principal stress (σ_3) is horizontal. Under such conditions, the faults are formed at high angles (more than 45°) with vertical component of movement parallel to σ_1. The hanging wall moves down relative to the footwall. The sense of movement indicates contraction along σ_1 which thus indicates *Z* strain axis, and extension along σ_3 which is horizontal and thus indicates *X* strain axis (Fig. 9.62a). The latter implies extension of the crust. Normal faults are thus formed in areas of crustal extension. The intersection line of the two fault planes is parallel to σ_2 which is the *Y* strain axis, and it remains unchanged assuming plane strain conditions of deformation. Normal faults have been described in detail in Chap. 10.

(b) **Thrust faults** are formed when the minimum principal stress (σ_3) is vertical and the maximum principal stress (σ_1) is horizontal. Under such conditions, the faults are formed at low angles (less than 45°). As a result, the hanging wall moves up relative to the footwall. The two thrust faults thus formed intersect at a line parallel to σ_2 which is horizontal (Fig. 9.62b).
Since σ_1 is horizontal and σ_3 is vertical, it implies that the thrust movement causes horizontal shortening and vertical extension. As such, thrust faults are formed under conditions of intense crustal compression. Further, since σ_3 is vertical, such faults are more likely to form at shallower levels of the crust. As with normal faults, the intersection line of the two fault planes is parallel to σ_2 which is the *Y* strain axis, and it remains unchanged assuming plane strain conditions of deformation. Thrust faults have been described in detail in Chap. 11.

(c) **Strike-slip faults** are formed when the maximum principal stress (σ_1) and the minimum principal stress (σ_3) are horizontal. Under such conditions, the two faults formed intersect along σ_2 which is vertical (Fig. 9.62c). The two faults thus formed are vertical. Since σ_1 (and also σ_3) are horizontal, the sense of movement of the faults is horizontal. Strike-slip faults are therefore vertical faults involving horizontal sense of movement of the associated blocks. Strike-slip faults have been described in detail in Chap. 12.

The above three types of faults can also be expressed on the basis of the Mohr-Coulomb theory of shear fracture as follows (Jaeger et al. 2007, p. 420):

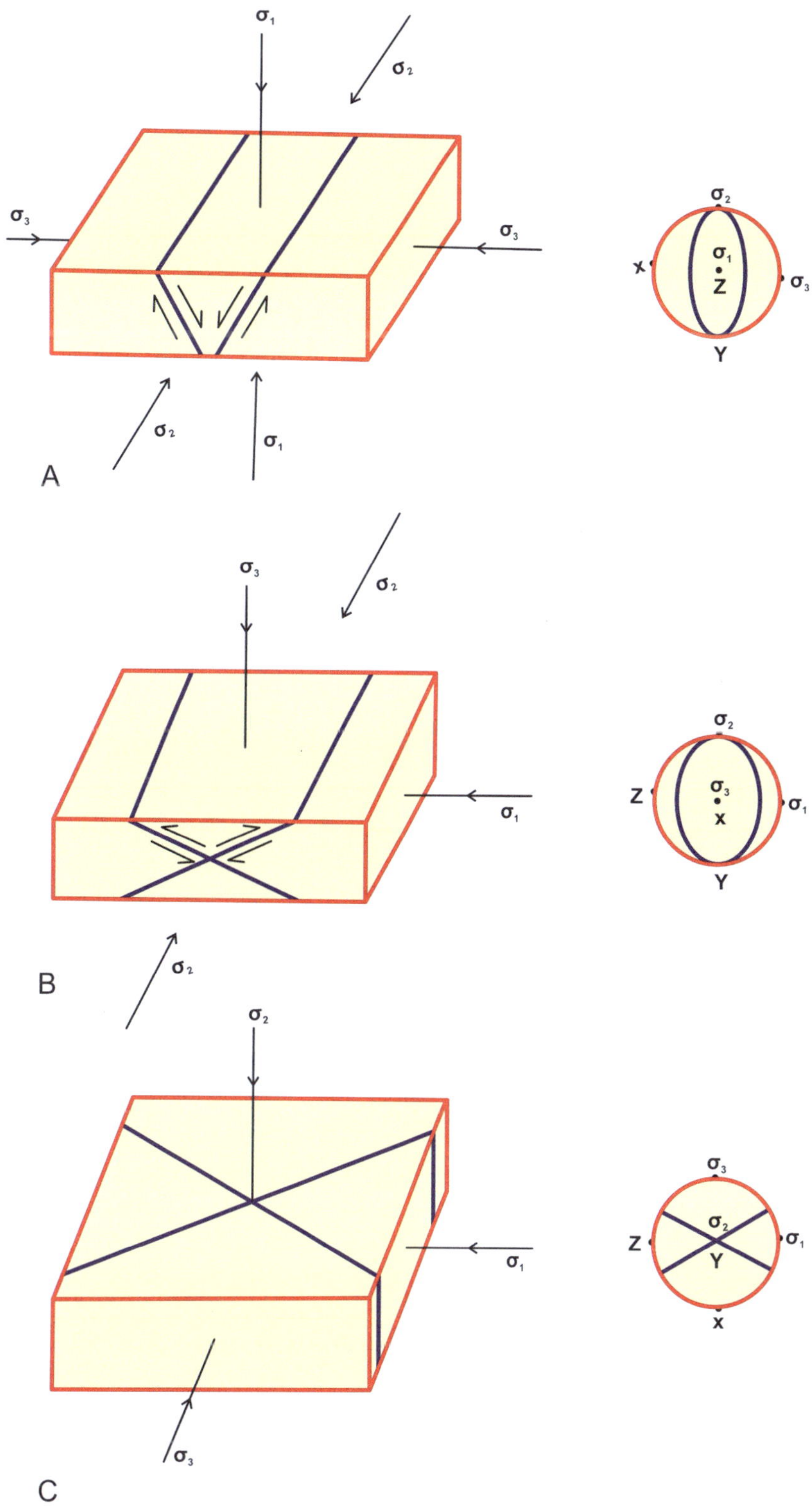

Fig. 9.62 Classification of faults based on the orientation of stress axes. (**a**) Normal fault. (**b**) Thrust fault. (**c**) Strike-slip fault. See text for details

Thrust faulting: If the vertical stress is the *least* principal stress, failure may occur on either of the two planes (Fig. 9.63a), inclined at an angle $\psi < 45^{\circ}$ to the horizontal.

Normal faulting: If the vertical stress is the *greatest* principal stress, failure may occur on either of the two planes (Fig. 9.63b), inclined at an angle $\psi < 45^{\circ}$ to the vertical.

Strike-slip faulting: If the vertical stress is the *intermediate* principal stress, failure may occur on either of the two vertical planes (Fig. 9.63c), inclined at an angle $\psi < 45^{\circ}$ to the σ_1 direction.

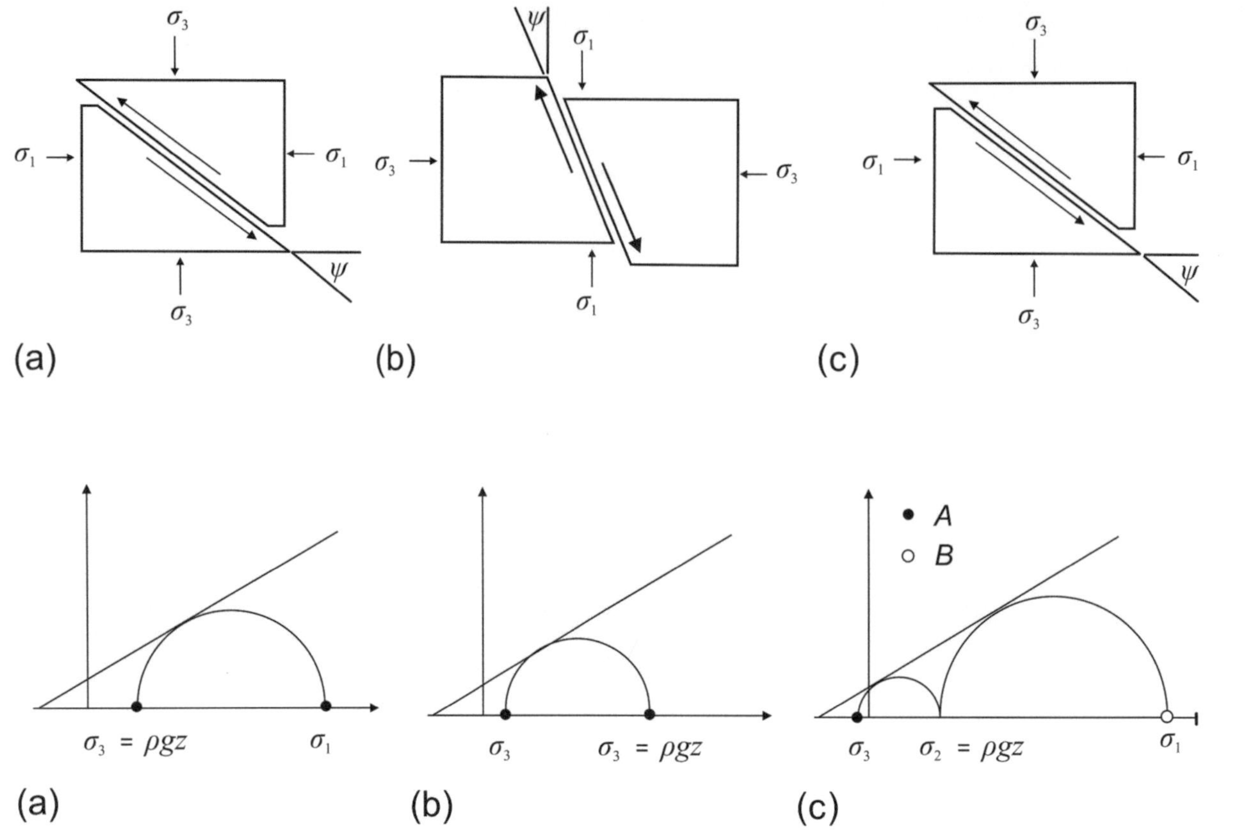

Fig. 9.63 The Andersonian types of faults as shown in two dimensions along with their Mohr circle analysis according to the Mohr-Coulomb failure criterion. (Reproduced from Jaeger JC, Cook NGW, Zimmerman RW, 2007, Fundamentals of Rock Mechanics, fourth edition, Figs. 14.1 and 14.2 with permission from John Wiley & Sons under Request ID: 600067224)

In the above cases, the corresponding Mohr circle is shown below each of the diagrams.

9.8.3 Hafner's Theory

Anderson's fault mechanism assumes that the fault planes should be straight. This is not always so with natural faults. Faults occasionally show curved outlines on all scales. Hafner (1951) had presented an explanation for faults that are curved in shape. He considered a simple two-dimensional rectangular block subjected to variable compressive horizontal stress (σ_h) on its sides together with a vertical stress (σ_v) at its lower part to counter the gravitational stresses. In order to keep the block in equilibrium, shear stresses must be applied at its bottom as well as at its sides. The block is now obviously free from any shear stress at its free upper surface. It is worth mentioning here that the earth's surface is also believed to be free from shear stresses. As such, the above-mentioned block can be assumed to represent a portion of the earth's crust.

Assuming that the earth's crust is heterogeneous, the block selected for analysis here is also heterogeneous. As such, the block shows variation of stresses from point to point, and this constitutes a *stress field* in the block. Further, the shear stress which is zero on the upper surface increases towards the bottom. The orientation of principal stress will thus show variation in different points of the block. A continuous line joining the adjacent points with different orientation of principal stress is called a *stress trajectory*. The stress trajectories and the predicted faults have been schematically shown in Fig. 9.64. According to Hafner, the inclination of each stress trajectory will change from surface to depth, thus showing a curved pattern. The faults thus formed under such conditions will accordingly be curved. The curvature of the predicted faults shows two possible directions of faulting (Twiss and Moores 2007, p. 257): one concave upward (Fig. 9.64a) that is reminiscent of many listric thrust faults, and the other concave downward (Fig. 9.64b) that is reminiscent of faults along some basement uplifts. The faults thus formed constitute a conjugate set of normal faults.

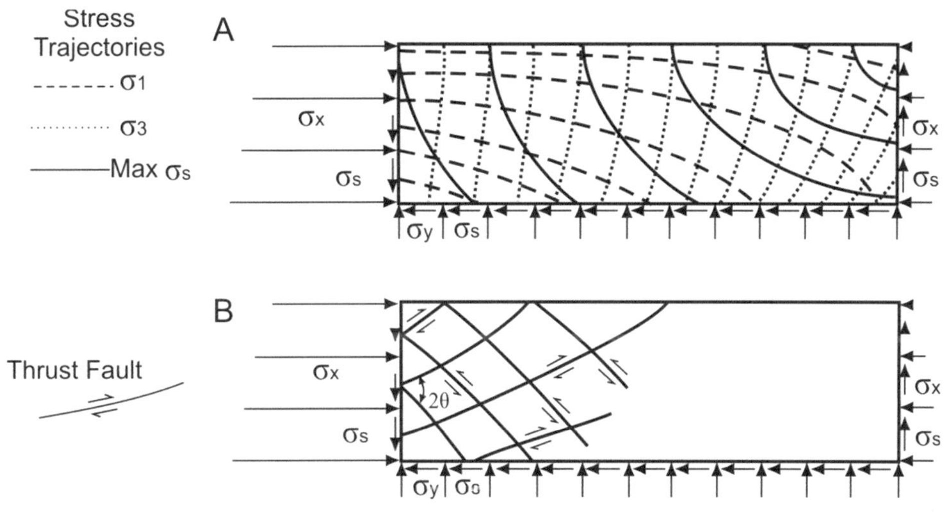

Fig. 9.64 Hafner's theory of fault mechanism assumes that most fault planes formed by horizontal compressive stresses are curved because the earth's crust is heterogeneous. (**a**) Stress trajectories for different stresses are curved. (**b**) Since the inclination of each stress trajectory changes from surface to depth, the faults thus formed should be curved. (After Hafner 1951, Fig. 6. Reproduced with permission from Geological Society of America)

9.8.4 Seismic Faulting

Study of earthquakes enhanced our understanding on the relationship between faulting and earthquake formation. The Wellington Earthquake of New Zealand of 1855 was possibly the first to establish the relationship between earthquakes and faulting. Our knowledge on this aspect enhanced further after the study of the great California Earthquake of 1906. A length of 435 km of the San Andreas fault had undergone displacement up to 6 m in places, and the western block of the fault moved northward relative to the eastern, and during the earthquake, the points adjacent to the fault underwent largest displacements while those located at 20–30 km away moved 1.4 m parallel to the fault (Suppe 1985, p. 296). On the basis of this earthquake, H. F. Reid developed a model of earthquakes that is known as *elastic rebound theory*. According to this theory, when at a point the release of stored elastic stresses exceeds the static friction, an earthquake is developed at that point. This theory is applicable for most of the shallow earthquakes and especially explains movement plan along coastal California.

Faulting and earthquakes are important manifestations of the earth's brittle behaviour to deforming stresses, and both these processes are believed to be interrelated. Seismologists commonly believe that earthquakes occur along faults. Faults of the shallower levels of the crust commonly generate earthquakes, and, conversely, formation of earthquakes generally creates faults. The width and complexity of surface rupture zone produced due to an earthquake depend largely on the mode of faulting, i.e. whether reverse, normal, strike-slip or oblique (Sibson 2003). The relationship between an earthquake and a fault is known as *seismic faulting* (see also Chap. 16).

9.8.5 Role of Friction in Fault Mechanics

Friction is the resistance to relative movement of two solid bodies. Whether any two solid bodies in contact would slide past each other depends upon friction. Friction constitutes an important factor in fault formation. Friction acts as a resistive force acting parallel to the direction of resultant motion. Two basic laws of friction were established for the first time by Leonardo da Vinci whose results were rediscovered by Guillaume Amontons who brought into light the two laws to the French Royal Academy of Science in 1699 (Suppe 1985 p. 289). *Amontons' first law* states that the frictional force required for slip along the fracture plane is directly proportional to the force acting perpendicular to the fracture plane. If F_f is the frictional force to F_n, the normal force then is

$$F_f = \mu F_n \tag{9.4}$$

Here, μ is a constant known as *coefficient of friction*, which remains constant for any two materials under given conditions and is different for any other pairs of materials. *Amontons' second law* states that the frictional force is independent of the area of contact between the solid bodies.

Amontons' laws of friction were further extended by Coulomb according to which the kinetic friction is almost independent of the speed of sliding. This is also known as *Amontons' third law* of friction.

Box 9.4 Friction in the Context of Faults

Friction is a force that obstructs two solid bodies to slide past. If the two bodies are in static position on a horizontal surface, they can slide only if the frictional force is overcome by the application of some horizontal

(continued)

Box 9.4 (continued)

force. If, on the other hand, the two bodies are on an inclined plane, the condition of rest or sliding of the body depends upon the interplay of its weight (gravity), friction between the body and the surface and angle of inclination of the surface.

Byerlee (1978) has shown experimentally that at normal effective stresses ($\sigma_n < 0.2$ GPa), the maximum frictional stress (σ_f) is given by

$$\sigma_f = 0.85\,\sigma_n \quad (9.5)$$

This relationship is called *Byerlee's law of rock friction*. It shows that for most rocks of the upper crust, the coefficient of friction in Amontons' laws is independent of the rock types. Byerlee has shown that at higher values of effective normal stress, i.e. between 0.2 GPa and 2 GPa,

$$\sigma_f = 50\ \text{MPa} + 0.6\sigma_n \quad (9.6)$$

The condition shown in (9.6) generally exists at greater depths.

Minerals of the earth's crust show different frictional behaviours. Gouge material occurring at fault surfaces obeys Byerlee's law, while the silicate minerals montmorillonite, vermiculite and illite show much lower friction than Byerlee's law and platy minerals such as chlorite, kaolinite, halloysite and serpentine show normal frictional properties (Suppe 1985).

Amontons' laws throw significant light to our understanding of relative motion of two solid bodies that are in frictional contact. The frictional force is of two types: *static friction* is the tangential force necessary to initiate sliding, and *kinetic friction* is the tangential force necessary to maintain sliding. For solid material, friction is believed to be the result of roughness of two surfaces. This aspect is explained in *Bowden's theory* (*in* Suppe 1985 p. 290) according to which two surfaces may not fit well even if they are well finished; there will be void at a number of places, and these voids are joined together at protuberances called *asperities*. If the motion has to take place, the asperities have to undergo deformation. Presence of normal stress at the points of contact will be high, and this will make the two surfaces adhere with each other; this is called *adhesion*. Water reduces the normal stress, which in turn reduces adhesion and thus promotes motion.

9.9 Significance of Faults

9.9.1 Academic Significance

- Faults are manifestations of brittle deformation of rocks. This implies that the rocks may have deformed in upper crustal layers, up to depths of 8–10 km.
- Faults are commonly formed in regions of extensional tectonics. As such, their presence indicates that the crust has undergone extension (in contrast to compression that forms folds).
- Analysis of faults helps indicate the directions of the forces that have formed the faults.
- Since faults are anisotropies in rocks, their presence indicates that the rock is mechanically weak. Study of faults therefore helps in selection of suitable rocks and building materials for civic constructions and engineering projects.
- Presence of faults also throws light on the depth where the rocks have deformed and then brought on to the surface of the earth by erosion or by some other processes.

9.9.2 Economic Significance

- Faults are important structures from economic point of view. Of all the structures in rocks, it is perhaps the faults that have more economic significance than other structures.
- Faults are occasionally the sites for metallic mineralization. The reason is that faults act as channels for flow of fluids including mineralized fluids. The mineralized fluids flow upwards, i.e. towards low-pressure zones, where they crystallize in the form of metallic mineral deposits on reaching low-pressure zones.
- Fault rocks around some large faults often contain economic mineral deposits. In such areas, the mineralizing fluids flowing up along the large fault enter into the fissures or open spaces of the adjacent fault rocks where they form mineral deposits called *lodes*.
- In petroliferous areas, faults constitute suitable sites for localization of petroleum and natural gas. Here again, the faults act as channels for movement of petroleum.
- Faults commonly act as barriers that obstruct the flow of petroleum in subsurface oilfields. As such, faults sometimes contain sizeable quantities of petroleum and gas.
- In terrains with hard rocks, sometimes groundwater is localized along faults. Water below the ground surface exists at high pressure. Once a fault is present nearby, groundwater rushes to the fault zone and moves up where it is trapped under suitable conditions, e.g. when an impervious rock is present upwards. As such, presence

of subsurface faults is always an important aspect during groundwater exploration.

- Faults that emerge on to the ground surface are occasionally water bearing or are associated with springs. In some areas, a row of springs is developed along a fault line.

9.10 Summary

- A *fault* is an elongated zone of high shear stress along which the adjacent rocks have been offset. A fault is therefore a discontinuity or anisotropy showing visible slip or displacement of rock masses parallel to the fault plane.
- A fault may be a discrete plane or may include several faults to constitute a fault zone.
- Since a fault is a planar feature, its parts and geometrical parameters resemble those of a plane.
- Faults can be classified on the basis of their geometry and genesis. The geometrical classification is more useful and practical than the genetic one.
- The geometrical classification of faults as described in this chapter is based on translational movement, rotational movement, fault association and slip on fault planes.
- Faults are recognized in field by a variety of ways. Some common methods of recognition of faults—mainly on the basis of the features characteristic of fault surface, geological effects of faults on strata and geomorphologic criteria—are described in the chapter.
- A fault after its formation commonly grows in length as well as in width due to opening of fractures followed by their joining with the adjacent shear joints with progressive deformation. Without progressive deformation, a fault will not grow and will become an inactive fault instead.
- Fault mechanics requires some specific conditions or criteria for the formation of faults. A common criterion is Coulomb criterion of failure, which states that the shear stress is the sum of the cohesive strength of a material and the coefficient of internal friction multiplied by the normal stress. The criterion gives a threshold value of shear stress at which a fracture is developed.
- Anderson's theory, assuming that faults are related to the principal stress axes and the shear stress, suggests three possibilities of fault formation, i.e. normal faults, thrust faults and strike-slip faults.
- Hafner's theory assumes that the shear stress which is zero on the upper surface of the earth increases with depth. As such, the stress trajectories will change from surface to depth with a curved pattern. The faults thus formed should accordingly be curved instead of being straight as predicted in the Anderson's theory.
- Faults have been shown to be related to earthquakes and vice versa. This is also known as *seismic faulting*. On this basis, the *elastic rebound theory* has been developed according to which when at a point the release of stored elastic stresses exceeds the static friction, an earthquake is developed at that point.

Questions

1. What are the differences between a fault surface and a fault trace?
2. What do you mean by separation of a fault? What is the difference between geographic separation and stratigraphic separation?
3. What is the difference between rake and plunge of a fault?
4. Are dip faults and dip-slip faults the same? If not, explain the difference between the two.
5. How would you distinguish between a strike fault and a strike-slip fault?
6. What is a half graben? How does it differ from a graben?
7. Are steep scarps always indicative of faults? Provide an explanation for this.
8. What are active faults? What is the difference between an active fault and a potentially active fault?
9. Write down the differences between the Anderson's and Hafner's theories of fault mechanics.
10. Give the significance of faults.

10 Extensional Regime and Normal Faults

A. R. Bhattacharya, *Structural Geology*, Springer Textbooks in Earth Sciences, Geography and Environment,
https://doi.org/10.1007/978-3-030-80795-5_10

Abstract

Extensional regime refers to a state of tectonism developed by the extension of a layer or a part of the crust. Extension means lengthening and is caused by tension (in contrast to compression that causes shortening). Our knowledge on extensional tectonism increased in the 1980s when several normal faults of the Basin and Range Province of the western USA were reinterpreted as low-angle extensional structures, called *detachment structures* that have accommodated tens of kilometres of displacement. Extensional regime commonly develops along divergent or constructive plate boundaries and intra-plate regions where extensional regime may develop locally by the formation of rifts, domino faults, grabens and growth faults. Models of extensional tectonism are discussed in this chapter. Normal faults are the most common structures to cause extension. A *normal fault* is one in which the hanging wall has gone down, dominantly by dip-slip movement, relative to the footwall. Because of a downward movement, this fault is also called a *gravity fault*. Other structures developed in extensional regime include domino faults, detachment faults, growth faults, metamorphic core complexes, rifts, ring faults and calderas. Normal faults bear great academic and economic significance.

Keywords

Extensional regime · Normal faults · Detachment faults · Domino faults · Growth faults · Metamorphic core complexes · Rifts

10.1 Introduction

As yet, our information on the nature of deformation and the structures produced in extensional regime is much less than that of the contractional regime. Our knowledge on extensional tectonics vastly increased in the 1980s when several normal faults, including high-angle ones, and shear zones of the Basin and Range Province of the western USA were reinterpreted as low-angle extensional structures. Several now-inactive normal faults that are almost flat or very lowly inclined have accommodated several tens of kilometres of displacement (Wernicke 1981, 1985; Hamilton 1987). These faults are called *detachment faults*. Later on, low-angle detachment faults have been recognized from several orogenic provinces of the world, and consequently our information on extensional structures multiplied in recent years.

10.2 Extensional Regime

10.2.1 What Is Extensional Regime?

Extensional regime broadly means a state of tectonism developed by the extension of a layer or a part of the crust. Technically, extension means lengthening caused by tension which is in contrast to compression that causes shortening. Normal faults are the most common structures to cause extension. But strike-slip faults can also cause extension of a layer. The word extensional regime thus sometimes creates confusion. In order to remove this confusion, the word extension is used when the lengthening of a layer takes place in the direction of displacement of a fault irrespective of the orientation of the layer. This implies that extension is perpendicular to the strike of a dip-slip normal fault. As such, normal faults constitute the most suitable structure that can create extensional regime because strike-slip faults cannot create such type of displacement. Thus, when we talk of extensional regime, we generally mean a system in which normal faults play their role. For this reason, normal faults are sometimes called *extensional faults*. Further, in order that there is no volume change, the horizontal extension is accompanied by a vertical thinning of the rock mass.

10.2.2 Mechanisms of Extensional Faulting

Wernicke and Burchfiel (1982) identified two broad mechanisms of formation of extensional structures and thus two broad classes of extensional faults: (a) Those which produce extension accompanied by rotation of layers as well as of the faults: This class of faults shows two possible geometries: planar and listric. The separation of two blocks on a curved surface causes extension. (b) Those which produce extension without rotation of geological features.

Wernicke and Burchfiel (1982) suggested that low-angle faults may be present at different structural levels, and as such the crust can be considered to show an imbricate stack of allochthonous slices with both rotational and non-rotational fault blocks in their hanging walls.

Experiments with analogue models (McClay and Ellis 1987) suggest that listric faults produce considerable rotation of hanging wall blocks, while in rollover structures collapse of the crestal region of the fold produces a second-order graben structure. The authors further suggest that fault nucleation (i.e. initiation of new fault planes) may occur in two ways: footwall collapse (a minor localized event) and hanging wall nucleation (a major event that dominates the fault sequences).

On the crustal scale, extensional processes are distributed depth-wise. Hamilton (1987) suggested that during extension, the brittle upper-crust blocks rotate and separate. The middle-crust lenses slide apart along ductile shear zones. Lower crust flattens pervasively. The crust thus shows a distributed pattern of extensional tectonics with depth.

10.2.3 Geological Environments for Extensional Regime

Extensional regimes are commonly developed in two different geological environments: (i) Divergent or constructive plate boundaries where crustal mass is created: Imbalances caused in different parts of the additional crustal mass thus created commonly develop extensional environment. (ii) Intra-plate regions where extensional regimes may develop locally by the formation of structures namely rifts, domino faults, grabens and growth faults. Passive continental margins are considered to be the most potential regions of large-magnitude extension (Wernicke and Burchfiel 1982).

10.2.4 Models of Extensional Tectonics

As yet, there are two important models of extensional tectonics: the crustal thinning model and the low-angle detachment fault model.

10.2.4.1 Crustal Thinning Model

This model involves pure shear homogeneous deformation. McKenzie (1978) proposed a model of extension of the continental lithosphere, which is therefore called the *McKenzie model*. According to this model (Fig. 10.1), within a vertical column, the continental lithosphere is uniformly and rapidly stretched. Stretching causes lithospheric thinning, which, in turn, causes subsidence and corresponding uplift or bulging of a part of the lithosphere. The lithosphere will thus accommodate vertical thinning, referred to as the *stretching factor* (β), which is given by the ratio of the length of a deformed line to its undeformed equivalent. The stretching causes isostatic compensation that results in subsidence of the upper surface of the lithosphere and consequently a rise of the Moho. This initial subsidence (S_i), that constitutes the depth of the newly developed basin, is given by

$$S_i = d(1 - 1/\beta)$$

where d is the initial thickness of the crust and lithosphere incorporating the densities of mantle, crust and new basin fill, the temperature at the base of the lithosphere and the coefficient of thermal expansion for mantle and crust. The vertical thinning caused due to stretching forms stretched basins, which on later sagging forms rifts due to faulting.

The McKenzie model assumes the initial stretch and subsidence to be geologically instantaneous.

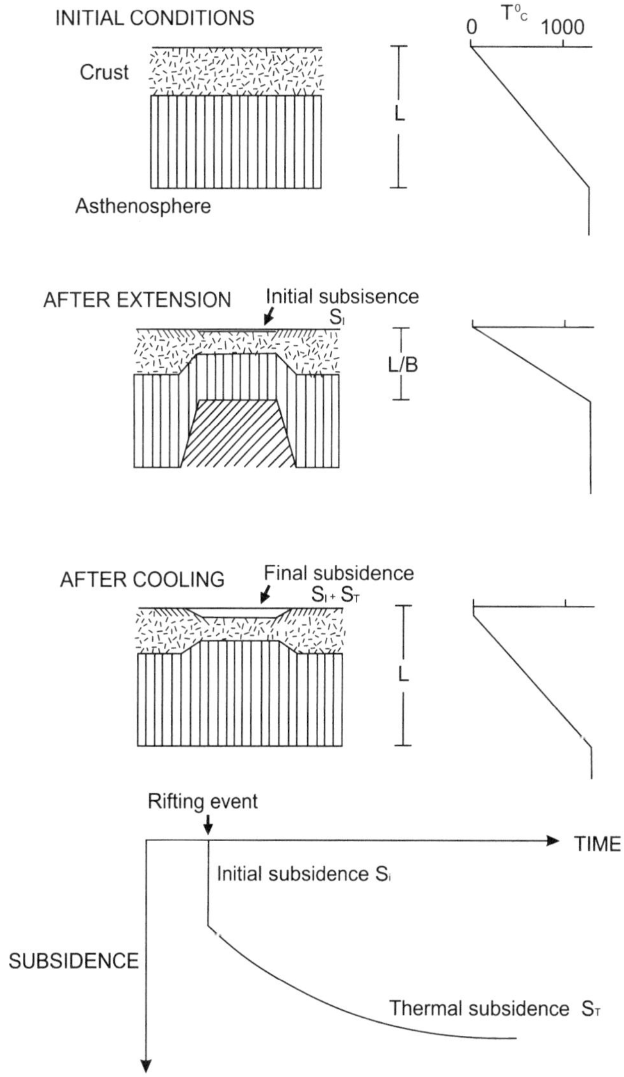

Fig. 10.1 McKenzie model to explain lithospheric extension. The model envisages that the crust (stippled) and mantle lithosphere (vertical lines) stretch simultaneously by a factor β. A stretched basin is thus developed below which lithosphere is replaced by hotter asthenosphere (diagonal lines); this raises the geothermal gradient within the lithosphere. This is followed by extension causing an initial subsidence of the basin floor, S_i. The resulting cooling of the thermal anomaly causes further subsidence which decays exponentially, S_T. This converts the uprisen asthenosphere to mantle lithosphere. (Reproduced from Roberts and Yielding 1994, Fig. 11.1 with permission from Elsevier Senior Copyrights Coordinator, Edlington, U.K. Submission ID: 1198387)

10.2.4.2 Low-Angle Detachment Fault Model

Crustal extension may also occur by asymmetric extension that involves simple shear. This model is also known as the *Wernicke model* (Wernicke 1985; Lister and Davis 1989) and takes into account a low-angle detachment fault that cuts across the entire lithosphere. In the light of their work in the Basin and Range Province, Lister and Davis (1989) are of the opinion that initially small extensions are accommodated by steep, seismically active faults that, on approaching the brittle-ductile transition, detach into an aseismic shear zone which produces low-angle faults that cut through the upper crust and the pre-existing steeper faults. The shear zone is thus domed and becomes inactive, but displacement is then taken up by the higher-up structures in the form of a *low-angle detachment fault.* To sum up, the Wernicke model is asymmetric extension involving simple shear, while the McKenzie model is a symmetric extension involving pure shear.

10.2.5 Extensional Tectonics in the Himalaya

The Himalayan mountain chain is believed to have formed by the collision of northward-moving Indian plate with the southern margin of the Asian plate. Although the Himalaya is a classic example of crustal deformation under contractional regime, extensional tectonics has also lately been identified to have played its role during the tectonic development of the mountain. Burg et al. (1984) reported the existence of E-W striking, gently (15°–30°) N-dipping, normal faults of regional scale in and northward of the Higher Himalaya. They separate the unmetamorphosed to weakly metamorphosed Ordovician rocks lying above the high-grade metamorphic rocks of the Higher Himalaya. The mildly deformed hanging wall of these faults persistently shows downward northerly displacement. The authors interpreted normal faulting as backsliding on an older thrust fault and suggested that this might be due to gravitational effect because of the high topographic relief of the mountain.

The above-mentioned significant observations of Burg et al. (1984) were interpreted by Royden and Burchfiel (1987), revealing that some facts have significant bearing on Himalayan tectonics: (i) The E-W striking, gently dipping normal faults (Fig. 10.2) are possibly Miocene or earliest

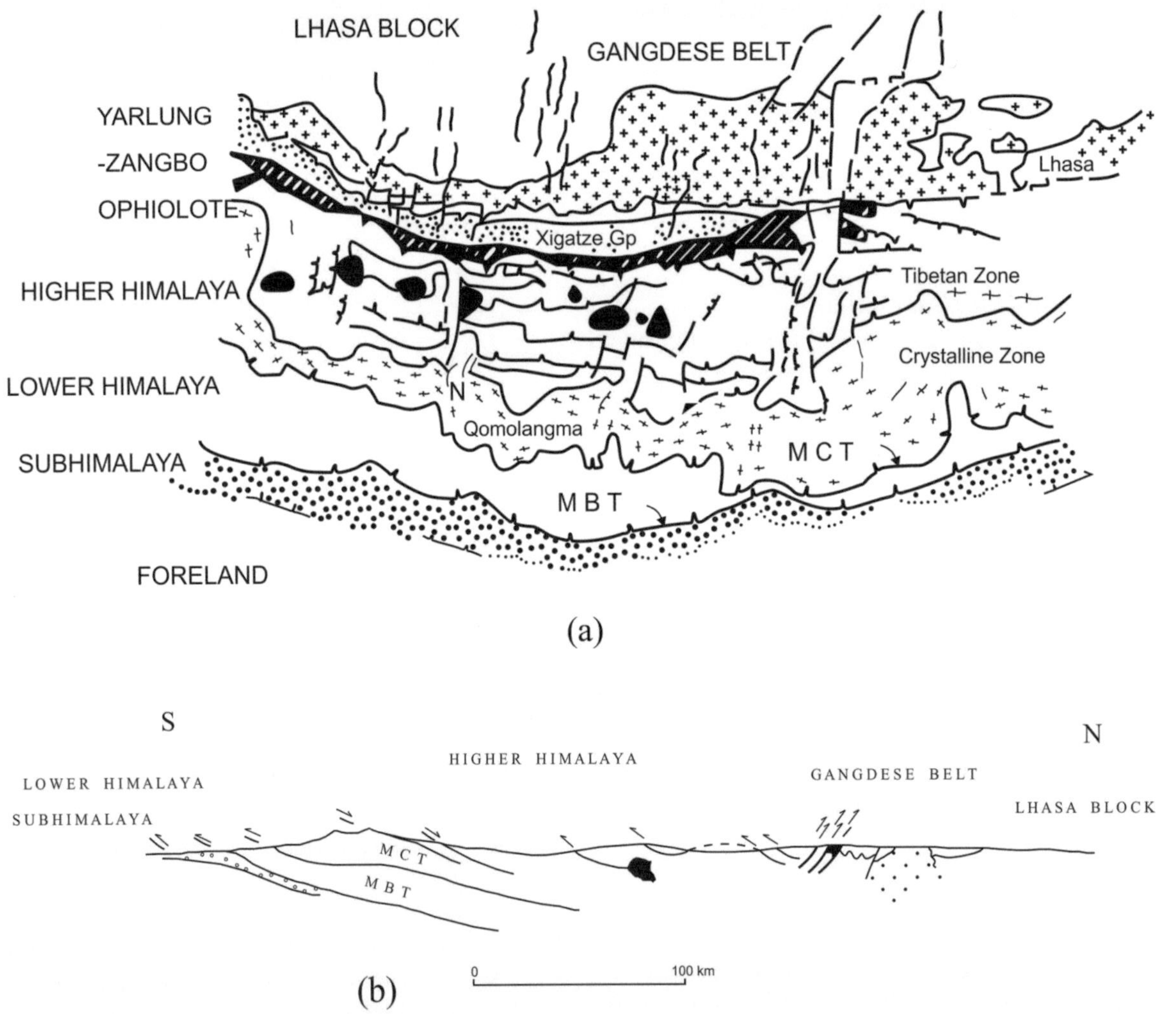

Fig. 10.2 (**a**) Generalized map showing tectonic elements of central Himalayan-southern Tibet region. *G* Guzuo; *MCT* Main Central Thrust; *MBT* Main Boundary Thrust; *N* Nyalam. (**b**) Generalized section across the Qomolangma area showing the position of the N-dipping normal faults and the S-vergent MCT and MBT and the N-vergent back-thrusts. (Reproduced from Royden and Burchfiel 1987, Geological Society of London Spl. Publ. 26, pp. 611–619. Fig. 2)

Pleistocene extensional features showing several tens of kilometres of downward northerly displacement. (ii) Large changes in the topographic elevations of the region may trigger sub-horizontal extensional stresses at the edge of a high-standing plateau in a direction parallel to that of regional compression. The sub-horizontal extensional stresses thus generated appear to be confined to upper several tens of kilometres of the crust only and probably do not extend deep into the lithosphere. (iii) The N-S crustal extension at shallow crustal levels of High Himalaya and southern Tibet can be mechanically consistent with contemporaneous N-S shortening of the mountain.

In the light of the above observations from the Himalaya, it appears that conditions for extensional tectonics can also develop in contractional regimes such as orogenic belts where the extensional stresses may operate in the direction of compression. The extensional stresses appear to have developed during convergence and are confined to the upper crustal levels only, and as such they do not indicate extension of the entire lithosphere.

10.3 Normal Faults

A normal fault (Fig. 10.3) is one in which the hanging wall has gone down relative to the footwall. The fault involves a dominant dip-slip component. In most cases, it is not sure whether only the hanging wall has gone down. There may be several possibilities. For example, both the blocks may have moved down, but the hanging wall has moved more than the footwall, or the footwall may have remained stationary while the hanging wall has moved up (Billings 1972, p. 244). Likewise, some other schemes of movements of the two blocks may be possible. Therefore, it is essential to consider the 'relative' movement of the two blocks, and in the case of normal faults the hanging wall should have a dominantly downward dip-slip component relative to the footwall.

Due to the typical pattern of movement of the two blocks, a normal fault creates horizontal extension in a direction perpendicular to the strike of the fault. In other words, the crust is lengthened during normal faulting, and this causes vertical thinning of the crust. Normal faults thus commonly occur in regions of *extensional tectonics* and are also called *extensional faults*. Extensional faults are able to produce large displacements measurable up to tens of kilometres or even more. Laboratory experiments on the formation of normal faults indicate that when two brittle layers of different shear strength undergo simultaneous extension, it is possible that a normal fault segment will be produced in one layer, and a tensile fracture in the other (Mandl 2000, p. 217).

For normal faults, the angle of inclination of the fault plane is important. In general, the dip of gravity faults (normal faults) may vary from nearly horizontal to vertical, but dips greater than 45 degrees are more common than dips less than 45 degrees (Billings 1972, p. 245). Normal faults are thus identified mainly on the basis of the relative sense of movement of the two blocks involved in faulting.

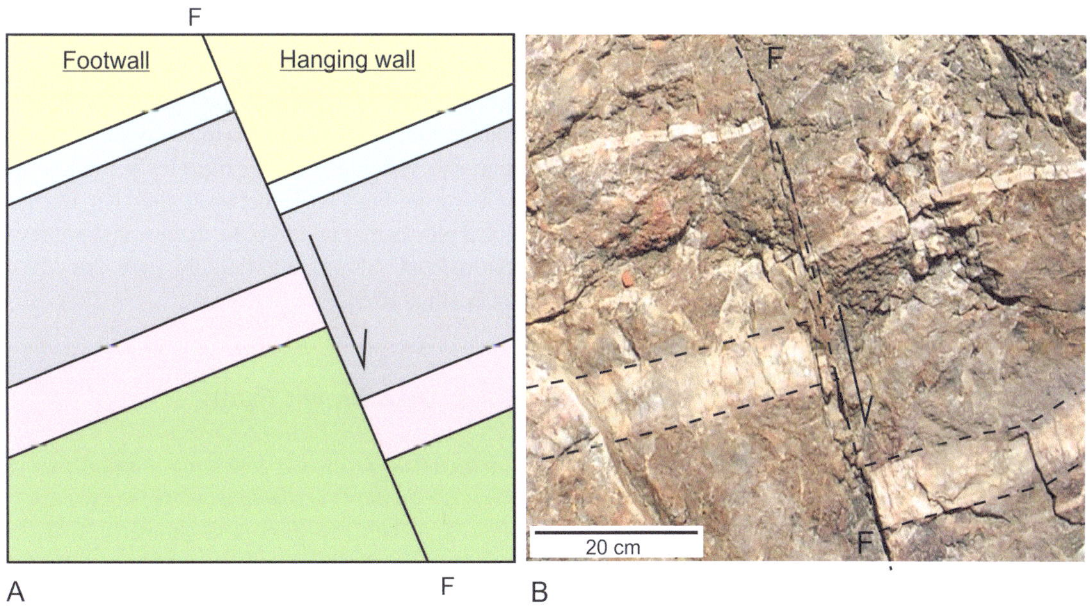

Fig. 10.3 Diagrammatic sketch (**a**) and field photograph (**b**) of normal fault. Note that the hanging wall has gone down relative to the footwall

Box 10.1 How to Identify the Downthrown Block of a Normal Fault?

A normal fault involves downward movement of hanging wall relative to the footwall. The fault thus divides a rock mass into two blocks such that a set of beds of one block has gone down relative to the other block along the fault surface. In the outcrop, a geologist has to identify the downthrown side (DTS) and the upthrown side (UTS) of the fault, which is not so easy simply by looking at the fault. We give you a simple tip.

Figure 10.4 is a diagrammatic sketch of an outcrop, also called a field section, showing a fault FF that separates the rock mass into two blocks. Note that every bed on one side of the block is offset on the other block along the fault such that the continuity of every bed is broken along the fault. Now concentrate on any block, say, the left-hand block. From base to top, number the beds as 1 (base), 2, 3 . . . 0.7 (top). This is in fact a stratigraphic sequence (assuming that the lithologic sequence of the area is not regionally inverted by any large-scale folding). Mark the same numbers to the correlative counterparts (beds) of the right-hand block. Now select any point, say A, on the fault plane. Here, bed 5 of the left-hand block is in contact with bed 7 of the right-hand block, i.e. younger bed occurs on the right-hand block. Repeat the process at, say, B. Here, bed 3 of the left-hand block is in contact

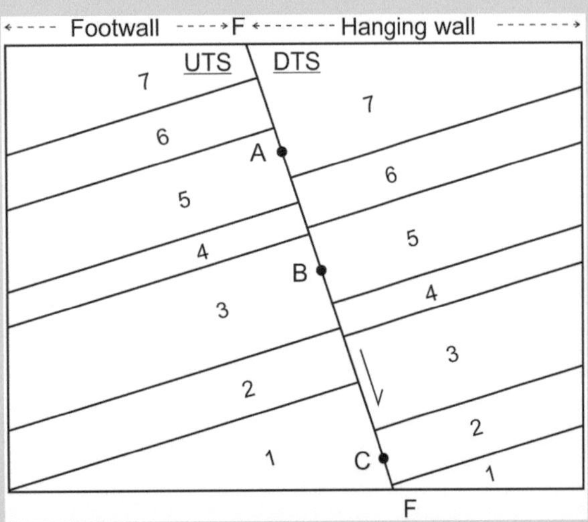

Fig. 10.4 Diagrammatic sketch to show how the downthrown side (DTS) of a normal fault is identified. The beds on the DTS are always younger than the corresponding beds of the upthrown side (UTS). The dip of the fault also needs to be known for ascertaining hanging wall and footwall. A, B and C are reference points to compare the relative age of the beds on either side of the fault

Box 10.1 (continued)

with bed 5 on the other side, i.e. younger bed again on the right-hand block. A similar result comes out at C too. All the three tests indicate that the right-hand block is the downthrown block of the fault.

The above test thus gives a thumb rule for identifying the downthrown side of a normal fault: *younger beds always occur on the downthrown block of a normal fault.*

It may be noted that there are also some other indicators of sense of movement of normal faults such as drag folds. But these are rather rare structures. In rocks, most beds associated with a normal fault occur as flat surfaces. Therefore, the method of numbering the beds and comparing them with their correlative counterparts on the other block, as described above, works as a foolproof method.

10.4 Types of Normal Faults

10.4.1 Domino Faults

Domino faults are formed by parallel planar faults for which both bedding and faults rotate (Groshong 2006, p. 329). The amount of rotation may be large, 30°–60° or more. Rotation of fault blocks involves faulting in a bookshelf or domino manner. Such rotated faults thus constitute a *domino model* or *bookshelf model* (Fig. 10.5) (Axen 1988; Groshong 2006). In the domino model, faulting of the rigid blocks occurs simultaneously, and all the faults are parallel to each other and produce similar effect (offset) to the involved strata. The fault plane and bedding undergo rigid body rotation keeping the angle (α in Fig. 10.5) between the two unchanged. The crustal blocks that undergo the extensional process involving simultaneous tilting and faulting give rise to *tilted fault blocks* (Fig. 10.5).

10.4.2 Detachment Faults

A *detachment fault* is *a low-angle normal fault associated with large amounts of displacement of the younger, overlying rocks*. The fault shows curved geometry in its section and becomes steeper towards the surface, thus showing a listric geometry (Fig. 10.6). *A listric normal fault is a curved fault (concave upwards), which may be divided into high-angle normal fault, medium-angle normal fault and bedding plane or sole fault segments* (Dennis et al. 1981). 'Listric' is a general term that actually refers to the 'geometry' of the

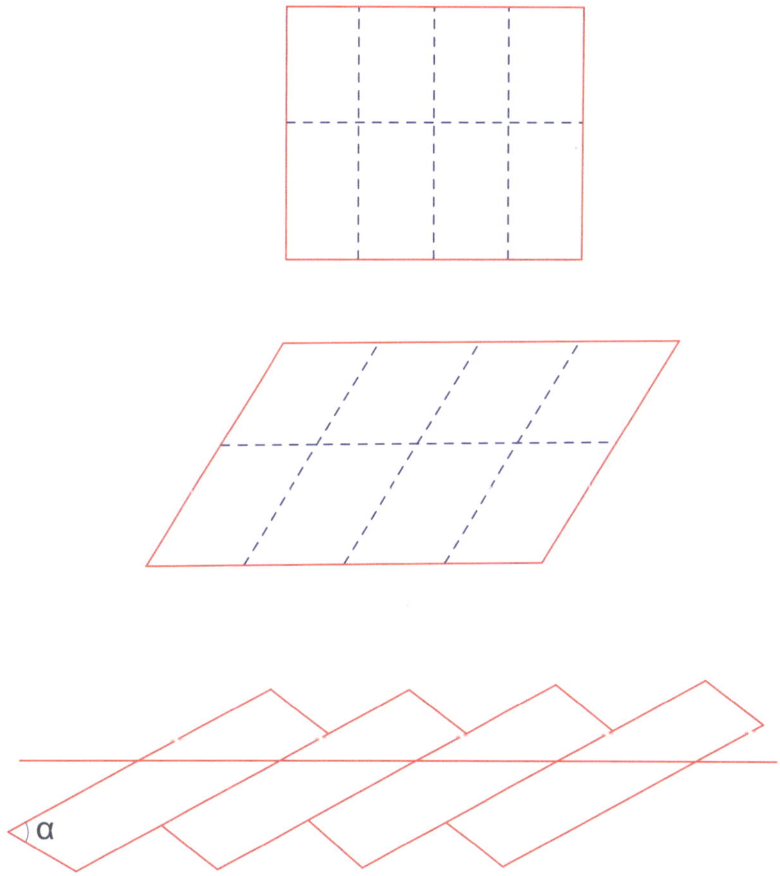

Fig. 10.5 Simplified diagram to show the domino model for development of faults. A few rigid blocks of rock masses (upper) when subjected to rotation (centre) involve faulting in a bookshelf or domino manner (lower). The inclined blocks are called tilted fault blocks

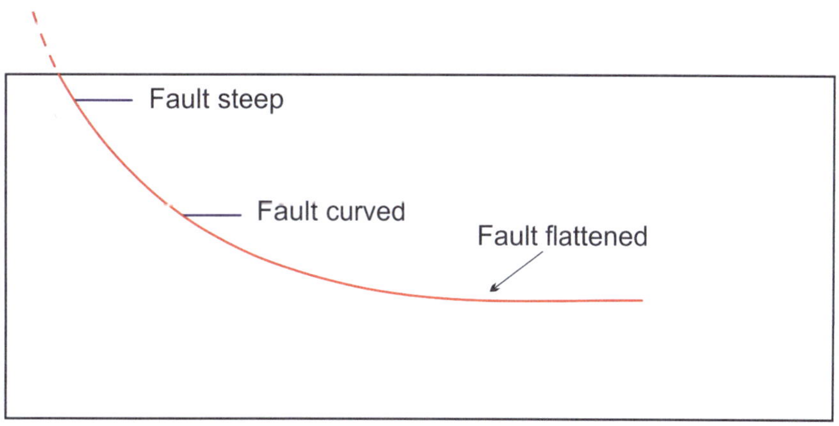

Fig. 10.6 Listric normal fault. The fault is steeper at the surface but flattens at the base

fault. Thus, they show a concave surface upwards. The 'detachment' of the overlying rocks commonly occurs due to the presence of some weak rock such as shale that acts as a lubricant to facilitate movement. Detachment faults occur in regions of extensional tectonics such as the Basin and Range Province and the Gulf Coast region of the USA.

Detachment faults are commonly associated with *roll-over structures* (Fig. 10.7) formed when the hanging wall slips down against the footwall. In order to fill the gap caused due to extension, the hanging wall rotates down the concave fault surface, thus forming an anticline called *rollover anticline*. This also gives rise to *reverse drag* on the fault surface.

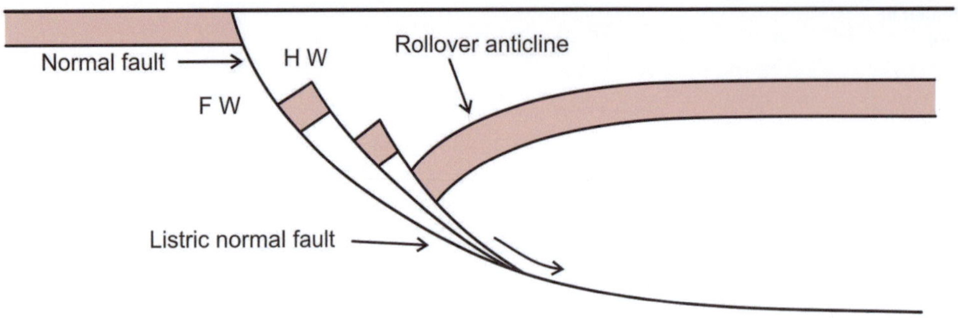

Fig. 10.7 Growth fault. It is formed when deformation is concurrent with sedimentation. The thickness of the sedimentary sequence in the downthrown block (right-hand side) is more than its correlative parts in the upthrown block (left-hand side). The hanging wall has slid down giving rise to reverse drag on the fault surface, thus forming the rollover anticline

10.4.3 Growth Faults

Growth faults (Fig. 10.7) are formed when deformation occurs during sedimentation. A growth fault moves during deposition and controls the thickness of the deposits on both sides of the fault (Groshong 2006, p. 204). These are typically normal faults. Since displacement on these faults occurs synchronous with sedimentation, these are called growth faults. During the formation of a growth fault, both blocks have subsided simultaneously, but the downthrown block has subsided faster than the upthrown block. That is why the sediment thickness of the downthrown block is always more than that of the upthrown block. In the light of seismic data, most growth faults are listric. In most growth faults, the downthrown block occurs towards the basin, and in such cases the growth faults are also called *down-to-basin faults.*

Growth faults commonly occur in two, may be more, different geological settings (Chapman 1983, p. 23): in rift-faulted basins in which continued basement faulting gave rise to the formation of growth faults, and in regressive sequences in which the upper sandy part develops growth faults that die out downwards and also upwards.

10.4.4 Metamorphic Core Complexes

Metamorphic core complexes are regions of large-scale crustal extension where low-angle normal faults (= detachment faults) have brought high-grade, deep-seated metamorphic rocks on to the surface. The detachment faults separate two different types of rocks: the lower unit (footwall) shows metamorphic rocks with evidences of ductile deformation, while the upper unit (hanging wall) exposes sedimentary rocks showing evidences of ductile-brittle deformation. The upper unit shows slicing of the rock units with the development of listric normal faults that join together with the detachment fault at depth. The detachment plane shows evidences of sliding and brecciation and in the process causes long transport of the upper unit, which may reach tens of kilometres or even more.

Metamorphic core complexes have been extensively reported from most of the areas of Cordillera of western North America. Subsequently, these have been reported from several other parts of the world. Metamorphic core complexes have been studied in detail in the Basin and Range Province (Fig. 10.8) of Nevada and Arizona, and the Scandinavian Caledonides (Wernicke and Burchfiel 1982; Lister and Davis 1989). The Basin and Range Province is characterized by extensional structures (Hamilton 1987). The Great Basin and the region N of the Snake River Plain are affected by Oligocene to Quaternary extension that has been highly variable in space, time and amount. For example, the pre-Oligocene width is almost doubled by extension, while the Great Basin has undergone late-Cenozoic extension by at least 65%. The crust is cut by detachment faults with underlying mid-crustal and overlying upper-crustal rocks.

The geological-structural setting of the metamorphic core complexes was a matter of debate in the 1970s. The typical association of low-angle normal faults had posed some mechanical problems mainly since normal faults are believed to be high-angle faults. Different models were subsequently put forward to explain this paradox, especially in the 1980s. Part of the problem also includes whether the faults are originally low-angled or high-angled. The existence of low-angle faults has been explained either due to exhumation processes or due to rotation of original high-angle faults.

10.4.5 Rifts

A *rift* is a linear zone where the earth's crust has been pulled apart. As the crust is pulled apart, a narrow linear zone goes down producing a graben or a half graben. Rifts are thus characterized by their geometry comprising a central down-faulted block, called graben or half graben that is surrounded by two uplifted blocks in case of a graben, or only one

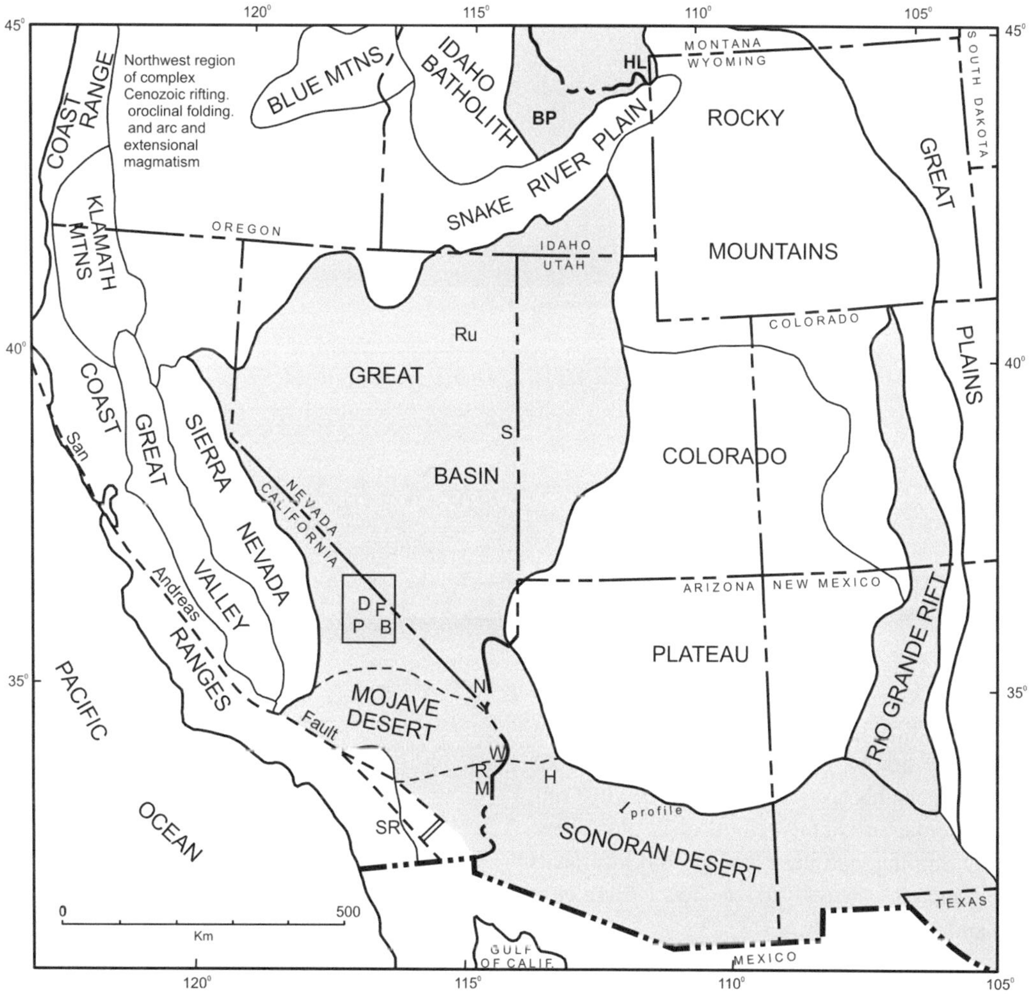

Fig. 10.8 Basin and Range Province of extensional faulting (shaded areas) as shown in an index map of a part of western USA. (Reproduced from Hamilton 1987, Continental Extensional Tectonics, Geological Society Special Publication 28, Fig. 1)

uplifted block in case of a half graben. Rifts are typically formed in regions of extensional tectonism, and the structures are associated with crustal thinning.

Rifts occur on the continents as well as on the ocean floor. An important rift is the *Sinai rift* formed due to eastern movement of Saudi Arabia and western movement of Egypt from the northern part of the Red Sea, which is a crustal scale graben. As the rift is widening, the Gulf of Aqaba and the Gulf of Suez are widening, while the Sinai Peninsula (centre) is getting progressively detached from its parent continent.

On the continents, a rift constitutes a *rift valley*, while on the ocean floor it is represented by the *mid-oceanic ridge*. Since rift valleys are depressions, they constitute sites of sediment deposition or volcanic activity. Several continental rift zones are associated with active volcanism.

Rifts are associated with several geomorphic features. In most cases, the down-faulted block accumulates huge quantity of water and thus forms *rift lakes*. Well-known examples of rift lakes include Titicaca, and several others in East Africa; the Lake Superior and the Lake Michigan in North America; and Baikal in central Europe. Volcanic rocks are also commonly present at the centre of many rifts, and some of them are active volcanoes. Rifts are occasionally associated with magmatic activity that takes place as a consequence of thinning of the lithosphere at the early stage of rift formation. At several places, rifts constitute mountain ranges called *Rift Mountains* such as those of East Africa.

In the continents, the process of rifting is able to break up the associated portion of the crust. Rifting may sometimes not reach to the point of breaking the crust; in such cases, it is called a *failed rift*.

10.4.6 Ring Faults and Calderas

Ring faults are a set of normal faults that occur in a concentric shape in a volcanic terrain (Fig. 10.9). After a volcanic eruption has taken place, a cavity is sometimes formed in

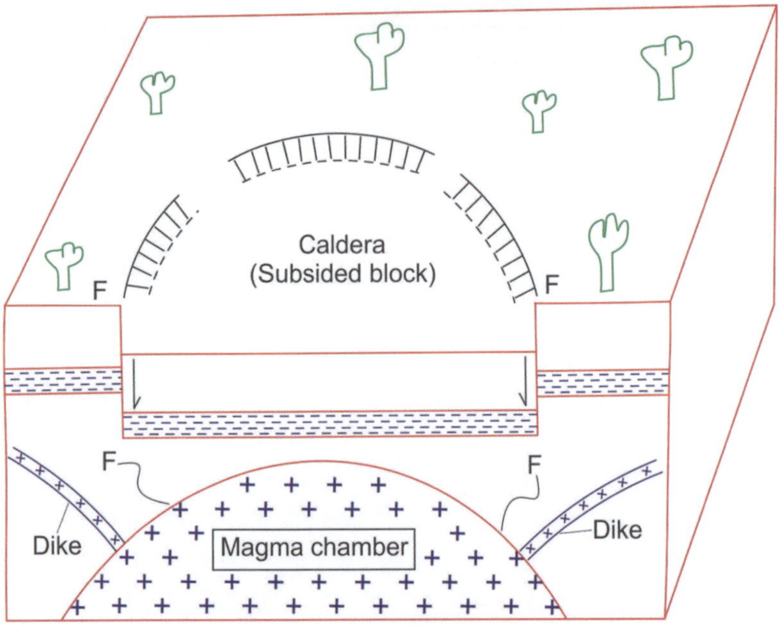

Fig. 10.9 Ring faults and calderas. See text for details

which the surface rocks collapse into the magma chamber along vertical/near-vertical normal faults. The structure thus formed, including the normal faults and the collapsed rock mass into the magma chamber, is called a *caldera*. Most ring faults are either circular or slightly elliptical in plan view, vertical or steeply dipping, and have vertical displacements from several hundred metres to a few kilometres (Gudmundsson and Nilsen 2006).

On the surface, ring faults are generally not continuous though the overall shape of the faults is concentric. The reason is that the initial displacement caused by a fault is not sufficient to bring down the bulk of the surface rocks for which a new fault needs to be developed to take up the load of the falling rock mass. The process of formation of new faults thus continues till the surface rocks involved in the process fully collapse into the cavity created by the eruption. At the base of the cavity where the surface rocks come to rest, the individual ring faults join together to constitute a single concentric fault.

The set of concentric normal faults formed due to volcanic eruption often constitute channel ways for emplacement of volcanic mass in the form of dikes, thus forming *ring dikes* (Fig. 10.10). However, this process may be repeated with time. According to Billings (1972, p. 360), if the central block subsides several times, a number of concentric ring dikes will form. The successive concentric sets are exposed only after erosion. Therefore, ring faults and ring dikes may occur both as a single concentric structure and with a few concentric structures one within another (Fig. 10.10).

Formation of ring faults is genetically related to volcanic eruption with accompanying collapse of surface rocks. Numerical models of Gudmundsson and Nilsen (2006, p. 83) suggest that a ring fault (and a ring dike) is most likely to form, in layered as well as non-layered host rocks, in a local stress field generated by a shallow sill-like chamber in a volcanic field subject to doming, tension or both. The diameter of the chamber must be much smaller than the diameter of the volcanic field subject to doming; in many of the numerical models, the volcanic field diameter is three to five times that of the chamber diameter. The models further suggest that for a 20-km-wide layered volcanic field, either tension or tension combined with doming may result in ring-fault formation, while for a 40-km-wide layered volcanic field, tension is not necessary; doming alone is sufficient to trigger ring fault formation.

Fig. 10.10 Ring dikes. See text for details

Ring faults are also believed to form as a result of the impact of *meteorites*. The term *bolide* is often used for large meteorites. Recent space researches show that bolides are common features on the surfaces of Moon, Mars and Venus.

10.5 Significance of Normal Faults

10.5.1 Academic Significance

- Presence of normal faults indicates that the rocks had undergone brittle deformation under upper crustal conditions, i.e. up to a depth of about 10 km.
- In some cases, depth of formation of the faults can also be estimated.
- Normal faults are manifestation of extensional tectonics. As such, their formation indicates thinning of the crust. Their presence thus has implications for interpreting the crustal conditions during the geological past.

10.5.2 Economic Significance

- Normal faults may form conduits or channels for fluid flow.
- Normal faults sometimes act as barriers to obstruct fluid flow, thus constituting sites for accumulation of sizeable quantities of hydrocarbon and gas.
- Since normal faults act as channels for mineralizing fluids, they are occasionally associated with several types of metallic deposits.
- In hard rock terrains, normal faults sometimes help in groundwater exploration. The reason is that groundwater exists at high pressure below the surface. Due to the presence of subsurface normal faults, the nearby groundwater is sucked and thus accumulates in large quantities along the fault that can be detected and then exploited by groundwater exploration methods.

10.6 Summary

- *Extensional regime* broadly means a state of tectonism developed by the extension of a layer or a part of the crust. Deformation under this regime causes lengthening of a part of the crust.
- Normal faults are the most common structures to cause extension.
- In the 1980s, several normal faults, including high-angle ones and shear zones, were reinterpreted as low-angle extensional structures. These are called *detachment faults* that can accommodate several tens of kilometres of displacement.
- Extensional regimes commonly develop in two different geological environments: (i) divergent or constructive plate boundaries where crustal mass is created and (ii) intra-plate regions where extensional regimes may develop locally by the formation of structures such as rifts, domino faults, grabens and growth faults. Passive continental margins are considered to be the most potential regions of large-magnitude extension.
- A *normal fault* is a dip-slip fault in which the hanging wall has gone down relative to the footwall. Because of a dominantly downward dip-slip component, such faults are also called *gravity faults*.
- The typical movement plan associated with a normal fault creates horizontal extension in a direction perpendicular to the strike of the fault, thus causing lengthening and vertical thinning of the units involved.
- Depending upon the geometry, normal faults are of various types: domino faults, detachment faults (including listric faults), growth faults, metamorphic core complexes, rifts, ring faults and calderas.
- Normal faults bear academic as well as economic significance.

Questions

1. What do you mean by extensional regime?
2. Discuss the geological conditions where extensional regime is commonly developed.
3. Describe how to identify the downthrown block of a normal fault.
4. What is the domino model of faults?
5. What do you mean by tilted fault blocks?
6. What are detachment faults? Describe their characteristics and implications for crustal deformation.
7. What is a rollover structure? How does this structure form?
8. Describe the formation of growth faults.
9. What is a rift? Explain its formation.
10. What are ring faults? How are these formed?

11 Contractional Regime and Thrust Faults

A. R. Bhattacharya, *Structural Geology*, Springer Textbooks in Earth Sciences, Geography and Environment,
https://doi.org/10.1007/978-3-030-80795-5_11

Abstract

Rocks travel? More than a century ago, a common man could not believe when a structural geologist said that in some mountain belts rocks have travelled for several, even tens of, kilometres in the geological past. Geological studies of several mountain belts revealed the occurrence of older rocks over younger ones, and the source area of the older rocks could be found at some distances ranging from a few kilometres to tens of kilometres. That rocks can travel along some weak planes, called *thrust faults*, was thus established. A *thrust fault*, also called a *reverse fault*, is one in which the hanging wall has moved up relative to the footwall along a surface that has low inclination (commonly less than 45°). Thrust faults are able to transport rocks to long distances.

Thrusts and folds are characteristic structures of *contractional regime*, which is one in which a layer or a part of the earth's crust is shortened. Various articulations of thrusts, known as *thrust geometry*, have come in a big way to help extract hydrocarbons and economic minerals from the earth's interior. Thrust geometry is nowadays a highly sought-after aspect of thrust faults. This chapter takes the readers to explore contractional regime, various aspects of thrusts, thrust geometries, genesis of thrust faults and a special structure called *salt diapir*.

Keywords

Contractional regime · Thick-skinned and thin-skinned deformation · Crystalline thrusts · Thrust faults · Reverse faults · Nappe · Klippe · Window · Upthrust · Overthrust · Underthrust · Thrust geometry · Thrust mechanics · Diapirs and salt domes

11.1 Introduction

Contractional regime is one in which a layer or a part of the earth's crust is shortened. Structures such as thrust faults, reverse faults and folds are commonly formed in such regimes. Folds are relatively more common structures of contractional regime that involve shortening of a layer in the direction of compressive stress. Contractional structures have drawn attention of geologists for over a century. Study of thrust faults has provided satisfactory solutions to several problems, especially where older rocks have been found to rest over younger ones, such as in the Alps and the Himalaya. Thrust faults have also drawn special attention because they show a variety of geometries, called thrust geometry, as described in this chapter. Study of thrust geometry, apart from others, has been found to be an important tool in hydrocarbon exploration because certain types of thrust geometries form suitable traps for hydrocarbon and gas.

This chapter is devoted to a description of contractional regime, thrusts and thrust-related structures, thrust geometry, genesis of thrusts and salt diapirs.

11.2 Deformation Styles of Contractional Regime

Large-scale deformation style of contractional regime is broadly twofold as encountered in orogenic belts: thick skinned and thin skinned.

11.2.1 Thick-Skinned Deformation

Thick-skinned deformation involves basement rocks. Crustal shortening is produced by thrust faults that cut both basement and cover rocks. This type of deformation leads to the formation of nappes that propagate from the root zone, thus giving rise to thrust sheets that are common features of orogenic belts. Many thrust faults continue to deeper levels of the crust in the form of shear zones. Thick-skinned deformation commonly affects the entire crust. The contraction produced is relatively smaller.

11.2.2 Thin-Skinned Deformation

Thin-skinned deformation refers to deformation that affects the upper layers of cover rocks. Thrust faults involved in this type of deformation do not reach the lower layers and the basement. The sedimentary cover is detached from the basement along a decollement composed of mechanically weak rocks (Boyer and Elliot 1982; Butler 1982). Thin-skinned deformation requires occurrence of sedimentary layers and presence of anisotropy containing mechanically weak rocks such as shale or salt along which large displacement can occur. Contraction produced by thin-skinned deformation is larger than that produced by thick-skinned deformation. Thin-skinned deformation is characteristic of many fold and thrust belts.

11.3 Crystalline Thrusts

By definition, a thrust can transport any type of rock. If they transport metamorphic and/or igneous rocks, they are called *crystalline thrusts*. Crystalline thrusts are commonly associated with long transport of rock leading to the formation of nappes and thrust sheets.

According to Hatcher and Hooper (1992), crystalline thrust sheets form as products of A- or B-subduction processes, and they may form in one mode and evolve into another during progressive deformation and transport. Their detachment occurs along zones of original mechanical weakness, thermally induced anisotropy, and by variation in rates of ductile flow. The authors identified two types of crystalline thrust sheets—type C and type F—as described below.

Type C thrust sheets are large (megathrust) sheets that are internally brittle slabs of intact crust (composite basement) that detach within the thermally softened brittle-ductile transition (BDT). After formation, they behave as thin-skinned thrust sheets. Type C megathrust sheets form by continent-continent or arc-continent collision accompanying A-subduction.

Type F thrust sheets are fold-related, lobe-shaped thrust sheets that form below or within the BDT (i.e. middle to lower crust under upper greenschist to granulite facies conditions) by attenuation of the common limb between antiforms and synforms in passive- or flexural-flow folding. They form as recumbent structures by anisotropy caused due to differences in flow (strain) rate in rocks undergoing uniform penetrative ductile flow. In plan view, they show a lobate shape due to a strong component of penetrative inhomogeneous simple shear producing strongly non-cylindrical to sheath folds in the direction of transport. Type F thrust sheets form via A- or B-subduction below the BDT.

Formation of both type C and F thrust sheets and their emplacement create *crustal thickening* that occurs more by subduction and collision than by normal thickening. Crustal thickening is also produced by an additional phenomenon of formation of crustal duplexes. Duplexes form in crustal rocks where faults at the base of type C thrust sheets can no longer propagate along the BDT, and therefore they ramp into the upper crust or, if available, into the foreland platform sequence (Hatcher and Hooper 1992).

11.4 Thrust Faults

Thrust faults have remained a paradox in structural geology for a long time. In the nineteenth century, a common geologist took it funny when a structural geologist said that rocks have travelled for several, or even tens of, kilometres in some mountain belts in the geological past. As a matter of fact, statement like this became a reality with time. Structural architecture of several mountain belts reveals the occurrence of older rocks over younger ones. When traced, the source area of the older rocks could be found at some distances ranging from a few kilometres to tens of kilometres. The surface separating the two contrasting rock sequences is commonly sub-horizontal, and the stratigraphy of the rocks does not show a normal order of sequence of the two blocks. Long transport of rock masses thus became a common belief. All this established the existence of thrust faults that are capable of transporting rock masses to long distances. Existence of thrust faults involving large-scale transport of rock masses was later on demonstrated in several thrust belts of the world such as the Canadian Rockies, Appalachians of the USA, Himalaya, Swiss Alps, NW Highland of Scotland and Scandinavian Caledonides.

A *thrust fault*, also called a *reverse fault*, is one in which *the hanging wall has moved up relative to the footwall* (Fig. 11.1). Although the movement along both these faults is predominantly dip-slip, the distinction between the two terms thrust fault and reverse fault is not always clear. The main difference is that of angle of inclination of the thrust plane. Some workers use the term thrust fault if the dip of the

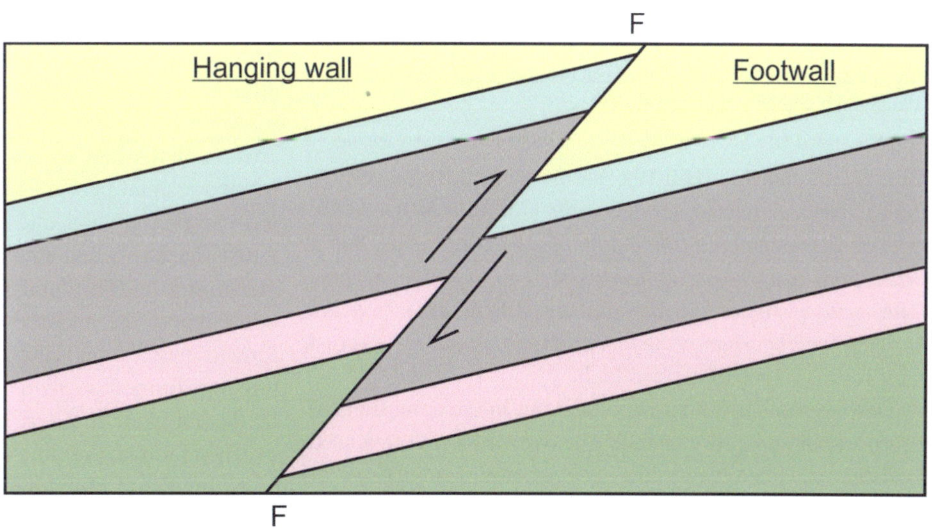

Fig. 11.1 Diagrammatic sketch of a thrust fault

Fig. 11.2 A thrust fault with thrust plane dipping less than 45°. (Photograph courtesy Gautam K Dinkar)

Fig. 11.3 A reverse fault with thrust plane dipping more than 45°. (Photograph courtesy Narendra K. Verma)

fault is less than 45° (Hill 1947); otherwise, the term reverse fault is used. *A reverse fault is a dip-slip fault in which the relative displacement of the hanging wall is upwards*, and *a thrust fault is a low-angle (commonly less than 45°) reverse fault, with a significant dip-slip component, in which the hanging wall overhangs the footwall* (Allaby 2008). A thrust fault is thus a type of reverse fault.

Despite some differences pointed out in the definitions of a thrust fault and a reverse fault, the two terms are often used loosely and interchangeably. For example, adjectives or prefixes such as *low-angle*, *moderate-angle* and *high-angle* have been used in the context of thrust faults. Further, some thrust faults may have developed as *sub-horizontal* or *low-angle* surfaces but became steeper due to later deformations. Such thrusts have been described as *moderate-angle thrusts* or *steep thrusts*, though with this attitude the thrust could have been equally described as a reverse fault. Without going into further discussion on this aspect, we shall use the term thrust fault for a fault with dip up to 45° with the hanging wall having moved up (Fig. 11.2) and reverse fault if the dip is more than 45° (Fig. 11.3).

11.5 Thrust Terminology

In thrust belts, thrust faults are associated with some specific structures. The zone or region from where the thrusts originate is called *root zone* from where the rock masses move in the form of *nappe*. A terrain showing nappe structure implies the presence of two different units of rock masses, an *allochthonous* unit that has moved far distances and has come to rest over an *autochthonous* unit that has not moved and is attached to its original place. If the movement of the upper unit involves small distances, it is called a *para-autochthonous* unit.

After emplacement of a thrust, the area is subjected to erosion that often removes a large part of a thrust sheet leaving a small part intact. The portion of a thrust sheet that has escaped erosion and is thus a remnant of the original thrust sheet is called a *klippe* (Fig. 11.4) (*klippe* is German, meaning a 'cliff'; plural *klippen*). Erosion of a nappe may often expose the underlying footwall rocks. The 'hole' thus formed enables the footwall rocks to be seen, and the structure thus formed is called a *window*. Since the closed outcrop of the footwall rocks is bounded on all sides against the overlying thrust, the structure is also called a *tectonic window* to distinguish it from a normal *window* which is formed due to erosional processes.

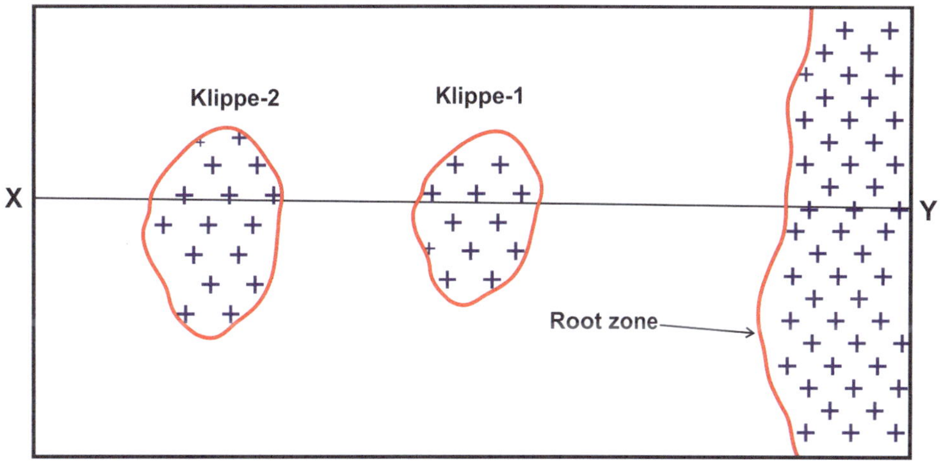

Fig. 11.4 Diagrammatic sketch of a klippe. The rock mass on the right-hand side (marked +) has been transported from the root zone towards X. After emplacement of the thrust sheet towards X, erosion has removed a major part of the thrust sheet leaving only two smaller outcrops; each is called a klippe. Since in the area two such outcrops are seen, there are two klippen, klippe-1 and klippe-2, in this area

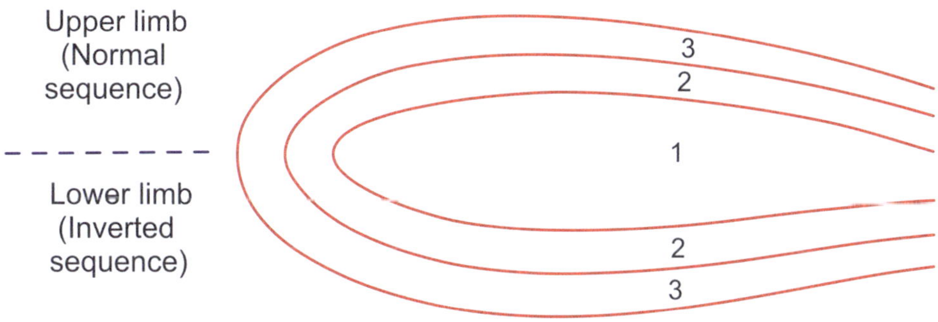

Fig. 11.5 A fold nappe in the form of a recumbent fold. Note that the beds of the upper limb show normal sequence, while those of the lower limb show inverted sequence (bed number 1 is older followed by 2 and 3)

Fig. 11.6 Imbricate zone in the Tethyan rocks of Kashmir Himalaya, India. Loc.: South of Pahalgam. (Photograph by the author)

The rock layers of a nappe may undergo large-scale recumbent folding that may show stratigraphic inversion of the lower limb. The structure thus formed is called a *fold nappe*. A fold nappe may be subjected to recumbent folding causing inversion of the lower limb as indicated by the numbering of beds in Fig. 11.5.

Thrust faults may occur in a group. In such cases, a larger thrust is associated with a few smaller thrusts, called *imbricate thrusts* (Fig. 11.6) that are formed in a subparallel set. Imbricate thrusts form due to branching from a basal decollement or the sole thrust (the lowest thrust of a thrust system) during thrust propagation. Imbricate thrusts are generally directed towards the foreland, but movement in opposite direction may also take place. A set of imbricate thrusts arranged in a parallel fashion give rise to an *imbricate zone* or *schuppen zone*.

11.6 Thrusts and Nappes

11.6.1 Definitions

Although the two terms *thrusts* and *nappes* are deeply entrenched in the literature, they often create confusion and as such demand clarification. This is so much so that the two terms were the subject of a special discussion at an international conference on "Thrust and Nappe Tectonics" held in 1979 at Imperial College, London. The conference report (published as edited volume by K.R. McClay and N.J. Price in 1981) brought out definitions and meanings of *thrusts* and *nappes* together with some other structures that are commonly associated with thrusts and nappes. We present below the following definitions from the above report. In this book, we shall follow these definitions as far as possible unless otherwise mentioned.

Thrust or *overthrust* (synonyms): *A surface of displacement with predominantly low dip (observed or inferred) along which rocks have been displaced for more than 5 km horizontally.* A *thrust sheet is the tectonic unit overlying a thrust*. *Nappe* is *a large, essentially coherent allochthonous, sheet-like tectonic unit that has moved a distance several times its thickness and in excess of 5 km along a predominantly sub-horizontal floor. Klippe is a small detached and isolated portion of nappe or thrust sheet.* The term *contraction fault* is used for *all classes of reverse faults, thrust faults and overthrusts*. It is defined as *a fault which shortens an arbitrary datum plane* (normally bedding). Conversely, an *extension fault is a fault which extends an arbitrary datum plane*. A *thrust fault is a map-scale contraction fault*. A *thrust nappe is an allochthonous tectonic sheet which has moved along a thrust fault*. A thrust nappe is diagrammatically shown in Fig. 11.7. A *fold nappe is an allochthonous tectonic unit which exhibits large-scale stratigraphic inversion and may have initiated from large recumbent folds. The underlying limbs of these folds may be sheared out into thrust faults*.

Although the above definitions convey the actual meaning of the structures, some inconsistencies still exist in the literature. Further, some modifications have also been added to some structures. For example, Cooper (1981) maintains that *thrust sheets are bodies of rock underlain by thrust surfaces and within which no significant stratigraphic repetition due to large-scale folding occurs*.

In addition to the above-mentioned terminology, one more structure is frequently used in geology, i.e. *underthrust*, which is *a thrust fault in which the footwall has been the active element* (Billings 1972, p. 198). However, the term underthrust is nowadays commonly used in the context of plate motion. For example, the Indian plate is underthrust below the Asian plate.

Box 11.1 Rocks Travel! But How Long?

Rocks travel! Sounds funny to any non-geologist!! In fact, it was a matter of disbelief even to geologists about a 100 years ago. As early as 1885, Lapworth had suggested that the metamorphic basement rocks of the NW Highlands of Scotland had travelled to far distances, up to 100 km, during Caledonian orogenic movements along the low-angle Moine thrust to rest over the Cambrian sediments. This view was perhaps not well taken by structural geologists of that time.

Much later, the work of Bailey (1935) had demonstrated that thrust can transport rock masses to larger distances. Subsequently, large displacements along thrust faults have been reported from several orogenic belts of the world. For example, in the Himalayan region, the crystalline rocks resting over the younger sedimentary belt of the Lesser Himalaya have been suggested (Heim and Gansser 1939; Gansser 1964) to have travelled southwards along the Main Central Thrust (MCT) from their root zone at the Higher Himalaya. Considering the movement of the MCT up to the southern margin of the Lesser Himalayan sedimentary belt in western Nepal, the displacement of this thrust should have been about 80 km (author's estimate from the data of DeCelles et al. 2001, p. 502). In the Maritime Alps, Fallot and Faure-Muret (1949), and later on Graham (1981), estimated

(continued)

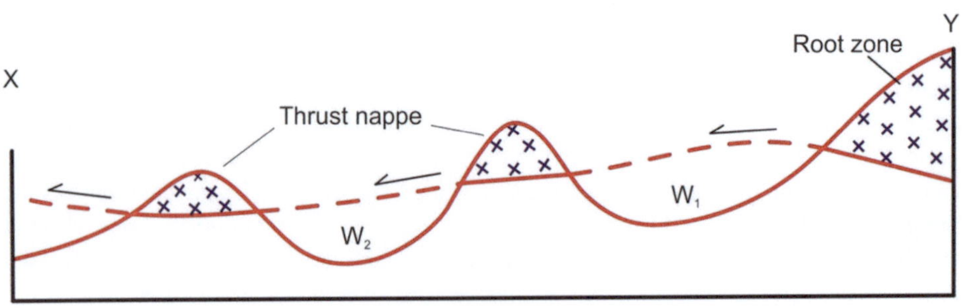

Fig. 11.7 Thrust nappe. W_1 and W_2 are two windows formed due to erosion of the overthrust mass of crystalline rocks (marked +). (This section approximately corresponds to Fig. 11.4)

Box 11.1 (continued)
20 km translation of the cover rocks. The Appalachian Orogen in North America shows imprints of three major deformational-thermal events (Hatcher 1981): the Taconic (Ordovician-Silurian), Acadian (late Devonian) and Alleghanian (Permian), each event involving large-scale west-directed horizontal transport of thrust nappes ranging from tens to hundreds of kilometres in different parts of the orogen. The Blue-Ridge-Inner Piedmont mega-nappe has undergone at least 225 km of horizontal transport (Hatcher 1981), while others such as the Chief Mountain of Montana have undergone a displacement of at least 40 km (Hatcher 1995, p. 200). The Moine thrust of NW Scotland has been estimated to show a total displacement of 40–50 km across the thrust belt (Ramsay 1969; Coward and Kim 1981).

All the above data are just a few examples to show that thrusts of orogenic belts can transport rock masses to large distances. Many such examples are available for different orogenic belts of the world.

That rocks travel nowadays seems to be an established fact. Techniques/methods have been developed that enable the estimation of large-scale transport of rock masses. Undoubtedly, this aspect of structural geology appears fascinating!

11.6.2 The Himalaya: A Storehouse of Thrust and Nappe Structures

The Himalayan mountain chain, like other orogenic mountain belts, shows an excellent example of large-scale thrusts and nappes. The Himalaya represents a classic example of intracontinental collisional belt. The mountain chain has been subdivided (Gansser 1964) into four major lithotectonic subdivisions, which from S to N are (i) Outer Himalaya (broadly includes the molassic Siwalik Supergroup of Mio-Pliocene ages, and its equivalents), demarcated in the north by the Main Boundary Thrust from the (ii) Lesser Himalaya (a thick pile of highly folded sedimentary units with a few outcrops of older crystalline rocks), separated in the north by the Main Central Thrust from the (iii) Greater or Higher Himalaya (exposing a massive, north-dipping pile of crystalline-metamorphic rocks that constitute the Central Crystalline Zone) with a fault contact against (iv) the Tethyan Himalaya (a thick pile of sedimentary rocks of Cambrian to Lower Eocene ages).

The rocks of the mountain chain are characterized by the presence of numerous thrusts that have transported rocks up to tens of kilometres. The Main Central Thrust (MCT) is a prominent geotectonic element that has transported a part of the crystalline-metamorphic rocks of the Higher Himalaya to rest over the younger sedimentary belt of the Lesser Himalaya (Fig. 11.8) during orogenic movements. The elliptical shaped crystalline outcrops of the Lesser Himalaya are

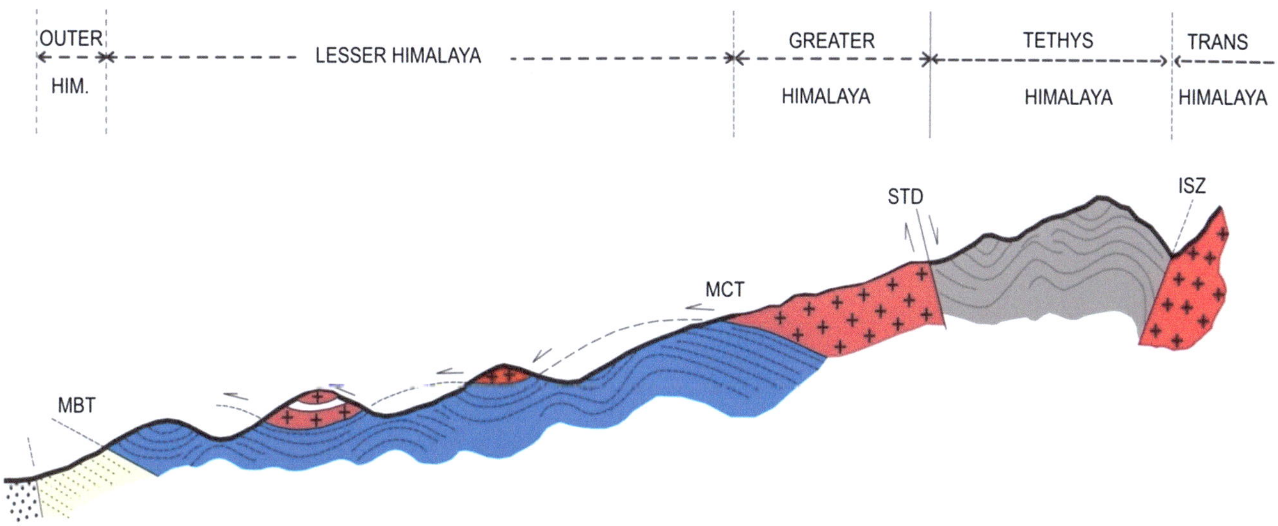

Fig. 11.8 Generalized structure of the Himalaya in the Kumaun sector that shows all the four major lithotectonic subdivisions of the mountain belt as indicated on top of the cross section. The structure shown within each subdivision is representative and highly generalized. Note that a portion of the crystalline-metamorphic rocks of the Greater Himalaya (red colour with +) has moved southwards along the Main Central Thrust (MCT) to rest over the younger sedimentary belt of the Lesser Himalaya in the form of thrust sheets. After erosion, only two remnants of the crystalline unit are exposed, each constituting a klippe. *ISZ* Indus Suture Zone; *MBT* Main Boundary Thrust; *STD* South Tibetan Detachment. See text for further details. (Based on Bhattacharya 1987, Fig. 2 with permission from Elsevier Senior Copyrights Coordinator, Edlington, U.K. Submission ID: 1193123)

remnants of these transported crystalline rocks that in field are seen physically overlying the younger sedimentary rocks. As such, according to Heim and Gansser (1939) and Gansser (1964), the crystalline thrusts of the Lesser Himalaya may be considered as southward extension of the MCT, and the former and the latter may be one and the same. Several workers (e.g. Gansser 1964; Searle 1986) have shown well-developed thrust and nappe structures in various sectors of the Himalaya.

To sum up, large thrusts are common features of orogenic mountain belts where individual thrusts extend for several kilometres. Most thrusts bring two rock units of different ages. As such, identification and precise location of thrusts are important in establishing the geology of an area or region.

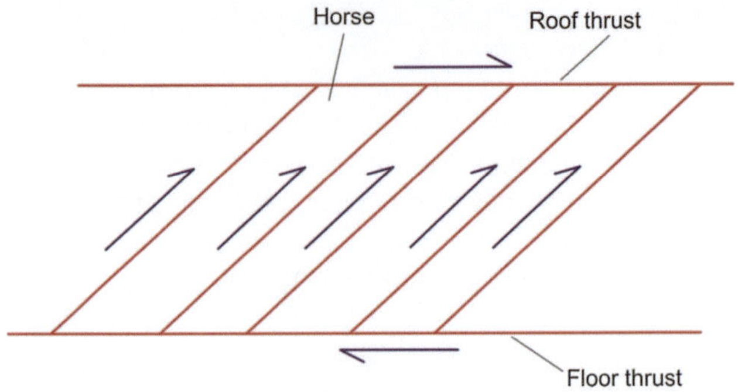

Fig. 11.9 Sketch of a duplex showing a set of parallel thrusts that are bounded by two detachment thrusts, a floor thrust at the base and a roof thrust at the top

Fig. 11.10 Outcrop photograph of a duplex developed in sandstone beds of Kumaun Lesser Himalaya, north of Nainital, India. (Photograph by the author)

11.7 Thrust Geometry

Thrusts show a variety of geometrical shapes that are commonly called thrust geometry. Our knowledge on the various geometries shown by thrusts had greatly increased since the 1960s when seismic profiles became a handy tool for search of hydrocarbon. Geologists engaged in exploration of hydrocarbon and economic minerals thus got a shot in arms that advanced knowledge on thrust geometry. Study of thrust geometry has today become an important aspect of structural geology. Works of Dahlstrom (1969, 1970), Boyer and Elliot (1982), Butler (1982, 1987), Suppe (1983, 1985), Hossack (1983), Boyer (1986), Mitra (1986, 1990), McClay (1992) and Butler et al. (2019) significantly advanced our knowledge on thrust geometry as highlighted below.

11.8 Types of Thrust Geometry

Since no unanimous scheme of classifying thrust geometry is available, we have grouped thrust geometry in three major categories: (i) planar thrust geometry, (ii) thrust geometry as related to folds and (iii) thrust geometry as related to stratigraphic sequence.

11.8.1 Planar Thrust Geometry

In this category, we include thrusts that show planar geometry. A *bedding thrust* is one that follows a bedding plane. As such, it is sometimes difficult to identify the presence of a bedding thrust. For this, one must carefully look for suitable evidences that indicate slippage of rocks past each other such as the presence of breccias or slickensides. A *listric fault* (Figs. 10.5 and 10.6), as an example, progressively becomes shallower at depth and ultimately becomes parallel to the bedding planes where the fault can be described as a bedding thrust. A *duplex* (Figs. 11.9 and 11.10) or a *thrust duplex* is a structure containing a set of parallel thrusts that are bounded by two detachment thrusts; one at the base is called a *floor thrust*, and the one at the top is called a *roof thrust*. The individual mass or slice surrounded on all sides by thrusts is called a *horse*. The imbricate thrusts and the horses are all oriented parallel to each other and thus show the same sense of movement.

Sometimes, thrusts may constitute a group of thrusts that are piled-up upon each other. The group of tilted thrusts that include the horses is called an *imbricate structure* or a *schuppen structure*. It is to be noted that a duplex is identified by the presence of both floor and roof thrusts, and if a roof thrust is not identified the structure should be named as an *imbricate stack* (Butler 1982). The individual imbricate thrusts that flatten towards the floor thrust and steepen towards the roof thrust constitute an *imbricate fan* (Figs. 11.11 and 11.12) (Boyer and Elliot 1982). Each unit (horse) of the fan is called an imbricate. An imbricate fan thus takes up a curved shape showing steeper dips at higher levels. The bounding thrusts or the master thrusts of a duplex can meet together at one end or at both. If the maximum slip is associated with the frontal thrust, it is called *leading edge imbricate thrust* (Fig. 11.11). If the maximum slip is associated with the rear thrust, it is called *trailing edge imbricate thrust* (Fig. 11.12).

Imbricate thrusts are commonly noticed in seismic profiles in subsurface rocks (Fig. 11.13). Individual horses within the imbricate fan may undergo later deformation or rotation that commonly follows different styles (Figs. 11.14 and 11.15).

A thrust may follow a bedding plane for some distance and then can cut the up-section rocks and then may follow another bedding plane. Repetition of this process produces a *ramp and flat structure* (Fig. 11.16). The high-angle portion of the thrust is called a *ramp*, while the portion of the thrust that follows the bedding plane is called a *flat*. The structure thus gives a staircase pattern to the host rocks. Due to up-section movement, a ramp can take older rocks to ride over the younger rocks. The ramp and flat geometry commonly occurs in rocks showing alternation of strong and weak beds. The flat follows a weak bed, while the ramp cuts through a relatively strong bed (Fig. 11.17). The horizontal segment followed by the flat is generally constituted of rocks such as shale, rock salt and some other weak rocks that promote easy propagation of the thrust. The segment followed by ramps is generally constituted of strong rocks that are also called *strut* and includes massive sandstone or massive limestone. The dip angle of ramps varies between 30° and 50°.

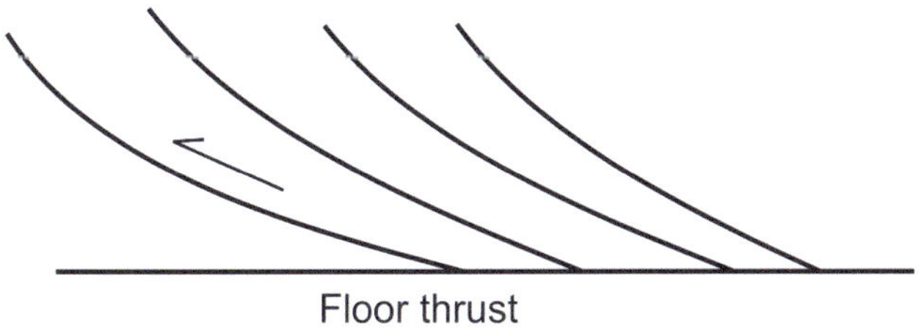

Fig. 11.11 Leading edge imbricate fan. The maximum slip is associated with the frontal thrust

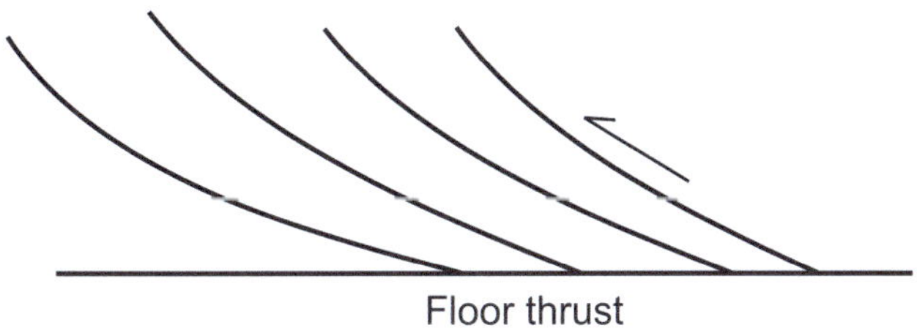

Fig. 11.12 Trailing edge imbricate fan. The maximum slip is associated with the rear thrust

A ramp shows three types of movement (Fig. 11.18) with respect to the displacement direction of the main fault. A *frontal ramp* is one in which the thrust moves forward and up-section from one horizontal surface to another. A frontal ramp intersects the main fault in a direction perpendicular to the direction of displacement of the main fault. In an *oblique ramp*, the direction of movement is at an angle with the direction of displacement of the main fault. In a *lateral ramp*, the strike of movement is parallel to the displacement direction of the main fault, and therefore it has a dominant strike-slip displacement.

An *emergent thrust* (Fig. 11.19) is one that reaches the ground surface. A *blind thrust* (Fig. 11.20) on the other hand is a propagating thrust fault that loses displacement and terminates up-section by transferring its shortening by developing a fold at the fault's tip. The thrust thus terminates within the sedimentary layers and never reaches the surface.

Fig. 11.13 Imbricate structure as noticed in seismic profile in subsurface rocks of Arabian Sea, western continental shelf, India. The structure is related to Decollement-2. (Seismic section with kind permission from Oil and Natural Gas Corporation, India)

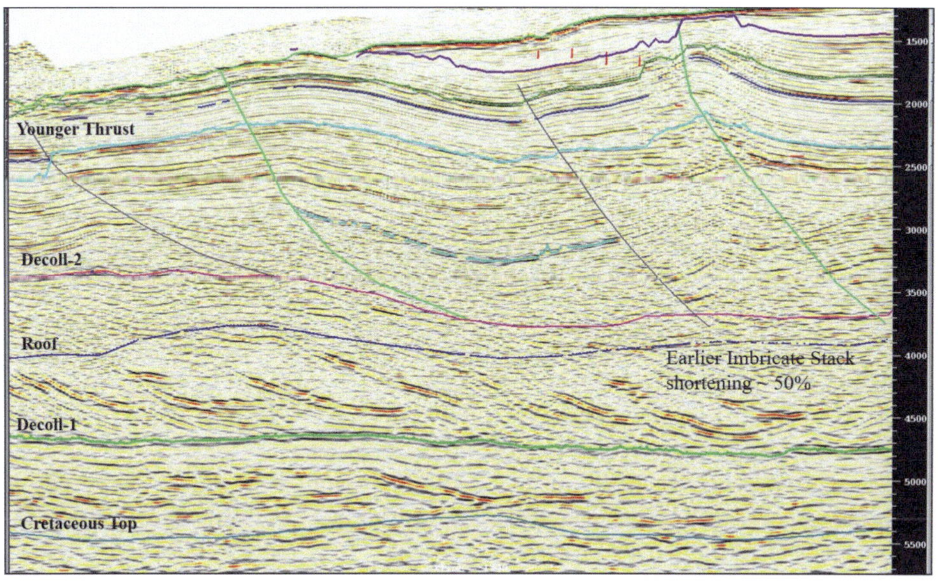

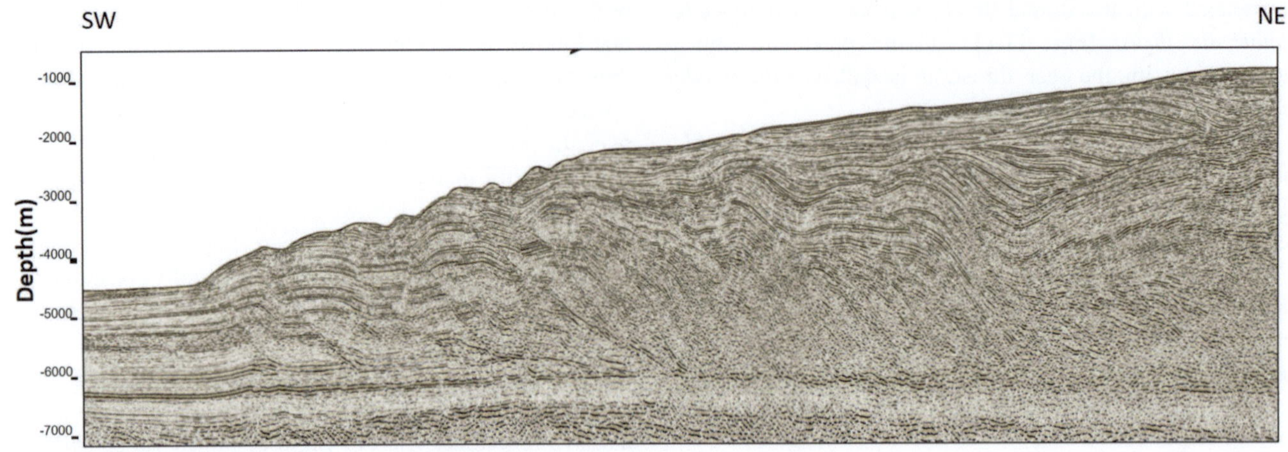

Fig. 11.14 Seismic line across accretionary prism, Outer Mumbai Shelf, Western Offshore Basin, India, showing imbricate thrusting and rotation. (Seismic section with kind permission from Oil and Natural Gas Corporation, India)

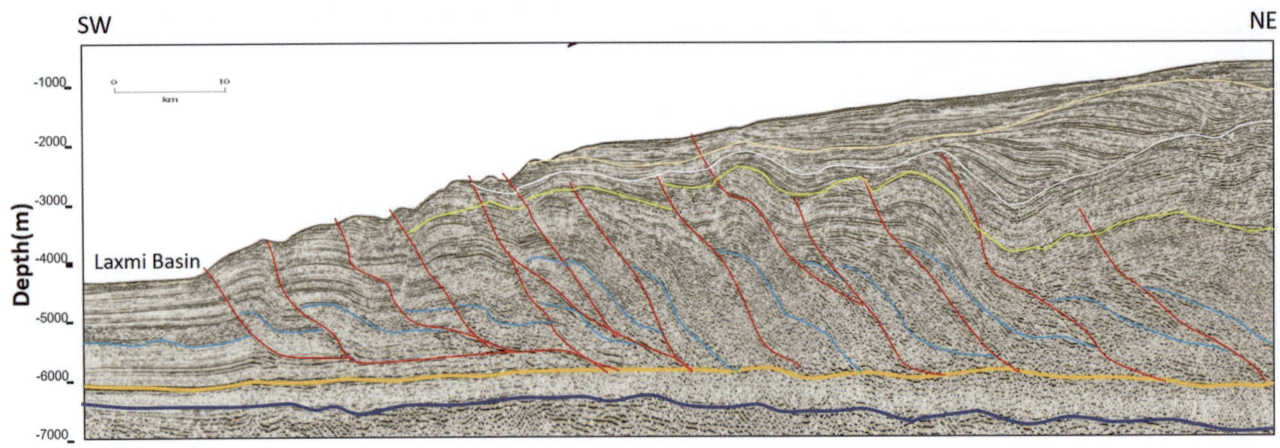

Fig. 11.15 Interpretation of the seismic section presented in Fig. 11.14. Note that individual horses have preserved imprints of later deformation that shows different styles in different horses. (Seismic section with kind permission from Oil and Natural Gas Corporation, India)

Fig. 11.16 Diagrammatic sketch of a ramp and flat structure

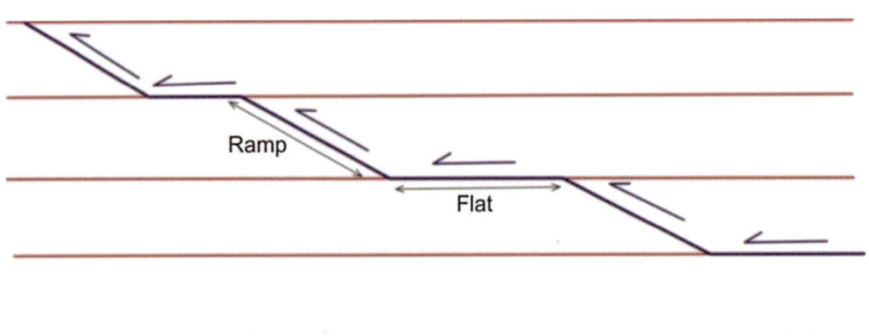

Most faults end up with a set of branches at its tip. A *splay* (Figs. 11.21 and 11.22) is a branch of a fault developed at a branch point. A splay is constituted of a single fault or a set of faults that are subsidiary to the main fault. If the fault is not exposed on the surface, the tip is described as *blind*.

11.8.2 Thrust Geometry as Related to Folds

Thrusts are invariably associated with folds in various fashions, thus forming various thrust geometries. A *break-thrust* (Fig. 11.23) forms when the connecting limb of an

Fig. 11.17 Field photo showing ramp and flat structure in the Siwalik rocks of the Garhwal Outer Himalaya. Loc.: Haridwar, India

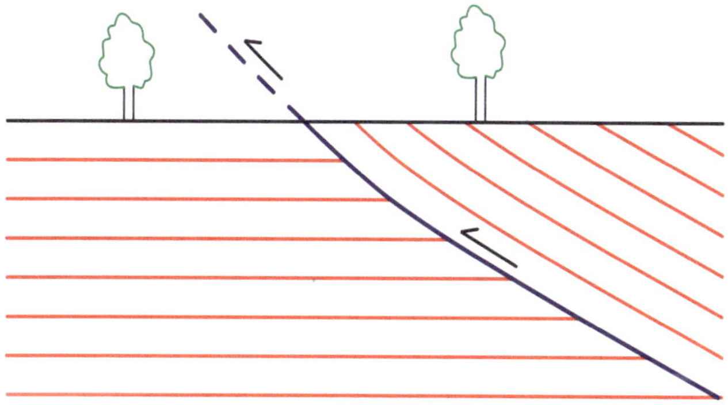

Fig. 11.19 Diagram of an emergent thrust. The thrust reaches the ground surface

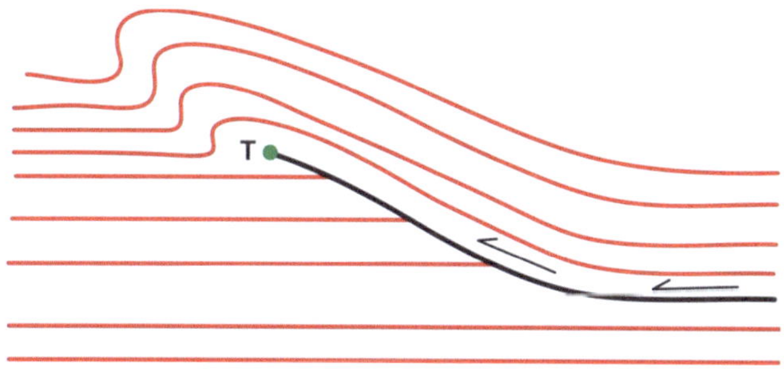

Fig. 11.20 Diagrammatic sketch of a blind thrust in which a propagating thrust fault loses slip and terminates up-section by transferring its shortening by developing a fold at the fault's tip (T) beyond which the fault does not move. The thrust thus never reaches the surface

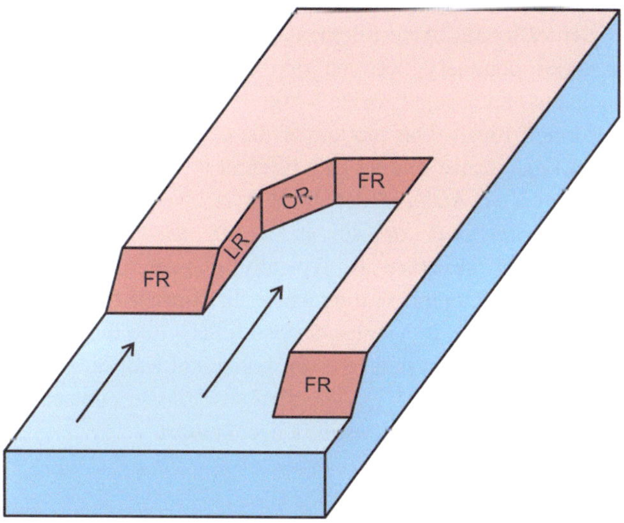

Fig. 11.18 Frontal, oblique and lateral ramps. *FR* frontal ramp; *LR* lateral ramp; *OR* oblique ramp. See text for details

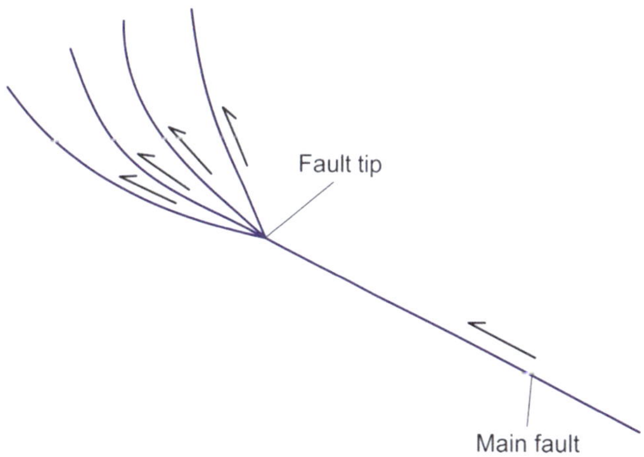

Fig. 11.21 Sketch of a splay structure developed at the fault tip of a fault

anticline-syncline pair is faulted such that the hanging wall anticline is overthrust while the footwall syncline is preserved (Hatcher 1995, p. 208). Break-thrusts are common structures in foreland fold-thrust belts of several mountain belts. A break-thrust may also affect the limbs of the associated folds. If the thrust cuts across the forelimb of a fold, it is called a *forelimb thrust* (Figs. 11.24 and 11.25), and if it cuts the backlimb of a fold it is called a *backlimb thrust* (Fig. 11.26). Further, break-thrusts are commonly associated with open to close, often overturned folds that have been named as *break-thrust folds* (Fischer et al. 1992).

Fig. 11.22 Field photograph of a splay structure developed in limestone beds of Garhwal Lesser Himalaya. Loc.: South of Devprayag, India. (Photograph by the author)

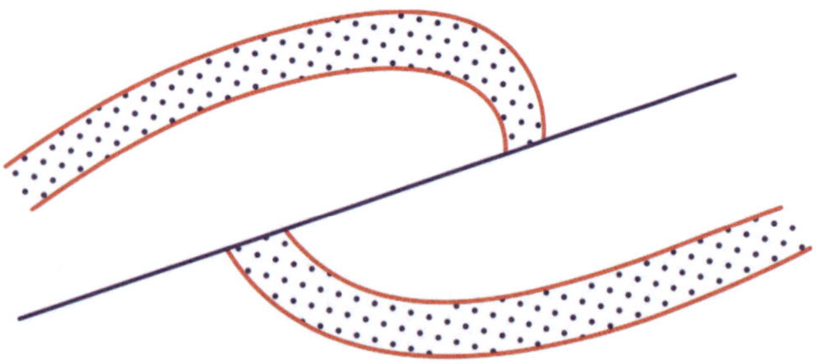

Fig. 11.23 Break-thrust formed due to faulting of an anticline-syncline pair. The hanging wall anticline is overthrust, while the footwall syncline is preserved

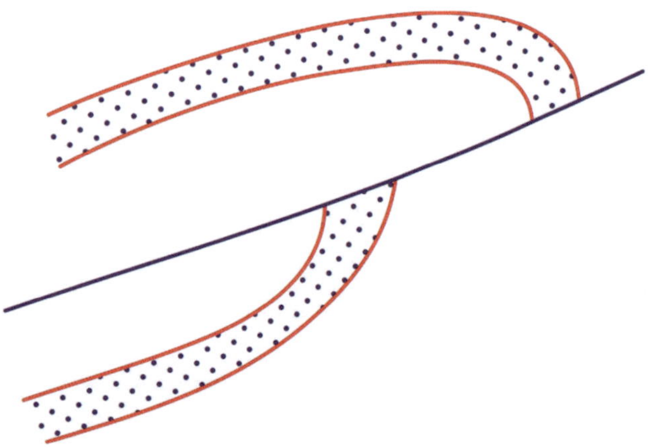

Fig. 11.24 Sketch of a forelimb thrust

Under certain circumstances, folding and thrust propagation may go together. A *stretch thrust* (Figs. 11.27 and 11.28) is thus developed when the overturned limb of a tight isoclinal fold is stretched and sheared. A thrust fault that got folded during some later folding episode is called a *folded thrust* or *thrust fold* (Fig. 11.29). Folded thrusts commonly occur in orogenic belts where they are formed due to folding of the thrust planes during later regional folding of the associated rocks.

If the fault surface is curved, the associated bed will undergo rotation in the hanging wall and will produce a fold called *ramp anticline* (Figs. 11.30 and 11.31).

A *fault-propagation fold* (Fig. 11.32) is one in which a propagating thrust fault loses slip and terminates up-section by transferring its shortening to a fold developing at the fault's tip (Mitra 1990). A fold is thus formed due to deformation at the end or tip point of the fault where it encounters a flat. A characteristic feature of fault-propagation folds is that the thrust fault loses displacement up-section (Dahlstrom 1969). Formation of fault-propagation fold requires continuous folding of beds at the tip of the fault (Mitra 1990).

Fault-bend fold (Fig. 11.33) was introduced by Suppe (1983) for a fold formed by bending of a fault block as it rides over a non-planar fault surface. According to him, if the rocks are layered, they may fold in response to riding over a bend in a fault. The fold, an anticline, is formed at the hanging wall due to the presence of a ramp (Figs. 11.33 and 11.34).

Sometimes, the anticline is accompanied by a complementary syncline (Fig. 11.35).

The term *decollement* (French, meaning to unglue or to separate) is used for a horizontal or low-angle thrust along which the strata get partly detached from, and slide over, the underneath ones. A decollement (Fig. 11.36) thus constitutes a major boundary between two blocks; the lower block includes undeformed rocks, while the upper block includes deformed rocks. The motion of the detached strata (upper block) may cause folding and different types of deformation structures. As such, a decollement is also called a *basal detachment fault* or simply *basal decollement*. The decollement structure is typically shown by the Jura Mountains of central Europe where the Jurassic rocks (limestone) are folded in a series of open folds while the underlying Triassic rocks (salts) remain horizontal and undisturbed (Laubscher 1972, 1977). The Jurassic-Triassic contact is a prominent horizontal thrust or a decollement.

11.8.3 Thrust Geometry as Related to Stratigraphic Sequence

Instead of their formation as a single or solitary structure, thrusts sometimes constitute a complex structure with more than one thrust developed in a sequence. The individual thrusts commonly affect a stratigraphic sequence of beds. Under this category, we consider the following types of thrusts:

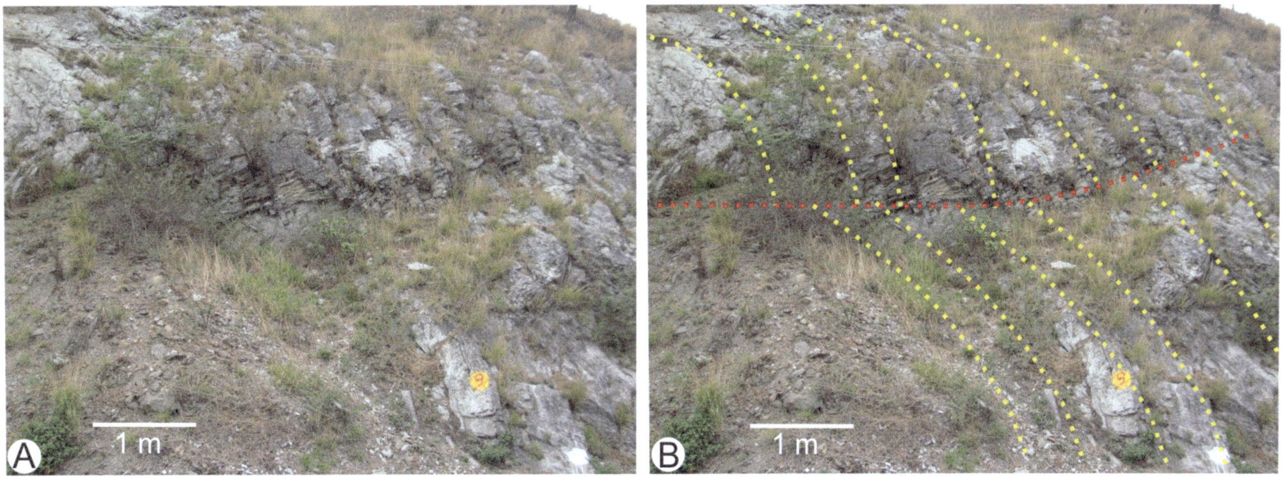

Fig. 11.25 (**a**) Field photograph of a forelimb thrust developed in limestone beds of Krol Group, Kumaun Lesser Himalaya. Loc.: NW of Bhimtal, India. (**b**) Elucidation of the structure. (Photograph by the author)

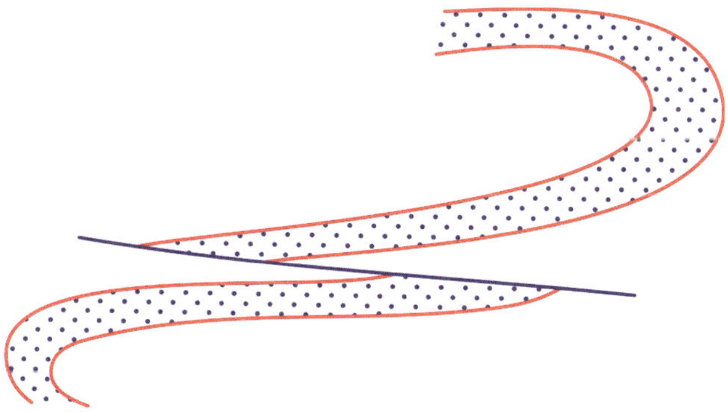

Fig. 11.26 Sketch of a backlimb thrust

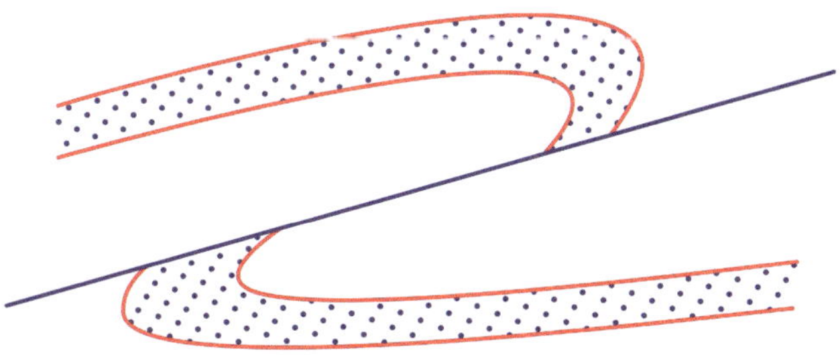

Fig. 11.27 Sketch of a stretch thrust

A *normal-sequence thrust* or *in-sequence thrust* is formed due to outward propagation of thrusts from the core of a mountain chain in a sequence (Fig. 11.37). The outward thrusts thus formed are progressively younger in sequence. An *out-of-sequence thrust* is formed when the normal sequence of outward propagation of thrusts is broken (Fig. 11.38). A *breaching thrust* (Fig. 11.39) (Butler 1987) is one that cuts across all the early-formed thrusts and is therefore a younger thrust.

A *piggyback thrust* (Fig. 11.40) is a complex structure formed when newer thrusts continue to form in a sequence in the footwall of the first-formed thrust (Butler 1982). The newer thrusts are formed in the direction of the foreland and continue to carry the early-formed thrusts on their back, thus giving a 'piggyback' geometry, and therefore the highest thrust will represent the earliest displacements while the lower ones the last displacements (Dahlstrom 1970). In the structure, therefore, the frontal (youngest) thrust is active while the earlier thrusts that it carried piggyback are inactive. In an *overstep* (Fig. 11.41) (Butler 1982), on the other hand, a new thrust surface is developed in the hanging wall of an older thrust. Thrust propagation takes place towards the hinterland in a sense opposite to the transport direction. The higher thrust will represent the later movements across the array of faults (Butler 1982). An overstep can thus be considered reverse to a piggyback structure.

Generally, in a thrust system, the associated thrusts show the same vergence. But occasionally, some of the associated thrusts show displacement in reverse direction with that of the main thrust; these are called *back-thrusts* (Fig. 11.42). The foreland fold and thrust belts of orogenic mountain belts commonly show the development of back-thrusts. If a back-thrust meets or truncates an earlier thrust, the area bounded by the two thrusts converging upwards is called a *triangle zone* (Elliott 1981) (Fig. 11.43). Sometimes, the structure undergoes later deformation and rotation (Fig. 11.44); in such cases, the structure deviates from its ideal geometry.

The hanging wall block created between a back-thrust and the frontal ramp is called a *pop-up structure* (Elliott 1981) (Figs. 11.45 and 11.46). Back-thrusts and pop-up structures are commonly noticed in thrust belts. The term pop-up structure is commonly used in the context of orogenic mountain

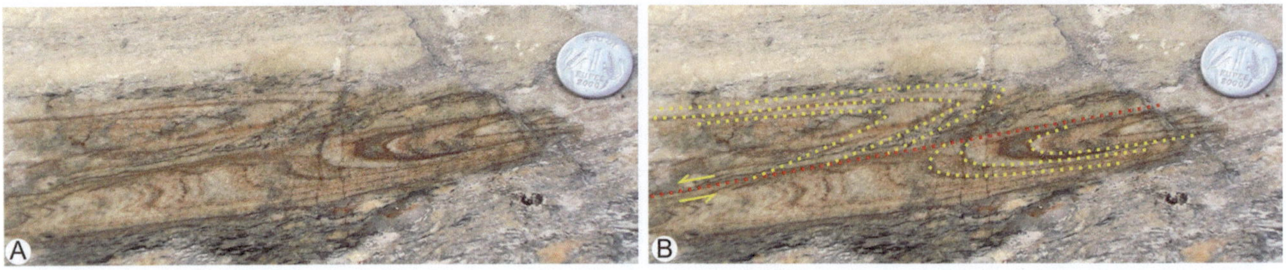

Fig. 11.28 (**a**) Field photograph of a stretch thrust developed in quartz mylonite in the Main Central Thrust zone of Garhwal Higher Himalaya, India. Loc.: North of Helong. (**b**) Elucidation of the structure. (Photograph by the author)

Fig. 11.29 A folded thrust developed in sandstone of Krol Group of Kumaun Lesser Himalaya, India. Loc.: Chamba, near Rishikesh, India. (Photograph by the author)

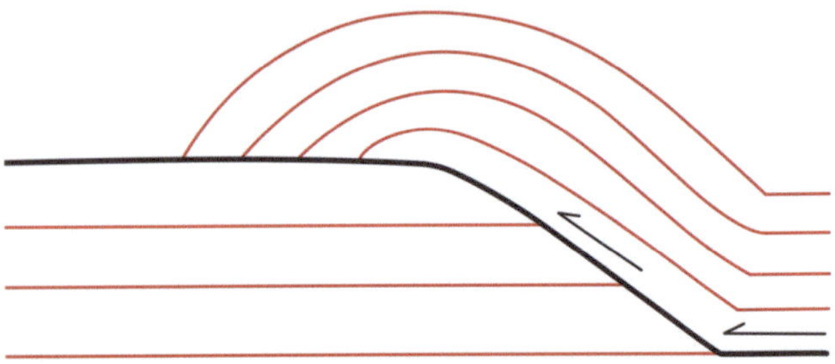

Fig. 11.30 Diagrammatic sketch of a ramp anticline

belts as in the Himalaya where it is developed at the collision zone.

11.9 Trishear

Trishear is a type of shear proposed by Erslev (1991) as an alternative model to explain folding in front of propagating thrusts in fault-propagation folds, for which the previous models used kink band geometries. Erslev pointed out that kink band kinematics cannot explain the curved fold surfaces and complex strain patterns in natural and experimental fault-propagation folds. During their formation, fault-propagation

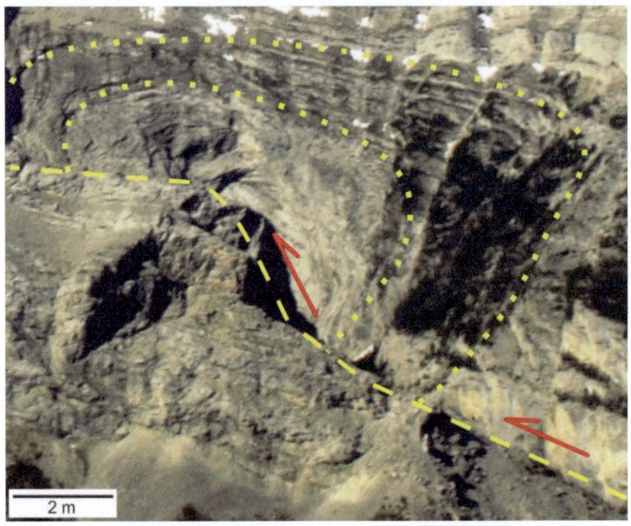

Fig. 11.31 Aerial photograph of a ramp anticline in the rocks of the Canadian Rockies in the Alberta region. A portion of the fault surface is curved over which the beds have rotated to produce the anticline. (Photograph courtesy Narendra K. Verma)

fold hinges tighten and converge downward, thus forming a triangular zone of penetrative deformation that is focused on the tip of the propagating fault. This downward convergence of deformation in fault-propagation folds has been modelled

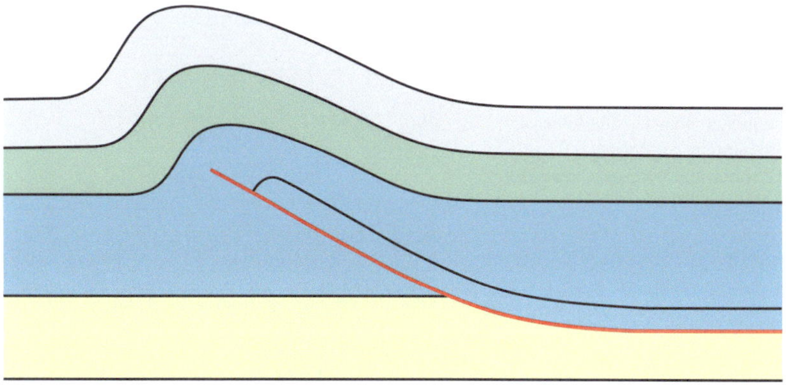

Fig. 11.32 Sketch of a fault-propagation fold

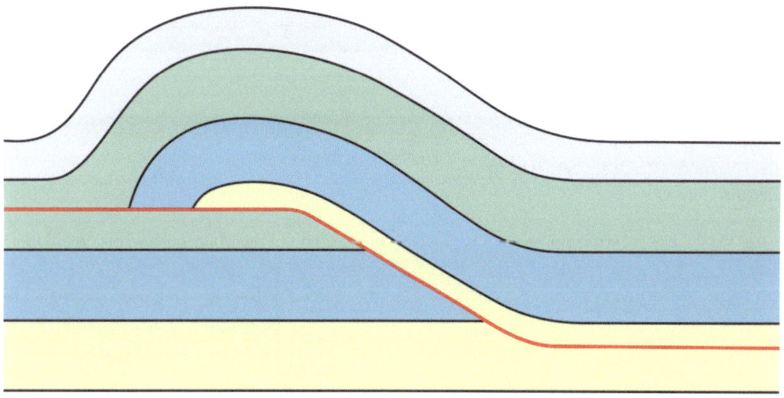

Fig. 11.33 Sketch of a fault-bend fold. See text for details

Fig. 11.34 Fault-bend fold developed in a thrust belt in the Andes Mountains, Santa Cruz Province, Patagonia, Argentina. (Photograph courtesy Narendra K. Verma)

as triangular shear zones, called as 'trishear', which is defined by Erslev as *distributed, strain-compatible shear in a triangular (in profile) shear zone*. Trishear resembles simple shear in a tabular shear zone with the main difference that area balance in a triangular shear zone requires curved displacement oblique to the fault-slip direction.

Analytical models of triangular shear-zone folding, as presented by Erslev (1991), assume that displacements within a triangular shear zone can be approximated by defining a set of initially fault-normal tie lines between the sides of a plane-strain shear zone (Fig. 11.47). These tie lines rotate with an increment of deformation; the rotation will be more for those

Fig. 11.35 Fault-bend fold containing an anticline and a complementary syncline developed in the Mesozoic sequence near Upsala Glacier, Patagonia, Argentina. (Photograph courtesy Narendra K. Verma)

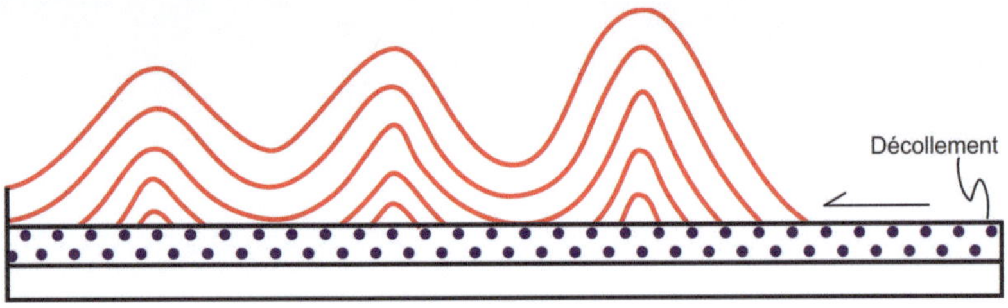

Fig. 11.36 Sketch of a decollement structure. See text for details

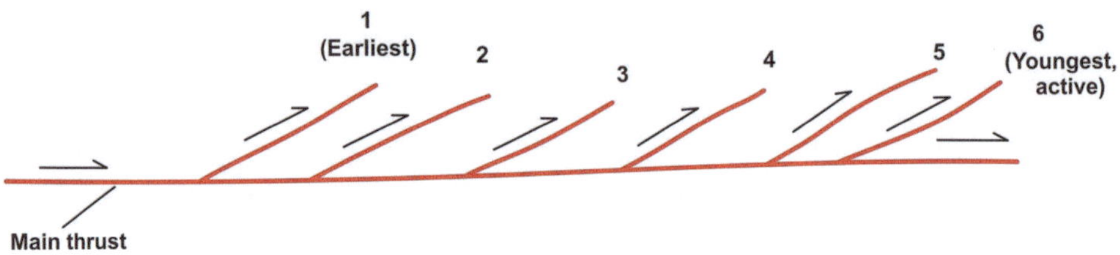

Fig. 11.37 Normal-sequence or in-sequence thrust. Thrusts sequentially form from the core of the mountain chain so that the younger thrusts progressively form in the direction of transport

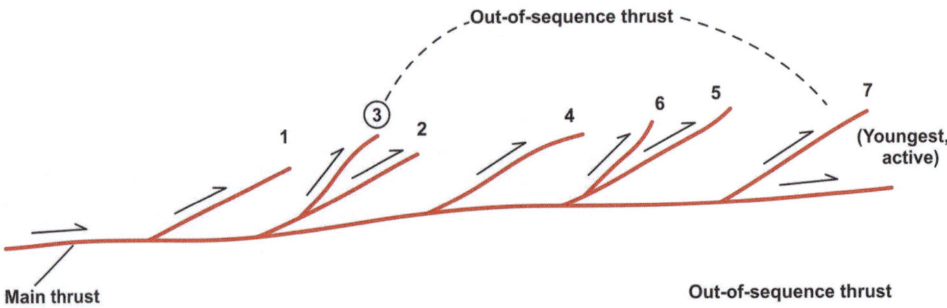

Fig. 11.38 Out-of-sequence thrust. The sequential outward formation of thrusts is broken so that one (thrust number 3) thrust breaks the normal sequence to come at the outer end

closer to the fault because the shear zone narrows towards the fault. If the shear zone is asymmetrically distributed either to the hanging wall (Fig. 11.47a) or to the footwall (Fig. 11.47b), a change in volume must occur in the triangular shear zone. In case of symmetric shear zones (Fig. 11.47c) that show equal involvement of the hanging wall and footwall, each polygon defined by adjacent tie lines and the connecting sides of the shear zone show equal volume during displacement.

Erslev further explains that if all motion trajectories are assumed to be parallel to fault slip, distributed simple shear can move each particle from an initial tie line to a deformed tie line. However, this situation results in volume imbalance, because due to simple shear the base of the polygons expands while the top of the polygons contracts. In order to maintain volume balance in the entire triangular shear zone, a component of motion oblique to the fault must bring material from the hanging-wall side to the footwall side of the shear zone.

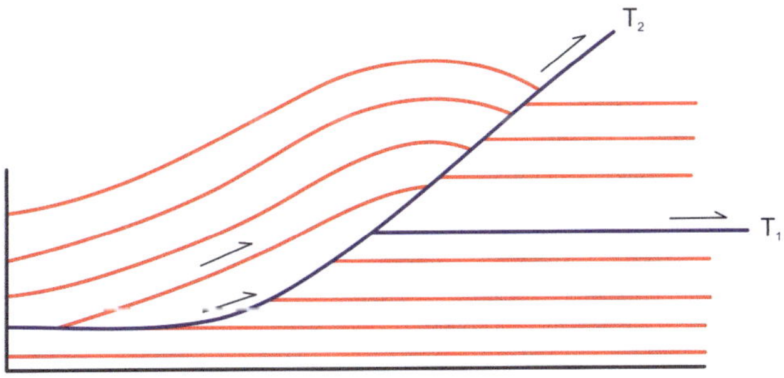

Fig. 11.39 Sketch of a breaching thrust. T_2 is the breaching thrust that cuts across an earlier thrust T_1

According to Erslev, this variant on simple shear is the trishear as defined above.

A general geometric feature of trishear deformation zone is that because strain is distributed through the triangular region, folds have curved hinges, and beds undergo stretching and thinning (Van der Pluijm and Marshak 2004, p. 462).

11.10 Models of Thrust Formation

11.10.1 Background

Thrust mechanics broadly means mechanical aspects of thrust formation. It deals with the stress-strain considerations necessary for initiation and motion of rock masses in the form of thrust sheets or overthrusts. Since thrusts occur under various geologic settings and are developed in diversified rocks of all ages, their formation has been explained in

Fig. 11.40 Piggyback thrust structure. See text for details

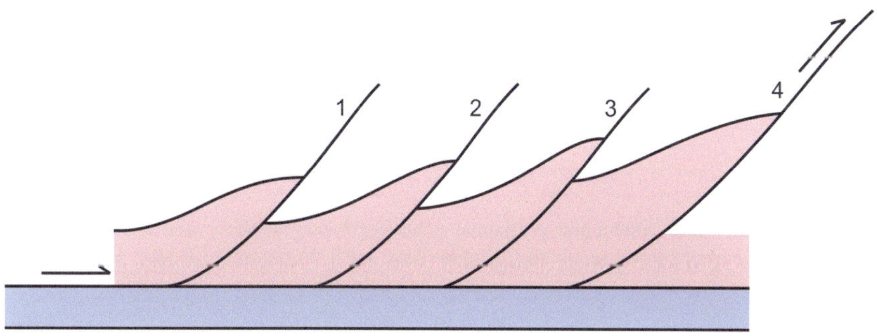

Fig. 11.41 Overstep structure. See text for details

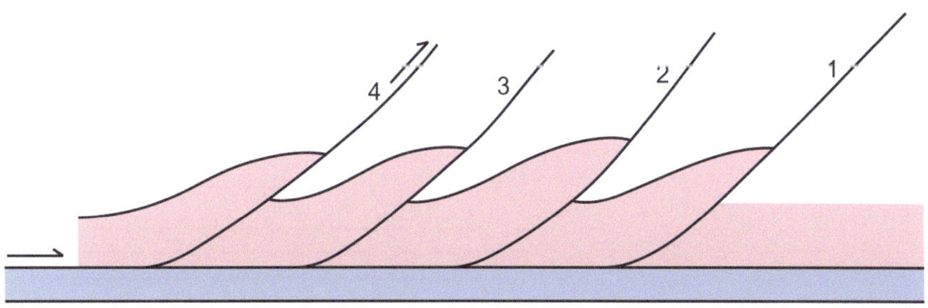

Fig. 11.42 Sketch of backthrusts. See text for details

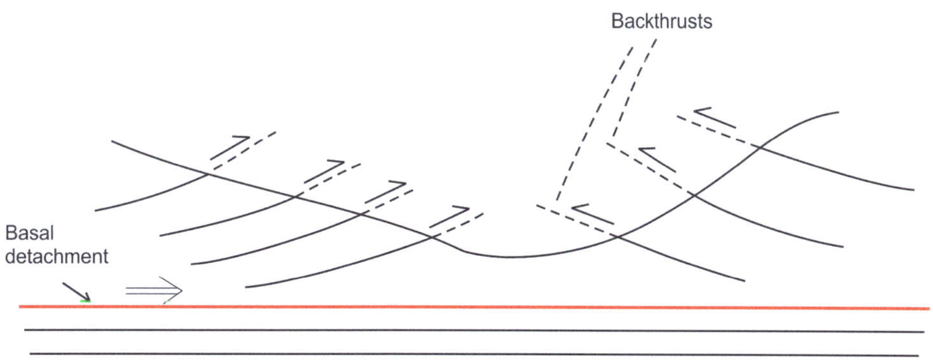

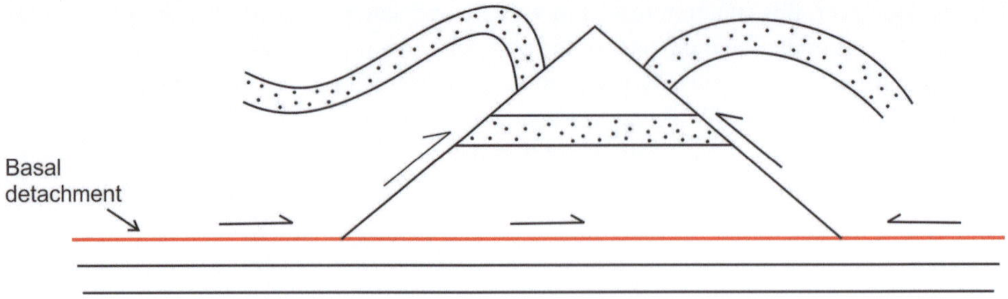

Fig. 11.43 Sketch of a triangle zone. See text for details

Fig. 11.44 Structure resembling a triangle zone developed in the turbidite sequence exposed along the Cinque Terre cliff, Vernazza, Alps, Italy. The structure has undergone rotation and later deformations and that is why it deviates from the ideal type. (This is the so-called Friday Folds of the Italian Alps). (Photograph courtesy Narendra K. Verma)

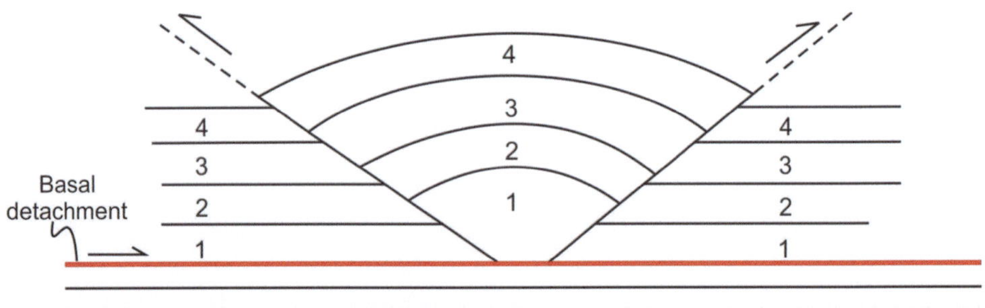

Fig. 11.45 Sketch of a pop-up structure. See text for details

various ways, and no unanimous opinion has as yet been put forward for their formation. Abode of most thrusts are in the foreland regions of orogenic belts where they occur in large numbers. Thrust formation in such regions is related to the processes of the particular orogenic belts. Most workers explained thrust mechanics by developing models that take into account field situations or experimental demonstration by analogue materials and sometimes by a combination of the two. In general, it is held that there are two dominant motive forces that produce thrusts: compression and gravity. However, both these forces are not independent as some components/inputs of one are needed in the other. Some common mechanics of thrust formation are discussed below.

11.10.2 Compressional Models

Compressional models are based on the assumption that compression is the motive force for thrust formation. According to Anderson (1951), thrust faults are formed at low angles (less than 45°) under conditions when the maximum principal stress σ_1 is horizontal and the minimum

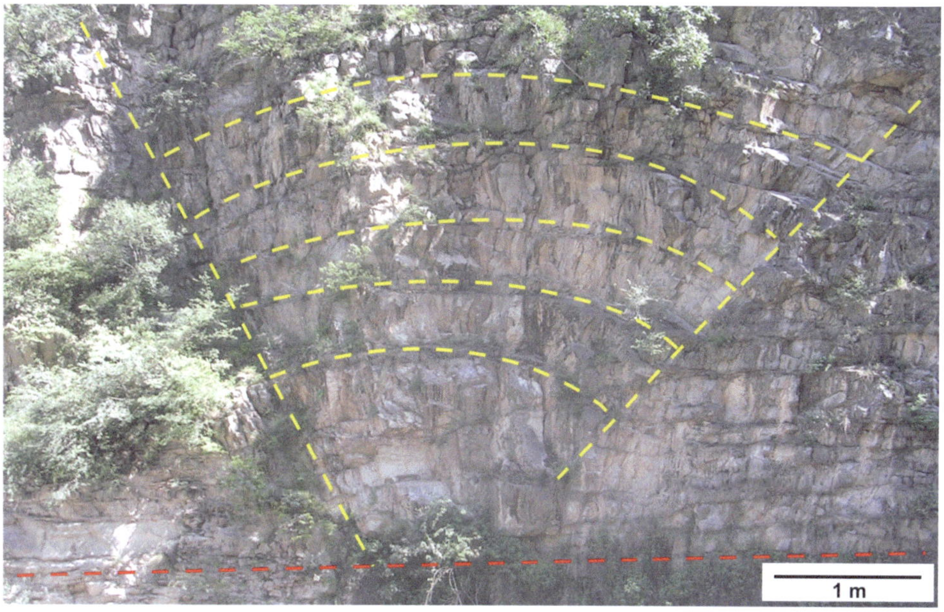

Fig. 11.46 Photograph of a pop-up structure developed in sandstone beds of Siwalik Supergroup of Outer Himalaya, India. Loc.: Near Haridwar. (Photograph by the author)

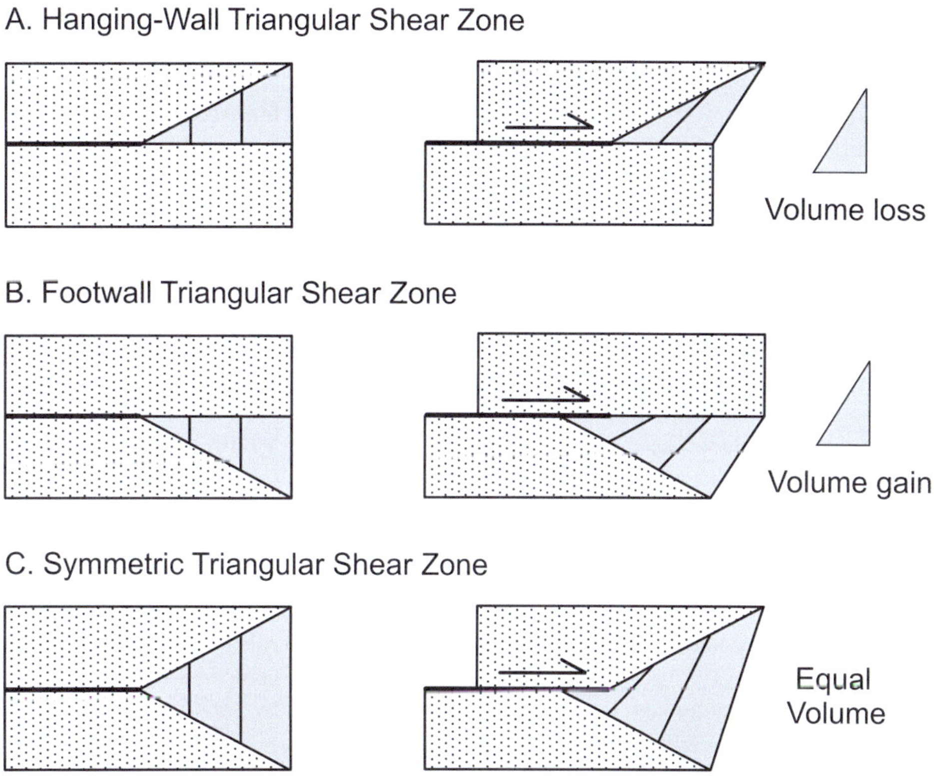

Fig. 11.47 Triangular shear-zone (trishear) folding showing geometric end members. See text for details. (Reproduced from Erslev 1991, Fig. 2 with permission from Geological Society of America)

principal stress σ_3 is vertical. This enables the hanging wall to move up relative to the footwall. Further, σ_1 being horizontal and σ_3 being vertical, the thrust movement causes horizontal shortening and vertical extension. This implies that thrust faults should form under conditions of intense crustal compression, and, since σ_3 is vertical, such faults are more likely to form at shallower levels of the crust.

Of all the models, that of Hubbert and Rubey (1959) has been described as an efficient one to explain the formation of thrusts in a satisfactory way. Their theory is based on the fact that fluid pressure significantly affects the strength and frictional resistance of rocks. They highlighted the role of fluid pressure in pushing a thrust mass from behind as well as sliding a thrust sheet down an inclined plane due to gravity. Presence of pore fluids, and thus fluid pressure of rocks, facilitates motion of rocks along planes of weakness much more than when the rocks are dry. Dry rocks on the other hand undergo fracturing, and this inhibits motion of the rocks. With fluid pressure, a rock has to spend less energy than a rock without pore fluids. Fluid reduces the effective normal stress and makes a rock mass buoyant, which in turn reduces the frictional resistance along fault planes, thus promoting the motion of the rock mass. Thrust faults thus commonly occur in zones of high fluid pressure and low frictional resistance. All this explains the existence of large thrusts with long transport in several orogenic belts.

Elliott (1976a, b) suggested that the displacement of a thrust is linearly related to the length of its outcrop trace and that a major part of the total energy is dissipated within the rock mass constituting the thrust sheet. According to him, the energy balance of a thrust sheet and partitioning of the energy among the various deformational processes are important factors that control initiation and motion of a thrust sheet. A horizontal compressive push that usually refers to the compressive crustal stresses is needed to initiate a thrust mass. In order that a thrust sheet is able to move, the compressive surface forces applied by the hinterland must be transmitted to the entire body of the thrust sheet. Thrust sheet thus acts as a 'stress guide'. If a thrust sheet does not have sufficient strength to serve as a stress guide, surface forces applied at the boundary would be dissipated within the thrust sheet. This would restrain any movement to the thrust sheet. This automatically calls for an alternative force, i.e. gravity, to move a thrust sheet. Elliott thus suggested that, although thrust faulting is commonly attributed to a horizontal compressive push, gravitational forces are necessary in the emplacement of large thrust sheets and that toe regions move under compressive forces transmitted by the main body of the thrust sheet, which itself moves by gravity.

Elliott also suggested that the mechanisms of deformation in the form of dissipation of energy in a thrust mass vary with its position (depth). He has shown that the top 5 km of the McConnell thrust of the Canadian Rockies is dominated by frictional sliding. At depths of more than 5 km, pressure solution slip becomes the dominant mechanism with signatures of cleavage and fibres. Most large thrusts are believed to have deformed by frictional sliding at their outer layers, which overlie a massive layer that mechanically behaves as a linearly viscous mass.

11.10.3 Gravitational Models

Orogenic movements commonly give rise to a well-developed topography to the affected region. Topography has been suggested to promote the movement of rock layers downhill. This implies that gravity is a dominant motive force to produce thrust faults. Several workers developed models for the formation of thrusts considering gravity as a motive force. In fact, compressional models also need to include gravity as a motive force to satisfactorily explain the formation of thrust sheets, especially when well-developed topography is present. Hubbert and Rubey (1959) suggested that high fluid pressure would reduce the high shear stress necessary to move a thrust sheet. As such, it is the body forces, i.e. gravity, that can move a thrust sheet. Elliott (1976a, b) also has shown that the main body of a thrust mass moves with gravity as a motive force. Gravity works in a variety of ways in the formation of thrusts mainly because of the well-developed topography of the hinterland during orogenic movements.

Thrust formation, dominantly by gravity, has been explained by four important models (see also Fossen 2016): gravitational gliding, gravitational collapse, gravitational extrusion and gravitational spreading.

11.10.3.1 Gravitational Gliding

This model is applicable for those orogenic belts where the thrust faults dip towards the foreland. Several thrust faults of the Alps were earlier mapped like this. However, the basal decollement was later on observed to always dip towards the hinterland. As such, the gliding model may not work with thrust belts. The foreland-directed dip of the Alps has been suggested to be an imprint of later deformations.

11.10.3.2 Gravitational Collapse

Under certain conditions, a part of the crust may become weak due to general heating or due to intrusion of warm magma. In contractional regime, this weak crust may collapse due to its own weight.

11.10.3.3 Gravitational Extrusion

Orogeny normally produces a wedge whose hinterland part is thicker and relatively more elevated. Once this stage is reached, gravity comes into play by pushing down—'extrusion' of—rock masses as thrust nappes that move towards, and perpendicular to, the foreland. That is why this model is called gravitational extrusion model. This process can be compared to glaciers whose downhill flow is also controlled by gravity.

11.10.3.4 Gravitational Spreading

Rock masses from the thick and highly elevated hinterland, as conceived in the extrusion model, may move sideways towards the foreland. This has been named as gravitational spreading model. The spreading is caused by the difference in surface elevation between the interior of the mountain belt and the craton (Elliott 1976a). Movement of thrust sheets is directed towards, i.e. perpendicular to the foreland. Tectonic push is relatively more along this (perpendicular) direction than those thrust sheets moving down sideways. Possibly, this is the reason that forelands are commonly curved. Curved foreland fronts of several orogenic belts may be ascribed to gravitational spreading. The curved front of the Himalayan foreland may also be viewed as an outcome of gravitational spreading of the Tibetan landmass in the form of thrust sheets that today constitutes the foreland fold-thrust belt.

11.10.4 Mixed Models

The models of thrust formation and the associated rock deformation described above represent rather ideal cases. Recent studies indicate that this is in fact not so for most thrust belts. Several workers have recently shown that thrust mechanism in many mountain belts operates due to the intervention of more than one mechanism and even ground configuration and overall geomorphology also play important roles.

Ballantyne (2002) and Bozzano et al. (2008) are of the opinion that lateral stress loading operates as the main mechanism to cause gravitational instabilities. Hungr et al. (2014) consider that in the case of rock-slope lateral spreading, the deformation contrast between a stiff rock mass and an underlying softer ductile material is essential, so that a free face can accommodate the horizontal movement. Di Luzio et al. (2004) and Della Seta et al. (2017) have shown that rock spreading can occur better due to fault-fold combinations such as the existence of faulted fold limbs.

Alfaro et al. (2019) carried out numerical modelling of the Eastern Betic Cordillera (Spain), which is a classic example of lateral spreading. They highlighted the role of slope tectonics and inherited structural elements for creep-driven slope deformations and suggested that there is a structural control due to folds and faults on the ongoing lateral spreading process. They concluded that the slope-scale gravity-induced processes work closely with the structural and geomorphological set-up.

11.10.5 Role of Detachment Geometries

Spitz et al. (2020) demonstrated that simulations with different and laterally varying detachment geometries cause a lateral transition from folding to overthrusting. They have shown that the Nadai strain, ε_s, generally increases from folding to overthrusting and that different lateral variations in geometry, such as oblique ramps, can generate different lateral finite strain variations. The workers thus highlighted the role of oblique ramps during overthrusting.

11.11 Diapirs and Salt Domes

11.11.1 Background

A *salt dome* is a dome-shaped structure formed by vertical intrusion of a column of salt into the overlying sedimentary strata (Fig. 11.48). The structure thus formed is called a *diapir*. The salt can intrude because its density is less than the surrounding rock. Vertical rise of magma from the magma chamber followed by its intrusion into the surrounding rocks is also similar to that of a salt dome, and therefore the term diapir is also extended to include magma. Diapirs can be constituted of ice, peat, evaporates (especially salt), mudstones and marls, occasionally sands, and some igneous rocks (Chapman 1983, p. 326). In the context of salt, the term salt diapir is commonly used.

By salt we commonly mean pure halite. But when salt occurs in sedimentary strata, it constitutes a component of evaporite that also contains anhydrite and gypsum. Clay minerals are also invariably present. The salt is impermeable, and therefore it constitutes a suitable site for dumping hazardous waste. The material above the diapir is a sedimentary sequence that is younger than the diapir material. On a horizontal section, a salt dome generally shows circular to elliptical structures. Invariably, diapir shows a cap of anhydrite or gypsum. The cap rocks of petroleum-bearing diapiric structures occasionally contain sulphur, which is probably biogenic (Chapman 1983, p. 327).

11.11.2 Geographical Distribution of Salt Domes

Salt domes occur in several parts of the world. The Middle East countries are known for the occurrence of several salt domes; well-known occurrences include those of Iran, Iraq, Oman and the United Arab Emirates. In the USA, they occur in the Gulf Coast region, Utah, Colorado and New Mexico. In northwest Europe, they occur in Norway, Northwest Germany and North Sea. They also occur in SW Russia and west central Africa.

11.11.3 Why Subsurface Salt Is Unstable?

Salt has a very low density and a very low viscosity. Combination of these two properties makes salt buoyant and mechanically a very weak rock with very low, or practically no, strength. In fact, salt is possibly the weakest of all

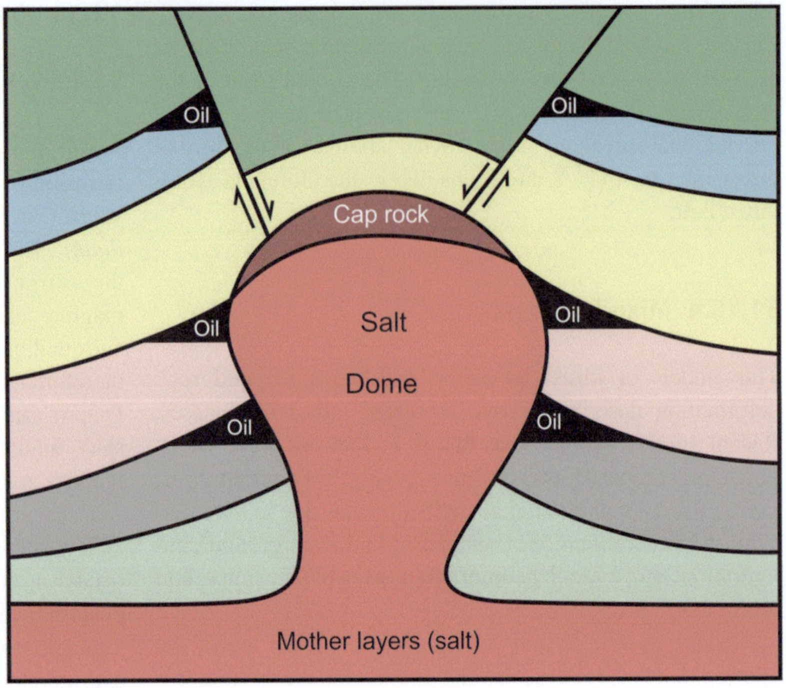

Fig. 11.48 A generalized diagram showing the structure of a salt dome. See text for details

sedimentary rocks (Jackson and Vendeville 1994). Further, because of these properties, salt produces a contrast in its rheological properties from the surrounding rocks that are mostly clastic siliceous. Being mechanically weak, the salt undergoes deformation and flows like a viscous material. If present in sedimentary strata, salt constitutes a mechanically weak layer to allow easy transport of rocks and is therefore capable to transport overlying rocks mainly in the form of detachments. Many decollement layers contain salt rocks. Being almost incompressible, salt does not deform by itself and thus maintains the tectonic transport as long as the stress system and other factors allow.

11.11.4 Formation of Salt Domes

Formation of salt dome practically means the movement of a salt mass from its substratum to upper layers of sediments. Although the density difference between the salt and the surrounding layers is important, movement is virtually triggered by gravitational cause. The latter includes variations in the density of the overburden as well as that of the salt mass. All these develop differential loading and thus a mechanical instability to the salt mass. Under such conditions, salt cannot remain stable in its own place for longer time and is driven up towards low-pressure zones by viscous flow due to its behaviour as a viscous fluid. The final shape of the salt, i.e. salt structure, depends strongly on the ambient tectonic conditions, i.e. whether contractional or extensional. Under contractional conditions, the salt is stretched vertically and shortened horizontally. Folds are commonly associated with such salt structures. Under extensional conditions, the salt assumes an opposite shape, i.e. stretched horizontally and thinned vertically. Faults are commonly associated with such salt structures. Most commonly, salt is intensively deformed and shows flow patterns (rather than folds), but faults are very rare (Chapman 1983, p. 327). This possibly suggests that extensional processes may play a subordinate role in the formation of salt diapirs.

Jackson et al. (1990) and Jackson and Talbot (1994) presented a model of how a salt dome behaves to sediment accumulation, which is highest at the flanks of the dome. With time, the load of sediments can cause lateral flowage of the underlying salt (dome). As a result, the squeezed-out salt progressively gets *welded* with lithological layers that originally had been stratigraphically separated by the salt layer. Thus, *salt welds* are formed. Increase of load causes increase of subsidence and sedimentation, leading more flow of salt. In this way, salt domes also affect, in addition to the structure, the local stratigraphy of the sedimentary layers.

Numerical modelling of two-dimensional finite deformation by Fuchs et al. (2015) shows that 'salt' material shows three different deformation regimes: (a) a squeezed-flow deformation regime, (b) a corner-flow deformation regime within the source layer and (c) a pure channel-flow deformation regime within the stem. During rise of the salt from its source layer, it follows different deformation paths and passes through different finite strain regimes until it reaches the surface where it spreads like a diapiric dome downslope as a *namakier* (i.e. salt glacier) (Sarkarinejad et al. 2018, p. 109).

In the absence of a decollement layer, the structural style developed includes narrow fold and thrust belt with consistent vergence.

11.11.5 Salt Tectonics

Salt tectonics refers to any tectonic deformation involving salt, or other evaporites, as a substratum or as a source layer, while *halokinesis* is a form of salt tectonics that is controlled entirely by gravity without significant contribution of lateral forces (Jackson and Talbot 1994, p.167). Salt tectonics refers to flow and tendency of salt to migrate when loaded by sedimentary overburden (Sarkarinejad et al. 2018, p. 111). Since salt is weak and buoyant, it behaves sympathetically to any tectonic environment it is subjected to. We present here excerpts of two major aspects of salt tectonics, i.e. extensional and contractional, from the work of Jackson and Talbot (1994).

11.11.5.1 Extensional Salt Tectonics

Extensional salt tectonics is developed where σ_1 is vertical and σ_3 is horizontal (or slope parallel) and less than the lithostatic stress. This type of salt tectonics is developed in regions of normal fault in the overburden such as the continental slopes. Under such conditions, the base of the salt initially remains undeformed. Due to thin-skinned extension, salt undergoes gravity spreading or gravity gliding. At advanced stages of extension, *raft tectonics* comes into play when deep, syndepositional grabens open and the intervening overburden separates into rafts that slide like block-glide landslides downslope on a decollement of thin salt. A *raft* is thus formed, which is a fault-block allochthonous overburden. After separation, this raft no longer rests on its original footwall; instead, it lies entirely on a decollement layer consisting of thin salt.

11.11.5.2 Contractional Salt Tectonics

Contractional salt tectonics is developed where either the minimum principal stress (σ_3) is vertical (thus creating shortening) or the intermediate principal stress (σ_2) is vertical (thus creating transpression). The regional shortening can be thin skinned (i.e. contraction of the cover) or it can involve basement. The pressure exerted by the salt decollement initiates decoupling of the overburden that results in the following: (a) more folding than thrusting; (b) absence of consistent vergence of folds and thrusts and, instead, common box folds and back-thrusts; and (c) development of extremely broad fold and thrust belts. In general, the structure is dominated by angular folds (box folds and chevron folds), thus reflecting internal deformation of anisotropic multilayers above a thin, ductile substratum.

11.11.6 Flow Mechanisms

Flow in a salt diapir mainly occurs by channel flow mechanism. The channel flow occurs within a salt layer by two major types of flow: *Poiseuille flow* that occurs when salt is loaded with overburden sediments of different thicknesses and it flows into a salt structure during the growth of a salt diapir, and *Couette flow* which is a shear-type salt channel flow with a linear velocity profile in the salt (Gemmer et al. 2005). In general, Couette flow includes simple shear within salt layer, and Poiseuille flow produces a high pure shear component in the centre of plan of the channel and high simple shear component along its margins (Sarkarinejad et al. 2018, p. 124). In a salt diapir, both pure shear and simple shear work in combinations.

Sarkarinejad et al. (2018) carried out kinematic analyses of channel flow in the Karmostaj salt diapir in the Zagros foreland fold belt, Iran. The diapir has reached the surface due to channel flow mechanism. The distribution of flow regimes within the salt diapir is the result of interaction of regional tectonics and salt diapirism. Microstructural studies reveal that various deformation mechanisms have operated in various parts of the diapir. Estimation of mean kinematic vorticity number (W_m) reveals involvement of 67.8% pure shear and 32.2% simple shear deformation. The authors thus conclude that the persistent flow of the diapir during extrusion was maintained due to a combination of pure shear and simple shear.

11.12 Significance of Thrust Faults and Salt Diapirs

11.12.1 Academic Significance

- Thrust faults have always drawn the attention of geologists from several angles. Their formation and geometry help us in understanding several geological processes. Since thrusts are formed under conditions of great compressional regime, they are commonly located in orogenic belts.
- Thrusts help in unravelling the directions along which crustal stresses were operative.
- Systematic geometric analyses of thrusts help in tracing the stages of deformation in an orogenic belt.
- Study of thrust geometry helps in unfolding the regional structure and in the genesis of folds and thrusts.

- Thrust geometry provides information to structural geologists about how and where the imposed strain on rocks is accommodated.
- The various thrust geometries shown by rocks are indication, directly or indirectly, of the mechanisms of deformation undergone by the rocks.
- Study of salt domes extends our knowledge on rock deformation from a special angle.

11.12.2 Economic Significance

- Thrusts have always drawn the attention of hydrocarbon geologists as well as economic geologists. The main reason is that thrusts serve as conduits of fluids. Several thrusts are known for their association with hydrocarbon and economic minerals including metallic minerals.
- Presence of a thrust system in a region of deformed rocks therefore draws the attention of petroleum exploration geologists, especially if a hydrocarbon reservoir is indicated by other surveys.
- Fluids of several kinds such as groundwater, oil, mineralized fluids and gas are always present in the upper crust. These fluids occur at high pressures at depth, and due to the presence of any nearby thrust, they reach a zone of low pressures where they can rest in physical equilibrium under the new conditions.
- The best hydrocarbon traps are provided by overlapping ramp anticlines of imbricate thrust systems because of (a) multiple stacking of the reservoir unit and (b) increased fracturing caused due to increased curvature of the overlapping sheets (Mitra 1986, 1990).
- While duplexes and imbricate thrust systems form some of the most complex hydrocarbon traps in overthrust belts, the independent ramp anticlines and fault-propagation folds constitute the simplest structural traps. Effective migration and trapping of hydrocarbons are favoured if the different thrust sheets of a duplex are connected by faults (Mitra 1986, 1990).
- Salt domes contain salt and some other evaporite minerals such as anhydrite and gypsum. The cap rocks of many salt diapirs are petroleum bearing, and they occasionally contain sulphur.

11.13 Summary

- *Contractional regime* is one in which a layer or a part of the earth's crust is shortened. Thrust faults and folds are commonly formed in such regimes.
- A *thrust fault*, also called a *reverse fault*, is one in which the hanging wall has moved up relative to the footwall. The term thrust fault is commonly used if the dip of the fault is less than 45°; otherwise, the term reverse fault is used.
- In thrust belts, thrusts are commonly associated with several types of structures. A *nappe* is a large, essentially coherent allochthonous, sheet-like tectonic unit that has moved a distance several times its thickness and in excess of 5 km along a predominantly sub-horizontal floor.
- A *thrust nappe* is an allochthonous tectonic sheet which has moved along a thrust fault. A *fold nappe* is an allochthonous tectonic unit which exhibits large-scale stratigraphic inversion and may have initiated from large recumbent folds. The underlying limbs of these folds may be sheared out into thrust faults.
- A *thrust sheet* is a tectonic unit overlying a thrust.
- Thrust faults show several types of geometries that have been grouped here under (i) planar thrust geometry, (ii) thrust geometry as related to folds and (iii) thrust geometry as related to stratigraphic sequence.
- Thrusts that have been grouped under planar thrust geometry include thrusts that behave as solid planes and include structures such as bedding thrust, duplex or thrust duplex, imbricate structure, ramp and flat structure, emergent thrust, blind thrust and splay.
- Thrusts as related to folds include geometries such as break-thrust, stretch thrust, thrust fold or folded thrust, fault-propagation fold, fault-bend fold and decollement.
- Thrusts as related to stratigraphic sequence include normal-sequence thrust, out-of-sequence thrust, breaching thrust, piggyback thrust, overstep, triangle zone and pop-up structure.
- Study of thrust geometry has been found to be of great importance in hydrocarbon exploration. Some of the thrust geometries form good traps for hydrocarbons.
- A *salt dome* is a dome-shaped structure formed by vertical intrusion of a column of salt into the overlying sedimentary strata. The salt dome structure is also described as a *diapir* because the vertical rise of magma from the magma chamber followed by its intrusion into the surrounding rocks is also similar to that of a salt dome.
- A very low density and a very low viscosity make salt a mechanically weak rock with practically no strength. All this makes a contrast in its rheological properties from the surrounding rocks, and this contrast makes the salt flow like a viscous material to allow easy transport of overlying rocks.
- Flow in a salt diapir mainly occurs by channel flow mechanism.

- Salt domes are associated with several types of economic minerals such as salt, gypsum, anhydrite and hydrocarbon.

Questions

1. What do you mean by contractional regime? Name the characteristic structures formed in this regime.
2. What is meant by thick-skinned deformation and thin-skinned deformation? Give the differences between the two.
3. Give a brief description of crystalline thrusts.
4. Differentiate between a nappe and a klippe.
5. What do you mean by a normal-sequence thrust and an out-of-sequence thrust?
6. How would you differentiate between leading edge and trailing edge imbricate structures?
7. What is a piggyback structure and an overstep? Give the difference between the two.
8. Briefly describe the role of gravity as a motive force to produce thrust faults.
9. Why do the hydrocarbon geologists take interest in the study of thrust geometry during their exploration work?
10. What is a salt dome? Describe how salt dome can rise up in the crust.

12 Strike-Slip Faults

A. R. Bhattacharya, *Structural Geology*, Springer Textbooks in Earth Sciences, Geography and Environment,
https://doi.org/10.1007/978-3-030-80795-5_12

Abstract

A *strike-slip fault* is one that shows horizontal slip parallel to the strike of a vertical or subvertical fault plane. Strike-slip faults are often associated with large displacements. The fault plane in most cases extends for great depth, and as such most of them show brittle behaviour on the upper part and ductile behaviour at depth. The trace of the fault may show curvatures where it shows some structural complexities. Due to motion along the strike-slip faults, the ground surface and the associated rock masses get rubbed against each other, thus producing earthquakes. In this chapter, we shall describe the various geometries and types of strike-slip faults and the structures associated with such faults.

Keywords

Strike-slip fault · Transform fault · Transcurrent fault · Wrench fault · Transfer fault · Transpression · Transtension · Flower structure · Terminations · Riedel shears

12.1 Introduction

A *strike-slip fault* (Fig. 12.1) is one that is associated with horizontal displacement parallel to the strike of a vertical or sub-vertical fault plane. As such, its geometric attributes are different from the two other major classes of faults, i.e. normal faults and thrust faults. After the identification of similar type of faults on the oceanic floor, the term strike-slip fault is generally restricted to continents only. Strike-slip faults are often of great lengths and extend for several tens, or even hundreds, of kilometres. Their counterparts on oceanic floor often extend for thousands of kilometres. Several strike-slip faults constitute boundaries of continental plates. The trace of these faults can be identified as a linear feature on the ground surface, and the larger ones can be seen in satellite images. Well-known examples of strike-slip faults include the San Andreas Fault of California, the Alpine Fault of New Zealand, the Great Glen Fault of Scotland and the Brevard Fault of the Appalachians.

The strike-slip fault plane is steep to vertical and in most cases extends for great depth. Therefore, many large-scale strike-slip faults show brittle behaviour on the upper part and ductile behaviour at depth (Freund 1974). The trace of the fault commonly looks straight and generally shows a simple geometry. The trace of strike-slip faults may also show curvatures (Fig. 12.2). The curved portions show some structural complexities in the vertical section of the fault. The strike-slip faults once formed cut across the pre-existing geological features or structures such as lithological contacts, folds and their axes, igneous rocks and veins.

Because of motion along the strike-slip faults, the ground and the associated rock masses get rubbed against each other, thus producing earthquakes. Several devastating earthquakes have been shown to be associated with strike-slip faults. Fault plane solutions in the Himalayan region reveal that several major earthquakes in the last century are associated with strike-slip movements. Many strike-slip faults are potential zones of earthquakes.

12.2 Motion of Strike-Slip Faults

In the Anderson (1951) model, a strike-slip fault is formed when the greatest principal stress axis is horizontal, the least principal stress axis is horizontal and the intermediate

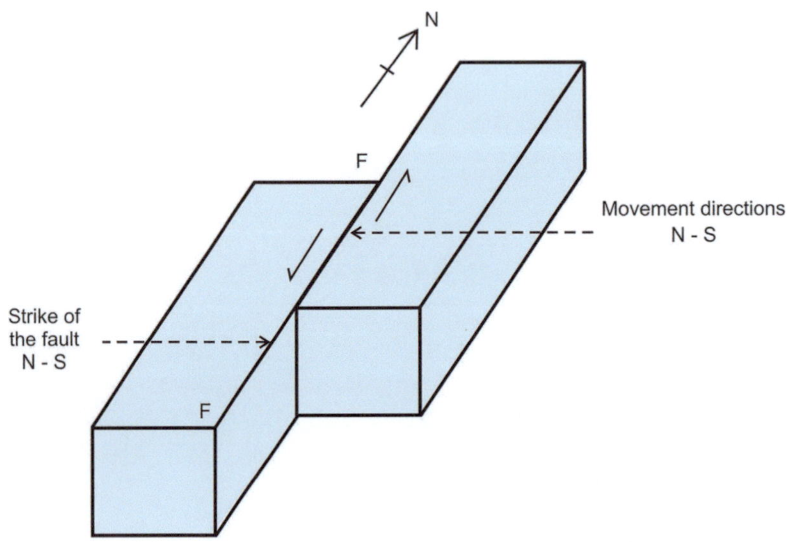

Fig. 12.1 Strike-slip fault. Note that the movement of blocks is parallel to the strike of the fault

principal stress axis is vertical. Under this condition, rupture occurs in a vertical direction when the differential stress, existing between the greatest and the least stresses, exceeds the strength of the rock. Strike-slip faults show horizontal movement in two opposite directions, i.e. right or left with respect to an observer. A *dextral* or *right-lateral strike-slip fault* (Fig. 12.3a) is one in which the block on the opposite side of an observer moves to the right. A *sinistral* or *left-lateral strike-slip fault* (Fig. 12.3b) is one in which the block on the opposite side of an observer moves to the left.

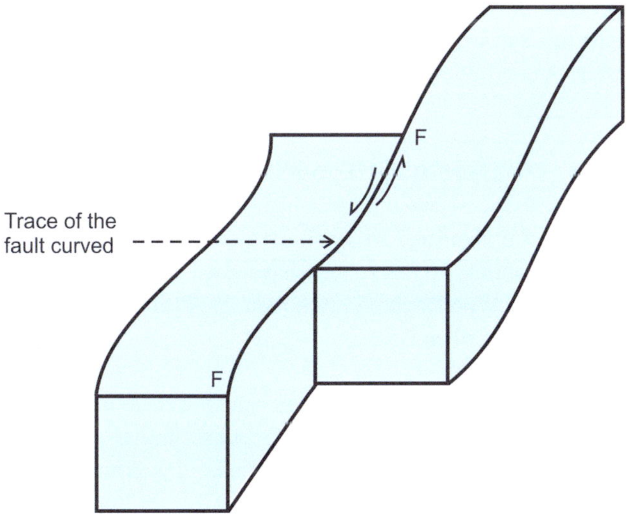

Fig. 12.2 Strike-slip fault showing curved trace of the fault

12.3 Types of Strike-Slip Faults

On the basis of the nature of displacement and the effects they produce in the surrounding rocks, several types of strike-slip faults have been identified of which some common types are described below.

12.3.1 Transform Fault

A *transform fault* (Fig. 12.4) is a strike-slip fault that cuts the whole lithosphere and fully accommodates the displacement across the plate boundary (Woodcock and Schubert 1994, p. 251). The associated plate may be oceanic or continental. The motion of the two blocks associated with a transform fault is horizontal and is either dextral or sinistral. Transform faults are boundaries of solid crustal plates, which always terminate at extensional or shortening structures, and as a rule are straight (Freund 1974, p. 94). They actually transfer the movement in such a way that lithosphere is neither created nor destroyed. As such, a transform fault is considered as a

Fig. 12.3 Strike-slip fault. (**a**) Dextral or right lateral. (**b**) Sinistral or left lateral

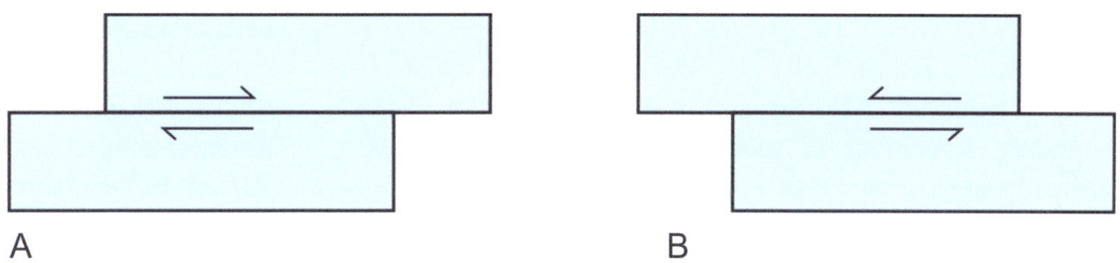

Fig. 12.4 Transform faults affecting diverging plate boundary (mid-oceanic ridge in this case). The portions AB and CD are transform faults. TC1, TC2 and TC3 are transcurrent faults

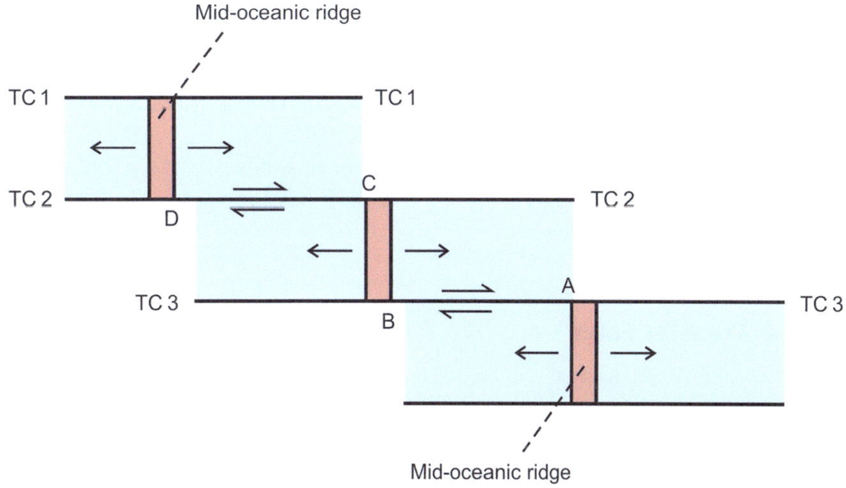

conservative plate boundary. A transform fault is believed to be an expression of accommodation of sea-floor spreading.

12.3.2 Transcurrent Fault

A *transcurrent fault* (Fig. 12.4) is a large-displacement strike-slip fault that cuts continental basement as well as sedimentary cover (Freund 1974; Woodcock and Schubert 1994). These faults commonly occur in groups (Fig. 12.4) and thus constitute a system of regional faults that affect a large segment of continental crust. Transcurrent faults usually terminate by splaying or bending (Freund 1974, p. 94). The longer a strike-slip fault, the larger is the associated displacement. In central Asia, several transcurrent faults have been described and are believed to result from northward movement of the Indian plate against the Asian plate (Molnar and Tapponier 1975).

The transcurrent faults and transform faults have both geometric and genetic differences (Freund 1974; Sylvester 1988). The transform faults occur at plate boundaries and show uniform deformation along its length, while the transcurrent faults show relatively more displacement in the middle portion. At the tips, the transform faults are joined with another plate boundary, while this is not so for transcurrent faults.

12.3.3 Wrench Fault and Tear Fault

These two types of faults are occasionally used synonymously with strike-slip faults that do not involve a plate boundary. According to Anderson (1951), a *wrench fault* has a dynamic rather than a kinematic significance and is one in which intermediate principal stress (σ_2) was vertical during deformation. Some workers (Moody and Hill 1956; Wicox et al. 1973; Biddle and Christie-Blick 1985) believe that wrench faults cut basement rocks and as such can be regarded synonymous to transcurrent faults.

A *tear fault* is a small-scale strike-slip or oblique-slip fault that runs across the strike of a contractional or extensional belt and is confined to the upper crust or to its sedimentary cover (Woodcock and Schubert 1994). In recent years, however, use of the terms wrench fault and tear fault has been rather limited in the literature.

12.3.4 Transfer Fault

In areas where a strike-slip fault ends and if some other fault with a different geometry is present nearby, the displacement of the former fault is transferred to the other fault. A *transfer fault* is one which transfers the displacement from one fault to another by strike-slip movement, though dip-slip movement is also possible. It is a vertical or near-vertical fault that can accommodate large amounts of displacement and can form in both extensional (Fig. 12.5a) and contractional (Fig. 12.5b) regimes. Occasionally, two or more normal faults may be joined at their end by faults with different orientations that are also called transfer faults. With these characters, the transfer faults bear similarities to tear faults, though the distinction between these two is not always clear. In general, the term transfer fault is used for any fault that transfers the displacement of one fault to another by strike-slip movement.

12.4 Classes of Strike-Slip Faults

Most of the identified strike-slip faults are large-size (continental) features. As such, they have been classified (Woodcock 1986; Woodcock and Schubert 1994) in plate tectonic setting as described below.

(a) **Ridge transform faults** are transform faults that join two oceanic ridge segments with similar spreading vectors. The faults thus maintain a constant length with time (Gilliland and Meyer 1976).
(b) **Boundary transform faults** separate plates that move parallel to their mutual boundary over lengths of hundreds of kilometres. Unlike ridge transform faults, the boundary transform faults usually join different plate boundaries (Gilliland and Meyer 1976) and sometimes show changes in length with time. These faults may persist for tens of millions of years and account for displacements of several hundreds of kilometres.
(c) **Trench-linked transform faults** or **strike-slip faults** occur at obliquely convergent plate boundaries. The oblique displacement kinematically resolves into two components: dip-slip on the subduction zone and strike-slip on a major fault parallel to, and some 100 km inboard from, the trench (Fitch 1972; Woodcock 1986). The term *trench-linked transform* is restricted to that type of trench-linked strike-slip fault that is discrete, whole-lithosphere fracture that separates a 'buffer' plate between itself and the trench.
(d) **Indent-linked strike-slip faults** form where a convergent plate boundary juxtaposes continental or arc crust on both plates (Woodcock 1986). The relative buoyancy of such crust causes collision and indentation of crust across the convergent zone.
(e) **Intra-plate strike-slip faults** occur away from the plate boundaries. They form due to intra-plate stresses that reactivate old faults that propagate these faults laterally and vertically into unfaulted rocks. Intra-plate faults

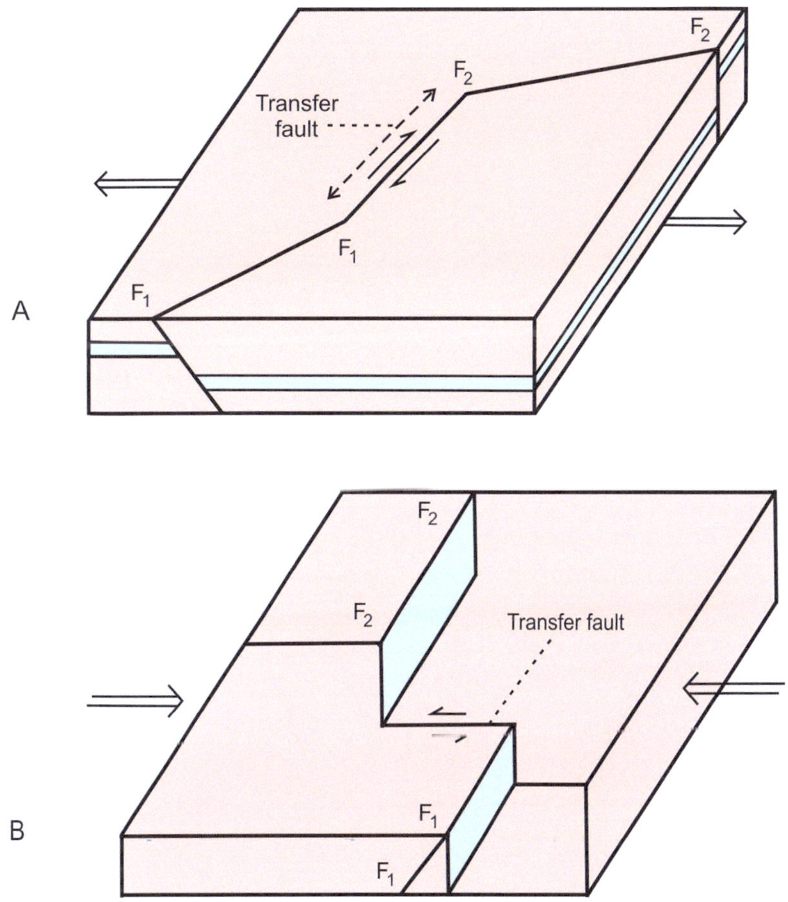

Fig. 12.5 Transfer fault. (**a**) Extensional regime. (**b**) Contractional regime. See text for details

show small displacements, but they may produce fault rocks and fault-surface features that obliterate earlier movement signatures.

12.5 Transpression and Transtension

Most strike-slip faults showing straight traces on the horizontal surface cause horizontal movement of two blocks along a vertical fault plane. As such, the displacement component of the fault is horizontal while the vertical component is negligible, and the two blocks undergo shearing parallel to the fault plane. However, if the traces of strike-slip faults show bends, the deformation pattern along the bends follows a different pattern. Along bends, the deformation can be either shortening (compression) or extension perpendicular to the fault plane. For these two types of deformation, Harland (1971) used the term *transpression* (Fig. 12.6) if the deformation at the bends involves shortening (compression) and *transtension* (Fig. 12.7) if the deformation involves extension.

Transpression is taken as a wrench or transcurrent shear accompanied by horizontal shortening across, and vertical lengthening along, the shear plane (Sanderson and Marchini 1984). Transpression occurs when the strike-slip component has a component of shortening.

This results in a combination of strike-slip movement and oblique compression and is believed to occur due to oblique convergence of a plate. Transpression thus involves a combination of pure shear and simple shear (Sanderson and Marchini 1984). Normal faults are formed at angles of more than 45°, while thrust faults are formed at angles of less than 45° (Fig. 12.6).

Transtension, on the other hand, occurs when the strike-slip component has a component of extension. This results in a combination of strike-slip movement and oblique tension. Normal faults are formed at angles of less than 45°, while thrust faults are formed at angles of more than 45° (Fig. 12.7). Kugler et al. (2019) suggested that orientations and heaves of faults can quantify transtension.

Sanderson and Marchini (1984) presented a model by considering a zone within which there is no volume change and which is laterally confined (i.e. there is no stretch along the zone leading to extrusion of material at its ends). Under such conditions, in order to conserve volume, the shortening across the zone results in a change in area that must be

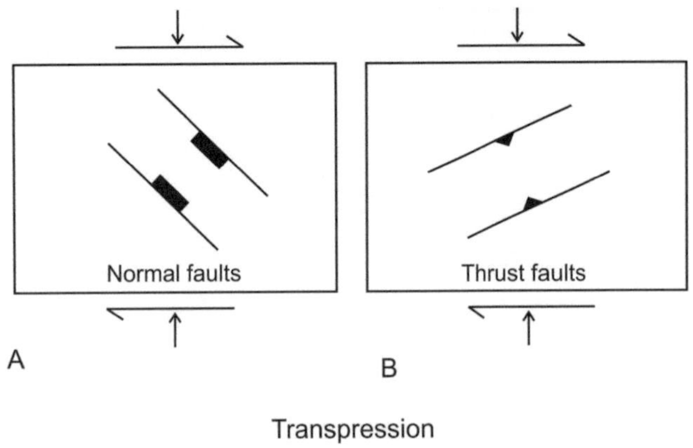

Fig. 12.6 Transpression. Note the orientations of normal faults (>45°) and thrust faults (<45°). See text for details

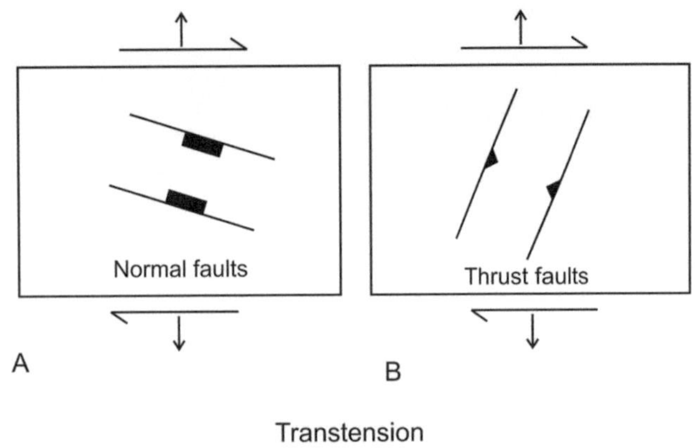

Fig. 12.7 Transtension. Note the orientations of normal faults (<45°) and thrust faults (>45°). See text for details

compensated by vertical thickening. They described transpression geometry by showing transformation of a unit cube by shortening parallel to *Y*-axis and shear parallel to *X*-axis such that the volume is conserved by lengthening parallel to *Z*-axis. Considering a component of shear along the zone, the deformation can be factorized into pure shear and simple shear components as follows:

$$D = \begin{pmatrix} 1 & \gamma & 0 \\ 0 & 1 & 0 \\ 0 & 0 & 1 \end{pmatrix} \begin{pmatrix} 1 & 0 & 0 \\ 0 & \alpha^{-1} & 0 \\ 0 & 0 & \alpha \end{pmatrix} = \begin{pmatrix} 1 & \alpha^{-1}\gamma & 0 \\ 0 & \alpha^{-1} & 0 \\ 0 & 0 & \alpha \end{pmatrix} \tag{12.1}$$

where α^{-1} denotes the shortening across the zone, α the vertical stretch and γ the shear strain parallel to the zone. In fact, α^{-1} is the ratio of the deformed to original width of the zone. The factorization (12.1) defines a strain in terms of two factors α and γ.

12.6 Structures Associated with Strike-Slip Faults

12.6.1 Fault Bend and Stepover

A *fault bend* is a curved part of a strike-slip fault that joins another fault generally with a different geometry. The segment of the fault that is subparallel to the regional slip vector is called a *straight* (Woodcock and Fischer 1986). The deformation within a bend is oblique due to relative motion of the two straight portions of the main fault. A *stepover* is the point where a strike-slip fault ends and the displacement is then taken up by another fault of the same orientation. Because of their specific geometries, bends and stepovers are areas of contraction and extension. A *contractional bend* or *restraining bend* (Fig. 12.8) is one within which the crust is shortened as a result of shearing by the main fault, thus producing contractional structures such as folds and thrust faults.

An *extensional bend* or *releasing bend* (Fig. 12.9) is one within which the crustal material is pulled apart due to the shear of the fault.

The bends and stepovers of large-scale strike-slip faults may show contraction or extension depending upon whether the fault is right lateral or left lateral and also whether on moving along the fault line one has to take a right turn or left turn to maintain the continuity of the fault. In bends and stepovers, extensional deformation pattern develops on the right-hand side of a right-lateral strike-slip fault and on the left-hand side of a left-lateral strike-slip fault; such bends are called releasing bends. Contractional deformation pattern develops, on the other hand, on the left-hand side of a right-lateral strike-slip fault and on the right-hand side of a left-lateral strike-slip fault; such bends are called restraining bends. Thus, a large-scale strike-slip fault may be associated with both contractional and extensional deformation of the crustal material along its bends and stepovers.

12.6.2 Flower Structure

A *flower structure* is a set of curved thrust faults as seen in the vertical section normal to the trace of the main fault. These are formed in the bends and stepovers of large strike-slip faults where deformation is accommodated by vertical movements leading to the formation of normal faults and thrust faults as well as folds. If the thrusts are concave upwards, the structure is called a *negative flower structure* or a *tulip structure* (Fig. 12.10a), and with convex structure it

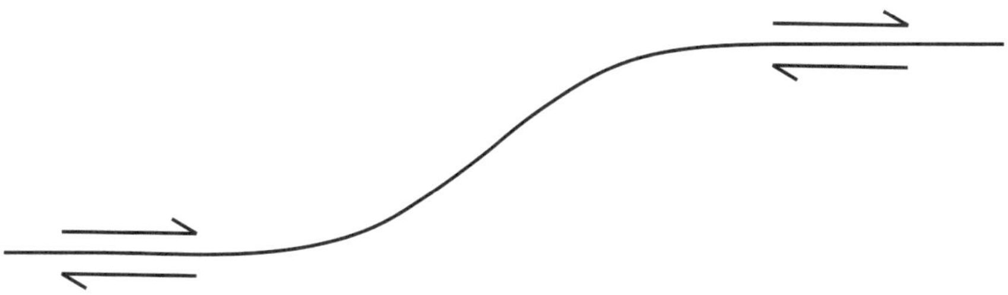

Fig. 12.8 Contractional bend at dextral strike-slip fault

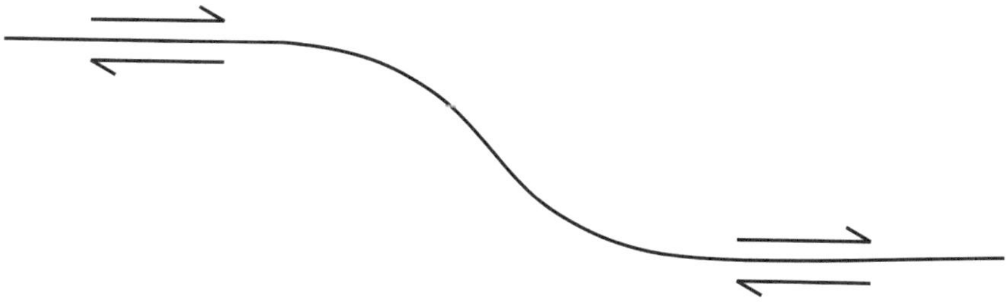

Fig. 12.9 Extensional bend at dextral strike-slip fault

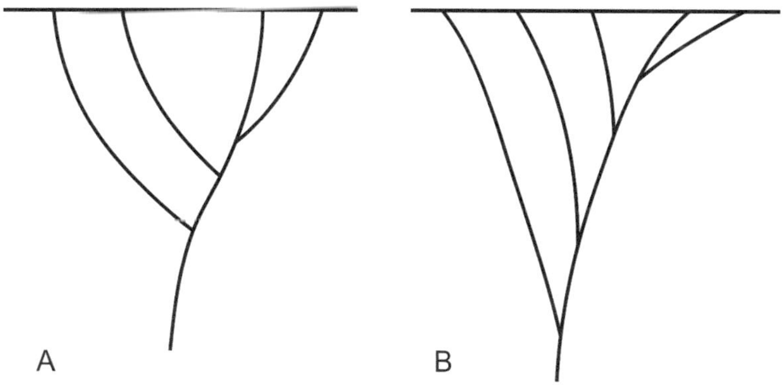

Fig. 12.10 Flower structure as seen in the vertical section normal to the trace of the main fault. (**a**) Negative flower structure, also called normal or tulip structure. The thrusts are concave upwards. (**b**) Positive flower structure, also called reverse or palm tree structure. The thrusts are convex upwards

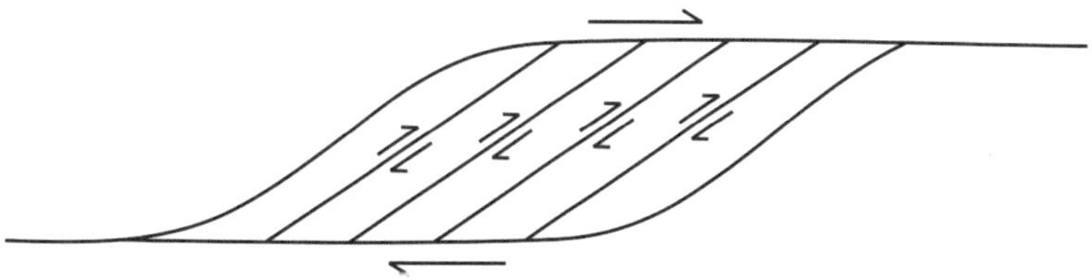

Fig. 12.11 Contractional strike-slip duplex developed at contractional bend

is called a *positive flower structure* or a *palm tree structure* (Fig. 12.10b) (Woodcock and Schubert 1994). These two types are also called *normal flower structure* and *reverse flower structure*, respectively.

12.6.3 Strike-Slip Duplex

Along bends of some strike-slip faults, the deformation bears a component across the bend that gives rise to a set of parallel thrusts or duplexes confined within the main fault and the structure thus formed is called a *strike-slip duplex*. As mentioned earlier, the deformation along a bend is either contractional or extensional. Accordingly, depending upon the nature of bend associated with a dextral or sinistral strike-slip fault, the strike-slip duplex developed can be either a *contractional strike-slip duplex* (Fig. 12.11) or an *extensional strike-slip duplex* (Fig. 12.12).

Contractional duplexes may form at restraining bends or offsets and extensional duplexes at releasing bends or offsets

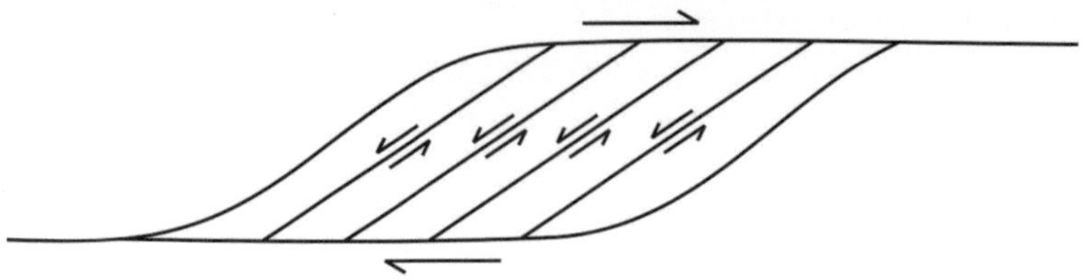

Fig. 12.12 Extensional strike-slip duplex developed at extensional bend

(Woodcock and Fischer 1986). In a vertical section, the contractional strike-slip duplex commonly develops a positive flower structure, while the extensional strike-slip duplex develops a negative flower structure.

12.6.4 Folds, Thrusts and Normal Faults

Strike-slip faults are commonly associated with folds and thrusts (Sylvester 1988). The latter structures are commonly disposed in an en echelon pattern and make an angle of about 135° with the strike-slip fault. These structures represent a component of horizontal shortening oblique to the fault and more or less perpendicular to the orientation of the folds and thrust faults (Twiss and Moores 2007, p. 139). Normal faults are also formed in the strike-slip fault in an en echelon pattern and are oriented about 45° with the main fault and thus almost perpendicular to the orientation of the folds and thrust faults. The normal faults represent a component of horizontal lengthening oblique to the main fault and thus more or less perpendicular to the shortening shown by the folds and thrust faults (Twiss and Moores 2007, p. 139).

12.6.5 Termination

A *termination* (Fig. 12.13) is the zone where a strike-slip fault ends or terminates with displacement gradually reaching to zero. This zone is associated with extensional or contractional deformation of the crust. In the extensional zone, the deformation is accommodated by the formation of splay and normal faults that form an imbricate fan. In the contractional zone, the deformation is accommodated by the formation of thrust faults and folds that form an imbricate fan. A strike-slip fault may also terminate by forming a fan of strike-slip splay faults called a *horsetail splay* (Fig. 12.13). The splay faults of a horsetail splay generally show a curvature with concavity towards the receding block.

Terminations are sometimes associated with plate margins. An example of termination is the Quetta-Chaman fault system of Pakistan that constitutes the western margin of the Indian plate (described below in detail). The termination is constituted of a sinistral fault system at its southern margin, where it swings westward and constitutes a series of thrust faults and folds (Fig. 12.15).

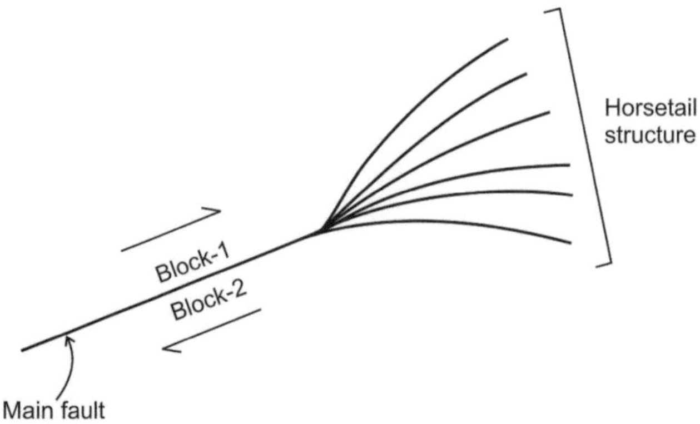

Fig. 12.13 Termination associated with a strike-slip fault. The main fault ends with the formation of splay faults that are called horsetail splay or horsetail structure. The concavity of the splay faults is directed towards the receding block (Block-2 in this case)

12.6.6 Riedel Shears

Of the various types of structures associated with strike-slip faults, some special types of shears called *Riedel shears* or *R shears* are important. These are named after the German structural geologist Riedel (1929) who performed experiments on the deformation of wet clay in a wooden box. He observed formation of a set of shear fractures in *en echelon* pattern when the clay layer was subjected to shearing. These shear fractures are formed at a low angle of about 15° with the main fault and show the same sense of shear as the main fault and are therefore synthetic to the main fault. These are called *R shears* (Fig. 12.14).

Formation of Riedel shears has been investigated by several workers through experiments and field studies (e.g. Tchalenko 1968, 1970; Mandl 1988). As a result, a few additional sets of brittle shears have also been found associated with Riedel shears as described below.

R′ shears is a set of subsidiary shears developed at about 75° with the main fault, and the sense of shear is opposite to that of the main fault. These are therefore antithetic to the main fault and are sometimes called as anti-Riedel shears. The two shears *R* and *R′* are thus conjugate making an angle of about 30°. *P shears* are formed with progressive deformation and are oriented symmetrically to the *R* shears with respect to the main fault. Since *P* shears are contractional, they accommodate fault-parallel shortening with progressive

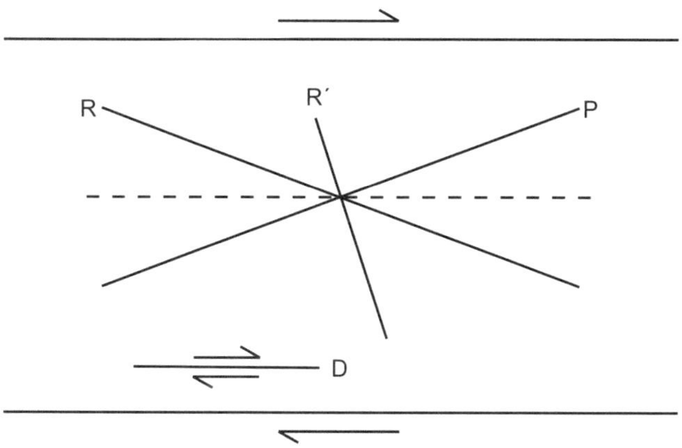

Fig. 12.14 Riedel shears. Note that the Riedel shears (R) are occasionally associated with a few more shears (R', P and D or Y) with varied orientations. See text for details

deformation. *D shears* or *Y shears* are synthetic shears subparallel to the main fault. All these Riedel shears may join with one another, thus forming an anastomosing network called *shear lenses*.

12.6.7 Pull-Apart Basins

The bends and stepovers are sites where the two ends of a strike-slip fault create extension or contraction under two contrasting modes of shear. In an extensional bend or releasing bend, the crustal material is pulled apart due to the shear of the fault, and the region under tension is bounded by faults of which two are the segments of the strike-slip fault showing primarily horizontal displacement, while the other two are bounded by normal faults showing vertical components of displacement. This causes the crust to subside, and the depression gets filled with water and sediment that form *pull-apart basins* or *sag ponds* (Burchfiel and Stewart 1966; Crowell 1974; Aydin and Nur 1982). Such basins are rhomb shaped or sigmoidal, and their size is variable and can extend up to several tens of kilometres.

12.7 Strike-Slip Faults on a Regional Perspective

Strike-slip faults, unlike several other structures, are manifested both on small or outcrop scale and on large scale. They constitute one of the largest single structures in rocks and are sometimes seen in satellite images. Some well-known areas/regions of strike-slip faults include the Alpine Fault of New Zealand, the Himalaya and Tibet, North Anatolia of Turkey, San Andreas Fault and White Mountains of California, the Brevard Fault of the Appalachians and the Great Glen Fault of Scotland. As an example, we briefly describe here the strike-slip fault systems of the Himalaya-Tibet region.

The *Himalayan mountain chain* shows two major zones of strike-slip faults (Fig. 12.15): one each to the east and to the west. All these faults have developed as a consequence of motion of the Indian plate. These are active faults and are thus associated with recurrent seismicity. Both the eastern and western fault systems are generally N-S trending. While the western fault system is left lateral, the eastern one is right lateral. Both the fault systems are characterized by their topographic expressions and as such can be seen in satellite images.

The eastern boundary of the Indian plate is demarcated by a N-S trending zone of right-lateral strike-slip faults that mostly affected the flysch deposits along the Arakan Yoma Line. The region is tectonically complex and seismically active. At the western boundary, the Indian plate is driven northwards along the northern extension of the Owen Fracture Zone of the Indian Ocean, represented by the Quetta-Chaman Fault, east of which a strike-slip fault system is developed. The northward motion of the Indian plate is facilitated by this zone of the left-lateral strike-slip faults.

Box 12.1 Strike-Slip System in a Craton

In recent years, vertical and sub-vertical shears and strike-slip shears have drawn attention especially in understanding continental growth and crustal evolution in several shield areas of the world (Burg 1999; Carreras 2001). The Bundelkhand craton of central India (Fig. 12.16) assumes great significance in understanding the crustal evolution of the Indian peninsula. A spectacular geomorphological-tectonic feature of the craton is the occurrence of quartz reefs that show a dominant NE-SW trending fabric. The reefs rise about 100–175 m above the ground surface and occur as vertical structures. They occur throughout the craton with an almost uniform density of occurrence (Fig. 12.17).

The quartz reefs of the Bundelkhand craton have been interpreted by various workers in different ways (see Bhattacharya and Singh 2013). However, our detailed study of the structural and tectonic fabrics relates them to a prominent NE-SW shear system of the craton. Our observations reveal that the quartz reefs show foliations that are vertical to sub-vertical. They are intimately associated with emplacement of granites. At several places, the quartz reefs show injection of hydrothermal veins followed by emplacement of quartz veins possibly at the end of the petrogenetic cycle. Occasionally, they are sheared with the development

(continued)

Box 12.1 (continued)
of S-C fabric. NE-SW trending mylonitic foliation is commonly developed in the reefs. Rocks (mylonites) fringing the reefs sometimes show emplacement of pink, fine-grained, non-foliated granite within the NE-SW trending mylonites. Locally, the surrounding mylonitic rocks show the development of NE-SW trending shears. All these rocks together constitute a complex that shows sinistral shear.

In the light of above observations, we (Bhattacharya and Singh 2013) have suggested that above rock types occupied the shear zones before introduction of the silica veins. This implies that the NE-SW trending shear system is an early development within which later events of acid magmatism, mineralization and silica emplacement took place. The quartz reefs thus represent strike-slip-dominated vertical to sub-vertical shear zones with dominantly sinistral sense of shear. The loci of the quartz reefs constitute the large-scale fractures developed as a consequence of extensional processes in this part of central India.

In addition to the above two zones of strike-slip faults, the region north of the Himalayan mountain chain shows several en echelon strike-slip faults. With the aid of analogue models, Tapponnier et al. (1986) proposed a polyphase extrusion model, in which several large blocks of the Asian lithosphere are successively pushed to the ESE by the northward movement of India. The strike-slip shear zones are formed as a

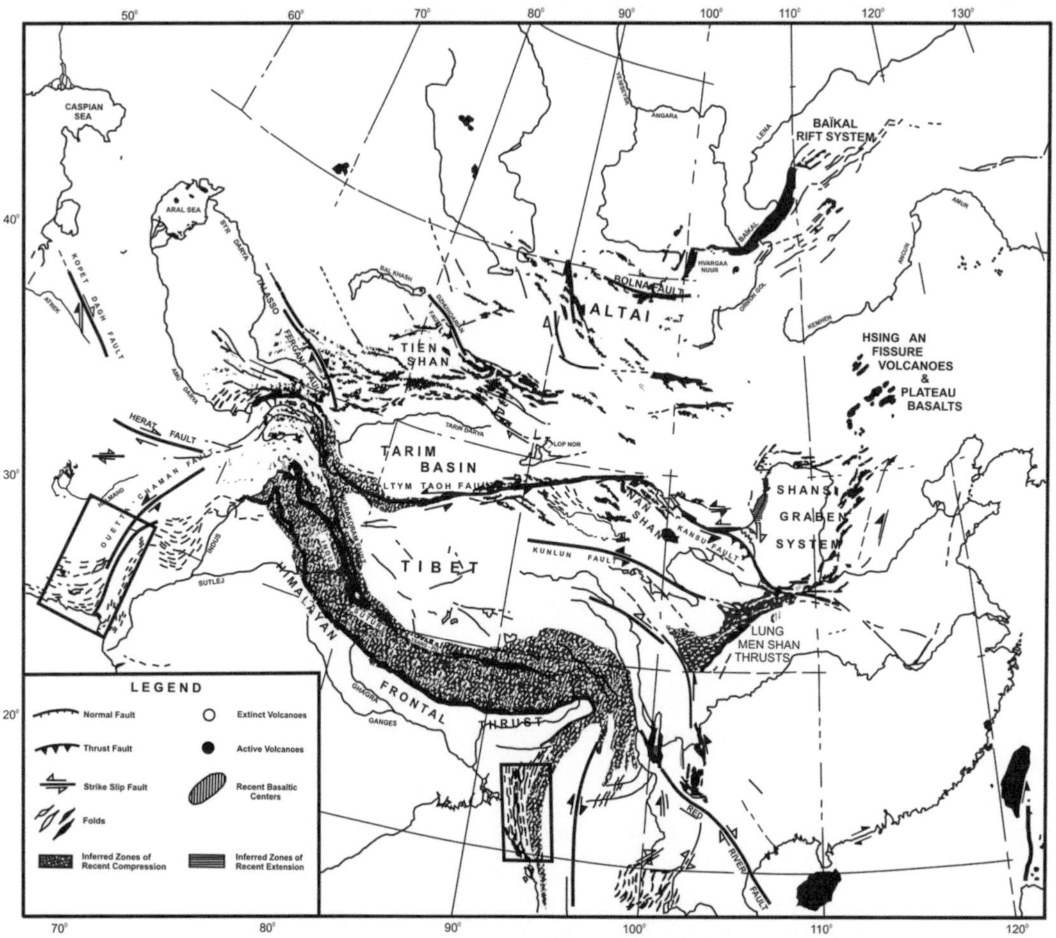

Fig. 12.15 Map showing regions of active tectonics in the Himalaya-Tibet region. Areas of active faults are shown. Major zones of strike-slip faulting have been indicated by rectangles: one along the eastern boundary along the Arakan-Yoma Line and the other along the western boundary of the Indian plate in Pakistan. (Reproduced from Tapponnier and Molnar 1977, Journal of Geophysical Research 82, Fig. 14, with permission from John Wiley & Sons - Books. Request ID: 600071602)

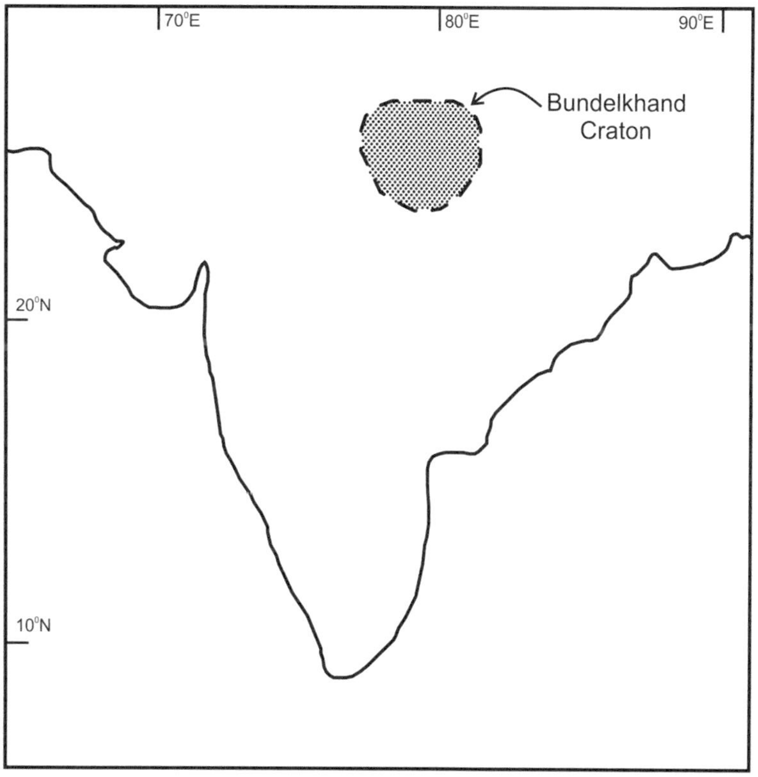

Fig. 12.16 Map showing the location of the Bundelkhand craton in the Indian peninsula. (Partially reproduced from Bhattacharya and Singh 2013, Fig. 1 (inset) with permission from Geological Society of India, Bengaluru)

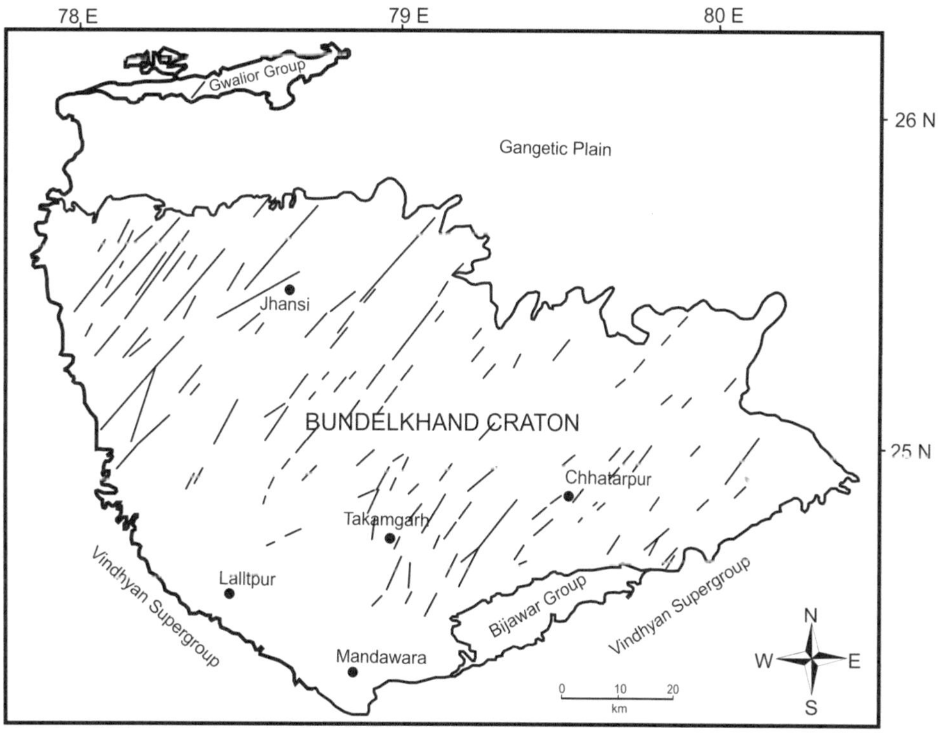

Fig. 12.17 Map showing the occurrences of NE-SW trending quartz reefs in the Bundelkhand craton. (Reproduced and simplified from Bhattacharya and Singh 2013, Fig. 1 with permission from Geological Society of India, Bengaluru)

result of northward penetration of India, which acts as a stiff indenter. The authors thus suggest that much of the underthrusting in the Himalayan region is primarily taken up by strike-slip faulting north of the collision belt. These en echelon, right-lateral, strike-slip faults provide an eastward displacement of the Tibetan plateau with respect to India.

One of the most classic examples of strike-slip faults is the San Andreas Fault in California, USA. The fault marks the boundary of the Atlantic and the American plates. This fault is active and as such is associated with recurrent seismicity.

12.8 Significance of Strike-Slip Faults

12.8.1 Academic Significance

- Since strike-slip faults are often associated with large, often tens of kilometres, displacement, they help in a better understanding of regional or continental tectonics.
- Large-size strike-slip faults have great depth below the surface. Study of their exposed parts often throws light on subsurface deformation patterns in the vicinity of faults.
- Occurrence of large-size strike-slip faults indicates brittle behaviour of the crust on the upper part and ductile behaviour at depth.
- Occurrence of numerous large-scale strike-slip faults on the ocean floor helps in understanding ocean-floor tectonics.
- Several large-scale strike-slip faults are associated with plate boundaries and mid-oceanic ridges. Study of such faults advances our knowledge on plate tectonic processes.

12.8.2 Applied Significance

- Motion of strike-slip faults causes rubbing of the associated rocks, thus producing earthquakes. Such faults therefore help identifying zones of active tectonics and thus locating potential zones of earthquakes.
- Active strike-slip faults are potential structures to produce earthquakes. Strike-slip faults often constitute a potential tool for understanding crustal seismicity.

12.9 Summary

- A *strike-slip fault* is one that is associated with horizontal displacement parallel to the strike of a vertical or steeply inclined fault plane. Strike-slip faults are often associated with large displacements of tens of, and even more than 100, kilometres. These large-scale structures are sometimes seen on satellite images.
- The fault plane generally extends to great depths below the ground surface. The trace of the fault is commonly straight but may also be curved.
- Strike-slip faults show horizontal movement in two opposite directions, i.e. right or left with respect to an observer. The former is called a *dextral* or *right-lateral strike-slip fault*, while the latter is called a *sinistral* or *left-lateral strike-slip fault*.
- Strike-slip faults are of several types: A *transform fault* is a strike-slip fault that cuts the whole lithosphere and fully accommodates the displacement across the plate boundary. It is always straight and constitutes the boundary of two crustal plates. A *transcurrent fault* is a strike-slip fault that has large displacement up to hundreds of kilometres and cuts continental basement as well as sedimentary cover. *Wrench fault* and *tear fault* are commonly used synonymously with strike-slip faults that do not involve a plate boundary. A *transfer fault* transfers the displacement from one fault to another by strike-slip movement, though dip-slip movement is also possible.
- Along bends of strike-slip faults, the deformation can be either shortening (compression) or extension perpendicular to the fault plane. The term *transpression* is used if the deformation at the bends involves shortening (compression) and *transtension* if the deformation involves extension.
- Strike-slip faults are sometimes associated with several types of structures such as fault bends or stepovers, flower structure, strike-slip duplex, folds, thrusts, normal faults, terminations as well as some special types of structures called *Riedel shears* or *R shears* that commonly show an en echelon pattern.
- Strike-slip faults bear great significance in geology, both of academic and applied nature.

Questions

1. What is a strike-slip fault? How does it differ from other types of faults?
2. Are both transform faults and transcurrent faults strike-slip faults? Give the similarities and dissimilarities of these faults.
3. What is a transfer fault? Under what situations it is formed?
4. What is meant by transpression and transtension? Explain the difference between the two with the help of diagrams.
5. Give the difference between a releasing bend and a restraining bend.

6. What is a flower structure? Describe with diagrams the various types of flower structures.
7. What are contractional duplex and extensional duplex? Write down the situations under which these structures are developed.
8. What are Riedel shears? Describe the various types of Riedel shears.
9. What are pull-apart basins? Highlight their crustal significance.
10. Name a few plate boundaries of the world that are associated with well-developed zones of strike-slip faults.

13 Joints and Fractures

A. R. Bhattacharya, *Structural Geology*, Springer Textbooks in Earth Sciences, Geography and Environment,
https://doi.org/10.1007/978-3-030-80795-5_13

Abstract

Rocks crack! Yes, they do, provided that they are brittle. A look on some outcrops occasionally reveals that the rocks are riddled with cracks that are called *joints* along which there has been little or no transverse displacement of rock. A related structure is *fracture*, which is a general term for any kind of break or discontinuity in rock. Yet another related structure is *shear fracture* in which wall-parallel displacement is discernible. Joints may be open or filled with some minerals. Joints and fractures filled with minerals are called *veins*. A variety of joint types have been identified on the basis of geometries shown by joints. Joint surfaces show several types of features that provide information on the direction of propagation and mode of formation of the joints. Joints form when the tensile stresses of a rock exceed its tensile strength, and they propagate where the component of maximum effective tensile stress is normal to the plane of the crack. *Joint mechanics* and causes of formation of joints are discussed in this chapter. *Fracture mechanics*, that concerns the study of stress concentrations caused by sharp-tipped flaws and the conditions for the propagation of these flaws, has also been discussed. Since joints make a rock mechanically weak, their study is important for engineering purposes and civic construction work. Further, since joints and fractures are avenues for passage of fluids, their study is also important for migration of petroleum, water and economic minerals.

Keywords

Joints · Fractures · Geometrical parameters of joints · Surface features of joints · Microcracks · Veins · Joint propagation · Lineaments · Joints and present-day stress field · Formation of joints · Neotectonic joints · Fracture mechanics

13.1 Introduction

A *joint* is a crack in rock, generally transverse to bedding, along which no appreciable shear displacement has occurred (Engelder and Geiser 1980). Joints are easily visible in the rock in the form of cracks. Though the displacement is not easily noticeable on the mesoscale, microscopic study clearly shows it. Joints are a few centimetres to hundreds of metres long structures whose faces can be seen only due to erosion and spalling of rocks.

Joints occur in all rocks of any origin. However, their origin and mode of formation vary with rock types. Joints are better preserved and noticed in sedimentary rocks where stronger/stiffer layers show relatively more development of joints. These occur in igneous rocks and especially in volcanic rocks. In the latter, joints form during cooling of the lava. Joints are also formed in unlithified sediments. Association of joints with folds, faults and other structures suggests their formation in a tectonic cycle, generally at later stages of the cycle. Joints also form during uplift and erosion of rocks of an area. With all these, joints constitute the most common structure in rocks.

Joints are a manifestation of brittle deformation of rocks. Their occurrence adds anisotropy to the host rock that, in turn, controls the mechanical properties of rock. In general, a strongly jointed rock is mechanically weak and, as such, is unable to support imposed load for long. Study of joints is therefore important to engineers for civic constructions. Further, since joints constitute an easy channel for migration of water and fluids, their presence strongly affects the hydraulic properties of rocks. Study of joints is therefore important for exploration of groundwater and hydrocarbons.

13.2 Joints, Fractures and Shear Fractures

Joints are planar tensile opening-mode fractures with little or no displacement parallel to the fracture plane (Narr and Suppe 1991). A joint is therefore an extension fracture. Joints occur in parallel sets (Fig. 13.1) and with a regular spacing from each other. These are commonly planar structures, but rarely these are curviplanar. Their opening is a manifestation of the movement caused when the tensile strength of the rock is exceeded. This movement takes place perpendicular to the joint surface and is so small that it is not visible in the outcrop. Occasionally, joints occur as gaps that are wide enough to be seen by naked eyes.

Joints and fractures are formed under varied geological conditions and are most commonly found in rocks located on or near the surface. It is because the frictional strength of a rock with many discontinuities is lower than the shear strength of an intact rock, and as stress increases, failure will happen on existing discontinuities to relieve the build-up of stress (Professor Terry Engelder, personal communication 2020). Joints also occur at depth.

Joints are commonly associated with *fractures* that are also breaks or discontinuities in rock. The two terms joint and fracture are often used interchangeably.

Another related structure is *shear fracture* in which wall-parallel displacement (Fig. 13.2) is discernible. Shear fractures are micro-faults along which shear movement has taken place parallel to the fracture plane. The fractures have accommodated displacement parallel to the fracture surfaces by small offsets. However, unlike faults, the shear fractures generally do not show features such as slickenlines or crystal

Fig. 13.1 Joints in rocks. One can notice that the entire rock mass appears 'broken' into a number of smaller pieces due to development of joints. The original rock mass now can no longer be considered as a cohesive mass

Fig. 13.2 Shear fractures in massive limestone occurring as planes of discontinuity that is associated with small shear displacement (shown by yellow dashes)

fibres. This sometimes makes confusion to distinguish shear fractures from joints. The small displacement or offset thus forms a general criterion to distinguish shear fractures from joints and even fractures. On a regional scale, shear fractures are less common than joints (Engelder 1987, p. 59).

In brief, joints show regular orientation and spacing; absence of these features may put the structure as fracture, not joints. Shear fractures show regular orientation and spacing; in the absence of these features, the structure can simply be described as fractures.

13.3 Understanding Joints and Fractures: From Laboratory Experiments

The terms joints and fractures often create confusion. For this, possibly the best solution, or the starting point, is to understand the nature of brittle deformation under laboratory conditions. Although laboratory experiments for understanding how rocks deform beyond their elastic limit have been done by several workers (e.g. Paterson 1978; Reches and Dieterich 1983; Duba et al. 1990), those performed by Griggs and Handin (1960) appear to throw better insights on this aspect.

Griggs and Handin (1960) performed laboratory experiments under conditions of triaxial compression and extension, and their results are summarized in Fig. 13.3. They used cylindrical specimens that are subject to a uniform compressive radial stress, the *confining pressure*, and an axial compressive stress. Depending upon the type of deformation shown by rocks, they classified the tests under two: *extension test* if the axial stress is the least compressive stress (σ_1) and *compression test* if the axial stress is the greatest compressive stress (σ_3). The results of Griggs and Handin (1960) have been presented by Pollard and Fletcher (2005, p. 334) in a simplified form as highlighted below.

In extension test, the rock specimens fail by the formation of an *extension fracture* (Fig. 13.3, case 1) formed at axial strains less than about 1%. The fracture is oriented perpendicular to the least compressive stress, and the relative motion

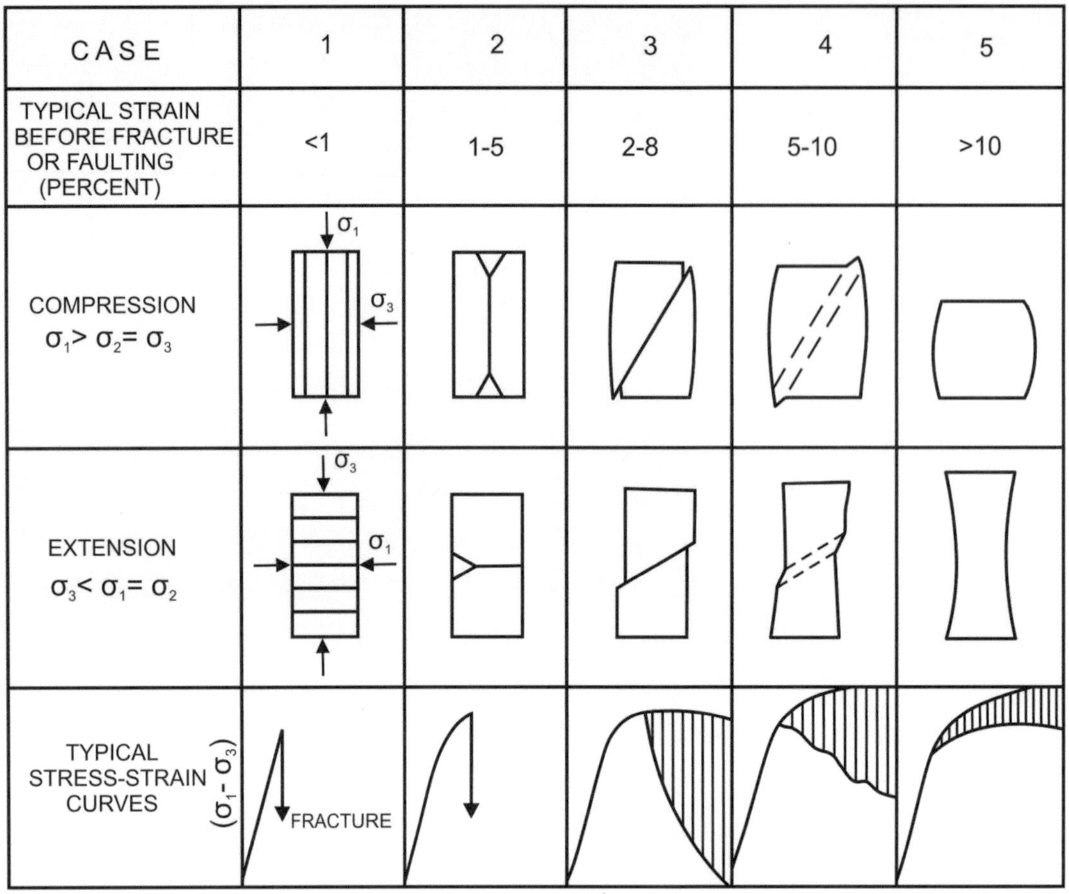

Fig. 13.3 Results of experimental deformation of rocks in the brittle field under laboratory conditions. See text for details. (Reproduced from Griggs and Handin 1960, Fig. 1 with permission from the Geological Society of America)

of the fracture surfaces is dominantly *opening*. The remaining tests are compression tests with increasing confining pressures. At low confining pressures, i.e. at strains 1–5%, *splitting fractures* (Fig. 13.3, case 2) are formed oriented parallel to the greatest compressive stress. Although the splitting fractures also show open mode, they develop wedge-shaped fractures near the ends of the specimens with dominantly shearing motion. At modest confining pressures, at strains 2–8%, *shear fractures* (Fig. 13.3, case 3) are formed at an acute angle to the greatest compressive stress. At still higher confining pressure, at strains 5–10%, the deformation is distributed across a *shear zone* (Fig. 13.3, case 4). At high confining pressures, at strains >10%, a *ductile state* (Fig. 13.3, case 5) is developed in which shearing is pervasive all through the specimen.

We learn from the above experiments that brittle deformation characteristically develops discrete fractures, while ductile deformation shows evidences of distributed flow. If the fracture has undergone substantial shear, it is usually classified as a fault; otherwise, it is denoted as a joint (Mandl 2000). Jaeger et al. (2007, p. 2) consider joints as cracks or fractures in rocks showing little or no transverse displacement, and faults (p. 3) as fracture surfaces on which a relative displacement has occurred transverse to the nominal plane of the fracture. Faults are thus geological fractures of rock for which there is a relative displacement of the rock on the two opposite faces, and as such they are shear fractures on both laboratory and geological scale (Jaeger et al. 2007, p. 419).

13.4 Types of Joints

Joints show a variety of geometries, and that is why a variety of joint types have been identified.

Systematic joints (Fig. 13.4) are planar surfaces that show regular spacing and parallel orientations within an area of observation. *Nonsystematic joints* (Figs. 13.4 and 13.5) show irregular geometry and terminate against older joints. A *joint zone* may consist of many individual joints closely enough spaced that when viewed from a distance the joints appear as an individual joint cutting the length of the outcrop (Engelder 1987). A *joint set* is a group of joints that are parallel or subparallel to each other. A *joint system* is formed by

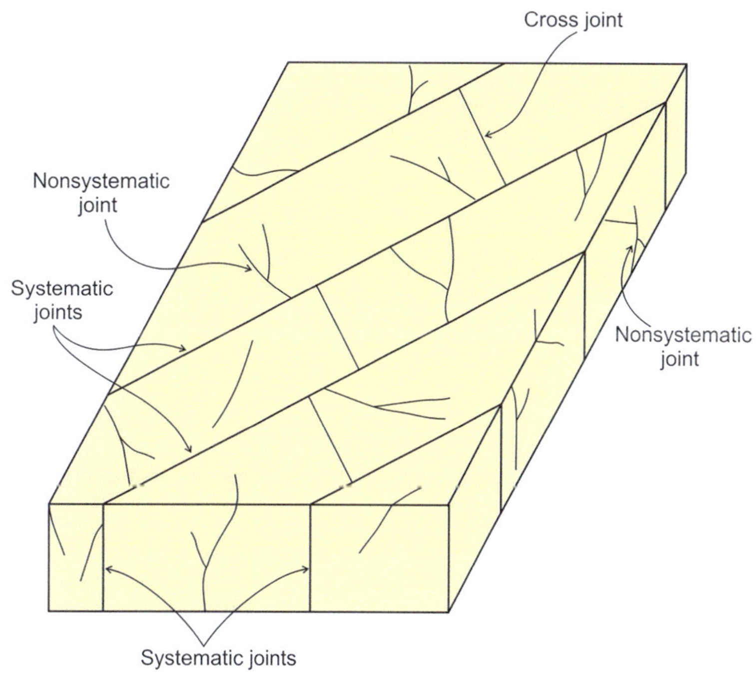

Fig. 13.4 Block diagram showing systematic, nonsystematic and cross joints

Fig. 13.5 Nonsystematic joints developed in the sandstone of Vindhyan Supergroup, central India. (Photograph courtesy Gautam K. Dinkar)

intersection of two or more sets of joints (Figs. 13.6 and 13.7), and the constant angle made by two sets of joints is called *dihedral angle*.

Single-layer joints are confined to one bed only, whereas *multiple-layer joints* cut across bedding planes (Dunne and Hancock 1994). The term single-layer joints can possibly be extended to include a particular type of rock also, instead of one bed only, such that joints are confined to one unit only (Fig. 13.8). *Faulted joints* (Cruikshank et al. 1991; Zhao and Johnson 1992) are those that have been subjected to shear such that the displacement can be detected (Fig. 13.9). Faulted joints can be identified in field by the offsetting of the pre-existing fabric or veins that have been subjected to shear such that the displacement can be detected.

En echelon veins are filled joints. They show an *en echelon* pattern (Fig. 13.10) when they are associated with shear zones. The veins sometimes constitute a shear zone with individual veins showing a sigmoidal pattern that gives the sense of shear. Due to the parallelism of veins and the conjugate shear zone, en echelon veins are believed to be

Fig. 13.6 Joint system showing two sets of systematically arranged joints in the quartzite of Bundelkhand craton, central India. Loc.: Khilarigaon. (Photograph by the author)

Fig. 13.7 Joint system showing two sets of systematically arranged joints in the Cretaceous-Paleogene turbidite sequence near Upsala Glacier in the Andes Mountains of Santa Cruz province, Patagonia, Argentina. The left to right trending joints are systematic joints, while the others constitute a cross-cutting set. Glacier-related striations can also be seen. (Photograph courtesy Narendra K. Verma)

Fig. 13.8 A shear zone developed in mélange in the Zanskar range, Ladakh, India, in which joints are developed in the shear zone unit only. (Photograph courtesy Narendra K. Verma)

shear cracks (Beach 1975). Synchronous propagation of all veins supports a shear origin for the parent vein (Engelder 1987).

Pinnate joints (Fig. 13.11) are a set of en echelon cracks propagating away from mesofaults at small angles (Hancock 1985; Hancock and Barka 1987). Engelder (1989) highlights the following characteristics of pinnate joints: (a) The acute angle between the pinnate joints and the host fracture closes (i.e. points) in the direction of displacement for the block containing the pinnate joints. (b) The small angle between the host fracture and pinnate joints suggests that these features are low-confining-pressure phenomena that show brittle behaviour between crack propagation and shear rupture at 25°–30° to σ_1. (c) Pinnate joints in association with host fractures can be used as a kinematic indicator of orientation and sense of slip on fractures of a region.

Filled joints are those filled with minerals (Fig. 13.12). The thickness of the fill is variable and may be seen in hand specimens or under microscope. Joints and fractures filled with minerals are called *veins*. The veins may range in size from outcrop scale to microscopic scale. Filling of veins may take place at higher as well as lower temperatures. Common minerals of veins include quartz, calcite, feldspar, zeolites as well as ore minerals. Joints are therefore of great significance

Fig. 13.9 Faulted joint showing offsetting of silica veins in the granite of Bundelkhand craton, central India. Width of the photograph is about 30 cm. (Photograph by the author)

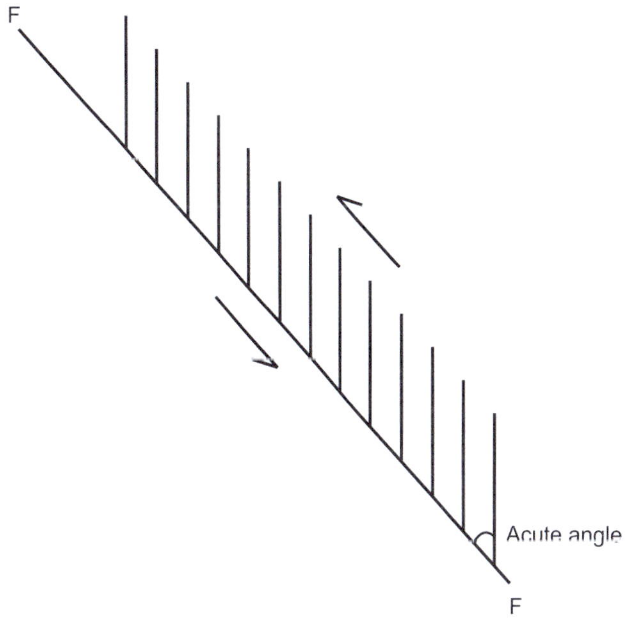

Fig. 13.11 Pinnate joints as related to a fault (FF)

Fig. 13.10 En echelon pattern shown by silica veins in limestone. Loc.: Near Hampteau Hotton, Belgium. (Photograph by the author)

Fig. 13.12 Filled joints. Joints if filled with secondary minerals are called veins. In the photograph, gneissic rocks show two sets of joints at high angles. Due to filling of secondary silica, these joints can be called silica veins. (Photograph by the author)

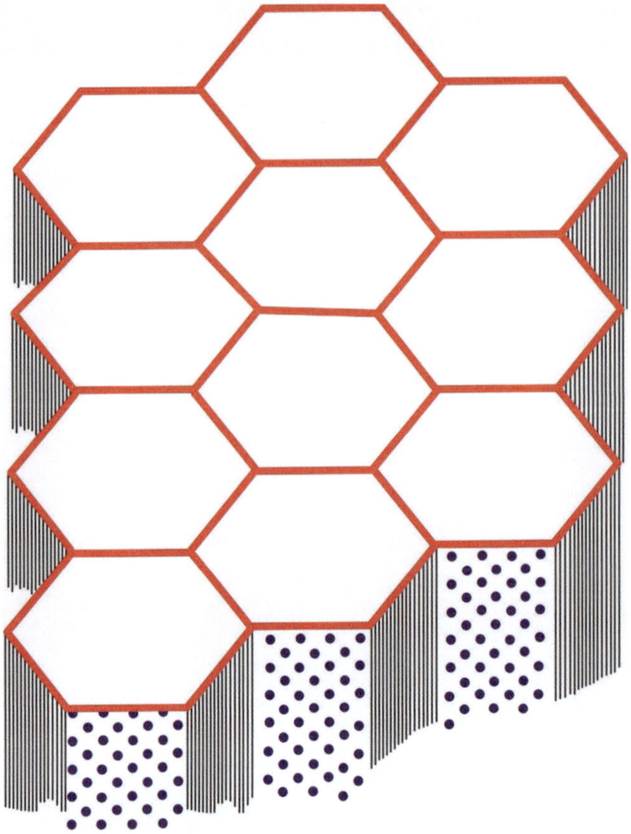

Fig. 13.13 Columnar joints. The individual joints look like vertical/sub-vertical columns that show hexagonal section under ideal conditions

to the economic geologists also. The term *dike* is used for fractures filled with igneous melt (magma). In the absence of any fill, i.e. with air or any present-day fluid or water, the joint is sometimes described as *fissure*.

Columnar joints are hexagonal in cross section (Fig. 13.13) with each face making an angle of 120°, though the columns may also be pentagonal. Columnar joints are commonly developed in basaltic rocks. After extrusion on to the surface of the earth, the flowing lava rapidly cools. During cooling, the column of the lava undergoes shrinkage due to release of gases. The cooling and contraction thus caused in the lava form vertical/sub-vertical columns. *Sheet joints* occur in more or less parallel sets generally with a curved geometry (Fig. 13.14) and affect the rock surface in such a way that the joints look like scales of an onion (Fig. 13.15). Along the sheets, the rock tends to peel off like onion scales. The rock surface thus looks curved at the edge where the joints are developed. Sheet joints are commonly developed in granitic rocks. Sheet joints are believed to form due to unloading caused by erosion or due to exfoliation. As such, these structures are also called *exfoliation joints*, or *sheeting*.

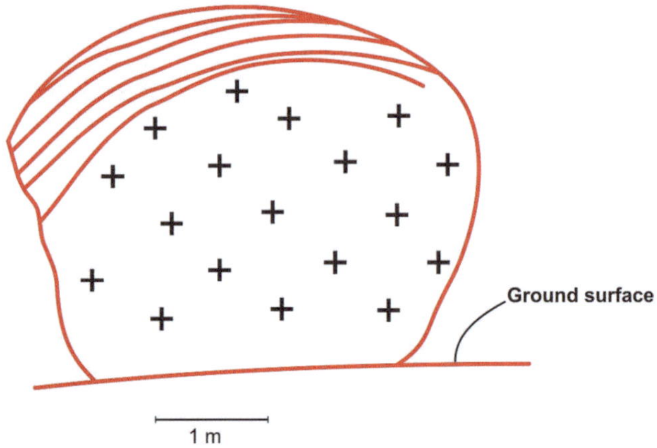

Fig. 13.14 Diagrammatic sketch of sheet joints or exfoliation joints. These joints are developed on the rock surface and look like onion scales occurring in parallel sets with a curved geometry

Fig. 13.15 Photograph of sheet joints developed in the rhyolite of Bundelkhand craton, central India. (Photograph by the author)

13.5 Geometrical Parameters of Joints

In addition to the above-mentioned geometrical aspects, joints are also studied with a few more geometrical parameters that provide information on the mode of formation of joints and fractures. These include scale, shape and spacing of joints.

13.5.1 Scale of Joints

Joints and fractures show a wide range of scales from a few centimetres to tens of metres, and some of them, called *master joints*, may even go up to a few kilometres. Fractures are sometimes so small that they are seen only under microscopes and are therefore called *microfractures*. In thin sections of rocks, the microfractures are generally easy to recognize.

13.5.2 Shape of Joints and Fractures

The shape of joints and fractures depends much on the lithology of the host rock. In rocks of homogeneous lithology such as granite, the individual fracture tends to be longer in extent. In rocks of varying lithologies, as in a sedimentary sequence, the individual fracture tends to be confined to a single bed with homogeneous lithology. In such cases, the fracture tends to extend parallel to the bedding or follow the lithological boundaries rather than across the bedding. Curved crack tips are commonly developed due to mechanical interaction between neighbouring cracks, while those showing straight path show little evidence of mechanical crack interaction (Olson and Pollard 1989).

There are various ways in which joint traces die out. A joint trace may hook and stop or it may hook forming a T-intersection with an adjoining joint developing from other direction. Also, a single joint trace may die out by forming a few en echelon fractures.

13.5.3 Joint Density and Joint Intensity

In addition to above, the following two parameters suggested by Dunne and Hancock (1994) have also been found useful. (a) *Joint density* is the total trace length of all joints in a unit area; this parameter gives an idea of the areal abundance of joints in a rock mass. (b) *Joint intensity* provides the total area of joint planes within a unit volume of rock, say 1 m^3; this parameter can be estimated after the joint density has been estimated on three exposed surfaces that are mutually perpendicular.

Mauldon et al. (2001) suggested a quantitative method for measuring joint (fracture) intensity, which is given as fracture length per unit area on the bedding surface. For this, bedding surface photographs are taken at each sampling site using digital circular scanlines. A digital circle of known radius is placed on each photograph. The number of intersections within this circle is noted down. The fracture/joint intensity for each sampling site is given by

$$I = n/(4r)$$

where I = estimated fracture intensity (mm^{-2}), n = number of fracture intersections within the digital and r = radius of circle. Fracture intensity thus estimated is given as fracture length per unit area on the bedding surface.

13.5.4 Fracture Spacing Index

Fractures of sedimentary rocks commonly show a constant ratio of layer thickness to joint spacing; this is called *fracture spacing index*. This index is approximately same in different rock types and in different structural locations over a substantial region (Narr and Suppe 1991).

Recently, Romano et al. (2020) studied fractures of a fault in the Majella Mountain in the central Apennines, Italy. They treated fault damage zones as a fractured volume and modelled as a fracture network. They distinguished volumes of the fault characterized by similar fracture properties and hence by similar hydraulic behaviour. They measured all the main fracture parameters such as spacing, orientations, sizes and shapes and observed that fracture intensity changes while moving away from the fault zone.

13.6 Microcracks

Microcracks are planar discontinuities that are too small to be seen in a hand specimen; their longest dimension is of the order of one to several grain diameters (about 100–1000 μm) with the small dimension of the order of 1 μm (Engelder 1987, p. 31). For igneous rocks, Simmons and Richter (1976) identify three general classes of microcracks: grain boundary cracks (located at the boundary between grains), inter-granular cracks (cracks cutting more than one grain) and intra-granular cracks (cracks contained within one grain). Photomicrographs showing these three types of cracks are presented here (Figs. 13.16, 13.17 and 13.18). Characteristic features of microcracks (Engelder 1987, p. 31) are outlined below.

Microcracks show two distribution patterns within the crust. One is within igneous rocks where the microcrack density is not related to local structures but to a pervasive

Fig. 13.16 Grain boundary cracks. A prominent crack runs from top left towards right-hand border. (Photomicrograph courtesy Amit K. Verma)

Fig. 13.17 Inter-granular cracks. A prominent crack can be seen diagonally from left to right. (Photomicrograph courtesy Amit K. Verma)

Fig. 13.18 Intra-granular cracks. A few such cracks can be seen disposed across the length of the porphyroclast. (Photomicrograph by the author)

process such as homogeneous cooling. The other type shows a distribution pattern that varies as a function of distance from local structures such as joints, shear fractures and larger faults. The tensile stresses necessary to propagate microcracks arise from either thermal or mechanical processes. The thermal stresses develop due to differential and incompatible thermal expansion or contraction between grains of different thermoelastic properties or between similar, but misaligned, anisotropic grains (Kranz 1983). Cracks of this type are commonly found within uplifted rocks because during burial and erosion the rocks undergo far larger thermal cycles. These cracks are therefore more likely to show uniform density within large volumes of rock. Microcracks that arise from mechanical causes are associated with local structures or discontinuities and therefore show a non-uniform distribution. Common abode of microcracks in the crust is within major fault zones all through the seismic portion of the crust.

The *driving stress for microcracking* is a microscopic horizontal tension developed in response to a vertical load acting through grains that are nearly unconfined by isobaric cooling (Nadan and Engelder 2009, p. 98).

Box 13.1 Recracking

Recracking is the phenomenon of growth of some cracks even after they have come to a halt. Some characteristic features of this phenomenon (Engelder 1987, p. 41) include the following: (a) Recracking may take three forms: intermittent growth, crack-seal growth and joint-zone growth. All these forms occur

Box 13.1 (continued)

on the mesoscopic scale. (b) Recracking is distinguished from *multiple cracking* that involves cracking in more than one orientation to form a regular pattern within a rock outcrop or over a large region. (c) Recracking indicates that stress magnitudes change in time. Tensile stresses, especially, cycle from low to high, thus causing a single crack to propagate further while a sealed crack to recrack. (d) The nature of stress cycling is believed to be due to changes in fluid pressure along joints.

The open microcracks, as described above, are sometimes filled with solutions released from the host grain or from the rock matrix; these are then called *healed microcracks*. The microcracks do not show tendency to heal in the absence of fluid but heal only when fluid is introduced by using the cement that is locally derived and transported by diffusion along the crack surface (Engelder 1987, p. 33).

13.7 Veins

Veins are joints that have been filled with a cement derived from hydrothermal processes including pressure solution (Engelder 1987, p. 42). Veins are commonly believed to form by accretionary processes during which narrow cracks propagate followed by the filling of the narrow open space with a cement, and this accretionary process is called the *crack-seal mechanism* (Ramsay 1980a) (see Chap. 16).

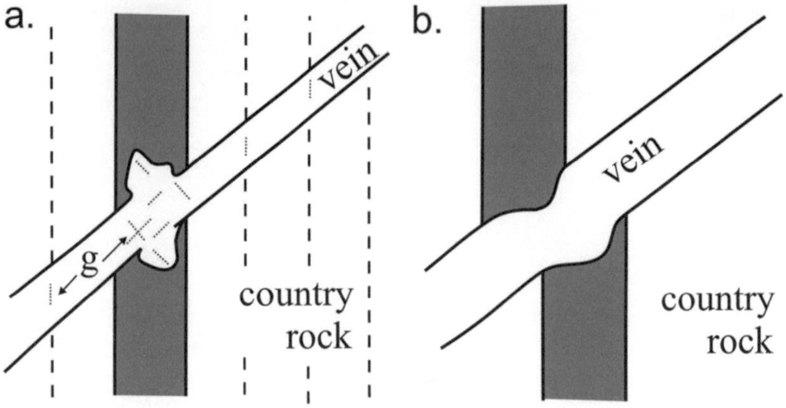

Fig. 13.19 Two types of veins: (a) replacement vein and (b) dilatant vein. See text for details. (Reproduced from Dunne and Hancock 1994, Fig. 5.1. In: Hancock PL (ed.) 1994. Continental Deformation. Pergamon Press, Oxford, with permission from Elsevier Science & Technology Journals. Request ID: 600072891)

Hydraulic fracturing is believed to be a mechanism for crack propagation, and the stress cycling involves effective stress where fluid pressure repeatedly exceeds the tensile strength of the vein material or vein-host rock boundary (Engelder 1987).

Dunne and Hancock (1994) identified two types of veins (Fig. 13.19): (a) **Replacement veins** contain precipitated minerals that chemically replace pre-existing country rock. (b) **Dilatant veins** contain minerals that were precipitated into pre-existing or propagating fractures. The replacement veins produce non-matching walls from country rocks with different chemical solubilities and do not offset the features in the country rock by preserving the 'ghost fabrics', while the dilatant veins have matching walls produced during fracture propagation, offset features in the country rock and lack ghost fabrics (Dunne and Hancock 1994, p. 101).

Bons et al. (2012) suggested that veins form due to formation of connected faults and fracture networks that allow circulation of fluids and healing of fractures by filling with mineral precipitates. Precipitation takes place when the fluid has reached favourable temperature and/or pressure conditions. Common minerals that constitute veins include quartz, calcite, ferruginous minerals, gypsum and feldspar.

13.8 Fracture Refraction

Refraction of structural markers such as cleavages and faults is commonly noticed in rocks when the structures propagate through layers of different competency. Likewise, microfractures also show the phenomenon of refraction, as recently described by Bose et al. (2020) in the meta-greywacke of Garhwal Lesser Himalaya, India. Their thin-section study reveals well-developed fracture refraction patterns showing microfractures that cut the flaky mineral-rich cleavage (c-) and porphyroclast-rich microlithon (m-) domains of the disjunctive foliation planes. The authors document shear-induced microfractures that get refracted at the boundaries between the cleavage and microlithon domains and have shown that higher competency contrast between c- and m-domains favours extension fractures over shear fractures, which develop more in the m-domains. The authors thus suggest that fracture refraction can serve as a powerful tool to quantify the domain-wise competence contrast.

13.9 Surface Features of Joints

Joint surfaces commonly show several types of features (Hodgson 1961; Bahat and Engelder 1984; Engelder 1985) that are formed under certain special conditions. Presence of these features provides information on the direction of propagation of joints and on the mode of formation of the extension fractures. Features on joint surfaces show a variety of shapes and sizes.

The surface features of joints (Fig. 13.20) broadly fall into two groups: (i) *Plumose structures* that are curved lines oriented parallel to the direction of joint propagation (Fig. 13.21): The plumes are smoothly curving lines that propagate up into the competent bed and then bend back to intersect the plastic bed boundary at an angle rather than propagating parallel to the bed boundary itself (Engelder 1987). (ii) *Arrest lines* that are curved lines oriented perpendicular to the direction of joint propagation: *Hesitation* and arrest are expressed as small rib marks or zones of increased surface topography that coincide with the outline or profile of the crack tip during a pause in the growth of a rupture (Lacazette and Engelder 1992). The plumes record joint rupture from initiation through propagation and arrest. Along the arrest lines, the fracture propagation is temporarily being halted by *rib marks*.

On the basis of the geometry of the plume axis, plumes can be of three types (Bahat and Engelder 1984; Engelder 1985): (i) straight or s-type plume, (ii) curving or c-type plume and (iii) rhythmic c-type plume. The s-type plumes are parallel to bedding, and horizontal where bedding is

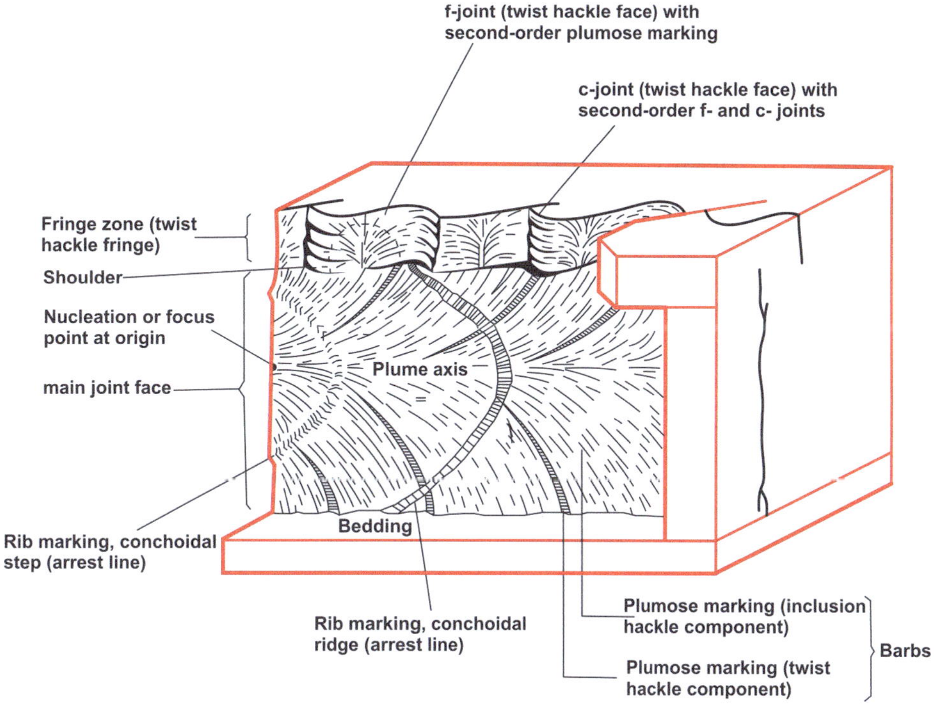

Fig. 13.20 Surface features of joints. See text for details. (Reproduced from Dunne and Hancock 1994, Fig. 5.25 with permission from Elsevier Copyrights Coordinator, Edlington, U.K. Submission ID: 1198165)

Fig. 13.21 Plumose structure developed in Mrakotin Granite, Czech Republic. (Photograph courtesy Professor Terry Engelder)

horizontal. These are relatively long and extend horizontally in both directions away from the initiation point. These initiate at layer boundaries and do not cross them. The c-type plumes are relatively short and propagate bilaterally. Their axes curve extensively. These plumes initiate at layer boundaries and may occasionally cross them. The rhythmic c-type plumes constitute a variation of the c-type plumes and consist of a series of barbs that fan repeatedly along a joint surface. These plumes propagate unilaterally (unlike the c-types that propagate bilaterally). The intensity of each fan pattern gradually increases until a convex perimeter is reached where the fan pattern disappears. The perimeter of the fan is convex towards the direction of propagation. The fan pattern is accentuated with increase of surface topography.

Fracture propagation is believed to be stopped by ripple marks, and in such cases these are called *Wallner lines* due to which few fan-shaped plumes (Fig. 13.22) are developed on the main joint surface.

13.10 Joint Propagation

A characteristic feature of joints is that they grow/propagate after their initiation. The concept that joints propagate at the end of several loading paths is based on the fact that many exposures contain joints in several orientations. Growth of joints occurs in the following manner (see Savalli and Engelder 2005): (a) On joint surfaces, joints initiate at and propagate away from a stress concentration point or initiation point (Fig. 13.23). (b) The orientation of plume lines on joint faces is a manifestation of the path taken by the crack tip line

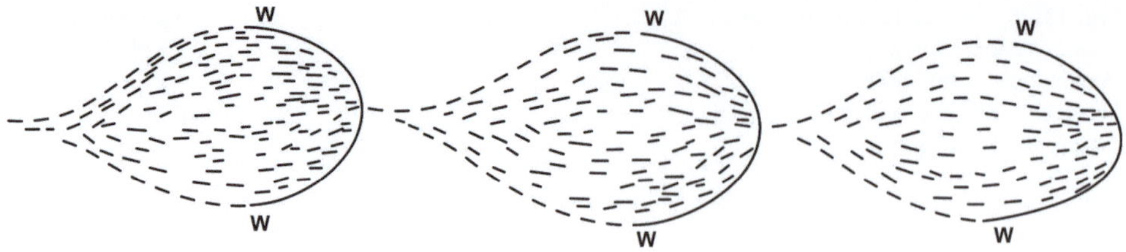

Fig. 13.22 Wallner lines (W) occurring as fan-shaped plumes on the main joint face of rocks

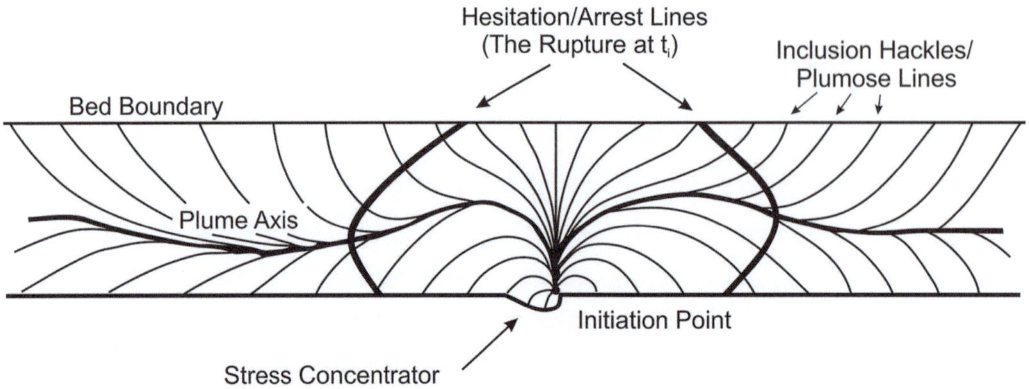

Fig. 13.23 Features on joint surfaces as related to growth of joints. See text for details. (Reproduced from Savalli and Engelder 2005, Fig. 1A with permission from Geological Society of America)

during rupture growth. (c) During primary growth in layered rocks, a rupture initiates either from the interior of a bed or from the bed boundary. From the initiation point, the rupture can propagate along either straight or curved trajectories as traced by the plume lines. (d) In layered, clastic sediments, joints grow through three stages, where velocity of the tip line, v_a, varies as a function of the crack-tip stress intensity, K_I. The *initial stage* of growth produces a rupture of approximately circular or elliptical shape that expands from an initial flaw with a velocity, v_a. Sometimes, this primary growth involves self-correction, i.e. return to a more stable circular shape from an elliptical rupture by redistribution of the crack-tip stress. A circular rupture front is the most stable geometry, and therefore the crack system favours evolution towards a circular (i.e. penny-shaped) geometry (Fig. 13.24). The *second stage* of growth occurs when the rupture tip line intersects both bedding interfaces and splits into two discontinuous segments that propagate simultaneously as a single, coherent rupture. The *third stage* is characterized by the detachment of the coherent rupture into one or more independently propagating, noncoherent tip lines. (e) During all the three stages of rupture growth, the K_I-dependent v_a points to subcritical propagation. *Subcritical crack growth mechanism* produces a distinct rupture pattern on joint surfaces. The rate of crack growth is a function of stress intensity at the crack tip. The subcritical stage ends up with sudden rise of growth, the critical growth, after which rupture growth dies out (see Savalli and Engelder 2005).

Fig. 13.24 Growth of a penny-shaped crack. The penny-shaped geometry, i.e. a circular rupture front, is the most stable geometry during the growth of a crack. Mrakotin Granite, Czech Republic (Photograph courtesy Professor Terry Engelder)

13.11 Joints as Related to Stresses

Joints form along fracture planes oriented parallel to the greatest compressive stress (σ_1) (Fig. 13.25) as well as along the intermediate compressive stress (σ_2) and perpendicular to the least compressive stress (σ_3) at the time of failure. With this kind of orientation, a joint can be considered as an extensional fracture and is one without any

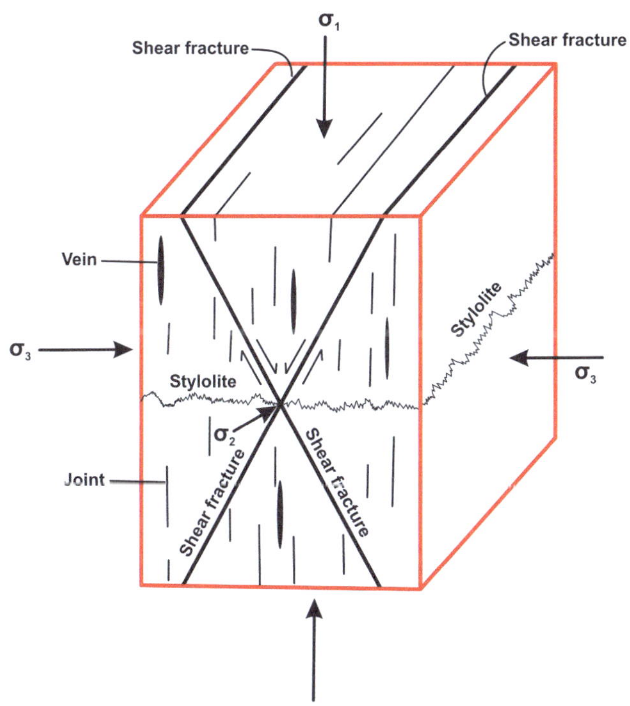

Fig. 13.25 Joints as related to the stress system. Block diagram to show the orientation of joints, shear fractures and stylolites in relation to the principal stresses σ_1, σ_2 and σ_3

component of shear. Extensional fracture may thus include joints and veins. Occasionally, contractional surfaces, sometimes called *anticracks*, are formed parallel to σ_3. These surfaces show displacements normal to their orientation and are commonly filled with fluids of secondary minerals. The zigzag structure (Fig. 13.25) thus formed is called *stylolite* that is commonly considered as an example of contractional structure.

13.12 Joints on a Larger Perspective (Lineaments)

Some joints, especially the master joints, may extend for several kilometres or even tens of kilometres. Such large-scale joints can therefore be noticed on aerial photographs or satellite images. As such, the master joints can also be considered as *lineaments*. However, use of the word lineament to refer to any master joint should be done with due care and reservations. A lineament, according to O'Leary et al. (1976), refers to *a linear feature of the earth's surface that can be recognized on maps or on aerial or satellite images and must be at least a few kilometres long*. Since lineament includes any large-scale linear feature, a master joint should therefore not be recognized on map or aerial photograph unless it is proved by field evidences. It may be mentioned that barring master joints, other types of joints are not large enough to be recognized on maps or aerial photographs.

13.13 Joints as Related to the Present-Day Stress Field

Since joints are stress indicators in some way or another, they may be related to the present-day or contemporary stress system of the earth. However, most joints reflect some earlier stress field(s) (Engelder and Geiser 1980; Younes and Engelder 1999). This may in turn provide important information on several aspects of earth science, e.g. the nature of active faults, predicting subsurface flow of water and oil, site selection for construction of large structures (dams, tunnels, etc.) as well as understanding the lithospheric processes. As yet, our knowledge on the relationship of joints with present-day stress field of the earth is meagre. In the light of available literature, there seems to be three different lines of thought on this subject as briefly described below:

1. *Hodgson's Hypothesis*: Hodgson (1961) opined that the orientation of joints in the upper sedimentary layers of the crust is controlled by stress in the underlying crystalline basement from where the joints propagate upwards to affect the otherwise unfractured sedimentary layers.
2. *Engelder's Hypothesis*: Engelder (1982) suggested that joints have probably not inherited from the crystalline basement rocks but may have formed under the influence of contemporary stress field. The Appalachian Plateau has more than one joint set, and only one is parallel to the contemporary tectonic stress field (Professor Terry Engelder, personal communication 2020). He cited the example of some regional joint patterns of eastern North America that may have formed in the present-day stress field without any input from the basement rocks. Engelder thus proposed that joints form perpendicular to the present-day σ_3.
3. *Hancock-Engelder Hypothesis*: Hancock and Engelder (1989) proposed that neotectonic joints generally form within the upper 0.5 km of the crust because unloading due to denudation and lateral relief consequent on uplift are prerequisites for their propagation. They have shown that late-formed joints have the characteristics of neotectonic joints that have propagated approximately parallel to directions of contemporary horizontal maximum stress; that is, the stress difference is low, and σ_3 is horizontal and tensile. Neotectonic joints in an area can therefore reflect the orientation of present-day tectonic stress field.

13.14 Joint Mechanics

13.14.1 Fundamental Principles

Joints result from the interplay of tensile stresses and tensile strength of a body or layer. The former acts from outside, while the latter is the strength of a rock that resists the tensile stresses. Joints form when the tensile stresses of a rock exceed its tensile strength. According to Engelder (1987), (a) microcracks and joints in rock propagate where the component of maximum effective tensile stress is normal to the plane of the crack; (b) joints propagate parallel to their own plane, which is the most stable orientation for crack propagation because it experiences maximum tensile stress; (c) many natural joint surfaces are not flat suggesting that crack propagation in rock deviates from its original plane; and (d) pore pressure and fluid pressure play an important role in the formation of joints. Pore pressure is the pressure of fluid present in pores and acts against all components of normal stress, while the fluid pressure, in the context of joints, is one that acts against the remote stress normal to the joint and is likely to be the least compressive stress.

Joint mechanics involves several underlying principles as highlighted below.

13.14.2 Linear Elastic Fracture Mechanics (LEFM)

Based on the theory of elasticity, the *linear elastic fracture mechanics* (LEFM) is primarily concerned with how a flaw responds to loading. Analysis of the stress distribution in the crack tip is done on the assumption that the theory of linear elasticity applies, i.e. any inelastic or non-linear elastic region is confined to a negligibly small zone at the crack tip, and there is no frictional loading on the crack surface (Atkinson 1987, p. 11). If brittle materials obey LEFM, it can be said that every brittle material has certain *fracture toughness* that resists it to fracturing. The parameter fracture toughness is independent of scale, specimen geometry and loading configuration (Engelder and Fischer 1996). Fractures (joints) develop when the stress intensity exceeds fracture toughness (Atkinson and Meredith 1987).

The basic tenets of LEFM, as outlined by Ingraffea (1987), are given below:

1. A *stress intensity factor*, K_1, is associated with a crack tip in a loaded structure.
2. A material has a critical stress intensity factor, K_{1c}, just as it has a critical tensile stress capacity, σ_T.
3. In the crack-front region, a material can behave inelastic and non-linear. However, as long as this region, the process zone, is small compared to the other characteristic dimensions of the structure, the criterion for crack advance is

$$K_1 = K_{1c} \tag{13.1}$$

4. The length of a crack extension is governed by the condition expressed in (13.1). The propagation of the crack will continue as long as this criterion holds, and it will stop when

$$K_1 < K_{1c}$$

13.14.3 Joint-Driving Mechanisms

Engelder and Fischer (1996) described a few *joint-driving mechanisms* that drive joints to release a distinct type of energy. The mechanisms are based on the Griffith energy-balance concept. According to Griffith (1920, 1924, *in*: Engelder and Fischer 1996), the equilibrium joint propagation requires the conservation of energy during propagation as expressed by what is known as the *Griffith energy-balance concept*:

$$\frac{dU_T}{d_c} = 0 \tag{13.2}$$

where U_T is the total energy of a rock-joint system and d_c is an increment of virtual joint extension.

The energy parameters of a rock-joint system include surface energy, U_S, which is the energy consumed by the newly created joint surface during joint propagation, and elastic strain energy, U_E, which is the energy of the solid rock. U_S and U_E together constitute the internal energy of the rock-joint system. During propagation of a joint, the rock may undergo elastic strain and so also a change in elastic strain energy. According to Engelder and Fischer (1996), for equilibrium, any change in the internal energy of a rock-joint system, ($U_S + U_E$), must be balanced by a change in potential energy of the system (U_W) such that

$$dU_S \pm dU_E = dU_W \tag{13.3}$$

The above equation provides the specific conditions necessary for conservation of energy during joint propagation as specified in Eq. (13.2).

Working on the concept of Griffith energy-balance concept, Engelder and Fischer (1996) outlined four joint-driving mechanisms: (a) *joint-normal stretching* occurs when layer-parallel extension keeps a slow pace or subcritical joint propagation; (b) *elastic contraction* occurs either during cooling of hot rocks or during unloading upon uplift and erosion: in both these cases, the joints are driven by the release of elastic strain energy; (c) *poroelastic contraction* occurs due to the presence of abundant pore fluid in rocks causing a fluid load that can drive either fast joint growth or subcritical crack propagation; and (d) *axial shortening* involves a jointing process called axial splitting that can drive joints by the axial shortening mechanism, which may involve either the release of potential energy of a load parallel to the joint-propagation direction or the release of elastic strain energy.

13.14.4 Joint-Driving Stress

Joint propagation requires what Engelder and Fischer (1996) described as a *joint-driving stress*. According to them, joint propagation is favoured only when joint-normal stress is tensile (tension has a +ve sign). The fluid within a joint that presses against the inside wall of the joint has a pressure P_i. Before the joints have propagated, $P_i = P_p$, where P_p is the pore pressure in the matrix of the rock. After the joints have propagated, $P_i \leq P_p$ (Engelder and Lacazette 1990).

During burial (the least horizontal stress, S_h, as is the case of vertical joints), the stress may compress the rock-joint system. This will inhibit joint propagation. Pore pressure, P_p, is then required to overcome the compressive stress of burial and thus produces effective tension (P_p has a +ve sign). Under such situations, the joint will propagate when $(S_h + P_i) > 0$. $S_h + P_i$ is called the *joint-driving stress* (Pollard and Segall 1987).

There are three different joint-driving stresses (Engelder and Fischer 1996): maximum horizontal stress S_H, minimum horizontal stress S_h and internal pore pressure P_i, where S_H operates in the joint-parallel direction and both S_h and P_i operate in the joint-normal direction.

13.14.5 Joint Loading Paths

The concept of loading paths is that joints propagate at the ends of several loading paths, and this is largely based on the fact that many exposures contain joints in several orientations (Engelder 1985, p. 461). The loading paths lead to failure of rock in tension during propagation of joints. Although several loading paths have been identified, Engelder (1985) described four loading paths that lead to joint propagation in sedimentary basins. These loading paths should be regarded as end members in the suite of all paths leading to tensile failure. The end members fall into two groups: hydraulic and tectonic joints, which propagate during burial or at the maximum depth of burial, and unloading and release joints, which propagate during uplift and erosion.

13.14.6 Folding

13.14.6.1 Conceptual Models

Field studies reveal that some joints are related to the geometry of folds and thus may owe their origin to the folding of rocks. Joints are reported to be associated with folds of various sizes, ranging from hand specimen to regional scales (Fig. 13.26).

Stearns (1968, in Bergbauer and Pollard 2004) proposed that fractures bear some definite geometric relationship with folds. He suggested that folded strata may show 11 fracture

Fig. 13.26 Development of joints (red lines) in the aerial photograph of a large anticlinal fold in the Zagros Mountains of Iran. The joints show dips towards the trace of the fold axial plane in the profile section and thus can be described as radial joints. (Photograph courtesy Narendra K. Verma)

orientations comprising 5 fracture sets that are systematically related to the fold axis and bedding. Bergbauer and Pollard (2004, p. 295) described these five sets as follows.

Set 1 forms due to a vertical intermediate stress and a greatest compressive stress parallel to the dip direction of the bed. This set has two orientations with dominant strike-slip and one extension fracture orientation. Set 2 forms when the least compressive stress is parallel to the dip direction. Set 3 fractures have two orientations of shear fractures and one orientation of extension fractures. Set 4 comprises two orientations of thrust faults. Set 5 fractures are interpreted as shear fractures resulting from bedding plane slip during folding. With these five sets, Stearns presented a conceptual model of 11 different fracture orientations developed within a folded bed.

Lisle (1992, 1994) suggested that the fracture density around a fold is directly related to the curvature of the fold.

Dunne and Hancock (1994, p. 116) suggested that joint sets that are symmetrically oriented within a fold are considered as extension or conjugate joints on the basis of whether they are normal or oblique to symmetry lines or planes, such as hinge lines, axial planes and layering.

Cosgrove and Ameen (2000) suggested some relationship between fold and associated fractures. According to them, fractures formed in association with buckle folding may be the result of the regional stress field or the local stresses generated as a result of buckling (e.g. extension in outer arc above the neutral surface and compression of the inner arc below).

13.14.6.2 Regional Studies

13.14.7 Example 1

Engelder and Geiser (1980) studied the orientation and development of joints of the Appalachian Plateau, New York. They have applied the basic principle that vertical joints propagate normal to the least principal stress and thus follow the trajectories of the stress field present at the time of propagation. They have shown that the three predominant joint sets on the Appalachian Plateau either cut across folds at high angles or strike subparallel to the fold axes.

13.14.8 Example 2

Bergbauer and Pollard (2004) studied joints in the sandstone beds of the Emigrant Gap anticline, Wyoming, USA. The structure is a doubly plunging fold. They identified two sets of joints within the sandstone beds: J_1, striking northwest, and J_2, striking northeast. Both sets are steeply dipping and mutually perpendicular. The younger set J_2 abuts against the younger set J_1. The authors are of the opinion that these two joint sets were developed prior to folding. Presence of the joints has facilitated folding to take place. During folding, the joints are formed parallel to the two pre-folding joint sets on the limbs, while on the hinge the fracture style depends upon the relative orientation of the hinge line and the pre-folding joints. The authors thus suggest that folding induced further propagation of the existing joints of both sets along strike and this led to infilling of new joints between the pre-folding joints, thus causing a dense network of joints in the anticline.

13.14.9 Example 3

On the basis of their study of Sheep Mountain anticline, Wyoming, which is a doubly plunging anticline, Savage et al. (2010) are of the opinion that if a fold grows laterally along its axis, early-formed fold-related joints may differ significantly in orientation from joints that form later. The authors carried out plate bending analysis that considers the displacement and stress fields associated with folding of isolated mechanical units. The soft layers, such as the shales of the Sheep Mountain, compartmentalize deformation, while the more competent layers deform if isolated.

In general, the development of joint patterns depends on whether the stresses were more effectively tensile on the top or bottom surface of the folded bed (Savage et al. 2010, p. 1468). Plate bending theory, according to the authors, thus predicts that joints initiate from the outer arc of the folded bed and that joint patterns resulting from stresses on the top and bottom of a bed will be at 90° from one another.

13.14.10 Example 4

Watkins et al. (2019) studied the Swift Reservoir Anticline (Fig. 13.27) in northwestern Montana to investigate the role of curvature in controlling fracture intensity and density. The fold has a steep forelimb dipping to the NE and shallow backlimb dipping to the SW. The structure has three main lithological units (Fig. 13.27): grain-supported dolomite, mud-supported dolomite and mud/grain-supported dolomite. The authors suggested the following: (a) Fracture intensity increases with increase in simple curvature bedding dip. (b) Although fracture intensity is influenced by structural factors such as bedding dip, fold curvature and structural position, lithology constitutes the main control on fracture intensity distribution across the fold structure. (c) Porosity and compositional factors constitute the main lithological controls on fracture intensity. (d) In general, the greatest controls on fracture intensity include porosity and quartz content, followed by structural factors such as simple curvature and structural position.

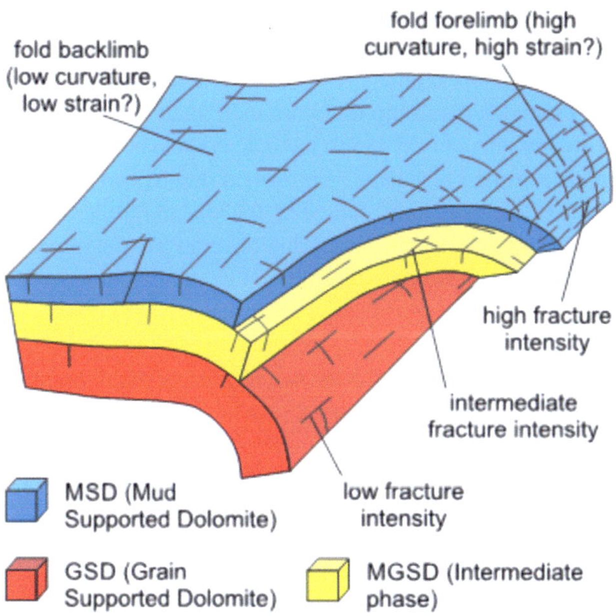

Fig. 13.27 Model showing variation of fracture intensity in the Swift Reservoir Anticline, Montana. Note that the fracture intensity increases with increasing fold curvature for a given lithology. Further, fracture intensity is much lower in grain-supported dolomites (GSD) showing higher percentages of quartz and porosity than the mud-supported dolomites (MSD), irrespective of fold curvature and structural position. (Reproduced from Watkins et al. 2019, Fig. 11. Courtesy Geological Society of London)

13.15 Causes of Joint Formation

13.15.1 Burial

Burial of a sedimentary pile, or addition of volcanic material, means increase of vertical gravitational loading in the subsurface rocks. This causes increase of pore fluid pressure in the underneath rocks in locations where the fluid cannot move vertically or sideways due to absence of any anisotropic plane such as bedding or joints. Pore fluid pressure is caused due to the presence of both groundwater and pore fluids trapped in the voids or spaces in the rocks. Pore fluids increase the tensile stress in the pre-existing cracks of rocks in a direction perpendicular to the greatest principal compressive stress along which the rocks tend to stretch. If the amount of tensile stress increases than that of the least principal compressive stress, the rock undergoes brittle failure by crack propagation. This process, called *hydraulic fracturing* or *hydrofracturing*, produces joints that are of various kinds such as systematic, nonsystematic, conjugate as well as joint sets as described earlier. Fractures formed by burial are mostly tensile fractures.

13.15.2 Uplift and Erosion

Uplift of an area is believed to be associated with formation of joints. Uplift causes removal of overburden due to erosion that in turn releases tensile stresses of rocks. The pre-existing state of stress of the underlying rocks is thus modified in three ways (Suppe 1985, p. 193) that commonly lead to horizontal tension: (i) Horizontal stretching through the geometry of uplift: Uplift commonly generates horizontal stretching of crust, especially at the crest of large anticlines. The bending of rocks thus caused produces joints under brittle crustal conditions. (ii) Expansion through the release of gravitational load: Uplift promotes erosion, thus releasing the gravitational load of rocks which in turn causes expansion of the compressed rocks. (iii) Contraction through cooling: Erosion causes a change in horizontal stresses due to cooling that depends on the rate of uplift, i.e. whether slow or rapid.

The combined effect of the above processes leads to new stresses mainly due to the release of gravitational load and thermal contraction. The change of the state of stress leads to horizontal extension of crust that, in turn, produces joints. Uplift and erosion commonly form tensile fractures in the form of joints parallel to the eroded surface such as sheet joints in granites and in some sedimentary rocks. Sheet joints are an example of axial cracks (extension fractures) driven by a high σ_1 parallel to the crack plane (Engelder and Fischer 1996).

Joints are commonly formed during and after the uplift of rocks. Examples of their formation in the early history of the rock are rare. Bahat et al. (2003) reported one such example, the Boršov granite of the South Bohemian Pluton exposed in the Boršov Quarry, southeast of Prague in the Czech Republic. According to the authors, the shape of the pluton could have controlled significant thermal and fluid pressure gradients, which, in turn, influenced the orientation of the joints. Such relationships most likely existed during the early history of the granite, before uplift, in a fluid-rich environment.

13.15.3 Thermal Contraction

Thermal contraction is a type of joint-driving mechanism (Engelder and Fischer 1996). Extrusion of volcanic lava on to the surface or near the surface of the earth commonly produces joints on cooling. When lava such as basalt is extruded on to the surface, it undergoes cooling which is faster in the top layers and progressively becomes slower at lower layers. As a result, the top layers undergo more contraction than the lower layers. This differential contraction produces tensile stresses more or less parallel to the layers. When the lava solidifies, the tensile stresses are released in the form of tensile fractures, also called *columnar joints* (Fig. 13.13), that are oriented perpendicular to the flow of lava. Initially, these joints are formed on the top layers that have solidified. With time, as the lower layers solidify, these joints propagate downwards. Columnar joints are also formed in dikes and sills emplaced at shallower levels of the crust.

Although columnar joints look like vertical columns, their geometric shapes are important. Because of space problem during contraction, the columns tend to occupy minimum space, thus forming six- or even eight-sided blocks. Under ideal or homogeneous conditions of cooling, accompanied by contraction, hexagonal columns are commonly formed. Since ideal conditions of contraction are not always met with, the columns depart from showing hexagonal sections by assuming any suitable polygonal shapes in horizontal sections before stabilizing. Joints formed by extrusive processes are thus characterized by their typical shapes in cross sections as described above.

13.15.4 Sheeting

Granitic rocks often show joints that occur as horizontal sets. These are sometimes also called *sheet joints* (Figs. 13.14 and 13.15). The rock thus gives the appearance of sedimentary layering. These joints are formed due to weathering, which causes removal of overburden. The underlying rocks thus expand with release of tensile stresses that produce joints parallel to the ground surface.

13.15.5 Tectonic Causes

Tectonic causes include a variety of geological situations under which the ambient tectonic stresses form joints in rocks. During deformational processes, for example, the orientation of stresses may change in the deforming system. This may set the platform anew for the formation of joints, or new joints if early-formed joints are present, by changing the horizontal components of stress. Thus, if (i) a horizontal stress becomes the least compressive stress and if a horizontal tensile stress is added, then vertical joints would form perpendicular to the tensile stress, and (ii) if the horizontal tensile stress becomes the greatest compressive stress, vertical joints would form parallel to the tensile stress (Twiss and Moores 2007, p. 246).

Development of joints is sometimes systematically related to the tectonic development of a region also. Engelder (1985) studied joint development in a sedimentary basin of the Appalachian Plateau in the Catskill Delta, USA. He observed four types of joints, viz. tectonic joints, hydraulic joints, unloading joints and release joints that are systematically related to the tectonic cycle of the region. *Tectonic joints* are formed at depth (up to 3 km) under the influence of high pore pressure caused due to tectonic compaction. *Hydraulic joints*, on the other hand, are formed due to compaction at depth (more than 3 km) caused by the load of the overburden. *Unloading joints* are formed due to removal of overburden during erosion. *Release joints*, like unloading joints, are also formed due to removal of overburden during erosion, but their orientation is largely determined by the presence of solution cleavage. In folded strata, release joints are commonly parallel to the fold axial planes. The unloading joints, on the other hand, are parallel neither to the fold axial planes nor to other tectonic fabrics of the associated rocks. The orientation of the unloading joints is however largely determined by the prevailing stress field.

13.16 Fracture Mechanics

13.16.1 What Is Fracture Mechanics?

Fracture mechanics concerns *the study of stress concentrations caused by sharp-tipped flaws and the conditions for the propagation of these flaws* (Atkinson 1987). In recent years, our knowledge on fracture mechanics had vastly increased due to experimental study of fracture mechanics of rocks. Most rocks and minerals contain flaws or defects, and this is virtually the starting point of fracture mechanics in which a defect is considered as a crack (Atkinson 1987). Some aspects of fracture mechanics that appear from the works of Atkinson (1987), Engelder (1987) and Younes and Engelder (1999) are described below.

13.16.2 Modes of Fracture Opening

Assuming a flat, sharp crack of zero thickness, there are three basic modes of loading (Fig. 13.28), i.e. by which a crack tip can be displaced and produce fracture: A: mode I, tensile or opening mode; B: mode II, in-plane shear or sliding mode; and C: mode III, anti-plane shear or tearing mode (Atkinson 1987). The leading edge of a fracture is called *tip line*, while the *tip* is the discrete point where the tip line is in contact with the rock surface. The three principal loading modes represent the configuration of the stresses at the tip line (Younes and Engelder 1999). A fracture will propagate only when the tensile stress developed due to loading exceeds the strength of the rock. If tensile stress is less than the strength of the rock, fracture propagation will stop in which case the propagation is described as *arrested*.

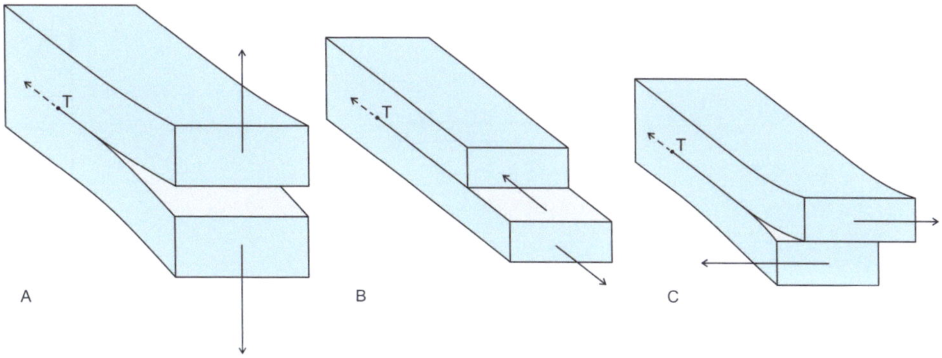

Fig. 13.28 Three fundamental modes of fracture: (**a**) mode I, tensile or opening mode; (**b**) mode II, in-plane shear or sliding mode; (**c**) mode III, anti-plane shear or tearing (scissoring) mode

Mode I loading occurs when a fracture propagates in its own plane under the influence of a tensile stress oriented perpendicular to the fracture plane. *Mode II* loading occurs due to the action of a shear couple directed parallel to the propagation direction and perpendicular to the tip line. *Mode III* loading occurs when the shear couple is perpendicular to the propagation direction and parallel to the tip line (Younes and Engelder 1999). Thus, mode I occurs with opening and has no shear couple, while modes II and III have shear couple parallel to the plane of the fracture surface, and no opening.

In addition to the above modes, sometimes *mixed mode of loading* occurs due to the combination of mode I and mode II loading, or mode I and mode III loading (Younes and Engelder 1999). Joints are commonly formed under mode I loading. The joint tip line is sometimes subjected to shear tractions causing the joint to deviate, thus forming hooks, abrupt kinks and twist hackles that are all manifestations of mixed mode loading at the joint tip line (Younes and Engelder 1999).

13.16.3 Stress Intensity

Stress intensity (Atkinson 1987, p. 6) concerns with the real forces applied to a crack tip, which will determine whether it will grow or remain stable. The stresses near the crack tip in a homogeneous, linear elastic medium are proportional to $r^{-1/2}$, where r is the distance measured from the crack. The coefficient of the $r^{-1/2}$ is called the stress intensity factor, K, which gives the magnitude of the crack-tip stress field for a particular mode in a homogeneous linear elastic material (Atkinson 1987).

13.16.4 Crack Extension Force

Crack extension force (Atkinson 1987, p. 7) is an alternative fracture mechanics approach to crack extension, other than crack-tip stresses, that examines the crack extension force, G, which is a global, not a local, parameter that includes contributions from all parts of the system (cracked specimen plus loading mechanism). The crack extension force is, in fact, the strain energy release rate that determines the loss of energy per unit of new crack separation area formed during an increment of crack extension. The parameter G is applicable to both stationary and running cracks (Atkinson 1987).

13.16.5 Crack Extension Laws

Crack extension laws in fracture mechanics are of two types (Atkinson 1987, p. 9):

(a) Equilibrium laws, according to which cracks may extend stably or unstably at some critical value of a fracture mechanics parameter
(b) Kinetic laws, according to which at certain subcritical values of fracture mechanics parameters, a crack can extend at a velocity that is function of the magnitude of the crack-driving force.

13.16.6 Microcracking by Process Zone

Process zone (Atkinson 1987, p. 13) is the *formation of a macrocrack extension zone by linking of microcracks in the non-linear zone*. It has been suggested (Lawn 1983, in: Atkinson 1987, p. 12) that brittle cracks are atomically sharp and propagate by the sequential rupture of bonds. Formation of process zone for rocks has been diagrammatically explained by Atkinson (1987) in Fig. 13.29. If blunt, machined notch is loaded to a rock, a few isolated microcracks are formed and the system behaves linear. On further loading, the intensity of microcracking increases and the crack-tip region now behaves non-linear.

On still further loading, the microcracks in this non-linear zone get linked resulting in the extension of the macrocrack. This process is called *process zone*, which is linked to the development of crack-tip microcracking.

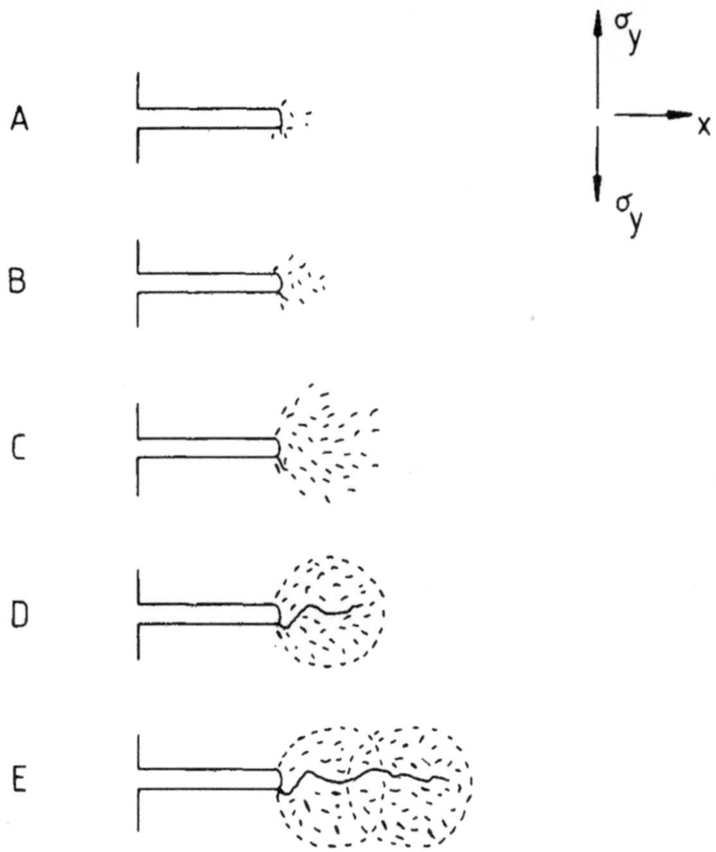

Fig. 13.29 Diagrams to show the development of a process zone and how it gives rise to macrocrack growth. In the present case, simple tensile deformation across the plane $y = 0$. The various stages **a**–**e** represent increase of stress by machined notch. (**a**) Machining develops a few isolated microcracks. (**b**) Isolated microcracks begin to develop but still showing linear elastic behaviour. (**c**) Microcracks develop with more intensity. Some microcracks begin to link up, and thus non-linear behaviour is observed. (**d**) The microcracks extend by link-up, and a zone of non-linear elasticity or process zone is developed within which the macrocrack extends by linking of microcracks. (**e**) Macrocrack extension is facilitated by migration of the process zone through the material ahead of the macrocrack tip. The microcracks thus allow extension of the macrocrack. (Reproduced from Atkinson 1987, Fig. 1.4 with permission from Elsevier Copyrights Coordinator, Edlington, U.K. Submission ID: 1198797)

13.16.7 Dynamic Fracture

The fractures as discussed above are in fact stationary cracks as far as propagation is concerned. According to Atkinson (1987, p. 16), in practice, running cracks can accelerate to significant fractions of the velocity of sound, and therefore an increase in crack-driving force beyond that is necessary to drive the crack at a limiting speed. This results in crack branching and formation of hackle on the crack surfaces. The limiting crack velocities, as observed in real solids, commonly range from 0.1 to 0.6 of the sound wave velocity (Irwin and de Wit 1983).

13.16.8 Shear Fracturing

Shear fractures develop as a plane of shear failure after prolonged microcracking (Lajtai 1971). Engelder (1987, p. 44) outlined some salient features of shear fracturing as follows: (a) Shear fracturing develops in the following steps: formation of individual microcracks, propagation and linking of these cracks, and then larger scale shear failure. (b) The latter, i.e. large-scale failure, may lead to formation of a zone of cataclastic, non-cohesive material, and the surrounding intact rock commonly contains microcracks. (c) Occurrence of microcracks at an acute angle to a joint is one of the surest signs of a shear origin for the fracture. (d) Shear fracturing can occur even under conditions of compressive applied stress in which case failure accompanies the propagation of microcracks. Under such conditions, the compressive stress is locally modified at the microcrack tips to become tensile, and this causes the microcracks to propagate further under mode I loading. With further compression, the microcracks are added until they start to link.

13.16.9 Griffith's Fracture Theory

In 1920, A.A. Griffith, an English aeronautical engineer, got perplexed to note that the stress needed to fracture bulk glass is about 100 MPa while that needed to break the atomic bonds of glass is about 10,000 MPa. He performed experiments on glass fibres to explain this anomalous result and observed that the fracture stress increases with decrease of fibre diameter. He suggested that fracture strength of brittle solids depends upon the presence of defects or *flaws* of microscopic sizes in the form of microcracks, dislocation planes, voids, inclusions and precipitates. Presence of flaws in a brittle solid lowers the *fracture strength*, which depends on the cohesive forces between the atoms as well as on the elastic properties of the solid. Flaws therefore constitute loci of *stress concentration* and are therefore called *stress concentrators*. An externally applied stress gets amplified at the tips of the microscopic flaws that are also called *stress raisers*. It is to be noted that mere presence of microcracks does not lower the fracture strength of a solid; it also depends upon the orientations and sizes of the microcracks.

The flaws play a significant role in fracture formation. An applied stress is amplified at any one of these flaws. Once the amplified or concentrated stress exceeds the cohesive strength of the solid, a crack becomes unstable and propagates to produce a fracture. This is called *Griffith's fracture theory*. These cracks are called *Griffith cracks* that are ideally assumed to be of elliptical shapes. The microcracks are planar discontinuities that are better seen in microscopes. Their length may range from 1 μm to about 1000 μm (Engelder 1987). In rocks, Griffith cracks are commonly represented by intra-granular, inter-granular and transgranular cracks. As a matter of fact, joints, faults and other anisotropy planes also constitute Griffith cracks on a much larger scale. Griffith proposed that these cracks are randomly oriented pre-existing cracks in brittle solids that fail by incremental propagation of such innumerable cracks.

Griffith considered the energy balance of an elastically solid body. When a crack propagates, the *elastic strain energy* is released. Once a fracture is created, new surfaces are formed that raise the *surface energy*. The propagation of a crack within a solid body produces two new surfaces along which the potential energy is raised because the body has to do work against the cohesive forces of the atoms on either side of the propagating crack. This causes an increase in surface energy. If the body is under stress, there is a decrease in potential energy because of release of stored elastic energy together with the work done by the externally applied stress. All this leads to an increase in surface energy caused due to a growing crack that creates new surfaces. The total potential energy per unit thickness of the body (U) is then given by

$$U = U_s + U_e \tag{13.4}$$

where U_s is the surface energy per unit thickness and U_e is the released elastic energy per unit thickness. Under such conditions, *Griffith's energy criterion* states that

$$U_e \geq U_s \tag{13.5}$$

The above criterion helps in predicting new crack growth at crack tips. Griffith extended the work of Inglis (1913) and thus proposed a fracture criterion following the concept of minimum total free energy of elastically solid bodies deformed by externally applied stresses.

Griffith's theory works well with brittle solids on both experimental and theoretical grounds. However, it does not stand properly for ductile materials, such as metals, which develop a plastic zone at the crack tip. With increasing load, the plastic zone increases in size with crack growth.

Griffith energy-balance concept has been further advanced by Engelder and Fischer (1996) by modelling joint propagation in the brittle crust with the help of two laboratory loading configurations that are distinguished on whether or not a loaded boundary moves as a joint grows. They observed that in the first case, i.e. when the loaded boundary undergoes displacement, the energy necessary for joint propagation comes from a remote stress, while in the second case, i.e. when the loaded boundary remains stationary, the energy for joint propagation develops upon release of elastic strain energy within the rock mass. These two genetic loading configurations suggest four common natural loading configurations: a joint-normal load, a thermoelastic load, a fluid load and an axial load. These four natural joint-loading configurations are ideal end members of a vast spectrum. The authors further suggest that each loading configuration triggers a different joint-driving mechanism, each of which is the release of energy through elastic strain and/or work. These four mechanisms for energy release, in other words, are joint-normal stretching, elastic contraction, poroelastic contraction under either a constant fluid drive or fluid decompression, and axial shortening, respectively, as described in Sect. 13.14.3.

13.17 Significance of Joints and Fractures

13.17.1 Academic Significance

- Joints and fractures are excellent examples of brittle deformation of rocks. As such, their study extends our knowledge on brittle processes of the upper crust.
- Since joints make a rock weak, their presence sometimes controls weathering processes.

- Study of joints also helps in understanding the surface/subsurface processes, mainly weathering of rocks.

13.17.2 Economic Significance

- Joints and fractures constitute important agents for fluid flow. As such, their occurrence is important in the exploration of hydrocarbons, economic minerals and groundwater.
- In the case of hydrocarbons, a rock containing well-developed joints may under suitable conditions constitute a trap rock provided that cap rocks are available.
- Mineralizing fluids at depth occur at high pressure and always have a tendency to move to zones of lower pressures. The fluids find it easy to pass through joints and fractures and move to zones of lower pressure, where they precipitate and form mineral deposits. Highly jointed rocks therefore always attract economic geologists.
- Fracture connectivity promotes better fluid flow, including hydrocarbon.
- The intensity of fracture networks also controls fluid flow.
- In rocky terrains, joints and fractures are important structures in groundwater exploration. These structures add permeability to the rocks that helps subsurface water to come up to the ground surface. In some terrains with hard rocks and scarcity of water, therefore, a modern method of groundwater exploration involves artificial production of joints and fractures in rocks by subsurface blasting.
- In general, occurrence of joints and fractures is considered as an important parameter for permeability of rocks to fluid flow and in turn to fluid storage properties of rocks. All these significantly help in evaluating aquifer characteristics in relation to migration of petroleum and gas.

13.17.3 Engineering Significance

- Joints severely affect engineering properties of rocks. In general, rocks containing joints cannot take up load as much as a rock of the same composition without joints. Before any engineering and mining work, joints should therefore be studied thoroughly.
- Since joints make a rock weak, their presence deserves serious considerations in civil engineering and mining work.
- Presence of joints strongly affects the mechanical properties of rock, and these are important in the design of structures such as dams and tunnels.
- Joints cause leakage of water and slippage of rocks. This is a serious problem especially in the construction of roads in hilly terrains.
- In general, the study of joints and fractures helps geologists and engineers in identifying zones of weakness in rocky terrains where large structures such as dams, bridges and tunnels are proposed to be constructed.
- In the light of above, the occurrence of joints thus helps in assigning the quality of rocks for construction of large structures.

13.17.4 Environmental Significance

- Since joints make a rock weak, jointed rocks make a terrain prone to environmental degradation.
- Rock showing intersecting joints is liable to give away pieces of smaller blocks of rocks from its hold, thus producing rock falls. This process becomes pronounced with higher inclination of the ground surface that promotes mass wasting and rock fall due to greater role of gravity. Such areas are environmentally unsafe.
- Joints thus play a significant role in the identification of zones of environmental degradation of an area. This in turn helps in a better planning for human habitation and civic constructions.

13.18 Summary

- *Joints* are cracks along which there has been little or no transverse displacement of rock.
- Joints are commonly associated with *fractures* that are also breaks or discontinuities in rock. The two terms joint and fracture are often used interchangeably. In a *shear fracture*, wall-parallel displacement is discernible.
- If the fracture has undergone substantial shear, it is usually classified as a fault; otherwise, it is denoted as a joint.
- *Microcracks* are planar discontinuities that are too small to be seen in a hand specimen; their longest dimension is of the order of one to several grain diameters (about 100–1000 μm) with the small dimension of the order of 1 μm.
- Joints and fractures occur on a wide range of scales from a few centimetres to a few kilometres.
- Joints and fractures filled with minerals are called *veins*.
- Fractures of sedimentary rocks commonly show a constant ratio of layer thickness to joint spacing; this is called *fracture spacing index*. This index is approximately same in different rock types and in different structural locations over a large area.

- Joint surfaces commonly show several types of features that show a variety of shapes and sizes. These features are formed under some special conditions and provide information on the direction of propagation of joints and on the mode of formation of the extension fractures.
- Since joints are stress indicators in some way or another, they may be related to the present-day or contemporary stress system of the earth. However, most joints reflect some earlier stress field(s).
- Joints are manifestation of brittle failure of rocks and form when the tensile stresses of a rock exceed its tensile strength. Joints in rock propagate where the component of maximum effective tensile stress is normal to the plane of the crack. The underlying principles of joint mechanics are highlighted in the text.
- Common causes of joint formation include burial, uplift and erosion, thermal contraction, sheeting and tectonic causes.
- *Fracture mechanics* concerns the study of stress concentrations caused by sharp-tipped flaws and the conditions for the propagation of these flaws.
- Fractures form when an externally applied stress exceeds the fracture strength of solids.
- Fracture opening occurs in three basic modes of loading: *mode I* (opening mode), *mode II* (sliding mode) and *mode III* (scissoring mode). These three modes are also called tensile, in-plane shear and anti-plane shear, respectively.
- Griffith's theory suggests that fracture strength of brittle solids depends upon the presence of defects or flaws of microscopic sizes in the form of microcracks, dislocation planes, voids, inclusions and precipitates. Presence of flaws lowers the fracture strength of solids.

Questions

1. Outline the differences between a joint and a fracture.
2. Why do joints generally occur at higher levels of the crust?
3. Is there any dependence of joint spacing on layer thickness of a bed? Discuss.
4. What are microcracks?
5. Describe some important features developed on joint surfaces.
6. Briefly describe joint propagation.
7. Are joints related to present-day stress field? Discuss.
8. Give some salient features of joint mechanics.
9. What is meant by fracture mechanics?
10. Give the significance of joints.

14 Foliation

A. R. Bhattacharya, *Structural Geology*, Springer Textbooks in Earth Sciences, Geography and Environment,
https://doi.org/10.1007/978-3-030-80795-5_14

Abstract

Foliation is a planar structure given by preferred orientation of minerals generally showing a platy or tabular habit. The preferred orientation is produced by deformation and is uniformly pervasive in a rock. Foliation is commonly developed in metamorphic rocks and includes cleavage, schistosity, gneissosity and gneissic banding. *Cleavage* or *rock cleavage* is a planar fabric along which rocks easily split or cleave into parallel or subparallel surfaces formed during metamorphism and deformation. Along the cleavage, a rock shows the ability to split or cleave into parallel or subparallel surfaces. Cleavage is generally developed in low-grade metamorphic rocks, while foliation is developed in rocks showing low- to high-grade metamorphism and is developed in all types of rocks. Foliation is assumed to represent the *XY* or the flattening plane of the strain ellipsoid. The foliation-forming processes can be considered to be those that give rise directly or indirectly to preferred orientation of components of a rock. In deformed rocks, study of foliation helps in tracing the deformation history.

Keywords

Foliation · Cleavage · Continuous cleavage · Spaced cleavage · Axial-plane foliation · Foliation fan · Foliation refraction · Transected foliation · Transposition foliation · Genesis of foliation

14.1 Introduction

Foliation (Latin *folium* means leaf) is a planar structure given by preferred orientation of minerals generally showing a platy or tabular habit. The preferred orientation is produced by deformation and is uniformly pervasive in a rock. Foliation is easily seen in outcrops as well as in the microscope.

Foliation is commonly developed in metamorphic rocks such as schists and phyllites in which case it is defined by parallel arrangement of platy, flaky or elongate minerals. The term foliation can also be used for sedimentary rocks because a bed also has a definite fabric formed due to sedimentary processes; this fabric is defined by colour banding, uniformity of grain size, texture and parallel alignment of micaceous minerals and clasts. In volcanic rocks, foliation is defined by flow banding and parallel alignment of platy minerals developed during the flow of magma. Foliation is thus used in a broader sense for any homogeneously distributed planar feature of rocks with a characteristic fabric. The fabric may be defined by parallel arrangement of elongate objects, flattened minerals, uniformity of grain size, colour banding or texture. Despite the fact that the above-mentioned features resemble foliation, the term foliation is used in structural geology for deformed rocks.

In the literature, terms like cleavage, schistosity, gneissosity and gneissic banding have also been used for the fabric shown by foliation. In this book, we shall use the term *foliation* to include all these types of structures.

14.2 Foliation and Cleavage

Cleavage or *rock cleavage* (*cleave* means to break) is a penetrative fabric along which rocks easily split or cleave into parallel or subparallel surfaces. Due to the presence of cleavage, the rock apparently does not lose its cohesion. Most geologists consider cleavage as a type of foliation though not necessarily both the terms carry the same meaning. For example, not all rocks break or tend to break along foliation, but this is so for cleavage. Because of this, property cleavage imparts a mechanical anisotropy to the rock while foliation necessarily does not. Cleavage is characteristic of rocks formed at low- to very-low-grade metamorphism while foliation develops over a broad spectrum ranging from low- to high-grade metamorphic rocks and also occurs in all types of rocks including those of sedimentary, igneous and metamorphic origins.

From above, it appears that both the terms foliation and cleavage enjoy independent status in structural geology. In this book, we shall however use the term foliation as a general term to include cleavage also. In a nongenetic, general sense, foliation includes all surfaces found in deformed metamorphic rocks (Hobbs et al. 1976, p. 213).

14.3 Foliation in Hand Specimens

A foliated rock shows two distinct domains (Figs. 14.1 and 14.2): foliation domain and microlithon. The *foliation domain* or *cleavage* domain is one in which the original fabric of the rock is significantly changed. The parallel fabric shown by the alignment of mineral constituents of a domain is parallel to the rock cleavage. The *microlithon* shows no or little alteration of the original fabric of the rock. These two domains alternate, and the spacing between them varies with different rocks.

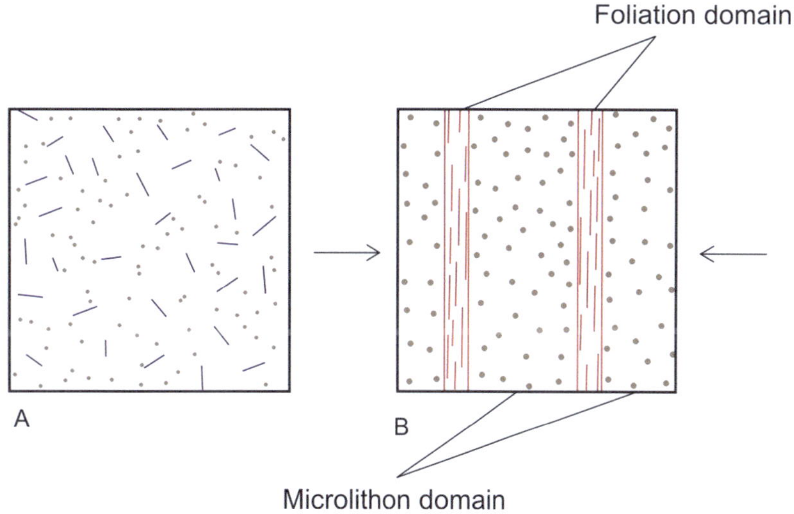

Fig. 14.1 (**a**) Sketch of an unfoliated rock. Note that the constituents of the rock do not show any systematic orientation. (**b**) Sketch of a foliated rock showing two distinct domains: foliation domain and microlithon domain. See text for details

Fig. 14.2 Photograph of a gneiss showing foliation in hand specimen with two distinct domains. The cleavage domain is shown by dark-coloured bands given by biotite, while the intervening space represents the microlithon domain. (Sample courtesy Geology Museum, University of Lucknow)

14.4 Foliation in Thin Sections

14.4.1 Some Common Forms of Foliation

In thin sections, foliation occurs in various forms. Some common forms are presented here: (1) Cleavage domain is almost equidistant from each other with respect to the microlithon domain (Fig. 14.3). (2) Cleavage domain is at unequal distances from each other with respect to the microlithon domain (Fig. 14.4). (3) Some minerals of the cleavage domain (e.g. micaceous minerals and quartz) are common in the microlithon domain and vice versa (Fig. 14.5). (4) Cleavage domains run almost straight on the scale of thin section (Fig. 14.6). (5) Cleavage domain appears wavy or shows gentle flexures or irregularities (Fig. 14.7); this is either due to swerving of the micaceous layers around some large grains or due to the effect of mild deformation suffered by the micaceous or the platy minerals at some later stages. (6) Cleavage domain is ill defined such that it is represented by only a few flaky minerals that may occur persistently or impersistently (Fig. 14.8).

14.4.2 Foliation in Some Common Rock Types

Characteristic features of foliation in thin sections of some common metamorphic rocks such as gneisses, schists and quartzites are briefly presented below.

In *gneisses* (Fig. 14.9), the microlithon domain is dominantly constituted of quartz and feldspar, also called the *QF domain* (Q for quartz and F for feldspar). This domain is generally a product of metamorphism of earlier bedding. Preferred orientation of these constituents is almost lacking. This domain may contain some platy minerals that show, but not always, preferred orientation. The foliation or cleavage

Fig. 14.3 Foliation in thin section. Note that the foliation domain is at almost equal distance from each other with respect to the microlithon domain

Fig. 14.4 A deformed rock in thin section in which the foliation domain is at unequal distances from each other with respect to the microlithon domain

domain is dominantly constituted of micaceous minerals, also called *M domain* (M for mica). The flakes of mica show preferred orientation by their longer faces.

In *schists* (Fig. 14.10), the microlithon domain is constituted of quartz, feldspar, calcite and a few non-platy or non-flaky minerals. Occasionally, mica and some flaky minerals are also present in the microlithon domains. The cleavage domain is dominated by several types of platy/flaky minerals such as mica, amphiboles, kyanite and chlorite. The layering pattern of the rock, as seen in hand specimens, is given by the difference of grain size and orientation patterns of components.

Fig. 14.5 A deformed rock showing some minerals of the cleavage domain (e.g. micaceous minerals and quartz) is common in the microlithon domain and vice versa

Fig. 14.6 In thin section, the cleavage domain runs almost straight on the scale of thin section

In schists, the composition of microlithon and cleavage domains as generalized above may show some variations. In some rocks, the cleavage domain consisting of clusters of micaceous minerals is almost equidistant from each other (Fig. 14.10), while in others (Fig. 14.11) it may not be equidistant. In the former rock type, the schistosity is distinctly visible to naked eyes, while in the latter it is not so. Further, the microlithon domain instead of containing quartz or feldspar or both may contain some other minerals such as calcite (Fig. 14.12). In altered basic rocks, the distinction between the microlithon and cleavage domains is not

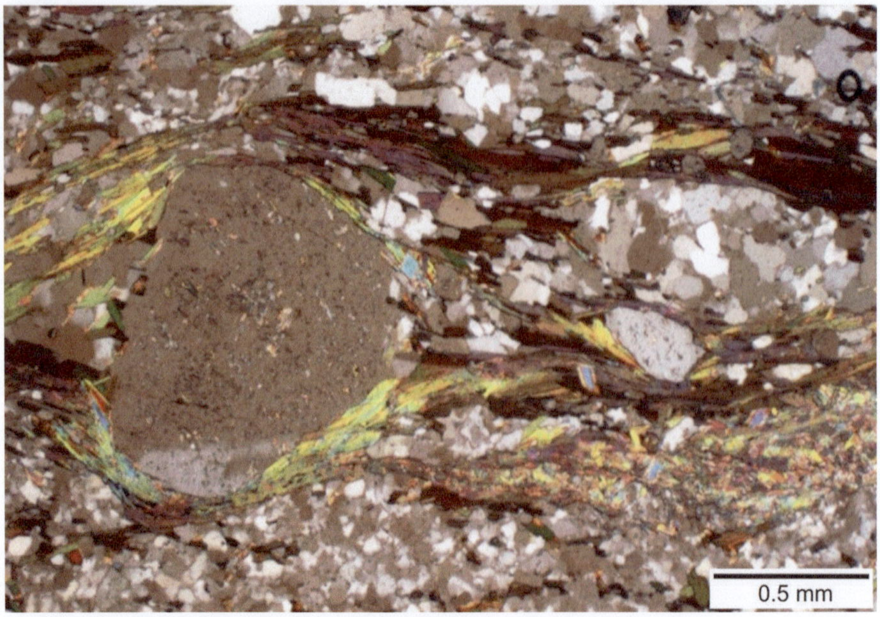

Fig. 14.7 The foliation domain in thin section of this rock appears wavy. The micaceous layers swerve around large grains of feldspar and quartz

Fig. 14.8 In thin section, the foliation domain of this rock is not well defined as can be seen by the presence of only a few flaky minerals that occur both persistently and impersistently

always clear as the schistosity is given by the preferred orientation of the platy minerals only (Fig. 14.13).

In *quartzites* (Fig. 14.14), the cleavage domain is given by highly deformed and stretched grains of quartz showing a preferred alignment. Occasionally, the presence of some flaky minerals in this domain accentuates the cleavage (Fig. 14.14). In the microlithon domain, the preferred orientation of quartz grains is feeble to lacking. In some highly deformed quartzites (Fig. 14.15), it is difficult to identify the cleavage and microlithon domains. It is only due to the presence of some mica flakes that the cleavage domain is identified.

14.5 Classification of Cleavage

Although the literature is aplenty with description of several types of cleavage, little work is available on the classification of cleavage. Modern classifications can be seen in the works of Dennis (1972), Gray (1977), Powell (1979) and Twiss and

Fig. 14.9 Foliation in gneiss as seen in thin section. The rock shows two distinct domains: microlithon domain (QF) with quartz and feldspar and mica domain (M)

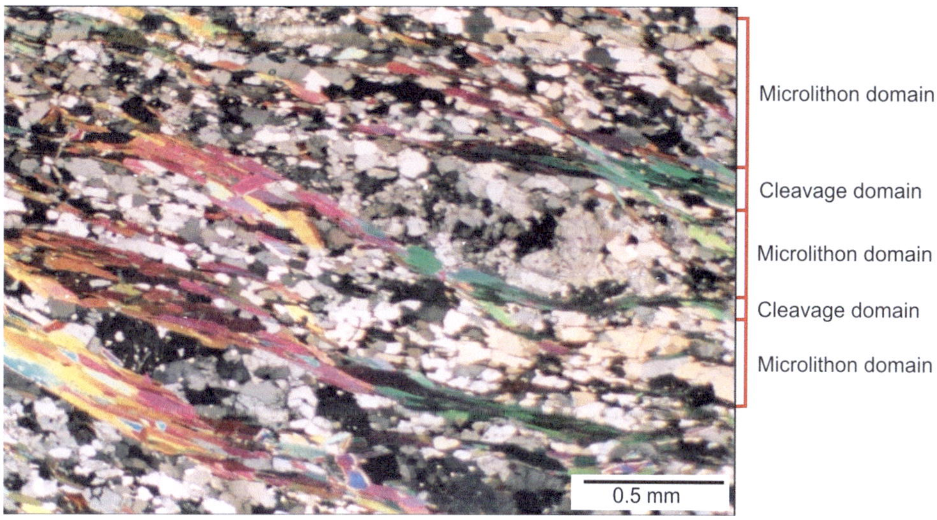

Fig. 14.10 A schist in thin section showing cleavage domain containing micaceous minerals is nearly equidistant from each other. The microlithon domain is constituted mainly of quartz and feldspar

Moores (2007), and all these reveal that there is no consensus in classifying the cleavage. Following Powell (1979), we consider here two major morphological subdivisions of rock cleavage: *continuous cleavage* and *spaced cleavage*.

14.5.1 Continuous Cleavage

Continuous cleavage is a pervasive structure characterized by distinct planar elements with spacing of less than 1 mm. As such, the rock develops a tendency to split into thin slices such as shown by slates. The cleavage domain is sometimes so dominant that microlithon is not visible. In continuous cleavage, therefore, the presence of microlithon is not that important. On the basis of grain size from finer to coarser, the continuous cleavage progressively becomes a slaty cleavage, phyllitic structure and schistosity.

Slaty cleavage is shown by slates (Fig. 14.16) that are characterized by planar structures given by micaceous layers commonly containing chlorite, muscovite or clayey matter. The micaceous layers are formed due to recrystallization of clay under very-low-grade metamorphic conditions and constitute the cleavage domains that are separated by microlithons containing an aggregate of small quartz and feldspar. Presence of slaty cleavage imparts a strong fissility character to the rock due to which it can be separated in thin layers. As such, slates are often used for roofing.

With increasing recrystallization, the grain size increases so that the slaty cleavage gives way to *phyllitic structure* as typically shown by phyllite (Fig. 14.17). Occurrence of tiny

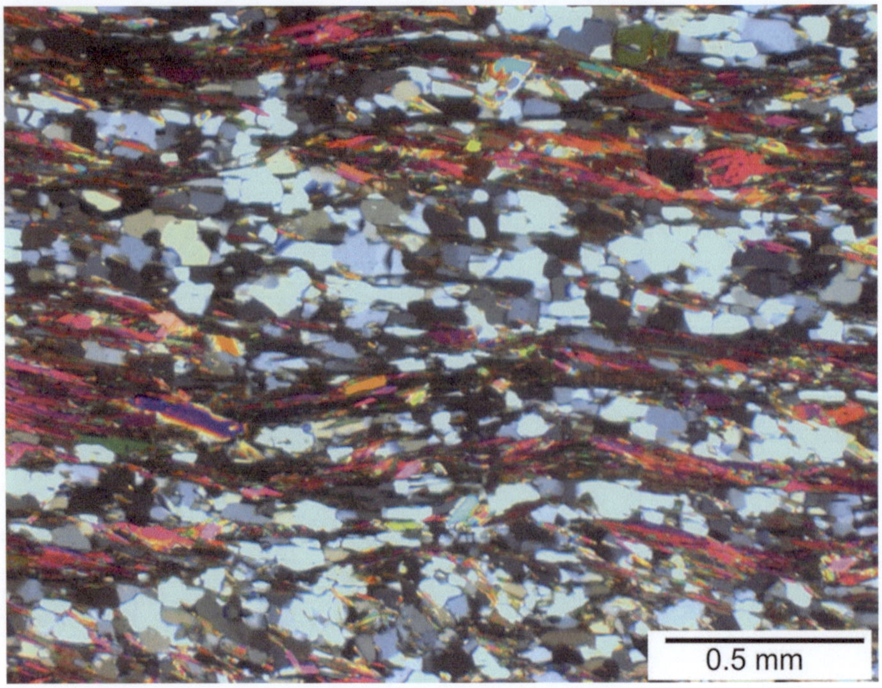

Fig. 14.11 Photomicrograph of a schist in which the cleavage domain containing micaceous minerals is not equidistant from each other

Fig. 14.12 Photomicrograph of a schist in which the microlithon domain is mainly constituted of calcite and some quartz, while the cleavage domain is constituted of micaceous minerals

grains of mica (commonly sericite) in layers occasionally gives a sheen to the rocks, and therefore the rock is sometimes called a *sheeny phyllite*. With further increase of metamorphism to medium and high grades, the grains grow larger and the rock now develops *schistosity* as shown by schists (Figs. 14.10, 14.11, 14.12 and 14.13). In such rocks, grains of quartz, feldspar, mica, garnet, amphiboles, etc. can be seen by naked eyes. Under conditions of high-grade metamorphism, the mineral assemblage of schists develops alternating bands of light- and dark-coloured minerals in which case the foliation is called *gneissosity* (Fig. 14.2) and the rock is called a gneiss. In thin sections (Fig. 14.9), the light-coloured bands show quartz, feldspar and an aggregate of quarto-feldspathic minerals, while the dark-coloured bands show micaceous minerals, amphiboles and pyroxenes.

Fig. 14.13 An altered basic rock in thin section in which the foliation is given by preferred orientation of platy minerals only; the microlithon and cleavage domains are not defined

Fig. 14.14 Foliation shown by quartzite. The cleavage domain is represented by stretched grains of quartz and some flaky minerals; the latter accentuate the cleavage. The microlithon domain is represented by deformed but feebly oriented grains of quartz

14.5.2 Spaced Cleavage

Spaced cleavage occurs in the form of domains with microlithons and cleavage. The spacing between the cleavage and microlithon domains is 1 mm or more. As such, spaced cleavage is visible in hand specimens. Spaced cleavage may show wide variations in spacing, mineral composition and preferred orientation of mineral constituents of individual domains. Spaced cleavage can be crenulations or disjunctive.

(a) *Crenulation cleavage* occurs in the form of small folds (when seen in hand specimens) or microfolds (when seen in microscopes) (Fig. 14.18) formed by folding/microfolding of a pre-existing cleavage. The new

Fig. 14.15 A highly deformed quartzite in which it is difficult to identify the cleavage and microlithon domains. However, the presence of some mica flakes defines the cleavage

Fig. 14.16 Slaty cleavage in slate

cleavage thus formed is given by nearly straight and parallel alignment of the axial planes of crenulations. The crenulations can be symmetric if the fold limbs are of equal length or asymmetric with unequal lengths of fold limbs. The fold hinges constitute the microlithons, while the cleavage domain is given by the parallel alignment of platy minerals of the limbs. Crenulation cleavage has also been described as *strain-slip cleavage* or *chevron folds* (see Chap. 8).

(b) *Disjunctive cleavage* (Latin *disjunctus* meaning 'disjoined' or 'detached') is commonly formed in previously unfoliated rocks such as mudstones or limestones as well as in some foliated rocks that cross-cut an earlier foliation. The disjunctive cleavage is characterized by distinct cleavage domains or seams and microlithons. Cleavage domains are relatively thinner (<1 mm) containing strongly aligned platy minerals, while the intervening microlithons are tabular to lenticular in which the platy minerals are either less abundant or randomly oriented.

14.6 Relative Chronology of Foliations

During a tectonic or metamorphic event, the early-formed planar surface in a rock generally changes its geometry to acquire a new one. In yet another event, the newly formed planar surface may again change its geometry. With this concept in mind, Bruno Sander, an Austrian geologist, proposed in 1930 to designate any planar surface in a deformed rock by S. He referred the earliest metamorphic planar surface as S_1. For planar surface of primary nature such as bedding or magmatic layering, he used the term

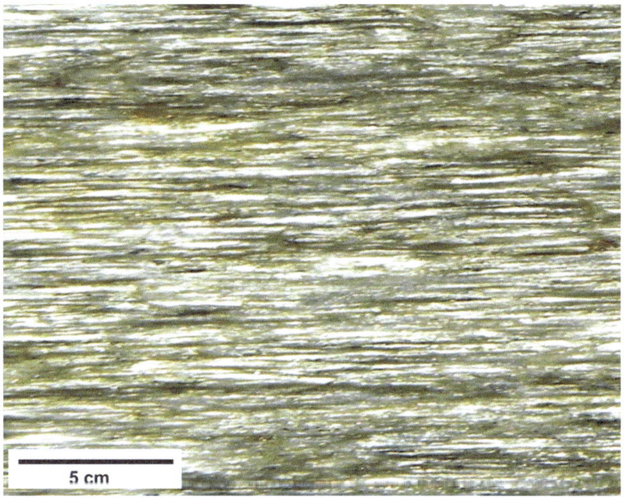

Fig. 14.17 Phyllitic structure in phyllite

S_0. In terrains of multiple deformations, the early-formed planar surface develops younger planar surfaces in successive deformation events for which he used the higher subscripts in a chronological order such as S_0, S_1 and S_2 (Fig. 14.19). The system of successive subscripts thus readily helps to understand the relative ages of different planar surfaces in multi-deformed rocks. Further, since folding also produces a new planar surface, the particular folding event can also be associated with the planar surface formed. For example, an axial-plane foliation (S_1) may form during the first folding event (F_1). The chronology of foliations such as S_1 and S_2 can thus be used to refer to the successive folding events such as F_1 and F_2. Likewise, this nomenclature system can also be used for successive linear structures such as L_1, L_2, ... and so also for successive deformational events D_1, D_2, ...

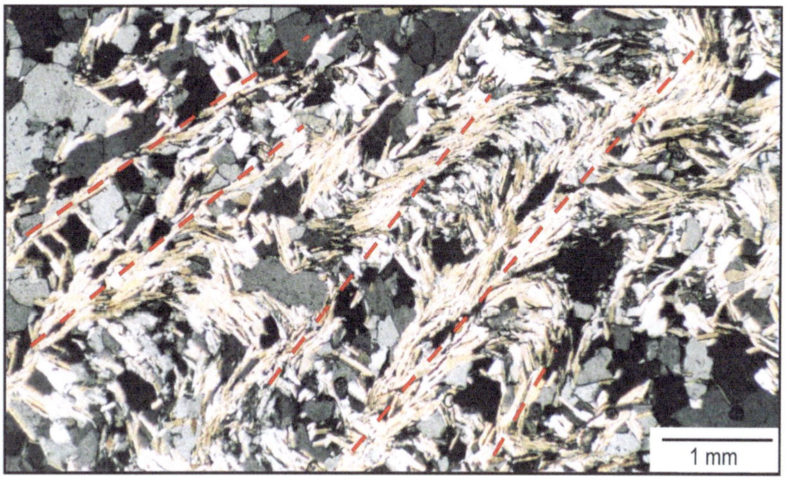

Fig. 14.18 Crenulation cleavage (shown by red dashes) as given by the chevron folds developed in a pre-existing cleavage

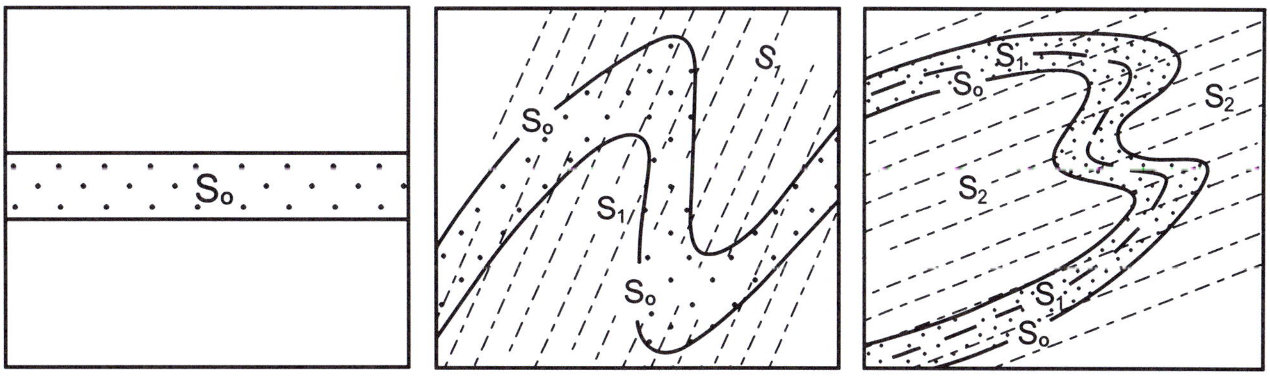

Fig. 14.19 Relative chronology of foliations. An earlier planar surface (S_o) is folded to give a new set of foliation (S_1). The limbs of the fold thus formed are again folded to give one more set of foliation (S_2)

14.7 Genesis of Foliation

14.7.1 Background

Foliation commonly forms normal, or near-normal, to the direction of maximum shortening and is assumed to represent the *XY* or the flattening plane of the strain ellipsoid. Foliation-forming processes can thus be considered to be those that give rise directly or indirectly to preferred orientation of components of a rock. The degree of orientation however may vary in rocks. With weak orientation, the rock may be said to show weak foliation, while with strong orientation, the rock may be said to show strong foliation. Although the formation of foliation has been discussed by several workers (e.g. Wilson 1961; Ramsay 1967; Billings 1972; Siddans 1972; Wood 1974; Gray 1977; Powell 1979; Ramsay and Huber 1983; Passchier and Trouw 2005; Twiss and Moores 2007; Hobbs and Ord 2015), as yet we do not have any unified opinion about the formation of foliation. On the basis of these works, we discuss below some common processes that develop foliations in rocks. These processes may operate singly or in combination with some other ones.

14.7.2 Role of Deformation

At the outset, the question arises as to whether deformation is necessary to form foliation, and if so then at what stage of deformation? Unfortunately, we have some limited information on this aspect. On the basis of the study of deformation of oolites from the South Mountain area of the Southern Appalachians, Cloos (1947) had estimated that a cleavage had appeared in the limestones when shortening in rocks had exceeded 30%. Likewise, Wood (1974) has shown that formation of slaty cleavage takes place when the rocks had been shortened by at least 50%. It thus seems that at least some deformation is necessary before a rock starts developing a foliation. However, some other factors can also develop foliation as discussed below.

14.7.3 An Overview of Foliation Formation

When we talk of formation of foliation, we mean how the preferred orientation of tabular, prismatic or acicular minerals of a rock was developed. In fact, the same logic can be applied for the formation of lineation when preferred orientation is given by minerals of acicular or prismatic habit. This implies that some of the processes overlap in both the cases; the difference is only of planar (foliation) and linear (lineation) features. In this context, Hobbs and Ord (2015, p. 313) presented three proposals for the formation of preferred mineral shape orientation:

1. The mineral grains form early in a deformation and are rotated during deformation to align with the principal plane or axis of strain [the Vernon (2004) model].
2. The mineral grains form to minimize the elastic or stored energy of the grain [the models of Kamb (1959), MacDonald (1960) and Brace (1960)].
3. Internal deformation by crystal slip during deformation aligns the grains with some kinetic framework.

With the above-mentioned three foliation-forming processes in mind, Hobbs and Ord (2015) however suggested a different model (mineral nucleation and mineral growth, described in Sect. 14.8 (14.8.11 and 14.8.12)). Some common foliation-forming processes are discussed below.

14.8 Foliation-Forming Processes

14.8.1 Folding

14.8.1.1 Can Folding Develop Foliation?

Folds are commonly developed in rocks that show well-developed layering, as in sedimentary rocks, or planar features shown by preferred orientation of minerals, as in metamorphic rocks. During folding of rocks, the otherwise randomly oriented constituent minerals are commonly aligned in a direction perpendicular to the greatest shortening direction. The realignment of the constituent minerals gives rise to foliation. In fact, the platy and flaky minerals easily undergo systematic realignment under the ambient stress field. The polygonal and equidimensional minerals however accommodate the stresses by occupying the available space. Some relation between folding and foliation formation thus appears implicit. Folding can develop some common types of foliation such as crenulation foliation, axial-plane foliation, foliation fan, foliation refraction and transacted foliation as described below.

14.8.1.2 Crenulation Foliation

Crenulation foliation is formed due to folding or microfolding of pre-existing foliation. Foliation is given by the axis of microfolds. Twiss and Moores (2007) suggested that symmetric crenulation foliation is formed due to shortening parallel to an early foliation S_1 (Fig. 14.20). As a result, this foliation is rotated towards low angles to form the axial surfaces of the new crenulations S_2. Asymmetric crenulation foliation is formed (Twiss and Moores 2007) by shortening of an early foliation S_1. The direction of greatest shortening is at low angle to S_1, thus producing asymmetric

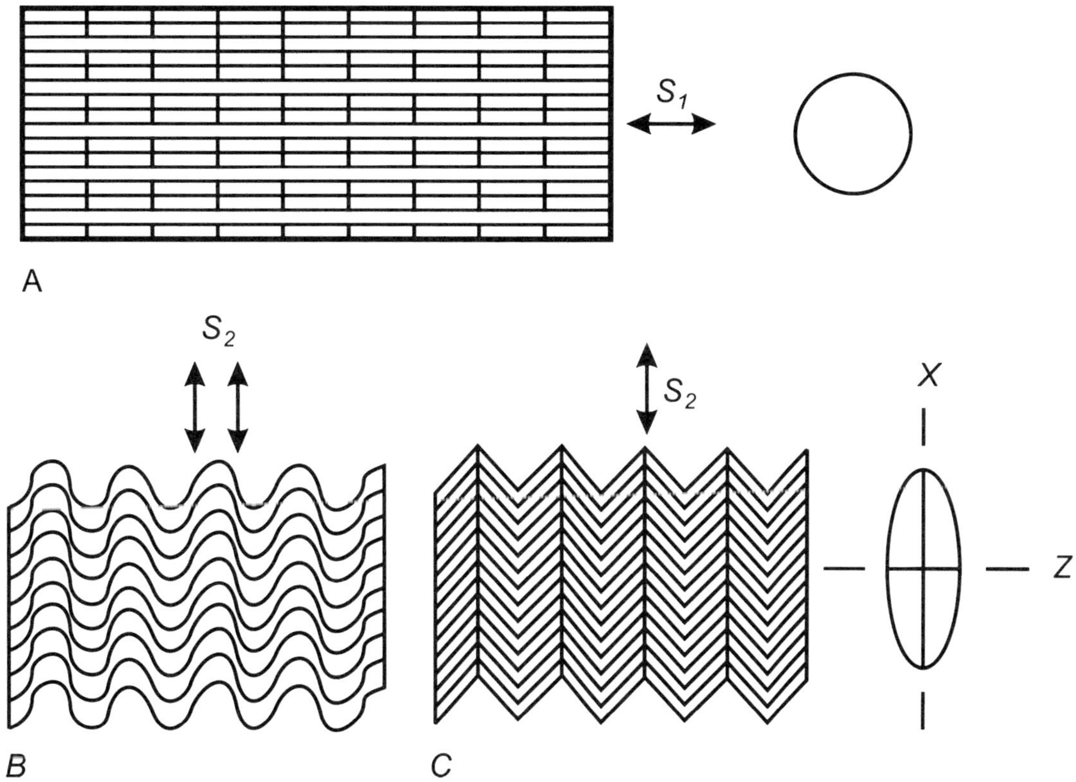

Fig. 14.20 Formation of symmetric crenulation foliation. See text for details. (Reproduced from RJ Twiss and EM Moores, 2007, Structural Geology, second ed. Fig. 14.2 with permission from Cambridge University Press under Licence number 52171308446)

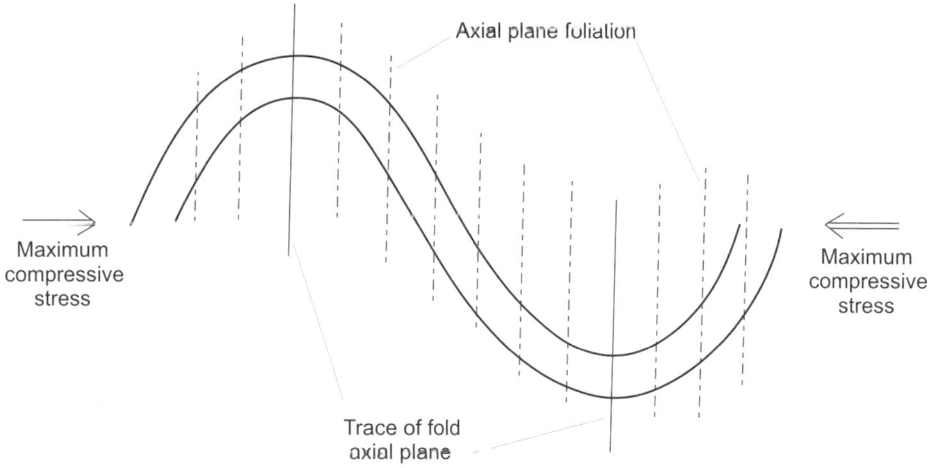

Fig. 14.21 Diagrammatic sketch of axial-plane foliation (dotted lines) developed parallel to the axial planes of the associated folds

crenulations. The axial surface of the crenulations gives the newly formed crenulation foliation S_2.

Asymmetric crenulation foliation can also develop (Twiss and Moores 2007) due to the formation of *shear bands* (Chap. 17) that are discrete bands of concentrated strain. A crenulation foliation is thus developed that is also called *shear band foliation*. An undeformed foliation may be subjected to simple shear within shear bands and thus can produce asymmetric crenulations. The shear bands now constitute the S_2 cleavage domains.

14.8.1.3 Axial-Plane Foliation

As described above, during folding of rocks, the pre-existing, randomly oriented constituent minerals are aligned in a direction perpendicular to the greatest shortening direction, which is commonly the direction parallel to the axial plane of the fold. This realignment of the constituent minerals gives rise to foliation called *axial-plane foliation* (Figs. 14.21 and 14.22). As such, the axial-plane foliation has always been given due significance in the study of folds. At the hinge of a fold, the bedding and the axial-plane foliation are at right angles to each other. Applying this relation, it is possible to locate the hinge zone of large-size folds.

The geometrical relationship of bedding and axial-plane foliation, also called *bedding-foliation relationship* (Fig. 14.23), is of great help in the case of overturned folds. In overturned folds, the axial-plane foliation changes its steepness relative to the bedding in the two limbs of the fold. Both the limbs dip in the same direction, but in the normal limb the cleavage is steeper than the bedding, while in the overturned limb the bedding is steeper than the cleavage, suggesting a synform in the opposite direction. Identification of the overturned limb also helps in establishing the normal stratigraphic order of beds in the fold.

In multilayered folds, the foliation sometimes does not maintain its continuity all through the fold. When there is alternation of competent and incompetent layers, the foliation gets refracted while passing from one layer to another. The overall geometry thus shown by the foliation in the fold is called *foliation refraction* (Figs. 14.24 and 14.25). Foliation refraction shows different orientations of cleavage in adjacent beds, and this is a reflection of the differences in the amount of shear the beds have undergone (Ramsay 1967; Van der Pluijm and Marshak 2004). Foliation refraction is sometimes used as an important marker to evaluate rheologic parameters for deformation of natural rocks. Czeck et al. (2019), for example, have shown that there is a linear relationship between strong (mineral) phase concentration and bedding/cleavage angle. As such, cleavage refraction directly

Fig. 14.22 Axial-plane foliation (shown by yellow dotted lines) developed parallel to the axial plane of the associated fold in sandstone beds of Southern Appalachians, near Rockwood, Tennessee, USA. The limbs of the fold also show foliation parallel to the axial plane of the fold. (Photograph by the author)

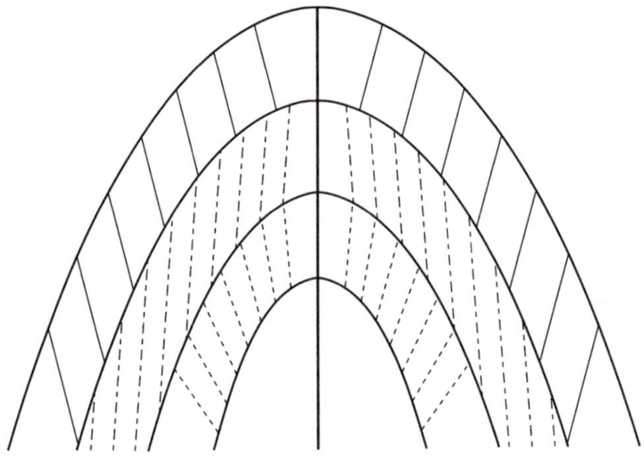

Fig. 14.24 Foliation refraction. Note that the foliations of a particular layer get refracted on entering the adjoining layers

Fig. 14.23 Bedding-foliation relationship. See text for details

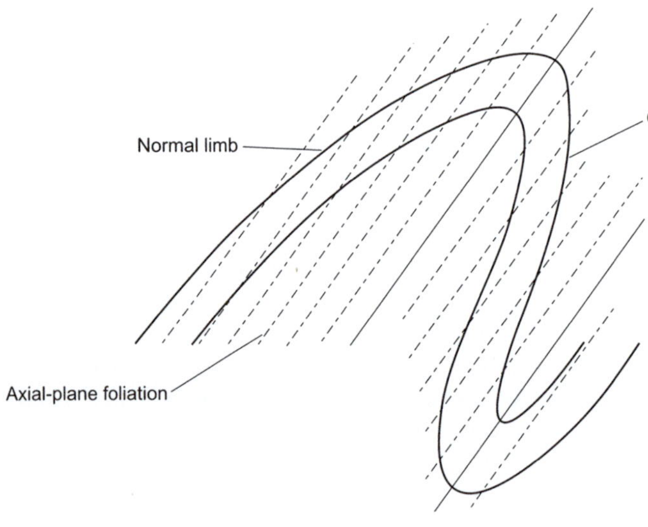

correlates to mineralogy. The authors thus suggest that cleavage refraction, or change in orientation of foliation, allows us to quantify effective viscosity ratios of the rock layers.

Foliation developed during folding sometimes assumes a fan shape in a fold, also called *foliation fan* (Fig. 14.26). The fan can be convergent or divergent with respect to the axial plane. The divergent fans diverge away from the core, while convergent foliation fans (Fig. 14.27) converge towards the core.

Although the axial-plane foliation is commonly parallel to the fold axial plane, in some cases, the foliation is developed at an angle with the fold axial plane. In other words, the foliation transects the fold axial plane. Foliation of this type is called *transected foliation* (Fig. 14.28), and the associated fold is called a *transected fold*.

We have thus shown that foliation commonly bears some definite relationship to folds. However, in the light of recent studies, some exceptions have been pointed out. Stephan et al. (2016), for example, while studying fold-cleavage relationship in the Rheno-Hercynian-Saxo-Thuringian boundary zone of Central European Variscides have observed different fold-cleavage types which, according to them, have resulted from a polyphase history with oblique convergence to pre-existing trends of mechanical anisotropies.

Fig. 14.25 Foliation refraction as developed in the slate of North Eifel, Germany. (Photograph by the author)

14.8.2 Ductile Shearing

14.8.2.1 Background

Rocks in ductile shear zones show foliations that are developed when the rocks are ductile. Ductile shear zones (Chap. 17) are developed due to localization of strong ductile shear strain at deeper levels and can develop in any type of rock. Under these conditions, rocks develop a shape fabric due to stretching of the constituent mineral grains. The preferred orientation of the flattened mineral constituents can be easily seen under microscope. Some common types of foliations developed in ductile shear zones are mylonitic foliation, transposition foliation and intrafolial folds, as described below, while another important structure, i.e. S-C structure, has been described in Chap. 17.

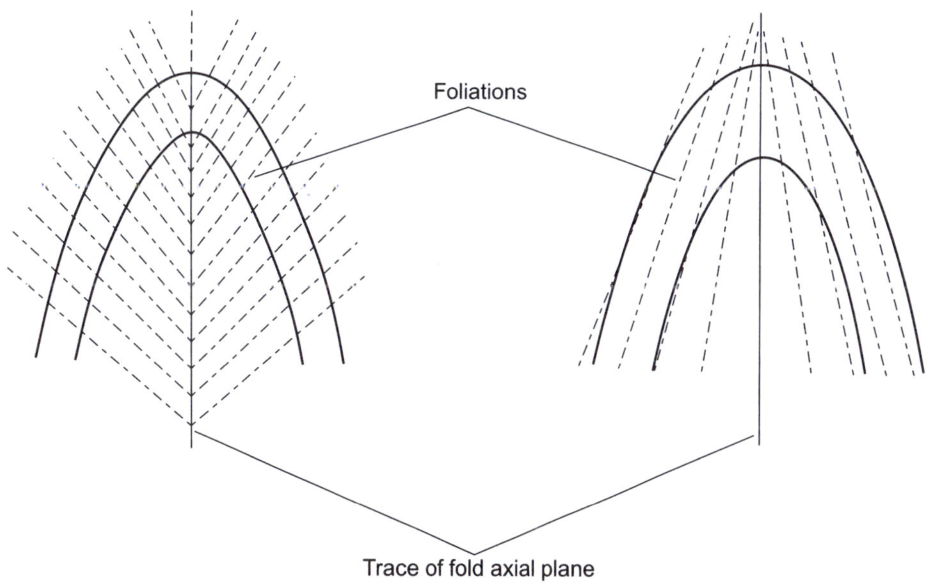

Fig. 14.26 Foliation fan associated with a fold. Left: Convergent foliation fan. Right: Divergent foliation fan

Fig. 14.27 Convergent foliation fan developed in a folded sandstone bed of the fold-thrust belt of the Southern Appalachian near Rockwood, Tennessee, USA. (Photograph by the author)

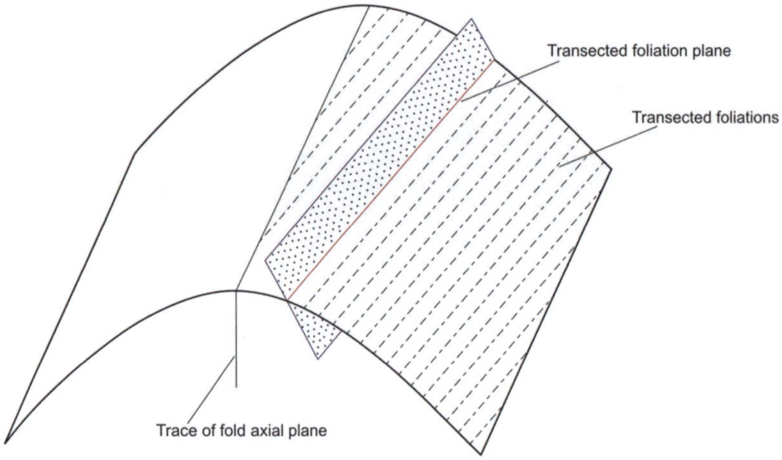

Fig. 14.28 Transected foliation. See text for details

14.8.2.2 Mylonitic Foliation

The *mylonitic foliation* is a characteristic of rocks called mylonite that is typically developed in ductile shear zones (Chap. 17). A mylonite is commonly formed by strain-softening processes. Grain size reduction is a common process in mylonites. (i) In *outcrops and hand specimens*, mylonite shows a fine-grained texture with alternating light- and dark-coloured bands (Fig. 14.29) that define the mylonitic foliation of the rock. Its mineral constituents are sometimes identified by hand lens also. The mylonitic foliation sometimes includes well-developed augens and boudins of quartzo-feldspathic material (Fig. 14.30).

(ii) In *thin sections*, mylonitic foliation occurs in several forms. In some mylonites, micaceous minerals occur in clusters (Fig. 14.31) whose alignment defines mylonitic foliation, while in others, mylonitic foliation is defined by preferred orientation of very thin layers of micaceous minerals (Fig. 14.32). Also, there are mylonites in which the mylonitic foliation is given by a strong orientation of all the major minerals, e.g. stretched quartz grains and micaceous minerals (Fig. 14.33).

14.8.2.3 Transposition Foliation

Transposition structures are formed due to stretching of an earlier structure under high strains. Parts of the structure thus change their original positions (transpose) to occupy some other positions. A fold (Fig. 14.34a) may be subjected to flattening (Fig. 14.34b) at high strains. With continued

Fig. 14.29 Field photograph of a quartz mylonite occurring in the Main Central Thrust of Garhwal Himalaya, India. Note that the rock is characterized by strong mylonitic foliation given by alternating light- and dark-coloured bands. (Photograph by the author)

Fig. 14.30 Field photograph of a mylonite from Southern Appalachians, USA, showing well-developed augens and boudins of quartzo-feldspathic material. The rock shows strong mylonitic foliation defined by alternating light- and dark-coloured bands. (Photograph by the author)

deformation, the limbs are broken, separated (transposed) and stretched into parallelism (Fig. 14.34c) that defines a new foliation. The hinge zone of some of the folds is thus isolated from the limbs; for such folds, the term *rootless folds* is also used. The rootless folds are recognized by isolated hinges of isoclinal folds with their axial surfaces parallel to the foliation.

In addition to folds, any earlier foliation or band may also undergo stretching to reorient into parallelism. Transposition foliation is commonly developed, but not necessarily, in ductile shear zones. An important factor for the formation of transposition foliation is shortening and consequent extension of an early structure, such as foliation or vein, under high strains such that parts of the early-formed structure are transposed to different positions.

14.8.3 Pressure Solution

Pressure solution is a process whereby crystals or crystal aggregates of a rock undergo dissolution at points of high compressive stress followed by redeposition at points of low

Fig. 14.31 Photomicrograph of a mylonite in which the foliation is given by strongly developed layers of micaceous minerals. (Photomicrograph by the author)

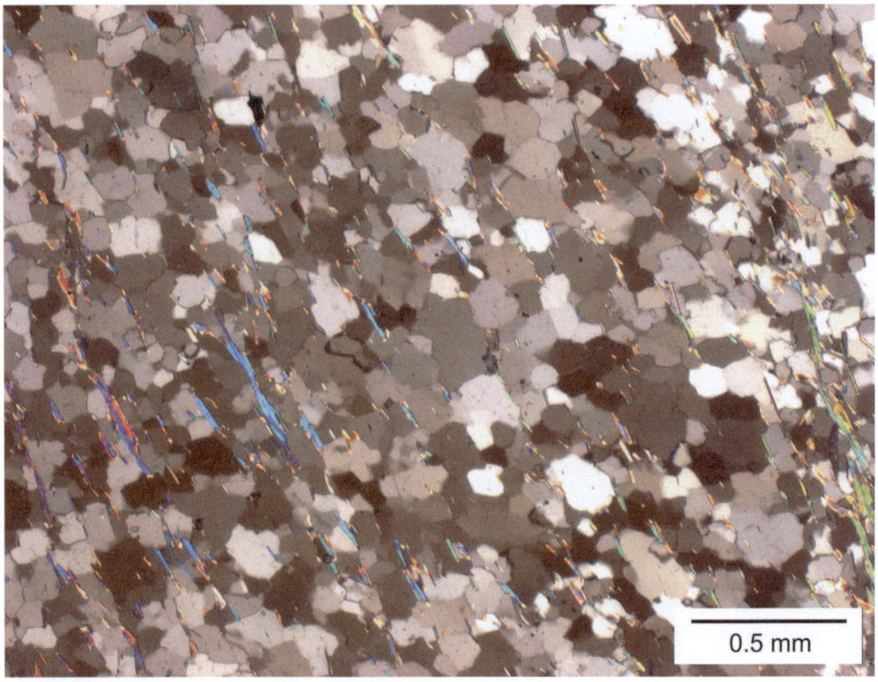

Fig. 14.32 Photomicrograph of a mylonite in which the foliation is given by very thin layers of micaceous minerals. (Photomicrograph by the author)

compressive stress. This process is also called *Riecke's principle*. It is based on the fact that the grains of a rock are at different stress levels and that there is a variation of solubility of the grains across the rock mass. Detailed studies (Williams 1972; Durney 1972, 1978; Elliott 1973; Gray and Durney 1979) suggest that pressure solution is an important mechanism for the formation of foliation in rocks as well as development of textures in metamorphic rocks.

The grains of a rock are under stress, which is compressive and non-hydrostatic. The grain contacts constitute areas of maximum stress and where grain dissolution may take place. The dissolved fluid may move to, and get deposited, in areas of low stress wherever available within the same rock. For example, signatures of pressure solution are commonly noticed in some rigid grains of garnet that show pressure shadows (Fig. 14.35) in the form of quartz grains aligned in the direction of extension.

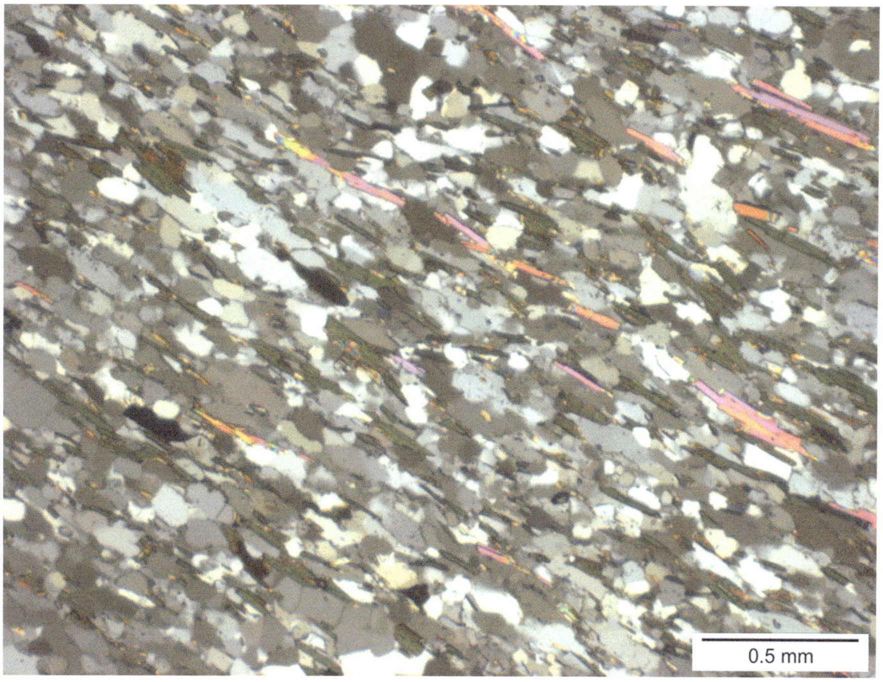

Fig. 14.33 Photomicrograph of a mylonite in which the foliation is given by stretched grains of quartz and micaceous minerals. (Photomicrograph by the author)

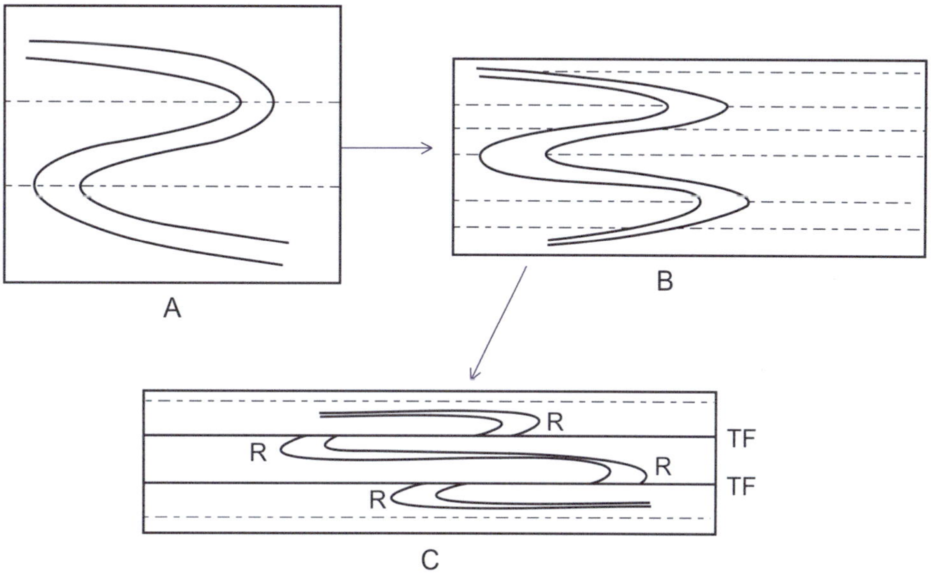

Fig. 14.34 Transposition foliation. (**a**, **b** and **c**) represent the various stages of formation of transposition foliation. *TF* transposition foliation; *R* rootless folds

Fig. 14.35 Signatures of pressure solution in the form of pressure shadows on either side of a garnet porphyroblast. The pressure shadows are developed in the extension direction. (Photograph by the author)

As a result of pressure solution, the fluid is diffused into the rock. The fluids may move short or long distances depending upon the available avenues. Commonly, they follow the grain boundaries, a process called *grain boundary diffusion*. In some cases, they may follow fractures, if present. Diffusion may also take place through the lattice of crystals as well.

In the *diffusion process*, quartz plays a major role because quartz grains have higher solubility than other minerals. As a result of flow of quartz, the flakes of mica are thrown into long stringers called *mica seams*. Continuation of this process leads to rotation of mica flakes of the mica seams towards the *XY*-plane of strain ellipsoid. This also explains the alternation of quartz-rich and mica-rich layers in many metamorphic rocks.

14.8.4 Grain Rotation

That mineral grains rotate during deformation of rocks is now an established fact. Rotation of garnets to give rise to snowball garnets in metamorphic rocks is a common example. This together with several other examples suggests that grain rotation is an important mechanism that produces foliations in deformed rocks. Three different models (Fig. 14.36) (see Twiss and Moores 2007) have been proposed to explain the formation of foliations as a consequence of grain rotation: the Jeffrey model, the March model and the Taylor-Bishop-Hill model.

The Jeffrey model (Fig. 14.36a) is applicable in cases where a mineral grain that behaves as a rigid object deforms in a ductile matrix under the influence of progressive simple shear. In such situations, the platy minerals develop a preferred orientation subparallel to the shear plane and perpendicular to the shear direction. The rotation of the platy mineral is slowest when it is parallel to the shear plane and fastest when it is perpendicular to the shear plane.

The March model (Fig. 14.36b) assumes that the randomly oriented platy grains of a rock act as passive markers and deform in the same way as the matrix. The platy grains tend to concentrate along the *XY*-plane and thus produce a foliation to the rock. With progressive homogeneous shear,

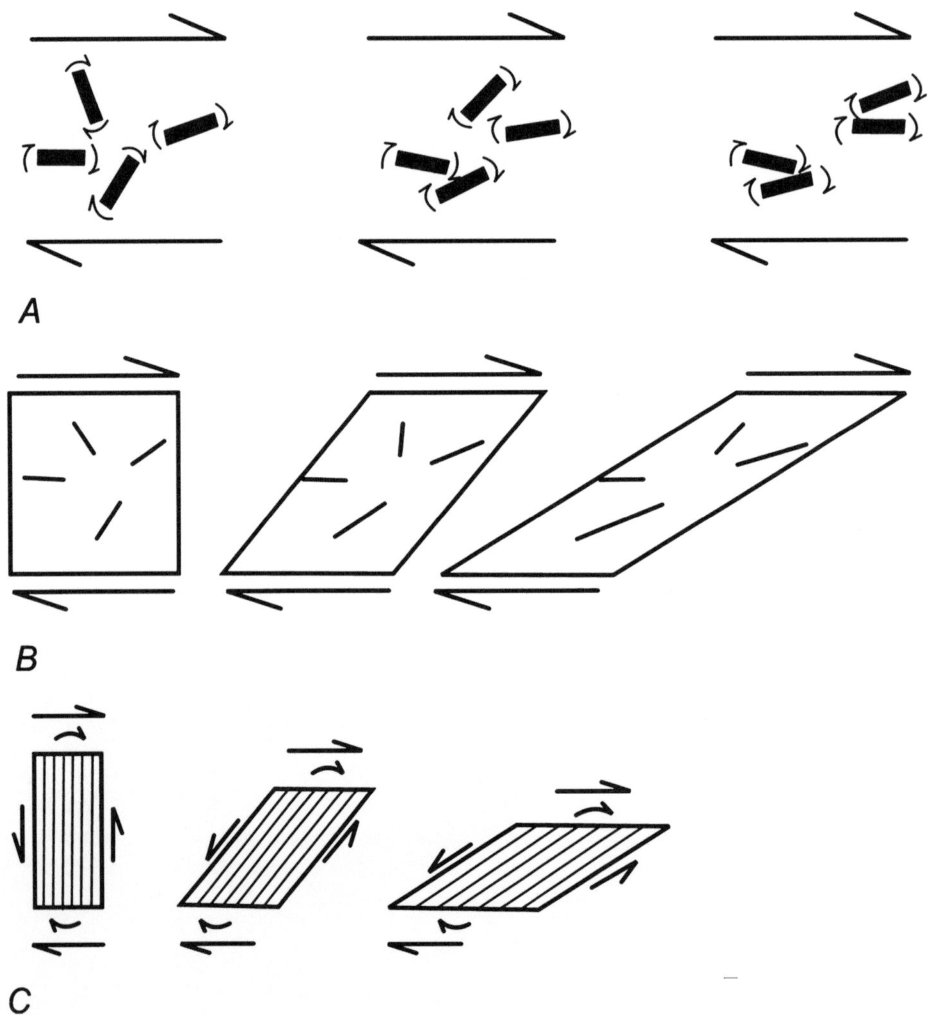

Fig. 14.36 Grain rotation as a mechanism for the formation of foliation under noncoaxial conditions. (**a**) The Jeffrey model. (**b**) The March model. (**c**) The Taylor-Bishop-Hill model. See text for details. (Reproduced from RJ Twiss and EM Moores, 2007, Structural Geology, second ed. Fig. 14.2 with permission from Cambridge University Press under Licence number 5217130843492)

the plane of flattening (*XY*) rotates towards the shear plane. Initially, i.e. at low strains, the angle between the platy minerals and the *XY*-plane is high. At advanced stages, i.e. at high strains, this angle becomes small when the foliation tends to become nearly parallel to the plane of flattening.

The Taylor-Bishop-Hill model (Fig. 14.36c) predicts that under noncoaxial shear condition, the crystallographic slip planes can rotate a grain. The rotation brings the crystal parallel to the plane of shear. The preferred orientation thus developed forms a foliation to the rock. This mechanism assumes that at very large strains, the crystals can rotate towards parallelism with the plane of flattening, but it does not take into consideration the externally applied shear.

14.8.5 Mineral Growth

It is believed that the formation of foliation in slates and schists is the result of syntectonic growth of minerals. The newly formed minerals tend to align with a preferred orientation under the ambient stress system, thus giving rise to a foliation to the rock. Some important aspects of foliation formation due to mineral growth, as arise from the work of a few workers (Williams 1972, 1977; Knipe and White 1977; Means 1977; Knipe 1981; Hobbs and Ord 2015), are highlighted below:

- In slates, the flakes of mica and chlorite grow in a direction parallel to the slaty cleavage during low-grade metamorphism.
- Major minerals of slates, such as quartz, chlorite and phengite, grow during metamorphic reactions.
- Preferred alignment of anisotropic minerals takes place under a stress field that is different from that of the surrounding minerals.
- In some slates, foliation is developed by crenulations of the bedding. In such rocks, while the flakes of phyllosilicates show parallelism with the axial planes of crenulations, those in the hinge zones show cross-cutting relations. All this suggests that the oriented growth of phyllosilicates is controlled by strain.
- In several metamorphic terrains, a gradual change from phyllites to schists is a common phenomenon. Also, the schists gradually grade from fine to coarser grained as the rock goes up from lower to higher grades of metamorphism. All this clearly indicates that phyllosilicates show a progressive growth in their size—a process called *mimetic crystallization*. The growth of phyllosilicates generally takes place along their length, and this accentuates the foliation.
- The foliation formation in slates and schists can thus be assigned, among others, to the growth of phyllosilicates under the combined action of deformation and metamorphism.
- Although mineral growth can satisfactorily explain the formation of foliation in slates and schists, the process ignores the role of some other processes that commonly go concurrently, such as grain rotation, shearing and pressure solution. As such, the foliation formation process(es) should be applied only to those cases where signatures of the latter processes are not observed.

In *ductile shear zones*, foliation formation owes much to the ductile flow of minerals. This is well demonstrated in mylonites and quartz-rich rocks of ductile shear zones all over the world. Ductile flow operates at moderate to high temperatures and leaves its signatures on mesoscopic, microscopic as well as lattice scales. Flow of silica under strong ductile deformation is a common feature of shear zones. The light-coloured silica layers alternate with those of dark-coloured minerals, and this imparts a strong foliation to the rock (Fig. 14.37).

At relatively lower temperatures, quartz grains in quartz mylonites are strongly flattened, thus forming quartz ribbons (Fig. 14.38). Quartz ribbons alternating with layers containing fine quartz produce a strong foliation to the rock that can be noticed both in hand specimens and in thin sections.

14.8.6 Dynamic Recrystallization

At moderate to high temperatures, the most common meso- to microstructure is *mylonitic foliation* formed due to grain-size reduction and consequent reorientation of mineral constituents in parallelism with shear direction. The grain-size reduction occurs due to *dynamic recrystallization* (see also Chap. 16), a process in which highly strained, large grains release their stored internal energy by the formation of smaller, strain-free grains. Formation of smaller grains makes the rock weak (strain softening). The outlines of the large grains may or may not be seen depending upon whether the grains have been completely converted to a mosaic of smaller grains or not. In cases where the rock has undergone high degree of dynamic recrystallization, it is sometimes difficult to identify the (mylonitic) foliation. Here, we present two examples of how to identify foliation in such rocks. In some cases, the outlines of the large, flattened grains can be identified (Fig. 14.39). In other cases, the boundary of the flattened grains is vaguely represented but can be easily demarcated due to the presence of some different minerals at its contact (Fig. 14.40). In both these cases, the direction of flattening gives the foliation. This also gives a mylonitic foliation to the rock.

Fig. 14.37 Polished section of a mylonitic gneiss showing foliation developed due to ductile flow of minerals. The flow bands are given by silica layers that alternate with layers containing dark-coloured minerals including garnet. All this produces a strong foliation to the rock. (Photograph by the author)

Fig. 14.38 Thin section of a quartz mylonite showing foliation due to the formation of a quartz ribbon. (Photomicrograph by the author)

At much higher temperatures, intra-crystalline processes operate causing lattice deformation. This imparts some specific crystallographic orientation (see Chap. 17, Sect. 17.7) of the minerals but not any visible foliation to the rock.

14.8.7 Progressive Shear Deformation

Due to progressive shear deformation, the original fabric of any pre-existing rock may be reoriented to form continuous foliation. During progressive simple or pure shear, the elongate and planar minerals undergo reorientation to form foliation irrespective of whether the rock initially had a random fabric or oriented fabric (Passchier and Trouw 2005, p. 82). Also, the originally equidimensional grains of a rock can develop foliation due to progressive simple or pure shear; in this case, the foliation developed is defined by grain shape preferred orientation.

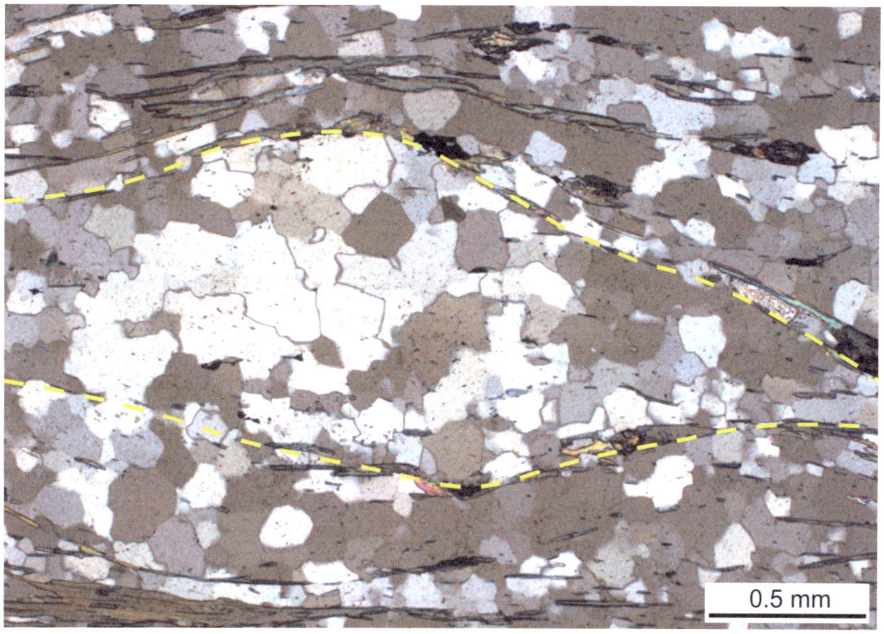

Fig. 14.39 Photomicrograph of a mylonite showing a large porphyroclast of quartz (traced with yellow dashes) that has undergone dynamic recrystallization by producing a mosaic of smaller strain-free grains. The boundary of the porphyroclast (running left-right) defines the direction of flattening of the grain which, consequently, gives the foliation. (Photomicrograph by the author)

Fig. 14.40 Photomicrograph of a mylonite showing a large porphyroclast of quartz that has undergone dynamic recrystallization by producing a mosaic of smaller strain-free grains. The boundary of the porphyroclast can be easily demarcated by the presence of some different minerals. The orientation of the long axis (left-right) of the porphyroclast thus identified defines the direction of flattening of the grain which, consequently, gives the foliation. (Photomicrograph by the author)

14.8.8 Metamorphism

Metamorphic processes are believed to be primary causes for the formation of foliation to a pre-existing sedimentary rock. The best example is *shale*. With increase of temperature and pressure, together with the influx of fluids, shale under conditions of prograde metamorphism progressively gives rise to slate, phyllite, schist and gneiss. Due to the role of pressure, the minerals tend to align in a direction perpendicular to the maximum compressive stress, while due to the role of temperature the rock undergoes mineralogical and geochemical changes. Influx of fluids acts as a catalyst to the system due to which shale is progressively converted to a *slate*, *phyllite* and *schist*. With further increase of temperature and pressure, the rock undergoes further compositional reconstitution such that there is segregation of light- and dark-coloured minerals in the form of alternating bands, thus producing *gneiss*. In all these cases, foliation is defined by the preferred orientation of fine-grained minerals in slate or phyllite, by coarse-grained minerals in schist and by alternating bands of light- and dark-coloured minerals in gneiss. Further, in the newly formed rocks, the term foliation is also called *slaty cleavage* for a slate, *schistosity* for a schist and *gneissosity* for a gneiss. To sum up, the prograde metamorphism can itself be assigned a mechanism of formation of foliation.

14.8.9 Metamorphic Differentiation

Metamorphic differentiation is the process by which minerals or mineral assemblages from an originally homogeneous rock get separated during metamorphism and accumulate as different layers. The segregated layers are commonly monomineralic and, together with the host rock, constitute a new metamorphic rock. Metamorphic differentiation is commonly observed in high-grade schists and gneisses. Development of gneissosity in gneissic rocks is often attributed to this process. Occurrence of veins and lenses of quartz or calcite is also believed to have developed due to this process. Due to later deformation, mainly shearing, these veins get disposed parallel to schistosity.

Box 14.1 Can Foliation Formation Reach a Steady-State Condition?

Rocks show a variety of foliations that are formed by various processes. Often, questions are raised on whether the foliation-forming processes are independent of each other or are dependent upon other related processes. Means (1981) opened up a discussion on whether foliation formation can ever reach a steady-state condition. Later on, Twiss and Moores (2007, p. 412) raised the question of whether foliation formation is an evolving process (cyclic) or it reaches a steady-state condition. We highlight here some excerpts from these works as this is an important aspect of both structural geology and metamorphic petrology.

Field evidences in tectonically disturbed areas indicate that an early foliation given by sedimentary or tectonic processes evolves into a crenulation foliation. The latter, with increase of metamorphism, may eventually form another type of foliation given by

Box 14.1 (continued)

compositional differentiation defined by cleavage domain and microlithons. With further increase of recrystallization, another type of foliation, i.e. continuous foliation, may develop. Yet again, the continuous foliation may give rise to crenulation foliation.

The above evidences suggest that foliation formation is possibly cyclic. If so, now the question arises as to whether the formation of foliation from one type to another is a continuous process, or a steady-state condition appears when this process stops once a particular type of foliation is formed. In such cases, a steady-state foliation can be visualized as one formed under the steady-state condition, i.e. when the flow (deformation) occurs under constant stress and constant strain-rate condition.

Although the concept of steady-state foliation seems to be a conceptual one, Ree (1991) produced experimentally a steady-state grain-shape foliation by deforming octachloropropane (OCP) by simple shear. He observed that the deformation history passed through foliation-strengthening and foliation-weakening processes before a steady-state foliation reached.

Assuming that rocks often reach steady-state flow during their deformation history, it is interesting whether formation of tectonic structures also follows suit or not. It is therefore too premature to say whether foliation formation is also cyclic or acquires a steady-state condition. This situation is something like putting the horse before the cart or the cart before the horse!

14.8.10 Crystal-Plastic Deformation

Crystal-plastic deformation involves deformation of lattice structures of minerals. It operates at higher temperatures by solid-state diffusion or dislocation creep. As a result, the mineral flattens along the *XY*-plane. The mechanism leaves signatures that are visible under a microscope such as deformation bands. Although the rock does not show any visible foliation on the mesoscopic or hand specimen scale, the mechanism may give rise to a preferred orientation to the minerals.

14.8.11 Mineral Nucleation

Hobbs and Ord (2015) suggested *mineral nucleation* as a prime mechanism for the development of mineral preferred orientation in metamorphic rocks. According to them

(p. 313), during metamorphism, new minerals nucleate and begin to grow once the affinity of the relevant mineral reaction becomes large enough. The basic principle is that during metamorphism, there is a competition between the energy of the growing volume of the new grain and the increasing energy of the surface energy. Once a critical grain size is reached, a decrease of Gibbs energy of the grain begins. At this stage, the grain is now free to grow. Anisotropy of the interfacial energies between the growing grain and the neighbouring grains may promote growth of a crystal in a solid medium.

14.8.12 Mineral Growth

Mineral growth, in addition to mineral nucleation, has been suggested by Hobbs and Ord (2015) as a prime mechanism for crystal shape preferred orientation during metamorphism. This is based on the observation that disconnections existing in many materials may constitute an important process of *grain growth*. The starting point is the step height that may be a single-unit cell of a crystal. In the presence of a stress field (Fig. 14.41), the grain boundary migration as controlled by the migration of steps (disconnections) constitutes a strong deformation mechanism as well as a preferred growth selection process. In general, the major driving forces for grain growth during a chemical reaction include (i) grain boundary migration driven by the chemical and/or thermally derived affinity of the reaction and (ii) motion of disconnections driven by the applied stress. According to the authors (p. 316), if the motion caused due to affinity is larger than that caused by stress, no shape preferred orientation develops. If, on the other hand, the motion due to stress is larger than that caused by affinity of the reaction, a crystal shape preferred orientation develops (Fig. 14.41).

Mineral growth has been suggested as a powerful mechanism for the development of preferred orientation of micas as seen in foliated rocks. Minerals showing other habits such as kyanite or sillimanite, on the other hand, may develop a mineral lineation.

14.9 Significance of Foliation

14.9.1 Academic Significance

- Study of foliation is important because it unravels the structure of a rock, especially in hand specimens and in thin sections.
- Foliation is related to structures of all sizes ranging from large scale to microscopic. As such, it provides a tool to understand the structure to geologists working in field as well as in the laboratory.
- Foliation bears strain significance because its orientation is always related to strain axes.
- In deformed rocks, careful study of foliation helps in tracing the deformation history especially when studied under microscope.

14.9.2 Economic Significance

- Foliations may serve as channels for passage of groundwater.
- During civil engineering and construction work, rocks showing strong foliation or no foliation help in rejecting/selecting suitable rocks or sites for construction.
- Presence of foliation sometimes makes a rock decorative in which case it bears economic values.

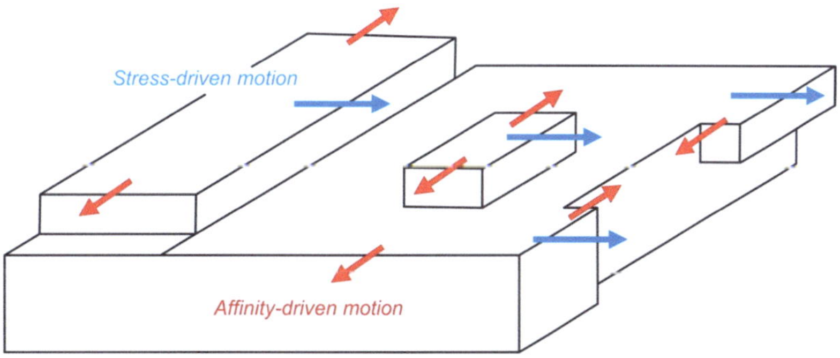

Fig. 14.41 Diagrammatic representation of the driving forces for grain growth during a chemical reaction. A grain shows two major types of driving forces: (**a**) grain boundary migration (red arrows) driven by the chemical and/or thermally derived affinity of the reaction and (**b**) motion of disconnections (blue arrows) driven by the applied stress. A crystal shape preferred orientation develops if the motion driven by stress (i.e. **b**) is larger than that driven by the affinity of the reaction. If the motion arising from affinity is larger than that driven by stress, no shape preferred orientation develops. (Reproduced from Hobbs and Ord 2015, Fig. 9.23 with permission from Elsevier Copyrights Coordinator, Edlington, U.K. Submission ID: 1198473)

14.10 Summary

- *Foliation* is a planar structure showing preferred orientation of minerals with a platy or tabular habit. The preferred orientation is produced by deformation and is uniformly pervasive in a rock.
- Foliation is developed in metamorphic rocks, and it includes cleavage, schistosity, gneissosity and gneissic banding.
- A foliated rock shows two distinct domains: (a) *foliation domain* or *cleavage domain* in which the original fabric of the rock is significantly changed and shows parallel fabric shown by the alignment of mineral constituents and (b) *microlithon domain* showing little or no alteration of the original fabric of the rock. These two domains alternate, and the spacing between them varies with different rocks.
- *Cleavage* or *rock cleavage* includes a set of fractures along closely spaced, parallel surfaces in a rock (usually low-grade metamorphic rock) formed by the alignment of various mineralogical and structural elements during metamorphism and deformation. Along the cleavage, a rock shows the ability to split or cleave into parallel or subparallel surfaces.
- Foliation commonly forms normal, or near-normal, to the direction of maximum shortening. Foliation-forming processes can be considered to be those that give rise directly or indirectly to preferred orientation of components of a rock.
- Common causes/processes that develop foliation in rocks include folding, ductile shearing, pressure solution, grain rotation, mineral growth, dynamic recrystallization, progressive shear deformation, metamorphism, metamorphic differentiation, crystal-plastic deformation, mineral nucleation and mineral growth.

Questions

1. What do you mean by foliation and cleavage? Is there any difference between the two?
2. What are foliation domain and microlithon? How would you distinguish between the two?
3. How would you distinguish a slate from a phyllite?
4. What is continuous cleavage? Name some common rocks showing this type of cleavage.
5. What is spaced cleavage? Give its types.
6. Describe with the aid of diagrams the various types of foliations as related to folds.
7. Describe some common types of foliation occurring in ductile shear zones.
8. Explain how pressure solution gives rise to formation of foliation.
9. Describe the formation of foliation due to metamorphism.
10. Give the significance of foliation.

15 Lineation

A. R. Bhattacharya, *Structural Geology*, Springer Textbooks in Earth Sciences, Geography and Environment,
https://doi.org/10.1007/978-3-030-80795-5_15

Abstract

Any fabric or orientation in a rock developed in a linear fashion as a result of tectonic deformation is called *lineation* or *linear structure*. Though lineation is commonly developed on the surface of rocks, it may also penetrate to a small extent, up to a few millimetres only. As such, lineation is broadly grouped as *non-penetrative lineation* such as slickensides, slickenlines and slickenfibres and *penetrative lineation* such as mineral lineation, crenulation lineation, intersection lineation, boudins, mullions, rods, pencil structure and pressure shadows. Lineation can form by processes that can be metamorphism dominated, deformation dominated or geometrically controlled, though it is difficult to draw any specific line of demarcation between these groups of processes. Study of lineation is important in structural geology as some lineation types constitute structural markers to relate smaller structures with the larger ones. Some other types, such as intersection lineation, given by a folded layer and an axial-plane foliation indicate the orientation of fold axis. This chapter describes the various types of non-penetrative and penetrative lineation, lineation as a tectonic fabric, genesis and significance of lineation.

Keywords

Lineation · Non-penetrative lineation · Penetrative lineation · L-tectonite · S-tectonite · LS-tectonite · Genesis of lineation

15.1 Introduction

Lineation or *linear structure* refers to any fabric or orientation developed in a linear fashion. It is a common feature noticed in deformed rocks and is commonly seen on the upper surfaces of rocks (Fig. 15.1). Lineation can develop in any rock of any origin. In some sedimentary rocks, for example, parallel orientation of grains or components of the rock may develop due to sedimentary processes such as flow of water. In an igneous rock, likewise, the mineral constituents in a flowing mass of magma may also be aligned in a particular direction, thus constituting a type of lineation. In metamorphic rocks, some minerals or mineral aggregates formed during metamorphic processes may also be aligned in a preferred direction because metamorphic processes involve both temperature and pressure.

The above examples indicate that any rock of any origin may develop some type of linear orientation of its fabric or its mineral constituents. All these examples, in fact, fall in two categories. In one category, the features are formed at the same time when the rock was formed such as seen in some sedimentary rocks and in some igneous rocks. The linear features developed in this category are grouped as *primary lineation*. In the other category, in which the linear features are formed after the formation of the rocks in the geological timescale, these are grouped as *secondary lineation*.

In common usage, primary lineation is considered to be restricted to sedimentary and igneous rocks while secondary lineation is considered to be restricted to deformed rocks or to metamorphic rocks in which the components of the rock have

Fig. 15.1 Lineation developed on the foliation surface of a quartzite exposed in the Børgefjell Window, central Scandinavian Caledonides, Norway, in a basement-cover horse. The lineation is parallel to the regional thrust transport direction. (Photograph courtesy Professor Reinhard Greiling)

undergone preferred orientation under the influence of stresses acting during tectonic deformation. In structural geology, we deal with secondary lineation only. Although the terms lineation and linear structure are used in the same sense, we shall use the term *lineation* only in this book.

15.2 Types of Lineation

Lineation was possibly first studied in detail by Cloos (1946) who described several types of lineations. Thereafter, the literature continued to be flooded with reports of such structures all over the world. As a result, lineation is now considered to be a common structure of deformed rocks.

Lineation occurs in rocks in a variety of ways and forms. As yet, no unanimous scheme of classification for lineation is available. Lineation can be a non-penetrative as well as a penetrative structure. Lineation can develop as a result of tectonic deformation as well as due to formation and consequent linear alignment of new minerals or mineral aggregates. Lineations are also observed in rocks deformed under plastic conditions as well as under brittle conditions.

In this work, we classify lineations under two broad categories: non-penetrative and penetrative.

15.3 Non-penetrative Lineation

Non-penetrative lineations are those that are noticed on the rock surfaces only. The most common occurrence of such lineations is along fractures where mineral growth takes place. The minerals thus developed are elongated in the direction of slip or movement along the fracture. Non-penetrative lineations are formed under brittle regime and are therefore commonly noticed in rocks deformed under upper crustal conditions. Common non-penetrative lineations include slickensides, slickenlines and slickenfibres as described below.

15.3.1 Slickensides

Slickensides are polished striated linear features that trend parallel to the fault motion. These are developed along the fault surface of all types of faults (Figs. 15.2 and 15.3) and may occur in any rock affected by a fault. The structure is formed due to frictional movement of two blocks along a fault and occurs as small grooves that rise on millimetre scales or even smaller and can be seen only when the upper block is not exposed. These grooves alternate with depressions oriented parallel to the adjacent grooves, and thus the structure is constituted of a set of parallel grooves and depressions (Fig. 15.3).

Slickensides usually constitute signatures of the last fault motion as the earlier slickensides may be overprinted by the last motion. Although slickensides indicate the direction of movement, the sense of slip of the fault is sometimes difficult to ascertain. This can be done only when the structure shows the presence of some steps or sudden depressions that are commonly oriented perpendicular to the slickenside grooves (Fig. 15.4). In such cases, the sense of slip of the fault is directed towards the descending side of the steps. However, one must be careful as there may also be anti-steps (Professor J.P. Burg, personal communication 2020). The depressions are commonly too small to be clearly seen, but one can 'feel' only when a finger is rubbed along the structure.

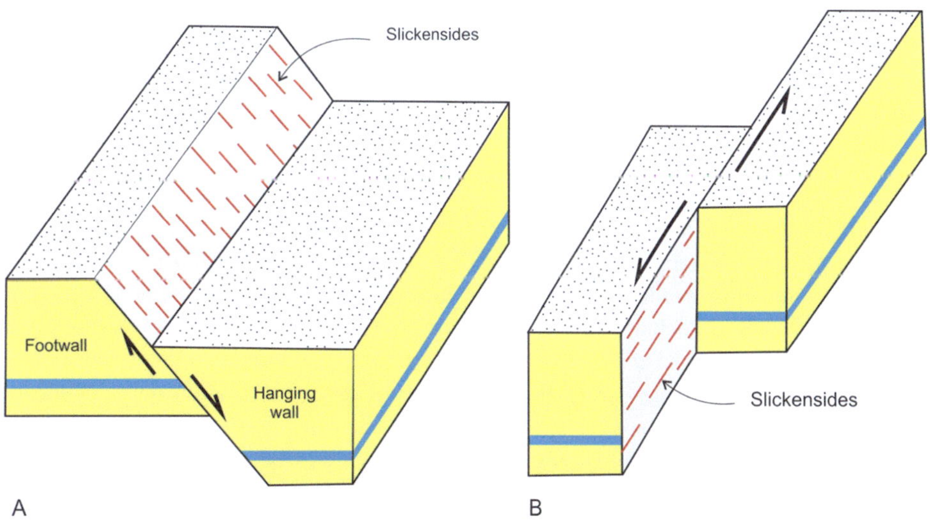

Fig. 15.2 Slickensides as developed along a fault surface. (**a**) Normal fault. (**b**) Strike-slip fault

Fig. 15.3 Field photograph of slickensides as they occur in a set of parallel grooves and depressions developed in the quartzite of Krol Group, south of Bhimtal, Kumaun Lesser Himalaya, India. (Photograph by the author)

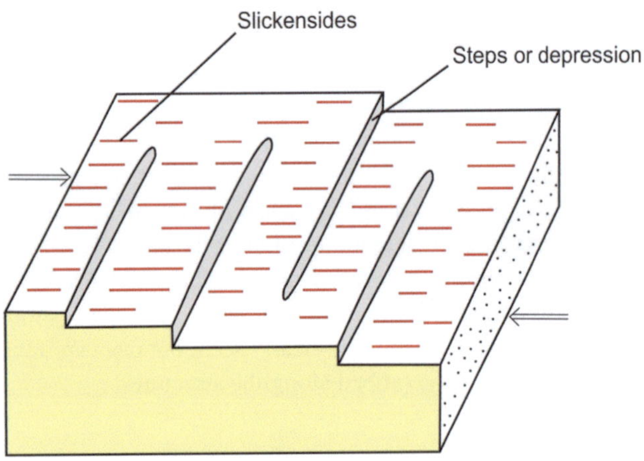

Fig. 15.4 Slickensides developed on a fault surface show steps or depressions that are oriented perpendicular to the slickenside grooves. The steps descend towards the right-hand side, which thus indicates the direction of slip of the fault

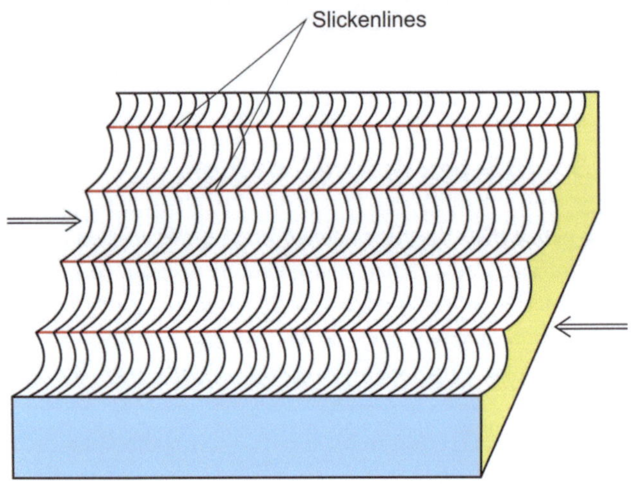

Fig. 15.5 Diagrammatic sketch of slickenlines as developed along a fault surface

15.3.2 Slickenlines

Slickenlines are the individual grooves of a slickenside that occur as linear scratches developed on a fault surface (Fig. 15.5). In other words, slickensides can be considered to be composed of, or include, slickenlines. Fault surfaces are not smooth but are rugged and show irregularities which, due to rubbing of two blocks along the fault surface, form grooves that are individually called slickenlines that can form by scratching of some components of either the hanging wall over the footwall or vice versa. Occurrence of slickenlines indicates the direction of fault movement, but deduction of the sense of movement is not always possible.

15.3.3 Slickenfibres

Slickenfibres are fibrous crystals formed during the frictional movement along a fault and are therefore oriented parallel to the direction of motion along a fault surface. The preferred orientation of the crystals usually marks the direction of extension during fault motion. Common minerals that can form slickenfibres include quartz, calcite, gypsum, chlorite or ferruginous minerals. When fully developed, the slickenfibres show a steplike pattern (Fig. 15.6), and the sense of movement is given by the direction in which the steps show down-dip pattern.

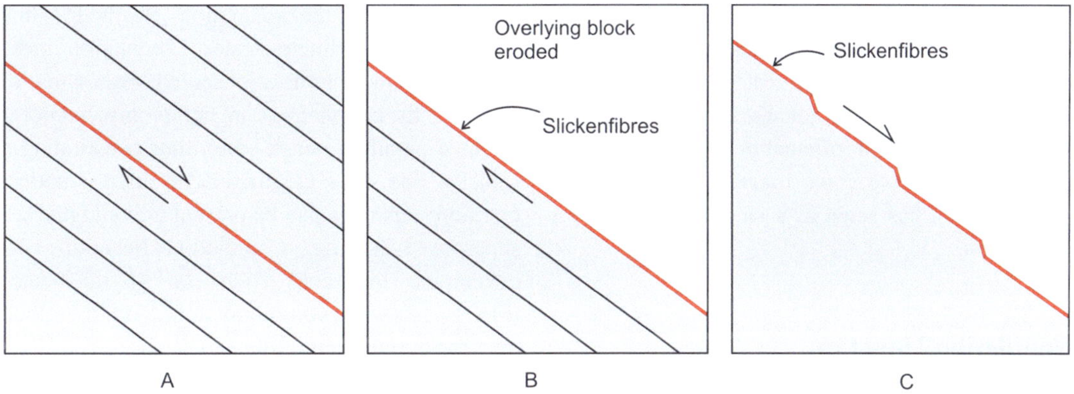

Fig. 15.6 Slickenfibres. (**a**) A fault (red line) develops between two layers. (**b**) The overlying block is eroded, thus exposing the slickenfibres. (**c**) The steplike pattern is shown by the slickenfibres, and the sense of movement is given by the direction in which the steps show down-dip pattern

Fig. 15.7 Lineation developed on the foliation surfaces in a mylonitic quartzo-feldspathic orthogneiss, which is subsequently folded with the foliation. Darker layers are relatively rich in biotite. The rocks are from the basal shear zone of the Seve Nappe, the major metamorphic nappe in the central Scandinavian Caledonides exposed on road to Stekenjokk mine, Sweden. (Photograph courtesy Professor Reinhard Greiling)

15.4 Penetrative Lineation

Penetrative lineations are those that extend within a rock. Most geologists consider penetrative lineations as mesoscopic structures. However, there is no precise scale to distinguish these two types of lineations (penetrative lineation and mesoscopic structure) as a few are reflected on microscale also, e.g. crenulations and rotated grains. Lineations themselves may undergo later deformation such as folding (Fig. 15.7). In this work, we include the following types of penetrative lineations.

15.4.1 Mineral Lineation

Mineral lineation is the preferred orientation of a single mineral grain or an aggregate of several minerals. The minerals generally have one axis much longer than the other two. As such, mineral lineation is commonly given by acicular, bladed, prismatic, fibrous and any linear shapes (Fig. 15.8) that can undergo a preferred alignment during deformation. The preferred orientation gives the direction of extension. Mineral lineation may occur in one or several sets (Figs. 15.9, 15.10 and 15.11).

Often, two or more sets are seen to be intersecting each other (Figs. 15.10 and 15.11). Common minerals that give rise to mineral lineations are hornblende, kyanite, sillimanite, chlorite, phyllosilicates and some mineral aggregates.

15.4.2 Lineation Given by Pebbles, Etc.

There are several cases when some components of a rock such as pebbles (Fig. 15.12), clusters of mineral grains, ooids and reduction spots get oriented in a parallel fashion during deformation. Preferred orientation of these components gives a lineation to the rock. We consider lineation given by all such components under one single category. In case of clusters of mineral grains, the individual grains in the clusters may not show any preferred orientation. In fact, during deformation, the grains have come together to form some clusters, each of which has acted as a single unit to give a preferred orientation.

15.4.3 Crenulation Lineation

Crenulation lineation occurs in the form of a set of minor folds or micro-folds of the pre-existing foliation (Fig. 15.13).

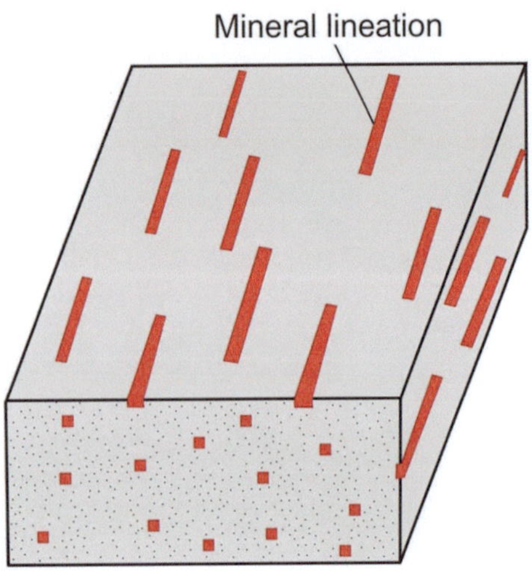

Fig. 15.8 Block diagram showing one set of mineral lineation

The amplitudes of the crenulations, i.e. the fold hinges, are on millimetre to centimetre scales. Crenulation lineation is also noticed in thin sections as observed under a microscope. The hinges of the crenulations are commonly, but not necessarily, oriented parallel to each other, thus constituting a lineation. Usually, one set of crenulation lineation is noticed in rocks, but more sets may also be present as conjugate sets or due to superposed folding. Crenulation lineation is commonly developed in rocks rich in phyllosilicate minerals (Fig. 15.14) and is commonly noticed in low-grade metamorphic rocks such as phyllite and schists.

15.4.4 Intersection Lineation

Since the intersection of any two planes gives a line, the intersection of any two planar features of a rock gives a linear feature, i.e. lineation. Thus, a variety of combination of intersection of two planar features of a rock is possible. Some common types of intersection lineations (Fig. 15.15) include the following: (a) Intersection of bedding and cleavage gives a lineation on the bedding surface (Fig. 15.15a) as well as intersection of bedding with the ground surface (Fig. 15.16). (b) Intersection of bedding and cleavage gives a lineation on the cleavage surface. Since the bed is seen on the cleavage surface as colour stripes, this type of lineation is also called *striping lineation* (Fig. 15.15b). (c) Intersection of axial plane of a fold with the limb of a fold produces a lineation (Figs. 15.15c,d and 15.17). Since the line thus formed also constitutes the axis of the fold, the latter (fold axis) is commonly considered as a lineation.

Fig. 15.9 Sericite schist showing one set of lineation. Lineation (L) is indicated by yellow line. (Sample courtesy Geology Museum, University of Lucknow)

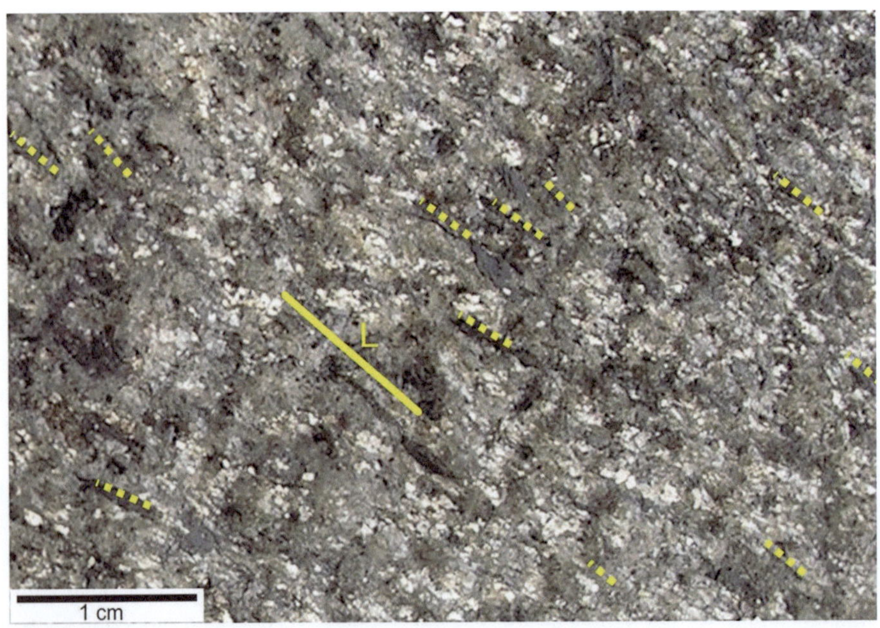

Fig. 15.10 A shale bed showing two sets of lineations (L_1, L_2) almost at right angles to each other. Lineation is indicated by yellow lines. (Sample courtesy Geology Museum, University of Lucknow)

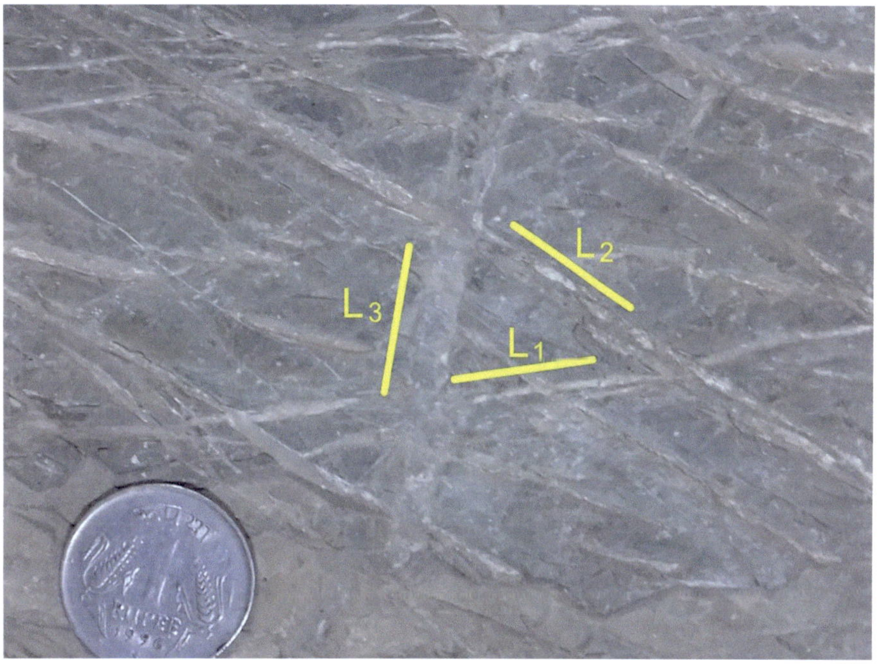

Fig. 15.11 Photograph showing three sets of lineations (L_1, L_2, L_3). Note that L_1 and L_3 lineations are weakly developed as compared to L_2. Lineation is indicated by yellow lines. (Sample courtesy Geology Museum, University of Lucknow)

15.4.5 Mullions

The term *mullion* is used for corrugated structures showing alternating convex and cusp surfaces developed at the interface of competent and incompetent layers (Fig. 15.18). Individual unit of the structure looks like a convex pillar with a flat base with cross section of a few centimetres and lengths measurable in several centimetres to a few metres. With this setting, the high-viscosity (competent) lower layer is able to buckle up into the low-viscosity (incompetent) upper layer, and by contrast the cusps protrude into the competent layer. The process thus produces a set of convex pillars alternating with cusps, and all these occur parallel to each other with a common flat base. Mullions are formed from the country rocks (Wilson 1953) and are commonly developed at the interface of quartzite and argillite or phyllite. These may occur associated with or without some other structures.

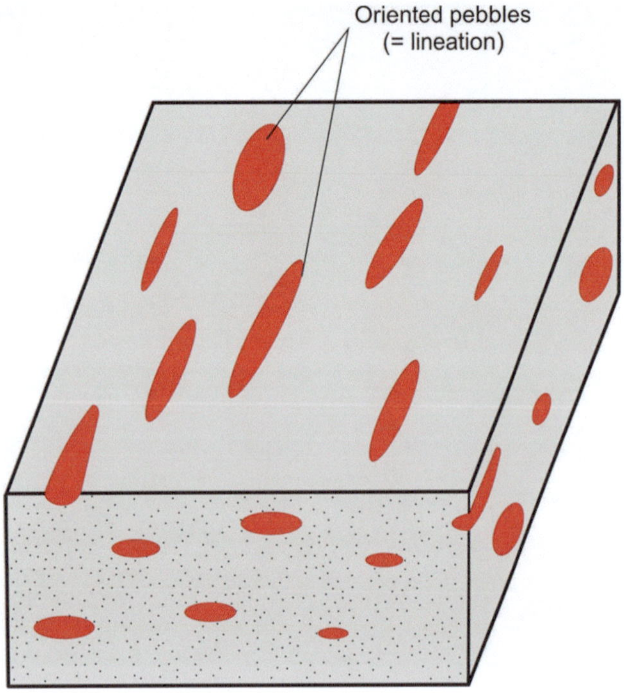

Fig. 15.12 Block diagram showing lineation given by pebbles

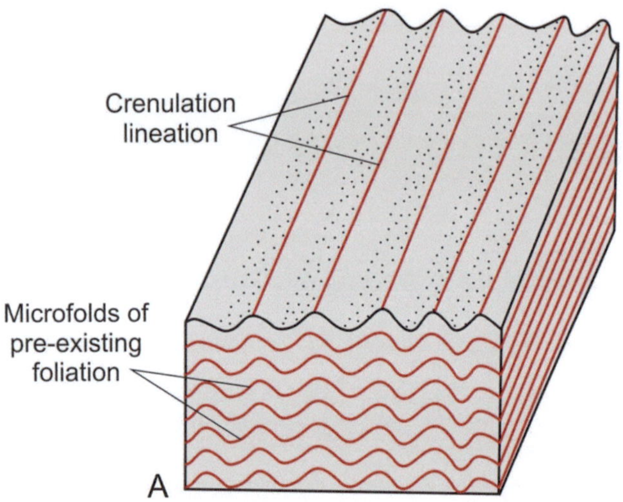

Fig. 15.13 Block diagram showing crenulation lineation

Wilson (1953, 1961) identified the following types of mullions: (a) *Fold mullions* include a set of narrow fold hinges developed in the competent layer, and their convexity is directed towards the incompetent layer. Formation of fold mullions could thus be related to layer-parallel shortening during deformation. Fold mullions are commonly noticed in sandstone alternating with shale (Fig. 15.19) or quartzite alternating with phyllite. (b) *Cleavage mullions* look similar to fold mullions and are commonly developed on the cleavage surface. (c) *Irregular mullions* are also cylindrical columns with an irregular cross section and are covered with a thin layer of mica. The columns commonly overlap with the neighbouring ones. Structures similar to irregular mullions are also sometimes noticed on the fracture surfaces in which case these have been named as *fault mullions* by Twiss and Moores (2007).

15.4.6 Rods

Rods, or *roddings*, are cylindrical columns (Figs. 15.20 and 15.21) composed of mineral aggregates that commonly occur in mica schists and gneisses. Generally, rods are monomineralic (quartz, calcite, feldspar, etc.), but in most cases these are constituted of irregular grains of quartz or vein quartz, and as such these are easily identified due to their white or light colour. Rods often look like mullions but are easily distinguished by their typical white or light colour. While mullions are composed of the same material as the host rock, rods are composed of different materials. Lengthwise, rods may extend for several tens of centimetres with a diameter of 1–2 cm. With their typical mode of occurrence, rods constitute a prominent lineation that represents the direction of extension during deformation. Rods are believed to be products of metamorphic segregation during deformation, or they may represent isolated fold hinges. When they occur in folded rocks, rods are parallel to the fold hinges.

15.4.7 Pencil Structure

The term *pencil structure* (Fig. 15.22) was initially used by Cloos (in: Engelder and Geiser 1979) to refer to the intersection of bedding and cleavage that facilitates the making of slate pencils. Pencil structure is commonly believed to form due to intersection of two cleavage planes, or a foliation plane and a cleavage. The intersection produces columns, i.e. pencil structures that may show rectangular or prismatic cross sections. The intersection thus produced promotes the rock to break along elongate columns. If the cleavage is an axial-plane cleavage, the lineation given by the pencil structure is nearly parallel to the associated fold axis. Pencil structure is generally developed in weakly deformed rocks or in weakly metamorphosed rocks.

15.4.8 Boudin

Boudins are sausage-shaped fragments of a competent rock layer that has been stretched parallel to layering. The term *boudinage* (Fig. 15.23) is commonly used to refer to the process of formation of boudins. Boudins are developed in a rock layer that behaves as a brittle substance or has very low

Fig. 15.14 Field photograph of crenulation lineation developed in the shale of Southern Appalachians, USA. (Photograph by the author)

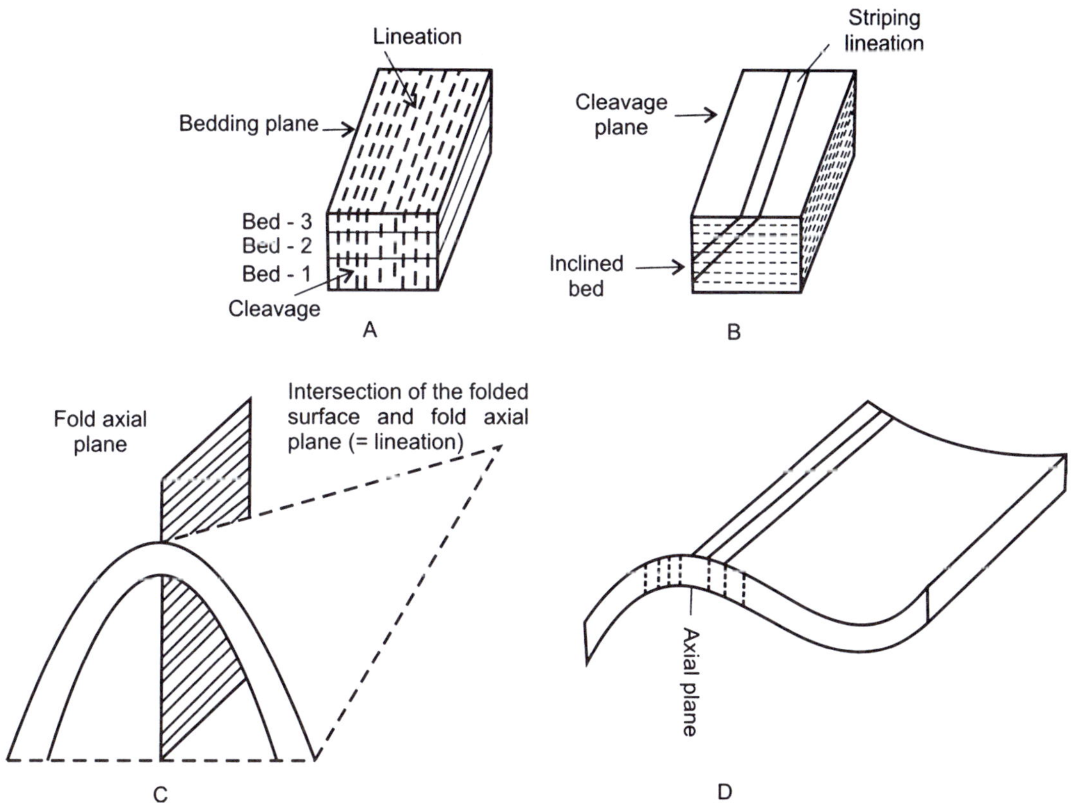

Fig. 15.15 Intersection lineations. (**a**) Intersection of bedding and cleavage gives a lineation on the bedding surface. (**b**) Intersection of bedding and cleavage gives a lineation on the cleavage surface, also called striping lineation. (**c**) Intersection of axial plane of a fold with the limb gives a lineation. (**d**) Lineations developed parallel to the axial plane of a fold

Fig. 15.16 Intersection lineation given by intersection of vertical shale beds with the ground surface. Loc.: Near Leh, Ladakh Himalaya, India. (Photograph courtesy Narendra K. Verma)

Fig. 15.17 Intersection lineation (indicated by the pen) formed due to intersection of axial plane of a fold with the folded surface

ductility while the surrounding layers have high ductility. During deformation, the brittle layer is thus pulled apart and is broken into several fragments during extension while the surrounding layers undergo ductile deformation. The individual boudins extend as linear objects and thus constitute a linear structure. Boudins are sometimes confused with *pinch-and-swell structures* (Fig. 15.24). The difference is that of continuity of the competent layer: in boudins, the continuity of the individual boudins is broken, while in pinch and swell the continuity is maintained.

Boudinage structure can be described with a few geometric elements (Fig. 15.23) (Wilson 1961; Goscombe and Passchier 2003; Goscombe et al. 2004; Twiss and Moores 2007): *Length* of a boudin or *boudin axis* is the maximum length of the structure. *Boudin line* is the line that separates two adjoining boudins. *Separation* is the distance between two adjacent boudins as measured in the transverse section. *Neck folds* or *scar folds* are the locally developed bends of the surrounding ductile layers within the zone of separation of two adjacent boudins. Neck folds are also sometimes described as *flanking structures* (Passchier and Trouw 2005). *Width* is the distance measured at the centre of the boudin along the layering. *Thickness* is the distance across the boudin. *Aspect ratio* is the ratio between the width and thickness of a boudin.

Boudins commonly form due to layer-parallel extension in one direction (Fig. 15.25). Boudins can also form due to layer-parallel extension in two directions (Fig. 15.26) that produce a number of three-dimensional segments called *chocolate boudinage* or *chocolate-tablet boudinage*. Often, the chocolate-tablet boudinage structures undergo faulting at some later stages such that the individual boudins show step faulting (Fig. 15.27).

Sometimes, strongly foliated rocks show the development of boudins even though there is no apparent contrast in competence between the segmented layer and the matrix. Such boudins are called *foliation boudins* that are distinguished from the normal boudins by their non-periodical occurrence and separation by fractures that are occasionally filled with secondary minerals.

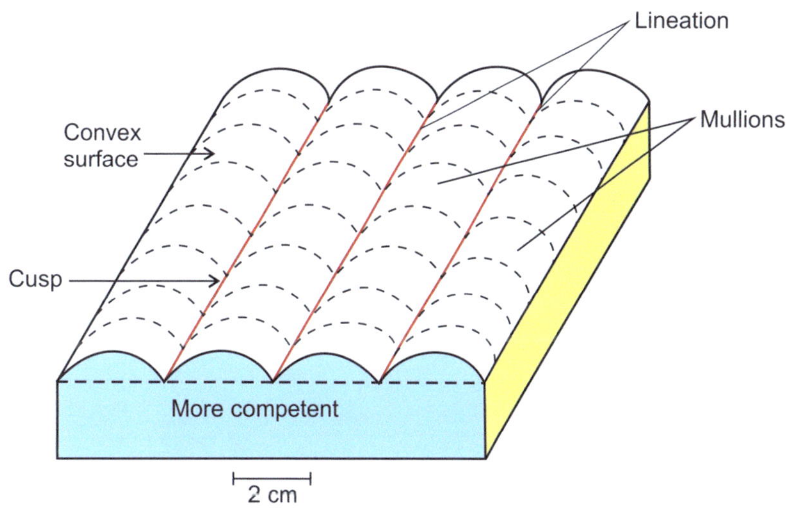

Fig. 15.18 Mullions as seen on the ground surface. See text for details

Boudins show a variety of shapes in transverse section. The different shapes reflect the degree of competence contrast between the boudinaged layer and the host. Recently, Zulauf et al. (2020) carried out scaled analogue models using non-linear viscous plasticine as rock analogue to investigate the influence of varying layer obliquity on the geometry of single-layer folds and boudins under bulk coaxial plane strain. They observed that boudins develop by combined necking and tensile fracturing at $\theta_{Z(i)} > 60°$. ($\theta_{Z(i)}$ is the initial angle between shortening axis and competent layer.) Because of high finite strain and related layer rotation, the boudins formed are symmetric. They suggest that the driving mechanism for boudinage formation is the transfer of stress from the matrix to the layer, and the thickness of the competent layer controls the geometry of boudins.

15.4.9 Stretching Lineation

The term *stretching* lineation is used in a broad sense to include lineations produced by stretching of minerals or mineral aggregates during deformation. This type of lineation is commonly developed in gneisses and in other metamorphic rocks in shear zones. Stretched recrystallized grains of quartz (Fig. 15.28), feldspar or quartzo-feldspathic aggregates commonly produce a penetrative lineation in mylonites. Highly stretched conglomerates also produce a penetrative lineation of this type. Rotation of fold hinges towards the direction of bulk shear in ductile shear zones (Bhattacharya and Siawal 1985; Verma and Bhattacharya 2015) also produces a stretching lineation that provides a shear sense marker.

15.4.10 Pressure Shadows

Pressure shadows are growth of some minerals such as quartz, mica and chlorite on opposite sides of a rigid, competent object such as garnet (Fig. 15.29), pyrite or magnetite. The growth of the minerals results in the formation of spindle-shaped objects that occur on the foliation surface, and their preferred alignment constitutes a lineation. The growth of the minerals in shadow zones is favoured by ductile deformation of the matrix materials.

Box 15.1 Lineations in Three-Dimensional Space

By lineation we mean any fabric or orientation developed in a linear fashion as a result of tectonic deformation. Pollard and Fletcher (2005, p. 77), however, considered that the geological contacts of two different geological units also constitute lineations as defined by the intersection of two geological surfaces that separate one volume of rocks from another. They cite three examples of Turner and Weiss (1963): (a) a fault that separates relatively young and undeformed sedimentary rocks from older and more deformed sedimentary or metamorphic rocks; (b) an igneous contact that separates older deformed sedimentary or metamorphic rocks from the younger rocks of an igneous intrusion; and (c) an angular unconformity that separates older sedimentary or metamorphic rocks from the overlying sedimentary strata. In all these cases, the contacts are three-dimensional surfaces that are planar or curved. Pollard and Fletcher consider these intersections as lineations in three-dimensional space.

Fig. 15.19 Mullions developed in sandstone alternating with shale. (**a**) Cross section. (**b**) A general view. Loc.: North Eifel, Germany. Chris Hilgers happily placed his hammer on the structure for scale. (Photographs by the author)

15.5 Lineation as a Tectonic Fabric

Lineations in whatsoever mode they appear, i.e. whether as non-penetrative or penetrative structures, definitely constitute an important tectonic fabric of rocks as they result from deformation. Being commonly associated with local structures, lineations are believed to have local significance. Some lineations, such as stretching lineation, have regional significance especially when they are associated with large-scale ductile shear zones. Therefore, when we talk of

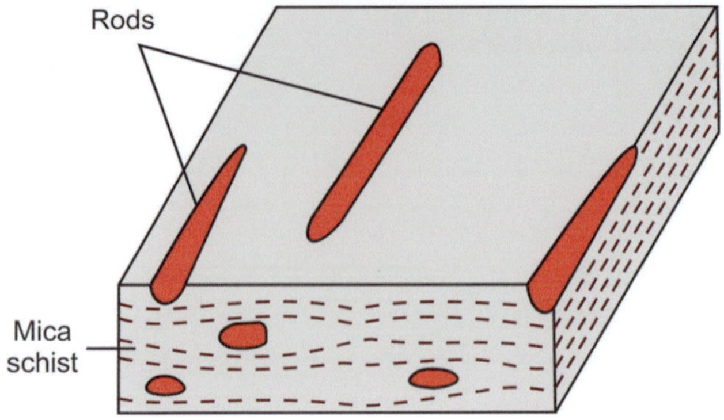

Fig. 15.20 Block diagram showing rods developed in a schistose rock

lineations as a fabric, it is better to refer to their mode of occurrence in a particular rock.

If a rock is dominated by lineation, it is called an *L-tectonite*. In such rocks, foliation is either absent or weak. L-tectonite suggests unidirectional stretching and is produced by both coaxial (pure shear) and noncoaxial (simple shear) deformation. These are penetrative structures formed under ductile conditions. An *S-tectonite* is a rock in which foliation is dominant, while lineation is either absent or very weak. S-tectonite suggests flattening and is produced by coaxial (pure shear) deformation. If a rock shows the development of both penetrative lineation and foliation, it is called an *LS-tectonite* (Fig. 15.30). An LS-tectonite is produced dominantly under shear deformation (noncoaxial) conditions. With the above definitions of L- and LS-tectonites, most geologists may not consider non-penetrative lineations such as slickensides or slickenfibres as lineations. The debate on this issue is still persistent. Nevertheless, the identification of L- and LS-tectonites makes lineation a tectonic fabric of deformed rocks.

The recognition of L- and LS-tectonites is highly qualitative, and there exists all gradation between these two types. Some geologists therefore assign some specific nomenclature on the basis of the dominant fabric. Thus, if L fabric dominates over S fabric, the rock may be assigned an $L > S$ fabric. With more dominance of L over S, use of $L \gg S$ is also sometimes used for the rock. Likewise, $S > L$ and $S \gg L$ are also in use. All these can be better expressed in a Flinn diagram (Fig. 15.31).

15.6 Genesis of Lineation

Lineation is formed under the influence of several processes, which may work singly or in combination of more than one. We consider four broad groups of processes for the formation of lineation: (i) metamorphism-dominated processes, (ii) deformation-dominated processes, (iii) geometrically controlled lineations and (iv) lineation as a consequence of

Fig. 15.21 Field photograph showing lineation given by rods. The rock shows mylonitic foliation and lineation in fine-grained gneisses at the base of the middle thrust units, central Scandinavian Caledonides, exposed at the Vojmån River, Sweden. The lineation is parallel to the regional thrust transport direction. Hammer handle (yellow colour) to the left provides scale. (Photograph courtesy Professor Reinhard Greiling)

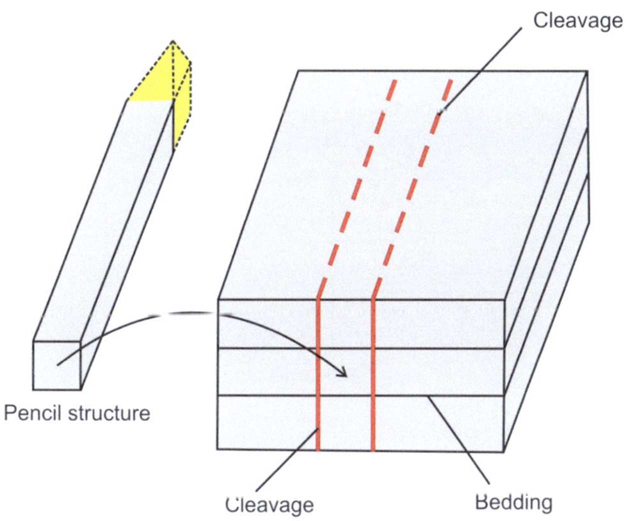

Fig. 15.22 Block diagram showing where pencil structure is developed

plate motion. It is however difficult to draw any specific line of demarcation between these four groups of processes. For example, during formation of any lineation under conditions of metamorphism, the role of deformation cannot be ruled out. Similarly, during deformation, the role of recrystallization (metamorphism) cannot be ruled out. The proposed grouping is therefore based on which of the major processes dominate in the formation of various types of lineation.

15.6.1 Metamorphism-Dominated Processes

During metamorphism, some new minerals are formed from a pre-existing rock due to geochemical and/or mineralogical reorganization. The rock may later on be subjected to layer-normal compressive stresses. These stresses are accommodated by alignment of the mineral constituents in a direction perpendicular to that of the imposed stresses. In metamorphic rocks, the preferred orientation of tabular, prismatic or acicular minerals is thus developed. In general, while minerals of tabular habit develop foliation to the rock, those of acicular or prismatic habit develop mineral lineation. Considering a foliation plane as a tectonic fabric, the lineations formed in the foliation plane get their longer axes oriented parallel to the direction of extension (X). The linear structure thus developed in the rock is called *mineral lineation*.

During metamorphism, some minerals, e.g. quartz, are segregated and are thus separated from the host rock. Under the influence of pressure, the segregated mass assumes a cylindrical shape. The cylindrical structure thus formed is called a *rod* or *rodding*. During folding, the rods tend to get reoriented parallel to the fold axis. As such, in regionally folded terrains, the orientation of rods commonly helps in identifying the fold axis. Further, in large-scale shear zones, since the linear features tend to align parallel to the direction of thrust transport, rods, if present, may also get aligned in the direction of thrust transport (Professor Reinhard Greiling, see Fig. 15.21). Likewise, mineral lineations are generally oriented parallel to the fold axis, excepting when the structure is a superposed fold.

15.6.2 Deformation-Dominated Processes

15.6.2.1 Deformation of Pre-existing Constituents

During deformation of rocks, the mineral constituents or a part of the rock mass tend to reorient or move in response to the deforming stress. In general, the constituents of a rock

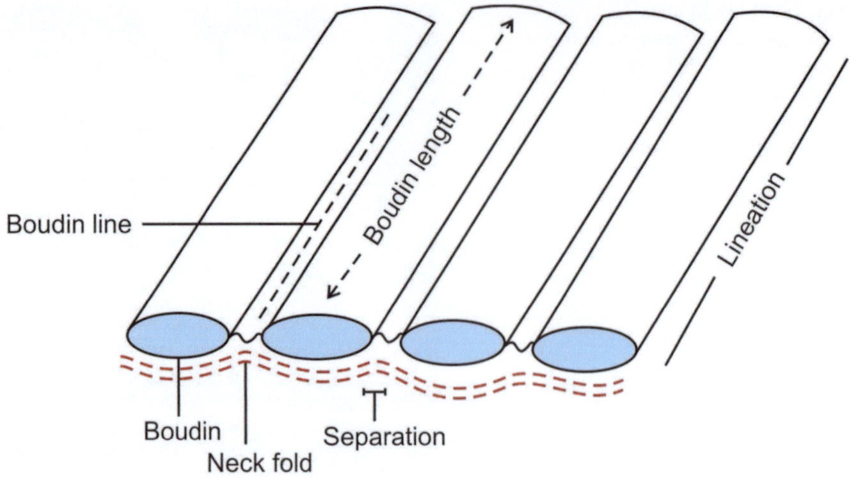

Fig. 15.23 Boudins formed due to fragmentation of the competent layer during extension so that the individual boudins are separated from each other

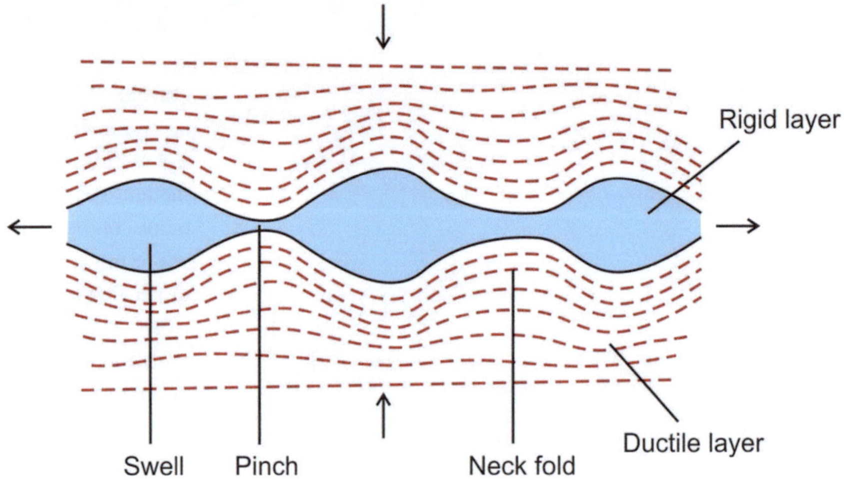

Fig. 15.24 Pinch-and-swell structure. Note that the continuity of the competent layer is maintained

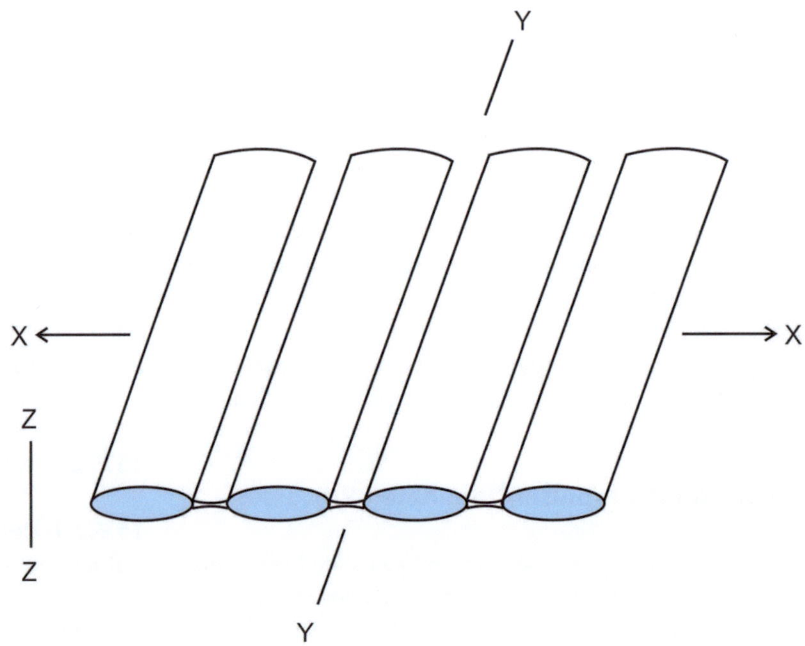

Fig. 15.25 Sketch of boudinages formed due to layer-parallel extension in one direction only (along Y-axis)

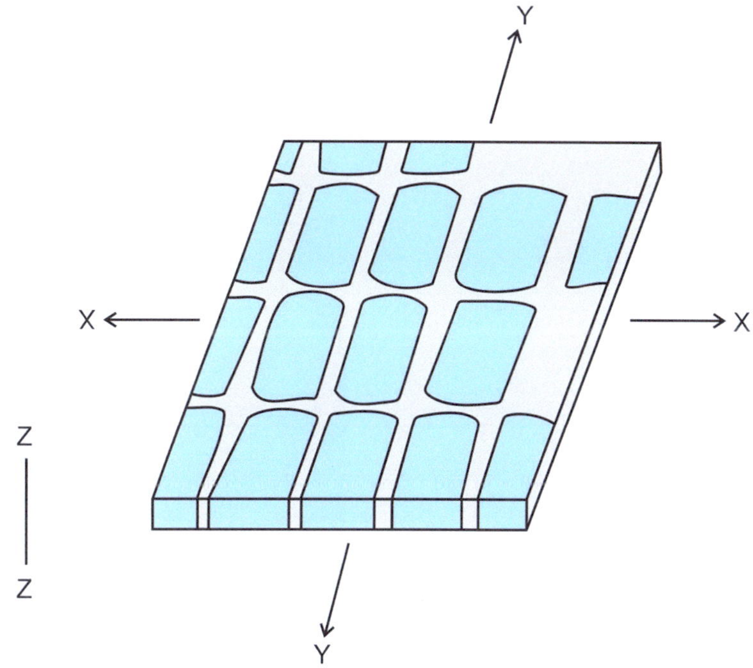

Fig. 15.26 Block diagram of chocolate-tablet boudinage. The boudins are formed due to layer-parallel extension in two directions (X and Y). The tectonic axes have been shown as a reference to their orientation

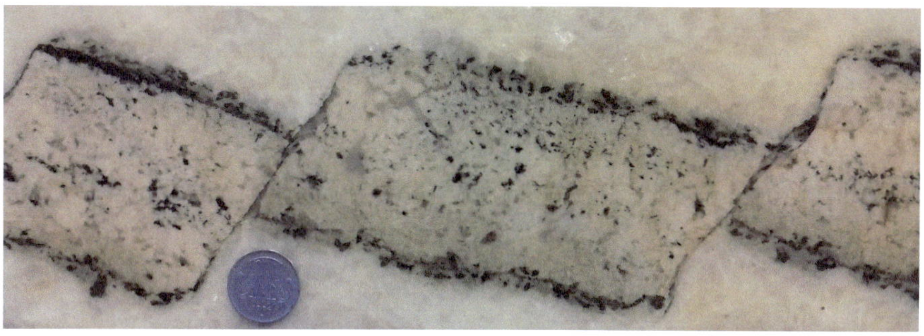

Fig. 15.27 A polished sample of marble from Greece showing chocolate-tablet boudinage. The individual chocolate boudins show step faulting. (Sample courtesy Professor J.L. Urai)

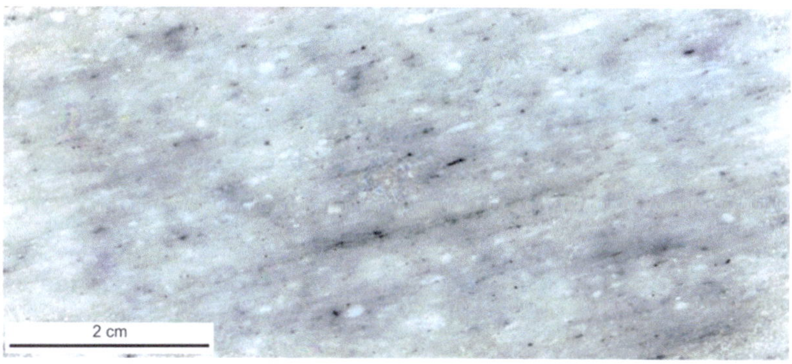

Fig. 15.28 Stretching lineation developed in quartzite of the Main Central Thrust zone near Joshimath, Garhwal Himalaya, India. (Photograph by the author)

tend to move in the direction of stretching, which in most cases is parallel to the *X*-direction of local strain ellipsoid. In the process, the rock develops a lineation. The type and orientation of the lineation depend upon the type of deformation and the structure thus formed. Likewise, some clastic sedimentary rocks contain pebbles that initially had elliptical shapes. During deformation of the rock, the long axes of the pebbles are generally aligned with a preferred orientation caused due to an imposed layer-normal compressive stress. The long axes of the pebbles when projected on to the *XY*-plane give a lineation to the rock. This lineation gives the direction of extension (*X*). The lineation thus formed is

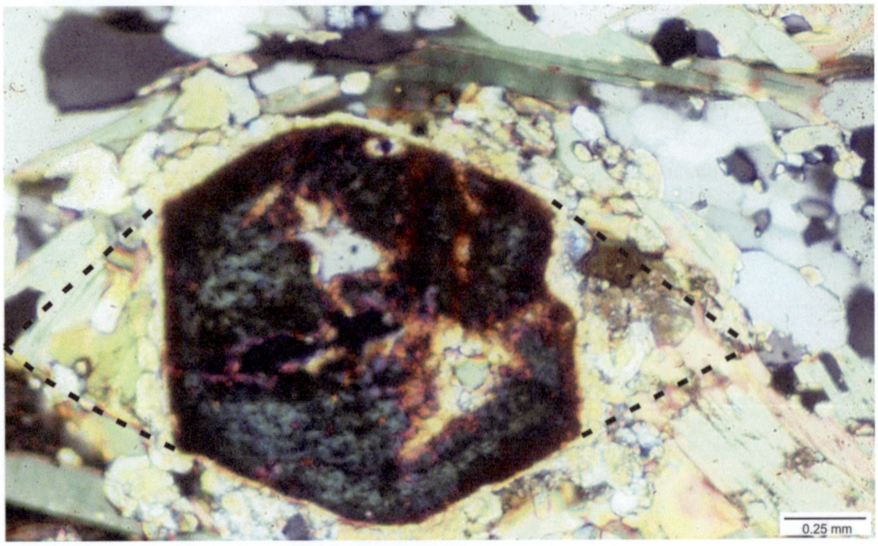

Fig. 15.29 Pressure shadows (shown by black dashes) developed on both sides of a garnet porphyroblast. (Photomicrograph by the author)

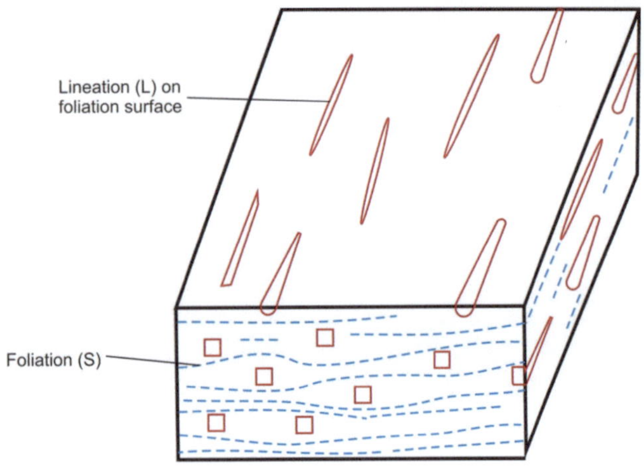

Fig. 15.30 A simplified diagram of an LS tectonite. If a rock is dominated by lineation (L) such that foliation is either absent or weak, the rock is called an L-tectonite. If a rock is dominated by foliation (S) such that lineation is either absent or weak, the rock is called an S-tectonite. With the development of both penetrative lineation and foliation, it is called an LS-tectonite

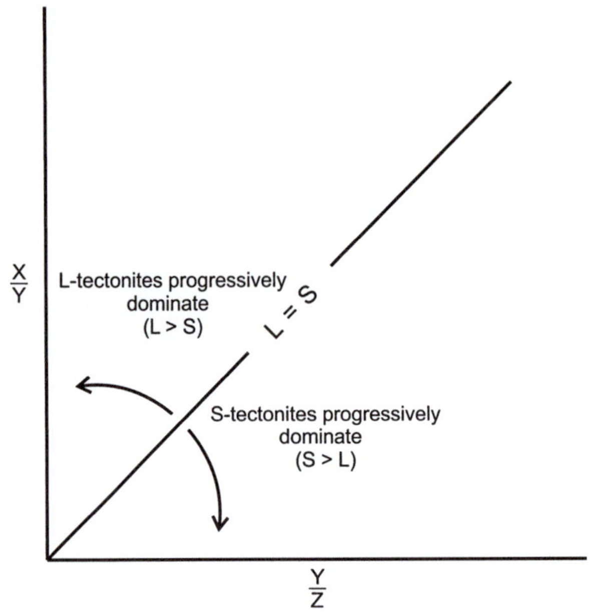

Fig. 15.31 LS-tectonite on a Flinn diagram

totally controlled by deformation, and there is no role of any metamorphism.

15.6.2.2 Folding

Rocks deformed by folding commonly show development of lineations. A fold may show more than one type of lineations in its different parts. This is because different parts of a fold, e.g. limbs, axial plane and hinge, behave differently to compressive stresses. For example, while the limbs undergo rotation, the axial-plane region undergoes stretching. As a result, the constituents of the rock contained in the fold show different types of movement plans. Also, during folding while some parts of the fold are under compressional field, others may be in extensional field.

Under pure shear conditions, the lineations tend to get reoriented perpendicular to the fold axis (Fig. 15.32a). Under simple shear conditions, as exist in large-scale ductile shear zones, rotation of limbs causes rotation of lineation that causes a progressive decrease in the angle between the lineation and the fold axis (*X*). With progressive simple shear, the pre-shear lineations formed at high angles with the fold axis rotate in the direction of bulk transport (Fig. 15.32b). At higher ductile strains, the lineations get reoriented parallel to the fold axis (Fig. 15.32c).

In general, folding may develop several types of lineations that are either developed during folding or the result of reorientation of some early-formed linear features during

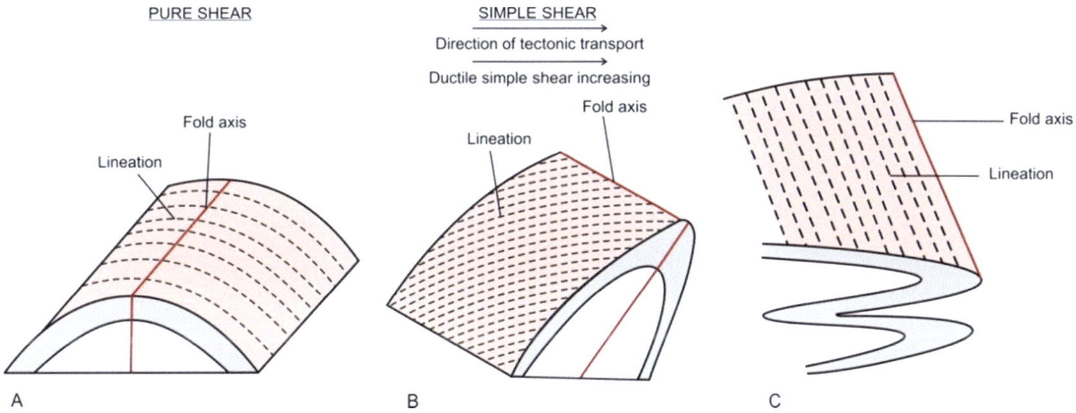

Fig. 15.32 Lineation as kinematic marker during folding. (**a**) Folding under pure shear conditions. The lineations tend to get reoriented perpendicular to the fold axis. (**b**) Folding under simple shear conditions. With progressive deformation, the lineations undergo reorientation towards the bulk tectonic transport. (**c**) Folding under simple shear conditions at higher strains that exist near the main thrust. The lineations tend to become parallel to the fold axis

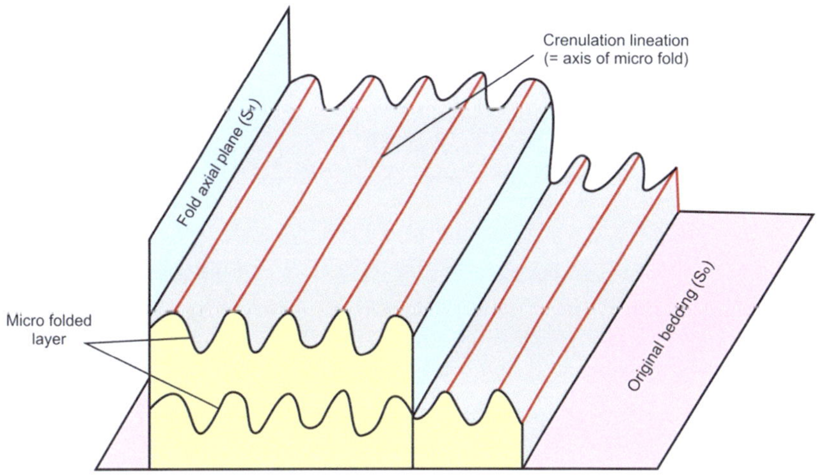

Fig. 15.33 Crenulation lineation as related to folds. The lineation is given by the axis of the micro-folds

folding. Some common types of lineations formed due to folding are described below:

– *Crenulation lineation* is a common type of lineation associated with folds. It is formed during small-scale folding or micro-folding of the foliation planes (Fig. 15.33). The lineation is given by the axis of the micro-folds and is formed due to stretching along the axis that is parallel to the fold axis.

Bell and Johnson (1992, p. 106) summarized the mechanisms for crenulation development under three categories:

(a) Crenulation nucleation by buckling, and volume loss by pressure solution (e.g. Gray 1979)

(b) Extensional crenulation in shear zones due to locally higher partitioned progressive shearing strain, but not necessarily any volume loss (e.g. Platt and Vissers 1980)

(c) Deformation partitioning and shear-strain controlled dissolution (Bell and Rubenach 1983; Bell and Cuff 1989), where the zones of progressive shearing become zones of differentiated crenulation cleavage development and zones of progressive shortening become hinges

In addition to crenulation lineation, a fold may undergo refolding due to which the early-formed lineation is sympathetically refolded. The new lineation thus formed is parallel to the refolded axis (Fig. 15.34).

– *Mullions* (Figs. 15.18 and 15.19) are sometimes formed during folding. In such cases, these are called

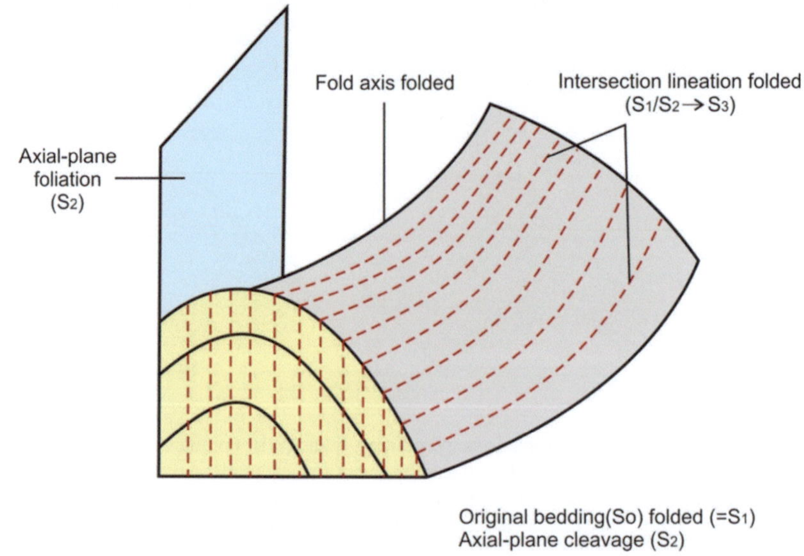

Fig. 15.34 Intersection lineation parallel to the refolded axis

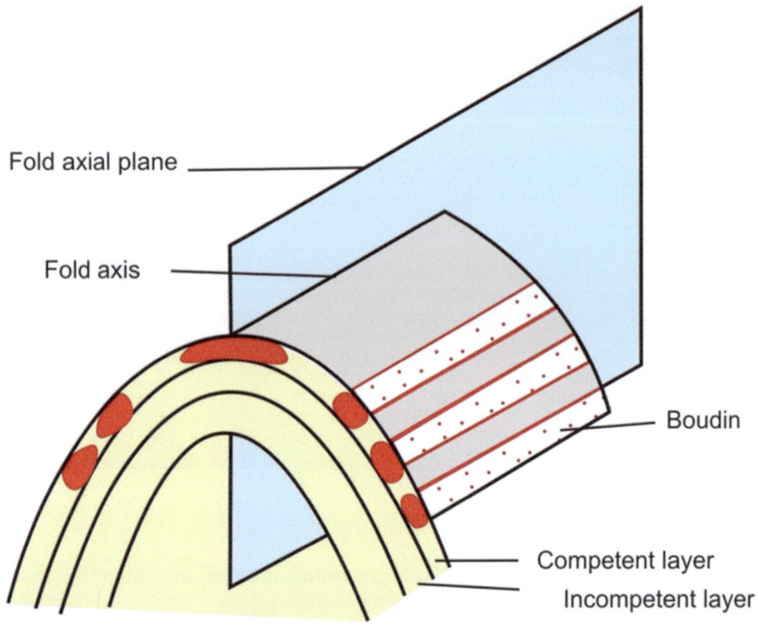

Fig. 15.35 Boudins as related to a fold. See text for details

fold mullions that are oriented parallel to the fold axes. Fold mullions thus help in identifying large-scale folds, especially their axes, if they are not clearly visible on outcrops.

- *Rods* commonly result from metamorphic segregation. However, if the formation of rod is followed by folding, the rods may be oriented parallel to the fold axis.
- *Boudins* are formed due to extension of a competent layer enclosed in an incompetent host (Fig. 15.23). These are commonly associated with, and genetically related to, the host folds. The boudins thus formed have their lengths parallel to the fold axis, while in profile section they appear as sausage-shaped bodies (Fig. 15.35).

Formation of boudins related to folding often leads to confusion. Although the boudin axis approximately parallels the fold axis, there are situations when the boudin axis is oblique to fold axis. Therefore, use of boudins to infer fold axis is to be done with caution, especially in terrains that have undergone multiphase deformations. In the light of experiments on folding of boudinaged layers, it has been shown (Sengupta 1983, p. 207) that a layer undergoing buckle folding may develop boudinage on the limbs at a comparatively late stage of folding when the limbs of the fold have rotated into the extension quadrant of strain; these are boudinaged folds.

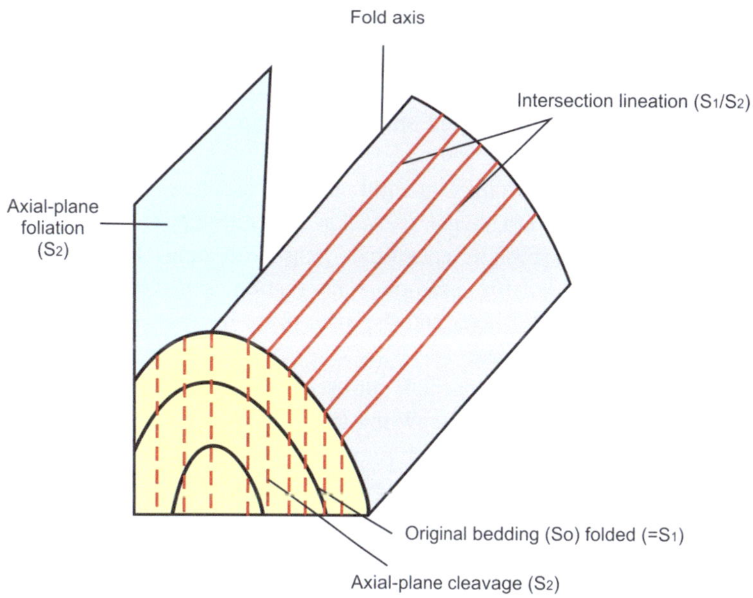

Fig. 15.36 Formation of intersection lineation due to intersection of an axial-plane cleavage with the bedding

15.6.2.3 Extensional Processes

Boudins may form in a layer when there is competence contrast between the boudinaged layer and the surrounding layers under both brittle and ductile conditions. Segmentation of a rigid layer may occur in extensional regime. The shape of the boudins formed depends upon the contrast in the ductility of the layer undergoing segmentation with that of the surrounding layers. If the ductility contrast is large, the boudins show sharp or angular edges. With decreasing contrast in ductility, the shape of boudins progressively becomes rounded as under ductile conditions.

15.6.2.4 Rotation

During deformation, some rigid minerals may rotate due to high competence contrast between the mineral and the softer matrix. Rotation of minerals may produce a penetrative lineation. Lineations in some schistose rocks owe their origin to rotation of some prismatic minerals such as hornblende.

15.6.2.5 Shear Deformation

During shear deformation, the original marker lines rotate towards the stretching direction, i.e. X-axis. At advanced stages of shear deformation, all these marker lines including the fold axes become nearly parallel to the X-direction and thus define a lineation.

15.6.3 Geometrically Controlled Lineation

Intersection lineation is formed due to intersection of two planar features (Fig. 15.15). As such, this type of lineation does not show any fabric that displays any linear orientation of constituents. Intersection lineations are invariably associated with folds. In a folded rock, a common type of lineation (= intersection lineation) is formed by the intersection of an axial-plane cleavage with the bedding (Fig. 15.36).

Intersection lineation neither indicates any physical or mechanical state of a rock, nor any process of deformation or metamorphism. Its presence however indicates that the rock has a history of its formation of metamorphic planes or some other planes of anisotropy such as bedding planes, foliations and cleavages.

15.6.4 Lineation as a Consequence of Plate Motion?

In recent years, the ultimate cause of formation of most of the tectonic structures in rocks has been assigned to the plate-related processes. With the same analogy, lineation too deserves a place in the list of plate-related tectonic structures. Unfortunately, this aspect has not yet been highlighted in textbooks. We cite here an example from the Himalaya. Shackleton and Ries (1984) have shown that the regionally consistent N-S trending stretching lineation observed in several lithotectonic subdivisions of the Himalaya may have developed due to northward movement of the Indian plate and its collision with the Asian plate.

Although the Himalayan N-S trending stretching lineation has been related to plate motion, it is rather premature to establish plate motion as a lineation-forming process for which more data/information from other collision belts of the world is needed.

15.7 Significance of Lineation

- Lineation is especially significant in understanding the geometry of the associated structures developed on all scales.
- Mullions are important structural markers as these are commonly related to smaller as well as larger structures.
- Intersection lineation is sometimes helpful in field. In cylindrical folds, for example, intersection lineation given by the folded layer and the axial-plane foliation is parallel to the fold axis.
- During civil engineering and construction work, rocks showing strong lineation draw the attention of engineers for suitability or otherwise for construction work.

15.8 Summary

- *Lineation* or *linear structure* is a general term used for any fabric or orientation developed in a linear fashion that has formed as a result of tectonic deformation.
- Lineation can be *non-penetrative* and *penetrative*. Common non-penetrative lineations include slickensides, slickenlines and slickenfibres. Common penetrative lineations include mineral lineation, crenulation lineation, intersection lineation, mullions, boudins, rods, pencil structures and pressure shadows.
- Since lineation is formed as a result of tectonic deformation, it constitutes a tectonic fabric. If the rock is dominated by a lineation, it is called an *L-tectonite*; if dominated by foliation, it is an *S-tectonite*; and with both lineation and foliation, it is an *LS-tectonite*.
- Lineation can form under the influence of several processes that may operate singly or in a combination of more than one. We have grouped the lineation-forming processes into four: metamorphism dominated, deformation dominated, geometrically controlled lineations and lineation as a consequence of plate motion.
- *Metamorphism-dominated processes* give rise to geochemical and/or mineralogical reorganization, thus giving rise to new minerals. Mineral lineation and rod or rodding are common examples.
- *Deformation-dominated processes* cause mineral constituents or part of a rock to reorient or move in response to the deforming stress. The constituents thus tend to move/reorient in the direction of stretching which in most cases is parallel to the *X*-direction of local strain ellipsoid. Crenulation lineation, mullions, boudins, slickensides, slickenlines and slickenfibres are common examples.
- *Geometry-controlled lineations* may form due to the intersection of two planar features, say bedding plane and cleavage, thus forming a lineation called intersection lineation.
- *Plate motion*, as exemplified from the Himalayan region, has also been suggested as a lineation-forming process. However, more data is needed from other collision belts of the world to establish this fact.

Questions

1. Explain what do you mean by lineation.
2. How do you distinguish between penetrative and non-penetrative lineations?
3. What is a pencil structure?
4. Differentiate between a boudin and a rod.
5. Describe the geometry and genesis of a chocolate-tablet boudinage with the help of sketches.
6. How is stretching lineation formed? Under which conditions this type of lineation is commonly formed?
7. What is crenulation lineation? How is it formed?
8. What do you mean by L-, S- and LS-tectonites?
9. Give examples of lineations formed due to faulting.
10. Describe the formation of lineation when metamorphism is a dominant process.

Part III

Wider Perspectives

16 Mechanisms of Rock Deformation

A. R. Bhattacharya, *Structural Geology*, Springer Textbooks in Earth Sciences, Geography and Environment,
https://doi.org/10.1007/978-3-030-80795-5_16

Abstract

The structures and microstructures of rocks that we see today are the culmination of several mechanisms that the rocks have undergone in the geological past. A *mechanism of rock deformation* broadly means the process that tends to accommodate large strains in rocks, thus leading to a stable structure/microstructure to the rock. The processes operate on different scales, but those operating on grain scales bring about significant changes to the structure and texture of rocks. The processes may operate singly or in combination with other ones. The various physico-mechanical conditions that prevailed during a particular mechanism may change with time and in turn may trigger another mechanism(s) to operate. This chapter is a survey of the various mechanisms of rock deformation that operate on microscale and mesoscale.

Keywords

Mechanisms of rock deformation · Recovery · Dynamic recrystallization · Crystal-plastic deformation · Diffusion creep · Dislocation creep · Cataclasis · Superplastic deformation · Strain softening · Crack-seal mechanism

16.1 Introduction

A *mechanism of rock deformation* broadly means the process that tends to accommodate large strains in rocks, thus resulting in a stable structure/microstructure to the rock. The processes operate on different scales alone or in combination with some other mechanisms until the accumulated strain is accommodated. They act under certain specific physico-mechanical conditions of rocks. A single mechanism may change with time depending upon the newly developed rheological state or structure/texture of the rock, giving way to some other mechanism(s) to operate. In many cases, the original texture of rocks is completely or largely changed. Remnant signatures/imprints of some earlier mechanism (s) often help deciphering the deformation history or strain path of the rock.

In structural geology, recognition and detailed understanding of mechanisms of deformation of rocks and minerals require field and optical microscopic study of rocks. In some cases, experimental study is also needed, especially to establish the flow laws that may have governed at the time of deformation.

16.2 Factors Controlling Mechanisms of Deformation

In natural rocks, any mechanism of deformation will operate only when certain necessary factor(s) are met with. These factors may not be so for other mechanisms and may be highly variable in nature and operate on different scales. Some common factors that control the mechanisms of rock deformation include (i) location of a rock inside the earth, (ii) temperature and pressure, (iii) fluids and gases present within the rocks and (iv) texture and grain size of the rocks. In addition to these, some other factors may also affect a rock, but their roles may not be that significant.

Recently, Gomez-Rivas et al. (2020) emphasized that of the several factors, temperature is the key parameter that determines what rock deformation processes are active and therefore how tectonic structures are formed. Also, it controls the migration of fluids and melt in the earth's crust that lead to phase transformations and changes in rock rheology. The authors highlight the following as to how temperature controls deformation mechanisms and processes: (i) Temperature and pressure, together with differential stress and strain rate, are the main parameters that determine how minerals and rocks deform. (ii) At low temperature and shallow crustal levels, deformation is normally dominated by brittle behaviour while ductile flow takes place at higher temperature (and higher effective confining pressure), and therefore generally at deeper levels. (iii) Rock deformation also depends on the mineral assemblage and spatial distribution of their components, for example grain size and grain size distribution, orientation of crystal lattice, porosity, permeability and presence and composition of interstitial fluids. (iv) Because of the above factors, brittle deformation is a temperature-independent process that strongly depends on pressure and presence of fluids. Brittle deformation operates by the nucleation and propagation of fractures as well as slip along existing discontinuities, e.g. fractures, bedding planes and faults. On the contrary, ductile flow (also called viscous or visco-plastic flow) is determined by thermally activated deformation mechanisms including dislocation glide and creep, twinning, kinking and diffusion creep, together with the associated processes such as intracrystalline recovery and dynamic recrystallization.

16.3 Classification of Mechanisms of Deformation

Deformation of rocks and minerals has been discussed by several workers (e.g. Hobbs et al. 1976; Nicolas and Poirier 1976; Schmid 1982; Poirier 1985; Nicolas 1987; Groshong 1988; Knipe 1989; Van der Pluijm and Marshak 1998; Vernon 2004; Gomez-Rivas et al. 2020; Schuck et al. 2020, and the references therein). The mechanisms of rock deformation have been classified in various ways such as brittle deformation and ductile deformation, scale-dependent mechanisms and temperature-dependent mechanisms. As such, no specific scheme exists on this topic. In this work, we have grouped the various deformation mechanisms into two: mechanisms on microscale and mechanisms on mesoscale.

16.4 Mechanisms on Microscale

16.4.1 Recovery

Recovery or *dynamic recovery* includes all those processes that attempt to return a deformed crystal to the undeformed state during which the stored energy of deformation decreases (Hobbs et al. 1976, p. 108; Gottstein and Mecking 1985, p. 193; Tullis and Yund 1985). The mechanism leads to a reduction of the dislocation density and the arrangement of the remaining dislocations in sub-boundaries perpendicular to the glide plane. Recovery operates well in crystals with high density of dislocations and point defects and leads to formation of a number of unstrained subgrains formed from the original distorted crystals that can be distinctly seen with different optical orientations from the neighbouring ones. Since during recovery the crystals release their energy by reducing the dislocation density, the mechanism operates at high temperature. Recovery is thus an accommodation mechanism operating at high temperature.

16.4.2 Dynamic Recrystallization

Dynamic recrystallization includes the processes by which a crystalline aggregate lowers its free energy during deformation by the formation of smaller (<0.01 mm), strain-free grains. It operates during syntectonic deformation when large, stressed grains tend to release their internal strain energy by the formation of smaller, strain-free grains (Fig. 16.1). With progressive deformation, the newly formed smaller grains continue to release their internal strain in a cyclic manner till the rock attains a stable, very fine (grain size <0.01 mm) microstructure. The mechanism thus leads to strain softening and formation of mylonites and as such influences the mechanical properties and deformation behaviour of rocks.

Dynamic recrystallization has been suggested as an important mechanism for the formation of mylonites. The process produces new microstructures and fabrics including lattice preferred orientation (LPO), which becomes stronger with progressive deformation. The *driving forces for dynamic recrystallization* originate due to decrease in the following four types of energy (Urai et al. 1986): (a) *intra-granular lattice defect energy* that is associated with the vacancies, dislocations and dislocation arrays within grains; (b) *grain-boundary energy* that provides a strong driving force for migration of boundaries with a small radius of curvature; (c) *chemical free energy* that exists where there is a small difference in composition between crystals of the same phase on either side of a migrating boundary; and (d) *elastic strain energy*, which is an external load-supporting energy localized in the vicinity of lattice defects and in grain corners or other asperities where "locked-in" stresses may exist.

Dynamic recrystallization, like recovery, is also an accommodation mechanism that operates at high temperature. However, the main difference is that while dynamic recrystallization leads to strain softening, recovery does not.

Dynamic recrystallization is a major mechanism of rock deformation that operates as long as a tectonic event persists and primarily depends upon two important processes, viz. grain-boundary migration and subgrain rotation. A few other processes are also believed to be associated, or run concurrent, with dynamic recrystallization such as migration recrystallization, rotation recrystallization, grain migration, polygonization and grain rotation. Role of the latter processes in dynamic recrystallization is variable and to what extent is not clearly known. Following Urai et al. (1986), we describe below some of the important mechanisms associated with dynamic recrystallization:

(a) *Grain-boundary migration*, also called *migration recrystallization*: It is a process of recrystallization whereby material from a grain that is being consumed enters the grain-boundary region and eventually recrystallizes on the lattice of a neighbouring grain that is growing. The mechanism is believed to be a conservative process involving no net gain or loss of material near a migrating boundary. Movement of a grain boundary takes place when the driving forces on both sides of the boundary are different and during migration the solute atoms prefer to stay with the boundary, thus exerting a dragging force on the boundary that reduces the net driving force on the boundary (Gottstein and Mecking 1985). During dynamic recrystallization, grain boundaries commonly migrate away from their

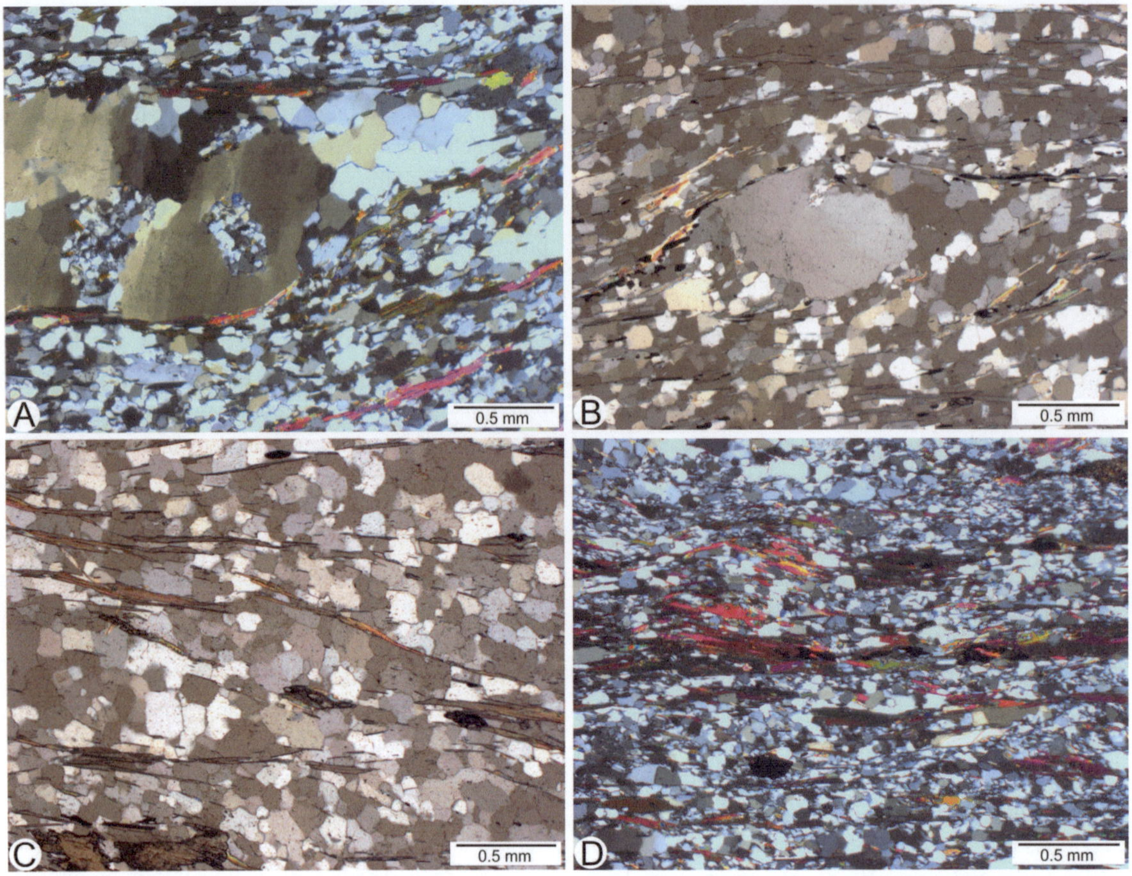

Fig. 16.1 Dynamic recrystallization. The photomicrographs show rocks that have undergone dynamic recrystallization to varying degrees. The process is not a sudden one but gradual. The micrographs selected here show different stages of dynamic recrystallization that ultimately change a strained large-quartz grain to a fine-grained rock. (**a**) A strained-quartz porphyroclast in a mylonite has undergone dynamic recrystallization which is only partial. As such, the strained portion looks cloudy (undulose extinction) in which faintly developed deformation bands/lamellae can also be seen. The smaller, clear grains represent products of dynamic recrystallization. (**b**) A large grain (outlines partly marked by the presence of a few dark-coloured fine-grained minerals) is selected here to show that the grain has released much of its internal strain by the formation of clear, unstrained, smaller grains excepting the central cloudy (strained) zone. (**c**) The dynamic recrystallization of a porphyroclast is at an advanced stage. The grain has released much of its internal strain by the formation of a large number of smaller, strain-free, grains. Mylonitic foliation is given by thin layers of muscovite-biotite. (**d**) A rock that has undergone almost complete dynamic recrystallization, thus forming a mylonite. (Photomicrographs by the author)

centres of curvature such that initially straight or gently curved boundaries may become wavy, lobate or sutured.

(b) *Subgrain rotation*: It is progressive misorientation of subgrains during recovery until high-angle grain boundaries are formed (Poirier and Nicolas 1975). A common evidence of subgrain rotation is the *core-and-mantle structure* (Fig. 16.2) where the core of host grain passes out transitionally into mantle with increasing subgrain formation and then into aggregates of recrystallized grains with similar size and orientations to the nearby subgrains. Subgrain rotation is considered as a common process associated with dynamic recrystallization.

(c) *Migration recrystallization*, also called *discontinuous recrystallization*: It is a mechanism due to which grain boundaries regularly undergo catastrophic jump in migration velocity.

(d) *Grain migration*: If in a deforming environment a grain loses material along some of its boundaries and gains material along some other parts, it may move through the material by a distance of the order of a grain diameter or more. This motion of grains through the material is called *grain migration*. The material enclosed within boundaries of the migrating grain is entirely different from the material originally enclosed and may persist indefinitely in dynamically recrystallizing materials.

(e) *Polygonization*: It is the arrangement of polygonal subgrains within larger grains (Fig. 16.3) as commonly observed by variations in extinction from adjacent grains caused by small difference in the orientation of

Fig. 16.2 Core-mantle structure preserved in a mylonitic gneiss of Main Central Thrust, Garhwal Himalaya, India. The core of the host grain constitutes a k-feldspar porphyroclast, which gradually passes out into a mantle of quartz subgrains and an aggregate of recrystallized subgrains. (Photomicrograph by the author)

Fig. 16.3 Polygonization as shown by the occurrence of polygonal subgrains within a large-quartz grain. The subgrains are formed from the original strained grain (outlines shown by yellow dashes) and are strain free

the subgrains (Hobbs et al. 1976, p. 108). The subgrains are formed from an original distorted crystal and are strain frcc.

16.4.3 Bulge Recrystallization

Bulge recrystallization or *bulging recrystallization* is a type of dynamic recrystallization that affects only the boundary region of porphyroclasts so that the resulting grain-boundary bulges may be separated from the old grains, forming relatively small recrystallized grains (Stipp et al. 2002). This mechanism is also called *strain-induced grain-boundary migration* since the process is envisaged to be driven by differences in stored strain energy from one place in an aggregate or grain to another (Hobbs and Ord 2015, p. 311). The bulges can separate from the initial grain and undergo grain-boundary migration as well as subgrain rotation, and therefore bulging recrystallization has been considered as a general recrystallization mechanism (Drury and Urai 1990). Four different models of bulging recrystallization have been proposed (Stipp et al. 2002):

1. A strain-free bulge is developed by local slow grain-boundary migration (Bailey and Hirsch 1962); separation of the bulge from the porphyroclast may be affected by microfracturing.
2. The bulge undergoes progressive subgrain rotation, causing a separation from the old grain by a bridging subgrain boundary (Means 1981; Urai et al. 1986).
3. A new high-angle boundary migrates from the bulge into the old grain (Tungatt and Humphreys 1984).
4. A subgrain at a grain boundary of a deformed grain bulges into the neighbouring grain (Drury et al. 1985).

16.4.4 Grain-Boundary Sliding

Grain-boundary sliding is a deformation mechanism that can accommodate very high strains when grains can slide past each other (Gomez-Rivas et al. 2020). The mechanism operates at high temperatures and low stresses and is assisted by the presence of fluids. Polycrystalline aggregates that are constituted of very-small-size grains are more susceptible to this mechanism.

16.4.5 Crystal Plastic Deformation

Crystal plastic deformation or *intracrystalline plasticity* is the accumulation of strain by processes that operate within a crystal such as glide movements (or intracrystalline gliding or intracrystalline slip). The main mechanism of crystal plastic deformation is the glide of dislocations in specific planes and along certain directions (termed slip systems) (Gomez-Rivas et al. 2020). It is an important mechanism of plastic deformation that causes permanent strain to a rock. The mechanism resembles *dislocation creep* in which the crystals accommodate shape changes by homogeneous simple shear due to which the dislocations move along some preferred crystallographic planes. This produces lattice preferred orientation (LPO) in crystals.

Box 16.1 Feldspar Recrystallization

Feldspar recrystallization is a common phenomenon in shear zones and can take place under both coaxial and noncoaxial deformations. Commonly, this phenomenon occurs synchronous with dynamic recrystallization of larger grains such as those of quartz. However, the course of dynamic recrystallization followed by quartz and feldspar is different. Quartz releases internal strain by progressive formation of smaller, strain-free, clear grains of quartz. Feldspar, on the other hand, releases internal strain by the formation of an aggregate of chlorite-biotite-muscovite and smaller strain-free, clear grains of quartz (Fig. 16.4). The pre-existing flaky minerals commonly swerve around the porphyroclast (A). In both the cases, the two tails are oriented parallel to the direction of extension (X) of the local strain ellipsoid. Because both quartz and feldspar grains release internal strain by giving rise to their by-products, feldspar recrystallization can also be considered as a process, or part, of dynamic recrystallization.

Crystal plastic deformation does not produce significant volume changes. Stress necessary to operate the mechanism is mainly utilized in moving the dislocations, i.e. intracrystalline gliding or slip, through the lattice

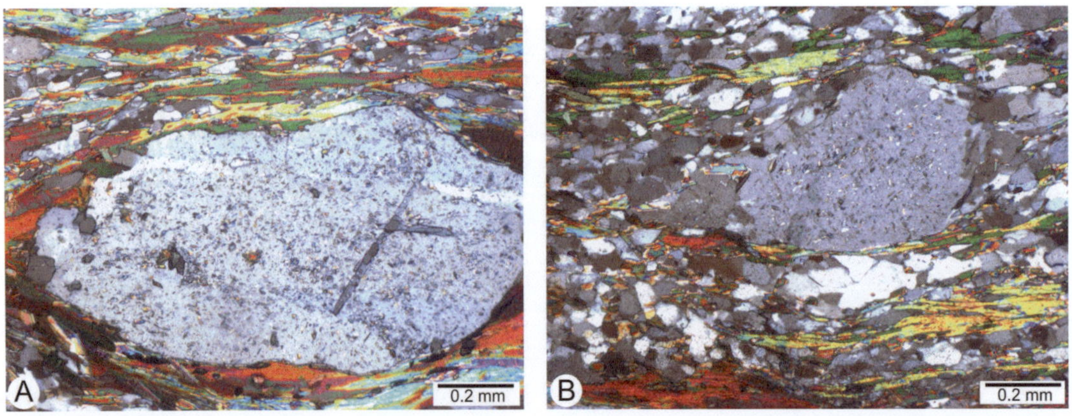

Fig. 16.4 Feldspar recrystallization. (**a**) A coaxially deformed k-feldspar releases its internal strain by the formation of an aggregate of chlorite-biotite-muscovite that envelopes the clast. (**b**) A noncoaxially deformed k-feldspar grain representing an advanced stage of dynamic recrystallization. Note that the grain has released much of its internal strain by the formation of an aggregate of chlorite-biotite-muscovite and smaller strain-free, clear grains of quartz. Note that the by-products of the porphyroclast tend to occupy the tails of the clast, which represents the direction of extension (X) of the local strain ellipsoid

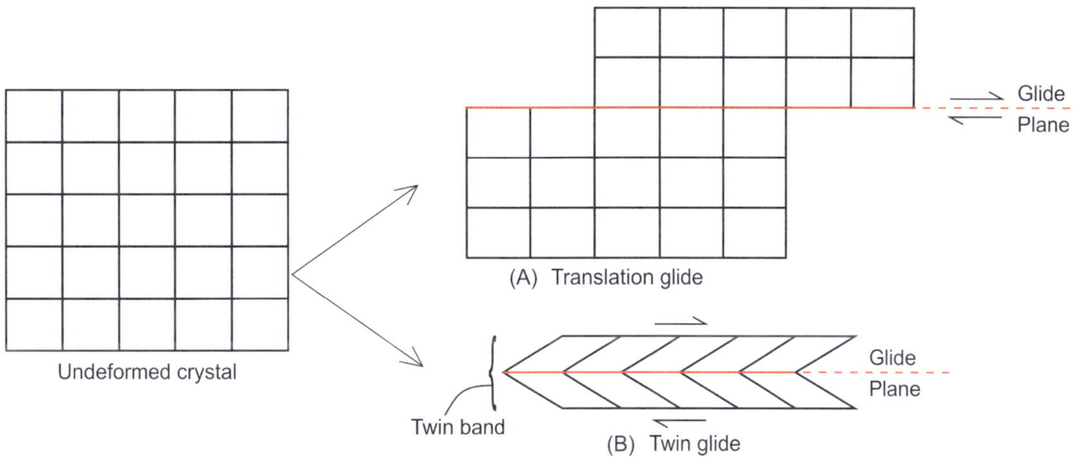

Fig. 16.5 Intracrystalline glide or slip occurs in two ways: translation glide (**a**) and twin glide (**b**). See text for details

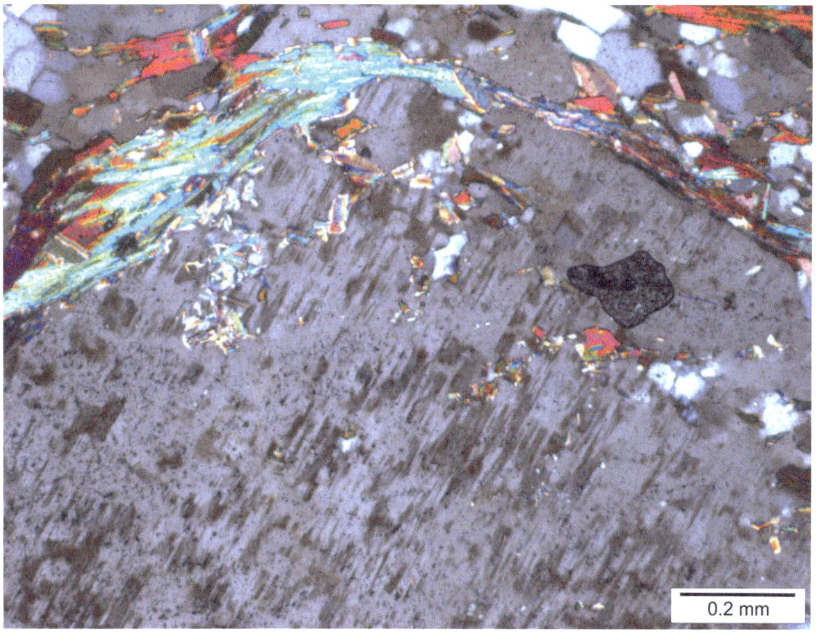

Fig. 16.6 A k-feldspar porphyroclast of a mylonitic gneiss has undergone crystal-plastic deformation by twin gliding forming a set of parallel twin lamellae. (Photomicrograph by the author)

structure. This is accomplished in two ways (Fig. 16.5): (a) *translation gliding* in which a crystal undergoes homogeneous simple shear such that the dislocations move along a pre-existing crystallographic plane and (b) *twin gliding* in which the dislocations slip along some fraction of the pre-existing distance between the crystallographic planes; this produces shear-induced twinning in crystals such as calcite, feldspar (Fig. 16.6) and quartz. Because of such movement patterns, translation gliding produces greater strain than twin gliding. Both these types of deformation involve homogeneous simple shear of a set of lattice points with respect to those of another set on other side of the glide plane.

Crystal plastic deformation is a common process in deformed rocks such as those in ductile shear zones. Although the various effects of this process are not easily visible in optical microscope, *undulose extinction* and *deformation lamellae* are commonly noticeable (Fig. 16.1a). Undulose extinction is an optical effect when a grain does not become extinct homogeneously at one instant. It occurs due to the presence of a large number of dislocations in a crystal lattice. Deformation lamellae, also known as *Fairbairn lamellae*, occur as a set of lamellae with preferred orientation that is distinguished by high optical relief.

The above evidences of intracrystalline microstructures such as undulose extinction and deformation lamellae disappear at elevated temperature due to recovery or recrystallization giving way to the development of lattice preferred orientation (Passchier and Trouw 2005).

16.4.6 Diffusion Creep

Diffusion creep is a mechanism of deformation of crystalline materials caused by diffusion of *vacancies* through the crystal lattice. No crystal has perfect lattice and contains some imperfections or defects in the form of (a) *point defects* such as vacancies or empty space, impurities and interstitials such as the presence of atom(s) and (b) *line defects* such as dislocations. Movement of point defects in the lattice of a crystal is broadly called *diffusion*. Diffusion creep is sensitive to temperature and occurs at high temperatures. The mechanism results in plastic deformation of rocks.

Occurrence of point defects enhances the internal energy and entropy of a crystal. If there is some difference in the energy levels within the crystal, the defects move by diffusion of atoms in the lattice. Vacancies are commonly localized along crystal faces due to tension and migrate in two major ways: (a) through the lattice, called the *Nabarro-Herring creep*, and (b) along the grain boundaries, called the *Coble creep*. Diffusion creep dominates at low stress and at small grain size (Chastel et al. 1993). At depths below a few kilometres where rocks deform by steady-state flow, diffusion-assisted dislocation creep dominates (Tsenn and Carter 1987).

16.4.7 Dislocation Creep

Dislocation creep is a mechanism of strain accommodation by dislocation glide of lattice defects through the crystal lattice. *Dislocations* are line defects that bind an area within a crystal where slip by an interatomic distance $|\mathbf{b}|$, where **b** is the Burgers vector, has taken place (Poirier 1991). The Burgers vector is contained in a glide plane. Crystals show two common types of dislocations: (a) *screw dislocation* when the Burgers vector is parallel to the dislocation and (b) *edge dislocation* when the Burgers vector is perpendicular. *Dislocation glide*, i.e. motion of dislocations, is the main strain-producing mechanism. A *glide plane* is a plane that separates two parts of a crystal affected by movement of dislocations. It contains the line along which dislocation takes place. Across a glide plane, which may be a crystallographic axis also, the atomic bonds are relatively weak. The number of glide planes and their orientations depend upon the crystal system, and therefore each mineral has its own glide planes that are different from others. During deformation, the orientation of the differential stress determines which of the glide planes should be active while others are not. Dislocation creep is highly sensitive to differential stress and results in plastic deformation of crystalline materials.

16.4.8 Grain-Boundary Pinning

In large-scale ductile shear zones, the progressive shear deformation is commonly associated with progressive microstructural changes that involve different types of deformational behaviour of the mineral phases of a rock. Under such situations, some specific type of interaction of two minerals, called *grain-boundary pinning*, may hinder grain-size reduction in the rocks of the shear zone. Pinning has been found to be a shear-related phenomenon and becomes pronounced with an increase of noncoaxial (rotational) ductile strain as traced towards the main thrust plane.

As yet, our knowledge on grain-boundary pinning is meagre. Herwegh and Jenni (2001) demonstrated that natural deformation in the carbonate mylonites bearing sheet silicates occurs via a complex interaction of granular flow and solution transfer processes and that the observed grain-size variability indicates basic differences in mass transfer behaviour between the individual layers, which might be related to differences in the fluid flux. Krabbendam et al. (2003) have shown that a dispersed second phase such as graphite in a naturally deforming rock can inhibit grain-boundary migration, stabilize the grain size and enhance grain-boundary sliding at the expense of dislocation flow. Herwegh and Berger (2004) suggest that the second-phase controlled microstructures predominantly deform via granular flow because pinning of calcite grain boundaries reduces the efficiency of dynamic recrystallization, thus favouring mass transfer processes and grain-boundary sliding. In contrast, according to them, the balance of grain-size reduction and growth by dynamic recrystallization maintains a steady-state grain size in microstructures that are only weakly affected by the second phase, thus promoting a predominance of dislocation creep.

Work on grain-boundary pinning and related processes has as yet focused mainly on mylonitic rocks containing graphite (Krabbendam et al. 2003) and calcite (Herwegh and Berger 2004). I carried out a detailed study of *quartz-mica pinning* and related grain-size stabilization in a crustal scale shear zone, the Main Central Thrust of the Himalaya. Outside the shear zone, quartz grains are large (ca. 0.5 mm and more) and show strain effects (undulose extinction, deformation bands, etc.). These rocks do not exhibit pinning of quartz boundaries due to the presence of the mica phases.

Within the shear zone, effects of pinning appear evident where quartz and mica grain occur together without any third phase (Fig. 16.7, upper part). It has been observed that in the mica-rich layers, quartz-mica pinning operates prominently, while in the mica-deficient layers large grains of strained quartz predominate (Fig. 16.7, lower part, Fig. 16.8) because pinning of grain boundaries is minimal. The role of mica in

Fig. 16.7 Quartz-mica pinning. Presence of mica with quartz (upper part) brings down the grain size of quartz and stabilizes with a common grain size. On the other hand, the region without mica (lower part) does not undergo any pinning, and as such the quartz grains maintain their original large size. (Photomicrograph by the author)

Fig. 16.8 Quartz-mica pinning. Additional demonstration of Fig. 16.7. One can notice that it is only the layer containing quartz and mica that undergoes pinning and grain-size reduction followed by grain-size stabilization. (Photomicrograph by the author)

the deforming regime progressively increases towards the trace of the main thrust, and therefore quartz grain-boundary pinning becomes more widespread (Fig. 16.9). In such cases, subgrain boundaries are progressively reactivated, apparently due to the increased role of fluids on approaching the main thrust. The thin films of fluid along the subgrain boundaries are interpreted to act as a lubricant that promotes mass transfer processes along the grain boundaries. Mica is thus able to move rather more easily into the mesh of relatively larger quartz grains. The enhanced movement of mica (Fig. 16.10) along the quartz boundaries promotes further pinning, which thus spreads to surrounding layers.

The above examples suggest that the presence of mica is necessary for the pinning process to operate. Since pinning initiates in a zone where the rocks start showing the effects of shear deformation, it appears that pinning is a shear-related phenomenon. Mica in the distal areas from the main thrust is represented by biotite while in the vicinity of the thrust by

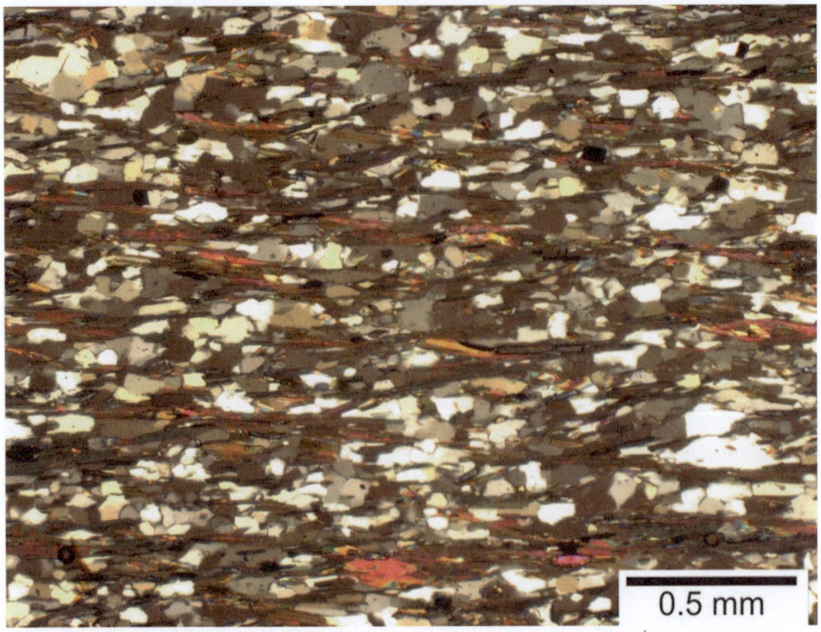

Fig. 16.9 Mica has played its role in pinning due to the presence of quartz grains, and both stabilize with a common, reduced grain size. (Photomicrograph by the author)

Fig. 16.10 An excellent example of quartz-mica pinning. Both the mineral phases have undergone grain-size reduction. The enhanced movement of mica along the quartz boundaries promotes enhanced pinning, which thus spreads to surrounding layers, and the process ultimately leads to grain-size stabilization. This rock is located along the trace of the Main Central Thrust of the Himalaya, where the effects of ductile shear strain are very high. (Photomicrograph by the author)

muscovite mainly due to progressive recrystallization of feldspar (Bhattacharya and Weber 2004). During shear deformation, both biotite and muscovite play equal roles in the pinning process. Therefore, in the context of pinning as related to grain-size reduction, both biotite and muscovite can be simply referred to as mica, and the process can therefore be generalized as *quartz-mica pinning*.

Since the presence of mica appears to alter the texture and microstructure of rocks in ductile shear zones, it is an open question whether grain-boundary pinning can be considered as a mechanism of rock deformation in microscale or not. I think it can be considered as a mechanism though more work is needed to establish this firmly.

16.5 Mechanisms on Mesoscale

16.5.1 Brittle Deformation Mechanisms

Brittle deformation takes place due to the presence of fractures or microfractures along which relative movement of fragments of grains occurs. *Fracturing* includes all the processes leading to formation of fractures, thus causing loss of cohesion of grains and creation of new fracture surfaces. Fractures may occur within the grains, along the grain boundaries or across the adjacent grains. Brittle deformation typically occurs at low temperatures and under low-grade metamorphism and is strongly controlled by fluid pressure (Gomez-Rivas et al. 2020). The nature of brittle deformation mechanisms thus varies accordingly. We consider here two common brittle deformation mechanisms: frictional grain-boundary sliding and cataclasis.

(a) Frictional grain-boundary sliding involves sliding of grains past the adjoining ones. Sliding takes place when the cohesion and friction between the grains are overcome. The grains thus become free to move and rotate as rigid objects without undergoing any internal deformation or crystal-plastic deformation. If cohesion between the grains is lost, the frictional grain-boundary sliding is known as *grain-boundary sliding* only, and this occurs at higher temperatures. Sliding and rotation of grains are favoured by flow of fluids occurring between the grains.
Frictional grain-boundary sliding operates at low confining pressure and at high fluid pressure that reduces the effective pressure, i.e. grain-to-grain contacts, and commonly occurs in slumping, faulting in unconsolidated sediments, fault gouges and cataclasites (Knipe 1989; Vernon 2004). In natural rocks, grain-boundary sliding can be recognized by the lack of variations in grain size, shape and internal structure in and out of the deforming zones (Knipe 1989).

(b) Cataclasis is basically a fracture process involving fragmentation of material, rotation, grain-boundary sliding, dilation and displacement along new surfaces created during deformation (Atkinson 1987; Knipe 1989). Cataclasis is a common process in faulting at high crustal levels and produces *gouges* and *breccias* (Wise et al. 1984; Sibson 1986). When microfracturing or microcracking occurs at moderately high confining pressures, the deformation is more pervasive and distributed leading to the formation of bands; this process is called cataclastic flow that may accommodate ductile deformation over broader areas. Rocks produced by cataclasis are called cataclasites that show no foliation (unless they are further deformed). The matrix typically is fine grained as compared to the host grains that are commonly angular. Cataclasites are sometimes associated with *pseudotachylite* (Fig. 9.37), which resembles a dark-coloured glassy rock disposed as veins and small intrusions in a fault rock. Pseudotachylites are formed during rapid seismic slip under dry conditions causing local melting and subsequent cooling (Sibson 1975).

16.5.2 Superplastic Deformation

Superplastic deformation, also called *superplasticity*, operates in very-fine-grained rocks (grain size less than 0.01 mm) in which grain-boundary sliding is the main mechanism supported by dislocation climb within the grains and diffusion (Boullier and Gueguen 1975). Under high strains, the grains slide past each other and rotate to accommodate the imposed strain without necking. Rotation produces high temperature, which is characteristic of this mechanism. The rock mass thus flows as a ductile material but does not develop noticeable lattice preferred orientation (LPO) despite being affected by large amounts of strain. Superplastic rocks commonly show minor changes in the shape of grains to accommodate the large imposed strain without developing noticeable microstructure. Superplastic deformation is a grain size-sensitive mechanism.

A superplastically deformed rock is sometimes called a '*superplastic' (SP) mylonite* which, like a mylonite, also shows a layered structure. While a mylonite shows crude layering formed by plastic flow, an SP mylonite shows well-developed layering by superplastic flow (Boullier and Gueguen 1975).

Superplastic deformation is common in metals. In natural rocks, the application of superplastic deformation should be done with caution. The main reason is that a superplastic rock resembles a mylonite. The very-fine-grained nature of a superplastic rock could also develop as a result of dynamic recrystallization. In general, the absence of well-developed LPO and very-fine-grained character of the rock are commonly taken as signatures of superplastic deformation.

16.5.3 Diffusive Mass Transfer

Diffusive mass transfer (DMT) (Rutter 1983; Etheridge et al. 1984; Knipe 1989) is a mechanism of *transfer of material from zones of high inter-granular normal stresses to zones of low normal stresses*. The mechanism causes volume loss to the rock due to removal of material. The driving force for this mechanism is the variation in chemical potential in a rock aggregate that depends upon (a) stress variation in the aggregate, (b) gradients of fluid pressure and (c) variations of internal strain energy of grains. DMT is more common in fine-grained rocks. Another favourable condition is low differential stress levels that do not promote crystal-plastic mechanisms to operate.

Common structures and microstructures produced by DMT include (a) development of a trail in oolitic and fossiliferous rocks formed due to removal of oolites and fossils during transfer of material, (b) development of veins in a rock (Fig. 16.11), (c) pitted surface in a pebbly rock due to removal of pebbles (Fig. 16.12) and (d) overgrowths and pressure shadows associated with some minerals. DMT is accomplished in three stages (Knipe 1989): (a) *Source mechanism* includes how the material enters a diffusion path. (b) *Migration or diffusion mechanism* involves transport of material along available transfer paths. (c) *Sink processes* include precipitation or deposition of the material in suitable sites where crystal growth is possible.

16.5.4 Pressure Solution

Pressure solution is the dissolution of material from a crystal or crystal aggregate from points of high pressure and its redeposition at points of low pressure. The mechanism and its importance in the deformation of crystals and rocks have been highlighted by several workers (e.g. Groshong 1975; Rutter 1976; Engelder and Marshak 1985; Dunne and Hancock 1994). Pressure solution operates (Gomez-Rivas et al. 2020, p. 7) by (a) dissolution of minerals in a fluid phase at grain boundaries, especially where these are in contact; (b) diffusion of the dissolved material to zones where stress is lower due to chemical potential gradients indicated by the stress difference; and (c) precipitation in zones of lower stress.

Pressure solution is basically a process of solution transfer. It can be intracrystalline, i.e. through a crystal

Fig. 16.11 Signatures of diffusive mass transfer. Photomicrograph showing the development of a vein in a rock. The vein may have moved from a point of high normal stress to that of a low normal stress. The host rock has accommodated the transfer of material of the vein by deformation along its contact with the moving mass. (Photomicrograph by the author)

Fig. 16.12 Signatures of diffusive mass transfer as represented by the pitted surface formed due to removal of pebbles in a fine-grained sandstone of Southern Appalachians, USA. Note that the pebbly layers have undergone folding during later deformation. The knife at the centre of the photograph gives the scale. (Photograph by the author)

Fig. 16.13 Photomicrograph showing evidences of intracrystalline pressure solution. The white linear patches mark the signature of passage of solution through the crystal. (Photomicrograph by the author)

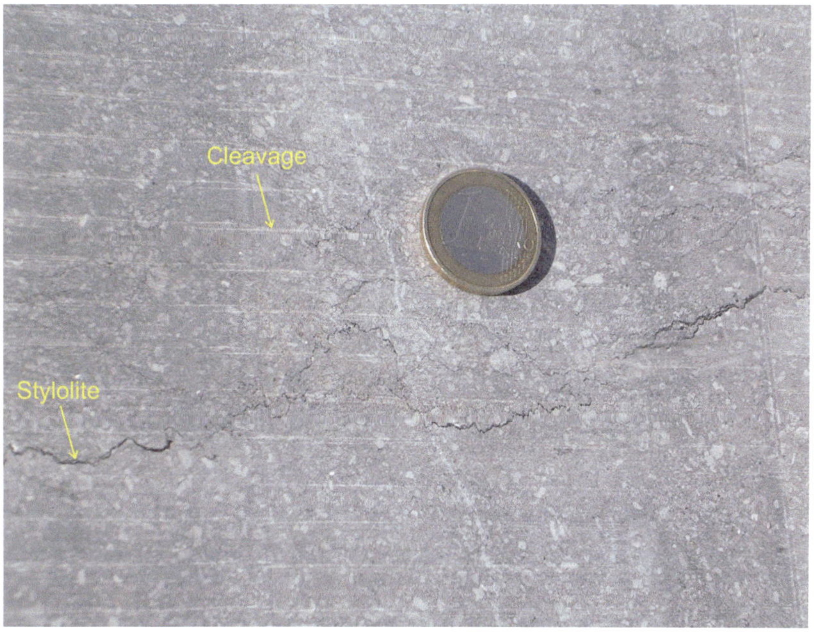

Fig. 16.14 Photograph of a limestone showing penetrative cleavage and stylolites. Both these structures are signatures of pressure solution (see text). Loc.: Near Hampteau Hotton, Belgium. (Photograph by the author)

(Fig. 16.13), or intercrystalline, i.e. along the grain boundaries.

Pressure solution is a common process operating from diagenesis to low- as well as high-grade metamorphic conditions of a rock during which a fluid phase is generally present. Since the process involves transfer of material from one part to another part of a rock, it can also be considered as a type of diffusive mass transfer. The only major difference is that diffusion by pressure solution takes place along a thin film of fluid along grain boundaries, and as such pressure solution is also sometimes called *wet diffusion*.

Pressure solution is a common mechanism of deformation of minerals and rocks. A variety of structures and textures are formed due to this process: (a) *Penetrative cleavage* (Fig. 16.14) in low-grade metamorphic rocks is believed to form due to transfer of solution from one point to another. (c) *Stylolites* (Fig. 16.14) are zigzag trails commonly observed in limestones in which they occur parallel as well as across the bedding. The zigzag structure of stylolites is due to local movement of the solution along the extension direction. The pressure-induced solution from these points moves through the network of grains of a rock followed by its precipitation at some other points, thus leaving zigzag trails. Stylolites developed at high angle to bedding are attributed to tectonic deformation of the host rock. (c) *Pitted pebbles* in some rocks are formed when soluble parts of the grains are dissolved in the high-stressed zones, thus leaving some pits or cavities in these dissolved points. Water, commonly present along the grain boundaries, dissolves a part of the grains and carries to some other points of low stress. (d) *Sutured grain boundaries* are developed due to movement of stress-induced solution along the grain boundaries of some rocks. The teeth-like minor trails thus developed across the grain boundaries represent extension. (e) *Pressure shadows* noticed on opposite sides of some rigid minerals such as magnetite and pyrite are also attributed to pressure solution. The minerals occurring in the pressure shadow zones are of different compositions from the host grain having formed due to solution transfer and consequent redeposition in the extension direction.

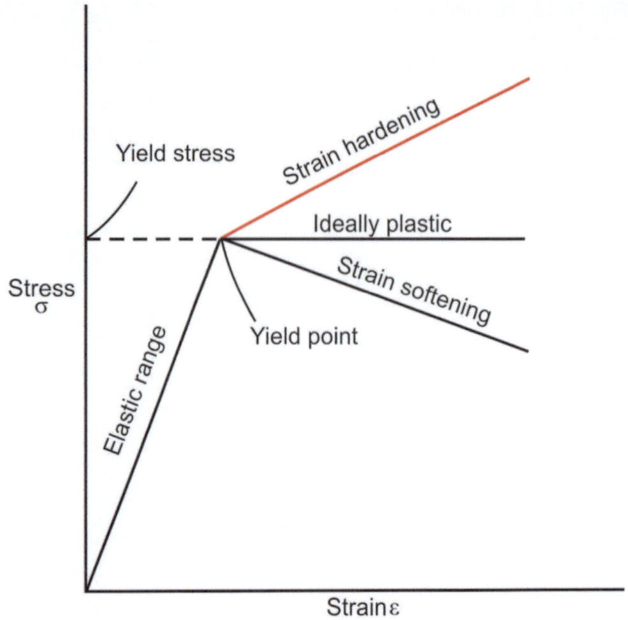

Fig. 16.15 Stress-strain graph to show strain hardening. After yield point, the graph may follow three paths: it may rise or move down or may run parallel to the strain axis. These three paths correspond to three different behaviours of the substance, i.e. strain hardening, strain softening and ideally plastic substance, respectively, in which case there is neither hardening nor softening

16.5.5 Strain Hardening and Strain Softening

Strain hardening, also called *work hardening*, is the mechanism of continuous increase of yield strength of a rock with progressive deformation. On loading, most rocks initially show an elastic behaviour during which stress remains proportional to strain (Hooke's law) as represented by the straight-line portion of the *stress-strain graph* (Fig. 16.15). If the rock has plastic property, the stresses continue to increase beyond the elastic limit and so the graph continues to rise. This rise of continuous stress to maintain further deformation is called strain hardening. The rock has now entered into plastic deformation stage. If the graph moves down till rupture, it is called *strain softening* (opposite of strain hardening) in which the substance lowers its stress level by dynamic recrystallization or dynamic recovery. Strain softening leads to the formation of shear zones and mylonites in narrow zones of intense shear deformation. And, if the stress-strain graph continues to move parallel to the strain axis, it represents an ideally plastic substance.

Strain hardening is an important mechanism of plastic deformation that results due to deformation at lattice scale. During strain hardening, the interatomic bonds break resulting in a rearrangement of atoms, thus causing modification of the original fabric of the rock. Strain hardening operates because of high dislocation density in the crystal lattices of solid substances. Increased stress is required to move these dislocations through the lattice. *This increase of stress is a measure of strain hardening.* Sometimes, the dislocations generate more rapidly than they can move through the crystal lattice, thus forming *dislocation tangles*. These tangles create obstruction to further generation of dislocations, thus resulting in strain hardening.

Before the onset of work hardening, the lattice of a crystal is defect free, i.e. free from dislocations, because of *annealing*, which is a process by which the lattice of a crystal is made defect free.

Dislocations are however not permanent defects of crystals; they decrease on heating a rock or even disappear at much higher temperature. As such, strain hardening reduces at higher temperature and disappears at much higher temperature.

16.5.6 Hydrolytic Weakening

Hydrolytic weakening is the mechanism that makes minerals and rocks weak or soft when exposed to water. A 'wet rock' is weak or soft than a 'dry rock'. The term 'water' is used to include water-related fluids. Water present in the upper crust significantly affects the physico-chemical properties of rocks, especially in the fault zones. Water is commonly present in the pore spaces of crustal rocks; this increases the pore pressure of rocks that influences the deformation behaviour of rocks and minerals.

Classic work of Griggs (1967) on hydrolytic weakening of quartz and other silicates brought into light several interesting aspects of this mechanism: (a) Rocks at high pressure and temperature become weak when brought into contact with water. (b) The weakening process is thermally activated, is reversible and is rate dependent. The rock deforms by intracrystalline glide. (c) The water-weakened quartz undergoes annealing that removes most of the dislocations and causes the formation of bubbles that are believed to be water bubbles formed by the reaction

$$Si(OH)_4 \rightarrow SiO_2 + 2H_2O$$

In this process, water migrates through the lattice. (d) Migration of water hydrolyses the strong Si–O–Si bonds that are replaced by weaker Si-OH-HO-Si. This promotes easy slip that enhances the motion of dislocations. In brief, hydrolytic weakening is the hydrolysis of Si-O bonds.

Pore fluids make the deformation behaviour of rocks to follow two different mechanisms (Duda and Renner 2013): (a) Weakening due to physico-chemical interactions between solids and fluids—this disturbs the stress level at which a rock can sustain permanent deformation and causes a reduction in the coefficient of friction (Lockner 1998). (b) Weakening due to drainage-related mechanisms—this controls the effective stress of a rock and causes the strain rate dependence of strength that significantly varies from dry to saturated rocks (Hubbert and Rubey 1959).

Hydrolytic weakening is a common mechanism operating in the rocks of the upper brittle crust and is especially effective in fault zones.

16.5.7 Crack-Seal Mechanism

The *crack-seal mechanism* was proposed by Ramsay (1980a). According to him, the fractures in a crystal propagate in discrete increments, and each incremental gap is filled up with new crystals in optical continuity with grains along the fracture walls (Fig. 16.16). The main evidence of crack-seal mechanism is the common occurrence, in many naturally deformed crustal rocks, of extension cracks and fissures that are filled with silicate or carbonate minerals. During each increment (opening) of the crack, the secondary minerals grow as fibres perpendicular to the walls of the crack. The material thus precipitated shows fibrous texture, and the fibres are oriented parallel to the direction of stretching. The crystals that make up the vein filling grow by a succession of 'crack-seal' increments due to which the elastically deforming rock fails by fracture. Subsequently, the walls of the open microcrack are sealed together by crystalline material, released by pressure solution in the rock matrix, followed by chemical transfer of this material into low-pressure fluid-filled space. Most of these extension veins are formed by an accretionary process resulting in the formation of a narrow fracture followed by filling of the open space by a crystalline material; this whole mechanism is called 'crack-seal'. The crack-seal mechanism may account for as much as 50% extension in some local situations (Ramsay 1980a). Trails of fluid inclusions commonly follow individual fracture openings. Evidences of crack-seal are therefore easily visible in light-coloured rocks such as sandstones, quartzites and limestones containing ferruginous inclusions.

16.5.8 Seismic Faulting

Sibson (1986, 1989) introduced and developed the concept that seismic activity can be considered as a process of rock deformation. According to him, geologists and seismologists have much to contribute to each other's understanding of fault processes. For example, much of the displacement occurring in the upper half of actively deforming continental crust is accomplished by earthquake faulting, and only the larger earthquakes within these regions contribute significantly to the displacement (Hyndman and Weichert 1983). Also, much of the high-level folding in the crust may also have taken place due to fault displacements, and, in some cases, incremental amplification of these folds is related to seismic slip episodes on faults (King and Vita-Finzi 1981; Stein and King 1984). Seismic activity may thus result in the formation of faults, folds and possibly some other structures. Sibson also extends the possibility of his concept to the self-similar behaviour as commonly applied to deformation structures, i.e. smaller structures mimic larger ones, especially to faults and fractures and possibly to complex structures such as thrust duplexes.

Sibson (1989) identified a *seismogenic regime* stating that seismic activity is largely restricted to the upper one-third to one-half of the deforming continental crust. From a mechanical viewpoint, the base of the regime represents a transition zone from unstable frictional faulting (FR) to quasi-plastic (QP) shearing flow in mylonitic rocks. The seismogenic regime is mechanically very important as a region of inferred peak shear resistance where larger earthquake ($M > 6$) ruptures occur. Majority of these earthquakes rupture in the upper crust. The base of the seismogenic zone, i.e. the FR/QP boundary, thus represents a zone of continuous and discontinuous shearing.

16.6 Summary

- A *mechanism of rock deformation* broadly means the process that tends to accommodate large strains in rocks, thus resulting in a stable structure/microstructure to the rock. The processes operate on different scales, but those

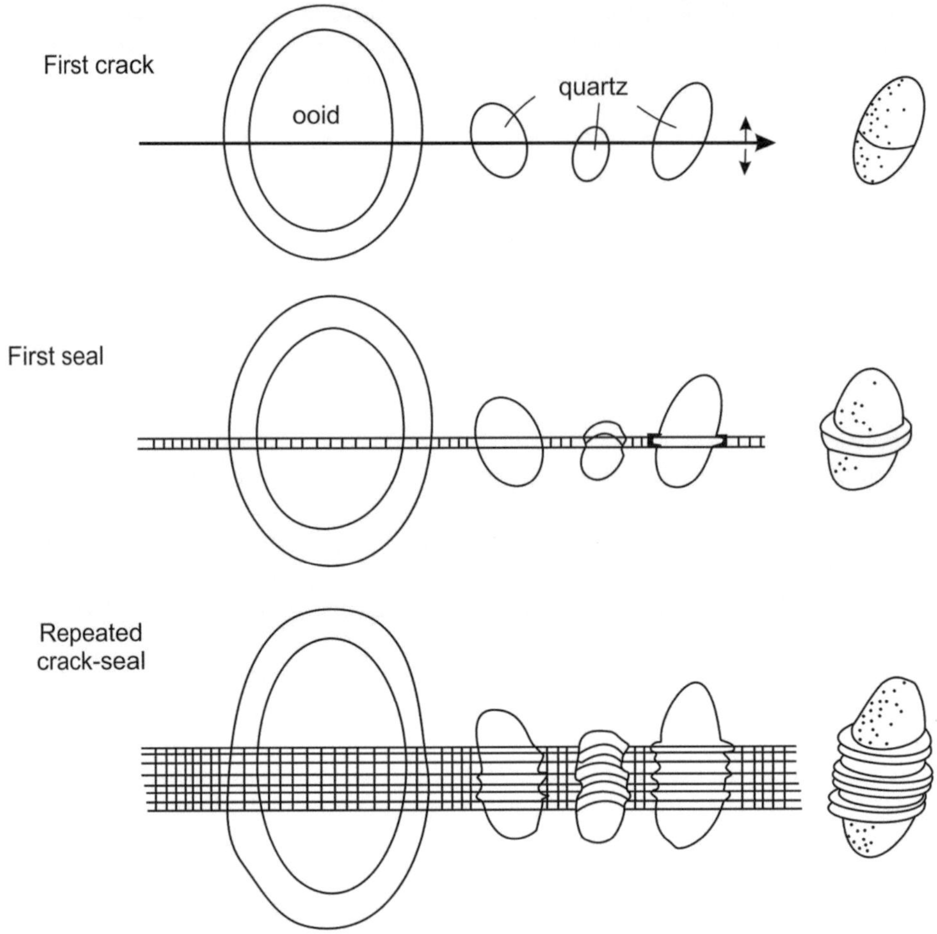

Fig. 16.16 Diagrammatic representation of the 'crack-seal' mechanism. See text for details. (Reproduced from Ramsay 1980a, Fig. 2 with permission from Springer Nature under Licence number 5217130012063)

operating on grain scales bring about significant changes to the structure and texture of rocks.

- The mechanisms act under certain specific physico-mechanical conditions of rocks. A particular mechanism may change with time depending upon the newly developed rheological state or structure/texture of the rock. This in turn may trigger some other mechanism(s) to operate.
- Common factors that control the mechanisms of rock deformation include (i) location of a rock inside the earth, (ii) temperature and pressure, (iii) fluids and gases present within the rocks and (iv) texture and grain size of the rocks.
- Temperature is believed to be the key parameter of rock deformation and formation of tectonic structures. It controls the movement of fluids and melt in the earth's crust that lead to phase transformations and changes in rock rheology.
- The various mechanisms that operate in natural rocks have been grouped here under two categories: mechanisms on microscale and those on mesoscale.
- Microscale mechanisms include recovery, dynamic recrystallization, bulge recrystallization, grain-boundary sliding, crystal-plastic deformation, diffusion creep, dislocation creep and grain-boundary pinning.
- Mesoscale mechanisms include brittle deformation mechanisms (including frictional grain-boundary sliding and cataclasis), superplastic deformation, diffusive mass transfer, pressure solution, strain hardening, strain softening, hydrolytic weakening, crack-seal mechanism and seismic faulting.
- Although most of the mechanisms are scale dependent, there are some that work on both micro- and mesoscales.

Questions

1. What do you mean by the mechanism of rock deformation?
2. What are the factors on which mechanisms of rock deformation depend?
3. Describe how dynamic recrystallization operates. Name some common changes this mechanism brings into the rocks.
4. How do you differentiate between dynamic recrystallization and recovery?

5. In how many ways the crystal-plastic deformation operates? Explain with diagrams.
6. Describe how cataclasis operates in rocks. What types of changes this mechanism brings into a rock?
7. What is superplastic deformation? Describe the conditions under which this mechanism operates.
8. What do you mean by pressure solution? Explain how the mechanism operates.
9. Outline the differences between strain hardening and strain softening?
10. Explain crack-seal mechanism with examples.

17 Shear Zones

A. R. Bhattacharya, *Structural Geology*, Springer Textbooks in Earth Sciences, Geography and Environment,
https://doi.org/10.1007/978-3-030-80795-5_17

Abstract

Shear zones are tabular bodies of rock that accommodate the bulk or whole deformation so that there is practically no or less deformation outside this zone. As such, they constitute anisotropy in a rock mass. An ideal shear zone is bounded by two parallel surfaces. Shear zones can occur in any rock type and can develop in various geologic settings, commonly in contractional, extensional and strike-slip settings. Most shallow-level faults continue at deeper levels where they form ductile shear zones containing rocks deformed under high temperature-pressure conditions and over a wide zone ranging up to tens of kilometres. Shear zones are characterized by the occurrence of some typical rocks because deformation pattern within a shear zone is different from that of the outside rocks. Common brittle shear zone rocks include *breccia, cataclasite* and *gouge* while those of ductile shear zones include *mylonite*. Large-scale ductile shear zones are characterized by progressive development of crystallographic preferred orientation (CPO) of their minerals, mainly quartz. Formation of shear zones depends upon the mechanisms that are able to localize strains in a narrow zone. Most shear zones grow in length and thickness over time. We take you through this chapter to explore the concept of shear zones, their geometry, rock types, classification, formation and significance.

Keywords

Shear zone · Brittle shear zone · Ductile shear zone · Breccia · Cataclasite · Mylonite · Mylonitic foliation · Shear heating · Geometrical softening

17.1 Introduction

Rocks of the earth's crust are lithologically heterogeneous, and the distribution of strain at different points is variable. There are zones where strain is localized much more than the surrounding. Such zones are called *shear zones. A shear zone is a zone in which strain is clearly higher than in the wall rock and whose margins are defined by a change in strain, typically seen by the rotation of pre-existing markers or formation of a new fabric* (Fossen and Cavalcante 2017, p. 435). Localization of shear strain occurs at all depth levels of the crust and therefore at all deformation regimes. Thus, strain localization produces *faults* at shallower levels of the crust where the rocks deform in the brittle regime and where strain is localized along a narrow, tabular zone with a measurable width. Faults often continue at depth where they enter the ductile regime of deformation where strain is generally distributed along a much wider zone. Faults in such zones are called *ductile shear zones*. For all practical purposes, faults and ductile shear zones can be considered under the category of *shear zones*. Shear zones thus occur all through the crust. Rocks within a shear zone are highly deformed, while those outside are less deformed or undeformed. A shear zone thus constitutes a zone of anisotropy in a rock mass and is able to accommodate differential movements in rocks of the crust as well as those of the mantle.

Shear zone may range in length from microscopic to several tens or hundreds of kilometres, and the width may also vary from microscopic to several kilometres. Shear zones can develop in any type of rock of any geologic age and under various geologic settings. Common geologic settings include contractional, extensional and strike-slip. However, the geometry of shear zones may vary in different geologic settings. For example, those developed in contractional and extensional settings are low angle, while in strike-slip setting these are steep to vertical. Our knowledge on shear zones has vastly increased since the 1980s. Since then, shear zones have drawn the attention of geologists all over the world.

17.2 Deformation Domains

A shear zone is associated with two contrasting domains: one where the bulk strain is concentrated and the other with less or practically no strain in rocks. Thus, the intensely deformed rock mass is separated by two planar surfaces at the margins (Fig. 17.1), each of which is called a *shear zone boundary* beyond which occur the *wall rocks* containing undeformed or less deformed rocks. *Shear plane* is parallel to the walls of a shear zone and contains the shear direction (the displacement vector) (Allaby 2008). Within a shear zone, the rocks deform by continuous but heterogeneous strain. However, other modes of deformation are also possible.

Box 17.1 Ductile Shear Zone-Brittle Fault Analogy

In orogenic belts, most large-scale ductile shear zones formed at deeper levels pass upwards as brittle shear zones in the upper crust. The upper crust is generally represented by cover sediments deposited over crystalline basement. Shortening of the basement results in an uplift of the internal part due to crustal thickening (Ramsay 1980b). As a result, the ductile shear zones move upwards to be transformed into ductile-brittle shear zones that with further rise become brittle shear zones in the upper crustal layers. In the sedimentary layers of the cover rocks, the shear zones move along

(continued)

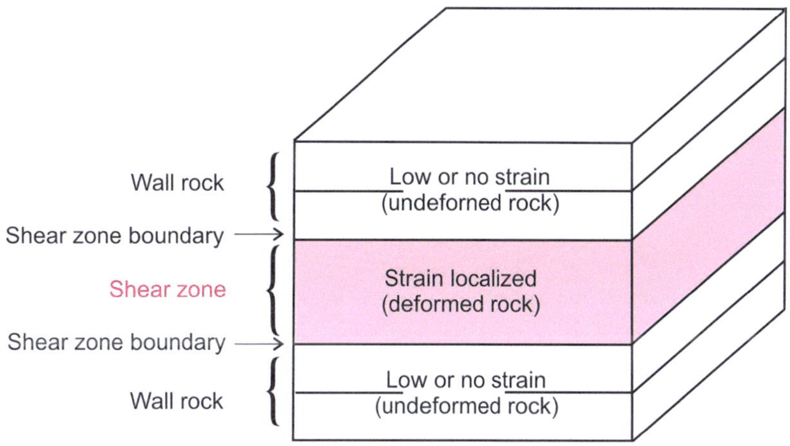

Fig. 17.1 Three-dimensional block diagram showing a shear zone, shear zone boundary and wall rock

Box 17.1 (continued)

the layering and step up to reach the surface as narrow, planar brittle shear zones. Thus, most large-scale ductile shear zones formed at deeper levels under high temperature-pressure conditions are represented as narrow, brittle shear zones on the surface.

From surface to depth, such faults represent variation of deformation mechanisms. At higher crustal levels, displacement along small fracture planes is dominant. This leads to *cataclastic flow* of rock masses. At deeper levels, especially below the brittle-ductile transition (10–15 km), fractures disappear because of higher temperature. This is the regime of ductile deformation where rocks deform at low strain rates dominantly by crystal plastic and dynamic recrystallization processes. Thus, there exists an analogy of ductile shear zones formed at deeper levels with brittle shear zones formed at shallower levels of the crust. The only major difference is that in the former strain localization and displacement occur by ductile deformation processes, while in the latter these occur by brittle deformation processes. Presence of a fault is in general an indication of near-surface deformation while that of a wide ductile shear zone is an indication of long exhumation of an orogenic belt. Most crustal scale shear zones are thus believed to have a narrow planar fault at the top and a wide zone of ductile deformation at the base.

17.3 Shear Zone and Finite Strain Axes

For a systematic study, let us consider a shear zone as a tabular body in the framework of finite strain axes *X*, *Y* and *Z* (Fig. 17.2). All the conditions of a shear zone are suitably met along the *XZ* section. It is therefore better to study a shear zone in the *XZ* sections (Fig. 17.2), which represents the

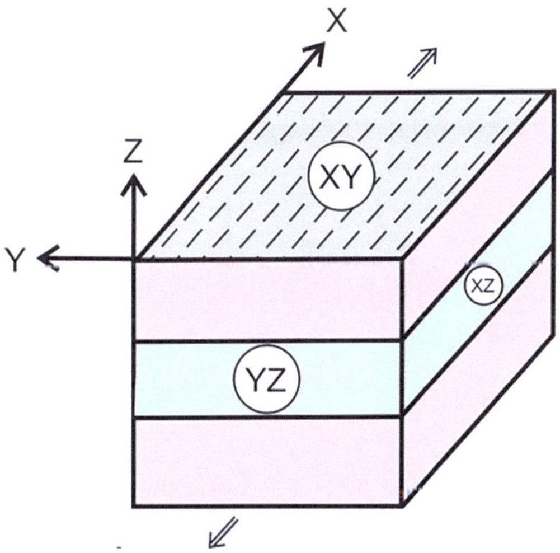

Fig. 17.2 Block diagram showing a shear zone in the framework of finite strain axes

optimum values of extension (*X*) and shortening (*Z*). The *XZ* section also represents the extension of particles parallel to *X*-plane and the corresponding shortening parallel to *Z*. Detailed anatomy of a simple shear zone is shown in Fig. 17.3 in which the various geometric elements are shown.

17.4 Strain Within a Shear Zone

In most shear zones, the deformation is continuous and therefore the marker planes maintain continuity within the shear zone. In an ideal shear zone, ductile strain progressively increases towards the centre of the shear zone (Fig. 17.4). Thus, between two (symmetrical) shear zone boundaries, shear strain shows a typical profile with the highest shear strain at the centre of the shear zone. Under the influence of simple shear, particles show progressive rotation together with sympathetic extension along *X* and shortening along *Z*, as better noticed in the *XZ*-plane.

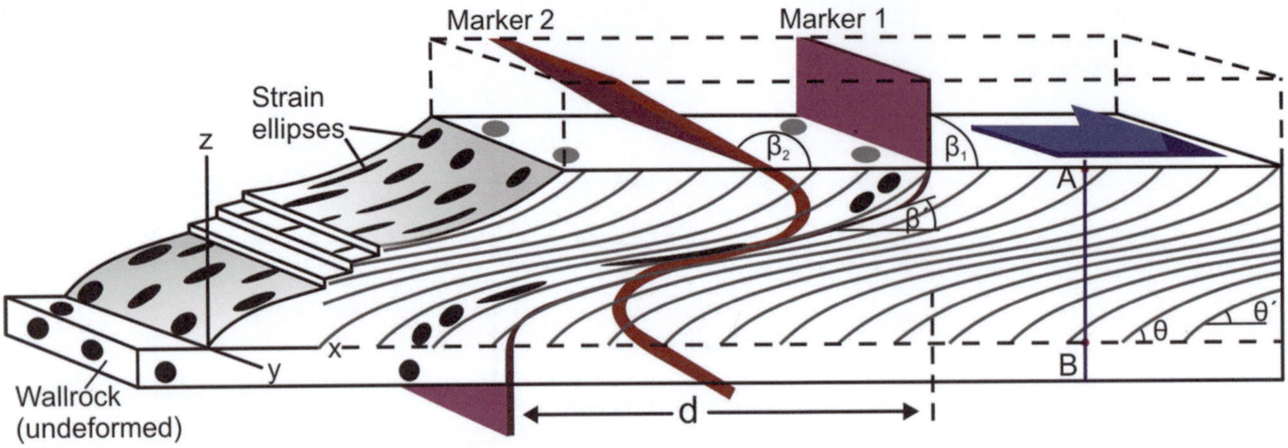

Fig. 17.3 Anatomy of a simple shear zone showing strain ellipses and the newly formed foliation. Note progressive changes in angular relations of marker lines (shown as curved lines) in different parts of the shear zone. (Reproduced from Fossen and Cavalcante 2017, Fig. 2a with permission from Elsevier Senior Copyrights Coordinator, Edlington, U.K. Submission ID: 1193222)

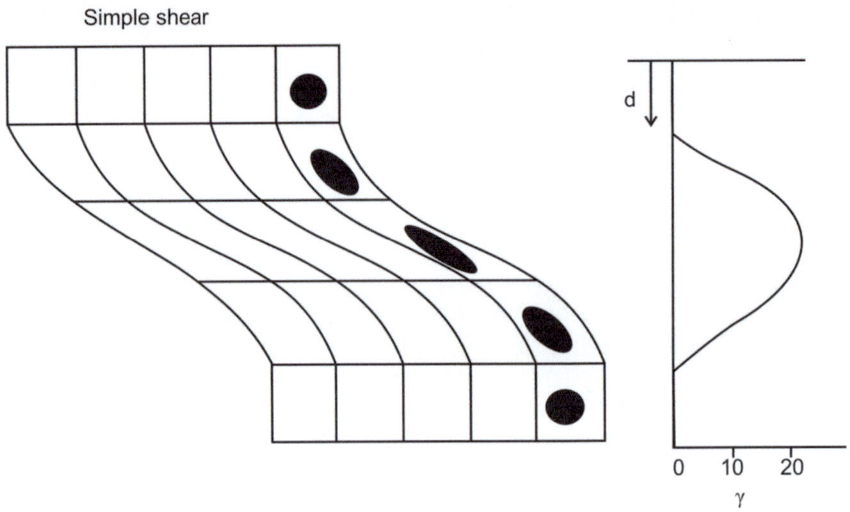

Fig. 17.4 Strain within a shear zone. In an ideal shear zone under simple shear conditions, the ductile strain on *XZ*-plane shows a progressive increase from shear zone boundary towards the centre of the shear zone, where it reaches its maximum value. This is accompanied by the rotation of particles in the direction of shear together with sympathetic extension along *X* and shortening along *Z*. The graph on the right-hand side represents the progressive increase of ductile strain in the shear zone. (Reproduced from Ramsay and Huber 1987, Fig. 26.4a with permission from Elsevier Science & Technology Journals. Request ID: 600072887)

What we have described above is an ideal case when the shear zone is symmetrical in which case the centre region shows the highest amount of strain. This is true for shear zone of all sizes, ranging from small outcrop to several kilometres. The large-scale or crustal scale shear zones however are generally not symmetric due to lithological and textural variations, coupled with the effects of later deformations. Even in such large-scale shear zones, the central part shows concentration of the highest strain, as we have demonstrated in Box 17.2.

Box 17.2 Strain Associated with Large Shear Zones

Most ductile shear zones are characterized by the concentration of large ductile strains. In Ramsay's (1980b) model, the strain shows a progressive increase towards the centre of the shear zone from both of its boundaries. This model is also reflected in field where the finite strain increases towards the main thrust. However, due to lithological and textural constraints, coupled with

(continued)

Box 17.2 (continued)
the effects of later deformations, this model does not follow ideally, and as such finite strain may show irregularities at several places within the shear zone.

We present here an example in which a crustal scale shear zone, the Main Central Thrust (MCT) of the Himalaya, shows progressive increase of fold-flattening strain towards the trace of the main thrust. The MCT is an important, E-W trending, geotectonic element of the Himalaya along which a part of the crystalline zone of the Greater Himalaya has been thrust over the younger sedimentary belt of the Lesser Himalaya (Gansser 1964). We have studied (Bhattacharya and Siawal 1985) an area of about 250 sq. km covering a 20 km strike length of the MCT in the Tejam-Girgaon area of eastern Kumaun Himalaya, India. Flattening strain of folds has been estimated by the method suggested by Ramsay (1962) along six N-S traverses across the strike of the rock formations as well as across the trace of the MCT, covering both the crystalline zone to the north (hanging wall) and the sedimentary belt to the south (footwall). The data on flattening strain have been presented for each traverse on a graph with respect to geographic distance from the MCT (Fig. 17.5). The strain data have also been presented on a geological map (Fig. 17.6) that shows the amount of strain at the locations where measured.

Measurement of strain and graphical representation of strain data of the area studied clearly indicate that (i) the trace of the MCT is characterized by the highest value of flattening strain: usually >90% in the crystalline rocks and >60% in the sedimentary rocks; (ii) the strain values gradually decrease away from the thrust within a tract of 5–6 km extent on either side of the MCT; and (iii) outside this tract, the values of strain reduce to normal values, i.e. up to about 45% in the crystalline rocks and up to about 40% in the sedimentary rocks (Fig. 17.5). The MCT thus appears to be the controlling factor responsible for the observed stronger flattening strain of the folds both in the hanging wall and in the footwall. The spatial distribution of strain (Fig. 17.6) shows that there is lateral continuity of the amount of flattening strain that enables construction of *strain contours* that gradually show an increasing pattern towards the trace of the main thrust.

The above example lucidly demonstrates that the centre of even large/crustal scale shear zone is also characterized by the concentration of highest strain and that Ramsay model holds good even for large-scale shear zones.

17.5 Classification of Shear Zones

Shear zones show a variety of shapes that are primarily a reflection of their mode of formation. Different workers have identified some specific parameters of shear zones from both geometric and genetic angles. As such, shear zones have been classified in several ways. Some common classifications are described below.

17.5.1 On the Basis of Geometry of Shear Zones

Ramsay (1980b) classified shear zones into three broad categories (Fig. 17.7) on the basis of their geometry that mainly depends on their modes of formation. A *brittle shear zone* (Fig. 17.7a) is formed by brittle deformation processes and is characterized by the presence of a fault that makes a marker line discontinuous (displaced) within the shear zone. Shear strain is mainly localized along the fault, while the remaining parts of the shear zone remain unstrained. A *brittle-ductile shear zone* (Fig. 17.7b) is formed when both brittle and ductile deformations operate together. A marker line shows the signature of rotational movement (curvature) as well as a discontinuity in the form of a fault. A brittle-ductile shear zone is sometimes marked by the presence of tension gashes. Depending upon the dominance of brittle or ductile processes, this type of shear zone can be named either brittle-ductile shear zone or ductile-brittle shear zone. A *ductile shear zone* (Fig. 17.d) is formed by ductile deformation processes. A marker line shows a sigmoidal shape indicating continuous deformation within the shear zone.

The three types of shear zones have received different names (Davis et al. 2012; Fossen 2016) though all lead to the same mode of formation in each case. While the brittle shear zone is also called *frictional shear zone*, the brittle-ductile shear zone is called *semi-brittle shear zone*. Some workers prefer to use the term *plastic shear zone* for a ductile shear zone as they assign plastic deformation processes for its formation. As a matter of fact, the ductile and plastic deformation processes should not be considered as one and the same even though both these processes can form this type of shear zone.

17.5.2 On the Basis of Kinematics of Deformation

The kinematics of deformation in the context of shear zones broadly means the relative movement of the two walls. The latter can be achieved by simple shear, pure shear, compaction or dilation as diagrammatically represented in Fig. 17.8.

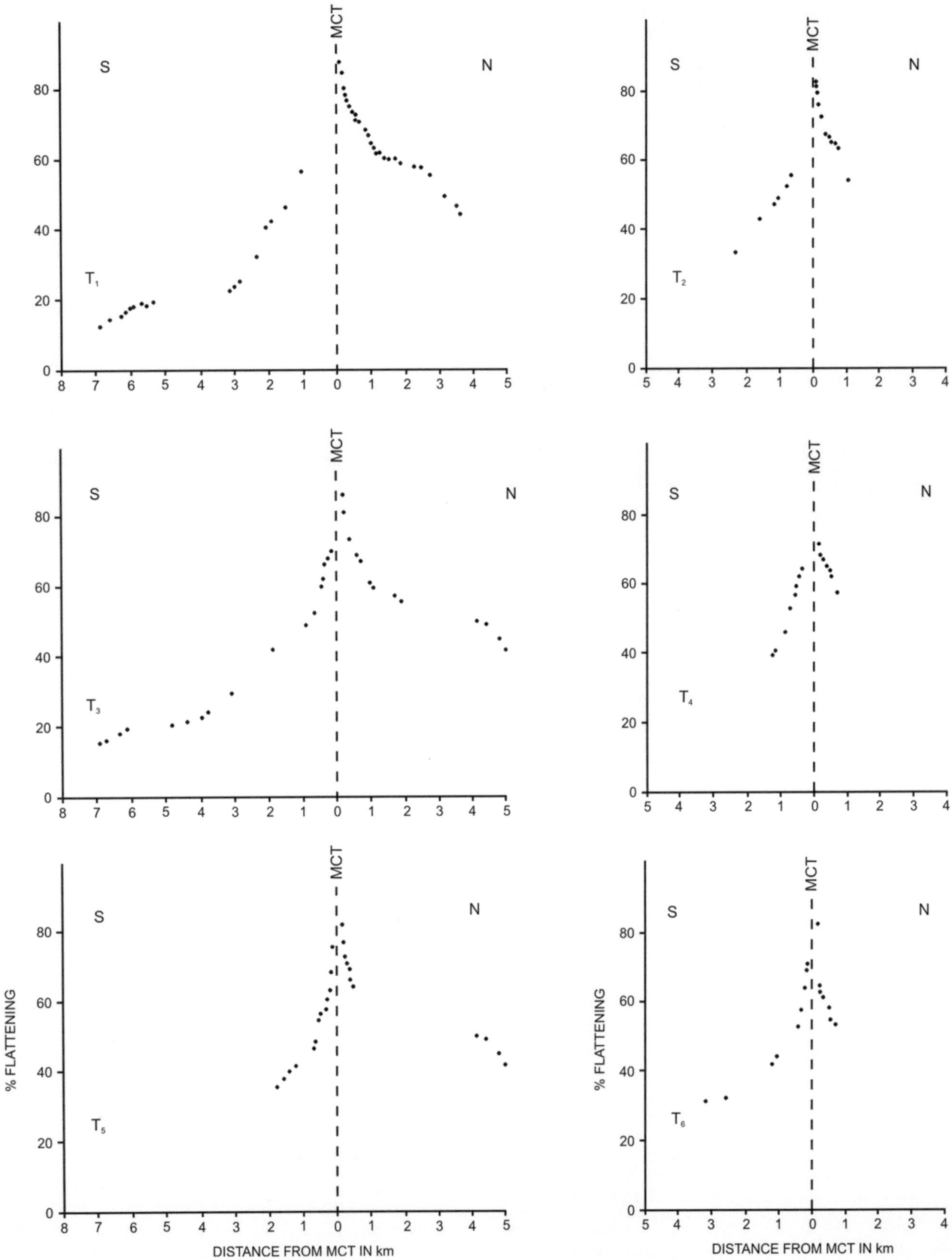

Fig. 17.5 Graphical representation of fold-flattening strain data for six traverses (T_1–T_6) in relation to distance from the MCT in the Tejam-Girgaon area, Kumaun Himalaya, India. Note that the amount of strain reaches its maximum value at the trace of the MCT in each traverse. (Reproduced from Bhattacharya and Siawal 1985, Fig. 6, with permission from KNGMG, the Netherlands)

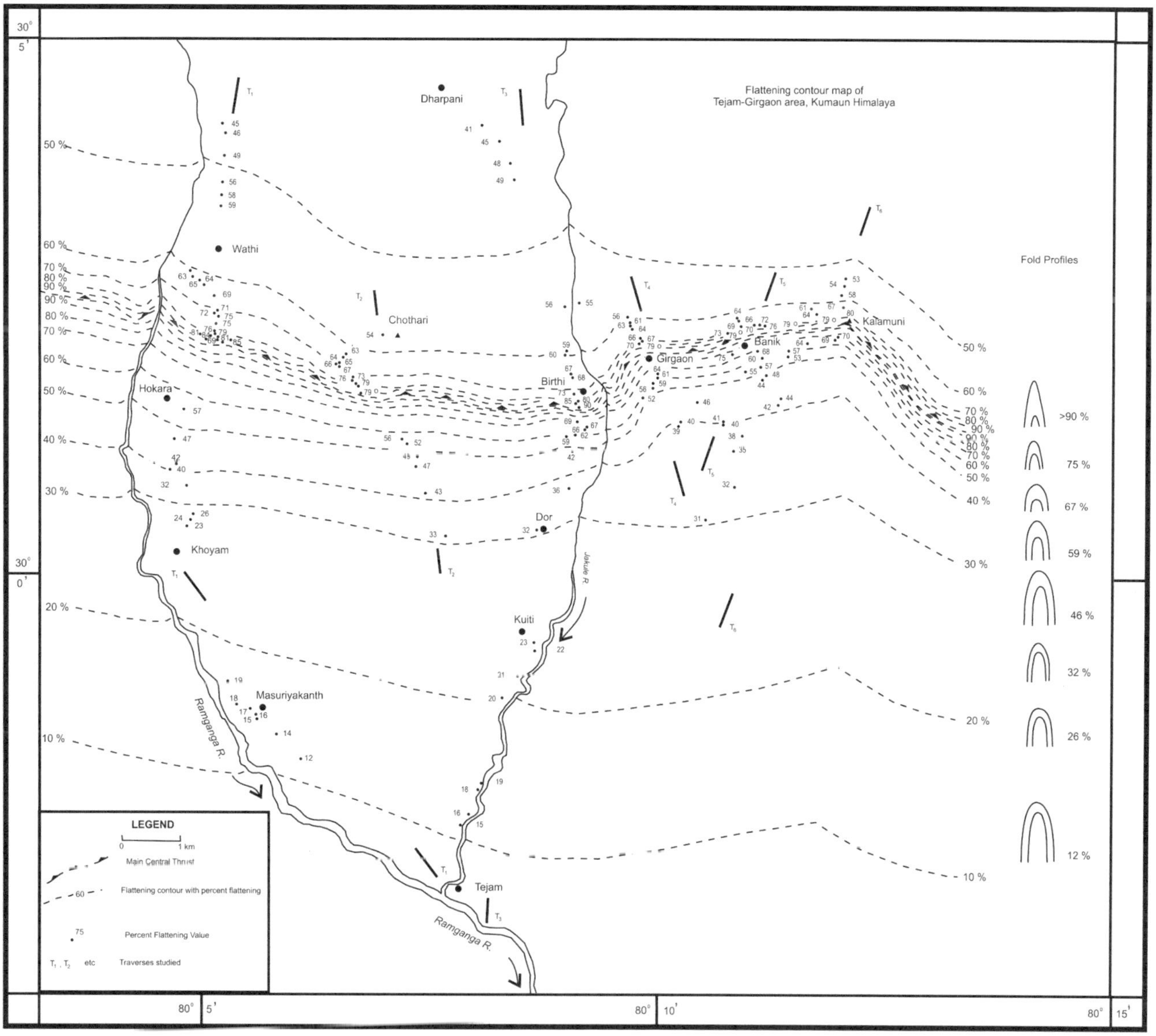

Fig. 17.6 Map showing spatial distribution of flattening strain data on either side of the MCT. Amount of strain is given at every point where measured. On the basis of spatial distribution of data, 'flattening contours' have been constructed. Note a progressive increase of strain data towards the MCT where the strain reaches its maximum value from either side. (Reproduced from Bhattacharya and Siawal 1985, Fig. 7, with permission from KNGMG, the Netherlands)

These four extreme types can thus be named as *simple shear zone*, *pure shear zone*, *compaction shear zone* and *dilation shear zone*. We assign these as *S-type*, *P-type*, *C-type* and *D-type* shear zones, respectively. The pure shear zones can be of compressional as well as of extensional types. Several intermediate types are also possible, e.g. *compactional simple shear zones*, *compactional pure shear zones*, *dilational simple shear zones* and *dilational pure shear zones* (see Fossen and Cavalcante 2017). Despite the possibility of formation of shear zones by these kinematic ways, most shear zones are by far believed to form by simple shear.

17.5.3 On the Basis of Microscale Deformation Mechanisms

On the basis of microscale deformation mechanisms, shear zones can be classified into two: (a) *plastic* (also called *crystal-plastic* or *viscous*) *shear zones* in which crystal-plastic mechanisms (dislocation creep and twinning) and diffusion dominate and (b) *frictional* (or *brittle*) *shear zones* in which brittle deformation mechanisms (grain fracturing, frictional sliding and grain rotation) dominate. Between these two end members, there are many shear zones that show

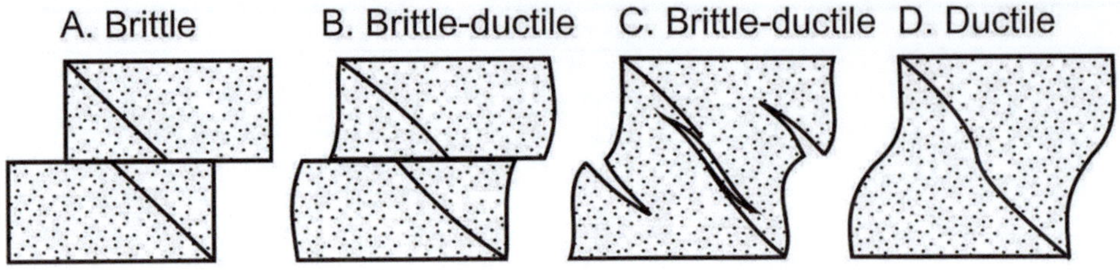

Fig. 17.7 Classification of shear zones by Ramsay on the basis of their geometry. (**a**) Brittle shear zones, (**b**, **c**) brittle-ductile shear zones and (**d**) ductile shear zones. (Reproduced from Ramsay and Huber 1987, Fig. 26.1 with permission from Elsevier Science & Technology Journals. Request ID: 600072887)

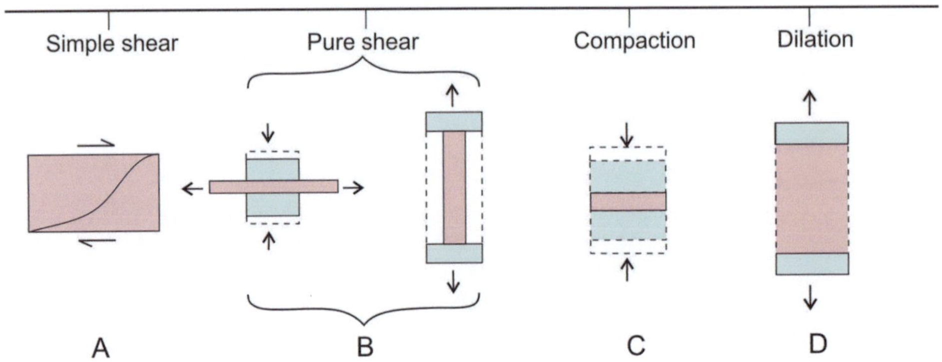

Fig. 17.8 Shear zones from the viewpoints of kinematics of deformation. Four ideal (extreme) cases of kinematics of deformation can be considered: simple shear, pure shear, compaction and dilation. These four extreme types can thus be considered to constitute simple shear zone (**a**), pure shear zone (**b**), compaction shear zone (**c**) and dilation shear zone (**d**). Pure shear zone can be of compressional and extensional types

components of both plastic and brittle (frictional) deformation mechanisms. For such shear zones, terms like *brittle-plastic shear zones* (Rutter 1986) or *frictional-viscous shear zones* (Stipp et al. 2002) or *frictional-plastic/brittle-viscous shear zones* (Fusseis and Handy 2008) have been used.

17.5.4 On the Basis of Relative Displacement of Rocks

Davis et al. (2012) classified shear zones on the basis of relative displacement of rocks on *opposite* sides of a shear zone, which gives the sense of shear of the shear zone. They identified four types of shear zones (Fig. 17.9): strike-slip, normal, reverse and oblique-slip. *Strike-slip shear zones* may be right-handed or left-handed (Fig. 17.9a,b). *Normal-slip shear zones* show hanging wall displacement *downward* relative to the footwall (Fig. 17.9c). *Reverse-* and *thrust-slip shear zones* show hanging wall displacement *upward* relative to the footwall (Fig. 17.9d). *Oblique shear zones* show components of both strike-slip and dip-slip. For sub-horizontal or variably dipping shear zones, the shear sense is described by the way the hanging wall moved, such as '*top to the west*' (Fig. 17.9e). For vertical shear zones with a dip-slip component of motion, terms like '*west-side up*' (Fig. 17.9f) or '*northeast-side down*' are used to convey the sense of shear.

17.5.5 Shear Zones Under Plate Tectonic Settings

The plate boundaries constitute suitable sites for the formation of shear zones. Considering this aspect, Davis et al. (2012) grouped shear zones under (a) continental collision-related shear zones, e.g. Himalayan thrust belt formed due to collision of India with Asia: these shear zones are associated with crustal shortening and thrust displacements due to which older, deeper rocks are brought at higher level and are then juxtaposed against younger rocks; (b) continental rifting-related shear zones, e.g. the African rift; and (c) transform faulting-related shear zones, e.g. seismically active strike-slip zones of the San Andreas Fault of California.

17.5.6 Ductile and Brittle Shear Zones

A simple classification of shear zones involves only two end members and is therefore called *ductile shear zones* and *brittle shear zones*. Ductile shear zones show a continuous

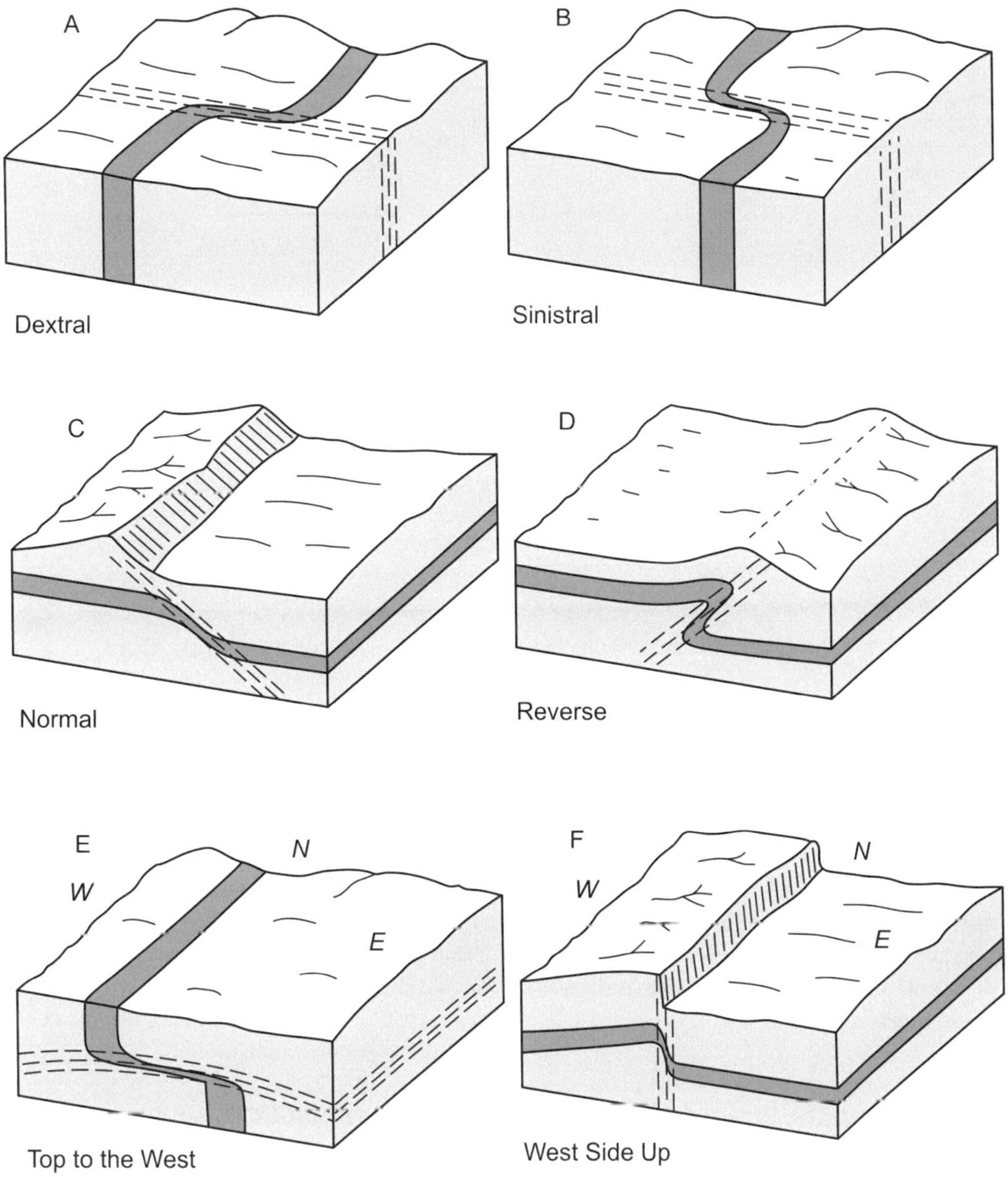

Fig. 17.9 Classification of shear zones on the basis of relative displacement of rocks on opposite sides of a shear zone. (Reproduced from GH Davis, JR Stephen, CF Kluth, 2012, Structural Geology of Rocks and Regions, third ed., Fig. 10.13 with permission from John Wiley & Sons. Request ID: 600067221)

displacement gradient across the zone, while brittle shear zones show displacement discontinuities. All plastic shear zones are ductile, but not all ductile shear zones are plastic; for example, sediments and sedimentary rocks can develop ductile shear zones in which the grain-scale deformation is purely frictional (Fossen and Cavalcante 2017, p. 436). A brittle shear zone is characterized by discontinuous deformation in which a continuous marker is displaced by a shear fracture. However, a *ductile-brittle shear zone* shows both continuous and discontinuous deformation.

17.5.7 On the Basis of Progressive Deformation or Flow

The coaxiality of deformation, which is described by the kinematic vorticity number W_k, has been used by Tikoff and Fossen (1993) to distinguish non-simple shear zones from simple shear zones. The authors relate W_k to simple and coaxial strain components because for steady-state flow W_k refers to an increment for the entire period of deformation. Thus, for *perfect shear zones*, $W_k = 1$. For most natural shear

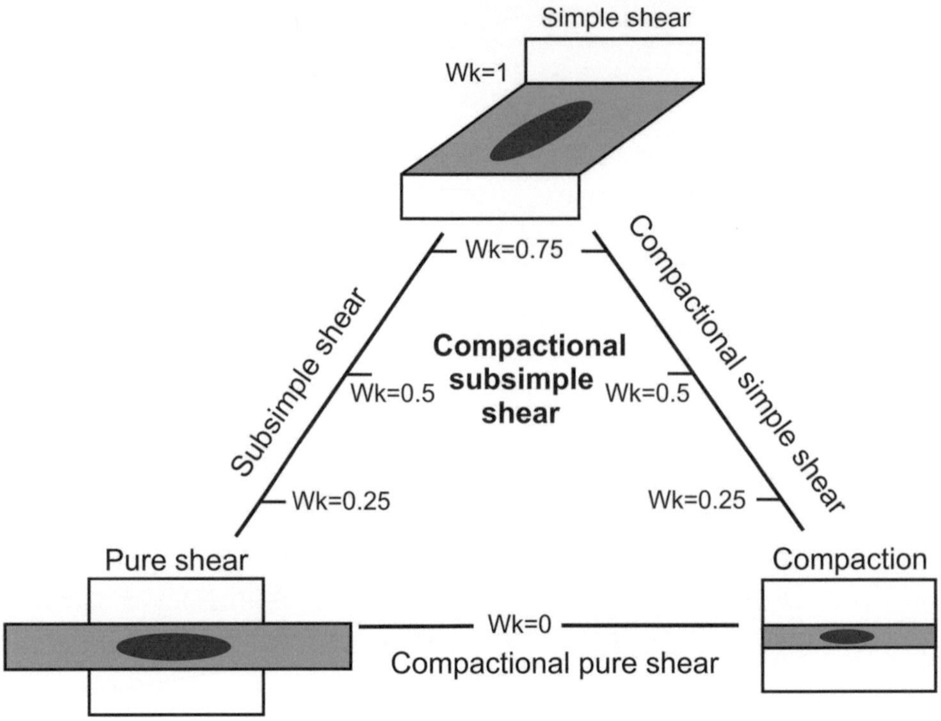

Fig. 17.10 Shear zones as classified on the basis of progressive deformation or flow. See text for details. (Reproduced from Fossen and Cavalcante 2017, Fig. 6 with permission from Elsevier Senior Copyrights Coordinator. Edlington, U.K. Submission ID: 1193222)

zones, W_k is lower and remains between 1 and 0. A *subsimple shear zone* shows deformations between simple and pure shear, and W_k shows values ranging from 0.6 to 1. All these have been represented by a triangular diagram (Fig. 17.10) that relates simple shear, pure shear and orthogonal compaction as end members. The diagram also defines the range of subsimple shear, compactional simple shear and compactional pure shear in terms of W_k.

17.5.8 Classification on the Basis of Shear Zone Profile

The marker planes within a shear zone and along the shear zone boundaries may show a variety of profiles. We identify six types of shear zones on the basis of shear zone profiles.

17.5.8.1 Shear Zone with Sigmoidal Foliation

The ductile shear zones are commonly characterized by sigmoidal foliation (Fig. 17.11) that is generally developed at an angle of about 45° with the shear zone walls.

This foliation gradually reduces its angle with gradual increase of shear strain towards the centre of the shear zone, where it shows maximum curvature and where the finite strain reaches the highest value. The foliation thus becomes sigmoidal within the shear zone. Adjoining foliations commonly show contrasting lithology that imparts a strong rheological contrast to promote shear deformation due to which the foliations easily yield to shear stress.

17.5.8.2 Shear Zone with Planar Foliation

Instead of showing a curved pattern, the foliations in some shear zones are disposed in a planar set (Fig. 17.12). The individual foliations are parallel to each other and make acute angles with the shear zone walls, where they give the sense of shear. The shear zone boundaries are parallel to each other. In ductile shear zones, the marker planes are given by mylonite layers, while in brittle shear zones, they are commonly given by a set of parallel joints.

17.5.8.3 Asymptotic Shear Zones

There are shear zones in which one of the boundaries is flat while the other is curved such that the shear zone looks asymptotic (Fig. 17.13). The foliation pattern within the shear zone is curved. However, the sense of shear shown by the foliation pattern remains unchanged.

17.5.8.4 Tapering Shear Zones

In some shear zones, the shear zone boundaries taper at one end (Fig. 17.14). The geometry of the foliation within the shear zone remains more or less uniform even near the termination.

17.5.8.5 Convergent Shear Zones

In some shear zones, the shear zone boundaries converge at one or both the ends (Fig. 17.15). The geometry of the foliation within the shear zone remains more or less uniform even near the termination.

Fig. 17.11 Sigmoidal shear zone

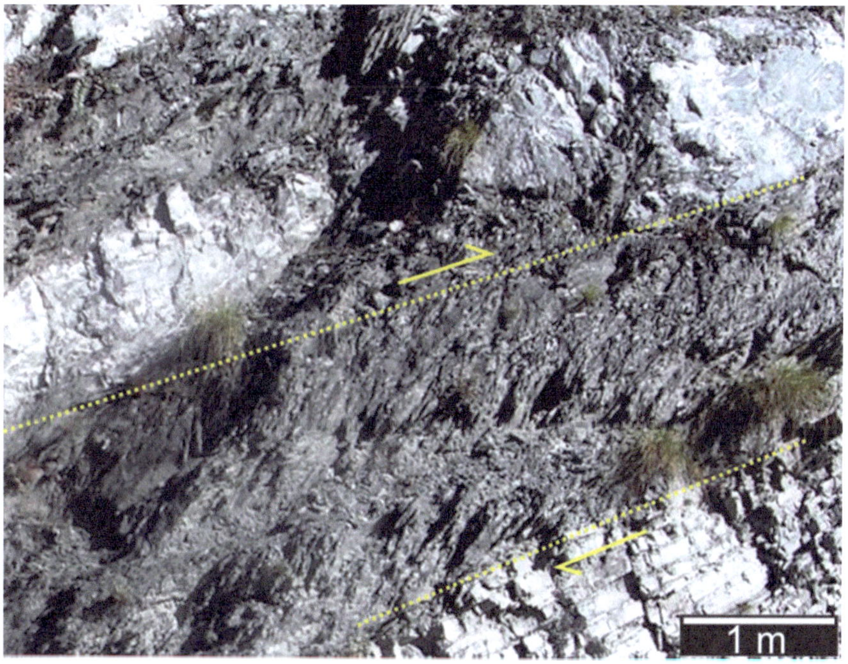

Fig. 17.12 Planar shear zone

17.5.8.6 Anastomosing Shear Zones

Occasionally, a number of foliations show anomalous patterns confined between two planar surfaces. Such types of shear zones are named here as *anastomosing shear zones* (Fig. 17.16) that are commonly noticed in terrains with highly metamorphosed rocks and migmatites. The rocks may have undergone either several events of deformation or deformation under highly ductile conditions within the shear zone. However, some other reasons are also possible for the formation of this type of shear zone.

17.5.9 Single and Multiple Shear Zones

Shear zones can occur as single or in groups. When more than one shear zone occurs together, the mode of their origin may be same or different, and they may have formed in one deformational event or more. We thus identify two types of shear zones: single shear zone and multiple shear zones.

Fig. 17.13 Asymptotic shear zone

Fig. 17.14 Tapering shear zone

Fig. 17.15 Convergent shear zone

17.5.9.1 Single Shear Zone

In an isotropic rock mass, ductile shear strain may be localized at a point. As a result, the foliation as defined by preferred orientation of flaky, platy and flattened minerals initially tends to make an angle of about 45° with the shear zone boundaries. With increase of strain, the foliation progressively rotates towards parallelism with the shear zone boundary. This produces a sigmoidal pattern (Fig. 17.17) of the foliation in the *XZ* section of the shear zone. The stretching of the flattened grains in the direction of extension (*X*) gives rise to a stretching lineation in the *XY*-plane. The acute angle that the foliation makes with the shear zone boundaries gives the sense of shear. In an ideal shear zone, the foliation shows maximum curvature at the centre of the shear zone where shear strain is the highest. The curvature of foliation is better manifested in rocks showing well-developed foliation such as metamorphic rocks and also if the deformation occurred under ductile conditions. The foliation under such conditions shows a smooth (curved) geometry that is indicative of a continuous strain gradient on either side of the central plane of the shear zone.

Fig. 17.16 Anastomosing shear zone

17.5.9.2 Multiple Shear Zones

Under conditions of local strain regime, several layers of rocks may develop shear zones at one place that are called *multiple shear zones*. Here, the shear zones occur in more or less parallel sets and thus constitute a *shear stack* (Fig. 17.18). Commonly, the individual shear zones in a shear stack show the same sense of shear, possibly suggesting their formation during one single event of ductile shear deformation. However, in some shear stacks, reversal of the sense of shear is also indicated possibly due to intervention of periods of reversal of shear.

17.6 Shear Zone Rocks

Shear zone rocks have been classified in several ways of which Sibson's (1977) classification of fault rocks seems to be more practical. Mylonite is the most common shear zone rock, while other rock types have been described in Chap. 9.

Mylonite is a foliated rock formed by strong ductile shearing (dominantly simple shear) and is characterized by grain-size reduction. Mylonite is nowadays believed to be an important rock type of ductile thrusts having formed at middle to lower levels of the crust, where high temperature-pressure regime favours grain-size reduction.

During ductile shear deformation, the foliation is partly or wholly overprinted by a newly formed foliation, the *mylonitic foliation* (Fig. 17.19) that can be observed in both hand specimens and under microscope. In outcrops (Figs. 14.29 and 14.30), mylonites occur as fine-grained rocks showing

Fig. 17.17 Single shear zone in quartzite of Garhwal Lesser Himalaya, near Nandprayag, India. (Photograph by the author)

Fig. 17.18 Multiple shear zones in the form of a stack of shear zones as developed in fine-grained sandstone beds of the Kumaun Lesser Himalaya, near Bhimtal, India. (Photograph by the author)

Fig. 17.19 Photograph of a slide of a mylonite. Mylonitic foliation, porphyroclasts and fine-grained matrix can be seen. (The slide was prepared at Aachen University of Technology, Germany, Courtesy Professor J.L. Urai)

bands that define mylonitic foliation. In thin sections (Figs. 17.20, 17.21, 17.22 and 17.23), one can see the fine-grained character of the rock showing distinct foliation planes and stretched porphyroclasts.

Like cataclasite, mylonite is also subdivided into protomylonite, mylonite and ultramylonite. Since the matrix of a mylonite is a product of grain-size reduction during crystal-plastic and diffusional processes, the grain size of matrix material has been given due importance in classifying mylonites that have thus been subdivided (Sibson 1977) into three types on the basis of proportion of matrix: (a) *protomylonite* (Fig. 17.20): matrix 10–50% of the rock, grain size of matrix material >0.5 mm; (b) *mylonite* or *orthomylonite* (Figs. 17.21 and 17.22): matrix 50–90%, grain size of matrix material <0.5 mm; and (c) *ultramylonite* (Fig. 17.23): matrix >90%, grain size of matrix material <0.1 mm.

17.7 Fabric Development in Shear Zones

Most large-scale shear zones are characterized by the development of typical fabric in deformed rocks. Quartz is the most common mineral of crustal rocks, and it is highly sensitive to crystal-plastic deformation during shear movements and thus develops its typical *c*- and *a*-axis

Fig. 17.20 Protomylonite with a few porphyroclasts. See text for details. (Photomicrograph courtesy Amit K. Verma)

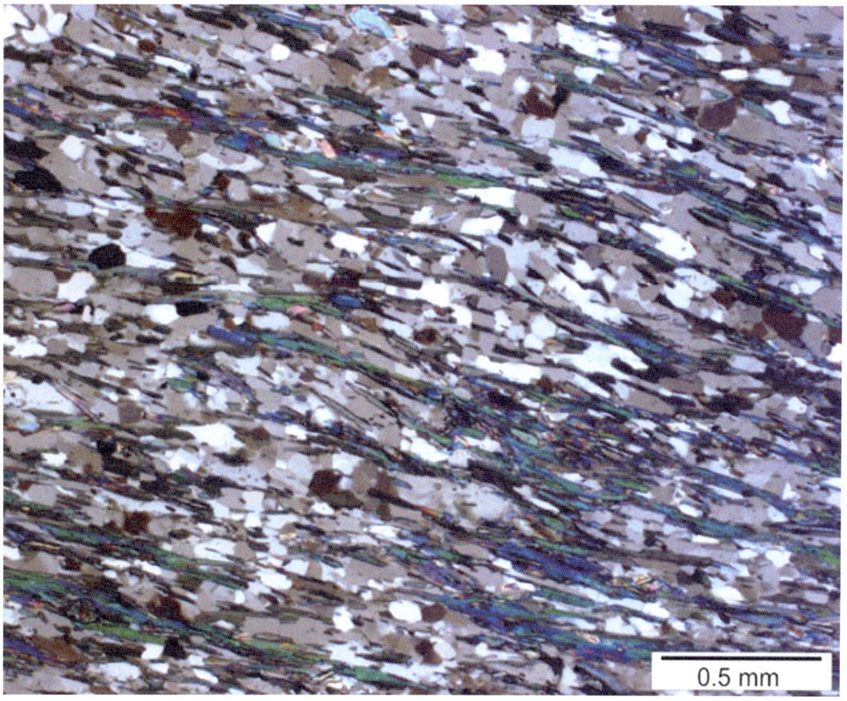

Fig. 17.21 Strongly foliated orthomylonite. See text for details. (Photomicrograph by the author)

fabrics. During progressive deformation, the crystallographic planes of minerals get activated. Activation is rather selective for individual minerals. In quartz, for example, activation of basal, rhomb and *a*-axis slip systems produces well-developed fabrics. These fabrics are developed during noncoaxial (rotational) deformation.

Quartz tends to show a sequential development of fabrics during shear deformation at large strains. Fabric studies help understand the flow mechanisms and deformation patterns in large shear zones. Since fabric development in quartz depends upon the dominant slip systems and the strain path or the kinematic framework, the crystallographic preferred orientation (CPO), also called lattice preferred orientation (LPO), of quartz has been found helpful in throwing light on the sense of shear, kinematics of deformation (flow kinematics) and slip systems operating during deformation (White 1973; Lister 1977; Lister and Williams 1979; Wenk et al. 1989; Herwegh and Handy 1998; Takeshita et al. 1999; Law 2014; Hunter et al. 2018; Rodrigues et al. 2019). The strength and intensity of quartz fabric development in relation

Fig. 17.22 Weakly foliated orthomylonite with a porphyroclast. See text for details. (Photomicrograph by the author)

Fig. 17.23 Ultramylonite. See text for details. (Photomicrograph courtesy Amit K. Verma)

to structural distance have been shown to constitute a useful proxy for relative finite strain magnitude and for delineating the spatial extents of shear zones (Knipe and Law 1987; Bhattacharya and Weber 2004; Starnes et al. 2019).

For *c*-axis measurement, oriented quartz-rich specimens are cut parallel to the stretching lineation and perpendicular to the foliation, thus representing the *XZ*-plane of the finite strain ellipsoid. Foliation represents the *XY*-plane, and lineation (*X*) that lies in this plane represents the direction of maximum finite elongation. Orientations of quartz *c*-axis are measured with the aid of an optical microscope fitted with universal stage. Texture goniometer and electron

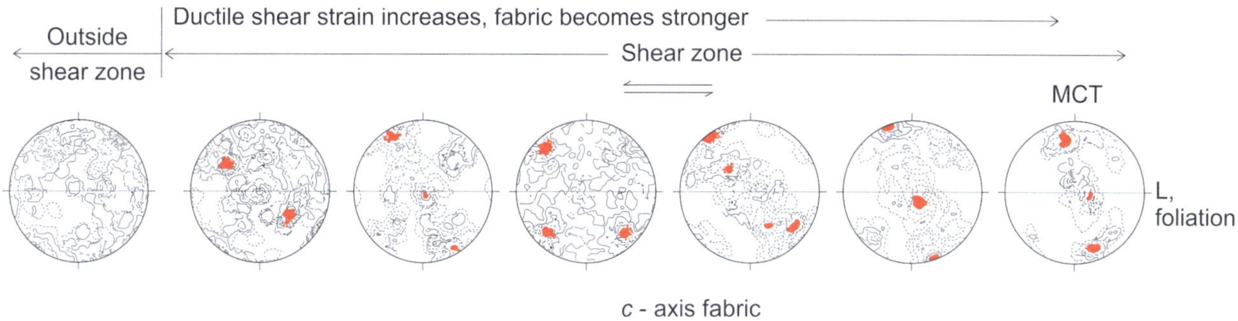

Fig. 17.24 Progressive development of quartz *c*-axis fabrics in a crustal scale ductile shear zone, the Main Central Thrust (MCT) of the Himalaya. Right to left is from south to north. The fabric diagrams are sequentially arranged from the MCT (south) northwards. The fabric diagrams are from the hanging wall (crystalline rocks) of the MCT. See text for details. (Reproduced with slight modification from Bhattacharya and Weber 2004, Fig. 13 with permission from Elsevier Copyrights Coordinator, Edlington, U.K. Submission ID: 1193137)

backscatter diffraction (EBSD) are also used for *c*-axis measurement. Fabric diagrams are obtained by plotting the *c*-axis data on an equal-area lower hemisphere Schmidt net with the help of a suitable program. In the fabric diagram, the *c*-axes tend to cluster along the shear-normal with progressive noncoaxial deformation.

The quartz *a*-axis fabric is obtained with the aid of X-ray texture goniometry. During noncoaxial deformation, the *a*-axis point maxima are more or less contained in the *XZ*-plane. In the fabric diagram, the *a*-axes preferentially align parallel to the macroscopic shear direction.

In shear zones, development of quartz CPO is dominantly influenced by the flow kinematics. Under coaxial plane strain conditions, quartz *c*-axis fabric typically shows an orthorhombic symmetry given by crossed girdle (Lister and Williams 1979; Law 1987; Law and Johnson 2010), while under dominant noncoaxial conditions, a monoclinic symmetry given by a single girdle develops oblique to the shear direction (Bouchez et al. 1983; Simpson and Schmidt 1983; Heilbronner and Tullis 2006). Development of quartz CPO of a crustal scale shear zone is shown in Fig. 17.24.

During progressive shearing and with higher strains, the CPO becomes progressively stronger and asymmetric as the main thrust is approached. The asymmetry gives the sense of shear for the shear zone in general. The Main Central Thrust (MCT) zone of the Himalaya is an excellent example of quartz *c*- and *a*-axis fabric development resulting from noncoaxial shear deformation (Bhattacharya and Weber 2003, 2004).

The MCT is a north-dipping shear zone along which a part of the deeply rooted Indian crust, now constituting the Central Crystalline Zone of the Higher Himalaya, was emplaced over the less metamorphosed sedimentary belt of the Lesser Himalaya during India-Asia plate collision. Rocks exposed over a 10–15 km wide tract are affected by strong ductile shear deformation associated with movements along the MCT. Dynamic recrystallization and crystal-plastic deformation progressively increase towards the trace of the MCT from both hanging wall (crystalline zone) and footwall (sedimentary belt). As such, the quartz *c*- (Fig. 17.24) and *a*-axis fabrics of quartz-rich tectonites become progressively stronger with the approach of the trace of the MCT. The persistent asymmetry of the *c*-axis fabrics indicates a sinistral sense of shear.

17.8 Shear Zone Formation

17.8.1 Background

As yet, a vast literature (e.g. Ramsay 1967; Ramsay and Graham 1970; Cobbold 1977; Ramsay and Allison 1979; Ramsay 1980b; Poirier 1980; Ramsay and Huber 1983; Segall and Simpson 1986; Hobbs et al. 1990; Passchier 1992; Austrheim and Boundy 1994; Means 1995; Hudleston 1999; Ingles et al. 1999; Pennacchioni 2005; Passchier and Trouw 2005; Pennacchioni and Mancktelow 2007; Fusseis and Handy 2008; Mittempergher et al. 2014; Hunter et al. 2018 and the references therein) has accumulated on various aspects of shear zone formation. As a result, a good deal of information is available on this aspect. In the light of the above works, and following a review on shear zones by Fossen and Cavalcante (2017), we consider some aspects of shear zone formation as described below.

17.8.2 Initiation of a Shear Zone

All shear zones develop in stages with time. Now the question arises as to how a shear zone initiates. All rocks and minerals contain flaws on micro- to mesoscale that are believed to constitute nucleation points. These flaws may be

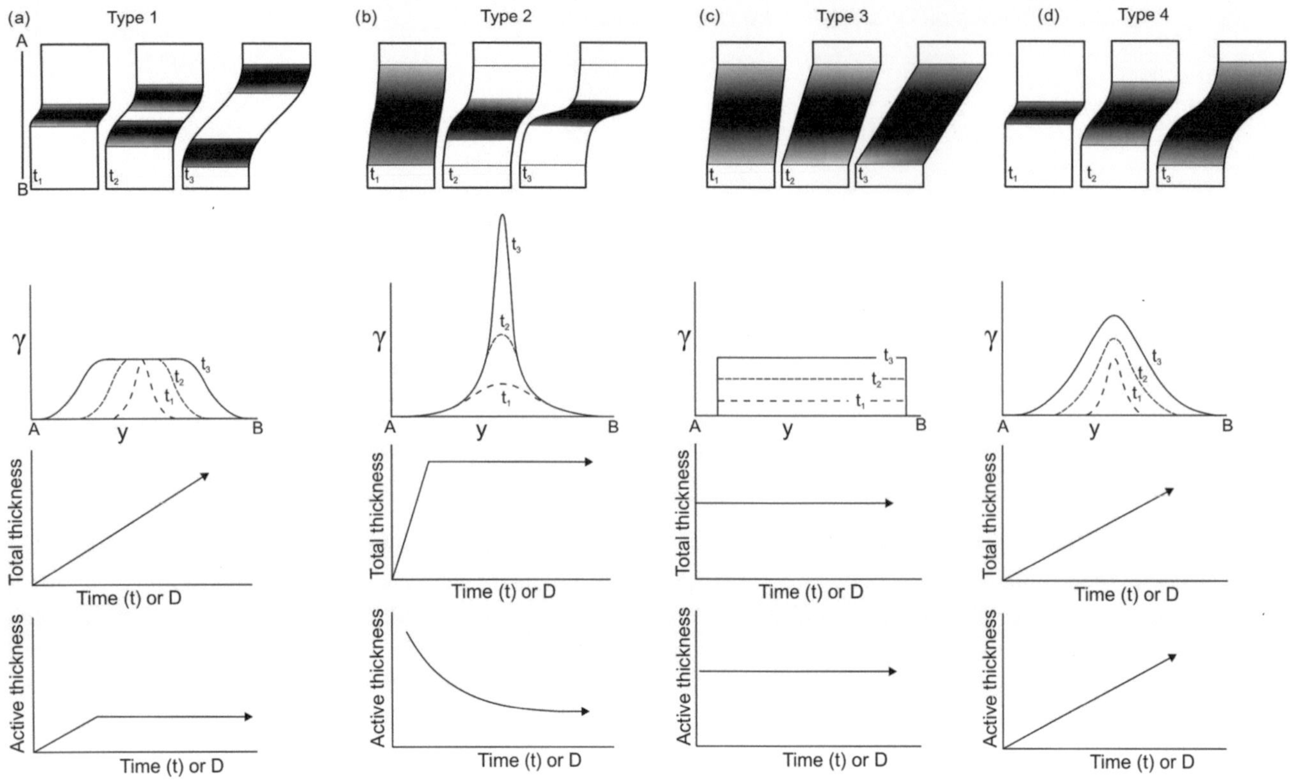

Fig. 17.25 Four ideal shear zones grouped on the basis of thickness and activity through time. The black-grey portions represent active parts of the shear zone. *Type 1* thickens with time leaving behind an inactive central part. *Type 2* shows increasing strain localization in the central part of the zone. *Type 3* has constant active thickness and is everywhere active at any given time. *Type 4* grows wider and remains everywhere active. (Reproduced from Fossen and Cavalcante 2017, Fig. 19 with permission from Elsevier Senior Copyrights Coordinator. Edlington, U.K. Submission ID: 1193222)

in the form of minor cracks or weak mineral phases. It is believed that shear zones can initiate on brittle fractures. With the activation of plastic deformation mechanisms, the shear zones undergo transformation to ductile deformation. These fractures act as avenues of fluid that weakens the rock and thus causes localization of plastic deformation.

17.8.3 Growth of a Shear Zone

Shear zones, like faults, grow in length with accumulation of displacement. The length of the initial fractures controls the length of the subsequent shear zones. As the lengths grow, they tend to connect other shear zones, thus forming composite systems or networks. It is believed that this growth is mainly by linkage, and tip propagation is not that significant in this process. Growth of shear zones by linkage has been demonstrated by Pennacchioni (2005) in his map of shear zones in a tonalitic rock from the Alps.

17.8.4 Evolution of Shear Zone Thickness

Theoretical models (Means 1984, 1995; Vitale and Mazzoli 2008) suggest that shear zone thickness evolves in relation to strain and accumulation of displacement. Fossen and Cavalcante (2017) proposed four idealized models, each of which produces different displacement profiles across the zone (Fig. 17.25). *Type 1* thickens with time as strain propagates into the walls, leaving behind an inactive central part. *Type 2* shear zone shows increasing strain localization in its central part, thus developing a characteristic bell type that evolves into a peak-type profile. Formation of types 1 and 2 can be explained by strain hardening and weakening, respectively. *Type 3* has constant active thickness and is everywhere active at any given time. *Type 4* grows thicker, while the whole shear zone remains active, and develops a bell-type profile that, unlike type 2, does not grow into a peak-type profile. In general, the shapes of the displacement profiles depend on the rate of strain hardening or softening, and on the kinematic vorticity number (W_k), and also whether the strain is plane or three-dimensional (Vitale and Mazzoli 2008; Fossen and Cavalcante 2017).

17.8.5 Strain-Softening Mechanisms

A shear zone, as we have explained above, is a zone of localized deformation such that the material outside this zone is apparently undeformed or less deformed. This implies that the strain concentrated in a shear zone should be able to soften the material so that it undergoes continuous ductile deformation without fracturing. In order to understand the formation of a shear zone, therefore, we need to know the strain-softening mechanisms operating in a shear zone. In general, strain softening mechanisms depend upon the thermal and mechanical state of the rock mass. As such, softening may occur in any part of a rock mass or in any segment of the crust. We consider here the following important mechanisms of strain softening: shear heating, geometric softening, and effects of quartz twinning.

17.8.5.1 Shear Heating

A hot rock, in contrast to a cold one, is weak and is mechanically soft. Under certain conditions, heat is sometimes developed in a rock mass. For example, presence of radioactive elements produces heat, which is dissipated in the surrounding rocks. Mechanical work done by rocks during deformation can also develop heat. If deformation is confined to narrow zones, dissipation of heat is relatively high that imparts some sort of viscosity to the rock. As a result, the rock softens and thus the shear stress tends to concentrate that enables the rock to undergo continuous ductile deformation. Softening thus developed may be assigned to *shear heating*.

17.8.5.2 Geometric Softening

Geometric softening is the progressive rotation of grain lattices towards parallelism with the shear plane during ductile deformation, and this is caused due to the rotation of lattice and grain shape (Etchecopar 1977). The rotation causes decrease of shear stress necessary to promote crystallographic slip. Alignment of lattices with the slip direction imparts a *lattice preferred orientation*. Geometric softening is common in minerals with well-developed slip systems, but the degree of softening depends upon how the initial fabric is oriented towards the shear plane.

17.8.5.3 Effects of Quartz Dauphiné Twinning on Strain Localization

McGinn et al. (2020) studied a network of exhumed shear zones of Fiordland, New Zealand. The shear zones expose middle and lower crust. The authors focused on a tonalite mylonite and a granodiorite mylonite both representing the middle crust. The granodiorite mylonite shows prevalent Dauphiné twinning, while the quartz in the tonalite mylonite is nearly untwinned. The authors demonstrate that Dauphiné twinning formed early in transpression and rendered the twinned grains less deformable relative to the untwinned grains over time. As such, the effectiveness of Dauphiné twinning in localizing strain is strongly influenced by its timing relative to the evolving shear zone kinematic deformation geometry.

17.9 Shear Zones on a Lithospheric Perspective

Poh et al. (2020), while studying deformation of Precambrian belts in compressive regimes, observed that the thermal state of the lithosphere strongly controls the distribution of the structural and metamorphic features in tectonic systems, especially ancient versus modern tectonics. They suggest that deformation in ancient tectonic system is distributed because of weak and hot lithosphere, while that of modern system is localized into prominent shear zones because of strong and cold lithosphere. Further, ancient deformation zones such as those observed in Precambrian accretionary orogens underwent much longer periods of deformation (>60 Ma) than in modern orogens, thus implying low average strain rates.

17.10 Significance of Shear Zones

17.10.1 Academic Significance

- Shear zones are tabular bodies of rock where bulk or whole deformation is accommodated so that there is less or practically no deformation outside this zone.
- The various strain markers of shear zones help in estimating the amount of deformation which, in turn, has implications for crustal deformation.
- Shear zones extend into deeper crust, sometimes into upper mantle also. As such, their study can help us understand the nature of deep crust and the geological-structural processes that take place therein.
- Most shear zones have developed over a long period of geological time. As such, the rocks of shear zone bear signatures of the various stages of deformation and the associated geological processes. Shear zones thus may constitute pages of geological history of the crust.
- Rocks of shear zones provide evidences of sense of shear. This enables us to know the directions in which stresses have been active during shear zone formation.

17.10.2 Economic Significance

- The movement of shear zones towards the upper crust together with concurrent deformation-geochemical processes is commonly associated with release of fluid systems, water, gases and mineralizing solutions in different parts of the shear zones. As such, shear zones are sometimes hunting grounds for mineral exploration.
- Careful study of shear sense markers may help trace the source areas of mineralizing fluids in the shear zones.

17.11 Summary

- A *shear zone* is one that is bounded by two parallel surfaces and within which the bulk strain is accommodated so that there is less or no deformation outside this zone.
- Shear zones constitute zones of anisotropy in a rock mass and are able to accommodate differential movements in rocks.
- Large faults commonly bear analogy to shear zones. A fault formed at shallower levels of the crust where the rocks deform in the brittle regime and strain is localized along a narrow, tabular zone has a narrow, measurable width. When the same fault continues at depth, it enters the ductile regime of deformation where strain is localized along a much wider zone in the form of a *ductile shear zone*.
- Shear zones show a variety of shapes that are primarily a reflection of their mode of formation. As such, they have been classified in several ways. The ideal shear zone (Ramsay type) is classified into three: *brittle*, *brittle-ductile* and *ductile shear zones*.
- Shear zones are characterized by the occurrence of some typical rocks because deformation within a shear zone is different from that of the outside rocks.
- Rocks of shear zones commonly show well-developed fabrics, called *crystallographic preferred orientation* (CPO), or *lattice preferred orientation* (LPO), formed during noncoaxial (rotational) deformation. The fabric thus developed depends upon the dominant slip systems and the strain path or the kinematic framework.
- Shear zones develop in stages over time. Starting from a flaw in the form of crack or weak mineral phases, a shear zone grows in length and width. As the length grows, a shear zone tends to connect other shear zones, thus forming composite systems or networks. It is believed that the growth of a shear zone is mainly by linkage.
- Since large strain is localized in ductile shear zones, the material undergoes continuous softening due to grain-size reduction without fracturing. Common strain-softening mechanisms are discussed in the chapter.
- Study of shear zones is important for understanding rock deformation under some special conditions of the crust.
- Shear zones are capable of transferring material from depth to the upper crust and sometimes even in the reverse direction. As a result, the fluid system, water, gases and mineralizing solutions can segregate economic minerals that are deposited in various parts of the shear zones. As such, shear zones are sometimes important from economic points of view.

Questions

1. Explain what do you mean by a shear zone.
2. What is the difference between a shear zone boundary and shear plane?
3. Describe the strain patterns in a shear zone.
4. On the basis of geometry, how would you differentiate between a brittle shear zone and a ductile shear zone?
5. What is a mylonite? What are its characteristic features?
6. Explain what do you mean by growth of a shear zone.
7. Why does quartz develop crystallographic preferred orientation during ductile shear deformation?
8. Explain the formation of shear zones by shear heating.
9. Discuss the statement 'Faults bear analogy to shear zones'.
10. Outline the significance of shear zones.

18 Shear-Sense Indicators

A. R. Bhattacharya, *Structural Geology*, Springer Textbooks in Earth Sciences, Geography and Environment,
https://doi.org/10.1007/978-3-030-80795-5_18

Abstract

A common practice in rocks deformed by shear deformation is to know the sense of shear. This is done by identification of some small-scale structures, called *shear-sense indicators* that provide the sense of shear of the shear zone. The indicators are developed on all scales ranging from mesoscopic to lattice scales and in practically all rock types. In this chapter, we have described some common shear-sense indicators developed in rocks deformed by both ductile and brittle deformations. Although shear-sense indicators indicate the sense of movement during deformation, studies reveal that the practice may sometimes lead to misleading results by providing opposite sense also. This phenomenon is called *opposite shear sense*.

Keywords

Shear-sense indicators · Ductile shear-sense indicators · Quartz CPO · Brittle shear-sense indicators · Opposite shear sense

18.1 Introduction

Shear zones usually contain a number of small-scale structures, called shear-sense indicators that provide the sense of shear of the shear zone. *A shear-sense indicator is a structure, resulting from progressive deformation, whose geometry is indicative of the progressive rotation of the finite strain axes with respect to the instantaneous stretching axes and/or the flow plane, at the scale of observation* (Hanmer and Passchier 1991). The shear-sense indicators are developed on all scales ranging from mesoscopic and microscopic to lattice scales and in practically all rock types. They are also developed in rocks deformed by both ductile and brittle deformation processes. This chapter is devoted to a description of common shear-sense indicators.

As indicated in Chap. 17, shear zones should be studied in *XZ* sections. However, in field, this is not always possible because the erosion line may cut the rock in any section. For studying shear-sense indicators, efforts should therefore be made to look for the *XZ* section or any section nearer to it. For studying in hand specimens, the sample may be cut in *XZ* section.

18.2 Ductile Shear-Sense Indicators

18.2.1 Sigmoidal Foliation

Sigmoidal foliation is characteristic of ductile shear zones. They show curvature at the centre of the shear zone and become tangential or parallel to the shear zone boundaries (Fig. 18.1) where they provide the sense of shear. The foliation shows maximum curvature towards the centre of the shear zone where the finite strain reaches the highest value.

18.2.2 Oblique Foliation

During ductile shear deformation, the pre-existing foliation undergoes realignment compatible with the ductile shear strain. In most metamorphic rocks, layers of biotite and quartz are arranged in planes that make acute angles with

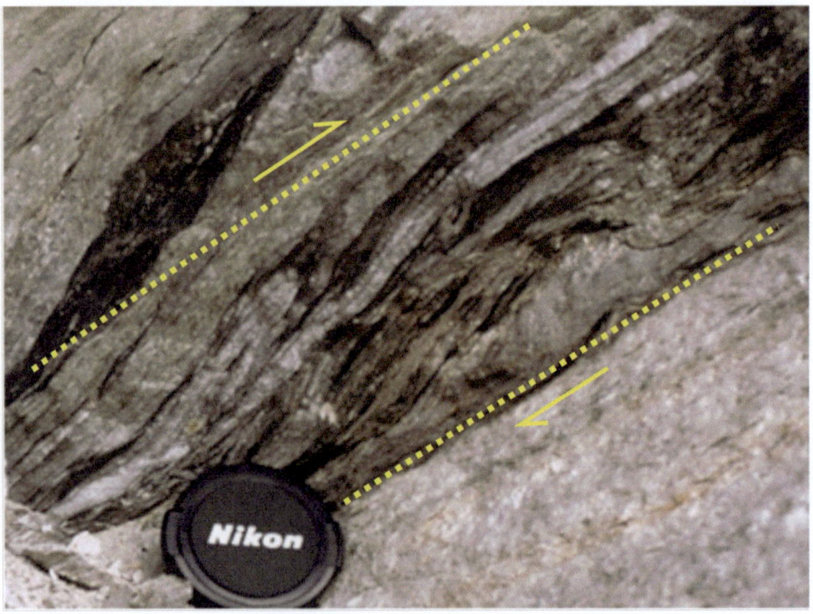

Fig. 18.1 Sigmoidal foliation in a quartz mylonite in the Main Central Thrust zone of Garhwal Himalaya, India. Loc.: Near Helong. (Photograph by the author)

Fig. 18.2 Oblique foliation given by layers of biotite and quartz in a gneissic rock of the Main Central Thrust zone of Garhwal Himalaya, India. Loc.: North of Joshimath

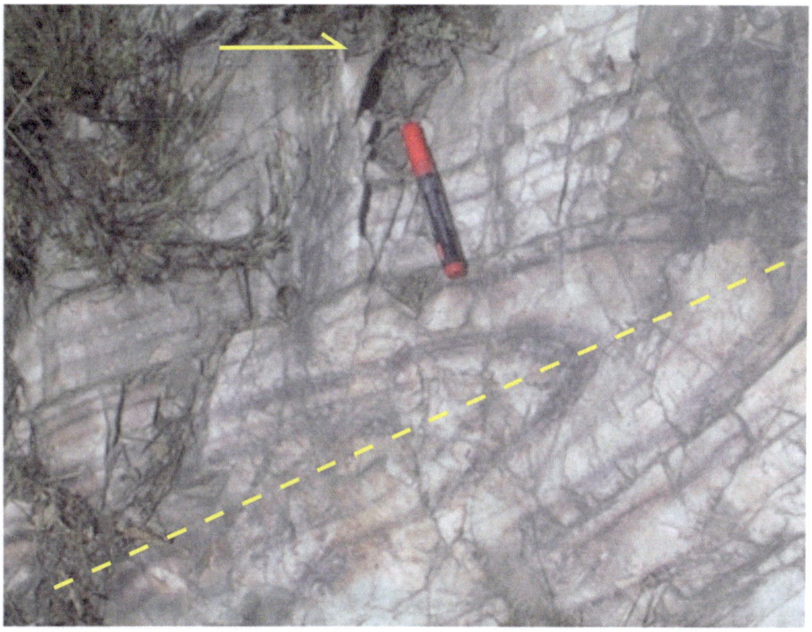

Fig. 18.3 Asymmetric isoclinal fold formed due to progressive simple shear deformation in the Main Central Thrust zone of Garhwal Himalaya, India. The axial plane of the fold (yellow dashes) has rotated clockwise suggesting dextral shear. Loc.: North of Helong

the shear zone walls (Fig. 18.2). The asymmetry thus produced gives the sense of shear for the shear zone.

18.2.3 Asymmetric Folds

The pre-existing folds often change their geometry due to rotational simple shear deformation within a shear zone. With progressive deformation, the fold hinge rotates towards the direction of stretching, or the *X*-axis, and thus the fold axial surfaces get inclined at an acute angle with the surrounding (mylonitic) foliation on *XZ* surface. The asymmetry and the acute angle give the sense of shear for the lithologic layer. The folds commonly tend to be isoclinal folds (Fig. 18.3) with their axes subparallel to the stretching lineation.

Folds are sometimes noticed in thin sections, and occasionally these are asymmetrically disposed (Fig. 18.4) with respect to the surrounding foliation in *XZ* sections. In such cases, the asymmetry of the axial plane of the microfolds gives the sense of shear.

Fig. 18.4 Asymmetric microfold (*XZ* section) developed in mica schist. The asymmetry of the axial plane gives the sense of shear, which is dextral in this case

Fig. 18.5 An oblique quartz lens has undergone rotation in response to simple shear, thus making a small angle with the surrounding foliation. The asymmetry thus developed gives the sense of shear. Loc.: South of Joshimath, Garhwal Greater Himalaya, India. (Photograph by the author)

18.2.4 Intrafolial Folds

Due to increase of rotational simple shear, some shear zone folds undergo progressive evolution such that one or both the limbs get highly stretched and are thus thinned and sometimes detached, thus giving rise to asymmetric *intrafolial folds* (Fig. 8.70). The acute angle of the axial surface of such folds gives the sense of shear. Intrafolial folds are reliable shear-sense indicators.

18.2.5 Oblique Lenses

Oblique quartz lenses (Fig. 18.5) that are occasionally S-shaped are commonly noticed on the *XZ* sections of mylonitic rocks of a shear zone. The quartz lenses may have earlier formed as a result of metamorphic differentiation. Under the influence of simple shear, they undergo rotation and thus make a small angle with the main foliation. The lengthwise orientation of these deformed lenses

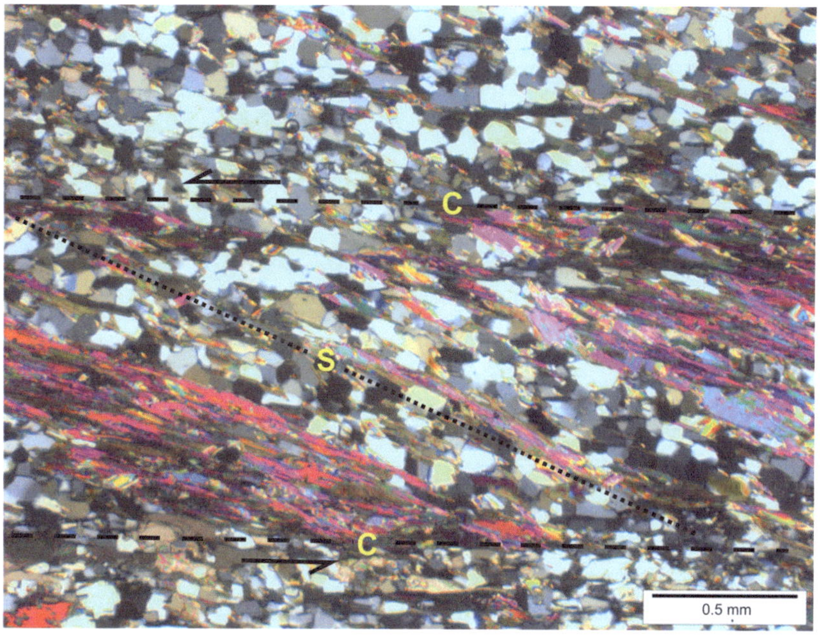

Fig. 18.6 S-C structures as noticed in thin section of a mylonitic rock. (Photomicrograph by the author)

constitutes the marker planes. The asymmetry of the quartz lenses gives the sense of shear. These structures are commonly developed in quartz-mica mylonites.

18.2.6 S-C Structures

S-C structures are developed on both microscopic and mesoscopic scales. During shear deformation, the pre-existing foliation, platy minerals and some grains undergo elongation in the direction of extension. A new foliation is thus formed that is inclined at acute angle with the shear zone boundary. Berthe′ et al. (1979) named this new foliation as *S-surface* (Fig. 18.6) (after the French word *schistosite*′, meaning schistosity). The S-surfaces are parallel to the *XY*-plane of the local strain ellipsoid. In a shear zone, the shear strain is considered parallel to the shear zone boundary. During shear deformation, the S-surface rotates towards the shear plane and at large strains becomes essentially parallel to the shear zone boundary (Ramsay and Graham 1970). The S-surface commonly becomes sigmoidal at the centre of the shear zone where shear strain is of highest value.

Berthe′ et al. (1979) recognized one more type of foliation in shear zones named as *C-surface* (Fig. 18.6) (after the French word *cisaillement*, meaning shear). It is in the form a narrow, local shear surface parallel to the shear zone boundary and is one where high shear strain is localized. The C-surface may occur as isolated or as a set of parallel shear zone boundaries. The foliation is parallel to the C-surface because at high shear strains the pre-existing foliation progressively rotates towards parallelism with the shear plane. A rock showing both S- and C-surfaces is called an *S-C tectonite*. In such rocks, the S- and C-surfaces define zones of low and high strains, respectively. With progressive rotational shear, new C-surfaces continue to form at this angle and are then rotated to become parallel to the shear zone boundary. Thus, in a shear zone, the C-surfaces constitute small-scale shear zones along which the shear zone foliation (S) curves into the C-surfaces. The acute angle between S- and C-surfaces gives the sense of shear for the shear zone.

Lister and Snoke (1984) identified two types of S-C mylonites: Type I S-C mylonites occur in quartzo-feldspathic rocks; the S-surfaces are defined by flattened grains of quartz and feldspar, while C-surfaces are defined by relatively finer, recrystallized subgrains of quartz. Type II S-C mylonites occur in quartz-mica rocks; the S-surfaces are given by the oblique foliation, while the C-surfaces are given by preferred orientation of mica.

Common association of S- and C-surfaces in ductile shear zones implies that both these structures have formed together (Berthe′ et al. 1979). S-surfaces rotate during deformation. All these strongly indicate that S-C structures are products of noncoaxial deformation that has affected the *earlier* foliation of the rock.

18.2.7 Shear Band Foliation

Rocks showing S-C foliation commonly show one more type of foliation called *shear band foliation* (Fig. 18.7) that occurs in the form of discrete, sometimes sigmoidal, bands oriented oblique to the C-surfaces. The bands are microscopic features

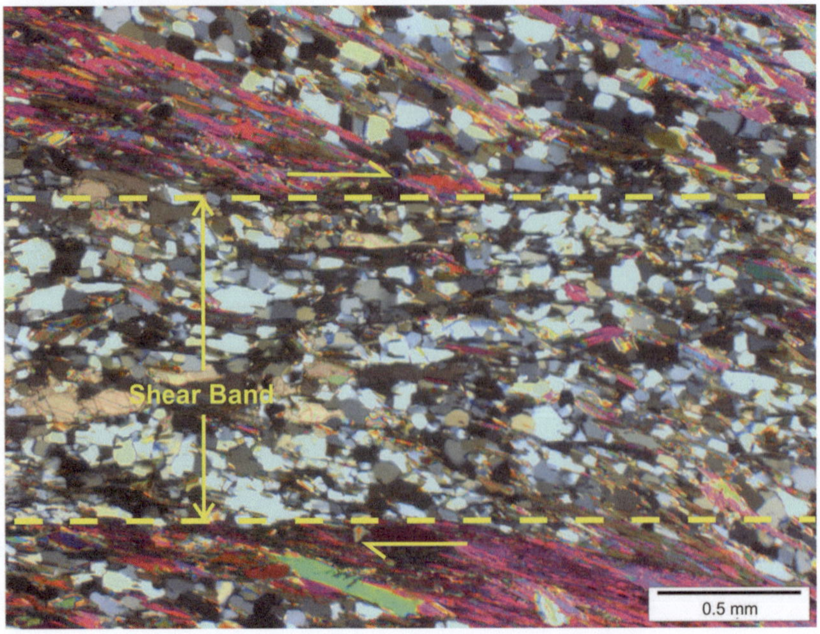

Fig. 18.7 Shear band foliation as noticed in thin section of a mylonitic rock. (Photomicrograph by the author)

but are also observed in hand specimens. Shear bands develop in rocks showing strong foliation and make an angle up to 45° with the earlier foliation.

Shear bands form as a single set where the shear parallel to the earlier foliation is large compared to shortening perpendicular to it, but if the shear component is small compared to the shortening component, a conjugate pair of bands is formed (Williams and Price 1990). If shear bands occur as a single set, they are inclined up to 45° with the main shear zone boundary, and this low angle provides the sense of shear for the shear zone. However, in the case of a conjugate shear band, the second set will provide an opposite sense of shear. In such cases, the shear bands are not considered as important shear-sense markers.

Shear band foliation is variously named as *shear band cleavage*, *C′-surface, extensional crenulation cleavage (ECC)* and *Riedel shear*. In the case of conjugate shear band, the second set is sometimes also called C'' or *C''-surface*.

18.2.8 Asymmetric Porphyroclasts

Most deformed rocks contain porphyroclasts that are relict of original rock or of any larger grain that has escaped ductile deformation. Quartz and feldspar constitute common porphyroclasts. The porphyroclasts are rigid objects that rotate in a finer grained matrix under noncoaxial conditions. The matrix is generally the product of dynamic recrystallization due to which large grains have been brought down to finer sizes during strong ductile deformation. With progressive ductile shear, the porphyroclasts undergo reaction with the matrix material on their outer surfaces, and thus an envelope of recrystallized material is formed. Such porphyroclasts are called *mantled porphyroclasts*. With increase of ductile shear strain, the mantled porphyroclasts develop 'tails' or 'wings' on their outer surfaces. The tails contain a mixture of smaller grains containing recrystallized material derived from the porphyroclasts as well as from the matrix material. With continued deformation, the tails undergo stretching into the matrix and thus become longer to finally become parallel to the foliation, though the porphyroclasts themselves do not necessarily stretch but rotate at their original positions (Fig. 4.16). The tails thus develop an asymmetry with the surrounding foliation, and this asymmetry provides the sense of shear.

For the purpose of determination of shear-sense, the porphyroclasts should be observed on surfaces that are parallel to the stretching lineation or mylonitic lineation and perpendicular to the main foliation. Porphyroclasts studied on such sections give maximum details of its shape and thus give reliable results. Passchier and Simpson (1986) described a simple method to identify the reference lines. Draw from the centre of the porphyroclasts a reference line parallel to C-surface (for thin sections) or parallel to the main foliation (for hand specimen or outcrop scale). The orientation of the tails then takes up different positions with reference to this reference line. This helps us in identifying four types of porphyroclasts: σ- and δ-types, φ-type (Passchier and Simpson 1986) and θ-type (Hooper and Hatcher 1988). The θ-type is not a shear-sense indicator. The porphyroclasts together with their tails constitute a *porphyroclast system*,

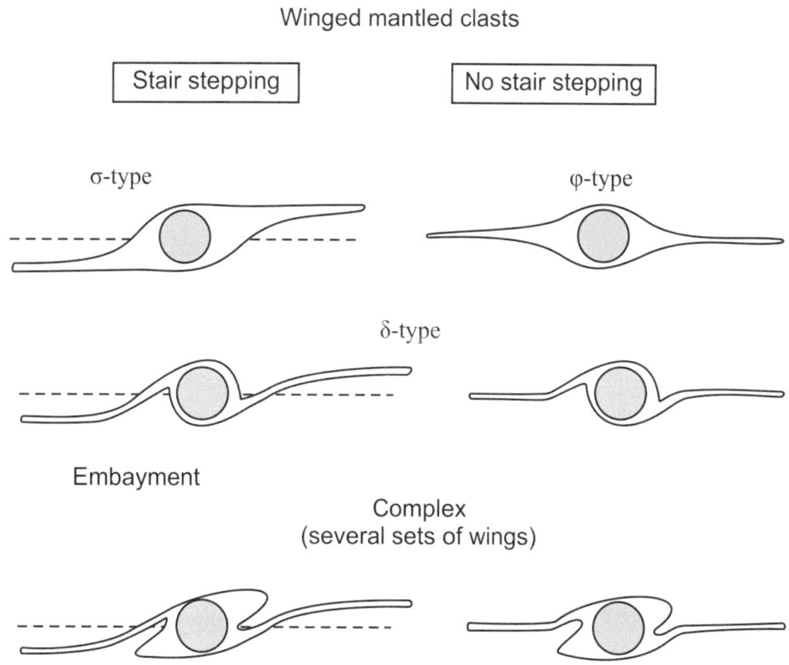

Fig. 18.8 Types of mantled porphyroclasts. The dotted line is a reference line with respect to which a dextral sense of shear is indicated. See text for details. (Reproduced from CW Passchier and RAJ Trouw, 2005, Microtectonics, second ed., Fig. 5.21 with permission from Springer Nature under Licence Number 5166460786777)

and an individual porphyroclast with tails is called a *winged mantle clast* (Fig. 18.8).

The σ- and δ-type porphyroclasts (Fig. 18.8) show 'stair-stepping' (Lister and Snoke 1984) of their tails with respect to the reference line, i.e. the reference line steps up in the direction of shear. In σ-type porphyroclasts, the tails are wedge shaped and give asymmetry at two different levels with respect to the reference line. The tails become thin at their ends and tend to be parallel to the reference line as well as to the foliation. In the δ-type porphyroclasts, the tails are curved and they cross the reference line close to the porphyroclast. The φ-types show symmetrical tails, while the θ-type porphyroclasts do not contain tails.

18.2.9 Quarter Structures

Mylonites occasionally contain porphyroclasts that are devoid of tails, and yet they indicate the sense of shear because of their asymmetric orientation with respect to the main foliation, if any. Such porphyroclasts sometimes show asymmetric microstructures in four quarters when considered in a framework of foliation and foliation-normal. Such asymmetric structures are named *quarter structures* by Hanmer and Passchier (1991).

Quarter structures may be of several types, of which Passchier and Trouw (2005) described the following three types (Fig. 18.9):

18.2.9.1 Quarter Folds

The microfolds developed in the extension quarter are called *quarter folds*. These folds form due to rotation of layering as they pass from the top of the porphyroclast during progressive deformation.

18.2.9.2 Quarter Mats

The concentration of mica layers around a porphyroclast in the shortening quarter is called *quarter mats*. These are believed to form when quartz is preferentially removed by solution transfer at locations where stress is concentrated around a porphyroclast.

18.2.9.3 Asymmetric Myrmekite

Myrmekite is occasionally located in the rim of K-feldspar crystals, if present in the shortening quarter. In such situations, the quartz lamellae in the myrmekite show an internal asymmetry that gives the shear sense.

18.2.10 Grain-Shape Foliation

During noncoaxial ductile deformation, occasionally some grains undergo grain-size reduction due to dynamic recrystallization in such a way that the subgrains thus formed are oriented in a parallel fashion with their long faces inclined to the main foliation (Fig. 18.10).

This preferred orientation of grains is called *grain-shape foliation* or *inclined subgrain fabric*. The asymmetry of the

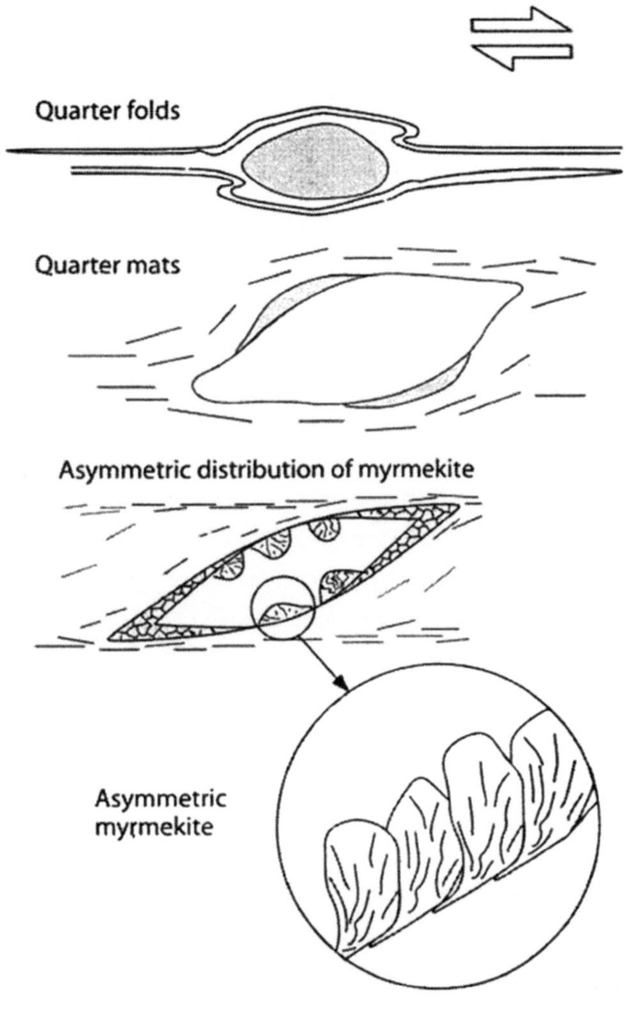

Fig. 18.9 Three types of quarter structures. Top: Quarter folds developed in the extension quarter. Middle: Quarter mats (upper diagram) given by concentration of mica layers around a porphyroclast in the shortening quarter. Lower: Myrmekite may also occur in the shortening quarter. The quartz lamellae in the myrmekite in the rim of K-feldspar show an internal asymmetry that gives the sense of shear. (Reproduced from CW Passchier and RAJ Trouw, 2005, Microtectonics, 2nd ed., Fig. 5.38 with permission from Springer Nature under Licence Number 5166460786777)

grains provides the sense of shear. This type of fabric is commonly developed in quartz ribbons after being subjected to dynamic recrystallization. The obliquity of recrystallized subgrains provides the sense of shear.

18.2.11 Mica Fish and Foliation Fish

In mica-rich mylonitic rocks, large, lens-shaped grains (porphyroclasts) of mica, called *mica fish*, commonly get oriented oblique to the main foliation or the C-surface (Lister and Snoke 1984). The obliquity of the mica fish with the foliation gives the sense of shear. Both the tips of the mica fish bend to become parallel to the shear direction, or the C-surface. As such, the orientation of the tips also suggests the sense of shear. Sometimes, the mica fish are connected to the adjoining one by trails of fine mica or matrix material (Fig. 18.11). The trails occasionally step up from one fish to another and thus give a sense of shear (Lister and Snoke 1984).

Apart from mica fish, some other minerals such as K-feldspar, plagioclase, tourmaline, kyanite and staurolite also commonly occur as lens-shaped single crystals in mylonites. These are collectively called *mineral fish* (Mukherjee 2011). Further, the micaceous layers in some mylonites locally show lozenge shapes. These are called *foliation fish*. The asymmetry of mineral fish and foliation fish against the main foliation or the adjoining C-surface can be used as a reliable kinematic marker.

18.2.12 Asymmetry of Boudins

Shear zone rocks commonly contain boudins within the lithologic layers. The boudins may show symmetric or asymmetric shapes (Fig. 18.12) with respect to the host foliation. They may also occur as fractured grains. Of these, the asymmetric boudins commonly serve as shear-sense indicators. The boudins take up various shapes that mainly depend upon the initial boudin neck orientation, i.e. whether normal or oblique to the long axis of the boudinaged grain (Fig. 15.23), the rigid or deformable property of boudins and the type of flow (Goscombe and Passchier 2003; Passchier and Trouw 2005). Shear-sense inferred from the asymmetry of boudins is of local significance.

18.2.13 Flanking Structures

Flanking structure is a *domain alongside a fault or vein where foliation is of a different orientation to that further away from the fault or vein* (Passchier and Trouw 2005, p. 311). The structure is in the form of deflections of foliation, called host element (HE), around a vein or fault, called cross-cutting element (CE) such that there is elevation, called lift, of the far-field HE above the cut-off at CE and a slip which is the displacement of the HE along the CE (Coelho et al. 2005; Passchier and Trouw 2005). The most common flanking structure is flanking fold (Fig. 18.13) and shear band foliation (described above). Flanking structures are formed due to emplacement of a cross-cutting element (Fig. 18.14) in a foliated rock. However, there are situations where the emplacement of CE may not cause any slip.

Several morphological types of flanking structures (FS) have been identified. According to Grasemann et al. (2003), the flanking structures can be of three types: *a*- and

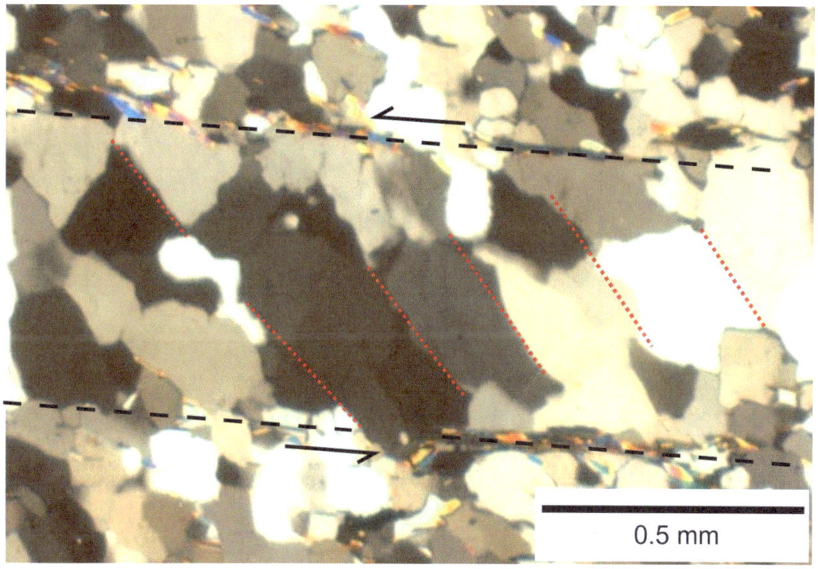

Fig. 18.10 Grain-shape foliation shown by inclined quartz subgrains. Note that the quartz subgrains show different stages of dynamic recrystallization. The cloudy ones are strained, while the clear ones are formed by release of internal strain from the previously strained grains by dynamic recrystallization. (Photomicrograph by the author)

Fig. 18.11 Photomicrograph of mica fish. The two mica fish are connected to the adjoining one by trails of fine mica or matrix material. The trails occasionally step up from one fish to another and thus give a sense of shear, which is dextral in this case. (Photomicrograph by the author)

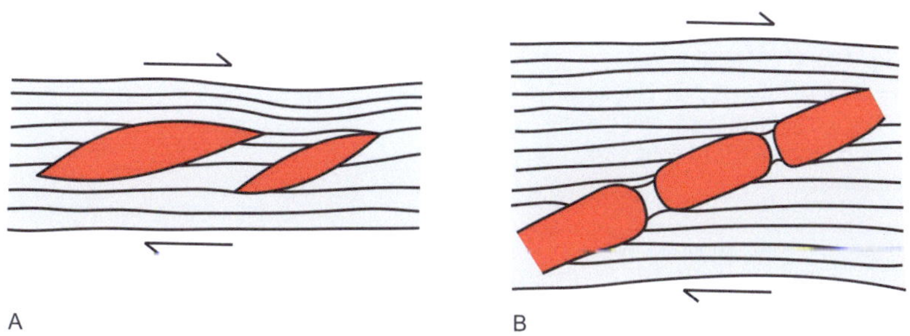

Fig. 18.12 Diagrammatic sketch of asymmetric boudins in section as they occur in lithologic layers. (**a**) Elliptical boudins. (**b**) Fractured boudin

s-type flanking folds, and *shear bands*, all of them show either a normal or a reverse drag of the HE near CE. The *a*-type flanking folds are counter-shearing, and *s*-type flanking folds are co-shearing with a contractional offset. The authors also suggested that flanking structures can form in homogeneous host rocks, without the need for pre-existing anisotropies or layering, provided that CEs already exist or can develop. Exner et al. (2004) suggested that *s*-type flanking folds can form in overall simple shear, in which there is no bulk stretch or shortening parallel to the shear zone boundary, and that these structures can be used as kinematic indicators in well-constrained natural structures.

Mukherjee (2014) reviewed FS and added two new classification schemes: One is based on the nature of the CE and whether HE penetrates it. The other takes into account all the potential combinations of drag/no drag and slip/no slip of the HE. Few key points from Mukherjee's (2014) review include the following: (a) In an FS, the style of drag is governed by only two factors: (i) the angular relation between the HE and the CE before shearing, and (ii) the relative magnitudes of throw and vertical separation resulting from slip of the HE. (b) During low, progressive bulk strains, FS may evolve from one type to another, e.g. from *a*- to *s*-type, but at higher strains intrafolial or sheath folds may develop where the CE and the HE are too close-spaced to be distinguished. (c) Several models of FS consider their developments in ductile shear zones of Newtonian and non-Newtonian rheologies bound by parallel boundaries.

The relationship between the HE containing marker horizons parallel to the shear zone boundaries and the CE—a planar slip surface such as fault, vein or dike inside the HE—may produce asymmetric structures useful as shear-sense indicators. However, such a use should be done with caution as a number of different kinematic histories can give rise to nearly identical FS morphologies (Passchier and Trouw 2005). If used, incorrectly, it can lead to severe misinterpretations (Dutta and Mukherjee 2019).

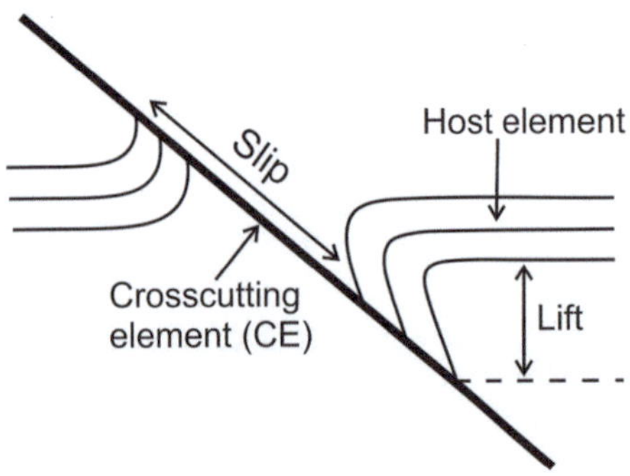

Fig. 18.13 Diagram of a flanking fold showing displacement (slip) of foliation (host element, HE) along a fault (cross-cutting element, CE). The lift is the elevation of the far-field HE above the cut-off at CE, and slip is the displacement of the HE along the CE

18.2.14 Mantled Structures Under Stick-and-Slip Conditions

Mulchrone and Mukherjee (2019) modelled the behaviour of rigid objects as being immersed in a linear Newtonian fluid with either a stick or a slip boundary condition. Stick boundary conditions are considered in theoretical work when the mantled structure is strongly bonded to the matrix. Slip boundary conditions are considered in analogue models when the mantled structure is both strongly bonded and loosely bonded (see Marques et al. 2014).

In the light of their field and laboratory studies, Mulchrone and Mukherjee found usefulness of the core-mantle structures in shear sense determination and suggest the following: (a) Asymmetry of orientation of objects relative to the shear direction may be problematic when used alone, particularly if stick boundary conditions prevail, but together with mantle structures there is less chance of confusion. (b) Only the geometry/inclination of the tails is used to deduce shear sense. (c) If most objects in an outcrop demonstrate a consistent synthetic orientation with respect to the direction of shear, then this is likely to support interpreting the sense of shear from this observation. (d) When the tail is not well developed then even at high strain, tails alone may not be a reliable shear-sense indicator.

Fig. 18.14 Slide showing flanking structures, NW Namibia. (Photograph with copyright under kind courtesy of Professor Cees Passchier)

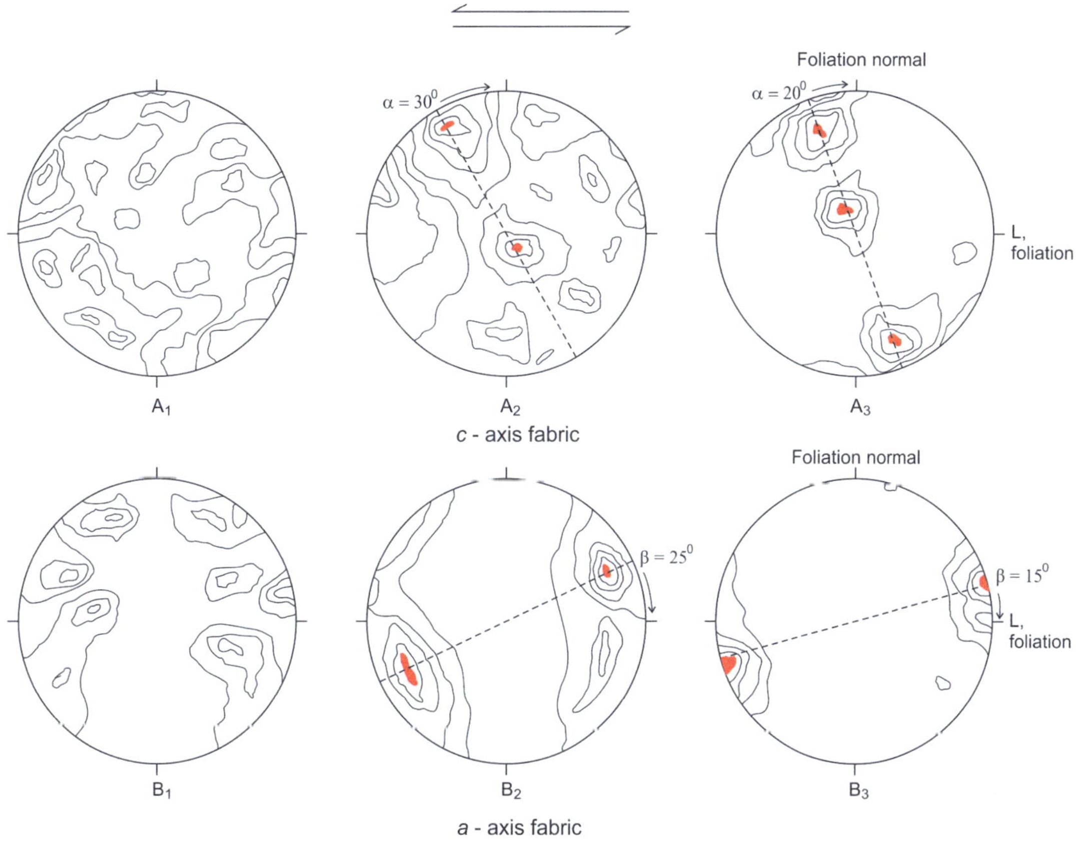

Fig. 18.15 Orientation of quartz *c*-axis (A_1 to A_3) and *a*-axis (B_1 to B_3) fabrics in undeformed and noncoaxially deformed quartz-rich rocks. A_1, B_1: Undeformed rocks showing random fabric. A_2, B_2. Moderately developed fabric. A_3, B_3: Strongly developed fabric. With progressive noncoaxial deformation, the symmetry plane (dashes) of *c*-axes rotates towards foliation-normal while that of *a*-axes rotates towards foliation. Shear direction is indicated at the top of the figure

18.2.15 Quartz CPO

In Chap. 17, we have described the development of quartz fabrics in shear zones. We shall now discuss the use of quartz *c*- and *a*-axis fabrics, also called *crystallographic preferred orientation* (CPO), as shear-sense indicators in quartz-rich mylonites in which CPO is developed as a result of rotation of the fabric during simple shear deformation.

18.2.15.1 Quartz *c*-Axis Fabric

Quartz *c*-axis fabric diagrams (Fig. 18.15A_1, A_2, A_3) are commonly of three types: random fabric, crossed girdle and single girdle. *Random fabric* (Fig. 18.15A_1) can be interpreted in several ways. Since it shows no definite fabric, it is possible that the rock does not show any visible effect of deformation. Another interpretation could be that the fabric may represent imprints of several deformations such that the imprint of any specific deformation cannot be identified. *Crossed girdle* may represent imprints of two deformations, each showing a single girdle of its own. Since the two girdles are generally at right angles to each other, it is possible that the sense of shear was reverse in the two deformation events. A crossed girdle may also represent a coaxial deformation. *Single girdle* (Fig. 18.15A_2, A_3), by analogy, may represent imprint of a strong noncoaxial deformation in which the quartz *c*-axis maxima, as represented by the single girdle, tend to rotate towards the foliation-normal. The quartz *c*-axis single girdle is considered aligned perpendicular to the shear zone boundary and is oblique to the foliation-normal, and the sense of obliquity of the girdle is consistent with the sense of shear (Bhattacharya and Weber 2004). For all practical purposes, single girdles are considered reliable shear-sense indicators.

18.2.15.2 Quartz *a*-Axis Fabric

Quartz *a*-axis fabric diagrams (Fig. 18.15B_1, B_2, B_3) have also been found useful as a reliable shear-sense indicator. Quartz *a*-axis fabric of quartz-rich tectonites is determined by X-ray texture goniometry. Quartz *a*-axis fabric diagrams, like those of quartz *c*-axis diagrams, are also of three types:

random fabric, crossed girdle and single girdle. Interpretation of quartz *a*-axis fabric diagrams should be done in the same way as for quartz *c*-axis fabric diagrams but in a perpendicular sense.

The fabric diagrams usually show point maxima of *a*-axes on either side of the flattening plane with maxima contained in the *XZ*-plane (Bhattacharya and Weber 2004). Rocks deformed by noncoaxial shear show an asymmetry of the point maxima with respect to the foliation plane. This asymmetry is consistent with the sense of shear. With increasing ductile shear strain, the *a*-axis maxima, as opposed to quartz *c*-axis fabrics, tend to fall closer to the foliation plane. This reduction of angle is related to higher strain (Etchecopar 1977; Bouchez and Duval 1982; Jessel and Lister 1990). Strong *a*-axis fabric suggests that *a*-direction acts as the most significant crystal slip direction during shearing event (Bhattacharya and Weber 2004).

Box 18.1 Strain-Insensitive Fabric

In deformed rocks, the fabric elements commonly take up the prevailing strain in respect of both finite strain and sense of shear. This is generally better reflected in microstructures than in hand specimens or in outcrops. Rocks that take up the prevailing finite strain and develop the fabric accordingly are said to show *strain-sensitive fabric*. However, there are situations when a part of the fabric or some components of the deformed rock do not accept the prevailing strain and instead show some deviations. Such rocks are said to show *strain-insensitive fabric*. The latter is commonly exemplified in ductile shear zones where the rocks undergo progressive deformation. The foliation of the strain-insensitive fabric is usually at an angle with the finite strain ellipse of the host rock. During subsequent stages of (dynamic) recrystallization, the newly formed grains progressively become strain free. As such, they may assume different shapes, say equant, elongated, polygonal, etc., depending upon the instantaneous strain and not the finite strain. The rock can thus be said to be the product of different states of strain. In the fabric of such rocks, therefore, the orientation of grains may not tell the overall or the general story. Therefore, one must be cautious of the strain-insensitive fabrics, especially in mylonitic rocks.

An example of strain-insensitive fabric is shown in Fig. 18.16 showing a large porphyroclast of quartz with imprints of at least three deformations, and in each case the geometry of the grains is different (see details in the caption of the figure).

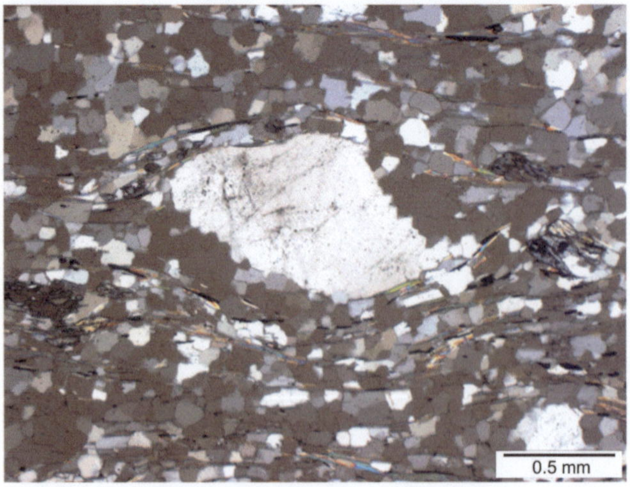

Fig. 18.16 The photomicrograph shows a large-quartz porphyroclast (at the centre of the photo) that has undergone partial dynamic recrystallization. The foliation outside the porphyroclast shows well-developed mylonitic foliation (running left-right). The porphyroclast shows three distinct zones showing signatures of three distinct stages of deformation: an outer zone showing a few dynamically recrystallized smaller, strain-free grains of quartz; a middle zone that shows feeble effects of dynamic recrystallization, i.e. grain-size reduction is not prominent; a central zone or core showing an approximately elliptical area of a strained, cloudy/undulose quartz grain. The central zone has preserved imprints of a dextral noncoaxial deformation as shown by the trails of fluid inclusions. At the same time, the strained grain itself is linearly (lengthwise) oriented, thus reflecting a sinistral sense of shear. The notable point is that though the host porphyroclast has undergone different stages of syntectonic deformation, why it is only the central zone that has preserved imprints of a dextral shear as given by the trails of fluid inclusions and a sinistral shear as shown by the linear orientation of the grain? All these may be ascribed to preferential acceptance of strain, and as such this may be considered as an excellent example of strain-insensitive fabric. (Photomicrograph by the author)

18.3 Brittle Shear-Sense Indicators

Rocks deformed by brittle deformation processes show relatively poor development of shear-sense indicators as compared to those deformed by ductile processes. Some common indicators of brittle deformation processes such as slickensides, slickenlines and slickenfibres are described in Chaps. 9 and 15. In addition, some other types of structures are described in this section.

18.3.1 'V'- Pull-Apart Structures

In many ductile shear zones, although the overall deformation remains under ductile conditions, some feldspar porphyroclasts, if present, occasionally deform by brittle

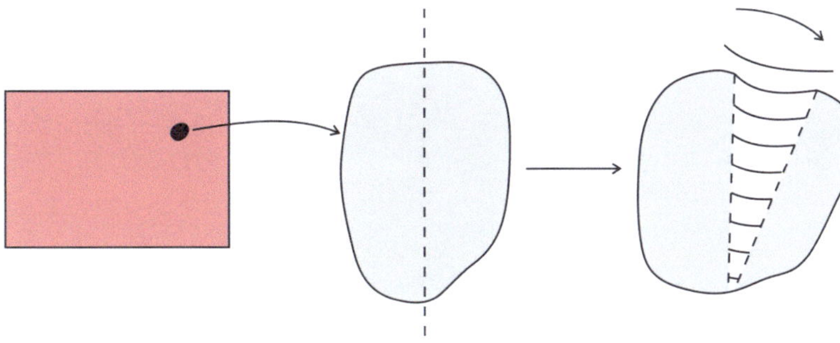

Fig. 18.17 Diagrammatic representation of 'V'-pull-apart structures. (**a**) Representation of thin section of a rock. The black dot represents the location of a porphyroclast shown in **b**. (**b**) The porphyroclast is going to develop a brittle fracture along the line shown. On rotation of the separated part, a 'V'-pull-apart structure is formed as shown in **c**. (**c**) Formation of a pull-apart structure. The surrounding foliation develops a deflection near the fragment that has rotated at the top of V. The matrix material or a part of the surrounding foliation thus filled in the V-shape space may sag down towards the closure of 'V'

Fig. 18.18 Tension gashes developed in limestone. Loc.: Near Hampteau Hotton, Belgium. (Photograph by the author)

fracturing. Hippert (1993) described a special type of pull-apart structure, called *'V'- pull-apart structures* (Fig. 18.17) that are developed due to opening of fractures at the rim of feldspar porphyroclasts (Fig. 18.17). The V-shape feature occasionally shows an asymmetry with the host foliation and can thus be used as a shear-sense indicator. The V-shape space, commonly filled with quartz, matrix material or a part of the surrounding foliation, may suck down towards the closure of 'V'. V-pull-apart structures have been described from thin sections.

18.3.2 Tension Gashes

Tension gashes are tension fractures in rocks. Most of these are shear-related structures. The fractures open up with progressive shear. The openings are later on filled by fluids that precipitate as secondary minerals such as quartz and calcite. These can therefore be described as extensional veins. With progressive simple shear, more gashes or arrays tend to form in an en echelon pattern (Fig. 18.18) between the shear zone boundaries. The opening of the fractures is maximum at the centre of the shear zone where shear strain attains the highest value. Under rotational shear conditions, tension gashes initially form at about 30° with the shear zone boundary. With progressive simple shear, rotation becomes stronger at the centre of the shear zone where the gashes bend and thus look sigmoidal in shape and the acute angle they make with the shear zone boundary gives the sense of shear. Tension gashes are commonly noticed in ductile-brittle shear zones.

18.3.3 Sheared Joints

Joints typically form due to brittle deformation processes. Sometimes, a set of joints (Fig. 18.19) are affected by shear

Fig. 18.19 A set of sheared joints developed in quartzite of the Main Central Thrust zone exposed south of Helong, Garhwal Himalaya, India. (Photograph by the author)

movements after their formation. In such cases, the acute angle made by the joints with the shear zone boundaries gives the sense of shear.

18.3.4 Duplex

A *duplex* (Chap. 11) is a set of parallel thrusts bounded by a *floor thrust* at the base and a *roof thrust* at the top (Fig. 11.9). Since parallel orientation of the thrusts is caused by simple shear, the structure constitutes a shear-sense indicator.

18.3.5 Domino Structure

Domino structures (Chap. 10) constitute a set of parallel planar faults formed by rotation of both bedding and faults (Fig. 10.5). Rotation is caused by simple shear. Faulting of the rigid blocks occurs simultaneously and causes similar offset and same amount of rotation. The structure in general constitutes a shear-sense indicator.

18.4 Opposite Shear Sense

Although shear-sense indicators constitute a useful tool to indicate the sense of movement during deformation, studies reveal that the practice may sometimes lead to misleading results. De Paor and Eisenstadt (1987) were possibly the first to present stratigraphic evidence for fault reversal from a sedimentary basin of Ellesmere Island and suggested fault reversals to be the norm rather than the exception in orogenic belts.

Bell and Johnson (1992), while investigating microstructural relationships in major movement zones in metamorphic rocks, pointed out numerous problems with current concepts of shear-sense criteria and their application. According to them, conflicts in shear-sense criteria in foliated metamorphic rocks stem from one or more of the following reasons: (i) differences in interpretation of how a particular geometry used as a shear-sense criterion formed; (ii) conflicts where the rocks have a complex deformation/metamorphic history, and the conflicting shear-sense criteria formed during different parts of the history; and (iii) deformation being in part, or in bulk, coaxial, leading to conflicting criteria formed during the deformation.

According to Bell and Johnson, the direction of apparent shearing commonly conflicts from one criterion to another, implying that the interpretations of shear sense along foliations from some mesoscale and microscale criteria have been erroneous. Different criteria in the same rock commonly give conflicting results. They suggested a new approach to interpret shear-sense by the use of strain fields that resolve conflicts in mesoscopic and microscopic criteria, thus providing a method for determining coherent shear-sense histories for any foliated metamorphic rock and for determining the shear-sense on the last foliation developed in a rock. The method uses geometries developed around competent heterogeneities such as quartz pebbles, pegmatite pods, veins, porphyroclasts, porphyroblasts and breccia clasts.

Hippertt and Tohver (1999) observed that most shear zones contain minor proportion of domains (thin section to outcrop scale) that show an opposing shear sense—named as *reverse shear*—relative to the overall kinematic framework. One therefore needs to be careful in interpreting opposing shear senses in the same shear zone as indicating superimposed deformation events. The authors are of the opinion that reverse shear is a universal feature in most shear zones and that most shear zones depart from ideal simple shear strain paths. They suggest that a kinematic analysis based on any indicator must always be statistical. Reverse shear most commonly occurs at low to moderate levels of finite strain, where a compositional tectonic banding is generally present.

Menegon and Pennacchioni (2010) reported that quartz CPO often shows both antithetic and synthetic orientation with respect to the bulk shear direction. Kilian et al. (2011) demonstrated that the obliquity of the long axes of the quartz grains to the shear plane is not consistent with the bulk shear sense.

From above, together with many other reports, it appears that shear-sense markers often give conflicting sense of shear for a rock body at the portion of observation. Recently, Dutta and Mukherjee (2019) reviewed this aspect from the viewpoints of genesis, global occurrences and numerical simulations. They prefer to use the words *opposite shear sense* (OSS) for the conflicting shear-sense indicators. They presented a catalogue of 59 reports of OSS in ductile and brittle regimes from 56 locations globally, from collisional, extensional and strike-slip tectonic settings besides a few from cratons. The authors presented several examples where shear-sense indicators create confusion or conflicting shear sense. For example, sigmoidal foliations are reliable indicators of shear sense in a ductile shear zone. But the presence of a deformed pre-existing foliation in the same rock body can be misleading and therefore should be avoided for interpretation. Some important results of the work of Dutta and Mukherjee (2019) on OSS are highlighted below:

(i) OSS can develop due to single or multiple deformation phase(s).

(ii) Seven major mechanisms have been put forward by various workers (see references in Dutta and Mukherjee 2019): (a) interaction between fracture and bedding interface, (b) viscosity heterogeneity, (c) strain rate contrast, (d) back rotation, (e) rigid block rotation and mutual interaction, (f) folding and (g) stretching faults. These suggested mechanisms are based on experimental results, theoretical concepts, numerical simulations, analogue experiments, and microstructural and field studies. As such, some of these are subject to verification.

(iii) Majority of the OSS owe their origin to multiple deformation phases during (a) inversion tectonics/fault reactivation, (b) exhumation of subducted crustal slices, (c) overturning or tilting of faults or lithosequences and (d) orogenic collapse and isostatic adjustment.

(iv) Multiply deformed terrains, which have undergone orthogonal switching of the principal stress axes (both local and regional), are more likely to exhibit OSS.

(v) Tectonic inversion of sedimentary basins is a major cause of producing OSS.

18.5 Summary

- Shear zones commonly contain a number of small-scale structures, called *shear-sense indicators* that provide the sense of shear of the shear zone.
- The shear-sense indicators are developed on all scales ranging from mesoscopic and microscopic to lattice scales and in practically all rock types. Some common shear-sense indicators for rocks deformed by ductile and brittle processes are described in this chapter.
- Although shear-sense indicators constitute a useful tool to indicate the sense of movement during deformation, their study sometimes leads to misleading results. In a rock body, the shear sense deduced from one indicator may show the reverse sense by another indicator. This has been variously named as reverse shear, kinematic reversal, shear-sense reversal, conflicting shear sense and opposite shear sense.
- Causes of shear-sense reversal suggested by various workers have been presented in the chapter.

Questions

1. What is meant by a shear-sense indicator?
2. What is sigmoidal foliation? How does it give the sense of shear?
3. Describe S-C structures and their role as shear-sense indicator.
4. Describe various types of porphyroclasts. Give their significance as shear-sense indicators with the aid of sketches.
5. What are quarter structures? Describe how they serve as shear-sense indicators.
6. What are mica fish and foliation fish? How do they differ?
7. Describe how quartz *c*- and *a*-axis fabrics are helpful as shear-sense indicators.
8. Describe how tension gashes can be used as a shear-sense indicator.
9. What is a domino structure? Why is this structure used as a shear-sense indicator?
10. What is meant by opposite shear sense? Discuss some causes that lead to the development of opposite shear sense in a rock body.

19 Deformation and Metamorphism

A. R. Bhattacharya, *Structural Geology*, Springer Textbooks in Earth Sciences, Geography and Environment,
https://doi.org/10.1007/978-3-030-80795-5_19

Abstract

It is believed that most metamorphic processes go together with deformational processes. While metamorphic processes involve several physico-chemical factors, operate in a complicated way and produce new minerals or alter some pre-existing minerals, the deformational processes provide suitable stresses that reorient the newly formed minerals and mineral assemblages to give rise to new fabric or microstructure to the rock. In other words, deformation promotes metamorphism and vice versa. Deformation can accelerate reactions in several ways, while metamorphism can promote deformation in several ways. In this chapter, we have described some common deformation structures in metamorphic perspective. The significance of porphyroblasts has especially been highlighted in this context. Also, a discussion on relative timing of deformation and metamorphism has been presented to elucidate how these two processes interact.

Keywords

Deformation · Metamorphism · Deformation–metamorphism interaction · Porphyroblasts · Relative timing of deformation

19.1 Introduction

Deformational processes play a significant role during metamorphic processes of a rock. The latter operate when a pre-existing rock is subjected to the combined action of pressure, temperature and fluids. For metamorphism, the role of both pressure and temperature is necessary albeit to any extent. The metamorphic processes operate in a complicated way during which the role of several physico-chemical factors is required. However, in a broader perspective, while the chemical-mineralogical processes produce new minerals, or alter some of the pre-existing minerals, the deformational processes provide suitable stresses that reorient the newly formed minerals and mineral assemblages and thus give rise to new *fabric* or *microstructure* to the rock. The net result is the formation of a metamorphic rock. In this chapter, we have described some important structural aspects that go together with chemical-mineralogical changes, thus giving rise to the fabric of metamorphic rocks. The literature on this aspect is vast, and the reader is suggested to consult some important books such as Turner and Weiss (1963), Spry (1969), Hobbs et al. (1976), Yardley (1989), Vernon (2004), Passchier and Trouw (2005), Philpotts and Ague (2009), Winter (2010) and Hobbs and Ord (2015) and the references therein.

19.2 Deformation and Metamorphism Are Interactive

It is a fact that most metamorphic rocks show the effects of deformation to varying extents. As such, both processes are interactive. In this context, Yardley (1989, p. 175–176) summarized two important results, i.e. (i) deformation can promote metamorphism, and (ii) vice versa, as described below.

(i) **Deformation can accelerate reactions in the following ways**: (a) juxtaposing reactant mineral grains as they move past one another; (b) increasing the free energy of reactant grains by straining them or breaking them into smaller pieces; (c) providing possible sites for preferential nucleation on newly created fracture surfaces; and (d) enhancing diffusion through movement on grain boundaries, increasing the area of grain boundary available and causing influx of fluid.

(ii) **Metamorphism can promote deformation in the following ways**: (a) Movement of fluid into zones of deformation can result in reactions involving hydration or metasomatism that would not otherwise have taken place. (b) Some mineralogical changes, mainly addition of K and removal of Ca, are associated with the formation of shear zones (e.g. NW Scotland); these transformations suggest both hydration and metasomatic changes. (c) Retrograde hydration reactions occur in zones that have also been affected by late-stage deformation, suggesting that the two processes assist one another. (d) Metamorphic reactions can reduce the strength of rocks in several ways so that even under high-grade metamorphic conditions the applied stresses must exceed the high finite strength of rocks to enable them to undergo plastic deformation.

19.3 Deformation Structures in Metamorphic Perspectives

We now describe some common deformation structures whose formation involves the role of metamorphism in some way or the other.

19.3.1 Foliation

Although *foliation* develops during metamorphism, it can also form by shearing that involves concentration of strain and consequent deformation in a rock mass between two parallel surfaces. The movements caused by shear are mainly accommodated by the formation of foliation. Presence of

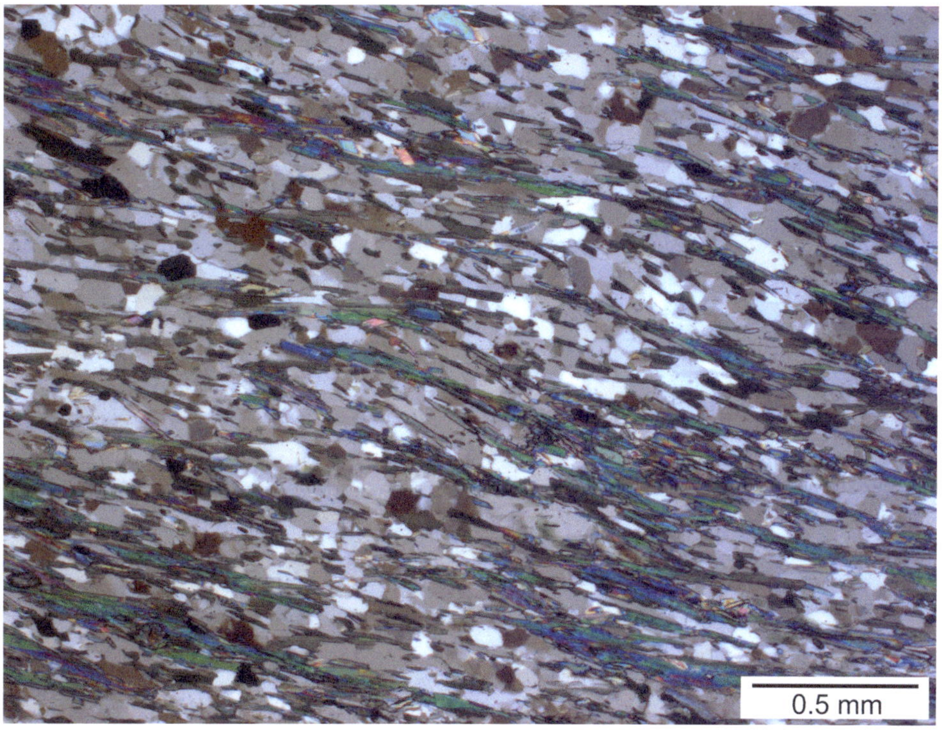

Fig. 19.1 Foliation developed due to the presence of micaceous minerals that facilitated strain-induced deformation

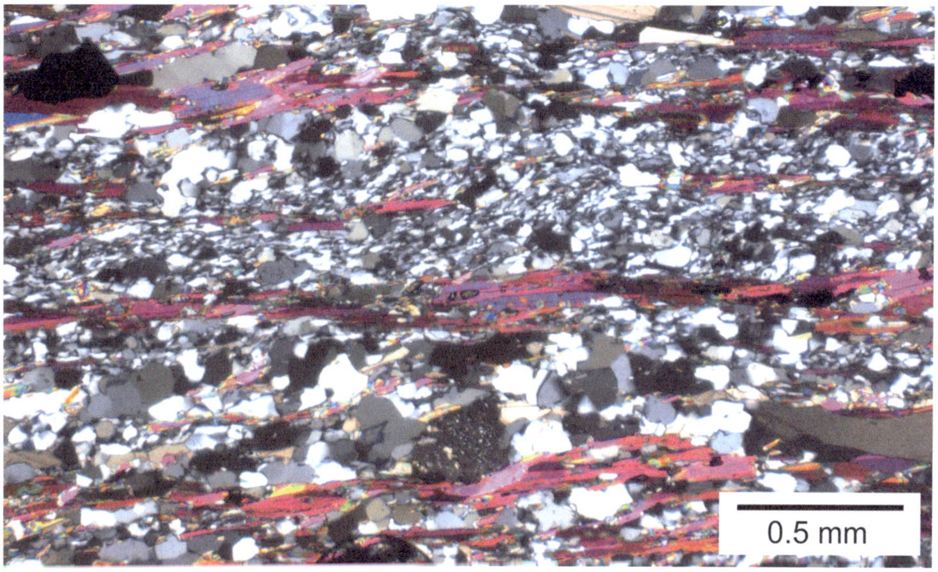

Fig. 19.2 Foliation developed due to preferred orientation of micaceous minerals during deformation

micaceous minerals (Fig. 19.1) facilitates strain-induced deformation. Formation of foliation due to preferred orientation of micaceous minerals is commonplace (Fig. 19.2). In such cases, foliation is accentuated by the micaceous layers.

Pre-existing foliation of a rock may be subjected to further deformation, thus causing deformed foliation (Fig. 19.3).

19.3.2 Pressure Shadows

When porphyroblasts of regionally metamorphosed rocks undergo metamorphism and deformation, a zone of low pressure is developed on either side of the porphyroblast. Quartz and chlorite thus dissolved from zones of high pressure in the surrounding rock mass are redeposited in zones of low pressure, thus forming *pressure shadows* (Fig. 19.4).

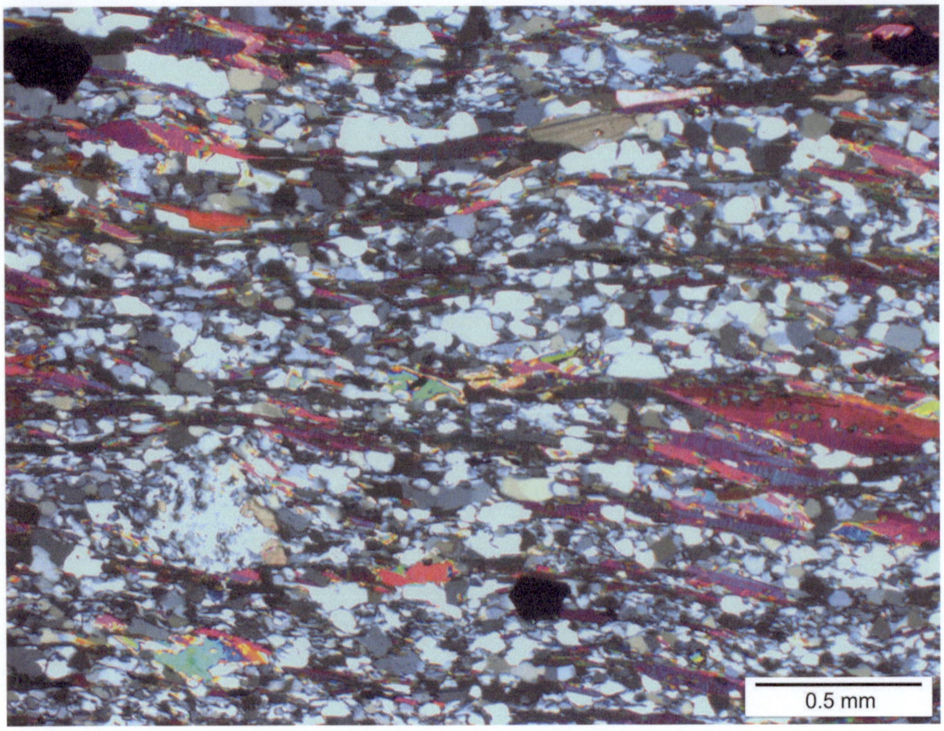

Fig. 19.3 Deformed foliation resulted from deformation of early-formed foliation

Fig. 19.4 During deformation, a zone of low pressure is developed on either side of a porphyroblast. Quartz and chlorite thus dissolved from zones of high pressure in the surrounding rock mass are redeposited in zones of low pressure

19.3.3 Crenulation Foliation

Folding of foliation in schistose rocks commonly produces small-scale folds, called *crenulations*, with wavelength of less than about 1 cm. These folds produce a lineation called *crenulation lineation* (Figs. 15.13 and 15.14), due to intersection of the later formed schistosity with that of first one. The curved pattern of the early schistosity as seen in cross section forms *crenulation cleavage* or *schistosity*.

Formation of crenulation cleavage, according to Philpotts and Ague (2009), is the result of interplay of deformation and metamorphic processes. According to them (p. 431), crenulation cleavage can result from successive episodes of metamorphism, but in many cases, it is produced by a single episode during which the orientation of the rock changes with respect to the stress field, possibly due to folding. During crenulation folding, the layers of quartz decrease in thickness, until they completely vanish in the main foliation plane of the crenulation cleavage. The crenulation folds are thus formed due to shortening of the rock normal to the crenulation cleavage planes as a result of pressure solution of quartz. Muscovite is concentrated in the crenulation cleavage as an insoluble residue.

19.3.4 Mineral Segregation

Several high-grade metamorphic rocks, mainly gneisses, show segregation of some monomineralic minerals such as quartz and calcite. These segregated layers follow similar orientation as that of the host rocks; that is, they are parallel to the foliation of the associated rocks. The structural relationship of these segregation layers is believed to be the result of certain thermodynamic readjustments. Under normal conditions, the grain boundaries between crystals of the same phase, or a structurally similar phase, are supposed to have a lower surface energy than those between totally dissimilar phases. Thus, the free energy of the system is reduced by segregation of the rock into layers that are dominated by one structural type (Yardley 1989, p. 170).

19.3.5 Veins

A characteristic feature of regionally metamorphic rocks is the occurrence of veins of minerals of which quartz is the commonest. The minerals of the veins are derived from the surrounding rocks. These syn-metamorphic veins are believed to be the result of extensional failure of the host rock caused by the development of pore fluid pressure formed due to progressive devolatilization reactions (Yardley 1989, p.176).

19.4 Porphyroblasts

Porphyroblasts are relatively large single crystals that have formed by metamorphic growth in a more fine-grained matrix (Passchier and Trouw 2005, p. 190). Most metamorphic rocks contain porphyroblasts, and such rocks are said to show a *porphyroblastic texture*. Since porphyroblasts are one of the most characteristic features of metamorphic rocks, they have been studied by several workers (e.g. Zwart 1962; Spry 1969; Schoneveld 1977; Bell 1985; Yardley 1989; Passchier et al. 1992; Vernon 2004; Passchier and Trouw 2005; Philpotts and Ague 2009; Hobbs and Ord 2015). In the light of these works, we present below some important aspects of porphyroblasts:

(i) Porphyroblasts are larger and stronger than the matrix material. They commonly occur in polymineralic rocks. The reason is that in polymineralic rocks, the number of nuclei formed of each mineral is generally not the same and this leads to variation in grain size of the different minerals in rock (Philpotts and Ague 2009, p. 436).

(ii) The amount of nucleation sites and the rate at which the nuclei grow govern the distribution and size of porphyroblast in a metamorphic rock, and that is why where many suitable nucleation sites are available several porphyroblasts are formed and where only a few suitable sites are present isolated large porphyroblasts are formed (Fig. 19.5) (Passchier and Trouw 2005, p. 190).

(iii) Porphyroblasts show two main types of schistosity (Fig. 19.6): *internal schistosity* (S_i) and *external schistosity* (S_e), which is the schistosity of the surrounding rock. The internal schistosity is due to inclusions of other minerals incorporated during the growth of the porphyroblast, and in such cases the rock is said to show a *poikiloblastic texture*. The relationship between S_i and S_e constitutes an important factor for understanding the deformation and metamorphism undergone by a rock, some of which are shown and explained in Fig. 19.7.

(iv) In the context of porphyroblasts as indicators of the relationship between deformation and metamorphism, Vernon (2004, p.431) outlined three types of links between deformation and metamorphism:

   (a) *Environment-controlled* links result from changes in P-T fluid conditions that simultaneously affect both the growth of porphyroblasts and the foliation development.

   (b) *Deformation-controlled* links involve selected microstructural sites, such as crenulation fold hinges, being favourable sites for porphyroblast nucleation and growth of porphyroblasts.

   (c) *Reaction-controlled* links involve the production of water in dehydration reactions, which can affect the deformation and hence control the microstructures of developing foliations; this is a form of reaction weakening.

(v) In the light of the vast literature on porphyroblasts, there appears to be no consensus about the formation of porphyroblasts.

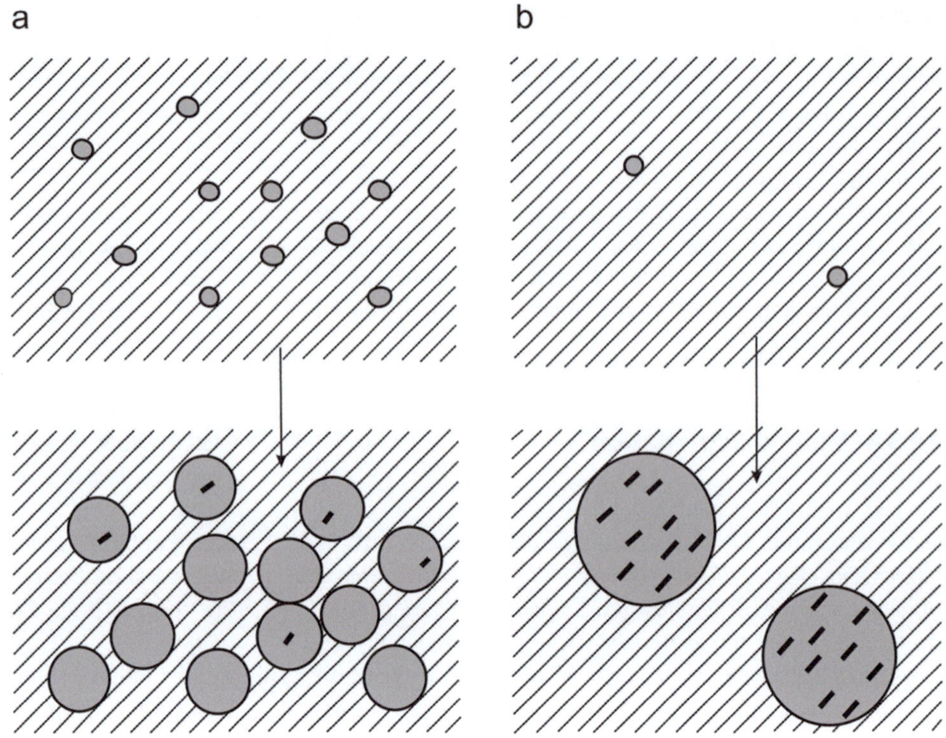

Fig. 19.5 The number of nucleation sites and the rate at which the nuclei grow govern the distribution and size of porphyroblast in a metamorphic rock, and that is why (**a**) where many suitable nucleation sites are available several porphyroblasts are formed and (**b**) where only a few suitable sites are present isolated large porphyroblasts are formed. (Reproduced from CW Passchier and RAJ Trouw, 2005, Microtectonics, 2nd ed., Fig. 7.2 with permission from Springer Nature under Licence Number 5166460142474)

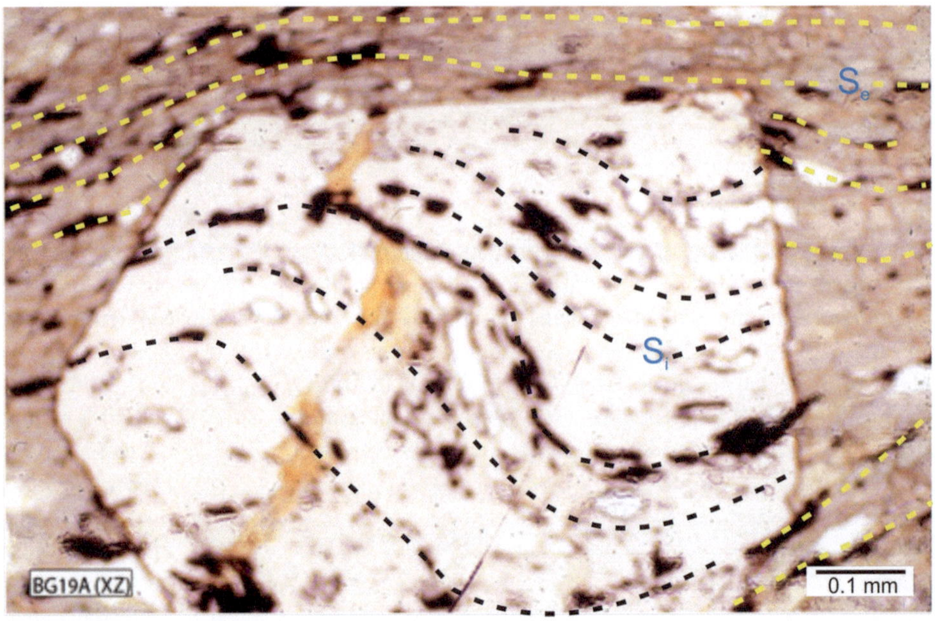

Fig. 19.6 A garnet porphyroblast showing trails of inclusions as internal schistosity (S_i) and an external schistosity (S_e). Rotation of garnet during deformation has caused rotation of the inclusion trails. The relationship between S_i and S_e constitutes an important signature for understanding deformation-metamorphism relationship. See text for details

19.5 Formation of Curved Trails in Porphyroblasts

From above, it is clear that the curved trails in porphyroblasts show varied geometric relations on which as yet there is no consensus on their mode of formation. The different views have been summarized by Vernon (2004) in four models that have been reorganized by Hobbs and Ord (2015) in three models as given below:

19.5.1 Model 1

This is the traditional model in which the porphyroblast rotates during growth and deformation. This promotes new

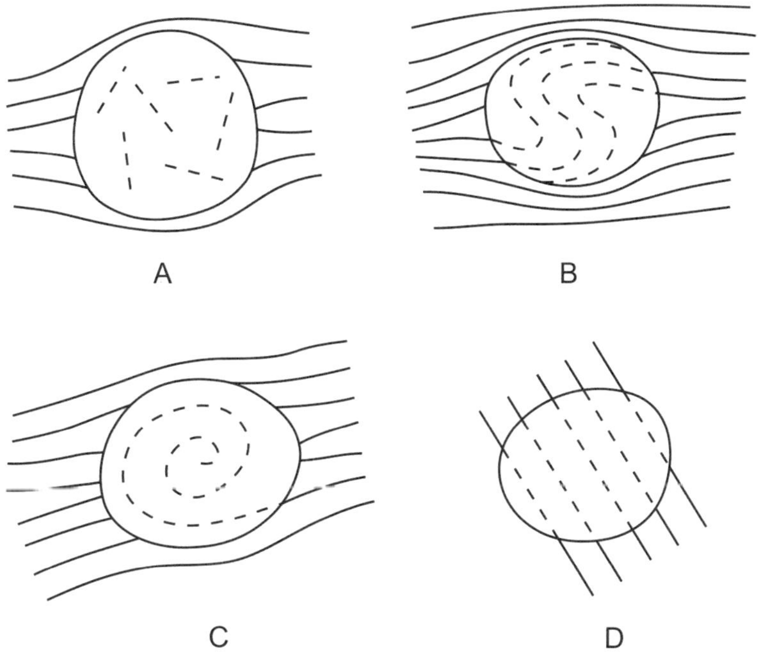

Fig. 19.7 Diagrammatic representation of different types of internal schistosity (S_i) of porphyroblasts in relation to the external schistosity (S_e) in metamorphic rocks. (**a**) Pre-tectonic porphyroblast: the internal schistosity (S_i) shows random pattern with respect to the external schistosity (S_e). (**b**, **c**) Syntectonic porphyroblast: Typical snowball garnet (**b**) and rotational garnet (**c**) showing spiral trails of inclusions (S_i). The folds developed within the porphyroblast (B) are called *helicitic folds*. (**d**) Post-tectonic porphyroblast: showing continuity of S_i with S_e

porphyroblast to overgrow by incorporating the surrounding fabric. This is analogous to the rotation of an ellipsoid in a shearing viscous medium. The rotations in this model consider only affine pure or simple shear deformation.

19.5.2 Model 2

This is the Ramsay (1962) model in which the porphyroblast does not rotate relative to the pre-growth reference coordinates while it is the matrix that rotates relative to the porphyroblast. The non-rotating porphyroblast then undergoes overgrowth and continues to incorporate the changing orientations of markers in the deforming matrix.

19.5.3 Model 3

In this model, the Bell model (Bell 1985; Bell and Johnson 1989, 1992; Bell et al. 1992), the porphyroblast is believed to be a composite entity developed during several episodes of growth and dissolution together with alternating periods of foliation development in the matrix. The composite porphyroblast does not rotate relative to the pre-growth reference axes, which are identified with a geographical frame. In the Bell model, the curved trails are inherited from the deformation of the matrix. Thus, the sense of shearing interpreted from the Bell model is opposite to that of the traditional model (a).

19.6 Relative Timing of Deformation

An important exercise of geologists in polydeformed metamorphic terrains is to work out the relative timing of deformation and metamorphism in a rock. Tremendous work has been carried out on this aspect in the last century. Most works concentrate on identifying three important stages to understand the structural-metamorphic processes: pre-tectonic crystallization, syntectonic crystallization and post-tectonic crystallization. Although field relations of rocks and the associated structures constitute an important tool, it is only the microscopic features that hold great potential to solve such problems.

Of the various criteria suggested for identifying the above-mentioned three stages, it is possibly the nature of internal schistosity (S_i) of porphyroblasts (Fig. 19.7) that alone is capable enough to solve the problem to a great extent. Earlier, Zwart (1962) had presented a basic three-tier scheme of formation of porphyroblasts in relation to deformation events in a metamorphic rock. This scheme has been modified by Passchier and Trouw (2005) by a four-tier scheme (Fig. 19.9) by the addition of an inter-tectonic stage. The latter is based on the inclusion patterns of porphyroblasts whose geometric relations with the surrounding rock suggest the relative age of mineral growth in relation to deformation.

Box 19.1 Porphyroblasts that Lack Inclusions

Porphyroblasts of metamorphic rocks invariably contain inclusions that constitute a source of information on several aspects pertaining to deformation as related to metamorphism and vice versa, e.g. in understanding the conditions of metamorphism and in throwing light on the timing relations between deformation and metamorphism, some of which are discussed in this chapter. However, there are also porphyroblasts that do not contain inclusions. In such cases, it is difficult to elicit information from them.

Passchier and Trouw (2005) outlined some geometric relations of such porphyroblasts with the surrounding foliation for identifying pre-, syn- and post-tectonic crystallization. Due to lack of suitable geometric relations in such cases, the methodology proposed by the authors is applicable to a few cases only as given below:

(i) If there is deflection of S_e or if strain shadows are present, porphyroblasts are pre-, inter- or syntectonic (Fig. 19.8a,b).
(ii) If there is no deflection of the external schistosity (S_e), porphyroblasts may be post-tectonic (Fig. 19.8c).
(iii) If large, isolated elongate mineral grains such as micas and amphiboles are oriented parallel to a foliation, and when lacking inclusion patterns, such minerals may be inter-, syn- or post-tectonic. Elongate porphyroblasts aligned with foliation are syntectonic.
(iv) If the porphyroblasts have rotated towards the foliation plane and may show slight but consistent obliquity with the foliation and internal deformation or replacement along the edges, then they may be inter-tectonic (Fig. 19.8b).
(v) If some of the crystals overgrew the foliation obliquely, then they may be post-tectonic porphyroblasts (Fig. 19.8d).

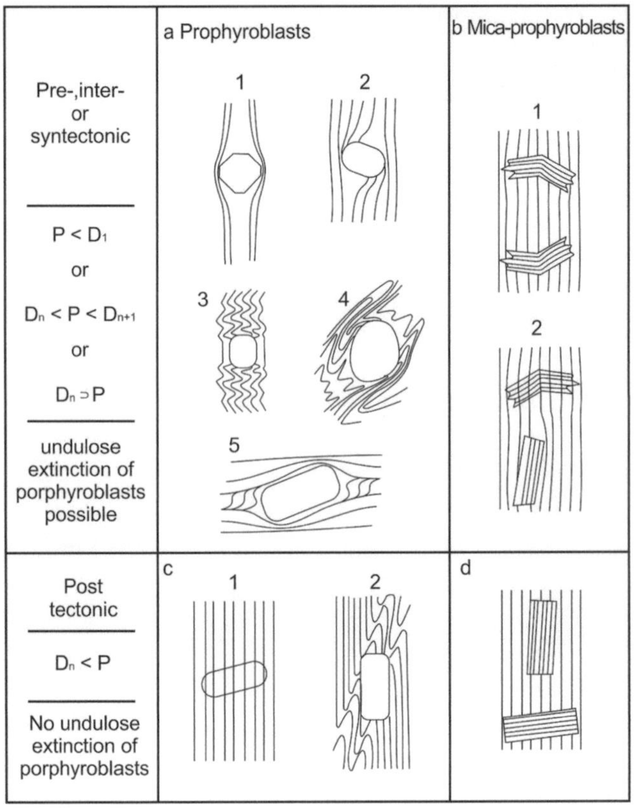

Fig. 19.8 How to elicit information for porphyroblasts if they do not contain inclusion trails? (Reproduced from CW Passchier and RAJ Trouw, 2005, Microtectonics, 2nd ed., Fig. 7.26 with permission from Springer Nature under Licence Number 5166460142474)

19.6.1 Pre-tectonic Crystallization

Pre-tectonic crystallization represents a stage of metamorphic history of a rock when the early-formed minerals are affected by later deformation. As a result, the earlier microstructure is overprinted by the development of a new microstructure. The latter can be identified by a number of features better seen in thin sections. Some of these include the following: (a) In porphyroblasts, the inclusions are randomly oriented (Fig. 19.9a,b) or may be surrounded by a matrix with polyphase deformation. (b) There is presence of undulose extinction, deformation bands (Fig. 19.10) and lamellae, pressure shadows, kink bands in phyllosilicate minerals and broken crystals across fractures. (c) Mylonitic texture is developed due to mylonitization of a metamorphic rock in which an earlier metamorphic texture is overprinted by the development of a strong foliation, i.e. *mylonitic foliation* (Figs. 17.18, 17.19, 17.20, and 17.21). White (1973) however considers a mylonitic texture as a syntectonic imprint.

19.6.2 Inter-tectonic Crystallization

Passchier and Trouw (2005) introduced an *inter-tectonic stage* for porphyroblasts that have grown over a secondary foliation and are surrounded by a matrix affected by a later deformation phase that has left no record in the porphyroblast (Fig. 19.9c,d).

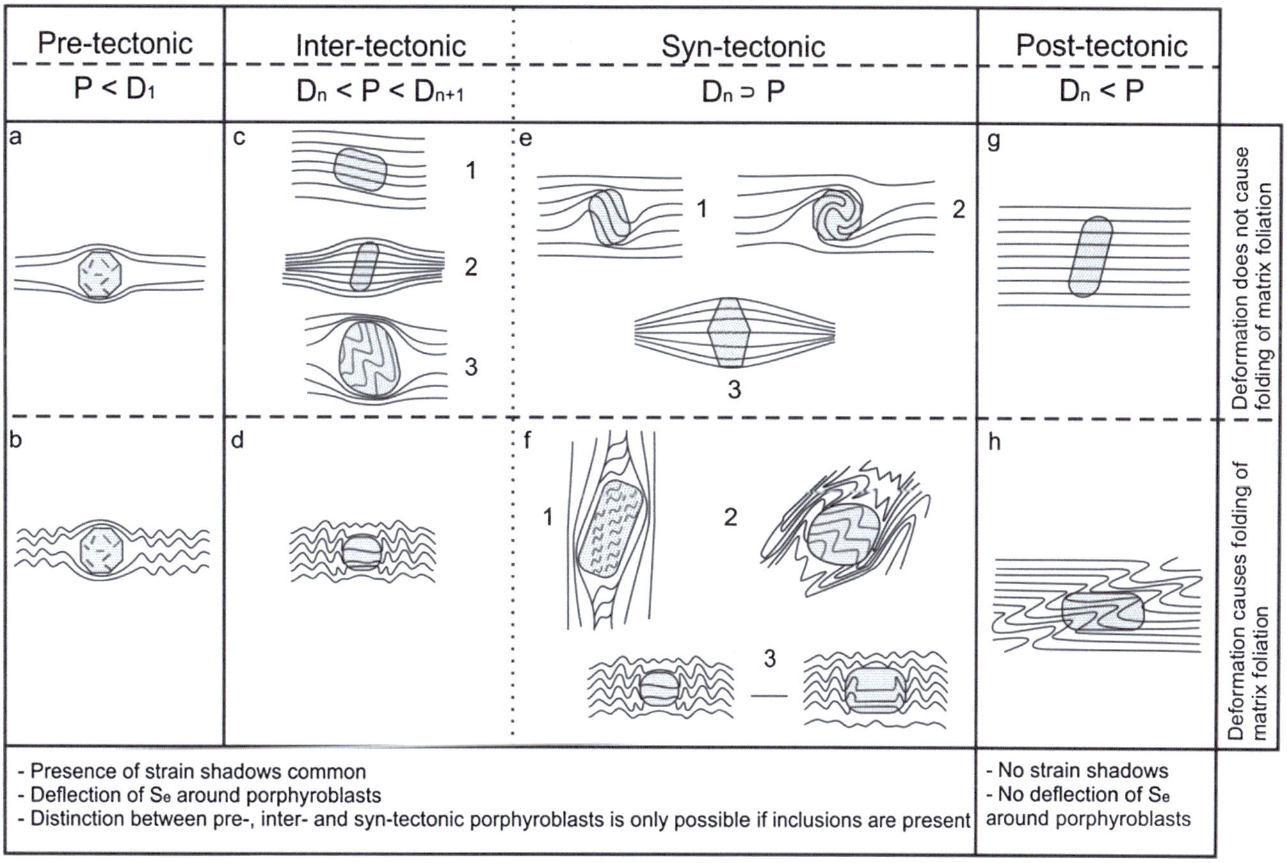

Fig. 19.9 Porphyroblasts are good indicators of pre-, inter-, syn- and post-tectonic crystallization in a metamorphic rock. P is porphyroblast and D is a deformation event. See text for details. (Reproduced from CW Passchier and RAJ Trouw, 2005, Microtectonics, 2nd ed., Fig. 7.9 with permission from Springer Nature under Licence Number 5166460142474)

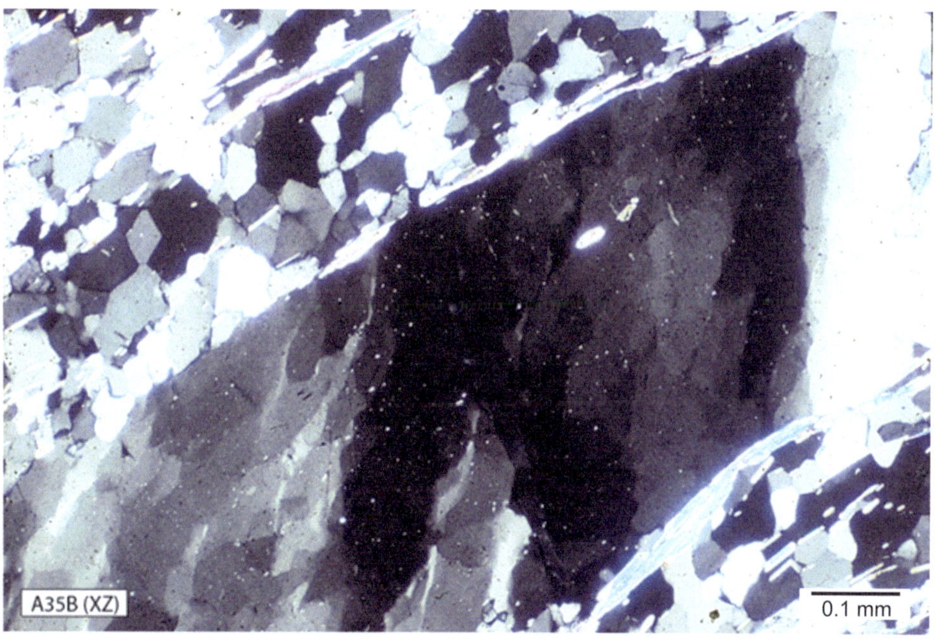

Fig. 19.10 Deformation bands and undulose extinction as indications of pre-tectonic crystallization. The bands developed after the formation of earlier (surrounding) foliation. (Photomicrograph by the author)

19.6.3 Syntectonic Crystallization

Syntectonic crystallization means deformation and crystallization of minerals going synchronously. Rocks showing syntectonic crystallization commonly occur in belts of orogenic metamorphism. Some common evidences of syntectonic crystallization include the following: (a) Pressure shadows. (b) Inclusion trails of porphyroblasts (Figs. 19.7b,c and 19.9e) that have grown during a single phase of deformation D_n: The inclusion patterns are generally curved in syntectonic porphyroblasts, while they are random or straight in pre- and inter-tectonic porphyroblasts (Passchier and Trouw 2005). The S_i can be symmetrically arranged in relation to S_e (Figs. 19.7b and 19.9e3,f3) or may show oblique S_i, sigmoidal or spiral geometry (Figs. 19.7c and 19.9e1,e2). The folds developed within the porphyroblast are called *helicitic folds* or *helicitic structures* (Figs. 19.7b and 19.9c3,f,h). (c) Development of continuous schistosity (Fig. 14.13). (d) Dynamic recrystallization (Fig. 16.1): Signatures of syntectonic crystallization are commonly noticed in dynamically recrystallized rocks of large-scale ductile shear zones for which we present the following two examples:

Example 1: A k-feldspar porphyroclast occurring in a mylonite (Fig. 19.11). The porphyroclast shows (a) an asymmetry with the surrounding mylonitic foliation; (b) feldspar recrystallization with the formation of muscovite and biotite that occupy the rim of the porphyroclast; (c) recrystallization into an assemblage of muscovite, biotite and quartz that are randomly spread within the porphyroclast; and (d) development of twin lamellae within the porphyroclast. The asymmetry of the porphyroclast and development of twin lamellae are signatures of deformation, while the other two reflect recrystallization. The porphyroclast thus appears to have preserved evidences of syntectonic crystallization.

Example 2: The photomicrograph (Fig. 19.12) shows a k-feldspar porphyroclast (left) in a mylonite and shows an asymmetry with the surrounding mylonitic foliation, thus showing a signature of noncoaxial ductile shear deformation. Formation of muscovite (and some biotite) occurring along the rim has taken place during feldspar recrystallization. An adjoining quartz porphyroclast (right) shows evidences of dynamic recrystallization in the form of grain-size reduction with the formation of new, unstrained grains along its tail to the left, while its central part still remains strained. Considering all these signatures together, the porphyroclasts appear to show evidences of syntectonic crystallization.

19.6.4 Post-tectonic Crystallization

Post-tectonic crystallization implies that metamorphism continued even after the deformation is over. It also includes intervention of any metamorphic phase after the end of deformation. This type of crystallization is not always easy to distinguish and is possible only when clear evidences of

Fig. 19.11 Photomicrograph showing a k-feldspar porphyroclast occurring in a mylonite. The porphyroclast shows evidences of syntectonic crystallization. See text for details. (Photomicrograph by the author)

Fig. 19.12 Photomicrograph showing a k-feldspar porphyroclast (left) occurring in a mylonite. The porphyroclast shows an asymmetry with the surrounding mylonitic foliation reflecting a sinistral sense of shear. Under noncoaxial ductile shear, it shows evidences of syntectonic crystallization as evidenced by the formation of muscovite (and some biotite) at its border. At the same time, an adjoining quartz porphyroclast (right) shows evidences of dynamic recrystallization, thus showing grain-size reduction by the formation of new, unstrained grains of quartz along its tail to the left, while its central part still remains strained. See text for details. (Photomicrograph by the author)

both deformation and crystallization are seen. In most cases, occurrence of a clear, unstrained porphyroclast cutting across the surrounding schistosity is considered to represent a post-tectonic crystallization. In post-tectonic porphyroblasts, deflection of S_e is absent (Figs. 19.7d and 19.9g,h), while it is common in pre-, inter- and syntectonic porphyroblasts. Further, the inclusion S_i is continuous with S_e (Figs. 19.7d and 19.9g) even if it is folded (Fig. 19.9h).

19.7 Summary

- Most metamorphic rocks bear imprints of deformation. The deformational processes are thus believed to go together with metamorphism under several circumstances.
- Deformation can accelerate reactions during metamorphism in several ways, and so also metamorphism can promote deformation in several ways. Deformation and metamorphism thus can be considered interactive to each other.
- Some common deformation structures in metamorphic rocks include foliation, crenulation foliation, crenulation lineation, pressure shadows, mineral lineation, mineral segregation and veins. Formation of all these structures includes the role of mineralizing fluids in several ways.
- *Porphyroblasts* are one of the most common structures of metamorphic rocks. These are relatively large single crystals that have formed by metamorphic growth in a more fine-grained matrix. Their study throws significant light on the structural and metamorphic history of rocks.
- Inclusions in porphyroblasts are highly helpful in tracing the deformation and metamorphic patterns of the rock, and thus in reconstructing the relative timing of mineral growth during deformation.
- In polydeformed metamorphic terrains, the relative timing of deformation can be considered in four stages: pre-, inter-, syn- and post-tectonic crystallization. In such studies, microscopic features of rocks constitute an important tool. Study of internal schistosity (S_i) and external schistosity (S_e) of porphyroblasts holds great potential in this exercise.

Questions

1. How can deformation influence metamorphic reactions?
2. How can metamorphism promote deformation?
3. Briefly explain the formation of pressure shadows.

4. What is a porphyroblast? Discuss the significance of porphyroblasts in the context of deformation-metamorphism relations.
5. What are S_i and S_e? Explain why their study is important.
6. Can porphyroblasts that do not contain inclusions help in understanding deformation as related to metamorphism and vice versa? Explain with figures.
7. Discuss some common structural-metamorphic processes during syntectonic crystallization.
8. How would you identify the presence of pre-tectonic crystallization in a rock?
9. What are helicitic folds? In what type of structural and/or metamorphic situations can these structures develop?
10. Briefly discuss 'deformation and metamorphism are interactive'.

20 Superposed Folds

A. R. Bhattacharya, *Structural Geology*, Springer Textbooks in Earth Sciences, Geography and Environment,
https://doi.org/10.1007/978-3-030-80795-5_20

Abstract

Superposed folds are complex folds formed due to superposition of an early-developed fold set by some later fold set(s). The resultant fold geometry is said to show a *fold interference pattern*. Superposed folds may form during a single deformation event or during different deformation events of a single orogeny. These folds commonly occur in most orogenic belts. The classical grouping of superposed folds is in two dimensions. However, some later workers have also considered three-dimensional grouping of superposed folds. Although our understanding of superposed folding is broadly field based, experimental studies of both single-layer and multilayer folds have thrown significant light on their mode of formation. Reconstruction of three-dimensional shapes of the superposed folds, and thus outlining the stages of evolution of the superposed folds, is sometimes difficult. However, the study of the reorientation of lineations and cleavages or the relations between folds and cleavage or lineation can help in distinguishing the earlier and later generations of folds. Recently, the concepts of *progressive deformation* and *deformation phase* are being considered in the context of fold superposition.

Keywords

Superposed folds · Fold interference pattern · Dome-basin pattern · Dome-crescent-mushroom pattern · Convergent-divergent pattern · Superposed buckling · Reorientation of lineations and cleavages · D-numbers · Progressive and multiphase deformation

20.1 Introduction

Folds occurring in most orogenic belts commonly show a complicated geometry due to refolding of their axial planes. Folds thus formed are called *superposed folds* that can be defined as *the complex folds formed due to superposition of an early-developed fold set by some later fold set(s)* (Ramsay 1967). Occurrence of superposed folds in an area is said to show a *fold interference pattern*, in which one can notice one or more axial planes in folded forms. Also, the limbs of the first fold set do not retain their original attitudes.

Superposed folds are better studied in field and better interpreted on structural maps of a region. Their study is important especially from the point of view that (a) they help in understanding the interactions of geometric factors in controlling the structure of strata at the earth's surface, and (b) the different types of superposed folds can be mapped in the field that helps in identifying and locating the major fold components of a region (Ramsay and Huber 1987, p. 499).

Some early workers (e.g. Reynolds and Holmes 1954; Sutton and Watson 1955; Weiss et al. 1955; Johnson 1956) had carried out a systematic study of superposed folds from the Scottish Highlands. Subsequently, such folds were recognized from several orogenic belts of the world. Our knowledge on superposed folds vastly increased due to the classic work of Ramsay (1967) and Ramsay and Huber (1987). Works of several other workers (e.g. Ghosh and Ramberg 1968; Park 1969; Thiessen and Means 1980; Watkinson 1981; Holdsworth 1990; Ghosh 1993; Faleiros et al. 2016; Neves et al. 2018; Mueller et al. 2020) have thrown significant light on various aspects of such folds.

20.2 Early Concepts on Superposed Folds

Superposed folds are commonly formed when an earlier fold is superposed by another folding event(s) during subsequent deformation phase(s). They may form during a single deformation event or during different deformation events of a single orogeny. Ramsay (1967) outlined the following types of deformation processes for the formation of superposed folds.

20.2.1 Crossing Orogenic Belts

Superposed folds may form in layered rocks due to successive deformations that are separated by a very large interval of time. A common example is when a folded basement complex undergoes a later orogenic deformation. In such cases, the strain patterns of the successive folding events are different from each other.

20.2.2 Successive Deformations in a Single Orogenic Cycle

Most terrains of superposed folding show a type of deformation pattern in which in one orogenic cycle the compressive deformations occur in separate pulses. This produces folds that are often of different styles due to different strain rates and physical state of rocks during fold formation. However, some common geometric features, say trend of axial surfaces, can still be present in the various sets of folds.

Fig. 20.1 Refolded fold in pink granitic gneiss of Bundelkhand craton, central India

20.2.3 Successive Folding During a Single Progressive Deformation

Sometimes, during a single progressive or continuous deformation, it is possible that there is a change in the orientation of layering with respect to the axes of incremental strain ellipsoid. This may cause oblique straining of any originally formed folds that, in turn, produces superposed folds with new orientations of the axial surfaces.

20.2.4 Synchronous Folding in Different Directions During One Deformation

There may be a situation when a competent layer is under the influence of a two-dimensional constrictive strain developed due to a three-dimensional constrictive strain. Under this condition, folding may occur in several directions synchronously. Synchronous folding can be distinguished from successive folding by studying the relationship of the axial surfaces of the two (or more) sets of folds (Ramsay 1967, p. 520).

Box 20.1 How to Identify Superposed Folds?

If an early fold is subjected to another generation of folding (or refolding), the resultant structure is called a superposed fold. Such folds are commonly developed in highly deformed rocks. During refolding, the geometric elements of the early fold are reoriented. Careful observation of the geometric elements of the superposed fold helps to distinguish the other generations of folds as briefly explained below.

Figure 20.1 shows a photograph of a refolded structure. Careful tracing of fold axes reveals an earlier F_1 fold superposed by another fold (F_2). The axes of the two folds are nearly perpendicular to each other.

Let us now consider a generalized case with the help of a diagrammatic sketch of a refolded fold. In Fig. 20.2 (upper), an early fold is shown in its refolded form. One can easily identify the trace of axial plane of the early fold (F_1) after it has been subjected to a second generation of folding (F_2). With the refolding pattern, one can also identify the geometric elements of the second fold, which in the present case is shown by the axial trace of the second fold (F_2) only. The axial plane of the second-generation (F_2) fold is at high angle to the first-generation fold (F_1).

Figure 20.2 (lower) shows another fold with a different orientation. With the above analogy, one can identify the geometric elements of the second generation of fold also.

20.3 Types of Superposed Folds

In terrains of superposed folding, the folds produced show a variety of three-dimensional forms. Ramsay (1967) identified three basic types (types 1, 2 and 3) of interference pattern of

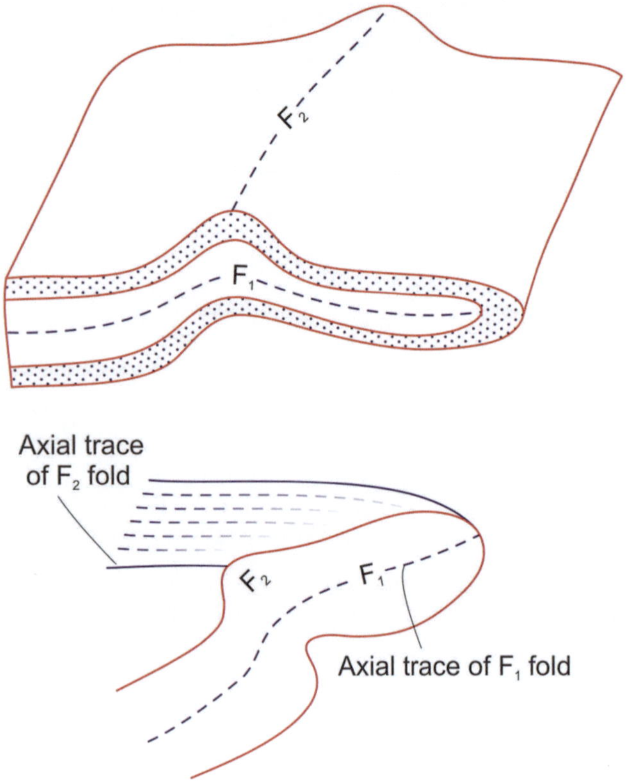

Fig. 20.2 Diagrammatic sketch of a refolded fold to illustrate the identification of different geometric elements of a second-generation fold

folds on the basis of flow direction of the superimposed movement for which the relative orientation of axial planes and fold axes are important components. Later on, the authors (Ramsay and Huber 1987) revised this scheme and identified four basic types of fold interference pattern (Fig. 20.3) named as type 0, 1, 2 and 3, as described below.

20.3.1 Type 0: Redundant Superposition

This type of superposition of two sets of folds does not produce any geometric feature characteristic of fold superposition (Fig. 20.4). As such, it has been named as type '0' or redundant superposition. The three-dimensional geometry thus produced is almost similar to the one produced due to single phase of deformation (Fig. 20.4). Superposition generally causes tightening of the resulting folds. The resulting geometry in all sections is identical to that of normal single-phase folds (Fig. 20.3a,c). Figure 20.3b shows the displacement pattern of the superposed second system when subjected to a heterogeneous simple shear with the shear plane orientation given by a_2 b_2. Figure 20.3c shows the final result of superposition. Figure 20.5 shows a field photograph closely resembling type 0.

20.3.2 Type 1: Dome-Basin Pattern

This type of superposition occurs when there is small angle between the first fold axial planes and the movement direction a_2 of the second folds. Under this condition, the differential shear of the second deformation does not produce any strong deflection of these axial planes from their original planar geometry. As a result, the limbs of the first folds are deflected forming new folds that plunge in directions away from or towards the first fold axial surface (Fig. 20.3). In the process, the planar geometry of the first axial surfaces is generally not strongly disturbed, but the second-generation movements will produce strong undulations of the hinges of the first folds (Fig. 20.6B). This results in the formation of culminations and depressions of the first folds. The structures thus developed are domes and basins (Fig. 20.3), and as such this type of superposition is said to constitute dome-basin patterns. Each dome is surrounded by four basins, and each basin is surrounded by four domes. This structure is also sometimes loosely called as *egg carton structure*. Figure 20.7 shows a field photograph closely resembling type 1.

20.3.3 Type 2: Dome-Crescent-Mushroom Pattern

In this type of superposed folds, the angle between the differential movement a_2 of the second phase and the first fold axial planes is high. During deformation, the first fold axial planes become strongly folded (Fig. 20.3, type 2C). Also, due to high angle between F_1 and b_2, the fold hinges of F_1 folds become strongly bowed (Figs. 20.8 and 20.9). In the resulting structure, the arching of the first fold limbs and axial surfaces in any one second fold bears the same sense with respect to the first axial plane (Fig. 20.3, type 2C). Due to upbowing and downbowing of the first fold hinges, dome-and-basin-like structures are formed and these are overturned with the same sense as that of the first folds.

20.3.4 Type 3: Convergent-Divergent Pattern

Type 3 superposed folds are formed when, like that of type 2, the differential movement direction a_2 of the second fold phase is at high angle to the first fold axial planes. However, unlike type 2, the b_2 direction is very close to the first fold hinge lines F_1. Under these conditions, the first fold axial surfaces become curved, but the first fold hinges are not significantly deflected. The fold axial directions of both first and second phases are subparallel. In the resulting structure, the limbs of the first folds will neither converge together nor diverge away from the first fold axial surfaces but will remain

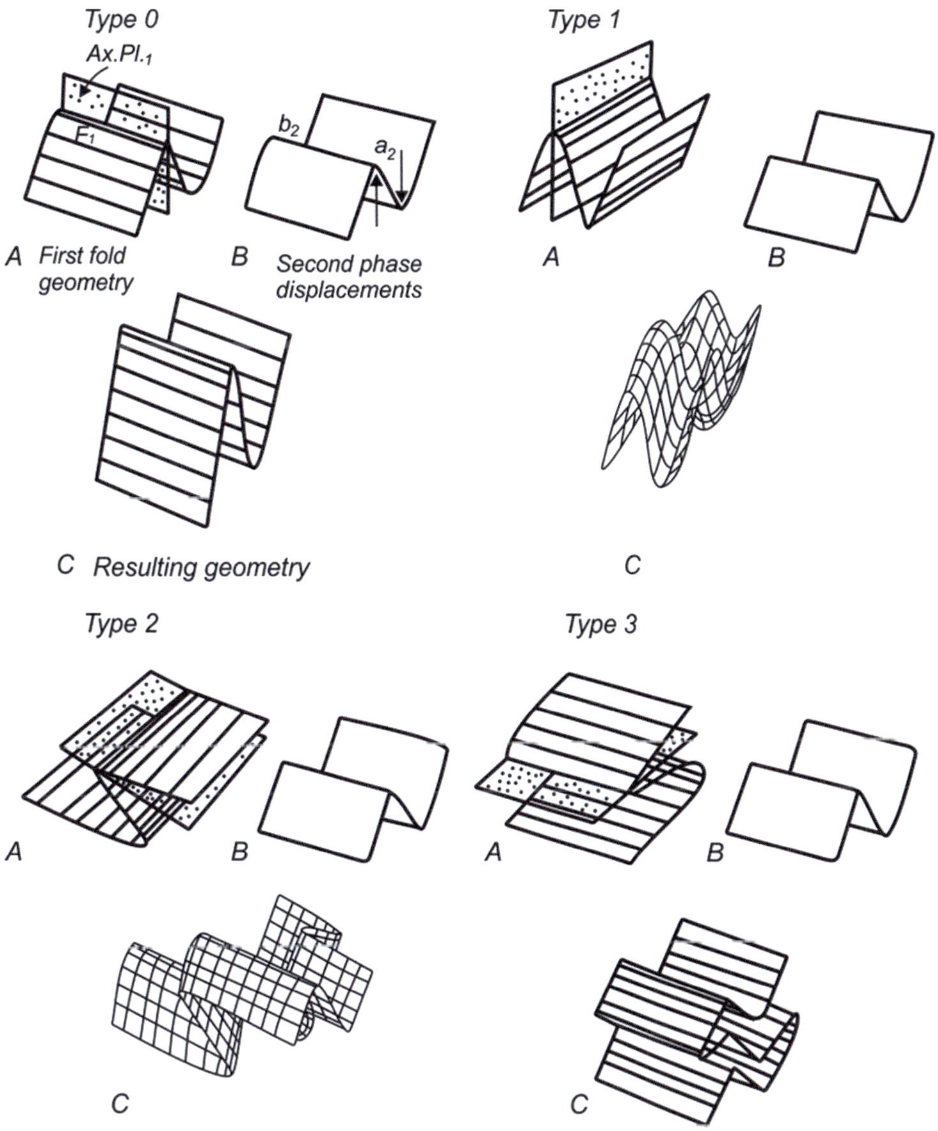

Fig. 20.3 Types of superposed folds. See text for details. (Reproduced from Ramsay and Huber 1987, Fig. 22.15 with permission from Elsevier Senior Copyrights Coordinator, Edlington, U.K. Submission ID: 1198150)

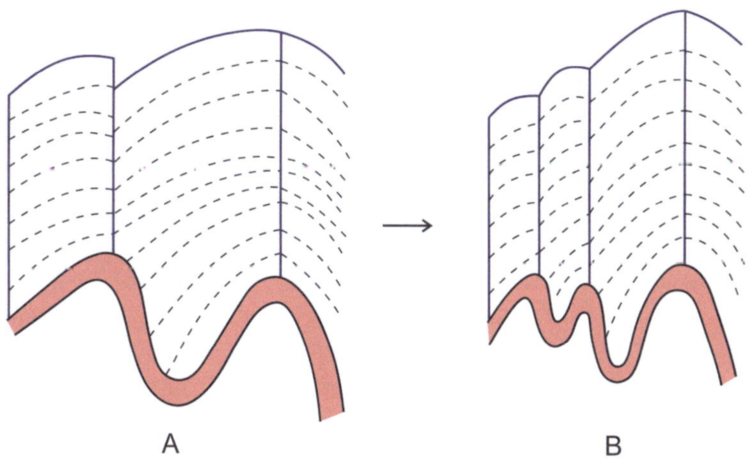

Fig. 20.4 Type 0 superposed folds. (**a**) An originally folded structure. (**b**) Resultant structure after refolding which does not produce any geometric feature characteristic of fold superposition

Fig. 20.5 Field photograph of folds that closely resemble type 0 superposed folds. (Photograph courtesy Professor T.K. Biswal)

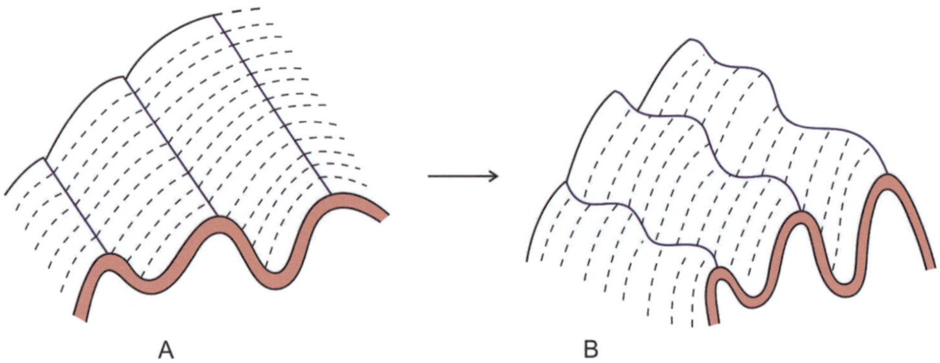

Fig. 20.6 Type 1 superposed folds. (**a**) An originally folded structure that is subjected to refolding with a small angle between the first fold axial planes and the movement direction a_2 of the second folds. (**b**) Under this condition, the differential shear of the second deformation does not produce any strong deflection of these axial planes from their original planar geometry but produces strong undulations of the hinges of the first folds, thus forming culminations and depressions of the first folds

as such even after refolding (Fig. 20.3, type 3). This type of superposition thus constitutes a *convergent-divergent pattern*.

20.4 Geometric Changes in Superposed Folds

Formation of superposed folds is the result of geometric changes undergone by the early-formed fold(s). Practically, every geometric element of an early fold takes part in the superposition process. However, the shape/geometry of the superposed fold is generally the result of some specific, not all, element(s) of the early-formed fold. Sometimes, only one element, say axial plane, is active, but in the process it involves/affects some other element(s) also. This aspect has been discussed by Ramsay (1967, pp. 538–546) in detail from which some key points are highlighted below:

(a) Superposed folds fall in two groups: those formed simultaneously during superposed deformation, and those that show geometric relations to the early-formed folds, and have undergone distortion due to later deformation.
(b) On refolding of a first-fold structure, the linear structures may show different orientations and commonly show some curved forms, while the planar structures undergo

Fig. 20.7 Folds closely representing dome-basin type of superposition. (Photograph courtesy Professor T.K. Biswal)

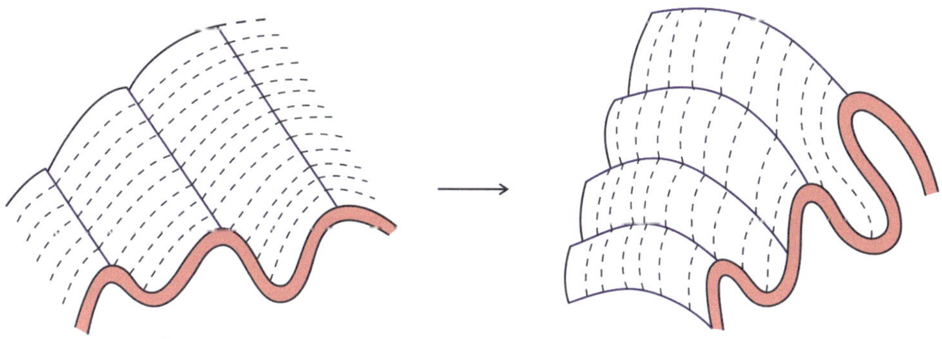

Fig. 20.8 Type 2 superposed folds. (**a**) An originally folded structure that is subjected to refolding such that the angle between the differential movement a_2 of the second phase and the first fold axial planes is high. (**b**) The resultant structure thus produced. See text for details

Fig. 20.9 Folds closely representing type 2 superposition. Note that the angle between the first fold axial plane (upper) and that of the second fold (lower) is high. As a result, during second folding, the hinges of first folds become bowed that results in arching of the first fold limbs. (Photograph courtesy Professor T.K. Biswal)

modification. The smaller first folds that are refolded on the limbs of the new folds are much tighter than their original shapes, while those that have undergone refolding near the hinge zones of the new folds show more open forms than their original shapes.

(c) The orientation of the axial planes of the new folds generally shows a constant geometry, while other surfaces and linear directions may show changes.

(d) The symmetric or asymmetric nature of the superposed folds depends on whether the axial planes of the new folds are subperpendicular to the first-fold limb, or if they make some other angle with the fold limb.

(e) The variation of the axial pattern of the new folds sometimes depends on the degree of tightness and size of hinge zone of the early folds.

(f) If the surfaces within the early folds show a considerable variation of orientations, the axial directions of the folds formed by refolding of this structure will show considerable variation.

(g) If the early folds are isoclinal, or nearly so and show narrow hinge zones, then the superposed structure will have only one new-fold axial direction.

(h) In areas of superposed folding, the amplitude of the new folds generally shows changes from locality to locality possibly due to variation in the intensity of the deformation during superposition. If the flow direction is oriented at a high angle to the surfaces being folded, the superposed folds thus formed have a larger amplitude than where this angle is small.

(i) To sum up, in regions of superposed folding, the various geometric elements of the first-fold structures undergo distortion to take up new orientations depending upon the mechanism and intensity of the new deformations.

20.5 Buckling of Superposed Folds

Although the understanding of superposed folding is broadly a field-based study of a region, experimental studies have thrown significant light on their genesis and mode of formation. In recent years, experiments on superposed buckling have been carried out by several workers (e.g. Ghosh and Ramberg 1968; Skjernaa 1975; Watkinson 1981; Ghosh et al. 1992). These studies broadly indicate (Ghosh 1993, p. 340) that (a) the pattern of superposed buckling depends upon the shapes of early folds, and (b) the profile shapes of folds are more varied in a multilayer than in a single layer, and therefore the refolding modes of multilayered rocks are more varied. Superposed buckling of a single layer and that of a multilayer are briefly described below.

20.5.1 Single-Layer Superposed Buckling

In the light of his study of superposed buckling of a single competent layer embedded in an incompetent layer, Ghosh (1993) suggested that the early folds may be refolded in any of the following four modes (Fig. 20.10).

20.5.1.1 First Mode of Superposed Buckling

This mode of buckle folding occurs when the early fold (F_1) is very gentle and has a small curvature of the layer at the hinge zone or with a large interlimb angle (greater than about 135°). The superposed folds (F_2) thus formed are approximately of the same size as F_1, and the interference produces a dome-and-basin pattern (Fig. 20.10a).

20.5.1.2 Second Mode of Superposed Buckling

If the first folds have moderate tightness (interlimb angle about 135° to 90°), the superposed buckling is inhibited. Under this condition, unlike the first mode of dome-and-basin pattern, a set of F_2 folds are formed that are much smaller in size than the F_1 folds (Fig. 20.10b). The smaller (F_2) folds ride over the hinges of the larger F_1 folds.

20.5.1.3 Third Mode of Superposed Buckling

This mode of superposed buckling occurs when the F_1 folds initially are moderately tight (interlimb angle lesser than about 90°). The resulting structures include the early and the late folds that have approximately similar arc lengths, but they neither produce dome-and-basin structure nor do the F_2 folds ride over the F_1 hinges (Fig. 20.10c). If the structure is oriented sub-horizontal, then the F_2 folds show plunge in opposite directions on the two limbs of the early (F_1) fold, and an antiformal F_2 on one limb of F_1 passes to a synformal F_2 on the other limb.

20.5.1.4 Fourth Mode of Superposed Buckling

This mode operates if the initial F_1 fold is very tight or isoclinal. During second deformation, the hinge and axial surface of the F_1 fold are folded with a shortening along the F_1 hinge line (Fig. 20.10d).

20.5.2 Multilayer Superposed Buckling

A multilayer undergoes superposed buckling almost in the same manner as a single layer; that is, it can show all the four modes of superposed folding as shown by a single layer as described above. However, the multilayer shows a modified second mode in which (a) the arc length of F_2 folds is much smaller (much less than one-fourth) than that of F_1, and (b) the small F_2 folds are overprinted on the long limbs of narrow-hinged F_1 folds such that the F_2 folds do not ride over their hinges.

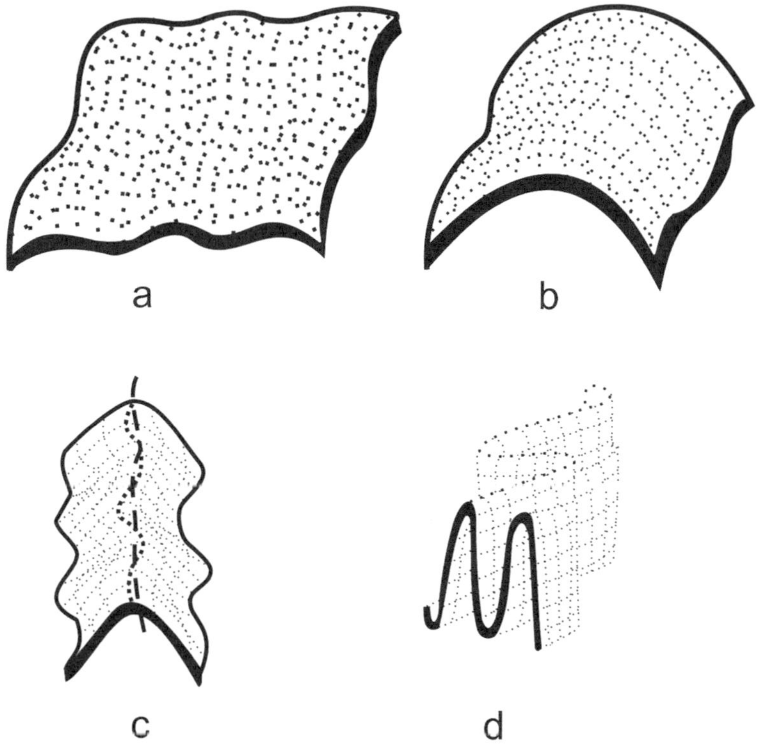

Fig. 20.10 Superposed buckling of a single competent layer embedded in an incompetent layer suggests that the early folds may be refolded in any of the four modes. See text for details. (Reproduced from Ghosh 1993, Fig. 15.11 with permission from Elsevier Copyrights Coordinator, Edlington, U.K. Submission ID: 1198160)

20.6 Classification of Thiessen and Means

Thiessen and Means (1980) modified Ramsay's (1967) classification of fold interference patterns by using angular parameters. They identified an angle, γ (gamma), between the axes of the first folds and the pole to the axial planes of the second folds. This, together with an angle β (beta), as used by Ramsay, provides a more natural basis for the classification of three-dimensional fold interference patterns. According to the authors, the two angles, β and γ, are sufficient for classification of three-dimensional interference patterns as presented in Fig. 20.11 by the use of the β-γ pair. *Types* 0, 1 and 3 *patterns* practically refer to those of Ramsay (1967). *Type 2 patterns*, however, form due to combinations of α, β and γ.

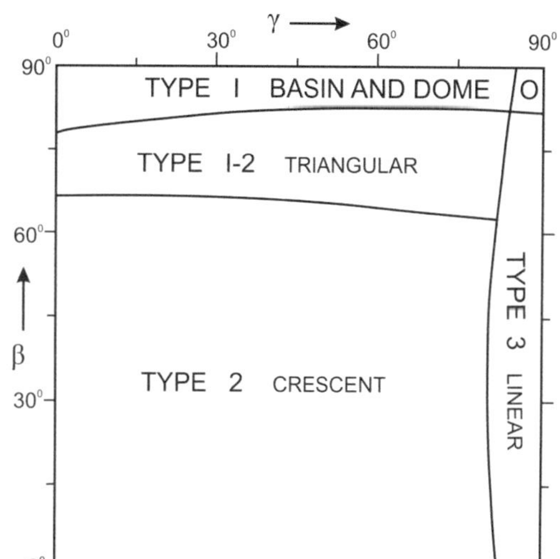

Fig. 20.11 Classification of fold interference patterns by Thiessen and Means (1980). Diagram showing regions of various fold types on β-γ projection. See text for details. (Reproduced from Thiessen and Means 1980, Fig. 8, with permission from Elsevier Copyrights Coordinator, Edlington, U.K. Submission ID: 1193147)

20.7 Classification of Grasemann and Others

Grasemann et al. (2004) considered graphical quantification of the interaxial angles of refold structures and thus identified six different end members of three-dimensional refold structures. The reference axes of refold structures are the fold axis, the pole to the axial plane and the shear direction in the axial plane. Assuming orthorhombic shear folds, the reference axes are defined by a three-dimensional Cartesian coordinate system. According to the authors, the relative orientations of the initial and superposed reference axes can be considered as a rotation about the origin by an orthogonal direction cosine matrix **L**:

$$\mathbf{L} = \begin{pmatrix} l_{11} & l_{12} & l_{13} \\ l_{21} & l_{22} & l_{23} \\ l_{31} & l_{32} & l_{33} \end{pmatrix}, \tag{20.1}$$

where

$$\begin{aligned} \cos^{-1} l_{11} &= \angle x_1' x_1;\ \cos^{-1} l_{12} = \angle x_1' x_2;\ \cos^{-1} l_{13} = \angle x_1' x_3 \\ \cos^{-1} l_{21} &= \angle x_2' x_1;\ \cos^{-1} l_{22} = \angle x_2' x_2;\ \cos^{-1} l_{23} = \angle x_2' x_3 \\ \cos^{-1} l_{31} &= \angle x_3' x_1;\ \cos^{-1} l_{32} = \angle x_3' x_2;\ \cos^{-1} l_{33} = \angle x_3' x_3 \end{aligned}$$

Since the symmetry of the folds is assumed to be orthorhombic, l_{ij} can take values between 0 and 1. By definition, the direction cosine requires that

$$\mathbf{L}_{ij}\mathbf{L}_{ij} = \mathbf{L}_{ji}\mathbf{L}_{ji} = 1 \tag{20.2}$$

As such, **L** can be replaced by a reduced two-dimensional direction cosine matrix **L***:

$$\mathbf{L}^* = \begin{pmatrix} l_{11} & l_{12} \\ l_{21} & l_{22} \end{pmatrix}, \tag{20.3}$$

where, because the angles are measured from one line to two orthogonal lines,

$$0 \le \mathbf{L}_{ij}\mathbf{L}_{ij} = \mathbf{L}_{ji}\mathbf{L}_{ji} \le 1 \tag{20.4}$$

The authors thus extended the well-established terminology of fold superposition classifying types 1–3 to three counterpart types 0_1–0_3 (Fig. 20.12), which are simply derived by 90° rotation of the superposed fold around its fold axis.

20.8 Reorientation of Lineations and Cleavages

A major objective in terrains of superposed folding is to identify the various fold generations. For this, the presence of lineations and cleavages helps in understanding the overprinting relations (Fig. 20.13) of folds, which in turn helps in reconstructing the three-dimensional shape of the superposed folds and outlining the stages of evolution of the superposed folds. In most cases, the three-dimensional forms of superposed folds are not exposed. In such cases, reorientation of lineations and cleavages helps in distinguishing the earlier and later generations of folds. Geometric relations between folds and cleavage or lineation can be explained by two simple examples (see Ghosh 1993, p. 354). In the first case (Fig. 20.13a), we consider a fold that shows strong lineation (L_1) parallel to the hinge lines of an early fold (F_1).

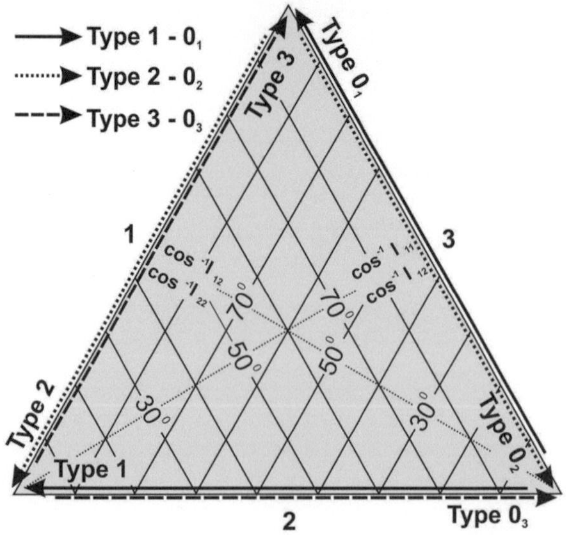

Fig. 20.12 Classification of refold structures by Grasemann et al. (2004) as represented by vector triangle plot showing graphical quantification of the interaxial angles of refold structures. Each side of the triangle represents two of the possible six end members as vectors with opposite polarity. See text for details. (Reproduced from Grasemann et al. 2004, Fig. 3, with permission from the Journal of Geology, the University of Chicago)

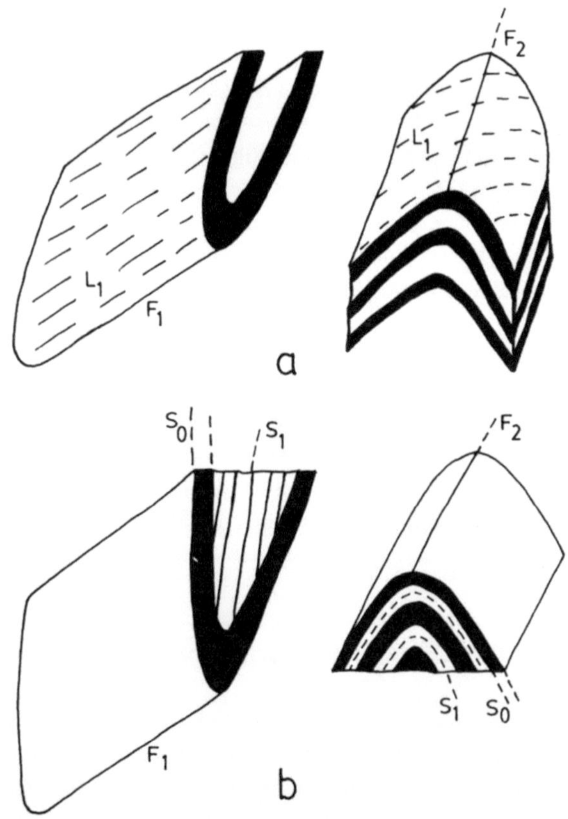

Fig. 20.13 Overprinting relations of lineations and cleavages in superposed folds. See text for details. (Reproduced from Ghosh 1993, Fig. 15.26 with permission from Elsevier Copyrights Coordinator, Edlington, U.K. Submission ID: 1198163)

This lineation will be deformed if the early fold is deformed by another fold generation (F_2), suggesting that the early fold has been deformed by another generation. Likewise, we consider another fold (Fig. 20.13b) in which an early cleavage (S_1) is parallel to the axial plane of the early fold (F_1) on its bedding (S_O). This cleavage will be deformed if the early fold is deformed by another fold generation (F_2), suggesting that the early fold has been deformed by another generation.

The above relations shown by the lineation and cleavage with the associated folds indicate that there exist some definite relations of lineation and cleavage during deformation of an early host fold. As such, even if the fold hinges are not exposed, one can identify the superposed fold(s) by the overprinting relations as demonstrated above.

20.9 Deformation Phase

In the context of superposed folding, it is important to understand what is meant by a *deformation phase*. This aspect has been discussed by Passchier and Trouw (2005) and Fossen et al. (2019). Passchier and Trouw (2005, p. 309) defined a deformation phase as a *period of deformation during which a group of structures have formed, separated from other structures by overprinting criteria*. According to them (p. 3): (a) The later deformation phases may merge together or may be separated by time intervals during which there may be little or no deformation and metamorphic conditions and the stress field may have undergone a change; (b) a deformation phase consists of distinct periods of active deformation of rocks on a scale exceeding that of a single outcrop, possibly separated by time intervals with little or no deformation during which metamorphic conditions and orientation of the stress field may have changed. For a deformation phase to be defined, its structural expression should be identified and mapped at a regional scale (at least hundreds of square kilometres). Each phase of deformation results in a characteristic family of structures that can be recognized by correlation between outcrops based on style, geometry, orientation, etc.

Progressive deformation (Fossen et al. 2019) refers to a period of continuous (non-stop) deformation during which composite structures such as overprinting cleavages, refolded folds, and folds and fabrics may form typically with significant variation in style and orientation. As a result, overprinting relations form progressively, and the overprinting structures may all be of different ages, regardless of similarities or differences in style and orientation. Hence, while an F_2 fold has formed, or is forming in one location, an F_1 fold is forming in a nearby location. All these constitute progressive deformation.

20.10 D-Numbers

In areas where rocks have undergone polyphase deformation, development of new planar structures and linear structures is commonplace. Each newly developed structural feature is given a special name to distinguish it from the previously formed and newly formed features. Thus, assigning S_0 to the original stratification, the subsequently formed planar structures have been assigned S_1, S_2, etc. in a sequence. Likewise, the linear structures are assigned L_0, L_1, L_2, etc. in a sequence. For folds, a sequence of F_0, F_1, F_2, etc. also holds good.

The above analogy of nomenclature for sequential formation of structures was also followed for each episode of deformation leading to the formation of new structures by assigning the *D-numbers*. As such, the soft-sediment disturbances were designated as D_0, into which S_0, L_0 and F_0 were included, while the first episode of tectonic deformation was given the symbolic notation of D_1, which included S_1, L_1 and F_1 (Rast 1997, p. 426). This system of D-scaling has been found to be convenient for relating the various geometric properties of deformed rocks to a particular deformation event in areas where the rocks have undergone polyphase deformation.

20.11 Summary

- *Superposed folds* are the complex folds formed due to superposition of an early-formed fold set by some later fold set(s). Their occurrence in an area constitutes a *fold interference pattern*. Such folds are common in orogenic belts and show a complicated geometry.
- Superposed folds are better studied in field and better interpreted on structural maps of a region. Their study throws light in understanding the interactions of geometric factors in controlling the structure of strata at the earth's surface, and in locating the major fold components of a region.
- In superposed folds, the earliest formed fold(s) do not retain their original geometries. The interference pattern thus produced shows a variety of three-dimensional forms.
- Superposed folds may form during a single deformation event or during different deformation events of a single orogeny. Some common deformation processes of formation of superposed folds include (i) successive deformations that are separated by a very large interval of time, (ii) a type of deformation pattern in which in one orogenic cycle the compressive deformations occur in separate pulses, (iii) successive folding during a single progressive deformation and (iv) synchronous folding in different directions during one deformation.

- In terrains of superposed folding, four basic types of fold interference patterns are identified: (i) type 0: redundant superposition; (ii) type 1: dome-basin pattern; (iii) type 2: dome-crescent-mushroom pattern; (iv) type 3: convergent-divergent pattern.
- Experimental studies suggest that the pattern of superposed buckling depends upon the shapes of early folds and that the profile shapes of folds are more varied in a multilayer than in a single layer, and therefore the refolding modes of multilayered rocks are more varied.
- During superposed folding, geometric elements of folds reorient themselves depending upon the mode or pattern of superposition. Such elements are especially helpful in cases where the three-dimensional forms of superposed folds are not exposed. Lineations and cleavages, especially, help in understanding the overprinting relations of folds, which in turn helps in reconstructing the three-dimensional shape of the superposed folds and outlining the stages of evolution of the superposed folds.
- The classical models of two-dimensional fold superposition have been modified by later workers by considering three-dimensional aspects of superposed folds.

Questions

1. What are superposed folds and where do they generally occur?
2. Describe the various processes that commonly form superposed folds.
3. Why are type 0 superposed folds also called redundant superpositions?
4. Discuss the formation of dome-basin pattern of folds.
5. What is a dome-crescent-mushroom pattern? Describe how they are formed.
6. What is the main difference between the type 2 and type 3 superposed folds?
7. Give some important aspects revealed by the experimental studies on buckling of superposed folds.
8. What are the common structural elements that help in establishing the overprinting relations of superposed folds?
9. Explain the concepts of deformation phase and progressive deformation.
10. Discuss the significance of superposed folds.

Appendix A: Stereographic Projection

Stereographic projection is a convenient method of graphically representing planar and linear structures in three dimensions and the angular relations existing among them. The starting point is an imaginary sphere or a reference sphere for representing the attitude of planes and lines that are shown as projection onto the horizontal *equatorial plane*. Stereographic projections have long been used in mineralogy for representing the crystal faces of minerals by projections in the upper hemisphere. In structural geology, however, the use of such projections came later, possibly in the 1940s. In contrast to mineralogy, the projections in structural geology are made in the lower hemisphere because most structures in rocks are present below the earth's surface.

Stereographic net or **Stereonet**: The basic diagram on which the stereographic projections are made is called a *stereographic net* or *stereonet* (Figs. A1.1 and A1.2). It is a circular diagram in which a number of great circles at regular intervals between 0° and 90° are drawn. Each great circle represents a plane that has a N-S strike and dip towards either E or W. Likewise, small circles are also drawn at regular intervals between 0° and 90° with centres at north and south poles. Two types of stereonets are commonly used: (a) *Equal-angle net* or *Wulff net* (Fig. A1.1) that shows orientation of planes at 2° interval: Each great circle and small circle are thus located at 2° interval. This net is commonly used in crystallography. (b) *Equal-area net* or *Schmidt net* (Fig. A1.2) that shows orientation of planes at 2° interval but with constant area: Each great circle and small circle are thus located at 2° interval. The projection of the great circles and small circles on to the primitive circle is called a *stereogram*. Schmidt net is used in structural geology for solving various types of problems. The method is efficient as it can conveniently handle a large number of data with an equal-area bias in a relatively shorter time. The method thus helps in comparing the data from equal areas of the earth's surface, and this is not possible with the Wulff net.

Primitive circle: Let us consider a reference sphere AWBE as shown in Fig. A1.3. If we cut it along a horizontal plane WSEN passing through the centre of the sphere, the equatorial sphere, the equatorial plane that has a circular outline is called the *primitive circle*. In structural problems, we always consider any plane or line—whether inclined, horizontal or vertical—to pass through this circle. The intersection(s) made by the planes and lines on the primitive circle are called their *horizontal projection*. The primitive circle is further used to refer to the geographic directions by imagining two perpendicular lines N-S and E-W to represent north (N), south (S), east (E) and west (W) as shown in Fig. A1.3.

For all structural problems, the lower hemisphere WBE is used. We orient planes and lines with their respective attitudes in this lower hemisphere and project them to the topmost part or north pole of the sphere called *zenith* (point A in Fig. A1.4). Likewise, a line (PP') in the lower hemisphere is projected to the zenith that meets the primitive circle at P′ (Fig. A1.5). The angular relationship given by the projections of the line (PP') on the primitive circle gives the attitude of the line.

Great circle: Let us imagine a plane SGN (Fig. A1.6) in the lower hemisphere that passes through the centre (O) at the equatorial plane. The plane intersects the lower hemisphere at points 1, 2, 3, 4, etc. These points will trace a circular intersection called *great circle*. On joining these points with the north pole, we get another set of intersection points 1′ 2′, 3′, 4′, etc. that will trace a circular arc at the equatorial plane; this arc gives the trace of a great circle. Thus, a number of great circles are possible each with a specific value of its own. A set of projections of great circles at 30° interval are shown in Fig. A1.7.

Small circle: If a circular cone with its apex at the centre of equatorial plane is imagined, it will trace a circle at the lower hemisphere with its centre at the south pole. Likewise, a number of such cones at equal angles will trace a series of circles at equal angles, each with centre at the south pole. The process can be reversed also, i.e. cones intersecting the northern hemisphere. In this case also, a series of circles at equal angles are traced, each with centre at the north pole. Each of these circles is called a *small circle*. A set of projections of great circles at 30° interval are shown in Fig. A1.7.

Stereographic projection of a plane: Planes and lines show different types of projection in a reference sphere. Let us imagine a plane SGN (Fig. A1.6) that may represent any planar structure, say bedding, foliation, fold axial plane, joint,

A. R. Bhattacharya, *Structural Geology*, Springer Textbooks in Earth Sciences, Geography and Environment,
https://doi.org/10.1007/978-3-030-80795-5

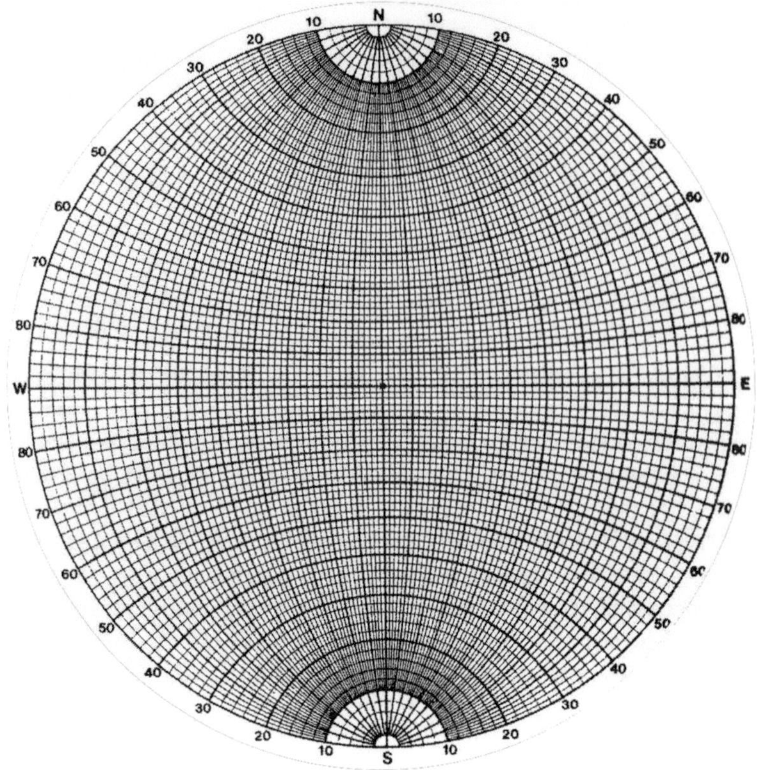

Fig. A1.1 Equal-angle net (Wulff net)

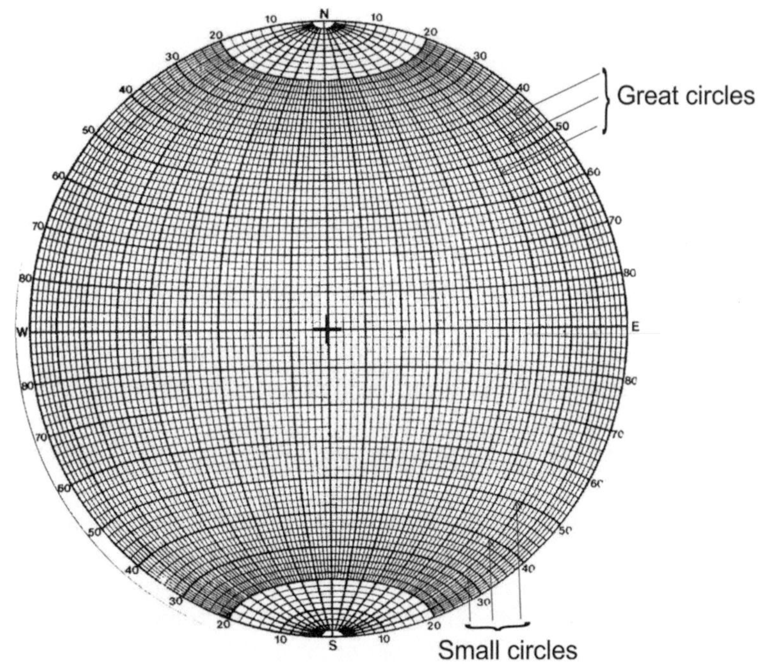

Fig. A1.2 Equal-area net (Schmidt net)

etc. If this plane lies on the equatorial plane, it is said to be horizontal, and if it lies along the vertical plane containing AB (Fig. A1.6), it is said to be vertical. For all positions between these two, the plane is said to be inclined, and the inclination increases from 0° at E to 90° at B. The points E and W in Fig. A1.6 can therefore be assigned 0° inclination and the centre O as 90°. For measurement of angles, the lines EO and WO can be equally divided between 0° and 90°.

The plane SGN in Fig. A1.6 intersects the centre at O and the lower hemisphere at G. Since the strike of the plane is north-south and the plane is situated towards E (east), the dip of the plane is towards east. The line joining G with the north

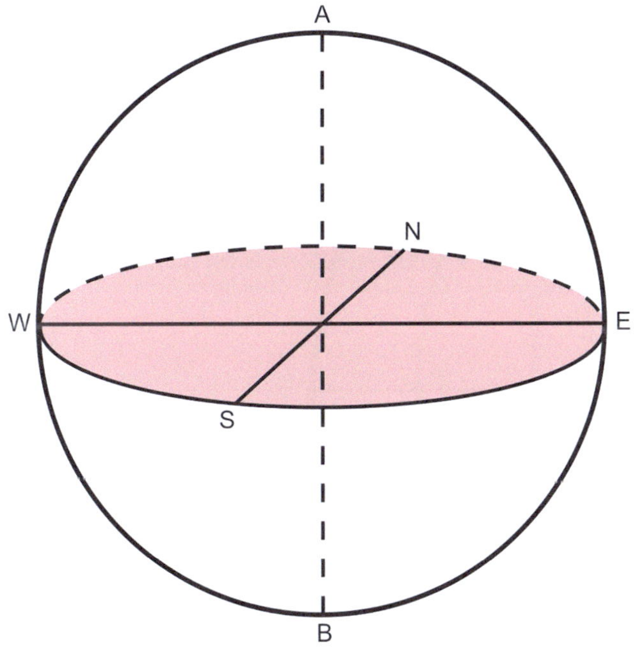

Fig. A1.3 Elements of stereographic projection. AWBE is a reference sphere, and WSEN is the primitive circle. WBE is the lower hemisphere, which is used for projecting structural data

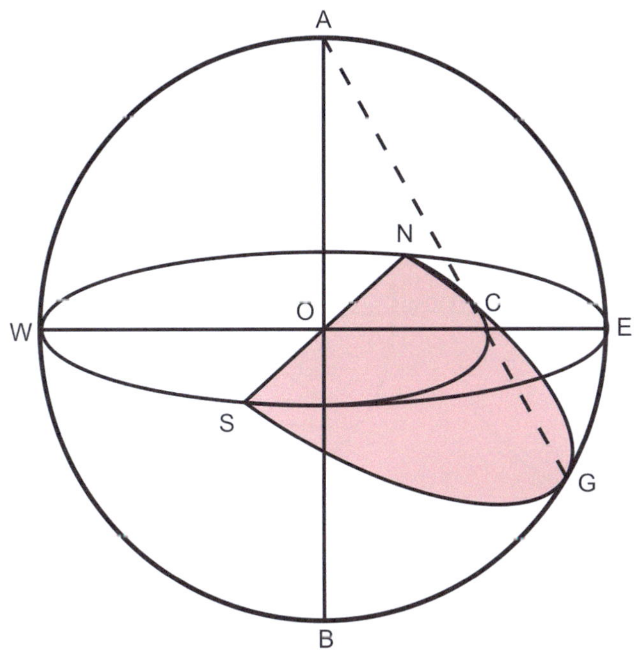

Fig. A1.4 Use of lower hemisphere of a stereograph for orienting a plane (SGN). The topmost part of the sphere (point A) is called the zenith

pole or zenith of the sphere intersects the equatorial plane at C. The plane SGN is thus represented by the girdle SCN as horizontal projection, and the distance CE equals the amount of dip.

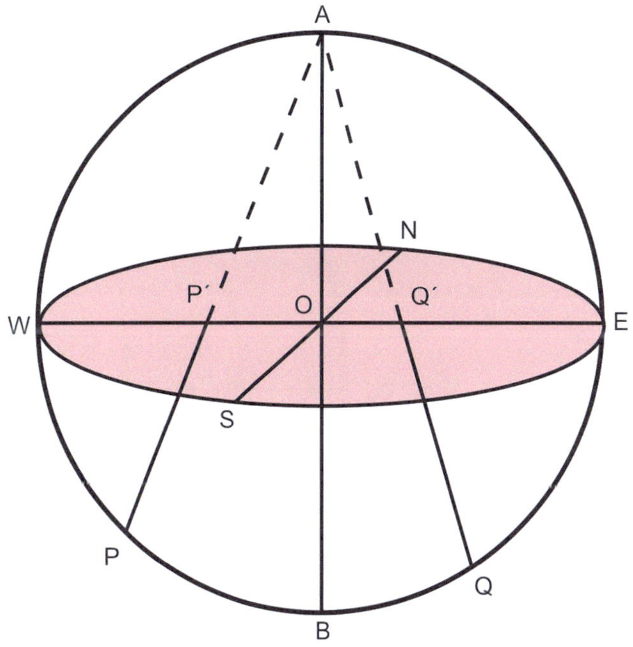

Fig. A1.5 Use of lower hemisphere of a stereograph for orienting a line (PP'). The topmost part of the sphere (point A) is called the zenith

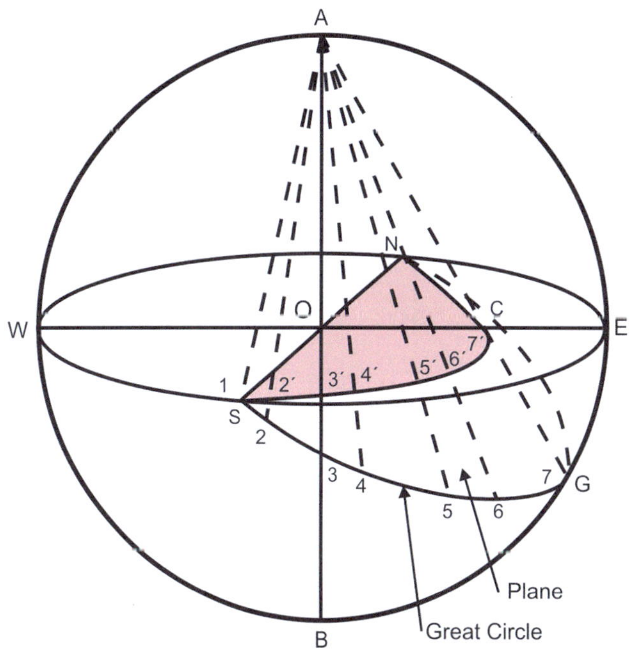

Fig. A1.6 Projection of a plane on the primitive circle. A plane SGN intersects the lower hemisphere at points 1, 2, 3, 4, etc. These points trace a circular intersection called great circle. Joining these points to the zenith (A) gives another set of points 1′ 2′, 3′, 4′, etc. that trace a circular arc at the equatorial plane; this arc gives the trace of a great circle

For use of the stereonet, it is convenient to attach it on a square cardboard. Take a tracing paper or overlay that covers the cardboard. Fix the overlay by a board pin at the centre of

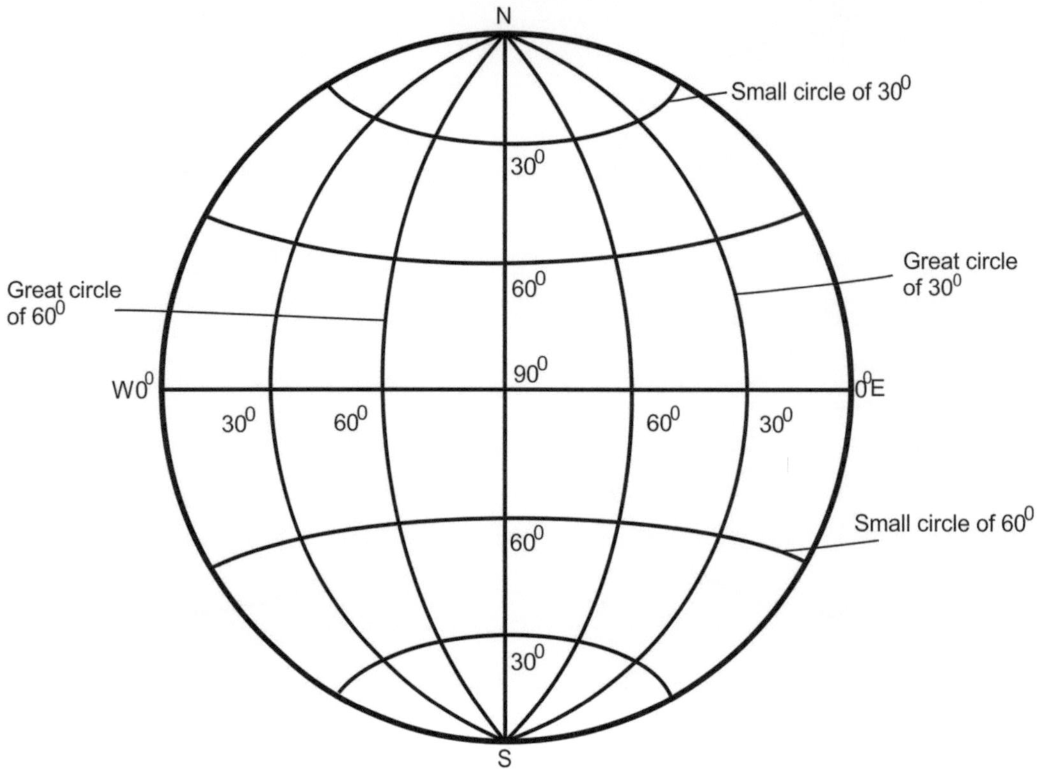

Fig. A1.7 Stereographic projection of great circles and small circles at 30° interval on the primitive circle

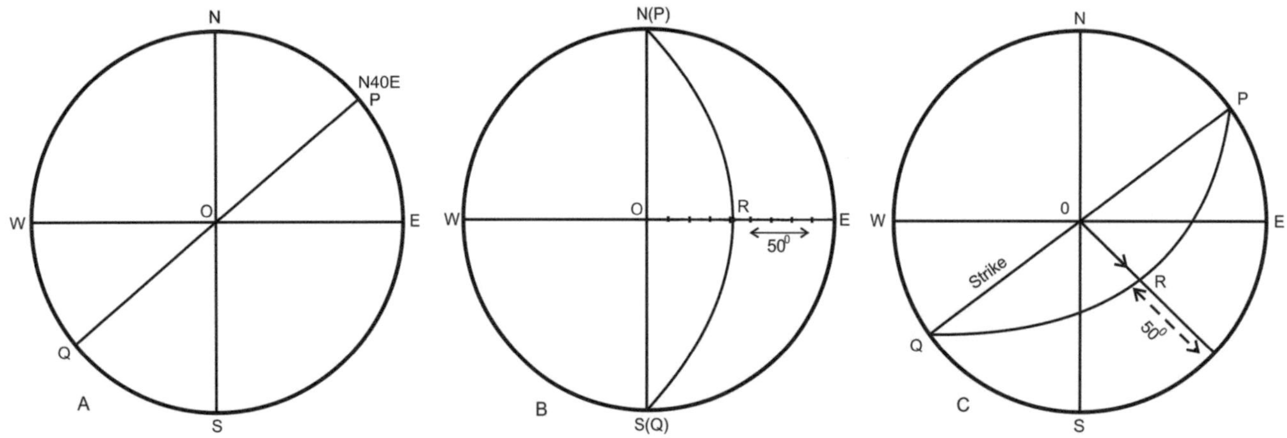

Fig. A1.8 Stereographic plotting of a plane. See text for details

the net. Trace the primitive circle by a pencil, and then mark north at the top of the net. Mark E, W and S also. With this, the net is now geographically oriented (Fig. A1.8). Let us now consider a plane that strikes N40°E and dips 50°SE. To obtain the stereographic projection of the plane, follow the steps given below:

1. Place the tracing paper over the stereonet, and fix it by a board pin at the centre O (Fig. A1.8a). Mark the directions N, S, E and W on the tracing paper.
2. Mark a point P at 40° east of N. Join P with the centre O, and then extend it to meet the periphery at Q. PQ is the strike of the plane (Fig. A1.8a).

3. Rotate PQ anticlockwise so that P falls on N. From E, move towards the centre by 50° (R); the distance ER is the amount of dip (Fig. A1.8b).
4. Rotate the line PQ clockwise so that it comes back to its original position (Fig. A1.8c). The girdle thus obtained gives the stereographic projection of the plane that strikes N40°E and dips 50°SE.

Pole diagram: The stereographic projection of a plane, as mentioned above, is represented by a girdle. In structural geology, one has to deal with a number of planar structures such as bedding, foliation and joints. Projection of a number of planar structures will make the stereonet highly crowded, and the reader will be confused. To avoid this, it is convenient to represent a plane by a point that is located at 90° from the girdle towards its convex side. The point thus obtained is called a *pole* (Fig. A1.9), and the stereogram is called a *pole diagram* or *π-diagram*. With the use of a pole diagram, a large number of planar structures can be conveniently plotted on a stereonet.

Projection of a line: Let us imagine a line PP' (Fig. A1.5) that may represent any linear structure, say, lineation, fold axis, plunge, etc. In stereographic projection, the attitude of this inclined line is obtained by joining it with the north pole of the sphere. The line intersects the equatorial plane at a point P′. In stereonet, a line, unlike a plane, is therefore represented by a point on the equatorial plane. The distance between W and P′ gives the amount of inclination of the line. Likewise, another inclined line at Q when joined with the north pole intersects the equatorial plane at Q'; the distance between E and Q' gives the amount of inclination of this line. All horizontal lines, on the other hand, will lie on the periphery of the equatorial plane (Fig. A1.10), while a vertical line will be represented by the centre, i.e. O.

Example: Let us now consider a line—that may represent any lineation or fold axis—plunging 30° towards S40°W. To obtain the stereographic projection of the line, follow the steps (Fig. A1.11) as given below:

1. Place the tracing paper over the stereonet, and fix it by a board pin at the centre (Fig. A1.11a). Mark the directions N, S, E and W on the tracing paper.
2. Mark a point P at 40° west of S (Fig. A1.11a).
3. Rotate the tracing paper anticlockwise so that P falls on S. From S, measure 30° to get a point R (Fig. A1.11b) that gives the amount of plunge.
4. Rotate the tracing paper clockwise so that P comes back to its original position (Fig. A1.11c). R gives the stereographic projection of the line that plunges 30° towards S40°W.

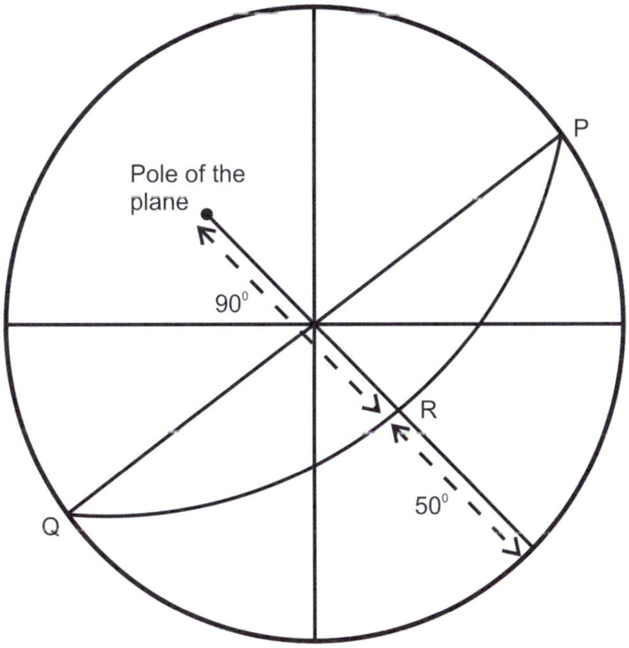

Fig. A1.9 Pole diagram. A plane shown by the girdle PRQ is represented by a point situated at 90° from this girdle

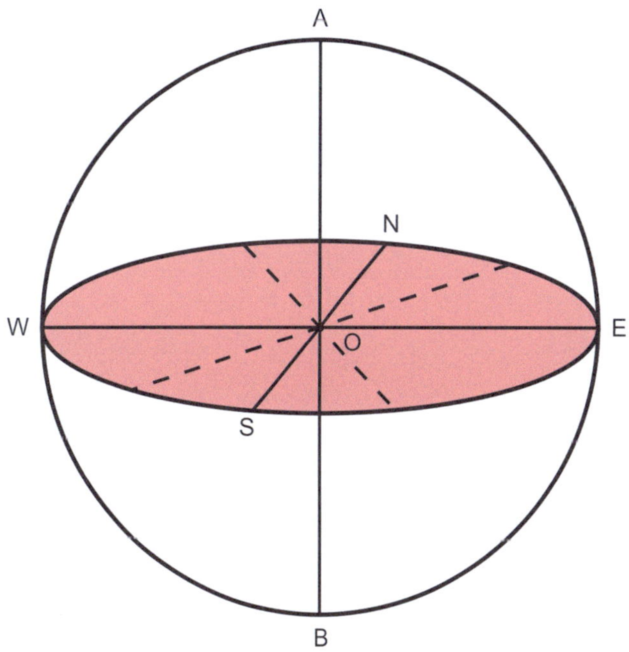

Fig. A1.10 In a stereonet, all horizontal lines plot along the periphery of the primitive circle. A vertical line is represented by the centre

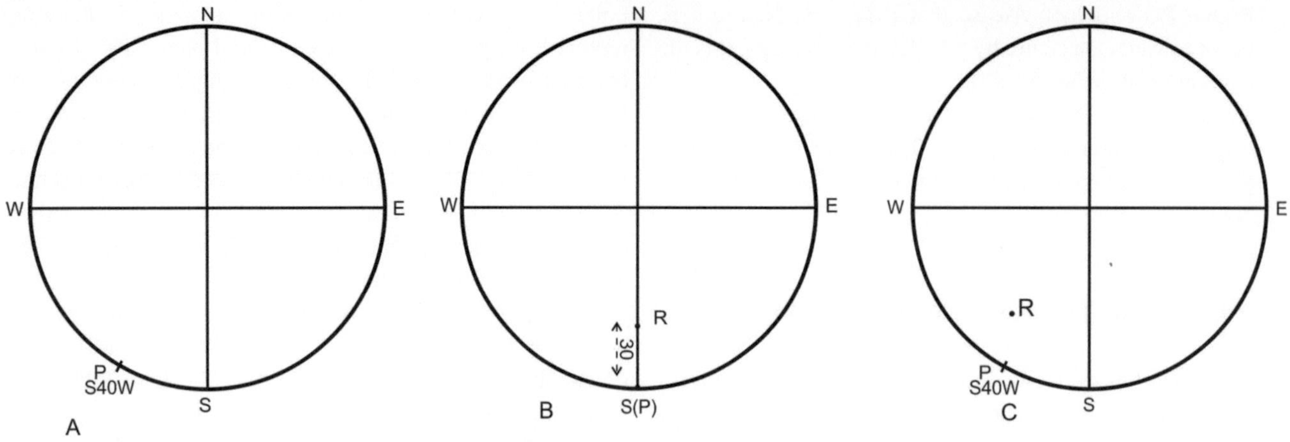

Fig. A1.11 Stereographic projection of a line. See text for details

Appendix B: Effect of Faults on Outcrops

Faults produce different types of effects on the outcrops as better seen on a geological map. The effects depend upon one or a combination of factors such as net-slip, dip-slip, dip and strike of the disrupted strata, strike of the fault and nature of the ground surface (undulating or flat) where observations are made. Further, while observing the effects of faults on outcrops, the geologist should remember one basic principle that younger beds occur on the downthrown block of a normal fault as we have demonstrated in Chap. 10 (Box 10.1). However, this is not so for strike-slip faults that cause lateral shifting of beds.

Some common effects of faults on outcrops are described below by considering the following effects:

1. Effect of vertical fault on horizontal strata
2. Effect of inclined fault on horizontal strata
3. Effect of strike-slip fault:
   - (i) Effect of vertical strike-slip fault on horizontal strata
   - (ii) Effect of inclined strike-slip fault on horizontal strata
   - (iii) Effect of inclined strike-slip fault on inclined strata
4. Effect of fault on folded strata:
   - (i) Effect of fault on anticline
   - (ii) Effect of fault on syncline
   - (iii) Effect of strike-slip fault on anticline
   - (iv) Effect of strike-slip fault on syncline

Effect of Vertical Fault on Horizontal Strata

A set of horizontal strata (Fig. A2.1a) may be affected by a vertical fault (Fig. A2.1b). After erosion, one can notice juxtaposition of different beds occurring at different heights or depths as the case may be (Fig. A2.1c). Apparent movement in the vertical section equals the net slip.

Effect of Inclined Fault on Horizontal Strata

Horizontal strata if affected by an inclined fault with dip-slip movement will cause, after erosion, apparent movement that is equal to net-slip in a vertical section (Fig. A2.2).

Effect of Strike-Slip Fault

Effect of Vertical Strike-Slip Fault on Horizontal Strata

In outcrops, the horizontally disposed strata occur one over another according to increasing height. Since in a strike-slip fault the displacement is directed horizontally, the fault (vertical in this case) will show no apparent effect on the outcrop pattern, i.e. the same bed will be exposed on either side of the fault (Fig. A2.3).

Effect of Inclined Strike-Slip Fault on Horizontal Strata

Horizontal strata if affected by an inclined strike-slip fault will show, after erosion, no apparent movement in a vertical section (Fig. A2.4). The net-slip equals the strike-slip. After erosion, a marker bed will appear continuous in a vertical section, thus apparently leaving no indication that the strata have been affected by a fault. In such cases, some other criteria should be followed to identify the presence of the fault.

Effect of Inclined Strike-Slip Fault on Inclined Strata

We consider here the effect of a strike-slip fault on inclined beds (Fig. A2.5a). The strike of the fault is perpendicular to that of the beds. After erosion, a marker bed shows strike-slip movement on the horizontal plane, while in a vertical section the beds apparently show dip-slip movement (Fig. A2.5b).

A. R. Bhattacharya, *Structural Geology*, Springer Textbooks in Earth Sciences, Geography and Environment,
https://doi.org/10.1007/978-3-030-80795-5

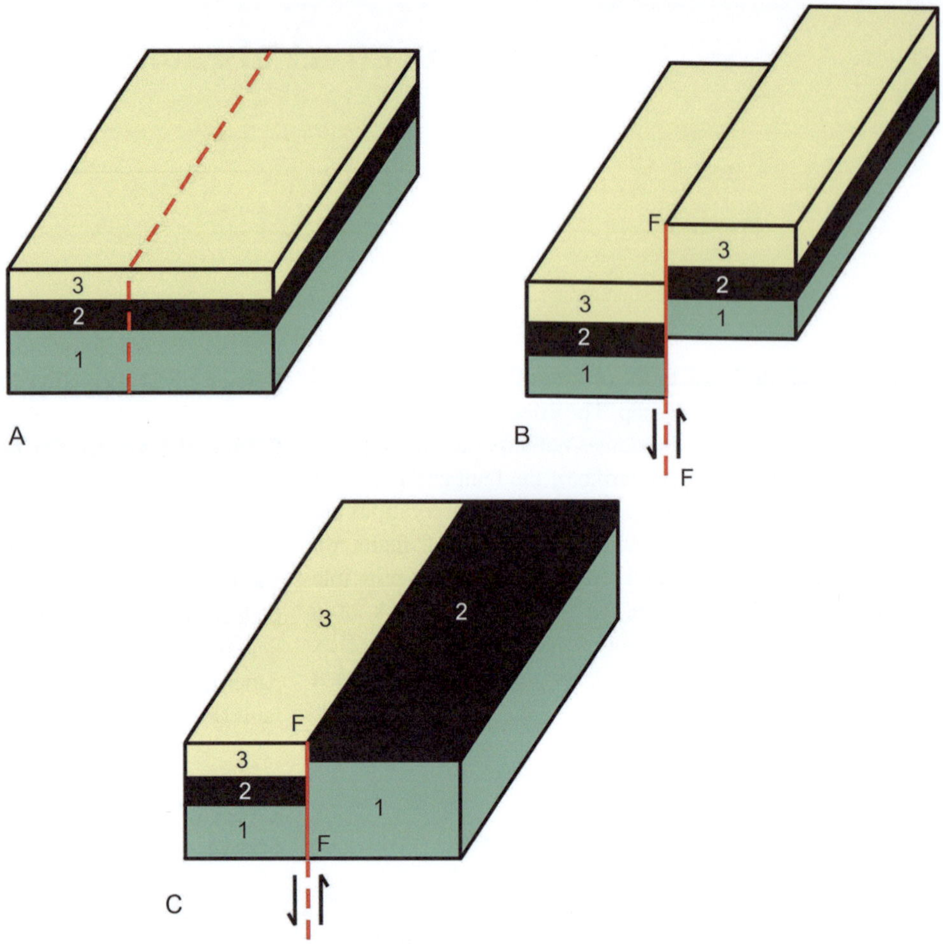

Fig. A2.1 Effect of a vertical fault on horizontal strata. (**a**) Block diagram to show the fault movement plan. (**b**) Block diagram showing displacement of beds due to faulting. (**c**) Final position of the faulted strata after erosion

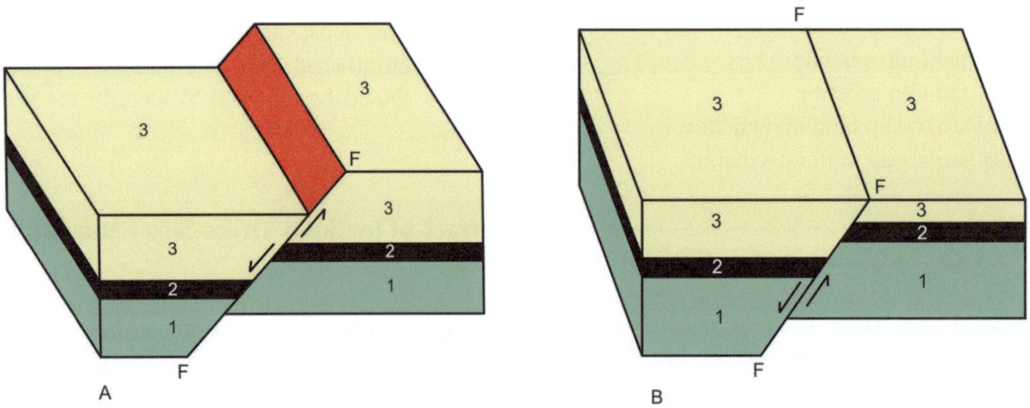

Fig. A2.2 Effect of an inclined fault (FF) on horizontal strata. (**a**) Block diagram showing movement of blocks after faulting. (**b**) After erosion. The fault face is coloured red

Effect of Fault on Folded Strata

In folded strata, the effects of fault on the outcrop patterns are different from the earlier examples; they produce widening or narrowing effects on the outcrops. Here, we consider the following two cases.

Effect of Fault on Anticline

A vertical fault that has affected the limbs (Fig. A2.6a,b), not the axial plane, of an anticline produces widening of strata on the upthrown block (Fig. A2.6c). Thus, widening of any outcrop across a fault can be used as a criterion to infer the upthrown block of an anticline.

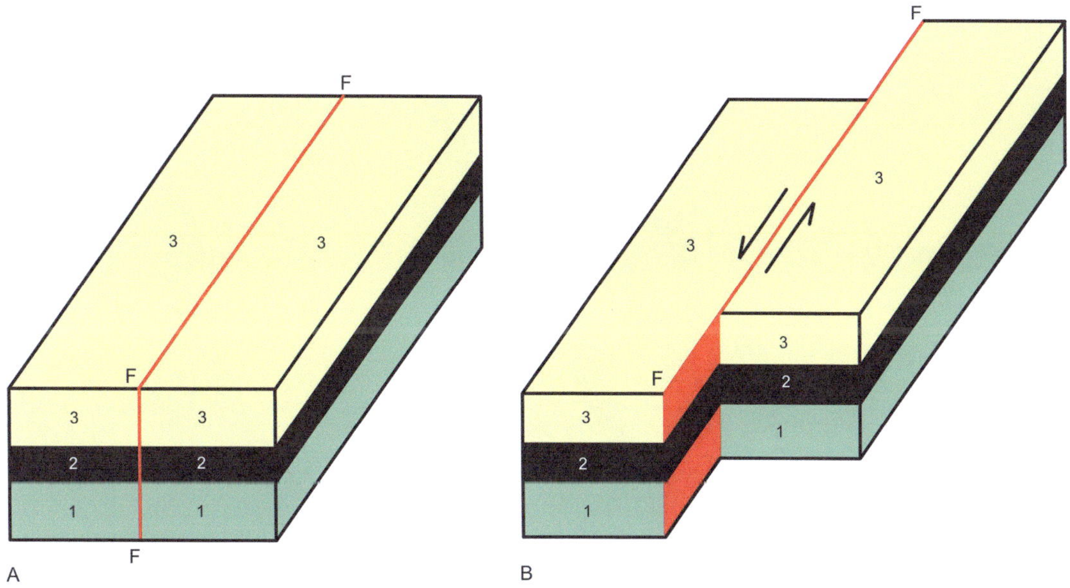

Fig. A2.3 Effect of strike-slip fault on horizontal strata. (**a**) Block diagram showing initial position of the strata. (**b**) Position after strike-slip faulting

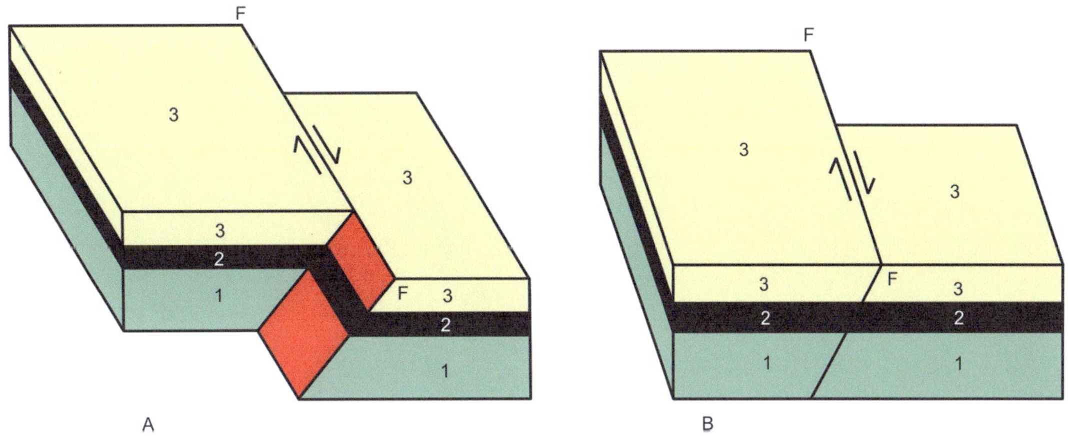

Fig. A2.4 Effect of an inclined fault on horizontal strata with strike-slip movement. (**a**) Block diagram showing movement of blocks after strike-slip faulting. The fault plane is coloured red within which the disposition of a marker bed (2) has been shown. (**b**) After erosion

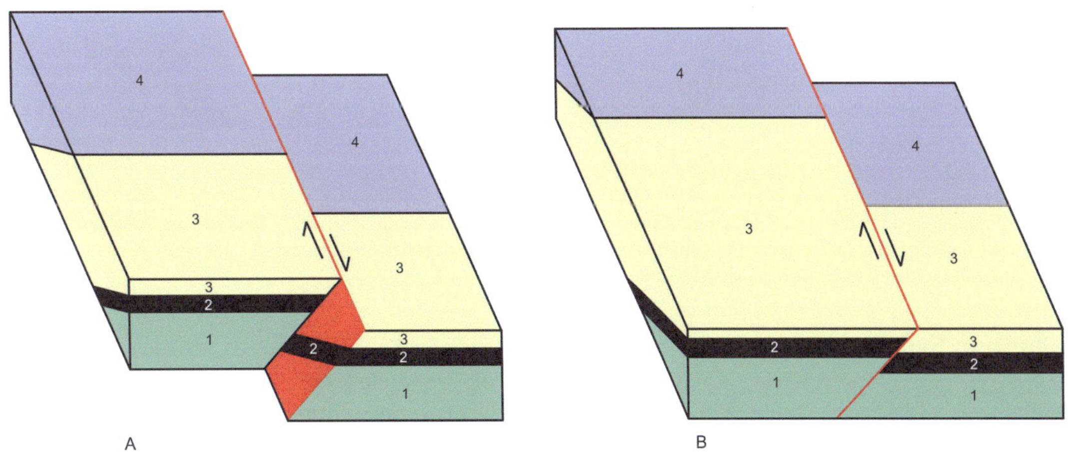

Fig. A2.5 Inclined strata affected by a strike-slip fault. (**a**) Block diagram showing movement of blocks due to strike-slip faulting. The fault plane is coloured red within which the disposition of a marker bed (2) has been shown. (**b**) After erosion

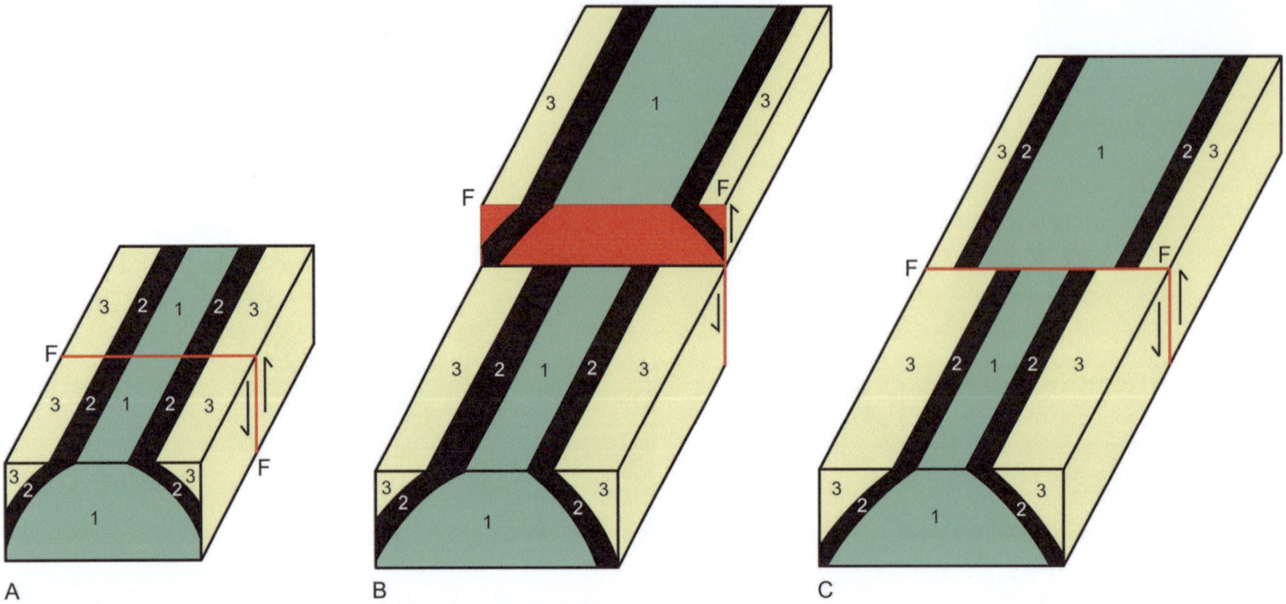

Fig. A2.6 Block diagrams to show the effect of fault on an anticline. (**a**) An anticline affected by a vertical fault at its limbs. (**b**) After faulting. The fault plane is coloured red within which the disposition of a marker bed (2) has been shown. (**c**) After erosion, widening of strata on the upthrown block takes place

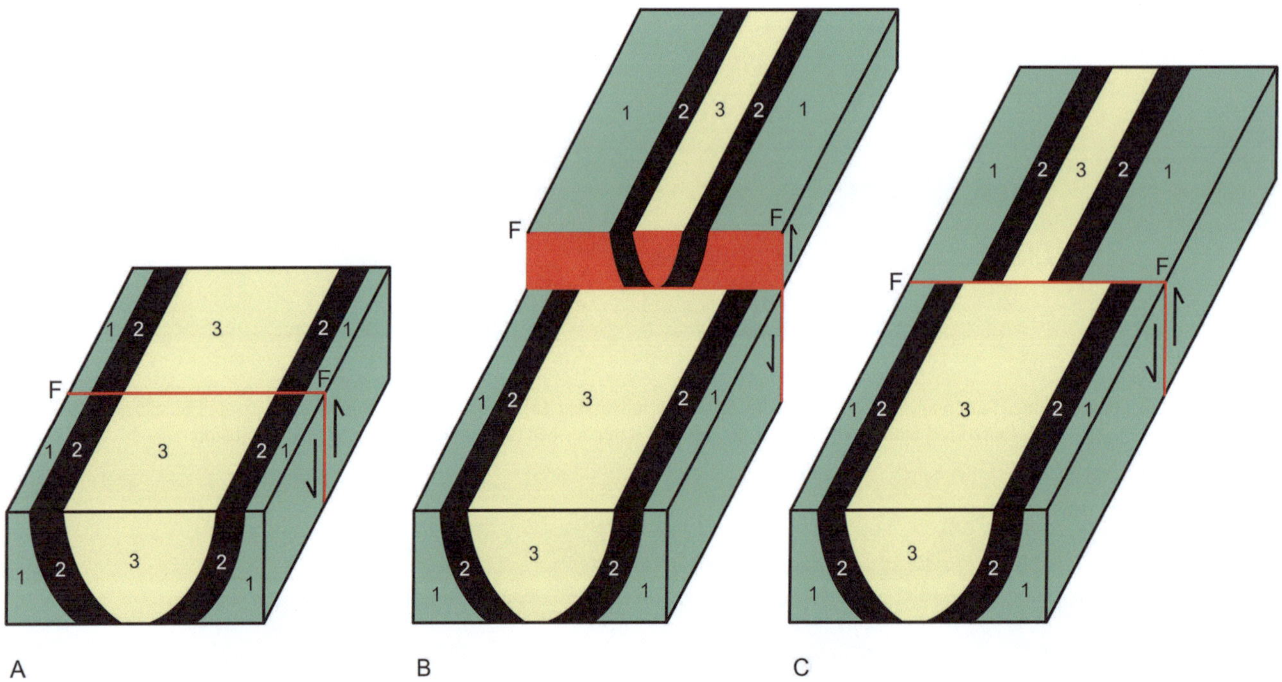

Fig. A2.7 Block diagrams to show the effect of fault on a syncline. (**a**) A syncline affected by a vertical fault at its limbs. (**b**) After faulting. The fault plane is coloured red within which the disposition of a marker bed (2) has been shown. (**c**) After erosion, narrowing of strata on the upthrown block takes place

Effect of Fault on Syncline

A vertical fault that has affected the limbs (Fig. A2.7a,b), not the axial plane, of a syncline produces narrowing of strata on the upthrown block (Fig. A2.7c). Thus, narrowing of any outcrop across a fault can be used as a criterion to infer the upthrown block of a syncline.

In the case of folded strata, we have considered the effects of dip-slip faults on the limbs only of folds. In such cases, the axial plane does not show any separation. If a fault affects the axial plane or if the fault has oblique slip movement, the trace of the axial plane will show separation on the outcrop, thus producing an effect of strike-slip movement on the geological map.

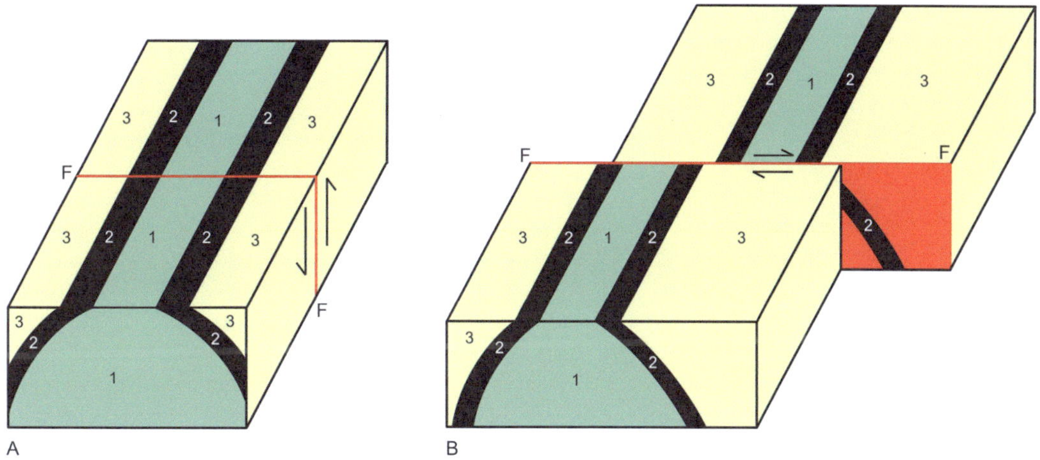

Fig. A2.8 Effect of strike-slip fault on an anticline. (**a**) Block diagram showing an anticline affected by a vertical fault at its limbs. (**b**) After strike-slip faulting. The fault plane is coloured red within which the disposition of a marker bed (2) has been shown

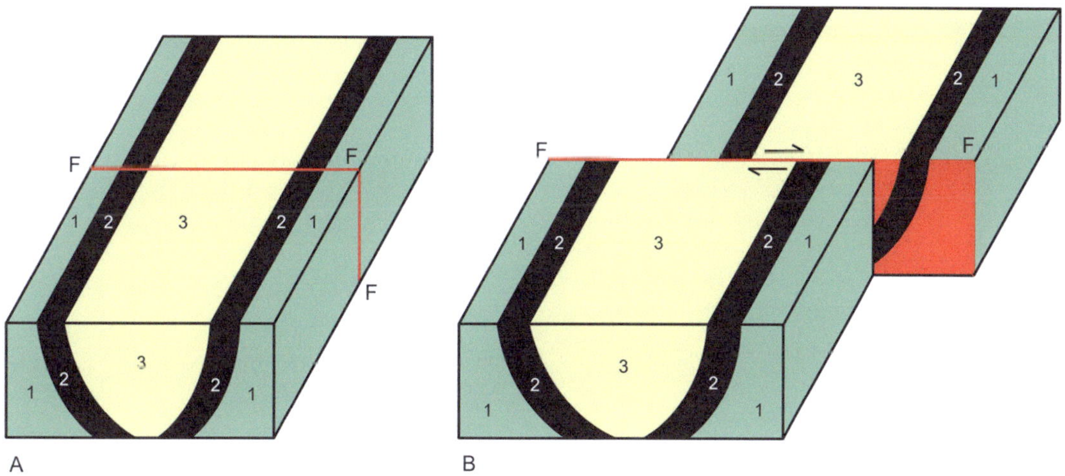

Fig. A2.9 Effect of strike-slip fault on a syncline. (**a**) Block diagram showing a syncline affected by a vertical fault at its limbs. (**b**) After strike-slip faulting. The fault plane is coloured red within which the disposition of a marker bed (2) has been shown

Effect of Strike-Slip Fault on Anticline

If an anticline is affected by a strike-slip fault, the set of beds will be shifted right or left of the fault depending upon whether the fault has caused a right-lateral or left-lateral effect. The width of the outcrops will not be changed. The only effect on outcrop will be offsetting of the strata to the right or left side of the observer (Fig. A2.8).

Effect of Strike-Slip Fault on Syncline

If a syncline, like anticline, is affected by a strike-slip fault, the set of beds will be shifted right or left of the fault depending upon whether the fault has caused a right-lateral or left-lateral effect. The width of the outcrops will not be changed. The only effect on outcrop will be offsetting of the strata to the right or left side of the observer (Fig. A2.9).

Concluding Remarks

To conclude, it can be said that faults produce a variety of effects on outcrops. As such, fieldwork in areas affected by faults should be carried out carefully. Understanding the effect of faults on outcrops is especially important where establishment of stratigraphy is to be done as well as in the exploration of minerals. Fossiliferous terrains, however, need to be carefully studied for the effect of faults.

Appendix C: Section Balancing

Construction of geologically *balanced cross sections* is a routine practice in geology. A balanced cross section is one in which the length and area of beds in the deformed state are the same as it were in the undeformed state. A *restored section* is one in which the rocks can be related to their undeformed state. Balanced cross sections predict that they are conceptually and geometrically correct. This practice is in fact a geometric test of checking cross-section interpretations prior to any applied aspect such as petroleum and mining exploration or engineering projects. This practice is of great significance in petroleum and gas exploration where high precision is the most important factor in the construction of geological cross sections as huge money is involved in exploration work.

Introduction

Construction of geologically balanced cross sections is a geometrical method that helps to understand, apart from others, the original undeformed state of the rocks from which the original cross-sectional area of sedimentary sequence can be calculated. Although it is a routine practice in geology, its importance becomes high especially in oil and gas exploration, where the structure of the rocks should be known precisely as huge money is involved in exploration work. Keeping in view the precision as the most important factor in the construction of geological cross sections, the concept of *balanced cross sections* grew in the second half of the last century. A balanced cross section refers to the rocks in the strained or deformed state and is one in which the length and area of beds in the deformed state are the same as it were in the undeformed state. A *restored section* is one in which the rocks can be related to their undeformed state. With time, section balancing gained momentum such that now it has become a powerful tool in structural geology. Development of seismic reflection methods gave a big boost to this method. For construction of balanced cross sections, data from surface, subsurface, seismic reflection techniques and other related methods are needed for accuracy.

The concept of balanced cross sections that was initially developed in the petroliferous regions of the Rocky Mountains (Hunt 1957; Bally et al. 1966) was later on successfully applied to unravel the structure of the crystalline thrust sheets of the Appalachians and the Basin and Range Province of NW USA (Allmendinger et al. 1983). In Europe, Laubscher (1962) applied this method to unravel some structural intricacies of the Jura Mountains, and later on he (Laubscher 1988, 1990) extended it to the Alps Mountains. Various aspects of the method have been explained in detail by several workers, and the reader is suggested to consult the work of Dahlstrom (1970), Hossack (1979), Hossack and Hancock (1983), Woodward et al. (1985), Woodward et al. (1986), Mitra and Namson (1989) and Groshong (1994, 2006).

Balancing of sections helps in calculating the depth of a decollement thrust and depth of detachment of a listric normal fault, in estimating the minimum contraction or extension across a deformed terrain and in unravelling the deep structure of mountain belts. The method also checks the validity of existing sections and thus adds accuracy in the preparation of suitable geological cross sections. As such, the method has been found highly useful during hydrocarbon exploration.

Some Generalizations

The classic work of Dahlstrom (1970) laid down a strong foundation for the construction of balanced cross sections. Some relevant excerpts from his work are presented below:

- In the initial stages of clastic deposition, voids (filled with fluids) practically outnumber mineral grains, thus making the density low. With progressive burial and compaction, density increases with depth.
- Volume change (reduction) during deformation is believed to be negligible. Since bed thickness does not change, the surface area of any bedding plane remains unchanged during deformation.

A. R. Bhattacharya, *Structural Geology*, Springer Textbooks in Earth Sciences, Geography and Environment,
https://doi.org/10.1007/978-3-030-80795-5

– Faults are usually curved in cross section rather than planar.
– Folding and thrust faulting are the two mechanisms that make a packet of rock shorter and thicker, and therefore these two mechanisms can be considered interchangeably.

A rule that can be derived from the law of conservation of volume is that: 'In adjacent cross sections the amount of "shortening" at a specific horizon between comparable reference lines must be nearly the same unless there is a tear fault between them' ('shortening' is the difference between the actual bed length and the horizontal distance it now occupies) (Dahlstrom 1970, p. 751).

The practice of section balancing implies that cross sections are used to convey predictions that they must be conceptually and geometrically correct. It is in fact a way of checking cross-section interpretations prior to any applied aspect such as petroleum and mining exploration or engineering projects. Thus, if a cross section passes the geometric tests, it is likely to be correct. This also implies that if a cross section does not pass the geometric tests, it could be incorrect.

Assumptions

Construction of a balanced and a restored cross section requires some assumptions to be followed: (i) the faulted beds should not show any gaps or overlaps in the stratigraphic sections; (ii) the rocks should not have undergone significant deformation along the strike of the rocks; (iii) in sedimentary sequences with faults, there should not be any material transfer across the fault surface; (vi) the geological section selected should be perpendicular to the strike of the rock units; (vii) in case of arcuate shape of outcrops, the section line selected should be along the highest curvature of the arc; (viii) there should not be repetition of beds due to faulting as it may change the original thickness of the affected beds in the vicinity of the concerned fault; and (ix) the ramps should move up-section in the direction of transport.

Needless to mention, the precision or accuracy of section balancing requires that *all* the above-mentioned preconditions should be taken into account. Experience indicates that all these conditions are generally not met with by the exploration geologists. Therefore, for precision and accuracy in section balancing, great caution should be observed, especially if the method is applied for locating hydrocarbon or any mineral deposit that involves huge money.

Methods of Balancing

There are two standard methods of section balancing: equal-area balancing and equal line-length balancing.

Equal-Area Balancing

Although this method has been in use for about a century, it remained riddled with several shortcomings. Mitra and Namson (1989) reviewed the existing assumptions, limitations and errors of this method and thus presented it in a modified form after verification from a number of fold and thrust belts. The equal-area balancing method as worked out by these workers is described below.

The equal-area balancing method is based on the principle that the total area of the beds remains unchanged before and after the deformation. The area-balancing method is used when the true stratigraphic thickness between the upper and lower horizons of the selected beds is known, and when it is possible to restore a rectangle for the selected beds. This ensures that it is possible to obtain the restored sections from the section in the deformed state. This method is particularly applicable to structures showing constant-volume, penetrative deformation resulting in variations in bed thickness and length.

The starting point of this method is the establishment of two pin lines at either end of the section in areas where there is no interbed slip (Woodward et al. 1985). The foreland pin line is taken as a fixed reference line for restoration of the complete section. For this, cross sections are commonly restored from the foreland to the hinterland. Pin lines are fixed in two different ways: undeformed rocks in the foreland for regional lines and axial planes of synclines for local sections.

Mitra and Namson (1989) proposed two main ways of equal-area balancing: (a) combination of equal-area restoration with line-length restoration of key beds that have not undergone significant thickness changes and (b) combination of key bed and area restoration.

Combination of Equal-Area and Key-Bed Balancing

This method (following Mitra and Namson 1989) can be applied to sections showing faulted rocks. The thrust sheets are restored sequentially from the foreland to the hinterland. All the thrusts are considered one by one such that for each thrust, the units in the hanging wall cut-off are matched against the corresponding units in the footwall cut-off.

Let us consider a deformed thrust sheet (W′X′Y′Z′) (Fig. A3.1a). Restoration of this thrust sheet by area balancing can be done by measuring the area (A_x) and the original

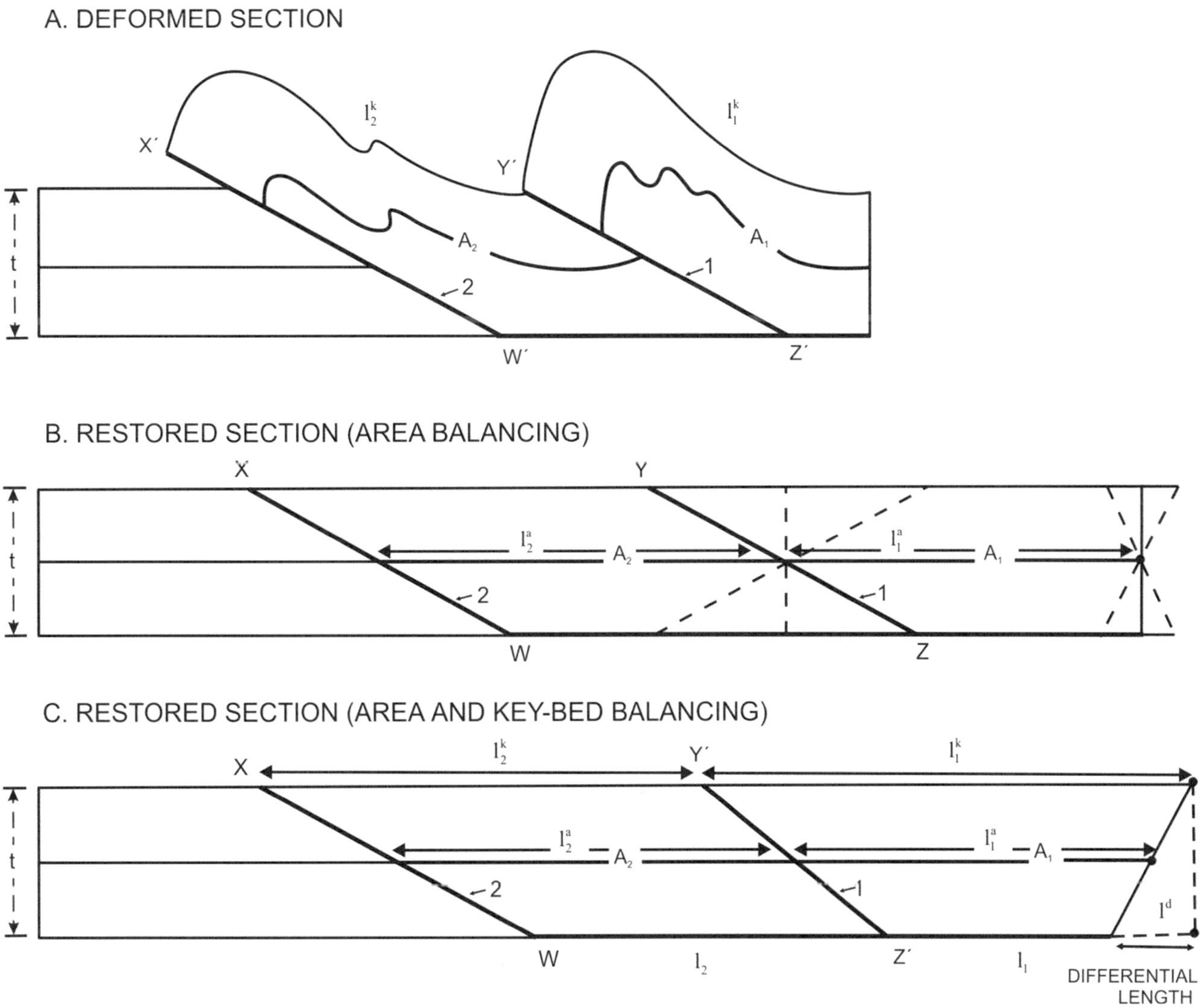

Fig. A3.1 (**a**) A cross section of deformed rock selected for restoration. (**b**) Restored section by area balancing. (**c**) Restored section by a combination of equal-area and key-bed balancing. (Reproduced from Mitra and Namson 1989, Fig. 1, with permission from American Journal of Science. Request ID: 600067196)

thickness(t) of the unit. Assuming plane strain, the restored average bed length can be estimated (Fig. A3.1b) from

$$l^a = A_x/t \tag{A3.1}$$

Here, only one 'average' bed length is estimated. As such, the lengths of the top and bottom of each unit are assumed to be equal to the average length, and therefore the shape of the undeformed sheet (WXYZ) is assumed to be a parallelogram (Fig. A3.1b).

Combination of Key Bed and Area Restoration

The above method gives the regional shortening in a section by comparing the deformed and undeformed lengths of individual sections (Fig. A3.1a,b), but it does not provide any information on whether the units are balanced or the fault trajectories in the final section are reasonable. As such, the area-balancing method alone does not constrain the orientations of the restored thrust geometries or the shortening profile. For this, the *method of combining the key bed and area restoration* is used for determining fault trajectories and thus ensuring proper balancing. The key-bed method involves identification of a key bed that deforms primarily by parallel folding and a minimum amount of penetrative deformation. This ensures accurate restoration using the line-length method.

In the combination of key-bed and area-restoration method, the most common case is one in which a thrust sheet with an initial trapezoid geometry WXY*Z* (Fig. A3.1c) is deformed to a final geometry shown by W′X′Y′Z′ (Fig. A3.1a). The average length $\left(l_2^3 = l_a\right)$ of the

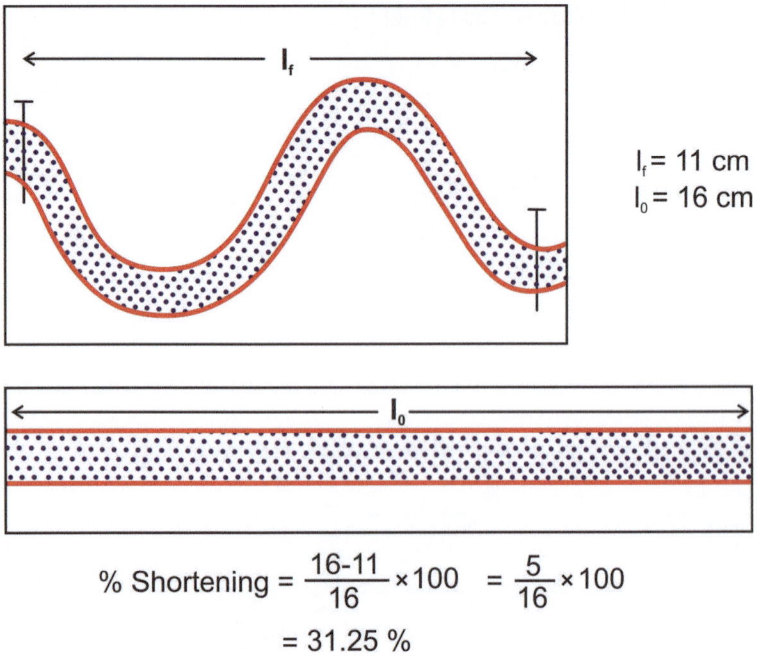

Fig. A3.2 Line-length balancing. A bed of original length l_o (lower diagram) is folded into a syncline and an anticline (upper diagram). In the deformed (folded) state, the length of the bed between two pin lines is l_f. The percent shortening is then given by $[(l_f - l_o)/l_o] \times 100$

first unit is obtained using Eq. (A3.1). Then, the length XY* or WZ* is assumed to be key unit, and therefore its length $(l_2^k = l_k)$ is unchanged during deformation. The other line length (l_2) is then obtained by using the equation

$$l_2 = 2(l_2^a) - l_2^k = 2A_2/t - l_2^k \qquad \text{(A3.2)}$$

In case the unit to be balanced shows a progressive variation of thickness from t^1 at one end to t^2 at the other, l_2 is obtained by the relationship

$$l_2 \approx 4A_2/(t' + t^2) - l_2^k \qquad \text{(A3.3)}$$

The restored lengths (l_k and l_2) can be used to restore thrust sheet 2 to its undeformed state. Repetition of this process for thrust sheet 1 will give the complete restored section. This completely balanced section will display reasonable fault geometries and approximately equal bed lengths in the undeformed state. If a regional cross section consisting of a large number of thrust sheets is selected, the thrusts are successively restored in this way from the foreland to the hinterland.

Line-Length Balancing

This method is based on the principle that the line lengths and thicknesses of the beds in their original sections and in their deformed sections should be equal. The method is described in the following steps:

Step 1: At the beginning, two vertical *pin lines* (Fig. A3.2) are chosen at each end of the section. With respect to the direction of transport, one would be the *leading pin line* (LPL) while the other the *trailing pin line* (TPL). The pin lines mark a point where there is no slip between the beds and the beds are in undisturbed state. The pin line is chosen at the end of a section where (a) maximum stratigraphic sequence is represented, (b) there is no slip of the beds and (c) the line is perpendicular to the dip direction and (d) lies along the axial plane of a fold, if any. From this point, a line should be selected across the strike of the beds along which there is neither any fault nor any visible deformation and the beds are represented by the complete stratigraphic sequence.

Step 2: Prepare a cross section with the help of surface and subsurface data.

Step 3: Measure the lengths of the beds. This can be done with the help of a string, and thus get the total length. This gives the original length of the beds (l_o) in their undeformed state. If the pin lines have been selected properly, the bed lengths should be equal at different vertical levels of the pin line.

Step 4: Measure the length of the section in the deformed state (from step 2). This gives the length of the beds (l_f) in deformed state. The shortening is then given by $[(l_f - l_o)/l_o] \times 100$.

Let us explain the line-length balancing method with an example of a folded bed. A segment of a folded layer has been selected between two pin lines as marked in Fig. A3.2. The length of this segment can be measured by moving a tag along the folded layer. This gives the original length (l_o) in the pre-deformed state. After deformation (folding), the length between the pin lines is measured; this gives the final length (l_f) of the folded layer. If

$$\begin{aligned} l_o &= 16 \text{ cm} \\ l_f &= 11 \text{ cm} \\ \%\text{Shortening} &= \{(l_o - l_f)/l_o\} \times 100 \\ &= \{(16 - 11)/16\} \times 100 \\ &= 31.25\% \end{aligned}$$

Thus, the deformation has caused 31.25% shortening of the original length of the bed.

Summary

- Construction of *balanced cross sections* is a routine practice in structural geology, especially in oil and gas exploration, as huge money is involved in such work. A balanced cross section is one in which the length and area of beds in the deformed state are the same as it were in the undeformed state. A *restored section* is prepared in which the rocks can be related to their undeformed state.
- Balanced cross sections predict that they are conceptually and geometrically correct. This practice is in fact a geometric test of checking cross-section interpretations prior to any applied aspect such as petroleum and mining exploration or engineering projects.

Glossary

A

Aberrant fold A fold that slightly deviates from the ideal cylindrical geometry.

Accelerating creep The rising part of the strain-time (creep) graph that represents rapid fatigue with time until the rock fails by rupture. The deformation takes place under constant stress.

Active faults Faults that show evidences of movement or activity as recorded within the historic or prehistoric periods or within the recent times as witnessed by man. Commonly, active faults are those that have shown activity during the last 10,000 years, i.e. during the Holocene Epoch.

Amontons' first law The frictional force required for slip along the fracture plane is directly proportional to the force acting perpendicular to the fracture plane.

Amontons' second law The frictional force is independent of the area of contact between the solid bodies.

Amontons' third law The kinetic friction is almost independent of the speed of sliding.

Anderson's theory Faults can be related to the principal stress axes and shear stress with the assumption that there cannot be any shear stress on the surface of the earth.

Angular shear See **shear strain**.

Anisotropic substances Substances whose properties vary with direction.

Anisotropy of magnetic susceptibility (AMS) The magnetic fabrics/subfabrics can be used for orientation studies of the constituent magnetic minerals of a rock. The AMS is believed to be related to the layer-parallel shortening of a rock. As such, this method finds great value in strain analysis.

Annealing The process or processes that involve removal of internal stress of a crystal such that its lattice is made defect free, i.e. free from dislocations.

Anticlinal theory A theory used by hydrocarbon geologists that hydrocarbons commonly accumulate in anticlines.

Anticline A fold that is convex upward towards the centre and the strata dip opposite to each other, and as such older rocks occur at the centre of curvature.

Anticlinorium A larger anticline that includes a number of smaller folds (anticlines and synclines) on its limbs.

Anticracks These are contractional surfaces formed parallel to σ_3 during formation of joints.

Antiform A fold that looks like an anticline but the age of the rocks at the core is not necessarily older or is not known.

Antiformal syncline A structure in which the strata dip away from the axis of the fold and the rocks at the core are younger; the structure can thus be described as a downward-facing syncline.

Antithetic fault A fault that dips oppositely to a larger fault with which it is associated.

Apparent dip It is the dip of a bed in a direction other than the true dip, and its amount is always less than the true dip. The amount of apparent dips of a bed may range from zero degree up to the amount of true dip.

Arrest line It is a line on a joint surface where joint propagation is temporarily stopped.

Asperities Irregularities or imperfections existing in a rock body.

Asymmetric fold A fold in which one limb dips more steeply than the other and therefore one limb is shorter than the other.

Asymmetric tensor A tensor with nine different components such as finite homogeneous strain.

Attitude It is a fundamental geometrical attribute that describes how a plane or a line is oriented on the surface of the earth.

Axial angle Angle that two limbs of a fold subtend at the hinge. The angle is formed by two lines joining hinge with the inflection points of the limb. (This parameter was introduced by Bhattacharya, 1992.)

Axial plane Plane formed by joining the hinge lines of successive layers of the same fold. Also called **axial surface**.

Axial surface See **axial plane**.

Axial trace It is the intersection of the fold axial plane with any surface such as fold surface or ground surface.

Axial-plane foliation The foliation that is parallel, or nearly parallel, to the fold axial plane.

A. R. Bhattacharya, *Structural Geology*, Springer Textbooks in Earth Sciences, Geography and Environment,
https://doi.org/10.1007/978-3-030-80795-5

B

Back-thrust Thrust that shows displacement in reverse direction with that of the main thrust.

Back-bearing Direction of an observer with respect to an object.

Backlimb The normal limb of an overturned fold; the beds of backlimb are disposed in normal position.

Balanced cross section A balanced cross section is one in which the length and area of beds in the deformed state are the same as it were in the undeformed state.

Basin See **structural basin**.

Bearing Direction of an object with respect to an observer.

Bedding fault A fault that is parallel to the adjacent beds.

Bending folds Folds formed by application of stresses transverse or inclined to layering. Bending constitutes a mechanism of folding.

Bending Application of stresses transverse or inclined to layering.

Biaxial stress A stress system that is represented in two directions: a normal stress and a shear stress. Also called **two-dimensional stress**.

Blind thrust A thrust that terminates within the sedimentary layers and never reaches the surface.

Bluntness The relative curvature of the hinge of a fold. It is defined by the ratio (b) between the radius of curvature (R_c) at fold closure and the radius of the circle (R_o) that is tangent to the limbs at the inflection points, i.e. $b = R_c/R_o$.

Body forces Forces that act upon a unit volume of the body and is given by the volume or size of the body. Such forces act on every point of the body and are therefore measured in three dimensions.

Bookshelf model See **domino model**.

Boudins, Boudinage Sausage-shaped fragments of a competent rock layer that have been stretched parallel to layering. The term boudinage is commonly used to refer to the process of formation of boudins.

Bowden's theory Two surfaces may not fit well even if they are well finished; there will be void at a number of places, and these voids are joined together at protuberances called asperities.

Box folds Folds in which the interlimb angles of the two folds meet at nearly 90° to each other and the hinge zones are rounded.

Breaching thrust A thrust that cuts across all the early-formed thrusts and is therefore a younger thrust.

Break-thrust A thrust that forms due to faulting of an anticline-syncline pair such that the hanging wall anticline is overthrust while the footwall syncline is preserved.

Breccia A fault rock dominantly containing more than 30% angular fragments.

Breddin graph A method for estimating strain mainly from deformed fossils that show relationship between angular shear and orientation of the strain ellipse.

Brittle deformation It involves failure and change in the shape of a rock by the formation of fractures or cracks along which the rock loses its cohesion. Brittle deformation is characteristic of rocks that are elastic.

Brittle fault A fault that has formed by brittle deformation mechanisms.

Brittle shear zone A shear zone formed by brittle deformation processes and is characterized by the presence of a fault or tension gashes. Shear strain is mainly localized along the fault while the remaining parts of the shear zone remain unstrained.

Brittle-ductile shear zone A shear zone formed by both brittle and ductile deformation processes. A marker line shows the signature of rotational movement (curvature) as well as a discontinuity in the form of a fault or tension gashes.

Brittle-ductile transition (BDT) The transitional zone between upper and lower crust where all the physico-mechanical properties of rocks are gradational depth-wise. In general, the BDT occurs at depths between 10 and 15 km, though it varies depending upon the thermal gradient and fluid content that broadly control mineralogical phase changes. In continental regions of high thermal gradients, the BDT occurs at relatively shallower levels, up to depths of 8–12 km, while in regions of low thermal gradients, it occurs up to depths of 15 km.

Buckling A process of formation of folds by the application of stresses parallel to layering of rocks.

Bulge recrystallization or bulging recrystallization A type of dynamic recrystallization that affects only the boundary region of porphyroclasts such that the resulting grain-boundary bulges are separated from the old grains, thus forming relatively small recrystallized grains.

Bulk deformation Change of volume and shape of a body during deformation.

Burgers vector It is the interatomic distance by which slip has taken place in a crystal.

Byerlee's law The linear correlation between confining pressure (given by depth) and strength of rock.

C

C′-surface Also called **shear-band foliation**.

Caldera It is a concentric structure formed by collapse, along concentric normal faults, of rock mass into a magma chamber after a volcanic eruption.

Cataclasis Generally, it is a fracture process in faulting involving fragmentation of material, rotation, grain-boundary sliding and displacement along new surfaces created during deformation. The process is common at high crustal levels and results in grain-size reduction. Cataclasis operates at low temperature and low confining pressure.

Cataclasite It is a rock produced by cataclasis. Texturally, it is a fine-grained, non-foliated rock formed by brittle deformation and is characterized by fracturing of grains, grain separation and rotation. The matrix typically is fine grained as compared to the host grains that are commonly angular.

Cataclastic flow A rock when reduced to fine grained by brittle deformation processes behaves as if formed by ductile processes that make it flow to produce bands; this process is called cataclastic flow.

Centre-to-centre method It is a method of estimating tectonic strain by measuring the distances and angles between a particular grain and its nearest neighbours.

Characteristic wavelength See **dominant wavelength**.

Chevron folds Folds with straight limbs of equal length and with angular hinges.

Chocolate boudinage Square to rectangular shaped boudins formed due to layer-parallel extension in two directions. Also called **chocolate-tablet boudinage**.

Chocolate-tablet boudinage See **chocolate boudinage.**

Class 1 folds Folds in which the dip isogons are convergent towards the inner arc and thus the curvature of the inner arc always exceeds that of the outer arc.

Class 2 folds Folds in which the dip isogons are parallel to the axial surface and thus the curvature of the inner arc is equal to that of the outer arc.

Class 3 folds Folds in which the dip isogons are divergent towards the inner arc and thus the curvature of the inner arc is always less than that of the outer arc.

Cleavage domain It is the domain in which the original fabric of the rock is significantly changed. This domain is characterized by parallel fabric shown by the alignment of mineral constituents that are parallel to the rock cleavage. Also called **foliation domain**.

Cleavage mullions Mullions developed on the cleavage surface.

Cleavage A planar fabric along which rocks easily split or cleave into parallel or subparallel surfaces formed during metamorphism and deformation. Along the cleavage, a rock shows the ability to split or cleave into parallel or subparallel surfaces. Also called **rock cleavage**.

Clinometer compass A geological compass that mainly measures directions and dips of beds.

Closed fold A fold with interlimb angle between 70° and 30°.

Coaxial deformation It is a type of deformation in which the material lines do not change their angular relations though their lengths have changed.

Coaxial strain path A body is said to have undergone a coaxial strain path if during progressive deformation the principal strain axes of the body coincide with the principal axes of incremental strain. In other words, the strain axes of the body continue to remain parallel as the body takes up different shapes with progressive deformation.

Coble creep Migration of vacancies along the grain boundaries.

Coefficient of friction Ratio of the frictional force to the normal stress of a material.

Competent bed Bed that allows stresses to pass through and, when stresses exceed the strength of the bed, undergoes deformation.

Compressive stress Stress that acts perpendicular to a plane and if the body is in equilibrium, it means that an equal and opposite stress must have been active. As a result, the body gets shortened.

Confining pressure Pressure that is exerted equally from all sides of a body. In the earth's crust, the confining pressure is the lithostatic pressure developed due to the load of the overlying rocks.

Congruous folds Smaller folds whose axes are parallel/subparallel to that of the larger fold.

Conical fold A noncylindrical fold that can be generated by rotating a line in a circular path from one end while keeping the other end fixed. The structure thus generated is cone shaped with convergent hinge lines.

Conjugate fold A combination of two asymmetric folds in which the axial planes dip in opposite directions and thus converge.

Conjugate kink folds Folds in which the hinge zones are angular and fold limbs are straight formed by the development of kink bands.

Constitutive equation Mathematical relation existing among intrinsic parameters (rigidity, elasticity and viscosity) and extrinsic parameters (temperature, pressure and time) of a material. Also called **rheological equation**.

Constitutive law Relationship between strain rate and stress.

Continuous cleavage A pervasive structure characterized by distinct planar elements with spacing of less than 1 mm. The rock thus develops a tendency to split into thin slices such as shown by slates.

Continuum mechanics Concept that considers a rock as a continuous medium that is free from asperities, microfractures, cracks or some other forms of discontinuities. Such a material deforms uniformly in all directions within its boundaries.

Contractional bend See **restraining bend**.

Contractional regime A deformation regime in which a layer or a part of the earth's crust is shortened. Structures such as thrust faults, reverse faults and folds are commonly formed in such regimes.

Contractional strike-slip duplex Structure formed when the bend of a strike-slip fault is associated with a dextral strike-slip fault.

Core-and-mantle structure A microscopic structure in which the core of host grain passes out transitionally into mantle with increasing subgrain formation and then into aggregates of recrystallized grains with similar size and with orientations to the nearby subgrains.

Coulomb criterion of failure Shear stress τ_s is the sum of the cohesive strength of a material (τ_0) and the coefficient of internal friction (μ) multiplied by the normal stress (σ_n), i.e. $\tau_s = \tau_0 + \mu\ \sigma_n$

Couple Two equal forces acting on a body in opposite directions in the same plane but not along the same line.

Crack-seal mechanism The mechanism whereby fractures in a crystal propagate in discrete increments and each incremental gap is filled up with new crystals in optical continuity with grains along the fracture walls.

Creep It is a time-dependent deformation that takes place under constant stress conditions well below the rupture strength of a rock. Also called **time-dependent deformation**.

Crenulation cleavage A planar structure (foliation) occurring in the form of small folds (when seen in hand specimens) or micro-folds (when seen in microscopes) formed by folding/micro-folding of a pre-existing cleavage.

Crenulation lineation Lineation that occurs in the form of a set of minor folds or micro-folds of the pre-existing foliation. The hinges of the crenulations are commonly, but not necessarily, oriented parallel to each other and thus constitute a lineation.

Crystal defects No crystal has a perfect lattice and contains some imperfections or defects in the form of (i) *point defects* such as vacancies or empty space, impurities and interstitials such as the presence of atom(s) and (ii) *line defects* such as dislocations.

Crystalline thrust Thrust that transports metamorphic and/or igneous rocks. Crystalline thrusts are commonly associated with long transport of rock leading to the formation of nappes and thrust sheets.

Crystallographic preferred orientation (CPO) Arrangement of lattices (atoms, ions or molecules) in a crystal. Also called **lattice preferred orientation (LPO)**.

Crystal-plastic deformation Deformation of the crystal lattice of a mineral. Also called **intracrystalline deformation**.

C-surface (French word *cisaillement*, meaning shear) Foliation in shear zones in the form of narrow, local shear surface parallel to the shear zone boundary and is one in which high shear strain is localized. The C-surface may occur as isolated or as a set of parallel shear zone boundaries.

Cylindrical fold A fold that shows the geometry of a cylinder that can be assumed to have formed when a line of fixed orientation, which is the fold axis, moves parallel to itself such that the fold surface thus formed contains a set of parallel lines. These lines are actually the hinge lines of the fold at the respective points on the fold profile. A cylindrical fold is thus characterized by parallel hinge lines that are parallel to the fold axis.

D

D shears Synthetic shears subparallel to the associated main fault. Also called **Y shears**.

Decollement Horizontal or low-angle thrust along which the strata get partly detached from, and slide over, the underneath ones. A decollement thus constitutes a major boundary between two blocks, a lower block with undeformed rocks and an upper block with deformed rocks.

Deformation history See **strain history**.

Deformation map See **deformation mechanism map**.

Deformation mechanism map A graph that represents the dominant deformation mechanism that has operated at various stages of deformation of a (polycrystalline) rock under a particular set of conditions. A deformation mechanism map broadly represents a relation between stress and temperature in a graph. The concept does not take into account brittle fracturing whose effect is superimposed, and thus ignored, by imposing sufficiently large hydrostatic confining pressure during experiment. Also called **deformation map**.

Deformation path See **strain path**.

Deformation structures Structures developed in rocks as a result of tectonic activity.

Deformation The change in the original shape, size or volume of a rock mass caused by external stresses.

Detachment fault A flat or low-angle normal fault associated with large amounts of displacement of the overlying rocks.

Deviatoric stress Component of a stress that expresses the difference between a normal stress and the mean stress. It can be both a two-dimensional stress and a three-dimensional stress. Deviatoric stresses tend to change the shape of a body, which is then said to have undergone distortion.

Diapir A dome-shaped structure formed by vertical intrusion of a column of salt or magma into the overlying sedimentary strata. In the context of salt, the term **salt diapir** is commonly used.

Differential stress Difference between any two stresses in a plane.

Diffusion creep Mechanism of deformation of crystalline materials caused by diffusion of vacancies through the crystal lattice.

Diffusive mass transfer Mechanism of transfer of material from zones of high inter-granular normal stresses to zones of low normal stresses. The mechanism causes volume loss to the rock due to removal of material.

Dilation Change in the volume of a body.

Dip fault The trace of the fault that is parallel to the direction of dip of the adjacent beds.

Dip isogons Lines joining the points of equal dip of two successive layers of a fold.

Dip Amount of inclination of a bed with the horizontal plane as measured in a vertical plane.

Dip-slip fault A fault in which the net slip is along the direction of the dip of the fault; the strike-slip component is thus zero.

Dip-slip The slip or displacement between two points along the dip of a fault plane.

Direct stress See **normal stress**.

Direction Orientation of an object with respect to north. On the surface of the earth, north has been taken as a key reference for all direction systems.

Discontinuous recrystallization See **migration recrystallization**.

Disharmonic folds Folds in which the different layers show different fold shapes. Disharmonic folds do not show any systematic geometric relations with the adjoining folds.

Dislocation creep Mechanism of strain accommodation by dislocation glide of lattice defects through the crystal lattice.

Dislocation glide Motion of dislocations.

Distortion Change in the shape of a body while the volume remains unchanged.

D-numbers The system of assigning a special name to each newly developed structural feature in areas of polyphase deformation. Each episode of deformation leads to formation of new structures. The sequential development of structures is assigned D-numbers such as D_1 and D_2.

Dome A special type of fold in which the strata dip away from the centre in all directions.

Dome-and-basin pattern A structure developed due to superposition of the first set of folds of an area leading to formation of domes and basins. Each dome is surrounded by four basins, and each basin is surrounded by four domes.

Dominant wavelength It is the wavelength of one of the several folds that initiates buckle folding when a thin competent layer is subjected to layer-parallel compressive strain. Also called **characteristic wavelength**.

Domino faults Parallel planar faults formed by rotation of both bedding and faults.

Domino model A system of parallel planar faults in which both bedding and faults rotate by 30° to 60° or more. Faulting of the rigid blocks occurs simultaneously, and all the faults are parallel to each other and produce similar effect (offset) to the involved strata. Also called **bookshelf model**.

Domino structure A system of parallel planar faults oriented in a domino manner.

Dormant fault A fault formed in a certain geologic period after which its movement or displacement may have been locked, i.e. it has not shown any evidence of its movement or activity after its formation.

Down-to-basin fault Growth fault in which the downthrown block occurs towards the basin.

Drag folds Folds that occur along a fault and are formed due to motion along the fault. Drag folds are typically asymmetric and thus indicate relative movement of two blocks along a fault.

Drape folds A flexure or open fold of the sedimentary layers that conforms to the surface geometry or configuration developed due to the presence of an underneath fault.

Ductile deformation Failure of a rock by plastic flow before rupture.

Ductile flow Movement (flow) of rock constituents that have been subjected to grain-size reduction by dynamic recrystallization during ductile shearing.

Ductile shear zones Shear zones formed by ductile deformation. Commonly associated with mylonites and several ductile deformation features.

Ductility Property of a rock to fail by plastic flow before rupture.

Duplex A set of parallel thrusts bounded by a floor thrust at the base and a roof thrust at the top.

Dynamic analysis Analysis of field data that helps in throwing light on the forces that have caused stress in a rock.

Dynamic recovery See **recovery**.

Dynamic recrystallization Process by which a strained grain or a crystalline aggregate releases its internal stored energy by the formation of smaller, strain-free grains. It operates during syntectonic deformation.

Dynamics of deformation Study of the stresses acting within a rock body as well as on its boundaries.

E

Edge dislocation It is a type of dislocation in crystals when the Burgers vector is perpendicular to the dislocation. (See *Burgers vector*.)

Effective stress Principal stresses (considered positive for compressive stresses) minus the pore pressure.

Elastic deformation Type of deformation when a rock returns to its original shape and volume after the applied stress is removed.

Elastic limit The highest point in the stress-strain diagram (Hooke's law) where a body ruptures.

Elastic rebound theory When at a point inside the earth the release of stored elastic stresses exceeds the static friction, an earthquake is developed at that point. This theory is applicable for most of the shallow earthquakes.

Elastic A material is said to be elastic if it returns to its original shape or size after the removal of the applied stress.

Elastica A fold in which the interlimb angle is negative.

Electron backscatter diffraction (EBSD) A technique for the study of microstructure of polycrystalline materials. Basically, it is a scanning electron microscopy (SEM) using X-ray diffraction (XRD).

Emergent thrust A thrust that reaches the ground surface.

Enveloping surface A surface formed by joining tangents to the crests and troughs of a set of folds.

Extension fracture Fracture in which the broken parts of a rock show movement across the fracture surface.

Extensional bend See **releasing bend**.

Extensional crenulation cleavage See **shear-band foliation**.

Extensional fault A fault in which the horizontal component of displacement caused by faulting extends the crust by tension. Extensional faults can produce large displacements measurable up to tens of kilometres or even more. Normal faults are the most common structures to cause extension and are therefore commonly called extensional faults.

Extensional regime A state of tectonism developed by extension of a layer or a part of the crust.

Extensional strike-slip duplex Structure formed when the bend of a strike-slip fault is associated with a sinistral strike-slip fault.

Extensional tectonics A state of tectonism developed by extension of a layer or a part of the crust. Normal faults are the most common structures to cause extension.

External schistosity See ***S_e*fabric**.

Extrinsic parameters External factors such as temperature, pressure and time in the context of rheology of a material.

F

Fabric Orientation of minerals in a rock.

Failed rift A rift formed when the process of rifting may not reach to the point of breaking the crust.

Fault bend fold A fold formed by bending of a fault block as it rides over a non-planar fault surface. An anticline is thus formed at the hanging wall due to the presence of a fault.

Fault blocks The two parts of a rock mass separated by a fault plane.

Fault breccia Breccia associated with a fault.

Fault cut During drilling operation, a borehole may encounter a fault; this point is called a fault cut.

Fault damage zone Material of deformed wall rocks around a fault surface formed during slip along the fault.

Fault gouge A fine-grained rock that is formed when the original rock is fully pulverized as a result of rubbing during faulting.

Fault line See **fault trace**.

Fault plane The plane or surface along which the actual movement associated with a fault takes place. Also called **fault surface**.

Fault-propagation fold A fold formed when a propagating thrust fault loses slip and terminates up-section by transferring its shortening to a fold developing at the fault's tip.

Fault scarp A scarp, i.e. a sudden rise of topography as exhibited by a planar slope, associated with an active fault or a rejuvenated fault.

Fault surface See **fault plane**.

Fault trace The line formed by the intersection of a fault with the ground surface. Also called **fault line**.

Fault zone A zone in which a fault that includes multiple faults occurring together.

Fault Fracture of high shear stress along which two blocks of rock have moved past each other. A discontinuity or anisotropy caused by shear displacement of rock masses parallel to the fault plane.

Faulted joints Joints that have been subjected to shear such that the displacement can be detected. Identified in field by the offsetting of the pre-existing fabric or veins that have been subjected to shear such that the displacement can be detected.

Filled joints Joints filled with minerals.

Finite strain The amount of strain estimated from a deformed rock at the time of observation.

First-order folds A large-scale or regional scale fold that has developed smaller folds in its limbs.

Flanking structure It is a deflection of planar or linear fabric elements in a rock alongside a cross-cutting object such as a vein, fracture or burrow.

Flattening strain Strain in folds developed due to superposition of homogeneous strain on the early-formed parallel buckle folds, thus leading to flattening of their hinge zones, reduction of interlimb angle and attenuation of fold limbs.

Flaws Defects of microscopic sizes present in brittle solids in the form of microcracks, dislocation planes, voids, inclusions and precipitates.

Flinn diagram Diagram that shows how an initially spherical object takes up different ellipsoidal shapes during progressive deformation. The diagram thus helps identify different states of strain during progressive deformation with a plot of axial ratios X/Y and Y/Z that are, respectively, represented by two parameters a and b. The method assumes constant volume during deformation.

Flow fold A fold that does not show any systematic relationship among its geometric parameters.

Flow law The general form of mathematical expression that relates differential stress, temperature and strain rate.

Flower structure A set of curved thrust faults developed in the vertical section of some strike-slip faults. These are formed in the bends and stepovers of large strike-slip faults, where deformation is accommodated by vertical movements giving rise to the formation of normal faults and thrust faults as well as folds.

Fluid A substance that shows constant rate of flow even under small stress.

Fold belts Occurrence of folds in regionally deformed terrains of orogenic provinces.

Fold interference pattern Superposed folds in which one can notice one or more axial planes in folded forms.

Fold mullions A set of narrow fold hinges developed in the competent layer, and their convexity is directed towards the incompetent layer. Formation of fold mullions could thus be related to layer-parallel shortening during deformation. Fold mullions are commonly noticed in sandstone alternating with shale or quartzite alternating with phyllite.

Fold nappe Structure formed when the rock layers of a nappe undergo large-scale recumbent folding such that the lower limb may show stratigraphic inversion.

Fold profile Form of a fold surface as seen on a plane normal to the hinge.

Fold system Folds occurring in a group.

Fold train Folds occurring adjacent to one another; the individual folds may show the same or different geometry.

Fold A bend formed as a result of permanent deformation of originally planar surfaces of rock layers.

Folded thrust See **thrust fold**.

Foliation domain See **cleavage domain**.

Foliation fish Lozenge-shaped micaceous layers locally occurring in some mylonites.

Foliation refraction Refraction of foliation while passing from one layer to another in multilayered folds showing alternation of competent and incompetent layers. Foliation refraction shows different orientations of foliation/cleavage in adjacent beds.

Foliation Planar structure given by preferred orientation of minerals generally showing platy or tabular habit. The preferred orientation is produced by deformation and is uniformly pervasive in a rock. Foliation is commonly developed in metamorphic rocks and includes cleavage, schistosity, gneissosity and gneissic banding.

Footwall The block below an inclined fault plane.

Force That which changes or tends to change the state of rest or of uniform motion of a body.

Forced fold Fold that has to take up the shape of the local ground configuration already developed by a fault in the basement rocks.

Foreland The frontal part of an orogenic belt where thin-skinned deformation dominates.

Forelimb The overturned limb of an overturned fold; the beds are disposed in an inverted position.

Fourier's law of heat conduction It provides a relation between heat flow and temperature gradient inside the earth. The law is based on the belief that conduction is the main mechanism for transfer of heat in the lithosphere and that thermal properties of rocks are a function of depth.

Fracture mechanics Study of stress concentrations caused by sharp-tipped flaws in a rock mass and the conditions for the propagation of these flaws.

Fracture General term for any kind of break or discontinuity in rock.

Friction Resistance to relative movement of two solid bodies.

Frictional grain-boundary sliding Sliding of grains past the adjoining ones. Sliding takes place when the cohesion and friction between the grains are overcome. As a result, the grains become free to move and rotate as rigid objects without undergoing any internal deformation or crystal-plastic deformation.

Fry method It is a method for estimating strain in deformed rocks. A set of points are plotted such that a central vacancy is produced whose shape and orientation are proportional to the shape and orientation of the strain ellipse of the deformed rock.

G

Geological compass Fundamental field instrument that mainly measures directions and dips of beds.

Geometric softening It is a mechanism of formation of shear zones by strain localization due to progressive rotation of grain lattices towards parallelism with the shear plane during ductile deformation, and this is caused by rotation of lattice and grain shape. The rotation causes decrease of shear stress necessary to promote crystallographic slip.

Glide plane A plane that separates two parts of a crystal affected by movement of dislocations. It contains the line along which dislocation takes place.

Gneiss A high-grade metamorphic rock showing alternating bands of light- and dark-coloured minerals. In thin sections, the light-coloured bands show quartz, feldspar and an aggregate of quartzo-feldspathic minerals, while the dark-coloured bands show micaceous minerals, amphiboles and pyroxenes.

Gneissosity A type of foliation given by alternating bands of light- and dark-coloured minerals.

Gouge A pulverized rock containing less than 30% fragments and thus appearing powdery in outcrops. Gouge is commonly formed due to fault motion.

Grain-boundary diffusion Movement (diffusion) of fluids formed due to pressure solution along grain boundaries.

Grain-boundary migration It is a process of recrystallization in which material from a grain that is being consumed enters the grain-boundary region and later on recrystallizes on the lattice of a neighbouring growing grain. The process does not involve any gain or loss of material near a migrating boundary. As such, the mechanism is believed to be a conservative process.

Grain-boundary pinning Process by which the presence of some particular mineral hinders the growth of some other mineral as observed under microscope. For example, in a quartz-mica-rich rock, the presence of mica may hinder the growth of quartz grains, so that the process ultimately leads to a stable microstructure to the rock in which both the minerals stabilize with uniform grain sizes. Pinning is a shear-related phenomenon.

Grain-boundary sliding Deformation mechanism that can accommodate very high strains that enable grains slide past each other. The mechanism operates at high temperatures and low stresses and is assisted by the presence of fluids.

Grain migration Motion of grains through the host material in a deforming environment in which a grain loses material along some of its boundaries and gains material along some other parts. The grain may move through the material by a distance of the order of a grain diameter or more.

Grain rotation Rotation of mineral grains during deformation.

Grain-shape foliation Foliation given by subgrains (formed due to dynamic recrystallization) oriented in a parallel fashion with their long faces inclined to the main foliation. Also called **inclined subgrain fabric**.

Gravitational extrusion A model to explain the formation of thrust faults in orogenic belts by production of a wedge whose hinterland part is thicker and relatively more elevated. Once this stage is reached, gravity comes into play by pushing down—'extrusion' of—rock masses as thrust nappes that move towards, and perpendicular to, the foreland.

Gravitational gliding A model to explain the formation of thrust faults in orogenic belts where thrust faults dip towards the foreland.

Gravitational spreading A model to explain the formation of thrust faults in orogenic belts, where it is assumed that rock masses from the thick and highly elevated hinterland may move sideways towards the foreland. The spreading of rock masses is caused by the difference in surface elevation between the interior of the mountain belt and the craton.

Gravity faults See **normal faults**.

Griffith's cracks Defects or flaws of microscopic sizes in brittle solids in the form of microcracks, dislocation planes, voids, inclusions and precipitates. These cracks are ideally assumed to be of elliptical shapes.

Griffith's fracture theory Fracture strength of brittle solids depends upon the presence of defects or flaws of microscopic sizes in the form of microcracks, dislocation planes, voids, inclusions and precipitates. Presence of flaws in a brittle solid decreases the fracture strength, which depends upon the cohesive forces between the atoms as well as on the elastic properties of the solid.

Growth faults Normal faults in which the thickness of the sedimentary sequence in the downthrown block is more than its corresponding parts in the upthrown block. These faults are formed when deformation is synchronous with sedimentation.

Growth structures Structures whose formation is related to sediment accumulation.

H

Hade The angle between a fault plane and the vertical plane.

Hafner's theory It states that faults occasionally show curved outlines on all scales.

Half graben A fault trough formed in the downthrown block of a graben when only a single fault has been active such that the sedimentary wedge is thicker along the fault.

Halokinesis Tectonic deformation of salt as entirely controlled by gravity without significant contribution of lateral forces.

Hanging wall The block that rests above an inclined fault plane.

Harmonic folds Folds in which the fold shape does not change within one lithic unit and is repeated a few times along the axial region, i.e. at every half-wavelength.

Hartman's rule The maximum principal stress (σ_1) bisects the angle between the conjugate shear fractures.

Healed microcracks Microcracks that are filled with solutions released from the host grain or from the rock matrix.

Heave The displacement component of an inclined fault along the horizontal.

Helicitic folds Folds developed within the porphyroblast of a metamorphic rock.

Heterogeneous deformation Deformation by the development of fracture without undergoing any flowage. Also called **inhomogeneous deformation**.

Heterogeneous strain A type of strain when deformation of a body has taken place in such a way that the reference lines and planes of the undeformed body have changed their geometric relations after deformation. Thus, the straight lines and the parallel lines of the undeformed body do not remain straight and parallel after deformation, and the bounding planar surfaces become curved. Also called **inhomogeneous strain**.

Hinge line The line on the fold surface that joins the points of highest curvature.

Hinge zone Part of a fold located near the hinge.

Hinge The point on the fold surface where the curvature is highest.

Hinterland The central part of an orogenic belt characterized by thick-skinned deformation, i.e. involvement of basement rocks and formation of crystalline thrusts.

Homocline A structure in which the strata dip persistently in one direction.

Homogeneous strain A type of strain when the changes in shape or size of a body have taken place in such a way that the reference lines and planes of the undeformed body do not change their geometric relations after deformation, such that the straight lines and the parallel lines of the undeformed body remain straight and parallel after deformation, and the bounding planar surfaces remain planar.

Homogeneous deformation Deformation by flowage without development of any fracture.

Hooke's law Strain of a material is directly proportional to the applied stress.

Hookean body A material that follows Hooke's law is called a Hookean body or an ideally elastic body; it ruptures within the elastic field.

Horizontal equivalent Length of a bed as projected on the horizontal surface.

Horizontal fold A fold showing plunge of its axis between 0° and 10°.

Horse The individual thrust mass or slice of a duplex surrounded on all sides by thrusts.

Horsetail splay A structure formed when a strike-slip fault terminates by forming a fan of strike-slip splay faults that generally show a curvature with concavity towards the receding block.

Horst and graben structure Normal faults occurring in a conjugate set in which individual faults on either side dip in opposite direction. The block that is uplifted between two oppositely dipping faults is called a *horst*, while the block that has been downthrown is called a *graben*.

Hydraulic joints Joints formed due to compaction at depth (more than 3 km) caused by the load of the overburden.

Hydrofracturing or hydraulic fracturing Formation of fractures by brittle failure by crack propagation under the presence of pore fluids in the underneath sedimentary rocks when the amount of tensile stress increases than that of the least principal compressive stress.

Hydrolytic weakening Mechanism that makes minerals and rocks weak or soft when exposed to water. A 'wet rock' is weak or soft than a 'dry rock'. The term 'water' is used to include water-related fluids.

Hydrostatic stress A state of stress when all the principal stresses are equal, i.e. $\sigma_1 = \sigma_2 = \sigma_3$.

Hyperbolic net A standard stereonet capable to yield valid results from initially uniform and imbricated fabrics by amalgamating the available methods of strain analysis. The net can simultaneously handle more than one method of strain analysis.

I

Ideally elastic material One in which the strain developed due to a load remains constant during loading but the strain developed in the body reduces to zero on removal of the load. Such a material recovers the total strain once the load is removed.

Ideally plastic material A material that has not undergone strain hardening. Also called **pure plastic material**.

Ideally viscous material See **Newtonian material**.

Imbricate fan A set of imbricate thrusts that are arranged in a parallel fashion and flatten towards the floor thrust and steepen towards the roof thrust. Characteristic of foreland fold-and-thrust belt.

Imbricate structure A structure formed by smaller thrusts that occur in a parallel set and join a master fault down dip. Also called **schuppen structure**.

Inactive faults Faults that have not shown activity during the past 1.65 million years.

Inclined fold A fold in which the axial surface is inclined.

Inclined subgrain fabric See **grain-shape foliation**.

Incremental strain It is finite strain of a rock that can be thought of as the cumulative effect of several deformation events, and the strain achieved during each deformation event is called incremental strain. The rock thus achieves final strain.

Infinitesimal strain Incremental strain that is infinitesimally small, i.e. below the finite limits.

Inflection point The point on the limb of a fold where the curvature reverses.

Inhomogeneous deformation See **heterogeneous deformation**.

Inhomogeneous strain See **heterogeneous strain**.

In-sequence thrust See **normal-sequence thrust**.

Interlimb angle Angle that the two limbs of a fold subtend at the hinge. It is also the angle formed by the two tangents drawn at the inflection point of the opposite limbs.

Internal schistosity See **S_i fabric**.

Intersection lineation Lineation given by the intersection of any two planar features of a rock.

Intracrystalline deformation See **crystal-plastic deformation**.

Intrafolial fold It is a tight to isoclinal fold with thinned, detached limbs. Such folds form due to intense deformation that causes transposition of the limbs.

Intrinsic parameters Factors or physical parameters such as rigidity, elasticity and viscosity in the context of rheology of a material.

Irregular mullions Cylindrical columns with an irregular cross section and are covered with a thin layer of mica. The columns commonly overlap with the neighbouring ones.

Isoclinal fold Fold with parallel limbs.

Isoclinal recumbent fold A fold whose limbs are parallel and axial surface is horizontal.

Isotropic substances Substances whose mechanical properties are the same in all directions.

J

Jeffrey model A model to explain the mechanisms of grain rotation wherein a mineral grain that behaves as a rigid object deforms in a ductile matrix under the influence of progressive simple shear.

Joint density Total trace length of all joints in a unit area. This parameter gives an idea of the areal abundance of joints in a rock mass.

Joint intensity The total area of joint planes within a unit volume of rock, say 1 m^3. This parameter can be estimated after the joint density on three exposed surfaces that are mutually perpendicular.

Joint set A group in which the joints are parallel or subparallel to each other.

Joint system It is formed by intersection of two or more sets of joints.

Joint zone Several closely spaced individual joints that when viewed from a distance look like an individual joint cutting the length of the outcrop.

Joint Crack along which there has been little or no transverse displacement of rock.

K

Kinematic analysis Analysis of geological data that gives an idea of quantitative deformation of a rock involving strain analysis to understand how the rock behaved to the deforming environment by changing its shape, size and orientation.

Kinematical vorticity number It is an expression of the degree of noncoaxiality, and thus the degree or amount of rotation of a body.

Kinematics Pertaining to the geometry of motion of rock bodies.

Kinetic friction Tangential force necessary to maintain sliding of a body.

Kink bands The steplike smaller limbs of kink folds.

Kink folds Folds with straight limbs and angular hinges but the limbs are of unequal lengths and therefore these are asymmetrical folds

Kinking Formation of kink folds.

Klippe The portion of a thrust sheet that has escaped erosion and is thus a remnant of the original thrust sheet.

L

Lattice preferred orientation See **crystallographic preferred orientation**.

Line defects The imperfections or defects in the lattice of a crystal in the form of dislocations.

Lineaments A linear feature of the earth's surface that can be recognized on maps or on aerial or satellite images and must be at least a few kilometres long.

Linear rheological model A substance that represents some ideal cases with respect to stress and strain rate.

Linear structures See **lineation**.

Lineation Any fabric or orientation developed in a linear fashion. It is a common feature noticed in deformed rocks and is generally seen on the upper surfaces of rocks. Also called **linear structure**.

Listric normal fault A normal fault that is steep at the surface but flattens at the base, thus showing concavity upwards.

Lithosphere The outer shell of the earth that responds to stress and undergoes deformation. It includes the crust and the uppermost rigid part of the mantle and is about 100 km thick.

Lithostatic stress Stress caused by the weight of the overlying burden in the earth's crust. Since rocks in the earth's crust are in equilibrium, i.e. physically stable, we can assume that at each point the stress is acting uniformly from all sides. In other words, the principal stresses are equal, all the principal stresses are compressive and there is no shear stress.

LS-tectonite A rock that shows both lineation and foliation. Generally produced by shear (noncoaxial) deformation.

L-tectonite A rock dominated by lineation such that the foliation is either absent or weak.

M

Mantled porphyroclasts Porphyroclasts which, with progressive shear, undergo reaction with the matrix material on their outer surfaces, thus forming an envelope of recrystallized material.

March model A model to explain the mechanism of grain rotation with the assumption that the randomly oriented platy grains of a rock act as passive markers and deform in the same way as the matrix. The platy grains tend to concentrate along the *XY*-plane and thus produce a foliation to the rock.

Master fault A large fault associated with a few smaller and parallel faults showing the same sense of displacement as the large fault.

Master joint Joint that extends up to a few kilometres.

M-domain In foliated rocks, the domain dominantly constituted of micaceous minerals (M for mica).

Mean stress In a system where all the principal stresses are unequal, i.e. $\sigma_1 \neq \sigma_2 \neq \sigma_3$, and all are compressive, we can imagine a mean stress, σ_m, which is given by $\sigma_m = (\sigma_1 + \sigma_2 + \sigma_3)/3$

Mechanism of rock deformation The process that tends to accommodate large strains in rocks, thus resulting in a stable structure or microstructure to the rock. The process may operate alone or in combination with other mechanisms until the accumulated strain is accommodated.

Metamorphic core complex Region of large-scale crustal extension where low-angle normal faults (= detachment faults) have brought high-grade, deep-seated metamorphic rocks with evidences of ductile deformation over sedimentary rocks showing evidences of ductile-brittle deformation.

Metamorphic differentiation Process by which minerals or mineral assemblages from an originally homogeneous rock get separated during metamorphism and accumulate as different layers that are commonly monomineralic.

M-fold See **symmetric fold**.

Mica fish Lens-shaped (porphyroclasts) of mica in mica-rich mylonitic rocks.

Mica seams Flakes of mica occurring as long stringers.

Microcracks Cracks or planar discontinuities in a rock that are visible under a microscope.

Microlithon A domain of foliated rocks that shows no or little alteration of the original fabric of the rock.

Micro-fault A fault that is visible under a microscope.

Microfractures Fractures that occur on microscopic or sub-microscopic scales.

Microstructure Structure of a rock as seen in microscope.

Migration recrystallization Mechanism due to which the grain boundaries of a crystalline aggregate regularly undergo catastrophic jump in migration velocity. Also called **discontinuous recrystallization**.

Mineral fish A general term used for lens-shaped (porphyroclasts) minerals such as K-feldspar, plagioclase, tourmaline, kyanite and staurolite.

Mineral lineation Preferred orientation of a single mineral grain or an aggregate of several minerals. The mineral generally has one axis much longer than the other two.

Mineral nucleation A process due to which new minerals nucleate during metamorphism and begin to grow once the affinity of the relevant mineral reaction becomes large enough. The mechanism develops mineral preferred orientation in metamorphic rocks.

Modulus of rigidity See **rigidity modulus**. Also called **shear modulus**.

Mohr diagram A diagram in which the normal stress σ_n and shear stress τ_s are represented by the abscissa and ordinate of a graph.

Mohr envelope Two tangents to Mohr circle. The tangents separate stable field from unstable field on the Mohr diagram.

Mohr strain diagram Representation of strain by a circle in which the reciprocal of quadratic elongation, λ' (= $1/\lambda$), is plotted as abscissa and modified shear strain, γ' (= γ/λ), as ordinate.

Mohr stress diagram A graphical method that represents in two dimensions the relationship between shear stress and normal stress. From the graph, it is possible to compute the values of shear stress, normal stress and angle of failure at the point when a crack is developed, i.e. when failure occurs.

Mohr-Coulomb criterion for shear failure Diagram formed by two tangents on a Mohr diagram. The tangents separate stable field from unstable field on the Mohr envelope. The region within the envelope describes the state of stable stress, while that outside the envelope is the state of unstable stress.

Monocline Local steepening of an otherwise uniformly dipping strata.

Mullions Corrugated structures showing alternating convex and cusp surfaces developed at the interface of competent and incompetent layers. Individual unit of the structure looks like a convex pillar with a flat base with cross section of a few centimetres and length measurable in several centimetres to a few metres.

Mylonite A foliated rock formed by strong ductile shearing (dominantly simple shear) and is characterized by grain-size reduction. It is a typical rock of ductile shear zones and is formed at middle to lower levels of the crust where high temperature-pressure regime favours grain-size reduction.

Mylonitic foliation Foliation developed due to partial or complete overprinting of a pre-existing foliation during ductile shear deformation. It is visible both in hand specimens and under microscope.

N

Nabarro-Herring creep Migration of vacancies through the crystal lattice.

Nappe A rock mass that has moved from a root zone along a thrust fault.

Negative flower structure See **normal flower structure**. Also called **tulip structure.**

Net-slip The total displacement of any point on a fault plane.

Neutral fold A fold that closes laterally; it may be an antiform or a synform.

Neutral surface A surface in a buckled competent layer that forms a boundary between the outer arc of the fold showing layer-parallel extension and the inner arc showing layer-parallel compression, while the surface itself does not show the effects of extension or compression.

Newtonian fluid Substance in which the rate of shear is directly proportional to shear stress.

Newtonian material A material in which strain remains constant with time (provided that the stress also remains constant). Also called **ideally viscous material**.

Noncoaxial deformation It is a type of deformation in which the material lines change their angular relations such that their lengths have changed.

Noncoaxial strain path A body is said to have undergone a noncoaxial strain path if during progressive deformation the principal strain axes of the body do not coincide with the principal axes of incremental strain. In other words, the strain axes of the body rotate as the body takes up different shapes with progressive deformation.

Noncongruous folds Smaller folds whose axes are not parallel/subparallel to that of the larger fold.

Non-cylindrical folds Folds that do not show cylindrical geometry. (See also **cylindrical folds**.)

Non-rigid body deformation Type of deformation of a body when the distance between any two points changes during motion.

Nonsystematic joints Joints that show irregular geometry and terminate against older joints.

Nontectonic structures Deformation structures formed in earth materials without being involved in tectonic processes.

Normal fault A fault in which the hanging wall has gone down relative to the footwall. Also called **gravity fault**.

Normal flower structure A type of flower structure in which the thrusts show concavity upwards. Also called **negative flower structure** and **tulip structure**.

Normal-sequence thrust A thrust formed due to outward propagation of thrusts from the core of a mountain chain in a sequence. The outward thrusts thus formed are progressively younger in the sequence. Also called **in-sequence thrust**.

Normal stress Stress acting perpendicular to a plane. If the body is in equilibrium, it means that an equal and opposite stress has been active. Also called **direct stress**.

Numerical modelling Mathematical simplification and/or application of computers of the structural database acquired from field, laboratory and experimental studies.

O

Oblique fault The strike of the fault is oblique to the strike of the adjacent beds.

Oblique foliation Foliation developed when a pre-existing foliation undergoes realignment compatible with the ductile shear strain.

Oblique-slip fault A fault in which the net-slip is parallel neither to the dip nor to the strike of the fault plane, and as such this fault has both a dip-slip component and a strike-slip component.

Opposite shear sense Conflicting shear sense given by the shear-sense indicators for a rock body at the portion of observation.

Orthomylonite A mylonite with a matrix 50–90% of the rock and grain size of matrix material <0.5 mm.

Out-of-sequence thrust A thrust formed when the normal sequence of outward propagation of thrusts is broken.

Overstep It is a complex structure in which a new thrust surface is developed in the hanging wall of an older thrust.

Thrust propagation takes place towards the hinterland in a sense opposite to the transport direction. The higher thrust will thus represent the later movements in the system of faults.

Overturned fold It is a fold in which the dip of both the limbs is in the same direction and the axial surface is inclined in such a way that the upper limb overrides the lower one. In such folds, the beds in the lower limb are disposed in an inverted position while those of the upper limb are in normal position.

P

Palaeopiezometers Features/signatures in rocks that help in estimating palaeostress.

Palaeopiezometry Estimation of palaeostress in a rock.

Palaeostress Stress that is locked in the rocks during their formation in the geological past.

Palinspastic map A map that shows the disposition of beds when they were originally deposited.

Palm tree structure See **reverse flower structure.** Also called **positive flower structure**.

Parallel folds Folds in which the layer thickness remains constant.

Parasitic folds Smaller folds developed in the limbs or hinge of a fold.

Passive folding Formation of folds when the layers simply undergo ductile flow as a passive marker without playing any significant part in deformation.

Passive folds Deformation of a rock layer by shear or flow across the layering.

Pencil structure Linear structure formed due to the intersection of two cleavage planes or a foliation plane and a cleavage. The intersection produces columns, called pencil structures, that may show rectangular or prismatic cross sections. The structure is commonly used to make slate pencils.

Penetrative lineation Lineation that extends within a rock.

Pericline Folds that assume elliptical or semi-elliptical shapes in plan view when a dome and basin close in both directions.

Phyllite A low-grade metamorphic rock with fine grains of mica.

Piggyback thrust It is a complex structure formed when newer thrusts continue to form in a sequence in the footwall of the first formed thrust. The newer thrusts are formed in the direction of the foreland and continue to carry the early-formed thrusts on their back, thus giving a 'piggyback' geometry. The highest thrust will represent the earliest displacements while the lower ones the last displacements.

Pinnate joints Set of en echelon cracks that propagate away from a fault at acute angles. Pinnate joints when associated with host fractures serve as a kinematic indicator slip on fractures of a region.

Pitch See **rake**.

Planar structures Rocks can be considered to constitute planes, and these planes contain structural features within them as well as on their surfaces. Planes in rocks are therefore called planar structures, such as bedding, fault planes, foliations, joints, dikes and sills.

Plane strain A state of deformation when extension along X is fully counterbalanced by shortening along Z so that there is no extension along Y, and therefore the deformation involves no volume change.

Plane stress A biaxial stress when two stresses, σ_1 and σ_2, act on a plane while the third, i.e. normal stress, σ_3, is zero. Since all the stresses act in one plane, the biaxial stress is also called plane stress.

Plastic deformation Deformation of a body that involves non-recoverable strain, i.e. the body shows permanent deformation. It involves deformation of the crystal lattice of the constituent minerals.

Plumose structures Curved lines oriented parallel to the direction of joint propagation as noticed on joint surfaces.

Plunge The angle subtended by a line on any plane with the horizontal plane.

Plunging fold A fold in which the hinge line is inclined to the horizontal plane.

Point defects The imperfections or defects in the lattice of a crystal in the form of vacancies or empty space, impurities and interstitials such as the presence of atom(s).

Polyclinal fold A group of adjacent folds with subparallel hinge lines; the fold set consists of more than two axial surfaces with different orientations.

Polygonization Occurrence of polygonal subgrains within larger grains as commonly observed by variations in extinction from adjacent grains caused by small difference in the orientation of the subgrains. The subgrains are formed from an original distorted crystal and are strain free.

Polyharmonic fold A large fold together with the included smaller folds of all sizes.

Pop-up structure The hanging wall block created between a back-thrust and the frontal ramp.

Pore fluid pressure Pressure caused by the presence of a fluid system in the pore spaces of rocks.

Porphyroblast Large single crystals that have grown during metamorphism in a relatively finer grained matrix.

Porphyroclasts Relict of some larger grains that have escaped ductile deformation as commonly noticed in most metamorphic rocks. The porphyroclasts are rigid objects that rotate in a finer grained matrix under noncoaxial conditions. Quartz and feldspar constitute common porphyroclasts.

Positive flower structure See **reverse flower structure**. Also called **palm tree structure**.

Post-tectonic crystallization It is a stage of metamorphism that continued even after the deformation is over.

Potentially active faults Faults that have shown their activity between about 1.65 million and 10,000 years, i.e. during the Pleistocene Epoch excluding the Holocene.

Power law At constant temperature, the strain is proportional to shear stress. It operates at low differential stresses and at high temperatures.

Pressure shadows Growth of some minerals such as quartz, mica and chlorite on opposite sides of a rigid, competent object such as garnet, pyrite or magnetite. The growth of the minerals results in the formation of spindle-shaped objects that occur on the foliation surface, and their preferred alignment constitutes a lineation.

Pressure solution Dissolution of material from a crystal or crystal aggregate from points of high pressure and its redeposition at points of low pressure.

Pre-tectonic crystallization A stage of metamorphic history of a rock when the early-formed minerals are affected by later deformation resulting in overprinting of the earlier microstructure by the development of a new microstructure.

Primary creep It is a type of time-dependent (creep) deformation that starts immediately after a rock has undergone the elastic response to the applied stresses. This is represented by the first straight-line portion of the strain-time graph.

Primary directions The directions that include north (0°, 360°), east (90°), south (180°) and west (270°) in the conventional 360° system.

Primary foliation Foliation as used for any primary feature that is homogeneously distributed as a planar feature of rocks with a characteristic fabric, e.g. a bed in a sedimentary rock or flow banding in the flow of magma.

Principal axes of stress Normal stresses acting parallel to each of the coordinate axes. The stress acting parallel to x-axis is designated as X, while the other two stresses are designated as Y and Z.

Principal stress Normal stress acting perpendicular to a face along which shear stress is zero.

Process zone Formation of a macrocrack extension zone by linking of microcracks.

Progressive deformation The process of deformation by which a body continues to change its shape and size in response to the imposed deformation.

Protomylonite A mylonite with a matrix 10–50% of the rock and grain size of matrix material >0.5 mm.

Pseudotachylite A glassy, very-fine-grained rock produced by high frictional heat along a fault causing partial melting and solidification due to sudden decrease of pressure. It is dark in colour and may contain fragments of wall rock. Formation of pseudotachylite is commonly linked with seismic event.

Ptygmatic folds Folds that show irregular geometry with respect to each other, and the thickness of the folded layer generally remains uniform.

Pull-apart basins Basin formed due to subsidence of the crust in an extensional bend or releasing bend of a strike-slip fault. Also called **sag ponds**.

Pumpelly's rule Small-scale structures generally depict large-scale structures formed at the same time.

Pure plastic material See **ideally plastic material**.

Pure shear A type of deformation in which the material lines of an undeformed rock do not change their angular relations. In other words, the orientations of the principal strain axes X, Y and Z have not changed even though their lengths have changed.

Q

QF domain In gneisses, the microlithon domain is dominantly constituted of quartz and feldspar, also called the QF domain (Q for quartz and F for feldspar).

Quarter structures Asymmetric microstructures shown by some porphyroclasts in four quarters when considered in a framework of foliation and foliation-normal.

Quartz ribbon An aggregate of strongly flattened quartz arranged in a band.

R

Rake The acute angle made by a linear structure on a plane with the strike of the plane. Also called **pitch**.

Ramp and flat structure Structure produced by a thrust by cutting up-section and then following a bedding plane. The high-angle portion of the thrust is called a *ramp*, while the portion of the thrust that follows the bedding plane is called a *flat*.

Ramp anticline An anticline formed on a curved fault surface due to rotation of the hanging wall.

Reclined fold A fold in which the axial plane dips between 10° and 80° and the hinge plunges down the dip of the axial plane.

Recovery Processes that attempt to return a deformed crystal to the undeformed state during which the stored energy of deformation decreases. Recovery operates well in crystals with high density of dislocations and point defects and leads to formation of a number of unstrained subgrains formed from the original distorted crystals that can be distinctly seen with different optical orientations from the neighbouring ones. Also called **dynamic recovery**.

Recumbent fold A fold in which axial surface is horizontal.

Release joints Joints formed due to removal of overburden during erosion, but their orientation is largely determined by the presence of solution cleavage.

Releasing bend A bend in a strike-slip fault within which the crustal material is pulled apart due to the shear of the fault.

Restored section A geological section in which the rocks can be related to their undeformed state.

Restraining bend One within which the crust is shortened as a result of shearing by the main fault, thus producing contractional structures such as folds and thrust faults. Also called **contractional bend**.

Reverse fault A fault in which the hanging wall has gone up relative to the footwall. The dip of the reverse fault is generally more than 45°.

Reverse flower structure A type of flower structure in which the thrusts show convexity upwards. Also called **palm tree structure** and **positive flower structure**.

R_f/φ method This is a method for the estimation of strain in deformed rocks. It is applicable to deformed rocks that contain originally elliptical objects such as oolites, pebbles and sand grains such that after deformation the original elliptical objects are again elliptical in shape. However, some marker planes formed during deformation such as rock cleavage, foliation or schistosity help in relating the new shape and orientation of these objects to the imposed strain.

Rheological equation See **constitutive equation**.

Rheological model Behaviour of a substance with respect to stress and strain rate.

Rheology Deformation and flow of rocks.

Riecke's principle Grains of a rock are at different stress levels, and thus there is a variation of solubility of the grains across the rock mass. As such, crystals or crystal aggregates of a rock undergo dissolution at points of high compressive stress followed by redeposition at points of low compressive stress.

Riedel shears or R shears A set of shear fractures formed at low angle of about 15° with the main strike-slip fault and that are therefore synthetic to the main fault.

Rift lakes Lakes formed at the downfaulted blocks of rifts.

Rift mountains Mountain ranges constituted of rifts.

Rift Linear zone where the earth's crust has been pulled apart. Rift is formed as a result of extensional tectonics that produces normal faults.

Rigid body deformation Type of deformation of a body involving translation and rotation such that the distance between any two points of the body remains unchanged and therefore there is no visible strain in the body.

Rigidity modulus Elastic constant given by the ratio of shear stress to shear strain. Also called **shear modulus** or **modulus of rigidity**.

Ring dikes A set of dikes formed due to volcanic eruption along circular to elliptical normal faults.

Ring faults A set of normal faults that occur in a concentric shape in a volcanic terrain. Ring faults are associated with collapse of rock mass after a volcanic eruption and have vertical displacements from several hundred metres to a few kilometres.

Rock cleavage See **cleavage**.

Rock mechanics Study of application of properties of materials such as stress, strain and strain rate.

Rodding Linear alignment of mineral aggregates, pebbles and other objects to give a rodlike appearance due to deformation.

Rods Cylindrical columns containing generally monomineralic (quartz or feldspar) aggregates or pebbles formed due to deformation.

Rollover anticline An anticline formed under conditions of extensional tectonics when the hanging wall of a detachment fault rotates down the concave fault surface.

Rollover structure A structure formed under conditions of extensional tectonics when the hanging wall of a detachment fault slips down against the footwall while the basal bed remains fixed.

Root zone The zone or region from where the thrusts originate.

Rootless folds Folds in which the original root of one or both the limbs is shifted to some other place during subsequent deformation. Also called **transposed folds**.

Rotational fault A fault that shows rotational movement.

R′ shears Antithetic shears developed at about 75° with the main fault. (See *Riedel shears*).

S

Sag ponds See **pull-apart basins**.

Salt diapir A dome-shaped structure formed by vertical intrusion of a column of salt into the overlying sedimentary strata.

Salt dome See **salt diapir**.

Salt tectonics Any tectonic deformation involving salt, or other evaporites, as a substratum or a source layer. It involves flow and tendency of salt to migrate under sedimentary overburden.

Scarp A sudden rise of topography as exhibited by a planar slope.

Schist A medium- to high-grade metamorphic rock containing grains of quartz, feldspar, mica, garnet, amphiboles, etc. that can be seen by naked eyes.

Schistosity Planar structure shown by schists in which grains of quartz, feldspar, mica, garnet, amphiboles, etc. can be seen by naked eyes.

Schuppen structure See **imbricate structure**.

Scissor fault A fault in which the displacement occurs across a point, called *hinge*, such that the displacement on both sides of the hinge is reversed. The amount of displacement at hinge is zero, while it increases from the hinge on either side.

Screw dislocation It is a type of dislocation in crystals when the Burgers vector is parallel to the dislocation. (See *Burgers vector*.)

S_e fabric It is the schistosity that surrounds a porphyroblast. Also called **external schistosity**.

Secondary creep It is a type of time-dependent (creep) deformation during which the strain rate or creep rate remains constant with time, and as such the strain and time of loading bear a linear relationship. This part of the strain-time curve represents plastic deformation.

Secondary directions The four directions given by the bisectors of the primary directions (N, S, E, W); these are NE, SE, SW and NW.

Second-order folds Folds developed on the flanks or limbs of a large-scale fold.

Sedimentary structures Features in rocks formed by sedimentary processes.

Seismic faulting The relationship between an earthquake and faulting.

Separation The displacement of a bed caused by faulting. The separation can be measured in any desired direction.

Shear-band foliation Foliation that occurs in the form of discrete, sometimes sigmoidal, bands oriented oblique to the C-surfaces. It is a microscopic feature but is also observed in hand specimens. Also called **extensional crenulation cleavage** (ECC), **shear-band cleavage** and **C′-surface**.

Shear fracture Fracture along which wall-parallel displacement is discernible.

Shear heating Heating produced in rocks due to the mechanical work done during shear deformation.

Shear joint Joint in which wall-parallel displacement is discernible.

Shear modulus See **rigidity modulus**. Also called **modulus of rigidity**.

Shear plane All planes other than the principal planes in a stress ellipsoid are shear planes because in such planes there is always a component of shear.

Shear-sense indicator A small-scale structure that provides the sense of shear of the associated shear zone.

Shear strain The tangent of the angular shear of a reference line caused by simple shear. Also called **angular shear**.

Shear stress Stress that acts in opposite directions but not along the same plane or line.

Shear zone It is a tabular body of rock that accommodates the bulk or whole deformation so that there is practically no or less deformation outside this zone.

Shear It is a couple that causes relative movement of rock material past one another.

Shear-band cleavage See **shear-band foliation**.

Sheath fold Non-cylindrical fold formed when the early-developed folds rotate during progressive simple shear. Due to rotation, the folds thus developed look elliptical or semi-elliptical in cross section and extend within the rock as tubes that are parallel to the direction of slip during shear movement.

Sheeny phyllite A phyllite that, due to the occurrence of tiny grains of mica in layers, gives sheen to the rock.

Sheet joint Joints that occur in more or less parallel sets generally with a curved geometry and affect the rock surface in such a way that the joints look like scales of an onion. Along the sheets, the rock tends to peel off like onion scales. These joints are formed due to weathering which causes removal of the overburden. Sheet joints are commonly shown by granitic rocks.

Sheet joints Joints that occur as horizontal sets, thus giving the appearance of sedimentary layering. Sheet joints are commonly shown by granitic rocks.

Sheeting The process of formation of sheet joints.

S_i fabric It is the fabric given by the inclusions of other minerals within a porphyroblast and has developed during the metamorphic growth of the porphyroblast. Also called **internal schistosity**.

Sigmoidal foliation Foliation that shows curvature at the centre of a shear zone and becomes tangential or parallel to the shear zone boundaries where it provides the sense of shear.

Similar folds Folds in which the curvature of the layers along the axial plane remains the same such that their shapes look similar in section.

Simple shear The type of deformation in which the material lines of an undeformed rock have changed their original angular relations. In other words, the orientations of the principal strain axes X, Y and Z have changed together with change in their lengths.

Slate A very-low-grade metamorphic rock characterized by planar structures given by micaceous layers commonly containing chlorite, muscovite or clayey matter. The micaceous layers are formed due to recrystallization of clay under very-low-grade metamorphic conditions and constitute the cleavage domains that are separated by microlithons containing an aggregate of small quartz and feldspar.

Slaty cleavage Pervasive planar structure given by micaceous layers commonly containing chlorite, muscovite or clayey matter. The micaceous layers are formed by recrystallization of clay under very-low-grade metamorphic conditions. Presence of slaty cleavage imparts a strong fissility character to the rock due to which it can be separated in thin layers. As such, slates are often used for roofing.

Slickenfibres Fibrous crystals formed during frictional movement along a fault and that are therefore oriented parallel to the direction of motion along a fault surface. The preferred orientation of the crystals usually marks the direction of extension during fault motion.

Slickenlines Individual grooves of a slickenside that occur as linear scratches developed on a fault surface. Slickenlines actually constitute the grooves of slickensides.

Slickensides Polished striated linear features that trend parallel to the fault motion. Slickensides are formed due to frictional movement during the motion of two blocks along a fault and occur as small grooves that rise on millimetre scales or even smaller.

Spaced cleavage A rock cleavage occurring in the form of domains with microlithons and cleavage. The spacing between the cleavage and microlithon domains is 1 mm or more. As such, spaced cleavage is visible in hand specimens.

Splay A branch of a fault developed at a branch point. A splay is constituted of a single fault or a set of faults that are subsidiary to the main fault.

Stable sliding mechanism It is a mechanism of growth of faults in which the fault maintains uninterrupted motion and as such build-up of stress does not take place.

Static friction The tangential force necessary to initiate sliding of a body.

Static recrystallization Recrystallization that takes place in the absence of any tectonic disturbance.

Steady-state creep Constant rate of flow under constant stress. Also called **steady-state flow**.

Steady-state flow See **steady-state creep**.

Step faults Faults arranged in steplike manner.

Stepover Point where a strike slip fault ends and the displacement is then taken up by another fault of the same orientation.

Stereographic plot A convenient method of graphically representing planar and linear structures in three dimensions and the angular relations existing among them.

Stick-slip mechanism It is a mechanism of growth of faults in which the fault shows sudden movement in response to accumulation of large stress. Release of stress causes a slip during which there is no motion.

Strain analysis Estimation of strain in deformed rocks.

Strain ellipse It describes the state of strain in two dimensions when a material circle changes its shape to an ellipse due to deformation by homogeneous strain. The ellipse can be referred to two mutually perpendicular axes X and Y. It is assumed that strain along one of these axes, X, is the greatest, while along the other the strain is the least.

Strain ellipsoids It describes the state of strain in three dimensions when a material sphere changes its shape to an ellipsoid due to deformation by homogeneous strain. The ellipsoid can be referred to three mutually perpendicular axes X, Y and Z. It is assumed that strain along one axis (X) is the greatest, while that along the other axes is the least (Z) and intermediate (Y), such that $X \geq Y \geq Z$.

Strain hardening The mechanism of continuous increase of yield strength of a rock with progressive deformation. If the rock has plastic property, this increase of continuous stress to maintain further deformation is called strain hardening. The rock thus enters into plastic deformation stage. Also called **work hardening**.

Strain history A sequence showing the various states of strain undergone by a rock starting from its undeformed state, as revealed by systematic analysis of strain in different parts of an area. Also called **deformation history**.

Strain marker Object in a deformed rock (seen in hand specimens or in thin sections) whose original shape or size is known.

Strain path The various states of strain that a body acquires from undeformed to deformed state during progressive deformation. Also called **deformation path**.

Strain rate Rate at which a rock changes its strain (i.e. deforms) with time.

Strain softening Softening of a rock in a shear zone due to strain concentration such that the material undergoes continuous ductile deformation without fracturing.

Strain tensor It is a three-dimensional quantity described by nine quantities such as finite homogeneous strain. A strain tensor is an asymmetric tensor.

Strain The amount of deformation that a body has undergone in response to an externally applied stress.

Stress concentrators Presence of flaws in a brittle solid lowers the fracture strength. Flaws therefore constitute loci of stress concentration and are called stress concentrators.

Stress ellipse If all the stresses acting on a body are represented by lines, the longest and shortest lines trace an ellipse called the stress ellipse.

Stress ellipsoid An imaginary ellipsoid that represents the state of three-dimensional stress such that the normal stresses acting upon three planes of a body are represented by the lengths of the axes of a triaxial ellipsoid and the shear stresses acting upon all the planes are zero.

Stress field A generalized state of stress in rocks of the earth's crust.

Stress history Change of stress conditions all through the long history of rocks of a sedimentary basin from burial to the present-day form.

Stress raiser The externally applied stress that gets amplified at the tips of the microscopic flaws of a body.

Stress tensor It is a three-dimensional quantity described by six quantities. A stress is a tensor of rank 2, and it is a symmetric tensor.

Stress trajectory A continuous line joining the adjacent points with different orientation of principal stress.

Stress Force per unit area of a body.

Stretch thrust A thrust developed when overturned limb of a tight isoclinal fold is stretched and sheared.

Stretching lineation Lineation produced by stretching of minerals or mineral aggregates during deformation. It is commonly developed in gneisses and in other metamorphic rocks in shear zones. Stretching lineation is commonly given by stretched recrystallized grains of quartz, feldspar or quartzo-feldspathic aggregates in mylonites.

Strike fault A fault that is parallel to the strike of the adjacent beds.

Strike or strike line The line formed by the intersection of a bed with the horizontal plane. A strike line always exists on the ground surface and is a straight line.

Strike-slip duplex A set of parallel thrusts or duplexes formed along bends of some strike-slip faults in which the deformation bears a component across the bend. The structure is confined within the main fault.

Strike-slip fault A fault that shows horizontal slip parallel to the strike of a vertical or sub-vertical fault plane; the dip-slip component is thus zero.

Strike-slip The slip or displacement along the strike of a fault plane.

Striping lineation Intersection of bedding and cleavage gives a lineation on the cleavage surface. Since the bed is seen on the cleavage surface as colour stripes, this type of lineation is also called striping lineation.

Structural basin A special type of syncline in which the strata dip towards the centre in all directions. Also called **basin**.

Structural geology A discipline of geology which encompasses the study of all aspects of deformation structures such as their geometry, field relations, geographic distribution, genesis and related aspects.

Structural map A map of an area showing the attitude of mesoscopic structural data on a geological map.

Structural terrace Structure formed when a bed includes a horizontal part between two inclined portions.

Stylolite Zigzag trails commonly observed in limestones in which they occur parallel as well as across the bedding. The zigzag structure is formed due to local movement of solution along the extension direction.

Subgrain rotation Progressive misorientation of subgrains during recovery until high-angle grain boundaries are formed. It is considered as a common process associated with dynamic recrystallization.

Superplastic deformation Deformation that operates in very-fine-grained rocks in which grain-boundary sliding is the main mechanism. Under high strains, the grains slide past each other and rotate to accommodate the imposed strain. Rotation produces high temperature, which is characteristic of this mechanism. The rock mass thus flows as a ductile material but does not develop noticeable lattice preferred orientation (LPO) despite being affected by large amounts of strain. Also called **superplasticity**.

Superplasticity See **superplastic deformation**.

Superposed folds Complex folds formed due to superposition of an early-developed fold set by some later fold set (s).

Supratenuous fold A fold in which the hinge is thin while the limbs are thicker.

Surface forces Forces that act upon the boundaries of a body. These forces act on a unit area of the body and are known as stress.

Symmetric fold A fold in which the axial plane is vertical and the two limbs dip at equal angles. Also called **M-fold**.

Symmetric tensor A tensor with six different components such as stress.

Syncline A fold that is concave towards the centre and the strata dip towards the centre of the structure from both sides, and as such younger rocks occur at the centre of curvature.

Synclinorium A larger syncline containing a number of smaller folds (synclines and anticlines) on its limbs.

Synform A fold that looks like a syncline but the age of the rocks at the core is not necessarily younger or is not known.

Synformal anticline A fold in which the strata dip towards the axis of the fold and the rocks at the core are older; the structure can thus be described as an upward-facing anticline.

Syntectonic crystallization Crystallization that is concurrent with tectonic deformation.

Syntectonic crystallization Refers to a state when deformation and crystallization of minerals go synchronously.

Synthetic fault A minor fault that has the same displacement sense and a similar orientation to a related major fault.

Synthetic strain markers Artificially prepared objects that mimic a natural strain marker, such as ooid or conglomerate. These objects are subjected to deformation under laboratory conditions, such as homogeneous/heterogeneous strain, pure shear and simple shear. An additional benefit of this method is that the synthetic objects can be packed and oriented in the same way as the natural markers in their outcrop position.

Systematic joints Planar surfaces (joints) that show regular spacing and parallel orientations within an area of observation.

S-fold An asymmetric, S-shaped fold in which alternate limbs are parallel and hinge is curved.

S-surface New planar structure developed from deformation of previously formed features. Thus, assigning S_0 to the original stratification, the subsequently formed planar structures are commonly assigned S_1, S_2, etc. in a sequence.

S-tectonite A rock in which foliation is dominant while lineation is either absent or very weak.

S-C structure It is a structure in rock showing S- and C-surfaces. S-surface is a new surface formed due to shear deformation of pre-existing foliation causing elongation of platy minerals and some grains in the direction of extension. C-surface is in the form of a narrow, local shear surface parallel to the shear zone boundary and is one where high shear strain is localized.

S-C tectonite A rock showing both S- and C-surfaces. In such rocks, the S- and C-surfaces define zones of low and high strains, respectively.

T

Taylor-Bishop-Hill model A model to explain the mechanism of grain rotation. It predicts that under noncoaxial shear condition the crystallographic slip planes can rotate a grain and that rotation brings the crystal parallel to the plane of shear. The preferred orientation thus developed forms a foliation to the rock.

Tear fault A small-scale strike-slip or oblique-slip fault. These faults are confined to the upper crust or to its sedimentary cover.

Tectonic joints Joints formed at depth (up to 3 km) under the influence of high pore pressure caused due to tectonic compaction.

Tectonic structures Deformation features of rocks caused by tectonic stresses.

Tectonics Study of deformation, bending, rupturing, movements and related processes of the lithosphere including the formation of mountains and ocean basins and evolution of the continents and of the crust.

Tectonite A deformed rock. A rock whose components have moved during deformation.

Tensile stress Stress that acts along a plane in opposite directions, and if the body is in equilibrium, it means that an equal and opposite stress must have been active. As a result, the body gets stretched.

Tension gashes Tension fractures in rocks filled by fluids that precipitate as secondary minerals such as quartz and calcite.

Tensor A tensor in general represents a set of numerical quantities that can describe the physical state of a material. It relates the various geometrical vectors that, when related to a coordinate system, constitute an organized frame for a set of physical properties of a body. Tensors thus constitute geometric objects that relate the vector and scalar quantities of an object and can be represented as a multidimensional array of numbers.

Termination A zone where a strike-slip fault ends or terminates with displacement gradually reaching to zero. A termination is associated with extensional or contractional deformation of the crust. In the extensional zone, the deformation is accommodated by the formation of splay and normal faults that form an imbricate fan, while in the contractional zone, the deformation is accommodated by the formation of thrust faults and folds that form an imbricate fan.

Tertiary creep It is a type of time-dependent (creep) deformation in which the rock shows rapid fatigue with time and with increasing creep until it fails by rupture. In the strain-time curve, it is represented by the rising part of the curve.

Tertiary directions A set of directions as a divider of the secondary directions (NE, SE, SW, NW), and can be written as NNE, ENE, ESE, SSE, SSW, WSW, WNW and NNW.

Thick-skinned deformation Deformation that involves basement rocks. This type of deformation is associated with crustal shortening produced by thrust faults that cut both basement and cover rocks leading to the formation of nappes that propagate from root zone, thus giving rise to thrust sheets that are common features of orogenic belts.

Thin-skinned deformation Deformation that affects the upper layers of cover rocks. This type of deformation requires occurrence of sedimentary layers and presence of anisotropy containing mechanically weak rocks such as shale or salt along which large displacement can occur. Thrust faults involved in this type of deformation do not reach the lower layers and the basement.

Third-order folds A set of smaller folds developed at the limbs of second-order folds.

Throw The vertical component of displacement of an inclined fault.

Thrust fault A fault in which the hanging wall has gone up relative to the footwall. The dip of the thrust fault is generally less than 45°.

Thrust fold A thrust fault that got folded during some later folding episode. Also called **folded thrust**.

Thrust geometry The variety of geometrical shapes shown by thrusts.

Thrust mechanics Mechanical aspects of thrust formation.

Thrust sheet A tectonic unit overlying a thrust.

Tilted fault blocks The crustal blocks that undergo the extensional process involving simultaneous tilting and faulting.

Time-dependent deformation See **creep**.

Torque It is a force that tends to rotate a body. A torque may increase or decrease the speed of rotation of the body. Since rotational forces seldom develop in rock masses of the earth, application of torque in structural geology is therefore negligible to absent.

Torsion Deformation of a body when it is twisted by applying a stress in one direction at one end while the other end remains either motionless or twisted in opposite direction.

Traction Distribution of forces acting on any specific surface of a solid substance, say a rock mass. The concept of traction appears to have not yet been widely used in structural geology, though there are several situations where the concept finds its utility.

Transcurrent fault It is a strike-slip fault that cuts continental basement as well as sedimentary cover and is associated with large displacement.

Transected fold A fold in which the fold axial plane and the foliation do not coincide. The fold is thus associated with a transected foliation.

Transected foliation A foliation developed at an angle with the fold axial plane.

Transfer fault A fault which transfers the displacement from one fault to another by strike-slip movement, though dip-slip movement is also possible. It is a vertical or near-vertical fault that can accommodate large amounts of displacement and can form in both extensional and contractional regimes.

Transform fault Strike-slip fault that cuts the whole lithosphere and forms boundaries of solid crustal plates. The plate may be oceanic or continental. The fault transfers the movement in such a way that lithosphere is neither created nor destroyed. As such, a transform fault is considered as a conservative plate boundary and is believed to accommodate sea-floor spreading.

Transient flow A type of flow in which the strain rate changes with time under constant stress.

Translation gliding Homogeneous simple shear of a crystal such that the dislocations move along a pre-existing crystallographic plane.

Translation Motion of a rigid body in which a straight line within the body remains parallel to itself throughout the motion and there is no change in its shape or volume and therefore it is no strain.

Transposed folds See **rootless folds**.

Transposition structure Structure formed due to stretching of an earlier structure under high strains. Parts of the structure thus change their original positions (transpose) to occupy some other positions.

Transpression Shortening or compression along bends of strike-slip faults.

Transtension Extension along bends of strike-slip faults.

Triangle zone The area bounded by two thrusts converging upwards. This situation occurs when a back thrust meets or truncates an earlier thrust.

Trishear It is a distributed, strain-compatible shear in a triangular shear zone (in profile). Trishear resembles simple shear in a tabular shear zone with the main difference that area balance in a triangular shear zone requires curved displacement oblique to the fault-slip direction. It is an alternative model to explain folding in front of propagating thrusts in fault-propagation folds, for which the previous models used kink band geometries.

True dip It is the amount of inclination of a bed with the horizontal plane as measured in a vertical plane at right angle to the strike, and it is the highest amount of inclination of a bed.

Tulip structure See **normal flower structure**. Also called **negative flower structure**.

Twin gliding Slip of dislocations of a crystal along some fraction of the pre-existing distance between the crystallographic planes; this produces shear-induced twinning in crystals as noticed in calcite, feldspar and quartz.

Two-dimensional stress See **biaxial stress**.

U

Ultramylonite A mylonite with matrix >90% of the rock and grain size of matrix material <0.1 mm.

Uniaxial compression A stress acting along one direction only, i.e. along its length. The stress causes shortening of the length. In this case, two of the three principal stresses are not acting and can therefore be considered to be of zero value.

Uniaxial stress Stress acting along one direction only, i.e. along the length of a body.

Unloading joints Joints formed due to removal of overburden during erosion.

Upright fold A fold in which the axial surface is vertical.

V

Veins Joints and fractures filled with minerals are called veins that may range in size from outcrop scale to microscopic scale.

Vergence The direction of overturning or leaning of the axial plane of a fold.

Vertical fold A fold in which the plunge of the fold hinge and dip of axial surface are nearly vertical.

Viscosity The drag that resists a fluid to flow.

Viscous deformation A type of deformation in which stress is proportional to strain rate implying that viscous deformation is a time-dependent deformation and that the rock accumulates strain with time. Viscous deformation occurs when rocks are at higher temperatures, near their melting points, when they behave as viscous materials, and as such some of the physical properties may be compared to those showing viscous behaviour.

Vorticity A parameter that expresses the average rate of rotation of all the material lines of a body under progressive deformation in relation to the coordinate axes.

W

Wallace-Bott's principle It states that slip along a fault plane should be parallel to the direction of the greatest shear stress. It gives the stress-slip relationship for a fault.

Wallner lines Lines that stop fracture propagation, thus developing fan-shaped plumes on joint surfaces.

Window Structure formed by erosion of a nappe such that the underlying footwall rocks are seen.

Work hardening See **strain hardening**.

Wrench fault This term is used synonymously with strike-slip fault that does not involve a plate boundary.

X

X-ray computed tomography A method in which one can view individual objects in three dimensions by the use of a photographic camera, an optical scanner, electron back-scattering image, X-ray maps, etc. or any combination of these.

Y

Y shears See **D shears**.

Young's modulus Elastic constant given by the ratio of normal stress to extension strain of a substance.

Z

Z-fold An asymmetric, Z-shaped fold in which alternate limbs are nearly parallel.

Zigzag fold Chevron folds on larger scale.

References

A

Alfaro P, Delgado J, Esposito C, Garćia Tortosa F, Marmoni GM, Martino S (2019) Time-dependent modelling of a mountain front retreat due to a fold-to-fault controlled lateral spreading. Tectonophysics 773:228233. https://doi.org/10.1016/j.tecto.2019.228233

Allaby M (2008) Oxford dictionary of earth sciences, 3rd edn. Oxford University Press, Oxford. 654p

Allmendinger RW, Sharp JW, Van Tish D, Sepa L, Brown L, Kaufman S, Oliver J (1983) Cenozoic and Mesozoic structure of the eastern Basin and Range Province, Utah, from COCORP seismic-reflection data. Geology 11:532–536

Allmendinger RW, Cardozo N, Fisher DM (2012) Structural geology algorithms: vectors and tensors. Cambridge University Press, New York. 302p

Anderson EM (1951) The dynamics of faulting and dyke formation with application to Britain, 2nd edn. Oliver & Boyd, Edinburgh. 206p

Angelier J (1994) Fault slip analysis and palaeostress reconstruction. In: Hancock PL (ed) Continental deformation. Pergamon Press, Oxford, pp 53–100

Arthaud F (1969) Méthode de détermination graphique des directions de raccourcissement, d'allongement et intermédiaire d'une population de failles. Bull Soc Géol Fr 7:729–737

Ashby MF (1972) A first report on deformation-mechanism maps. Acta Metall 20:887–897

Atkinson BK (1987) Fracture mechanics of rock. Academic, London

Atkinson BK, Meredith G (1987) The theory of subcritical crack growth with application to minerals and rocks. In: Atkinson BK (ed) Fracture mechanics of rocks. Academic, London, pp 111–166

Austrheim H, Boundy TM (1994) Pseudotachylytes generated during seismic faulting and eclogitization of the deep crust. Science 265:82–83

Axen GJ (1988) The geometry of planar domino-style normal faults above a dipping basal detachment. J Struct Geol 10:405–411

Aydin A, Nur A (1982) Evolution of pull-apart basins and their scale independence. Tectonics 1:91–105B

B

Bahat D, Engelder T (1984) Surface morphology on cross-fold joints of the Appalachian Plateau, New York and Pennsylvania. Tectonophysics 104:299–313

Bahat D, Bankwitz P, Bankwitz E (2003) Preuplift joints in granites: evidence for subcritical and postcritical fracture growth. Bull Geol Soc Am 115:148–165

Bailey EB (1935) Tectonic essays, mainly Alpine. Oxford University Press, Oxford, England

Bailey JE, Hirsch PB (1962) The recrystallisation process in some polycrystalline metals. Proc R Soc Lond A 267:11–30

Ballantyne CK (2002) Paraglacial geomorphology. Quat Sci Rev 21:1935–2017. https://doi.org/10.1016/S0277-3791(02)00005-7

Bally AW, Gordby PL, Stewart BA (1966) Structure, seismic data and orogenic evolution of southern Canadian Rocky Mountains. Can Petrol Geol Bull 14:337–381

Beach A (1975) The geometry of en echelon vein arrays. Tectonophysics 28:245–263

Bell TH (1985) Deformation partitioning and porphyroblast rotation in metamorphic rocks: a radial reinterpretation. J Metamorphic Geol 3:109–118

Bell TH, Cuff C (1989) Dissolution, solution transfer, diffusion versus fluid flow and volume loss during deformation/metamorphism. J Metamorphic Geol 7:425–447

Bell TH, Johnson SE (1989) Porphyroblast intrusion trails: the key to orogenesis. J Metamorphic Geol 3:109–118

Bell TH, Johnson SE (1992) Shear sense: a new approach that resolves conflicts between criteria in metamorphic rocks. J Metamorphic Geol 10:99–124

Bell TH, Rubenach MJ (1983) Sequential porphyroblast growth and crenulation cleavage development during progressive deformation. Tectonophysics 92:171–194

Bell TH, Johnson SE, Davis B, Forde A, Hayward N, Wilkins C (1992) Porphyroblast inclusion-trail orientation data: eppure non son girate. J metamorphic Geol 10:295–300

Bergbauer S, Pollard DD (2004) A new conceptual fold-fracture model including prefolding joints, based on the Emigrant Gap anticline, Wyoming. Bull Geol Soc Am 116:294–307

Berthé D, Choukroune P, Jegouza P (1979) Orthogneiss, mylonite and noncoaxial deformation of granites: the example of the South Armorican shear zone. J Struct Geol 1:31–42

Bhattacharya AR (1983) On the nature of the Main Boundary Fault in Kumaun-Garhwal Himalaya, with special reference to measurement of flattening in folds. Neues Jahrb Geol Paläont Mh 1983(4):193–204

Bhattacharya AR (1986) Wavelength-amplitude characteristics of polyphase folds in the Precambrian Bundelkhand Complex, India. Tectonophysics 128:121–125

Bhattacharya AR (1987) A 'ductile thrust' in the Himalaya. Tectonophysics 135:37–45

Bhattacharya AR (1992) A quantitative study of hinge thickness of natural folds: some implications for fold development. Tectonophysics 212:371–377

Bhattacharya AR (1997) Control of ductile strain and rheology on fold geometry: a mathematical perspective. Geoinformatics 8:143–148

Bhattacharya AR (2005) A classification of folds: role of axial angle and thickness ratio. Geoinformatics 16:27–34

Bhattacharya AR, Siawal A (1985) A phenomenon of unusual flattening in natural folds associated with a Himalayan thrust. Geol Mijnb 64:159–165

A. R. Bhattacharya, *Structural Geology*, Springer Textbooks in Earth Sciences, Geography and Environment,
https://doi.org/10.1007/978-3-030-80795-5

Bhattacharya AR, Singh SP (2013) Proterozoic crustal scale shearing in the Bundelkhand Massif with special reference to quartz reefs. J Geol Soc India 82:474–484

Bhattacharya AR, Verma AK (2020) Grain-shape controlled strain in quartz grains in high ductile flow regime: observations from the Main Central Thrust Zone of the Kumaun Himalaya, India. J Earth Syst Sci 129:213. https://doi.org/10.1007/s12040-020-01479-w

Bhattacharya AR, Weber K (2003) Effect of carbonate and phyllosilicate mineral phases on the quartz < a > axis fabrics of noncoaxially deformed rocks. GeoActa 2:109–116

Bhattacharya AR, Weber K (2004) Fabric development during shear deformation in the Main Central Thrust zone, NW-Himalaya, India. Tectonophysics 387:23–48

Biddle KT, Christie-Blick N (1985) Strike-slip deformation, basin formation, and sedimentation. Society of Economic Paleontologists and Mineralogists Special Publication 37, 386 p.

Billings MP (1972) Structural geology, 3rd edn. Prentice-Hall, Upper Saddle River, NJ. 606p

Biot MA (1957) Folding instability of a layered viscoelastic medium under compression. Proc R Soc Lond A 242:444–454

Biot MA (1961) Theory of folding of stratified viscoelastic media and its implications in tectonic and orogenesis. Bull Geol Soc Am 72:1595–1620

Biot MA (1964) Theory of internal buckling of a confined multilayered structure. Bull Geol Soc Am 75:563–568

Biot MA (1965) Mechanics of incremental deformation. Wiley, New York. 504p

Biot MA, Odé H, Roever WL (1961) Experimental verification of the theory of folding of stratified viscoelastic media. Bull Geol Soc Am 72:1621–1631

Bons PD, Elburg MA, Gomez-Rivas E (2012) A review of the formation of tectonic veins and their microstructures. J Struct Geol 43:33–62

Borradaile GJ, Henry B (1997) Tectonic applications of magnetic susceptibility and its anisotropy. Earth Sci Rev 42:49–93

Borradaile GJ, Jackson M (2010) Structural geology, petrofabrics and magnetic fabrics (AMS, AARM, AIRM). J Struct Geol 32:1519–1551

Bose N, Dutta D, Mukherjee S (2020) Refraction of micro-fractures due to shear-induced mechanical stratigraphy in a low-grade meta-sedimentary rock. J Struct Geol 133:103995. https://doi.org/10.1016/j.jsg.2020.103995

Bott MHP (1982) The interior of the earth: its structure, constitution and evolution. London, Edward Arnold. 403p

Bouchez JL, Lister GS, Nicolas A (1983) Fabric asymmetry and shear sense in movement zones. Geol Rundsch 72:401–419

Boullier AM, Gueguen Y (1975) SP-mylonites: origin of some mylonites by superplastic flow. Contrib Mineral Petrol 50:93–104

Boyer SE (1986) Styles of folding within thrust sheets: examples from the Appalachian and Rocky Mountains of the USA and Canada. J Struct Geol 8:325–339

Boyer SE, Elliot D (1982) Thrust systems. Bull Am Assoc Pet Geol 66:1196–1230

Bozzano F, Bretschneider A, Martino S (2008) Stress-strain history from the geological evolution of the Orvieto and Radicofani cliff slopes (Italy). Landslides 5:351–366. https://doi.org/10.1007/s10346-008-0127-2

Brace WF (1960) An extension of the Griffith theory of fracture to rocks. J Geophys Res 65:3477–3480

Breddin H (1956) Die tectonische Deformation der Fossilien in Rheinischen Schiefergebirge. Z Deutsch Geol Ges 106:227–305

Burchfiel BC, Stewart JH (1966) "Pull-apart" origin of the central segment of Death Valley, California. Bull Geol Soc Am 77:439–441

Burg JP (1999) Ductile structures and instabilities: their implication for Variscan tectonics in the Ardennes. Tectonophysics 309:1–25

Burg JP, Brunel M, Gapais D, Chen GM, Liu GH (1984) Deformation of leucogranites of the crystalline Main Central Sheet in southern Tibet (China). J Struct Geol 6:535–542

Butler RWH (1982) The terminology of structures in thrust belt. J Struct Geol 4:239–245

Butler RWH (1987) Thrust sequences. J Geol Soc Lond 144:619–634

Butler RWH, Bond CE, Cooper MA, Watkins H (2019) Fold-thrust structures - where have all the buckles gone? In: Bond CR, Lebit HD (eds) Folding and fracturing of rocks: 50 years of research since the seminal text book of J.G. Ramsay. Geological Society, London., Spec. Publications 487. https://doi.org/10.1144/SP487.7

Byerlee JD (1978) Friction of rocks. Pure Appl Geophys 116:615–626

C

Carreras J (2001) Zooming on Northern Cap de Creus shear zones. J Struct Geol 23:1457–1486

Carter NL, Avé Lallemant HG (1970) High temperature flow of dunite and peridotite. Bull Geol Soc Am 81:2181–2202

Chapman RE (1983) Petroleum geology. Elsevier Academic Press, Amsterdam. 415p

Chastel YB, Dawson PR, Wenk HR, Bennett K (1993) Anisotropic convection with implications for the upper mantle. J Geophys Res 98(B10):17757–17771

Cloetingh S, Wortel R (1986) Stress in the Indo-Australian plate. Tectonophysics 132:49–67

Cloos E (1946) Lineation: a critical review and annotated bibliography. Geol Soc Am Menoir 18:122

Cloos E (1947) Oölite deformation in the south mountain fold, Maryland. Bull Geol Soc Am 58:843–918

Cobbold PR (1977) Description and origin of banded deformation structures. II. Rheology and the growth of banded perturbations. Can J Earth Sci 14:2510–2523

Coelho S, Passchier CW, Grasemann B (2005) Geometric description of flanking structures. J Struct Geol 27:597–606

Cooper MA (1981) The internal geometry of nappes: criteria for models of emplacement. In: McClay KR, Price NJ (eds) Thrust and Nappe tectonics. Special Publication, vol 9. Geological Society, London, pp 225–234

Cosgrove JW, Ameen MS (2000) A comparison of the geometry, spatial organization and fracture patterns associated with forced folds and buckle folds. In: Cosgrove JW, Ameen MS (eds) Forced folds and fractures. Special Publication, vol 169. Geological Society, London, pp 7–21

Coward MP, Kim JH (1981) Strain within thrust sheets. In: McClay KR, Price NJ (eds) Thrust and Nappe tectonics. Special Publication, vol 9. Geological Society, London, pp 275–292

Cowie PA, Scholz CH (1992) Physical explanation for the displacement-length relationship of faults, using a post-yield fracture mechanics model. J Struct Geol 14(10):1133–1148

Crowell JC (1974) Origin of late Cenozoic basins in southern California. In: Dickinson WR (ed) Tectonics and sedimentation. Special Publication, vol 22. Society of Economic Paleontologist and Mineralogists, Tulsa, OK, pp 190–204

Cruikshank KM, Zhao G, Johnson AM (1991) Analysis of minor fractures associated with joints and faulted joints. J Struct Geol 13:865–886

Currie JB, Patnode HW, Trump RP (1962) Development of folds in the sedimentary strata. Bull Geol Soc Am 73:655–674

Czeck DM, Traut JT, Hudleston PJ (2019) Rheologic information determined from cleavage refraction in naturally deformed interlayered quartzites and phyllites. Special Publication. Geological Society, London. https://doi.org/10.1144/SP487-2018-49

D

Dahlstrom CDA (1969) Balanced cross sections. Can J Earth Sci 6:743–757

Dahlstrom CDA (1970) Structural geology in the eastern margin of the Canadian Rocky Mountains. Can Petrol Geol Bull 18:332–406

Davis GH, Stephen JR, Kluth CF (2012) Structural geology of rocks and regions, 3rd edn. Wiley, Hoboken, NJ

De Paor DG (1981) Strain analysis using deformed line distributions. Tectonophysics 73:T9–T14

De Paor DG (1988) $R_f/Ø_f$ strain analysis using an orientation net. J Struct Geol 10(4):323–333

De Paor DG (1996) Structural geology and personal computers. Pergamon, New York. 527p

De Paor DG, Eisenstadt G (1987) Stratigraphic and structural consequences of fault reversal: an example from the Franklinian Basin, Ellesmere Island. Geology 15:948–949

De Sitter LU (1958) Structural geology, 2nd edn. McGraw-Hill Book Co., New York. 551p

De Sitter LU (1964) Variation in tectonic style. Bull Can Petrol Geol 12:263–278

DeCelles PG, Robinson DM, Quade J, Ojha TP, Garzione CN, Copeland P, Upreti BN (2001) Stratigraphy, structure, and tectonic evolution of the Himalayan fold-thrust belt in western Nepal. Tectonics 20:487–509

Della Seta M, Esposito C, Marmoni GM, Martino S, Scarascia Mugnozza G, Troiani F (2017) Morpho-structural evolution of the valley-slope systems and related implications on slope-scale gravitational processes: new results from the Mt. Genzana case history (Central Apennines, Italy). Geomorphology 289:60–77. https://doi.org/10.1016/j.geomorph.2016.07.003

Dennis JG (1972) Structural geology. Ronald Press, New York. 532p

Dennis JG, Price RA, Sales JK, Hatcher R, Bally AW, Perry WJ, Laubscher HP, Williams RE, Elliott D, Norris DK, Hutton DW, Emmett T (1981) What is a thrust? What is a nappe? In: McClay KR, Price NJ (eds) Thrust and Nappe tectonics. Geological Society, London, pp 7–9

Di Luzio E, Saroli M, Esposito C, Bianchi Fasani G, Cavinato GP, Scarascia Mugnozza G (2004) Influence of structural framework on mountain slope deformation in the Maiella anticline (Central Apennines, Italy). Geomorphology 60:417–432. https://doi.org/10.1016/j.geomorph.2003.10.004

Dijkstra AH, Drury MR, Vissers RLM, Newman J (2002) On the role of melt-rock reaction in mantle shear zone formation in the Othris Peridotite Massif (Greece). J Struct Geol 24:1431–1450

Drury MR, Urai JL (1990) Deformation-related recrystallization processes. Tectonophysics 172:235–253

Drury MR, Humphreys FJ, White SH (1985) Large strain deformation studies using polycrystalline magnesium as rock analogue. Part II. Dynamic recrystallization mechanisms at high temperatures. Phys Earth Planet Inter 40:208–222

Duba AG, Durham WB, Handin JW, Wanget HF (1990) The Brittle-Ductile transition in rocks. Am Geophys Union 56:256

Duda M, Renner J (2013) The weakening effect of water on the brittle failure strength of sandstone. Geophys J Int 192:1091–1108

Dunne WM, Hancock PL (1994) Palaeostress analysis of small-scale brittle structures. In: Hancock PL (ed) Continental deformation. Pergamon Press, Oxford, pp 101–120

Dunnet D (1969) A technique for finite strain analysis using elliptical particles. Tectonophysics 7:117–136

Dunnet D, Siddans AWB (1971) Non-random sedimentary fabrics and their modification by strain. Tectonophysics 12:307–325

Durney DW (1972) Solution transfer, an important geological deformation mechanism. Nature 235:315–317

Durney DW (1978) Early theories and hypotheses on pressure-solution-redeposition. Geology 6:369–372

Dutta D, Mukherjee S (2019) Opposite shear senses: geneses, global occurrences, numerical simulations and a case study from the Indian western Himalaya. J Struct Geol 126:357–392

E

Elliott D (1970) Determination of finite strain and initial shape from deformed elliptical objects. Bull Geol Soc Am 81:2221–2236

Elliott D (1973) Diffusion flow laws in metamorphic rocks. Bull Geol Soc Am 84:2645–2664

Elliott D (1976a) The motion of thrust sheets. J Geophys Res 81(5):949–963

Elliott D (1976b) The energy balance and deformation mechanisms of thrust sheets. Philos Trans R Soc Lond A 283:289–312

Elliott D (1981) The strength of rocks in thrust sheets. EOS 62:397

Engelder T (1982) Is there a genetic relation between selected regional joints and contemporary stress within the lithosphere of North America. Tectonics 1:161–178

Engelder T (1985) Loading paths to joint propagation during a tectonic cycle: an example from the Appalachian Plateau, U.S.A. J Struct Geol 7:459–476

Engelder T (1987) Joints and shear fractures in rocks. In: Atkinson BK (ed) Fracture mechanism of rocks. Academic, London, pp 27–69

Engelder T (1989) Analysis of pinnate joints in the Mount Desert Island granite: Implications for postintrusion kinematics in the coastal volcanic belt, Maine. Geology 17:564–567

Engelder T (1993) Stress regimes in the lithosphere. Princeton University Press, Princeton

Engelder T, Fischer MP (1996) Loading configurations and driving mechanisms for joints based on the Griffith energy-balance concept. Tectonophysics 256:253–277

Engelder T, Geiser P (1979) The relationship between pencil cleavage and lateral shortening within the Devonioan section of the Appalachian Plateau, New York. Geology 7:460–464

Engelder T, Geiser P (1980) On the use of regional joint sets as trajectories of paleostress fields during the development of the Appalachian Plateau, New York. J Geophys Res 85(B11):6319–6341

Engelder T, Lacazette A (1990) Natural hydraulic fracturing. In: Barton N, Stephansson O (eds) Rock Joints. A.A. Balkema, Rotterdam, pp 35–44

Engelder T, Marshak S (1985) Development of cleavage in limestone of a fold thrust belt in eastern New York. J Struct Geol 7:345–359

Erslev EA (1988) Normalized center-to-center strain analysis of packed aggregates. J Struct Geol 10:201–209

Erslev EA (1991) Trishear fault-propagation folding. Geology 19:617–620

Erslev EA, Ge H (1990) Least-squares center-to-center and mean object ellipse fabric analysis. J Struct Geol 12:1047–1059

Etchecopar A (1977) A plane kinematic model of progressive deformation in polystyrene aggregate. Tectonophysics 39:121–139

Etheridge MA, Wall VJ, Cox SF, Vernon RH (1984) High fluid pressures during regional metamorphism and deformation - implications for mass transport and deformation mechanisms. J Geophys Res 89:4344–4358

Exner U, Mancktelow NS, Grasemann B (2004) Progressive development of s-type flanking folds in simple shear. J Struct Geol 26:2191–2201

F

Fagereng A, Biggs J (2019) New perspectives on 'geological strain rates' calculated from both naturally deformed and actively deforming rocks. J Struct Geol 125:100–110

Fallot P, Faure-Muret A (1949) Sur l'extension du de′collement de la se′rie de couverture subalpine. C R Seances Acad Sci Fr 228:616–619

Fischer MP, Woodward N, Mitchell M (1992) The kinematics of break-thrust folds. J Struct Geol 14:451–460

Fitch TJ (1972) Plate convergence, transcurrent faults, and internal deformation adjacent to Southeast Asia and the Western Pacific. J Geophys Res 77:4432–4460

Fleuty MJ (1964) The description of folds. Proc Geol Assoc 75:461–492

Flinn D (1962) On folding during three-dimensional progressive deformation. Quart J Geol Soc Lond 118:385–433

Forsyth D, Uyeda S (1975) On the relative importance of the driving forces of plate motion. Geophys J R Astron Soc 43:163–200

Fossen H (2016) Structural geology, 2nd edn. Cambridge University Press, Cambridge UK. 524p

Fossen H, Cavalcante GCG (2017) Shear zones – a review. Earth Sci Rev 171:434–455

Fossen H, Cavalcante GCG, Pinheiro RVL, Archanjo CJ (2019) Deformation – progressive or multiphase? J Struct Geol 125:82–99

Freund R (1974) Kinematics of transform faults. Tectonophysics 21:93–134

Frost HJ, Ashby MF (1973) A second report on deformation mechanism maps. Division of Engineering and Applied Physics Harvard University, Cambridge, MA. 108p

Frost HJ, Ashby MF (1982) Deformation mechanism maps. Pergamon Press, Oxford. 166p

Fry N (1979) Random point distributions and strain measurement in rocks. Tectonophysics 60:89–105

Fuchs L, Koyi H, Schmeling H (2015) Numerical modeling of the effect of composite rheology on internal deformation in down-built diapirs. Tectonophysics 632:111–122

Fusseis F, Handy MR (2008) Micromechanisms of shear zone propagation at the brittle-viscous transition. J Struct Geol 30:1242–1253

G

Gansser A (1964) Geology of the Himalayas. Interscience Publishers, London. 289p

Gemmer L, Beaumont C, Ings S (2005) Dynamic modelling of passive margin salt tectonics: effects of water loading, sediment properties and sedimentation patterns. Basin Res 16:199–218

Ghosh SK (1968) Experiments on buckling of multilayers which permit interlayer gliding. Tectonophysics 6:207–249

Ghosh SK (1993) Structural geology: fundamentals and modern development. Pergamon Press, Oxford. 598p

Ghosh SK, Ramberg H (1968) Buckling experiments on intersecting fold patterns. Tectonophysics 51:83–97

Ghosh SK, Ramberg H (1976) Reorientation of inclusions by combination of pure shear and simple shear. Tectonophysics 34:1–70

Ghosh SK, Mandal N, Khan D, Deb S (1992) Modes of superposed buckling in single layers controlled by initial tightness of early folds. J Struct Geol 14:381–394

Gilliland WN, Meyer GP (1976) Two classes of transform faults. Bull Geol Soc Am 87:1127–1130

Gomez-Rivas E, Butler RWH, Healy D, Alsop I (2020) From hot to cold - the temperature dependence on rock deformation processes: an introduction. J Struct Geol 132:103977. https://doi.org/10.1016/j.jsg.2020.103977

Goscombe BD, Passchier CW (2003) Asymmetrical boudins as shear sense indicators - an assessment from field data. J Struct Geol 25:575–589

Goscombe BD, Passchier CW, Hand M (2004) Boudinage classification: end-member boudin structures. J Struct Geol 26:739–763

Gottstein G, Mecking H (1985) Preferred orientation in deformed metals and rocks. In: Wenk H-R (ed) Introduction to modern texture analysis. Academic, London, pp 183–218

Graham RH (1981) Gravity sliding in the Maritime Alps. In: McClay KR, Price NJ (eds) Thrust and Nappe tectonics. Special Publication, vol 9. Geological Society, London, pp 335–355

Grasemann B, Stüwe K, Vannay JC (2003) Sense and non-sense of shear in flanking structures. J Struct Geol 25:19–34

Grasemann B, Wiesmayr G, Draganits E, Fusseis F (2004) Classification of Refold structures. J Geol 112:119–125

Gray DR (1977) Morphologic classification of crenulation cleavage. J Geol 85:229–235

Gray DR (1979) Geometry of crenulation-folds and their relationship to crenulations cleavage. J Struct Geol 1:187–205

Gray DR, Durney DW (1979) Investigation on mechanical significance of crenulation cleavage. Tectonophysics 58:35–79

Griffith AA (1920) The phenomena of rupture and flow in solids. Philos Trans R Soc Lond A 221:163–198

Griffith AA (1924) The theory of rupture. In: Proceeding of first international Congress of Applied Mechanics, Delft, pp 55–63

Griggs DT (1967) Hydrolytic weakening of quartz and other silicates. Geophys J R Astron Soc 14:19–31

Griggs DT, Handin J (1960) Observations on fracture and a hypothesis of earthquakes. In: Griggs DT, Handin J (eds) Rock deformation (A symposium), Mem Geol Soc Am, vol 79, pp 347–364

Groshong RH Jr (1975) Strain, fractures and pressure solution in natural single-layer folds. Bull Geol Soc Am 86:1363–1376

Groshong RH Jr (1988) Low-temperature deformation mechanisms and their interpretation. Bull Geol Soc Am 100:1329–1360

Groshong RH Jr (1994) Area balance, depth to detachment and strain in extension. Tectonics 13:1488–1497

Groshong RH Jr (2006) 3-D structural geology, a practical guide to quantitative surface and subsurface map interpretation, 2nd edn. Springer, Heidelberg. 400p

Gudmundsson A, Nilsen K (2006) Structures, models, and stress fields of ring faults in composite volcanoes. In: Natale GD, Kilburn C, Troise C (eds) Mechanisms of activity and unrest at large calderas. Special Publication, vol 269. Geological Society, London, pp 83–108

H

Hafner W (1951) Stress distributions and faulting. Bull Geol Soc Am 62:373–398

Hamilton W (1987) Crustal extension in the Basin and Range Province, southwestern United States. In: Coward MP, Dewey JF, Hanback PL (eds) Continental extensional tectonics, Geological Society Special Publication No. 28. Geological Society, London, pp 155–176

Hancock PL (1985) Brittle microtectonics: principles and practice. J Struct Geol 7:437–458

Hancock PL, Barka AA (1987) Kinematic indicators on active normal faults in Western Turkey. J Struct Geol 9:573–584

Hancock PL, Engelder T (1989) Neotectonic joints. Bull Geol Soc Am 101:1197–1208

Hanmer S, Passchier CW (1991) Shear sense indicators: a review. Geol Surv Can Pap 90:17–72

Harland WB (1971) Tectonic transpression in Caledonian Spitsbergen. Geol Mag 108:27–42

Hatcher RD Jr (1981) Thrusts and nappes in the North American Appalachian Orogen. In: McClay KR (ed) Thrust tectonics. Geological Society, London, pp 491–500

Hatcher RD Jr (1995) Structural geology: principles, concepts and problems, 2nd edn. Prentice and Hall, Englewood Cliffs, NJ

Hatcher RD Jr, Hooper RJ (1992) Evolution of crystalline thrust sheets in the internal parts of mountain chains. In: McClay KR (ed) Thrust tectonics. Geological Society, London, pp 217–233

Heard HC (1960) Transition from brittle fracture to ductile flow in Solenhofen limestone as a function of temperature, confining pressure and interstitial fluid pressure. In: Griggs DT, Handin J (eds) Rock deformation, Mem. Geol. Soc. Am., vol 79, pp 193–226

Heilbronner R (2000) Automatic grain boundary detection and grain size analysis using polarization micrographs or orientation images. J Struct Geol 22:969–981

Heilbronner R, Barrett S (2014) Image analysis in earth science microstructures and textures of earth materials. Springer, Heidelberg. 520p

Heilbronner R, Tullis J (2006) Evolution of c axis pole figures and grain size during dynamic recrystallization. J Geophys Res 111:1–19

Heim A, Gansser A (1939) Central Himalayas: geological observations of Swiss expedition, 1936. Mem Soc Helv Sci Nat 73(1):1–245

Herwegh M, Berger A (2004) Deformation mechanisms in second-phase affected microstructures and their energy balance. J Struct Geol 26:1483–1498

Herwegh M, Handy MR (1998) The origin of shape preferred orientations in mylonite: inferences from in-situ experiments on poly-crystalline norcamphor. J Struct Geol 20:681–694

Herwegh M, Jenni A (2001) Granular flow in polymineralic rocks bearing sheet silicates: new evidence from natural examples. Tectonophysics 332:309–320

Hill ML (1947) Classification of faults. Geol Bull Soc Am 31:1664–1673

Hippert JFM (1993) V-pull-apart microstructures: a new shear sense indicator. J Struct Geol 15:1393–1404

Hippertt JFM, Tohver E (1999) On the development of zones of reverse shearing in mylonitic rocks. J Struct Geol 21:1603–1614

Hobbs BE (2019) The development of structural geology and the historical context of the Journal of Structural Geology: a reflection by Bruce Hobbs. J Struct Geol 125:3–19

Hobbs BE, Ord A (2015) Structural geology: the mechanics of deforming metamorphic rocks, vol I. Elsevier, UK. 665p

Hobbs BE, Means WD, Williams PF (1976) An outline of structural geology. Wiley, New York. 571p

Hobbs BE, Mühlhaus HB, Ord A (1990) Instability, softening and localization of deformation. Geol Soc Lond Spec Pub 54:143–165

Hodgson RA (1961) Classification of structures on joint surfaces. Am J Sci 259:493–502

Hooper RJ, Hatcher RD Jr (1988) Mylonites from the Towaliga fault zone, central Georgia: products of heterogeneous non-coaxial deformation. Tectonophysics 152:1–17

Hossack JR (1968) Pebble deformation and thrusting in the Bygdin area (Southern Norway). Tectonophysics 5:315–339

Hossack JR (1979) The use of balanced cross sections in the calculation of orogenic contraction, a review. J Geol Soc Lond 136:705–711

Hossack JR (1983) A cross-section through the Scandinavian Caledonides constructed with the aid of branch-line maps. J Struct Geol 5:103–111

Hossack JR, Hancock PL (1983) Balanced cross sections and their geological significance. J Struct Geol 5:98–223

Hubbert MK, Rubey WW (1959) Role of fluid pressure in mechanics of overthrust faulting: I. Mechanics of fluid-filled porous solids and its application to overthrust faulting. Bull Geol Soc Am 70:115–166

Hudleston PJ (1973a) Fold morphology and some geometrical implications of theories of fold development. Tectonophysics 16:1–46

Hudleston PJ (1973b) An analysis of "Single-layer" folds developed experimentally in viscous media. Tectonophysics 16:189–214

Hudleston PJ (1986) Extracting information from folds in rocks. J Geol Educ 34:237–245

Hudleston P (1999) Strain compatibility and shear zones: is there a problem? J Struct Geol 21:923–932

Hudleston PJ, Stephansson O (1973) Layer shortening and fold development in the buckling of single layers. Tectonophysics 17:299–321

Hudleston PJ, Treagus SH (2010) Information from folds: a review. J Struct Geol 32:2042–2071

Hungr O, Leroueil S, Picarelli L (2014) The Varnes classification of landslide types, an update. Landslides 11:167–194

Hunt CW (1957) Planimetric equation: The Oil and Gas Journal. Penn. Well Publishing Company, Tulsa, pp 259–264

Hunter NJR, Weinberg RF, Wilson CJL, Law RD (2018) A new technique for quantifying symmetry and opening angles in *c*-axis pole figures: implications for interpreting the kinematic and thermal properties of rocks. J Struct Geol 112:1–6

Hyndman RD, Weichert DH (1983) Seismicity and rates of relative motion on the plate boundaries of Western North America. Geophys J Int 72:59–82

I

Ingles J, Lamouroux C, Soula JC, Guerrero N, Debat P (1999) Nucleation of ductile shear zones in a granodiorite under greenschist facies conditions, Néouvielle massif, Pyrenees, France. J Struct Geol 21:555–576

Inglis CE (1913) Stresses in a plate due to the presence of cracks and sharp corners: Transactions of the Institute of Naval Architects 55:219–230. In: Hatcher RD, Jr. (1995) Structural Geology: Principles, Concepts and Problems. 2nd ed., Prentice and Hall, Englewood Cliffs, New Jersey

Ingraffea AR (1987) Theory of crack initiation and propagation in rock. In: Atkinson BK (ed) Fracture mechanics of rock. Academic, London, pp 71–110

Irwin GR, de Wit R (1983) Introduction to fracture mechanics and its geophysical applications. In: Atkinson BK (ed) Fracture mechanics of rocks. Academic, London

J

Jackson MPA, Talbot CJ (1994) Advances in salt tectonics. In: Hancock PL (ed) Continental deformation. Pergamon Press, Oxford, pp 159–179

Jackson MPA, Vendeville BC (1994) Regional extension as a geologic trigger for diapirism. Bull Geol Soc Am 106:57–73

Jackson MPA, Cornelius RR, Craig CH, Gansser A, Stöcklin J, Talbot CJ (1990) Salt diapirs of the Great Kavir, Central Iran. Mem Geol Soc Am 177:139

Jaeger JC, Cook NGW, Zimmerman RW (2007) Fundamental of rock mechanics, 4th edn. Blackwell Publishing, Oxford. 475p

Jessel MW, Lister GS (1990) A simulation of the temperature dependence of quartz fabrics. In: Knipe RJ, Rutter EH (eds) Deformation mechanisms, rheology and tectonics. Special Publication, vol 54. Geological Society, London, pp 353–362

Johnson AM (1977) Styles of folding: mechanics and mechanisms of folding of natural elastic material. Elsevier, Amsterdam. 406p

K

Kamb WB (1959) Theory of preferred crystal orientation developed by crystallization under stress. J Geol 67:153–170

Kaninskaite I, Fisher QJ, Michei EAH (2020) Faults in tight limestones and dolostones in San Vito lo Capo, Sicily, Italy: internal architecture and petrophysical properties. J Struct Geol. https://doi.org/10.1016/j.jsg.2019.103970

Keller EA (2001) Environmental geology, 8th edn. Prentice-Hall, Upper Saddle River, NJ

Kilian R, Heilbronner R, Stünitz H (2011) Quartz microstructures and crystallographic preferred orientation: which shear sense do they indicate? J Struct Geol 33:1446–1466

Kim YS, Peacock DCP, Sanderson DJ (2004) Fault damage zones. J Struct Geol 26:503–517

King GCP, Vita-Finzi (1981) Active folding in the Algerian earthquake of 10 October 1980. Nature 292:22–26

Kirby SH (1983) Rheology of the lithosphere. Rev Geophys 21:1458–1487

Knipe RJ (1981) The interaction of deformation and metamorphism of slate. Tectonophysics 78:249–272

Knipe RJ (1989) Deformation mechanisms - recognition from natural tectonites. J Struct Geol 11:127–146

Knipe RJ, Law RD (1987) The influence of crystallographic orientation and grain boundary migration on microstructural and textural evolution in an S-C mylonite. Tectonophysics 135:155–169

Knipe RJ, White SH (1977) Microstructural variation of an axial plane cleavage around a fold - a H.V.E.M. study. Tectonophysics 39:355–380

Kohlstedt DL, Evans B, Mackwell SJ (1995) Strength of the lithosphere: constraints imposed by laboratory experiments. J Geophys Res 100 (B9):17587–17602

Krabbendam M, Urai JL, van Vliet IJ (2003) Grain size stabilization by dispersed graphite in a high-grade quartz mylonite: an example from Naxos (Greece). J Struct Geol 25:855–866

Kranz RL (1983) Microcracks in review. Tectonophysics 100:449–480

Kugler J, Waldron JWF, Durling PW (2019) Fault development in transtension, McCully gas field. Tectonophysics, New Brunswick, Canada. https://doi.org/10.1016/j.tecto.2019.228313

L

Lacazette A, Engelder T (1992) Fluid-driven cyclic propagation of a joint in the Ithaca siltstone, Appalachian Basin, New York. In: Evans B, Wong T-F (eds) Fault mechanics and transport properties of rocks. Academic, London, pp 297–324

Lahiri S, Rana V, Bhatt S, Mamtani MA (2020) Paleostress and statistical analysis using quartz veins from mineralized and non-mineralized zones: application for exploration targeting. J Struct Geol 133:104006. https://doi.org/10.1016/j.jsg.2020.104006

Lajtai EZ (1971) A theoretical and experimental evaluation of the Griffith theory of brittle fracture. Tectonophysics 11:129–156

Lan L, Hudleston P (1996) Rock rheology and sharpness of folds in single layers. J Struct Geol 18:925–931

Lapworth C (1885) The highland controversy in British Geology: its causes and consequences. Nature 8:558–559

Laubscher HP (1962) Die Zweiphasenhypothese der Jurafaltung. Ecolgae Geologicae Helvetiae 55:1–22

Laubscher HP (1972) Some overall aspects of Jura dynamics. Am J Sci 272:293–304

Laubscher HP (1977) Fold development in the Jura. Tectonophysics 37:337–362

Laubscher HP (1988) Material balance in Alpine orogeny. Geol Soc Am Bull 100:1313–1328

Laubscher HP (1990) The problem of the deep structure of the southern Alps: 3-D material balance considerations and regional consequences. Tectonophysics 176:103–121

Launeau P, Robin PYF (1996) Fabric analysis using the intercept method. Tectonophysics 267:91–119

Launeau P, Robin PYF (2005) Determination of fabric and strain ellipsoids from measured sectional ellipses—implementation and applications. J Struct Geol 27:2223–2233

Law RD (1987) Heterogeneous deformation and quartz crystallographic fabric transitions: natural examples from the Moine Thrust zone at the stack of Glencoul, Northern Assynt. J Struct Geol 9:819–833

Law RD (2014) Deformation thermometry based on quartz c-axis fabrics and recrystallization microstructures: a review. J Struct Geol 66:129–161

Law RD, Johnson MRW (2010) Microstructures and crystal fabrics of the Moine Thrust zone and Moine Nappe: history of research and changing tectonic interpretations. Geol Soc Lond Spec Pub 335:443–503

Lawn BR (1983) In: Atkinson BK (ed) Fracture mechanics of rock. Academic, London, p 12

Lisle RJ (1977) Estimation of the tectonic strain ratio from the mean shape of deformed elliptical objects. Geol Mijnb 56:140–144

Lisle RJ (1985) Geological strain analysis. Pergamon Press, New York. 102p

Lisle RJ (1992) Strain estimation from flattened buckle folds. J Struct Geol 14:369–371

Lisle RJ (1994) Palaeostrain analysis. In: Hancock (ed) Continental deformation. Pergamon Press, New York, pp 28–42

Lisle RJ (2010) Strain analysis from point fabric patterns: an objective variant of the Fry method. J Struct Geol 32:975–981

Lisle RJ, Bastida F, Aller J (2019) Measuring the research impact of the book *Folding and Fracturing of Rocks* by John G. Ramsay. In: Bond CE, Lebit HD (eds) Folding and fracturing of rocks: 50 years of research since the seminal text book of J.G. Ramsay, Special Publications, vol 487. Geological Society, London. https://doi.org/10.1144/SP487.8

Lister GS (1977) Discussion: Crossed girdle c-axis fabrics I: quartzites plastically deformed by plane strain and progressive simple shear. Tectonophysics 39:51–54

Lister GS, Davis GA (1989) The origin of metamorphic core complex and detachment faults formed during Tertiary continental extension in the northern Colorado River region U.S.A. J Struct Geol 11:65–94

Lister GS, Snoke AW (1984) S-C mylonites. J Struct Geol 6:617–638

Lister GS, Williams PF (1979) Fabric development in shear zone: theoretical controls and observed phenomena. J Struct Geol 1:283–297

Llorens MG (2019) Stress and strain evolution during single-layer folding under pure and simple shear. J Struct Geol 126:245–257

Lockner DA (1998) A generalized law for brittle deformation of Westerly granite. J Geophys Res 103(B3):5107–5123

M

MacDonald GJF (1960) Chapter 1: Orientation of anisotropic minerals in a stress field. Mem Geol Soc Am 79:1–8

Maltman A (1994) Prelithification deformation. In: Hancock (ed) Continental deformation. Pergamon Press, New York, pp 143–158

Mandl G (1988) Mechanics of tectonic faulting. Elsevier, Amsterdam. 407p

Mandl G (2000) Faulting in brittle rocks: an introduction to the mechanics of tectonic faults. Springer, Berlin. 434p

Marques FO, Mandal N, Taborda R, Antunes JV, Bose S (2014) The behavior of deformable and non-deformable inclusions in viscous flow. Earth Sci Rev 134:16–69

Mauldon M, Dunne WM Jr, Rohrbaugh MB (2001) Circular scanlines and circular windows: new tools for characterizing the geometry of fracture traces. J Strut Geol 23:247–258

McAdoo DC, Sandwell DT (1985) Folding of oceanic lithosphere. J Geophys Res 90:8563–8569

McCarthy DJ, Meere PA, Petronis MS (2015) A comparison of the effectiveness of clast based finite strain analysis techniques to AMS in sandstones from the Sevier Thrust Belt, Wyoming. Tectonophysics 639:68–81

McCarthy D, Meere P, Mulchrone K (2019) Determining finite strain: how far have we progressed? Special Publication. Geological Society, London. https://doi.org/10.1144/SP487-2018-62

McClay KR (1992) Thrust tectonics. Chapman and Hall, London

McClay KR, Ellis PG (1987) Geometries of extensional fault systems developed in model experiments. Geology 15:341–344

McClay KR, Price NJ (1981) Thrust and Nappe tectonics. Special Publication, vol 9. Geological Society, London

McGarth AG, Davison I (1995) Damage zone geometry around fault tips. J Struct Geol 17:1011–1024

McGinn C, Miranda EA, Hufford LJ (2020) The effects of quartz Dauphiné twinning on strain localization in a mid-crustal shear zone. J Struct Geol 134:103980. https://doi.org/10.1016/j.jsg.2020.103980

Mckenzie DP (1978) Some remarks on the development of sedimentary basins. Earth Planet Sci Lett 40:25–32

McNaught M (1994) Modifying the normalized fry method for aggregates of non-elliptical grains. J Struct Geol 16:493–503

Means WD (1976) Stress and strain. Springer, New York. 339p

Means WD (1977) Experimental contributions to the study of foliation in rocks: a review of research since 1960. Tectonophysics 39:329–355

Means WD (1981) The concept of steady-state foliation. Tectonophysics 78:179–199

Means WD (1984) Shear zones of types I and II and their significance for reconstruction of rock history. Geol Soc Am Abstr 16:50

Means WD (1995) Shear zones and rock history. Tectonophysics 247:157–160

Menegon L, Pennacchioni G (2010) Local shear zone pattern and bulk deformation in the Gran Paradiso metagranite (NW Italian Alps). Int J Earth Sci 99:1805–1825

Mercier JCC, Anderson DA, Carter NL (1977) Stress in the lithosphere: inference from the steady state flow of rocks. Pure Appl Geophys 115:199–226

Michibayashi K (1993) Syntectonic development of a strain-independent steady-state grain size during mylonitization. Tectonophysics 222:151–164

Milton NS (1980) Determination of the strain ellipsoid from measurement on any three sections. Tectonophysics 64:T19–T27

Mitra S (1986) Duplex structures and imbricate thrust system. Bull Am Assoc Petrol Geol 70:1087–1112

Mitra S (1990) Geometry and kinematic evolution of inversion structures. Bull Am Assoc Petrol Geol 77:1159–1191

Mitra S, Namson J (1989) Equal-area balancing. Am J Sci 289:563–599

Mittempergher S, Dallai L, Pennacchioni G, Renard F, Toro GD (2014) Origin of hydrous fluids at seismogenic depth: constraints from natural and experimental fault rocks. Earth Planet Sci Lett 385:97–109

Molnar P, Tapponier P (1975) Cenozoic tectonics of Asia: effect of a continental collision. Science 189:419–425

Moody JD, Hill MJ (1956) Wrench-fault tectonics. Bull Geol Soc Am 67:1207–1246

Mukherjee S (2011) Mineral fish: their morphological classification, usefulness as shear sense indicators and genesis. Int J Earth Sci 100:1303–1314

Mukherjee S (2014) Review of flanking structures in meso- and micro-scales. Geol Mag 151(6):957–974

Mulchrone KF (2013) Fitting the void: data boundaries, point distributions and strain analysis. J Struct Geol 46:22–33

Mulchrone KE, Mukherjee S (2019) Numerical modelling and comparison of the temporal evolution of mantle and tails surrounding rigid elliptical objects in simple shear regime under stick and slip boundary conditions. J Struct Geol 132:103968. https://doi.org/10.1016/j.jsg.2019.103968

Mulchrone KF, Talbot C (2016) Strain estimation in 3D by fitting linear and planar data to the March model. Tectonophysics 686:63–67

Mulchrone KF, O'Sullivan F, Meere PA (2003) Finite strain estimation using the mean radial length of elliptical objects with bootstrap confidence intervals. J Struct Geol 25:529–539

Mulchrone KF, Meere PA, Choudhury KR (2005) SAPE: a program for semi-automatic parameter extraction for strain analysis. J Struct Geol 27:2084–2098

Müller WH, Schmid SM, Briegel U (1981) Deformation experiments on anhydrite rocks of different grain size: rheology and microfabrics. Tectonophysics 78:527–543

N

Nabavi ST, Fossen H (2018) Fold geometry and folding - a review. Earth Sci Rev. https://doi.org/10.1016/j.earscirev.2021.103812

Nadan BJ, Engelder T (2009) Microcracks in New England granitoids: a record of thermoelastic relaxation during exhumation of intracontinental crust. Bull Geol Soc Am 121:80–90

Narr W, Suppe J (1991) Joint spacing in sedimentary rocks. J Struct Geol 11:1037–1048

Neves SP, Santos TAS, Medeiros PC, Amorim LQ, Casimiro DCG (2018) Interference fold patterns in regional unidirectional stress fields: a result of local kinematic interactions. J Struct Geol 115:304–310

Newman J (1994) The influence of grain size and grain size distribution on methods for estimating paleostresses from twinning in carbonates. J Struct Geol 16:1589–1601

Nicolas A (1987) Principles of rock deformation. D. Riedel Publishing, Dordretch, Holland. 201p

Nicolas A, Poirier JP (1976) Crystalline plasticity and solid state flow in metamorphic rocks. Wiley-Interscience, London. 444p

O

O'Leary DW, Freidman JD, Pohn HA (1976) Lineament, linear, lineation: some proposed new definitions for old terms. Bull Geol Soc Am 87:1463–1469

Oertel G (1996) Stress and deformation: a handbook on tensors in geology. Oxford University Press, Oxford. 305p

Olson J, Pollard DD (1989) Inferring paleostresses from natural fracture patterns: a new method. Geology 17:345–348

Ord A, Christie JM (1984) Flow stress from microstructure in mylonitic quartzites of the Moine Thrust zone, Assynt area, Scotland. J Struct Geol 6:639–654

P

Panozzo R (1984) Two dimensional strain from the orientation of lines in a plane. J Struct Geol 6:215–221

Panozzo HR (1992) The autocorrelation function: an image processing tool for fabric analysis. Tectonophysics 212:351–370

Park RG (1969) Structural correlation in metamorphic belts. Tectonophysics 7:323–338

Passchier CW (1992) Pseudotachylyte and the development of ultramylonite bands in the Saint-Barthélemy Massif, French Pyrenees. J Struct Geol 4:69–79

Passchier CW, Simpson C (1986) Porphyroclast systems as kinematic indicators. J Struct Geol 8:831–843

Passchier CW, Trouw RAJ (2005) Microtectonics, 2nd edn. Springer, Berlin. 366p

Passchier CW, Trouw RAJ, Zwart HJ, Vissers RLM (1992) Porphyroblast rotation: eppur si muove? Metamorphic Geol 10:283–294

Paterson MS (1978) Experimental rock deformation: the Brittle Field. Springer, New York

Paterson MS, Weiss LE (1966) Experimental deformation and folding in phyllite. Bull Geol Soc Am 77:343–374

Peacock DCP, Price SP, Whitham AG, Pickles CS (2000) The World's biggest relay ramp: Hold With Hope, NE Greenland. J Struct Geol 22:843–850

Pennacchioni G (2005) Control of the geometry of precursor brittle structures on the type of ductile shear zone in the Adamello tonalites, Southern Alps (Italy). J Struct Geol 27:627–644

Pennacchioni G, Mancktelow NS (2007) Nucleation and initial growth of a shear zone network within compositionally and structurally heterogeneous granitoids under amphibolite facies conditions. J Struct Geol 29:1757–1780

Philpotts AR, Ague JJ (2009) Principles of igneous and metamorphic petrology, 2nd edn. Cambridge University Press, Cambridge

Platt JP, Vissers RLM (1980) Extensional structures in anisotropic rocks. J Struct Geol 2:397–410

Poh J, Yamato P, Duretz T, Gapais D, Ledru P (2020) Precambrian deformation belts in compressive tectonic regimes: a numerical perspective. Tectonophysics 777. https://doi.org/10.1016/j.tecto.2020.228350

Poirier JP (1980) Shear localization and shear instability in materials in the ductile field. J Struct Geol 2:135–142

Poirier JP (1985) Creep of crystals: high-temperature deformation processes in metals, ceramics and minerals. Cambridge University of Press, Cambridge. 260p

Poirier JP (1991) Introduction to the physics of the earth's interior. Cambridge University Press, Cambridge

Poirier JP, Nicolas A (1975) Deformation-induced recrystallization and progressive mis-orientation of subgrain-boundaries, with special reference to mantle peridotites. J Geol 83:707–720

Pollard DD, Fletcher RC (2005) Fundamentals of structural geology. Cambridge University Press, New York. 500p

Pollard DD, Segall P (1987) Theoretical displacements and stresses near fractures in rock: with applications to faults, joints, veins, dikes, and solution surfaces. In: Atkinson BK (ed) Fracture mechanics of rock. Academic, London, pp 277–349

Powell CMA (1979) A morphological classification of rock cleavage. Tectonophysics 58:21–34

R

Ramberg H (1961) Relationship between concentric longitudinal strain and concentric shear strain during folding of homogeneous sheets of rocks. Am J Sci 259:382–390

Ramberg H (1963) Fluid dynamics of viscous buckling applicable to folding of layered rocks. Bull Am Assoc Pet Geol 47:484–505

Ramberg H (1964) Selective buckling of composite layer with contrasted rheological properties, a theory for the formation of several orders folds. Tectonophysics 1:307–341

Ramberg H (1967) Gravity, deformation and the earth's crust as studied by centrifuged models. Academic, London. 241p

Ramberg H (1975) Particle paths, displacement and progressive strain applicable to rocks. Tectonophysics 28:1–37

Ramberg H, Stephansson O (1964) Compression of floating elastic and viscous plates affected by gravity: a basis for discussing crustal buckling. Tectonophysics 1:101–120

Ramsay JG (1962) The geometry and mechanics of formation of "Similar" type folds. J Geol 70:309–327

Ramsay JG (1967) Folding and fracturing of rocks. McGraw-Hill Book Company, New York. 568p

Ramsay JG (1969) The measurement of strain and displacement in orogenic belts. Spec. Publ, vol 3. Geological Society, London, pp 43–79

Ramsay JG (1980a) The crack-seal mechanism of rock deformation. Nature 284:135–139

Ramsay JG (1980b) Shear zone geometry: a review. J Struct Geol 2:83–99

Ramsay JG, Allison I (1979) Structural analysis of shear zones in an alpinized Hercynian granite (Maggia Lappen, Pennine Zone, Central Alps). Schweizerische Mineralogisches und Petrographische Mitteilungen 59:251–279

Ramsay JG, Graham RH (1970) Strain variation in shear belts. Can J Earth Sci 7:786–813

Ramsay JG, Huber MI (1983) The techniques of modern structural geology. In: Strain analysis, vol 1. Academic, London. 307p

Ramsay JG, Huber MI (1987) The techniques of modern structural geology: folds and fractures. Academic, London. 400p

Ramsay DM, Sturt BA (1973) An analysis of noncylindrical and incongruous fold pattern from the Eo-Cambrian rocks of Söröy, northern Norway: I. Noncylindrical, incongruous and aberrant folding. Tectonophysics 18:81–107

Ranalli G (1984) Grain size distribution and flow stress in tectonites. J Struct Geol 6:443–447

Ranalli G (1987) Rheology of the earth: deformation and flow processes in geophysics and geodynamics. Allen and Unwin, London

Rast N (1997) Continuous and discontinuous events in orogenic deformation. In: Sengupta S (ed) Evolution of geological structures in micro- to macro-scales. Springer, New York, pp 425–441

Reches Z (1978) Analysis of faulting in three-dimensional strain fields. Tectonophysics 47:109–129

Reches Z, Dieterich JH (1983) Faulting of rocks in three-dimensional strain fields. I. Failure of rocks in polyaxial, Servo-Control experiments. Tectonophysics 95:111–132

Ree JH (1991) An experimental steady-state foliation. J Struct Geol 13:1001–1011

Reynolds D, Holmes A (1954) The superposition of Caledonoid folds on an earlier fold-system in the Dalradians of Malin Head Co. Donegal. Geol Mag 91:417–444

Reynolds SJ, Lister GS (1987) Structural aspects of fluid-rock interactions in detachment. Geology 15:362–366

Riedel W (1929) Zur Mechanik geologischer Brucherschei-nungen: ein Beitrag zum Problem der "Fiederspälten". Centralblatt für Mineralogie, Geologie, und Päleontologie, Part B 1929:354–368

Roberts A, Yielding G (1994) Continental extensional tectonics. In: Hancock PL (ed) Continental deformation. Pergamon Press, Oxford, pp 223–250

Robin PYF (2002) Determination of fabric and strain ellipsoids from measured sectional ellipses—theory. J Struct Geol 24:531–544

Robin PYF (2019) 'Strain probe': best-fitting a homogeneous strain to the motions of scattered points. J Struct Geol 124:211–224

Robin PYF, Charles CRJ (2015) Quantifying the three-dimensional shapes of spheroidal objects in rocks imaged by tomography. J Struct Geol 77:1–10

Rodrigues SWO, Martins-Ferreira MAC, Faleirs FM, Neto MDC, Yogi MTAG (2019) Deformation conditions and quartz c-axis fabric development along nappe boundaries: the Andrelandia Nappe System, Southern Brasilia Orogen (Brazil). Tectonophysics 766:283–301

Romano V, Bigi S, Carnevale F, Hyman JD, Karra S, Valocchi AJ, Tartarello MC, Battaglia M (2020) Hydraulic characterization of a fault zone from fracture distribution. J Strut Geol. https://doi.org/10.1016/j.jsg.2020.104036

Royden LH, Burchfiel BC (1987) Thin-skinned N-S extension within the convergent Himalayan region; Gravitational collapse of a Miocene topographic front. In: Coward MP, Dewey JF, Hanback PL (eds) Continental extensional tectonics, Geological Society Special Publication No. 28. Geological Society, London, pp 611–619

Rutter EH (1976) The kinetics of rock deformation by pressure solution. Philos Trans R Soc Lond Ser A 283:203–219

Rutter EH (1983) Pressure solution in nature, theory and experiment. J Geol Soc Lond 140:725–740

Rutter EH (1986) On the nomenclature of mode of failure transitions in rocks. Tectonophysics 122:381–387

S

Sander B (1930) Gefugekunde der Gesteine. Springer, Berlin. 352p

Sanderson DJ, Marchini WRD (1984) Transpression. J Struct Geol 6:449–458

Sarkarinejad K, Sarshar MA, Adineh S (2018) Structural, micro-structural and kinematic analyses of channel flow in the Karmostaj salt diapir in the Zagros foreland belt, Fars province, Iran. J Struct Geol 107:109–131

Savage HM, Shackleton JR, Cooke ML, Riedel JJ (2010) Insights into fold growth using fold-related joint patterns and mechanical Stratigraphy. J Strut Geol 32:1466–1475

Savalli L, Engelder T (2005) Mechanisms controlling rupture shape during subcritical growth of joints in layered rocks. Bull Geol Soc Am 117:436–449

Schmid SM (1982) Microfabric studies as indicators of deformation mechanisms and flow laws operative in mountain building. In: Hsu KJ (ed) Mountain building processes. Academic, London, pp 95–110

Scholz CH (1990) The mechanics of earthquakes and faulting. Cambridge University Press, Cambridge

Schoneveld C (1977) A study of some typical inclusion patterns in strongly paracrystalline-rotated garnets. Tectonophysics 39:453–471

Schuck B, Desbois G, Urai JL (2020) Grain-scale deformation mechanisms and evolution of porosity in experimentally deformed Boom Clay. J Struct Geol 130:103894. https://doi.org/10.1016/j.jsg.2019.103894

Searle MP (1986) Structural evolution and sequence of thrusting in the High Himalayan, Tibetan-Tethys and Indus suture zones of Zanskar and Ladakh, Western Himalaya. J Struct Geol 8:923–936

Segall P, Simpson C (1986) Nucleation of ductile shear zones on dilatant fractures. Geology 14:56–59

Sengupta S (1983) Folding of boudinaged layers. J Struct Geol 5:197–210

Shackleton RM, Ries AC (1984) The relation between regionally consistent stretching lineations and plate motions. J Struct Geol 6:111–117

Shimamoto T, Ikeda Y (1976) A simple algebraic method for strain estimation from deformed ellipsoidal objects. 1. Basic theory. Tectonophysics 36:315–337

Sibson RH (1975) Generation of pseudotachylite by ancient seismic faulting. Geophys J R Astron Soc 43:775–794

Sibson RH (1977) Fault rocks and fault mechanisms. J Geol Soc Lond 133:191–213

Sibson RH (1986) Earthquakes and rock deformation in crustal fault zones. Annu Rev Earth Planet Sci 14:75–149

Sibson RH (1989) Earthquake faulting as a structural process. J Struct Geol 11:1–14

Sibson RH (2003) Thickness of the seismic slip zone. Bull Seismol Soc Am 93:1169–1178

Siddans AWB (1972) Slaty cleavage, a review of research since 1815. Earth Sci Rev 8:205–232

Simmons G, Richter D (1976) Microcracks in rocks. In: Strens RGJ (ed) The physics and chemistry of minerals and rocks. Wiley-Interscience, New York, pp 105–137

Simón LJ (2019) Forty years of paleostress analysis: has it attained maturity? J Struct Geol 125:124–133

Simpson GDH (1998) Dehydration-related deformation during regional metamorphism, NW Sardinia, Italy. J Metamorphic Geol 16:457–472

Simpson C, Schmidt SM (1983) An evaluation of criteria to deduce the sense of movement in sheared rocks. Bull Geol Soc Am 94:1281–1288

Skjernaa L (1975) Experiments on superimposed buckle folding. Tectonophysics 27:255–270

Soares A, Dias R (2015) Fry and Rf/ϕ strain methods constraints and fold transection mechanisms in the NW Iberian Variscides. J Struct Geol 79:19–30

Spitz R, Schmalholz SM, Kaus BJP, Popov AA (2020) Quantification and visualization of finite strain in 3D viscous numerical models of folding and overthrusting. J Struct Geol 131:103945. https://doi.org/10.1016/j.jsg.2019.103945

Spry A (1969) Metamorphic textures. Pergamon Press, Oxford. 350p

Srivastava DC, Lisle RJ, Imran M, Kandpal R (1998) The kink-band triangle: a triangular plot for paleostress analysis from kink-bands. J Struct Geol 20:1579–1586

Starnes JK, Long SP, Gordon SM, Zhang J, Soignard E (2019) Using quartz fabric intensity parameters to delineate strain patterns across the Himalayan Main Central thrust. J Struct Geol. https://doi.org/10.1016/j.jsg.2019.103941

Stearns DW (1968) Certain aspects of fractures in naturally deformed rock. Bull Geol Soc Am 116:294–307

Stein RS, King GCP (1984) Seismic potential revealed by surface folding: 1983 Coalinga, California. Earthq Sci 224(4651):869–872

Stephan T, Kroner U, Hahn T, Hallas P, Heuse T (2016) Fold/cleavage relationships as indicator for late Variscan sinistral transpression at the Rheno-Hercynian-Saxo-Thuringian boundary zone, Central European Variscides. Tectonophysics 681:250–262

Stipp M, Stünitz H, Heilbronner R, Schmid SM (2002) The eastern Tonale fault zone: a 'natural laboratory' for crystal plastic deformation of quartz over a temperature range from 250° to 700° C. J Struct Geol 24:1861–1884

Suppe J (1983) Geometry and kinematic of fault-bend folding. Am J Sci 283:684–721

Suppe J (1985) Principles of structural geology. Prentice Hall, Englewood Cliffs, NJ

Sylvester AG (1988) Strike-slip faults. Bull Geol Soc Am 100:1666–1703

T

Takeshita T, Wenk HR, Lebensohn R (1999) Development of preferred orientation and microstructure in sheared quartzite; comparison of natural data and simulated results. Tectonophysics 312:133–155
Talbot CJ (1970) The minimum strain ellipsoid using quartz vein. Tectonophysics 9:47–76
Tanner PWG (1989) The flexural-slip mechanism. J Struct Geol 11:635–655
Tapponnier P, Molnar P (1977) Active faulting and tectonics in China. J Geophys Res 82:2905–2930
Tapponnier P, Peltzer G, Armijo R (1986) On the mechanics of the collision between India and Asia. Geol Soc Lond 19:113–157
Tchalenko JS (1968) The evolution of kink bands and the development of compression textures in sheared clays. Tectonophysics 6:159–174
Tchalenko JS (1970) Similarities between shear zones of different magnitudes. Bull Geol Soc Am 81:1625–1640
Ter Heege JH, Bresser JHP, Spiers CJ (2004) Composite flow laws for crystalline materials with log-normally distributed grain size: theory and application to olivine. J Struct Geol 26:1693–1705
Thiessen RL, Means WD (1980) Classification of fold interference patterns: a reexamination. J Struct Geol 2:311–316
Thissen CJ, Brandon MT (2015) An autocorrelation method for three-dimensional strain analysis. J Struct Geol 81:135–154
Tikoff B, Fossen H (1993) Simultaneous pure and simple shear: the unified deformation matrix. Tectonophysics 217:267–283
Truesdell C (1954) The kinematics of vorticity, Indiana Univ. Pub. Sci. Series no. 19. Indiana Univ. Pub., Bloomington
Tsenn MC, Carter NL (1987) Upper limits of power law creep of rocks. Tectonophysics 136:1–26
Tullis J, Yund RA (1985) Dynamic recrystallization of feldspar: a mechanism for ductile shear zone formation. Geology 13:238–241
Tungatt PD, Humphreys FJ (1984) The plastic deformation and dynamic recrystallization of polycrystalline sodium nitrate. Acta Metall 32:1625–1635
Turcotte DL, Schubert G (1982) Applications of continuum physics to geological problems. Wiley, New York. 450p
Turner FJ, Weiss LE (1963) Structural analysis of metamorphic tectonites. McGraw-Hill, New York. 545p
Twiss RJ (1977) Theory and applicability of a recrystallized grain size paleopiezometer. In: Wyss M (ed) Stress in the earth. Springer, Basel AG, pp 227–244
Twiss RJ, Moores EM (2007) Structural geology, 2nd edn. W. H. Freeman and Company, New York. 736p

U

Urai JL, Means WD, Lister GS (1986) Dynamic recrystallization of minerals. In: Heard HC, Hobbs BE (eds) Mineral and rock deformation: laboratory studies, the Paterson volume, Geophysics monograph, vol 36. Am. Geophys. Union, Washington, DC, pp 161–200

V

Van der Pluijm BA, Marshak S (2004) Earth structure: an introduction to structural geology and tectonics, 2nd edn. W. W. Norton and Company, New York, p 656
Verma AK, Bhattacharya AR (2015) Reorientation of lineation in the Central Crystalline Zone, Munsiari-Milam area of the Kumaun Greater Himalaya. J Earth Syst Sci 124:449–458
Vernon RH (2004) A practical guide to rock microstructure. Cambridge University Press, Cambridge. 665p
Vinta BSSR, Srivastava DC (2012) Rapid extraction of central vacancy by image-analysis of Fry plots. J Struct Geol 40:44–53
Vitale S, Mazzoli S (2008) Heterogeneous shear zone evolution: the role of shear strain hardening/softening. J Struct Geol 30:1383–1395

W

Waldron JWF, Wallace KD (2007) Objective fitting of ellipses in the centre-to-centre (Fry) method of strain analysis. J Struct Geol 29:1430–1444
Watkins H, Bond CE, Cawood AJ, Cooper MA, Warren MJ (2019) Fracture distribution on the Swift Reservoir Anticline, Montana: implications for structural and lithological controls on fracture intensity. In: Bond CE, Lebit HD (eds) Folding and fracturing of rocks: 50 years of research since the seminal text book of J.G. Ramsay. Spec. Publ., vol 487. Geological Society, London. https://doi.org/10.1144/SP487.9
Watkinson AJ (1981) Patterns of fold interference: influence of early fold shapes. J Struct Geol 3:19–23
Weiss LE, Mclntyre DB, Kursten M (1955) Contrasted styles of folding in the rocks of Ord Ban, Mid-Strathspey. Geol Mag 92:21–36
Welker AJ, Hogan JP, Eckert A, Tindall S, Liu C (2019) Conical folds – an artefact of using simple geometric shapes to describe a complex geologic structure. J Struct Geol 123:96–104
Wellman HW (1962) A graphical method for analysing fossil distortion caused by tectonic deformation. Geol Mag 99:348–352
Wenk HR, Canova G, Molinari A (1989) Viscoplastic modeling of texture development in quartzite. J Geophys Res 94:17895–17906
Wernicke B (1981) Low-angle normal fault in the Basin and Range Province; nappe tectonics in an extending orogen. Nature 291:645–648
Wernicke B (1985) Uniform normal-sense simple shear of continental lithosphere. Can J Earth Sci 22:108–125
Wernicke B, Burchfiel BC (1982) Modes of extensional tectonics. J Struct Geol 4:104–115
White S (1973) Syntectonic recrystallization and texture development in quartz. Nature 244:276–278
Whitten EHT (1966) Structural geology of folded rocks. Rank McNally & Co., Chicago. 680p
Wicox RE, Harding TP, Seely DR (1973) Basic wrench tectonics. Bull Am Assoc Petrol Geol 57:74–96
Williams PF (1972) Development of metamorphic layers and cleavage in low grade metamorphic rocks at Bermagui, Australia. Am J Sci 272:1–47
Williams PF (1977) Foliation: a review and discussion. Tectonophysics 39:305–328
Williams PF, Price GP (1990) Origin of kink bands and shear-band cleavage in shear zones: an experimental study. J Struct Geol 12:145–164
Wilson G (1953) Mullion and rodding structures in the Moine Series of Scotland. Proc Geol Assoc 64:118–151
Wilson G (1961) Tectonic significance of small-scale structures and their importance to the geologist in the field. Annals de la Societé Geologique de Belgique 84:424–548
Winter JD (2010) Principles of igneous and metamorphic petrology, 2nd edn. Pearson Education Inc., Upper Saddle River, NJ
Wise DU, Dunn DE, Engelder JT, Geiser PA, Hatcher RD Jr, Kish SA, Odom AL, Schamel S (1984) Fault-related rocks: suggestions for terminology. Geology 12:391–394

Wood DS (1974) Current views of the development of slaty cleavage. Annu Rev Earth Planet Sci 2:369–401

Woodcock NH (1986) The role of strike-slip fault systems at plate boundaries. Philos Trans R Soc Lond A317:13–29

Woodcock NH, Fischer M (1986) Strike-slip duplexes. J Struct Geol 8:725–735

Woodcock NH, Schubert C (1994) Continental strike slip tectonics. In: Hancock PL (ed) Continental deformation. Pergamon Press, New York, pp 251–263

Woodward NB, Boyer SE, Suppe J (1985) An outline of balanced cross-sections, Studies in geology, vol 11. University of Tennessee, Department of Geological Sciences, Knoxville. 170p

Woodward NB, Gray DR, Spears DB (1986) Including strain data in balanced cross-sections. J Struct Geol 8:313–324

Y

Yamaji A (2008) Theories of strain analysis from shape fabrics: a perspective using hyperbolic geometry. J Struct Geol 30:1451–1465

Yamaji A (2015) How tightly does calcite *e*-twin constrain stress? J Struct Geol 72:83–95

Yardley BWD (1986) Fluid migration and veining in the Connemara Schists, Ireland. In: Walther JV, Wood BJ (eds) Fluid-rock interactions during metamorphism. Springer, New York, pp 109–131

Yardley BWD (1989) An introduction to metamorphic petrology, Longman earth science series. Wiley, New York

Younes AI, Engelder T (1999) Fringe cracks: key structures for the interpretation of the progressive Alleghanian deformation of the Appalachian plateau. Bull Geol Soc Am 111:219–239

Z

Zhao G, Johnson A (1992) Sequence of deformations recorded in joints and faults, Arches National Park, Utah. J Struct Geol 14:225–236

Zoback ML, Zoback MD (1989) Tectonic stress field of the continental United States. Geol Soc Am Mem 172:523–539

Zulauf G, Zulauf J, Maul H (2017) Quantification of the geometrical parameters of non-cylindrical folds. J Struct Geol 100:120–129

Zulauf J, Zulauf G, Hattingen E (2020) Boudinage and two-stage folding of oblique single layers under coaxial plane strain: layer rotation around the axis of no change (Y). J Struct Geol 135:104023. https://doi.org/10.1016/j.jsg.2020.104023

Zwart HJ (1962) On the determination of polymetamorphic mineral associations and its application to the Bosost area (central Pyrenees). Geol Rundsch 52:38–65

Author Index

A. R. Bhattacharya, *Structural Geology*, Springer Textbooks in Earth Sciences, Geography and Environment,
https://doi.org/10.1007/978-3-030-80795-5

Subject Index

A. R. Bhattacharya, *Structural Geology*, Springer Textbooks in Earth Sciences, Geography and Environment,
https://doi.org/10.1007/978-3-030-80795-5

W

X

Y

Z